MW00838039

# Gourmet and Health-Promoting Specialty Oils

# Gourmet and Health-Promoting Specialty Oils

Editors

**Robert A. Moreau**
**Afaf Kamal-Eldin**

Urbana, Illinois

**AOCS Mission Statement**
To be a global forum to promote the exchange of ideas, information, and experience, to enhance personal excellence, and to provide high standards of quality among those with a professional interest in the science and technology of fats, oils, surfactants, and related materials.



AOCS Press, Urbana, IL 61802


**Library of Congress Cataloging-in-Publication Data**
Gourmet and health-promoting specialty oils / editors, Afaf Kamal-Eldin, Robert Moreau.
p. cm.
Includes bibliographical references and index.
ISBN 978-1-893997-97-4 (alk. paper)
1. Oils and fats, Edible. I. Kamal-Eldin, Afaf. II. Moreau, Robert.

TX407.O34G68 2008
641.3'385--dc22

2008055039

Printed in the United States of America.
13 12 11 10 09 5 4 3 2 1

The paper used in this book is acid-free and falls within the guidelines established to ensure permanence and durability.

# Contents

# Preface

It is currently recognized that the contribution of dietary lipids to our health and well-being is determined by their compositional factors.

Both the fatty acid profile (i.e. the relative levels of omega-9, omega-6, and omega-3 fatty acids) and a wide range of common and specific minor lipid components have been shown to influence the physiological functions of our bodies.

Upon invitation by the AOCS Press, we have edited this book, which contains 21 chapters that describe about 40 different oils and fats from plants, algae, fish, and milk. As much as possible, the chapters followed a similar outline describing their sources, extraction and processing, chemical components, and health effects. We hope that the book will be useful for many people involved in oils and fats applications.

We sincerely thank all authors for their contributions and for their friendly cooperation during the preparation of the book. The help and support given to us by the AOCS staff, especially Brock Peoples and Jodey Schonfeld, was essential for the completion of our task and is very much appreciated.

*Robert A. Moreau* and
*Afaf Kamal-Eldin*

# Introduction

**Robert A. Moreau[1] and Afaf Kamal-Eldin[2]**
[1]*Eastern Regional Research Center, Agricultural Research Service, United States Department of Agriculture, 600 East Mermaid Lane, Wyndmoor, Pennsylvania 19038;*
[2]*Department of Food Science, Swedish University of Agricultural Sciences, Box 7051, 750 07 Uppsala, Sweden*

## Gourmet Specialty Oils

The word gourmet is used to describe a connoisseur to the refined sensation of fine foods and drinks and a "gourmet food" is a food of the highest quality, perfectly prepared and artfully presented (http://www.epicurious.com/tools/fooddictionary). Gourmet vegetable oils are, thus, characterized by their aroma and taste, mainly resulting from the fact that the oils are not refined. The most popular gourmet oil is olive oil, with an annual international production of about 2.75 million tons (Table I.1). Other oils that are used as gourmet oils in fine restaurants include avocado oil and some tree nut oils. However, many of what we know today as gourmet oils retain their flavor and taste because they are produced at very low volumes and are not processed by conventional refining, bleaching, and deodorizing (RBD), which are processes routinely used to remove impurities and extend the shelf life of commodity vegetable oils consumed by humans. Common examples of commodity oils include palm, soybean, cottonseed, rapeseed/canola, sunflower, and peanut (McKevith 2005). The main commodity oils, their production volumes, producing countries, and compositional characteristics are also summarized in Table I.1. Interestingly, the unique gourmet flavors of some gourmet oils, such as butter, butter oil, ghee, sesame oil, and pumpkin seed oil, are generated either by toasting the seeds or by treating the oils at high temperatures.

Distinguishing specialty oils from commodity oils requires one to define certain terminologies. A recent review on "minor specialty oils" suggested that oils with an annual production of less than 1 million tons per year could be considered to be "minor" oils (Gunstone, 2006). We chose to include in this book three of the four oils listed at the bottom of Table I.1 (olive oil, sesame oil, and flaxseed oil) because each of them have definite gourmet or health-promoting properties, which will be described in detail in the respective chapters. We did not include corn oil, because most of the commercial corn oil produced worldwide is sold for commodity uses and is sold at commodity prices (Moreau, 2005) and not sold for gourmet and health-promoting

**Table I.1. Worldwide Annual Production and Compositional Characteristics of Major Edible Oils and Fats**

| Oil | Production (million tons)* | Main Producing Countries | Predominant Fatty Acid | Other Compositional Characteristics |
|---|---|---|---|---|
| Soybean | 33.6 | USA, Brazil, Argentina | 18:2 | High levels of γ- and δ-tocopherols |
| Palm | 31.4 | Malaysia, Indonesia | 18:2 | Oils contain carotenoids |
| Animal fats | 25.4 | Worldwide | saturated | Oils contain cholesterol |
| Rape/canola | 17.7 | EU, China, India, Canada | 18:1 or 22:1 | Meal contains glucosinolates |
| Sunflower | 12.4 | Former USSR, EU | 18:2 and 18:1 | Increasing levels of 18:1 |
| Peanut | 5.7 | India, China, USA | 18:2 and 18:1 | Proteins contain allergens |
| Cottonseed | 5.4 | China, USA | 18:2 | Meal may contain gossypol |
| Palm kernel | 3.8 | Malaysia, Indonesia | 12:0 | Low melting point |
| Coconut | 3.7 | Indonesia, Philippines | 12:0 | Low melting point |
| Olive | 2.8 | EU, Tunisia | 18:1 | Phenolic antioxidants |
| Corn | 2.5 | USA, China, Brazil | 18:2 | High phytosterols and γ-tocopherol |
| Sesame | 0.9 | India, China | 18:1/18:2 (ca 1:1) | Furofuran lignans |
| Linseed/Flaxseed | 0.8 | Canada, China, India, USA | 18:3 n-3 | Applications for drying oil and ALA |

*Estimated average for the period 2006-2010 (Gunstone 2005).

applications. Gourmet oils are alternatively described as virgin or cold-pressed oils. Virgin oils are defined as those obtained without the involvement of chemicals and without altering the nature of the oil; i.e., by pressing and expelling under mild heating (<50°C) and subsequent purification by water washing, settling, filtration, and centrifugation (FAO/WHO 1993). Application of excessive heating (>50°C), e.g. leading to destruction of vitamins and chlorophyll or other pigments compromises the virgin designation. Cold-pressed oils, the highest-grade virgin oils, which are sold at much higher prices than commodity oils, are considered the best options for cooking and salad dressings in fine restaurants and are offered as gifts to friends and relatives. These oils are recognized as high quality oils and particularly appreciated for their flavor, color, viscous flow, and healthy connotations. Virgin and cold-pressed oils are obtained from various plant parts that include seeds, nuts, and fruit mesocarps. Many, but not all, of gourmet and health-promoting oils are produced by cold-pressing, but other extraction technologies are also employed (Table I.2) and they will be discussed in detail in each of the following chapters.

A number of gourmet oils are discussed in this book (Table I.3). Avocado and hempseed oils are generally characterized by their green color imparted by chlorophyll, while the color of olive oils vary from green to greenish-yellow, depending on the concentration of chlorophylls and pheophytins. However, while olive oil has a slightly bitter/pungent taste due to its high content of different various phenolic compounds (see Chapter *Olive Oil*), avocado and hempseed oils have a milder taste (Chapter *Avocado Oil*). Hempseed oil should also not be heated to high temperatures for long periods of time because of its exceptional high content of polyunsaturated fatty acids which will oxidize and eventually form *trans*-fatty acids. The major tree nut oils include the oils of almonds, Brazil nuts, cashew nuts, hazelnuts, macadamia nuts, pecans, pine nuts, pistachio nuts, and walnuts. Tree nut oils originate from diverse botanical families and are considerably different in their respective chemical compositions. Like olive oil, some other oils are considered to be gourmet oils when they are used as virgin oils although they are also used in refined forms. Sesame seed oil is an example of such an oil and currently virgin rapeseed/canola oil is also sold as a speciality oil. As mentioned above, oils are usually considered to be gourmet oils because of the small amounts that are produced, and because they are often utilized without refining.

Most of the gourmet and health-promoting oils in Table I.3 are obtained from seeds where they serve as a storage form of energy and carbon, which are utilized when the seeds germinate. However, it is interesting to note that both olive oil and avocado oil are obtained from "fruits" (the botanical term for this tissue is the mesocarp) that surround the seeds. Olive and avocado oils are not utilized during germination and their only apparent function is to attract animals that eat the fruit and thereby assist in the dispersal of the seeds and thus ensure the continuation of the species. The pleasing taste and aroma of olive oil and avocado oil may help to attract animals to these

**Table I.2. Oil Extraction Processes for Specialty and Commodity Oils**

| Method | Advantages | Disadvantage | Oil refined, bleached, deodorized (RBD) | Examples |
|---|---|---|---|---|
| Cold pressing or centrifugation systems | Retains minor compounds like volatiles, phenolic compounds and chlorophyll | Low yields of oil | No | Virgin olive oil, Avocado oil, Hempseed oil |
| Supercritical fluid extraction ($CO_2$) | Is nontoxic and safer than hexane. No need to remove solvents from miscella or meal | Is more expensive and yields may be lower than with hexane extraction | Optional | Oat oil |
| Ethanol extraction | Solvent is less toxic and safer than hexane | Extracts non-lipids, more costly to remove solvent from miscella and meal | Yes | Corn kernel oil |
| Standard pressing | Simple and economical technology for large scale industrial production | Lower oil yield than hexane extraction, high temperatures cause some chemical changes to oil and meal | Yes | Many commodity oils |
| Hexane extraction | Low cost, high oil yields | Health and safety issues | Yes | Many commodity oils |
| Pre-pressing + hexane extraction | Good for seeds with >20% oil | Requires more equipment | Yes | Many commodity oils |
| Aqueous enzymatic extraction | A gentle, "green" technique | High cost of enzymes, oil yield is lower than hexane extraction | Optional | In development |

oil-rich fruits. The last three oils in Table I.3 are also not seed oils. Fungal oil, algal oil, and fish oil all function as a storage form of energy in these organisms. The oil in milk that is used to make butter, butter oil and ghee is an energy-rich and nutritious food for the offspring of cows and water buffalo.

Although one could technically classify butter, butter oil, and ghee as "gourmet fats" (because like other fats, they are solid at room temperature) and not "gourmet oils," we have included them in this book partly because of their widespread gourmet use in many parts of the world and partly because the term "butter oil" qualifies it to be a gourmet oil.

## Health-Promoting Specialty Oils

Although the fatty acid composition of vegetable oils is often the property that receives the most attention, a number of minor lipid-soluble components are known to have significant contribution to the nutritional properties of these oils. Vegetable oils (Tables I.1 & I.3) are convenient sources of the essential fatty acids, linoleic acid (18:2 n-6, also called 18:2 ω-6) and α-linolenic acid (18:3 n-3, also called 18:3 ω-3), and the other fatty acids derived from them (Fig. I.1). A common fatty acid abbreviation system will be used in this book – the abbreviation X:Y n-Z denotes a fatty acid that has X carbons, Y double bonds, and the first double bond is on the Z carbon, counting from the "tail" (noncarboxyl) end of the fatty acid. It is now acknowledged that an adequate balance of essential fatty acids is required to serve as precursors for the production of prostaglandins and leukotrienes, in order to provide optimal pro-inflammatory and anti-inflammatory status in the body. It is believed that humans increased their consumption of linoleic acid (18:2 n-6) after the agricultural revolution leading to increased levels of arachidonic acid (20:4 n-6) and the inflammatory products dinoprostone (or prostaglandin $E_2$) and a number of leukotrienes. It is advised that humans balance their consumption of n-6-rich fats by increasing the relative intake of other fatty acids including oleic (18:1 n-9), α-linolenic (18:3 n-3), γ-linolenic (18:3 n-6), eicosapentaenoic (EPA, 20:5 n-3) and docosahexaenoic acids (DHA, 22:6 n-3) (Simopoulos, 2004). However, recommendations given for EPA and DHA differ significantly between organizations (Table I.4).

Scientific interest in the health-promoting effects of *omega*-3 (also designated as n-3 or ω-3) over the omega-6 (also designated as n-6 or ω-6) fatty acids started in the late 1970s and has continued, since both are important. The low incidence of coronary heart disease (CHD) in Greenland Eskimos was attributed to their traditional diet of marine animals and fish (Bang et al, 1980). Besides CHD, health benefits of n-3 fatty acids extend to other diseases, such as diabetes, metabolic syndrome, skin problems, inflammation, immune disorders, cancers, Crohn's disease, and depression. Oils that are recognized for health-promoting properties include algal and fish oils that contain very long-chain fatty acids, particularly EPA and DHA. The oils of flaxseed, perilla, hempseed, camelina, some berry seeds and walnuts are good sources of

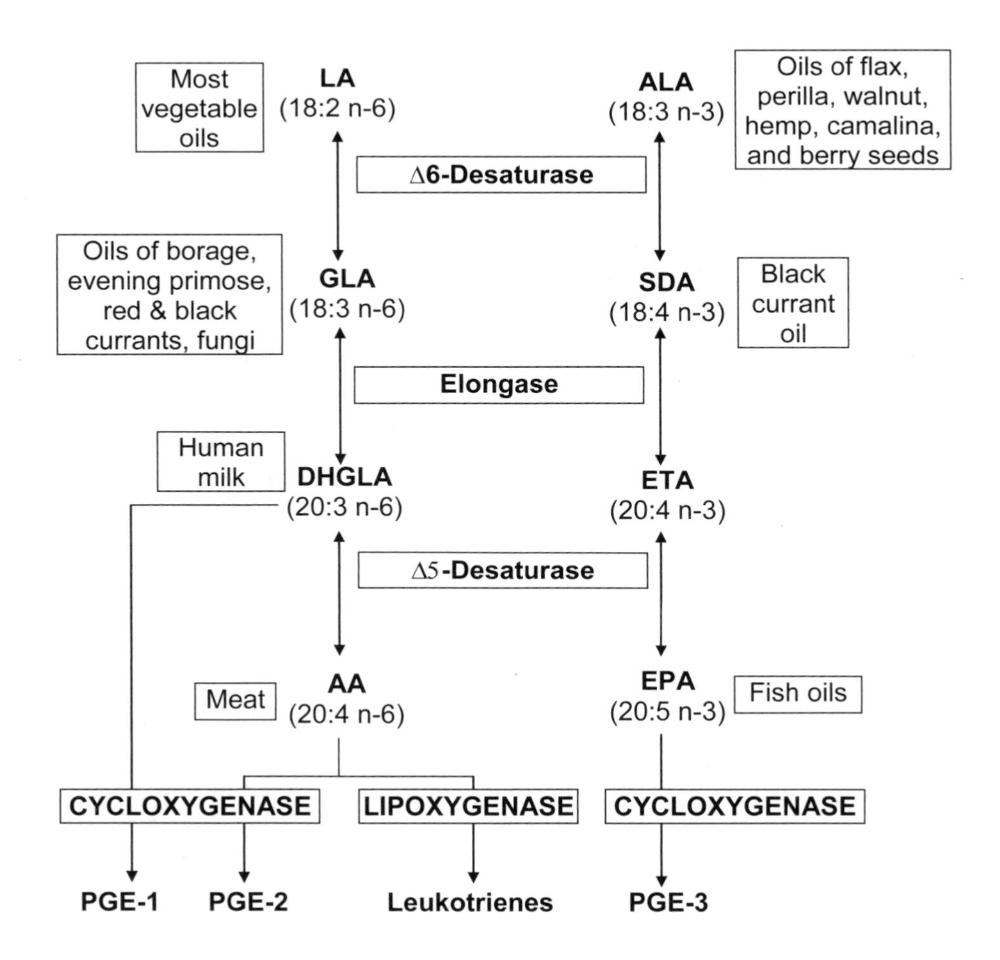


**Fig. I.1.** Dietary sources and metabolic conversions of essential fatty acids in humans. Abbreviations (http://en.wikipedia.org/wiki): LA (linoleic acid, 18:2 Δ9,12), GLA (gamma-linolenic acid, 18:3 Δ6,9,12), DHGLA (dihomo-gamma-linoleic acid, 20:3 Δ6,9,12), AA (arachidonic acid, Δ5,8,11,14), ALA (alpha-linolenic acid, 18:3 Δ9,12,15), SDA (stearidonic acid, 18:4 Δ6,9,12,15), ETA (eicosatetraenoic acid, 20:4 Δ8,11,14,17), EPA (eicosapenatenoic acid, 20:5 Δ5,8,11,14,17), PGE-1 (prostaglandin E1 or alprostadil), PGE-2 (prostaglandin E2 or dinoprostone), and PGE-3 (prostaglandin E3) .

**Table 1.3. Selected Gourmet and Health-Promoting Specialty Oils and Fats: Valuable Components, and Major Gourmet and Health-Promoting Properties.**

| Chapter | Oil | Most Abundant Fatty Acids | Most Valuable Unique Components | Gourmet Properties Reported | Major Proven and Potential Health-Promoting Properties |
|---|---|---|---|---|---|
| 1 | Olive oil (*Olea europaea*, Oleaceae) | 18:1>16:0>18:2 | 18:1, chlorophylls, squalene, phenolics, volatiles | Yes | CVD, inflammation |
| 2 | Avocado oil (*Persea americana*, Lauraceae) | 18:1>16:0>18:2 | 18:1, chlorophyll | Yes | CVD |
| 3 | Almond oil (*Prunus dulcis*, Rosaceae) | 18:1>18:2>16:0 | 18:1 | Yes | CVD |
| 3 | Brazil nut oil (*Bertholletia excelsa*, Lecythidaceae) | 18:2>18:1>16:0 | squalene | NR | NR* |
| 3 | Cashew nut oil (*Anacardium occidentale*, Anacardiaceae) | 18:2>18:1>16:0 | 18:1 | NR | NR* |
| 3 | Hazelnut oil (*Corylus avellana*, Betulaceae) | 18:1>18:2>16:0 | 18:1 | Yes | CVD |
| 3 | Macadamia nut oil (*Macadamia spp.*, Proteaceae) | 18:1>16:1>16:0 | 18:1 | Yes | NR* |
| 3 | Pecan nut oil (*Carya illinoinensis*, Juglandaceae) | 18:1>18:2>16:0 | 18:1 | Yes | NR* |
| 3 | Pine nut oil (*Pinus spp.*, Pinaceae) | 18:2>18:1>PLA | PLA | Yes | CVD |
| 3 | Pistachio nut oil (*Pistacia vera*, Anacardiaceae) | 18:1>18:2>16:0 | 18:1 | NR | NR* |
| 3 | Walnut oil (*Juglans spp.*, Juglandaceae) | 18:2>18:1>ALA | ALA | Yes | CVD |
| 4 | Flaxseed oil (*Linum usitatissimum*, Linaceae) | ALA>18:2>18:1 | ALA | Yes | CVD, inflammation |
| 4 | Perilla oil (*Perilla frutescens*, Lamiaceae) | ALA>18:2≥18:1 | ALA | NR | CVD, inflammation |
| 4 | Camelina oil (*Camelina sativa*, Brassicaceae) | ALA>18:2≥18:1 | ALA | NR | CVD, inflammation |
| 5 | Hempseed oil (*Cannabis sativa*, Cannabaceae) | 18:2>ALA>18:1 | ALA, GLA, SDA | Yes | CVD, anti-eczema, inflammation |
| 6 | Grapeseed oil (*Vitis spp.*, Vitaceae) | 18:2>18:1>16:0 | tocotrienols | Yes | NR |

*cont. on p. 8.*

**Table I.3., cont. Selected Gourmet and Health-Promoting Specialty Oils and Fats: Valuable Components, and Major Gourmet and Health-Promoting Properties.**

| Chapter | Oil | Most Abundant Fatty Acids | Most Valuable Unique Components | Gourmet Properties Reported | Major Proven and Potential Health-Promoting Properties |
|---|---|---|---|---|---|
| 6 | Gooseberry oil (*Ribes grossularia*, Grossulariaceae) | 18:2>GLA> ALA | GLA, ALA | NR | NR |
| 6 | Red currant oil (*Ribes rubrum*, Grossulariaceae) | 18:2> ALA>18:1 | GLA, ALA | NR | CVD |
| 6,7 | Black currant oil (*Ribes nigrum*, Grossulariaceae) | 18:2>GLA≥ ALA | GLA, ALA | NR | Immune function |
| 7 | Borage oil (*Borago officinalis*, Boraginaceae) | 18:2>GLA>18:1 | GLA | NR | Inflammation, blood pressure |
| 7 | Evening primrose oil (*Oenothera biennis,* Onagraceae) | 18:2>18:1>GLA | GLA | NR | Inflammation, CVD |
| 7 | Fungal oil (*Mucor javanicus* and others) | 18:1>GLA>18:2 | GLA | NR | CVD |
| 8 | Sesame oil (*Sesamum indicum*, Pedaliaceae) | 18:1≅18:2>16:0 | Lignans | Yes | CVD, inflammation |
| 9 | Niger seed oil (*Guizotia abyssinica*, Asteraceae) | 18:2>18:1≥16:0 | Vitamin K | NR | CVD, inflammation |
| 10 | Black cumin oil (*Nigella sativa*, Ranunculaceae) | 18:2>18:1>16:0 | Alkaloidal quinones | Yes | CVD, skin care |
| 11 | Camellia oil (*Camellia oleifera* and *C. sinensis*, Theaceae) | 18:1>18:2≥16:0 | 18:1, lignans | Yes | CVD, |
| 12 | Pumpkin seed oil (*Cucurbita pepo*, Cucurbitaceae) | 18:2>18:1>16:0 | volatiles | Yes | Prostate health |
| 13 | Wheat germ oil (*Triticum aestivum*, Poaceae) | 18:2>18:1≅16:0 | α-tocopherol | NR | CVD |
| 14 | Rice bran oil (*Oryza sativa*, Poaceae) | 18:1≥18:2>16:0 | γ-Oryzanols, tocotrienols | NR | CVD |
| 15 | Corn fiber oil (*Zea mays*, Poaceae) | 18:2>18:1>16:0 | Phytosterols | NR | CVD |
| 15 | Corn kernel oil (*Zea mays*, Poaceae) | 18:2>18:1>16:0 | Lutein and zeaxanthin | NR | Eye health |
| 16 | Oat oil (*Avena sativa*, Poaceae) | 18:1≥18:2>16:0 | Phenolic compounds | NR | Skin health, CVD |

**Table I.3., cont. Selected Gourmet and Health-Promoting Specialty Oils and Fats: Valuable Components, and Major Gourmet and Health-Promoting Properties.**

| Chapter | Oil | Most Abundant Fatty Acids | Most Valuable Unique Components | Gourmet Properties Reported | Major Proven and Potential Health-Promoting Properties |
|---|---|---|---|---|---|
| 17 | Barley oil (*Hordeum vulgare*, Poaceae) | 18:2>18:1≅16:0 | Tocotreinols | NR | Serum cholesterol |
| 18 | Parsley seed oil (*Petroselinum crispum*, Apiaceae) | 18:1>18:2>18:0 | 18:1, carotenoids | NR | CVD |
| 18 | Carrot seed oil (*Daucus carota*, Apiacaeae) | 18:1>18:2>16:0 | 18:1 | NR | CVD |
| 18 | Onion Seed oil (*Allium cepa*, Alliaceae) | 18:2>18:1>16:0 | tocopherols | NR | NR |
| 19 | Algal oils (example, *Crypthecodinium cohnii*, DHASCO® oil) | DHA>18:1v>14:0 | DHA, EPA, ARA | NR | CVD |
| 20 | Fish oils (example, *Brevoortia patronus*, Menhaden oil) | 16:0>18:1>EPA | EPA, DHA | NR | CVD, |
| 21 | Butter oil and Ghee (from *Bos spp.* and *Bubalus bubalis*) | 18:1≅16:0>18:0 | Short-chain FAs, flavor components, CLA | Yes | Anticancer (CLA) |

*Although health promoting properties have been demonstrated for these tree nuts, no health-promoting properties have been reported for the respective tree nut oils. Abbreviations: ALA, α-linolenic acid, 18:3 n-3; ARA, arachidonic acid 20:4 n-6; CVD, prevention of cardiovascular disease; DHA, docosahexanoic acid, 22:6 n-3; EPA, eicosapentanoic acid, 20:5 n-3; GLA, γ-linolenic acid, 18:3 n-6; PLA, pinolenic acid, 18:3 Δ5,19,12; SDA, stearidonic acid; 18:4 n-3: 18:1v, vaccenic acid, 18:1 n-7; NR, not reported

**Table I.4. Recommendations for total EPA and DHA Intake for Adults by Different Organizations**

| Organization | Recommendation (g/day) |
|---|---|
| American Heart Association | 0.5-1.0 |
| Belgium Health Council | 0.68 |
| British Nutrition Foundation Task Force | 1.0-1.5 |
| French Agency for Food Safety (AFSSA) | 0.10-0.12 |
| Institutes of Medicine (USA) | 0.27 |
| ISSFAL | 0.65 |
| UK Department of Health | 0.2 |
| WHO/FAO | 0.5 |

Sources: Givens and Gibbs (2008) and organization's websites.

another type of n-3 fatty acid, α-linolenic acid (ALA) (18:3 Δ9,12,15), which require more metabolic effort to produce EPA and DHA, but are still claimed to have many beneficial effects as a biologic precursor to long chain derivatives. SDA (stearidonic acid, 18:4 n-3), found in hempseed and some berry seed oils, requires less metabolic effort than ALA to produce EPA and DHA. The isomeric n-6 fatty acid, γ-linolenic acid (18:3 Δ6,9,12), which is found in borage, evening primrose, red and black currant, hempseed and fungal oils, may also promote health *via* the modulation of prostaglandin balance ($PGE_1$ *versus* $PGE_2$). A detailed discussion of the health issues of n-6 versus n-3 fatty acids was recently published (Lands, 2005).

As mentioned above, unrefined vegetable oils retain their non-acyl lipid constituents; these are also known as the unsaponifiable components. These include lipid-soluble vitamins, sterols, and other phenolic and aroma compounds. For example, the health benefits of olive oil are attributed to the complex effects of oleic acid, squalene, and a wide range of phenolic compounds. Other oils with characteristic minor lipid components include wheat germ oil (which contains α-tocopherol), grapeseed, rice bran, and barley oils (which contains tocotrienols), niger seed oil (which contains vitamin K), corn kernel oil and parsley seed oil (which contains carotenoids), rice bran oil (which contains γ-oryzanols), sesame and camellia oils (which contains lignans), and oat oil (which contains phenolic compounds).

## Chemical Composition and Oxidative Stability of Gourmet and Health-Promoting Oils

It is well known that the oxidative stability of oils and fats is primarily determined by their fatty acid composition with increased instability resulting from an increased degree of unsaturation. Thus, fish oils that contain high levels of EPA (eicosapentanoic acid, 20:5 n-3, *all-cis*-5,8,11,14,17-eicosa-5,8,11,14,17-pentaenoic acid) and DHA (docosapentenoic acid, 22:6 n-3, *all-cis*-docosa-4,7,10,13,16,19-hexaenoic acid) are extremely vulnerable to oxidation. Moderately vulnerable to oxidation are SDA (stearidonic acid:18:4 n-3), ALA (α-linolenic acid; 18:3(n-6), *all-cis*-9,12,15-octadecatrienoic acid, n-3) and GLA (γ-linolenic acid; 18:3 n-6, *all-cis*-6,9,12-octadecatrienoic acid, n-6)-containing oils. Primarily oleate-based oils are the most stable. In addition, the oxidative stability of fats and oils is significantly influenced by the presence of the minor anti- and pro-oxidant compounds. These components are especially important in the case of virgin and unrefined specialty oils.

## Authenticity and Adulteration

The gourmet and health-promoting oils are distinguished from the major commodity oils by the following characteristics:

- Gentle processing (gentle extraction or cold pressing and gentle or no refining)
- Unique flavor and/or aroma
- Unique health promoting properties
- Lower production amounts
- Higher prices

Because of the small amounts that are produced and their higher prices, adulteration of more expensive oils with cheaper oils is sometimes encountered. International institutions, such as the International Olive Council, are actively involved in preventing frauds and protecting olive oil consumers (IOC, 2006). The result is that olive oil is the strictest regulated edible oil in the world. Although the current challenges (addition of hazelnut oil or mild deodorized virgin olive oil) have their days numbered, other niches of adulteration seem to have appeared such as the authenticity of VOO (virgin olive oil) flavored with natural extracts, or the geographical traceability of virgin olive oil (García-González & Aparicio, 2006; Aparicio et al, 2007). With respect to the other edible oils, the detection of sesame oil in other oils can be performed colorimetrically by the Villavecchia or Baudouin tests, which are based on the reaction of sesamol with acidic furfural, assayed at 518 nm (Budowski et al, 1950). Adulteration of other vegetable oils with soybean oil can easily be identified by an increased level of α-tocopherol, and other vegetable oils may be identified by the presence of certain sterols or other unsaponifiable compounds.

## Allergic and Toxic Compounds

Although fats and oils are not normally considered to have allergic properties, some unrefined oils do contain small amounts of proteins and other compounds that may cause allergic reactions in some people. Several types of toxic compounds (including mercury, other heavy metals, pesticides, herbicides, PCBs, and dioxin) have also been detected in some edible oils, especially fish oils and fish oil capsules. These topics will be covered in detail in the individual chapters.

## Conclusions

The following twenty-one chapters of this book describe more than forty gourmet and health-promoting specialty oils (Table I.3). It should be noted that some oils that have been identified as "specialty" oils have unique properties that also make them valuable for industrial/non-edible applications (Gunstone, 2006) and these uses will not be covered in this volume. Although a number of books and book chapters have focused on individual lipids with gourmet and/or health-promoting properties, the approach of this book is to examine the natural oils themselves. Because some of these gourmet and health-promoting specialty oils contain colorful pigments or are obtained from interesting and beautiful plants, a series of color photos of many of these oils and plants materials appears in this book.

[†]Mention of trade names or commercial products in this publication is solely for the purpose of providing specific information and does not imply recommendation or endorsement by the U.S. Department of Agriculture.

### References

Aparicio, R.; R. Aparicio-Ruiz; D.L. García-González; Rapid methods for testing of oil authenticity: The case of olive oil. *Rapid Methods;* A. van Amerongen,D. Barug, M. Lauwars, Eds.; Wageningen Academic: Wageningen, Holland, 2007; pp. 163–188.

Bang, H.O.; J. Dyerberg; H.M. Sinclair; The composition of the Eskimo food in northwestern Greenland. *Am. J. Clin. Nutr.* **1980**; *33,* 2657-2661.

Budowski, P.; R.T. O'Connor; E.T. Field; Sesame Oil. IV. Determination of free and bound sesamol. *J. Am. Oil Chem. Soc.*, **1950,** *27,* 307-310.

FAO/WHO, Food and Agricultural Organization/World Health Organization of the United Nations. 1993. Joint FAO/WHO food standards programme: Codex Alimintarius Commission: Fats, oils and related products, Codex standard for named vegetable oils, Codex-STAN 210, volume 8, second edition (amended 2003, 2005), Rome, Italy.

García-González, D.L.; R. Aparicio; Olive Oil Authenticity: The Current Analytical Challenges. Lipid Technol. **2006**, *18*, 81-85

Givens, D.I.; R.A. Gibbs; Current intakes of EPA and DHA in European populations and the potential of animal-derived foods to increase them. *Proceedings of the Nutrition Society* **2008**, *67*, 273-280.

Gunstone, F.D.; Vegetable oils. *Bailey's Industrial Oil and Fat Products*, Vol. 1; 6$^{th}$ ed.; F. Shahidi, Ed.; Edible Oil and Fat Products: Chemistry, Properties, and Health Effects; John Wiley and

Sons: NY, 2005; pp 213-267.

Gunstone, F. D.; Minor Specialty Oils. *Nutraceutical and Specialty Lipids and their Co-Products*; Vol. 5; F. Shahidi, Ed.; Nutraceutical Science and Technology Series: Taylor and Francis; Boca Raton, 2006; pp 91-125.

IOC, International Olive Council. 2006. Trade Standard Applying to Olive Oil and Olive-Pomace Oil. COI/T.15/NC nº3/Rev. 2, Madrid Spain.

Lands, W.E.M., Fish, Omega-3 and Human Health, 2nd Edition, AOCS Press, USA, 220 pp, 2005.

McKevith, B; Nutritional aspects of oilseeds, British Nutrition Foundation. *Nutrition Bulletin* **2005**, *30*, 13–26.

Moreau, R.A., Corn Oil, F. Shahidi (ed), in "Bailey's Industrial Oil & Fat Products, Sixth Edition, Volume 2, Edible Oil & Fat Products: Edible Oils," Wiley-Interscience, Hoboken, pp 149-172, 2005.

Simopoulos, A.P.; Omega-6/omega-3 essential fatty acid ratio and chronic diseases. *Food Reviews International,* **2004,** *20*, 77-90.

# List of Color Illustrations

*This full-color photograph section has been included to show the wondrous variety of sources for the oils discussed in this volume.*

*Images courtesy of the chapter authors unless otherwise attributed.*

**Fig. 1.** Olive tree with fruit. See Chapter 1.

**Fig. 2.** Olives — from tree to toast. See Chapter 1.

Fig. 3. Olive oil. See Chapter 1.

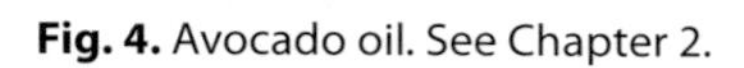

**Fig. 4.** Avocado oil. See Chapter 2.

**Fig. 5.** Brazil nut pod. See Chapter 3.

**Fig. 6.** Field of flax in bloom. See Chapter 4.

**Fig. 7.** Flaxseed. See Chapter 4. Image courtesy of the Flax Institute of Canada.

**Fig. 8.** Flaxseed oil and soft gels. See Chapter 4. Image courtesy of the Flax Institute of Canada.

**Fig. 9.** Hemp field. See Chapter 5.

**Fig. 10.** Hemp seed. See Chapter 5.

Fig. 11. Cranberry harvest. See Chapter 6. Image courtesy of the USDA-ARS.

Fig. 12. Evening primrose in bloom. See Chapter 7. Image courtesy of the USDA-ARS.

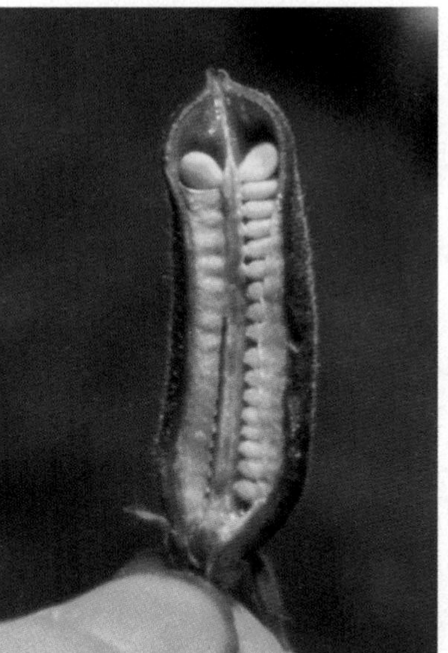

**Fig. 13.** Sesame flower and seed pods. See Chapter 8. Images courtesy of Prof. Jason Tzen.

**Fig. 14.** Niger seeds. See Chapter 9.

**Fig. 15.** *Nigella sativa*. See Chapter 10. Source: Wikimedia Commons.

Fig. 16. Camellia flower. See Chapter 11.

Fig. 17. Camellia fruit. See Chapter 11.

**Fig. 18.** Camellia seeds. See Chapter 11.

**Fig. 19.** Oil from Camellia seeds. See Chapter 11.

**Fig. 20.** Pumpkins in the field. See Chapter 12.

**Fig. 21.** Wheat kernels, wheat germ, unrefined wheat germ oil, and refined wheat germ oil (RBD). See Chapter 13.

Fig. 22. Rice in the field. See Chapter 14. Image courtesy of the USDA-ARS.

Fig. 23. Corn, corn fiber, and corn fiber oil. See Chapter 15.

**Fig. 24.** Oats and oat oil. See Chapter 16. Grain image courtesy of the USDA-ARS. Oat oil image courtesy of Ceapro, Inc.

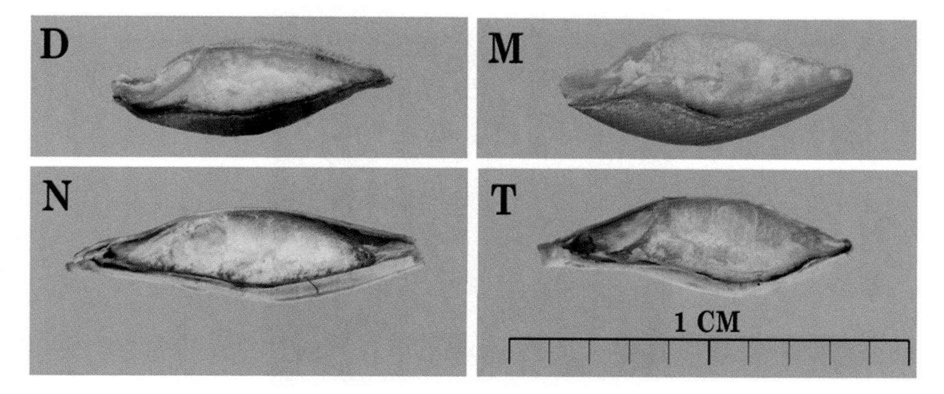


**Fig. 25.** Barley kernels. See Chapter 17.

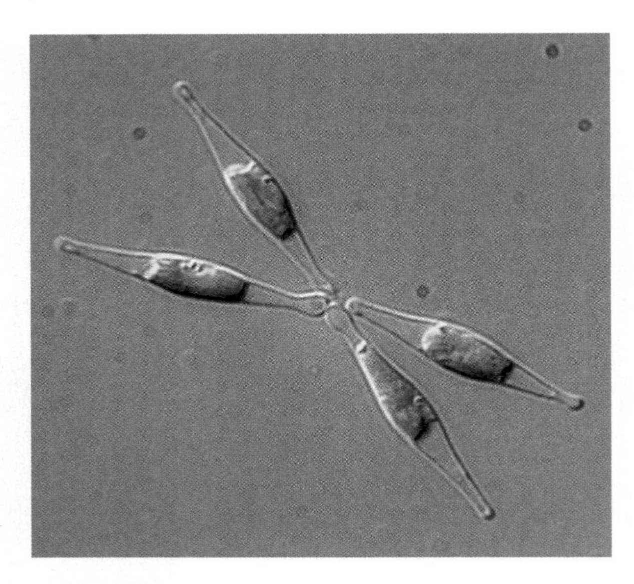

**Fig. 26.** *Phaeodactylum tricornutum*, an oil-rich microalgae. See Chapter 19. Image courtesy of Alessandra de Martino, and Chris Bowler, Stazione Zoologica, and Ecole Normale Supérieure.

**Fig. 27.** Carrot seed crop in bloom. See Chapter 18. Image courtesy of the USDA-ARS.

**Fig. 28.** School of mackerel, a fish used in the production of fish oil. See Chapter 20. Image courtesy of Dr. Jim Lyle.

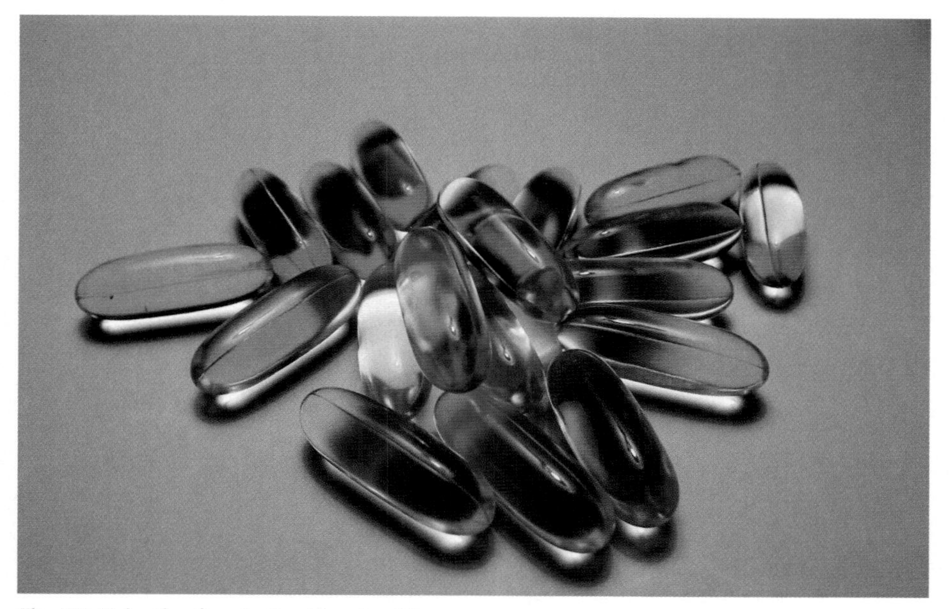

**Fig. 29.** Fish oil soft gels. See Chapter 20.

**Fig. 30.** Ghee: cool (left) and warm (right). See Chapter 21.

**Fig. 31.** Pure cow ghee. See Chapter 21.

# Olive Oil

**Diego L. García-González, Ramón Aparicio-Ruiz, and Ramón Aparicio**
*Instituto de la Grasa (CSIC), Padre García Tejero, 4, 41012, Sevilla, Spain*

## Introduction

The olive tree, one of the oldest known cultivated trees, is the symbol of friendship and peace; it also plays other social and religious roles described in Greek mythology and the Old Testament. More than 1275 autochthonous olive cultivars are identified and characterized (Bartolini et al., 1998) from an ancestor that is still subject to debate, but its origin is dated circa 5000 years ago though controversy also exists about its possible geographical origin in Mesopotamia. Nevertheless, Phoenicians and Greeks were responsible for the spread of the olive tree to Western regions where they traded.

The growth of olive oil cultivation was followed by the development of olive processing. The squeezing of olives in stone mortars, operated by early farmers, was not suitable for the increasing trade of olive oil in the Roman Empire. Romans decisively contributed to technological development with the milling crusher, which expedited the crushing process, and the wooden or iron manually activated screw press. They represented the major revolution in olive processing until Joseph Graham (1795) invented the hydraulic pressing system. The third revolution was in the second half of the twentieth century with the centrifugation system. The cost reduction of fully automated systems, which produce high-quality olive oils, the industrialization of agriculture, and the nutritional benefits of consuming olive oil abruptly increased olive oil demand and hence, olive oil production.

Nowadays, 600 million productive olive trees grow on the Earth and are spread out on 7 million ha, with pedoclimatic conditions such as those prevailing in the Mediterranean countries that account for not less than 97% of world production. The European Union is the major producer and also consumer, and surprisingly the number one exporter and the number two importer. Olive oil is marketed in accordance with the designations of the International Olive Council trade standards (IOC. 2006). Thus, virgin olive oil must be obtained by only mechanical or other physical means under conditions, particularly thermal, which do not lead to alterations in the oil. This oil has not undergone treatment other than washing, decantation, centrifugation, and filtration. Extra-virgin olive oil (EVOO) and virgin olive oil (VOO) are different edible grades of virgin olive oil. Lampante VOO is not fit for consumption and is intended for refining or for technical purposes. Refined olive oil (ROO) is the oil refined by methods that include neutralization, decolorization with bleaching earth, and deodorization. Olive oil is an edible blend of VOO and ROO. Olive–pomace oil is obtained by solvent extraction of the olive milling by-products; its triacylglycerol composition is similar to that of VOO, but some of the nonsaponifiable compounds (e.g., waxes) may differ significantly requiring the oil to be winterized before refining. Olive–pomace oil designation means a mixture refined olive-pomace oil with virgin olive oil. The result is that olive oil is the most controlled edible oil, overseen by numerous chemical standards and regulations (Table 1.1) that protect consumers against false copies.

The great diversity in the chemical profiles of olive oils is mainly a consequence of the numerous olive cultivars. In analyzing the constituents of olive fruit, moisture and oil content constitute 85–90% of pulp weight, while the rest is composed of organic matter and minerals. Pulp is rich in potassium as well as in major monosaccharides (e.g., xylose, galactose, mannitol, and glucose) and small amounts of organic acids. The skin, pulp, and seed contain different lipid fractions though the major components (acylglycerols) do not contribute much to olive oil characterization, but minor compounds account for 0.5–2.0% of the olive oil composition only. They consist of a series of compounds, such as sterols, alcohols, hydrocarbons, tocopherols, phenols, pigments, et cetera, and two hundred or so aromatic compounds. The large number of chemical compounds is the basis of olive oil authentication to protect against fraudulent blends and copies (Aparicio et al., 2007) as well as the determining of the VOO sensory quality by means of volatiles and phenols (Aparicio et al., 1996). Furthermore, the chemical composition—in particular the balance between mono- and polyunsaturated fatty acids and the content of phenols, tocopherols, and carotenes—explains the cardioprotective properties of the olive oil and its effects on several other pathways, including insulin sensitivity, blood pressure, inflammatory markers, and arterial wall function, among others.

## Processing

The primitive man who accidentally crushed fallen olives and noticed that they secreted oil was the first producer. Successive civilizations of the Mediterranean basin contributed to the technological development in olive oil processing (Di Giovacchino, 2000).

The modern concept of the processing system includes all the processes, from the harvesting of the olives to their crushing and extraction (Fig. 1.1). Olive picking is an important operation that contributes to VOO quality since its sensory quality depends on the health and ripeness of the harvested olives; olives infested by parasites or

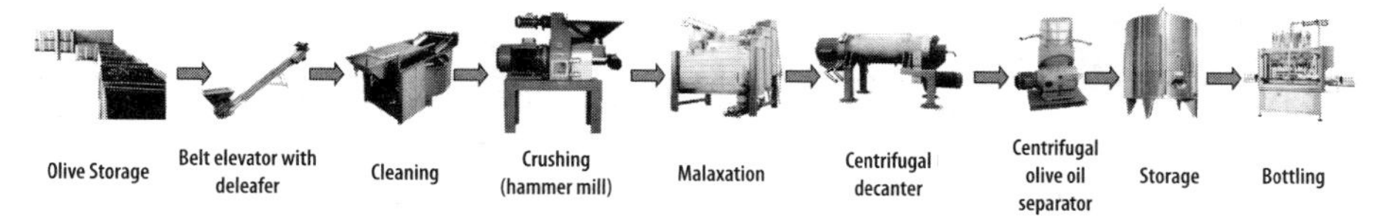


**Fig. 1.1.** Scheme of an automatic centrifugation system.

olives collected from the ground are processed to be refined. The olives are no longer hand-picked in most of the producing countries; manual or fully automated shakers are used. The percentage of olives dropped by the shakers is increasingly higher as the harvesting is delayed. However, the olive ripeness is also a determinant variable in the olive oil quality (Aparicio & Morales, 1998), and olive oil from overripened olives is characterized by a ripe flavor and a sweet taste. This oil is less cherished by consumers than an oil characterized with a slight bitter taste and a herbaceous scent. Thus, the harvesting time is a compromise between olive oil quality and harvesting efficiency.

The picked olives should be transported to olive mills in plastic cases with holes that allow air circulation to diminish the heat in the olives placed on the bottom layer; cases can vary in capacity from 25 kg to 300 kg. After transporting the olives, the olives are stored under optimal conditions from the moment of harvesting up to their processing in the mill. The time between both events should be less than two days without risking significant changes in the oil sensory profile.

The harvesting process does not prevent the presence of foreign materials with the olives that could be harmful to the machinery (e.g., stones) and modify the sensory attributes of the extracted olive oil. The processes of removing foreign materials and washing the olives are carried out as soon as the olives reach the olive mill. After they are weighed a sample of olives representing the whole batch is randomly taken for analytical analyses (humidity, free acidity, yield, sensory assessment). The washing step avoids the VOO being characterized with negative attributes such as muddy sediment and earthy, among others (Table 1.2).

Cleaned and washed batches of olives are then put on a moving belt and taken to the metallic crushers. They consist of a metallic body and a high-speed rotating piece of various shapes that throws the olives against a fixed or a mobile metal grating; hammers are the most common shape although other alternatives, toothed disks, cylinders, rollers, etcetera, are available. All the pieces are made of stainless steel. The olive crushing also affects the sensory profile of the resulting olive oil. A crusher with small holes in the grating gets high extraction yield, but the resulting oil is qualified with a bitter taste. The crushing process also activates various endogenous enzymes such as polyphenoloxidase and peroxidases that promote the oxidation of phenols. Furthermore, the metallic sensory attribute, an undesirable qualifier, may be detected in VOO if the inner parts of the crusher are covered with a thin layer of iron oxide that is solubilized by free fatty acids.

**Table 1.1. Olive Oil Trade Standards. Source: [a], IOC 2006**

| Designations | (1) | (2) | (3) | (4) | (5) | (6) | (7) |
|---|---|---|---|---|---|---|---|
| Extra Virgin Olive Oil | ≤0.05 | ≤0.05 | ≥1000 | ≤4.5 | ≤250 | ≤0.10 | ≤0.2 |
| Virgin Olive Oil | ≤0.05 | ≤0.05 | ≥1000 | ≤4.5 | ≤250 | ≤0.10 | ≤0.2 |
| Ordinary Virgin Olive Oil | ≤0.05 | ≤0.05 | ≥1000 | ≤4.5 | ≤250 | ≤0.10 | ≤0.2 |
| Lampante Virgin Olive Oil | ≤0.10 | ≤0.10 | ≥1000 | ≤4.5[a] | ≤300[a] | ≤0.50 | ≤0.3 |
| Refined Olive Oil | ≤0.20 | ≤0.30 | ≥1000 | ≤4.5 | ≤350 | - | ≤0.3 |
| Olive Oil | ≤0.20 | ≤0.30 | ≥1000 | ≤4.5 | ≤350 | - | ≤0.3 |
| Crude Olive-Pomace Oil | >0.20 | >0.10 | >2500 | >4.5[a] | >350[a] | - | >0.6 |
| Refined Olive-Pomace Oil | >0.40 | >0.35 | >1800 | >4.5 | >350 | - | >0.5 |
| Olive-Pomace Oil | >0.40 | >0.35 | >1600 | >4.5 | >350 | - | >0.5 |

| Designations | (15) | (16) | (17) | (18) | (19) | (20) | (21) |
|---|---|---|---|---|---|---|---|
| Extra-Virgin Olive Oil | ≤3.0 | ≤0.1 | ≤0.1 | ≤0.5 | ≤0.5 | ≤0.1 | ≤4.0 |
| Virgin Olive Oil | ≤3.0 | ≤0.1 | ≤0.1 | ≤0.5 | ≤0.5 | ≤0.1 | ≤4.0 |
| Ordinary Virgin Olive Oil | ≤3.0 | ≤0.1 | ≤0.1 | ≤0.5 | ≤0.5 | ≤0.1 | ≤4.0 |
| Lampante Virgin Olive Oil | ≤3.0 | ≤0.1 | ≤0.2 | ≤0.5 | ≤0.5 | ≤0.1 | ≤4.0 |
| Refined Olive Oil | ≤3.0 | ≤0.1 | ≤0.05 | ≤0.5 | ≤0.5 | ≤0.1 | ≤4.0 |
| Olive Oil | ≤3.0 | ≤0.1 | ≤0.05 | ≤0.5 | ≤0.5 | ≤0.1 | ≤4.0 |
| Crude Olive-Pomace Oil | - | - | - | ≤0.5 | ≤0.5 | ≤0.1 | ≤4.0 |
| Refined Olive-Pomace Oil | ≤3.0 | ≤0.1 | ≤0.05 | ≤0.5 | ≤0.5 | ≤0.1 | ≤4.0 |
| Olive-Pomace Oil | ≤3.0 | ≤0.1 | ≤0.05 | ≤0.5 | ≤0.5 | ≤0.2 | ≤4.0 |

| Designations | (28) | (29) | (30) | (31) | (32) | (33) | (34) |
|---|---|---|---|---|---|---|---|
| Extra-Virgin Olive Oil | ≤0.05 | 7.5-20.0 | 0.3-3.5 | ≤0.3 | ≤0.3 | 0.5-5.0 | 55.0-83.0 |
| Virgin Olive Oil | ≤0.05 | 7.5-20.0 | 0.3-3.5 | ≤0.3 | ≤0.3 | 0.5-5.0 | 55.0-83.0 |
| Ordinary Virgin Olive Oil | ≤0.05 | 7.5-20.0 | 0.3-3.5 | ≤0.3 | ≤0.3 | 0.5-5.0 | 55.0-83.0 |
| Lampante Virgin Olive Oil | ≤0.05 | 7.5-20.0 | 0.3-3.5 | ≤0.3 | ≤0.3 | 0.5-5.0 | 55.0-83.0 |
| Refined Olive Oil | ≤0.05 | 7.5-20.0 | 0.3-3.5 | ≤0.3 | ≤0.3 | 0.5-5.0 | 55.0-83.0 |
| Olive Oil | ≤0.05 | 7.5-20.0 | 0.3-3.5 | ≤0.3 | ≤0.3 | 0.5-5.0 | 55.0-83.0 |
| Crude Olive-Pomace Oil | ≤0.05 | 7.5-20.0 | 0.3-3.5 | ≤0.3 | ≤0.3 | 0.5-5.0 | 55.0-83.0 |
| Refined Olive-Pomace Oil | ≤0.05 | 7.5-20.0 | 0.3-3.5 | ≤0.3 | ≤0.3 | 0.5-5.0 | 55.0-83.0 |
| Olive-Pomace Oil | ≤0.05 | 7.5-20.0 | 0.3-3.5 | ≤0.3 | ≤0.3 | 0.5-5.0 | 55.0-83.0 |

(1): Trans oleic fatty acids (%)
(2): Sum of *trans* linoleic & linolenic fatty acids (%)
(3): Total sterol content (mg/kg)
(4): Erythrodiol and uvaol content (% total sterols).
(5): Wax content: C40+C42+C44+C46 (mg/kg).
(6): Stigmastadiene content (mg/kg)
(7): Difference between the actual and theoretical ECN42 triacylglycerol content
(8): Content of 2-glyceryl monopalmitate; *b*, C16:0 ≤ 14.0% and 2P ≤ 0.9% or C16:0 > 14:0% and 2P ≤ 1.0%;,*C*, C16:0 >14.0% and 2P ≤ 0.9% or C16:0 > 14:0% and 2P ≤ 1.1%.
(9): Absorbency in ultra-violet at $K_{232}$
(10): Absorbency in ultra-violet at $K_{270}$
(11): Absorbency in ultra-violet ($\Delta K$)
(12): Free acidity (%m/m expressed in oleic acid)
(13): Peroxide value (in milleq. peroxide oxygen per kg/oil)
(14): Moisture and volatile matter (% m/m)
(15): Trace of iron (mg/kg)
(16): Trace of copper (mg/kg)
(17): Insoluble impurities in light petroleum (% m/m)
(18): $\Delta^7$-Stigmastenol (%)
(19): Cholesterol (%)

| (8) | (9) | (10) | (11) | (12) | (13) | (14) |
|---|---|---|---|---|---|---|
| *B* | ≤2.50[a] | ≤0.22 | ≤0.01 | ≤0.8 | ≤20 | ≤0.2 |
| *B* | ≤2.60[a] | ≤0.25 | ≤0.01 | ≤2.0 | ≤20 | ≤0.2 |
| *B* | - | ≤0.30[a] | ≤0.01 | ≤3.3 | ≤20 | ≤0.2 |
| *C* | - | - | - | >3.3 | no limit | ≤0.3 |
| *C* | - | ≤1.10 | ≤0.16 | ≤0.3 | ≤5 | ≤0.1 |
| *B* | - | ≤0.90 | ≤0.15 | ≤1.0 | ≤15 | ≤0.1 |
| ≤1.4% | - | - | - | no limit | no limit | ≤1.5 |
| ≤1.4% | - | ≤2.00 | ≤0.20 | ≤0.3 | ≤5 | ≤0.1 |
| ≤1.2% | - | ≤1.70 | ≤0.18 | ≤1.0 | ≤15 | ≤0.1 |

| (22) | (23) | (24) | (25) | (26) | (27) |
|---|---|---|---|---|---|
| <Camp | ≥93.0 | - | Me=0 | Me>0 | - |
| <Camp | ≥93.0 | - | 0<Me≤2.5 | Me>0 | - |
| <Camp | ≥93.0 | - | 2.5<Me≤6.0[a] | - | - |
| - | ≥93.0 | - | Me >6.0 | - | - |
| <Camp | ≥93.0 | acceptable | - | - | *D* |
| <Camp | ≥93.0 | good | - | - | *E* |
| - | ≥93.0 | - | - | - | - |
| <Camp | ≥93.0 | acceptable | - | - | *F* |
| <Camp | ≥93.0 | good | - | - | E |

| (35) | (36) | (37) | (38) | (39) | (40) |
|---|---|---|---|---|---|
| 3.5-21.0 | ≤1.0 | ≤0.6 | ≤0.4 | ≤0.2 | ≤0.2 |
| 3.5-21.0 | ≤1.0 | ≤0.6 | ≤0.4 | ≤0.2 | ≤0.2 |
| 3.5-21.0 | ≤1.0 | ≤0.6 | ≤0.4 | ≤0.2 | ≤0.2 |
| 3.5-21.0 | ≤1.0 | ≤0.6 | ≤0.4 | ≤0.2 | ≤0.2 |
| 3.5-21.0 | ≤1.0 | ≤0.6 | ≤0.4 | ≤0.2 | ≤0.2 |
| 3.5-21.0 | ≤1.0 | ≤0.6 | ≤0.4 | ≤0.2 | ≤0.2 |
| 3.5-21.0 | ≤1.0 | ≤0.6 | ≤0.4 | ≤0.2 | ≤0.2 |
| 3.5-21.0 | ≤1.0 | ≤0.6 | ≤0.4 | ≤0.2 | ≤0.2 |
| 3.5-21.0 | ≤1.0 | ≤0.6 | ≤0.4 | ≤0.3 | ≤0.2 |

(20): Brassicasterol (%)
(21): Campesterol (%); Camp = %Campesterol
(22): Stigmasterol (%)
(23): The value of β-Sitosterol is calculated as sum of: $\Delta^{5,23}$-Stigmastadienol + Clerosterol + β-Sitosterol + Sitostanol + $\Delta^{5}$-Avenasterol + $\Delta^{5,24}$-Stigmastadienol.
(24): Organoleptic characteristics: odor & taste
(25) Organoleptic assessment: median of defect
(26): Organoleptic assessment: median of the fruity attribute
(27): Organoleptic assessment: color; *D*, light and yellow; *E*, light and yellow to green; *F*, light, yellow to brownish yellow.
(28): Myristic acid(% m/m methylesters)
(29): Palmitic acid (% m/m methylesters)
(30): Palmitoleic acid (% m/m methylesters)
(31): Heptadecanoic acid (% m/m methylesters)
(32): Heptadecenoic acid (% m/m methylesters)
(33): Stearic acid (% m/m methylesters)
(34): Oleic acid (% m/m methylesters)
(35): Linoleic acid (% m/m methylesters)
(36): Linolenic acid (% m/m methylesters)
(37): Arachidic acid (% m/m methylesters)
(38): Eicosenoic acid (% m/m methylesters)
(39): Behenic acid (% m/m methylesters)
(40): Lignoceric acid (% m/m methylesters)

**Table 1.2. Main Sensory Defects Produced in the VOO Extraction Process. The Causes of the Defects and Possible Volatiles Responsible for Them**

| Process | Defect | Cause | Volatiles |
|---|---|---|---|
| Harvesting | Grubby<br>Hay-wood | Infestation by *Dacus oleae* or *prays*<br>Olives that are dried out. | Ethanol and acids |
| Washing | Muddy<br>Earth | Uncleaned ground picked olives | Heptan-2-ol |
| Olive storing | Fusty<br>Winey<br>Mustiness | Olives in an advanced stage of anaerobic fermentation.<br>Olives stored in piles for a long time<br>Olives with fungi and yeasts. | Pentanoic, Butanoic and Acetic acids, Butan-2-ol, 2-Methyl-butan-1-ol 1-Octen-3-ol, 1-Octen-3-one |
| Crushing | Metallic | Uncleaned crusher in first days crop | ---- |
| Mixing | Heated | Excessive temperature (T≥35°C) or prolonged heating time (t >60min) | Aldehydes (Pentanal. Hexanal, Nonanal) |
| Pressure | Mats ("esparto") | Uncleaned mats or diaphragms | Not described |
| Decantation | Muddy sediment<br>Vegetable water | Olive oil in contact with wastewater for a long time | Ethyl acetate, Ethanol |
| Centrifugation | Greasy | Uncleaned horizontal decanters | Not described |
| Olive oil storage in deposits | Rancid | Inadequate storing conditions | Hexanal, Nonanal & long chain aldehydes and acids |
| Olive oil storage in bottles | Cucumber | Olive oil hermetically stored in plastic bottles for a long time | 2,6-Nonadienal |

Source: Angerosa, 2002; Morales et al. 2005.

A metallic crusher can cause emulsions with a negative effect on yield. The malaxation operation diminishes this effect by merging the droplets produced by crushing the olives to increase the percentage of available oil. The malaxation process involves stirring the olive paste slowly which aids in the coalescence of small drops into large ones and favors the breakage of the unbroken cells containing oil. The malaxators are two or three cylindrical vats intandem, each one with a rotating helix with several wings that mix the paste at low speed (15–20 rpm) for 30–75 min. The vats, made of stainless steel, have double walls for circulation of heated water. Temperature and time are variables of the malaxation that are automatically controlled. The temperature of the paste should not exceed 30°C to avoid a change in oil color (from yellow-green to reddish), an increase of acidity, and destruction of the volatile compounds. An increase of the mixing time contributes to the loss of the phenolic content by oxidation. New malaxators with a controlled atmosphere of around 2% oxygen content allow the extraction of VOO with better sensory and nutritional qualities (Servili et al., 2003), if the temperature is inside the range of 25–27°C and the whole process lasts less than 60 min (Aparicio et al., 1994).

After this step, the main constituents of olive paste are liquids (olive oil and olive mill wastewater-OMW) and solids such as small pieces of kernel and tissues. The next step is the separation of olive oil from the other constituents. Modern olive mills extract VOO by means of centrifugation systems since they allow the obtaining of high-quality olive oils with less production cost. Centrifugation is a continuous process that separates olive oil from the other materials (water and solids) by Stoke's law which states the speed at which two nonmiscible liquids are separated under the centrifugal force. Separation is carried out inside a decanter, a cylindrical bowl with a similarly shaped screw with helical blades that gyrate at 3500–3600 rpm. The small difference between the speed at which the bowl and the inner screw rotate results in the movement of the solid (olive pomace) to one end of the centrifuge, while oil and water (the other two constituents) are moved to the other end. The addition of lukewarm water increases fluidity and helps the separation of liquid and solid phases by centrifugal force. The addition of water also increases the amount of OMW (97.2 L/100 kg olives) that has negative environmental effects due to its resistance to biodegradation. Thus, a new decanter is implemented in modern olive mills to avoid OMW contamination. It is the so-called two-phase decanter, in opposition to the three-phase decanter, that does not need the water addition.

This new decanter does not show significant differences compared to the three-phase decanter concerning quality parameters, purity criteria, and sensory assessment (Table 1.3). The only exception is the bitterness perception due to a noticeably higher amount of phenols as a consequence of the absence of water during the centrifugation process. The main disadvantage is the production of a slurry of olive pomace (with 60–70% of water) that cannot be managed easily. Irrespective of the process used for the oil extraction, a final centrifugation of the produced VOO is needed to remove water and small solids from the oil. This process is carried out in vertical centrifuges that rotate at a high speed (6000–7000 rpm).

The whole process produces VOO, which is stored in large containers to protect against oxidation, and formation of by-products. The by-products are OMW and/or olive pomace and, the less important, twigs and leaves. Olive pomace is extracted by solvent (usually hexane) after a previous drying step. Olive pomace from two-phase decanters has to be heated at high temperature due to the moisture content, and it increases the cost and also the risk of generating polycyclic aromatic hydrocarbons (PAH). Nonedible grades of olive oil, such as lampante VOO and olive pomace oil

**Table 1.3. Chemical and Sensory Information of the Products Resulting from the Processing of Olives by Three- and Two-Phase Decanters. Range of Values Corresponds to the Processing of Several Single Cultivars Processed by Both Decanters Simultaneously**

| Chemical & Physical information from several cultivars | Three-phase decanters | Two-phase decanters |
|---|---|---|
| Extraction yield | 83.5-87.7 | 84.2-88.1 |
| Virgin olive oil | | |
| Acidity | 0.2-0.4 | 0.2-0.4 |
| Induction time (h) | 20.2-15.8 | 18.6-15.6 |
| Peroxide value (meq $O_2$/kg oil) | 3.4-3.9 | 3.7-4.5 |
| Chlorophylls pigments (ppm) | 6.2-7.3 | 6.8-7.6 |
| σ-tocopherol | 147-286 | 167-320 |
| $K_{232}$ | 1.517-1.562 | 1.493-1.538 |
| $K_{270}$ | 0.098-0.112 | 0.093-0.114 |
| Total phenols (mg/L gallic acid) | 301-363 | 240-326 |
| Sensory assessment[1] | 71-78 | 73-80 |
| Fruity perception | 53-71 | 55-77 |
| Bitterness perception | 37-46 | 44-56 |
| Taste intensity | 52-57 | 58-63 |
| Odor intensity | 44-45 | 43-26 |
| Grass flavor | 67-79 | 38-76 |
| Volatiles[2] | | |
| $C_6$-alcohols | 188-10815 | 1093-2964 |
| $C_6$-aldehydes | 802-17173 | 6140-34213 |
| $C_6$-acetates | 47-668 | 134-364 |
| Oil in by-products | 2.6-2.9 | 2.3-2.5 |
| Wastewater | | |
| Quantity (L/100 kg olives) | 87.6-97.2 | 7.6-8.5 |
| Oil (kg/100 kg olives) | 1.05-1.27 | 0.09-0.15 |
| Dry residue (kg/100 kg olives) | 8.1-8.5 | 1.1-1.3 |
| Olive-pomace | | |
| Quantity (kg/100 kg olives) | 48.6-50.9 | 70.8-72.7 |
| Moisture (%) | 47-2-51.6 | 54.2-59.6 |
| Oil (kg/100 kg olives) | 1.51-1.63 | 2.24-2.37 |
| Dry pomace (kg/100 kg olives) | 22.1-24.9 | 29.6-32.8 |

[1] Sensory assessment with hedonic scale of 100 mm;
[2] According to the method described by Aparicio et al. (1996)

(Table 1.1), are intended for refining. The process of refining follows all the steps applied to other edible oils to remove acid, color, and odor: degumming, neutralization, winterization, bleaching, and deodorization. Refining causes: (i) losses of tocopherols, phenols, and squalene, (ii) of sterols (≤15%), (iii) of linear alcohols from waxes, and (iv) the formation of conjugated double bonds and geometrical isomers, etcetera.

## Edible and Nonedible Applications

The olive is a versatile foodstuff which has historical applications which include its use as an edible oil, first and foremost, and table olive; and also has other nonedible uses, such as soap making, cosmetics preparation, ointments, perfumes, lubrication, or lamp oils. Today, research on new environmentally friendly olive oil extraction systems focuses on the investigation of applications of the resulting by-products which can vary from identifying new bioactive compounds to biodiesel production (Fig. 1.2).

### Edible Applications

VOO with a rich and fruity flavor is best for uncooked dishes and for dressing salads, soups, vegetables, and cold meats. Light, refined olive oils are perfect for frying or

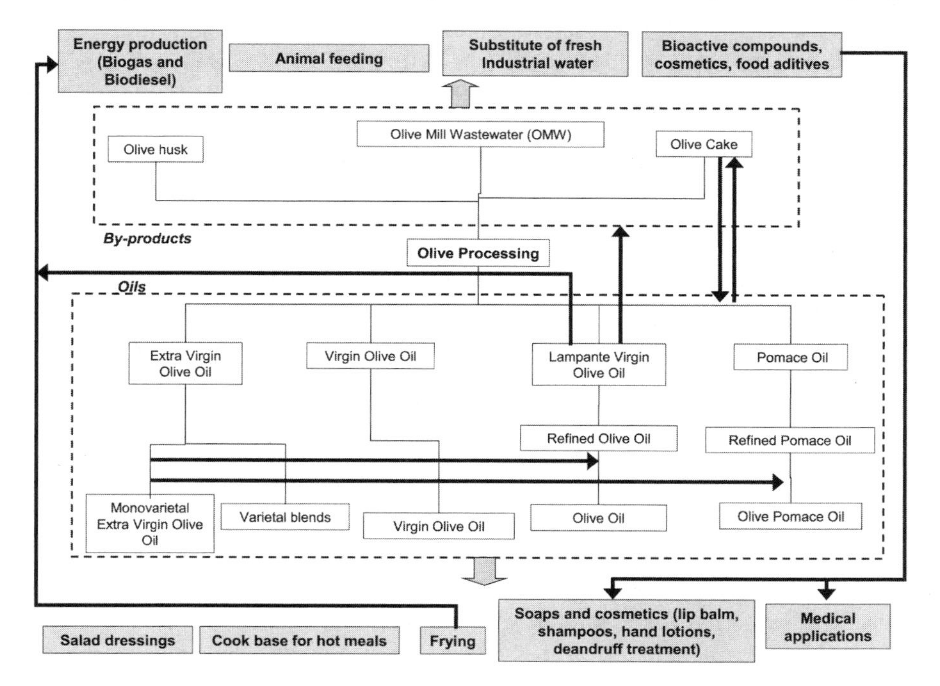


**Fig. 1.2.** Basic scheme of the main edible and nonedible applications of olive oil.

for making more delicate dishes such as mayonnaise in which the obtrusive flavor of EVOO would be overwhelming. Each recipe prescribes which category of olive oil should be used. The highest consumption rate in terms of kilograms per person per year is found in Greece (24 kg)—followed by Spain (13.1 kg) and Italy (12.5 kg)—with a sharp increase of 5–7% per year. The United States and Japan lead the consumption increase, 12% and 30%, respectively, among nonproducer countries (Mataix & Barbancho, 2006). In fact, the world olive oil demand has increased in the last decade due to the discovery of olive oil as an "everyday" cooking oil by new consumers who are abandoning the idea of buying olive oil as a gourmet oil to be used in selected meals only (García-Martínez, et al. 2002). However, standard olive oil needs to be repositioned and distanced from EVOO by giving customers enough cues to revise their opinions and consumption strategies. Furthermore, the awareness of consumers with a higher sensory quality and a clear geographical identity has encouraged producers to develop new products based on varietal VOOs, characterized by different sensory attributes and VOOs with a declared geographical origin (Protected Designation of Origin or PDO) that are controlled by a Regulatory Council when produced within the European Union. Today, olive oil authenticity, sold by PDOs and reputed sellers, is guaranteed, and the consumers' expectations of sensory quality are fulfilled when they buy a VOO bottle.

ROO and even VOO are used in fish canning (e.g., anchovy and tuna) because this combines the positive qualities of ω-3 polyunsaturated fatty acids and oleic acid from the coverage oil, which makes the result ideal for preventing cardiovascular diseases. In the case of VOO, a partitioning also exists toward the brine phase of major phenol compounds with all the advantages of these antioxidant compounds. On the other hand, producers recently released new products based on virgin or olive oil flavored with spices and herbs, from garlic to rosemary, and occasionally mixed with vinegar, ready to be used as a salad dressing or dip.

Deep- and pan-frying are the second most important applications of olive oil, an ancient culinary practice of the Mediterranean countries. The composition of olive oil, with a low content of saturated fatty acid and a high content of antioxidants (e.g., phenols), makes it an excellent oil to be applied in culinary processes that apply high temperatures, such as in frying and baking. After 10 hours of heating at 180°C, the total polar compounds of VOO can be lower than 25%, while other vegetable oils, such as sunflower oil or commercial blends intended for frying purposes, easily reach 29% of polar compounds (Kalantzakis et al., 2006). Furthermore, some varietal VOO (e.g., Picual) are even more resistant to thermoxidation, and after 25 hours of heating at 180°C, they only contain 15% of polar compounds (Brenes et al., 2002). Although some phenols (e.g., hydroxytyrosol and tyrosol-like substances) are dramatically decreased in the first hours of frying, other phenols (1-acetoxypinoresinol and pinoresinol) remain in the oil after 25 hours. Furthermore, high-stability olive oil has a low melting point that allows it to easily drain from fried foodstuffs, the other property appreciated in frying oils.

## Nonedible Applications

Nonedible applications of olive oil have existed since ancient times, and some of them remain today. Thus, olive oil is used in such preparations as lip balm, shampoo, and hand lotion, as well as in dry skin and dandruff treatments. Furthermore, olive oil contains antioxidant and anti-inflammatory compounds that make it suitable for topical applications in skin damage and dermatitis. Together with the traditional nonedible uses of olive oil, the reuse of by-products is being investigated to avoid environmental problems and also to reduce the high cost of olive oil production, thus increasing the competitiveness. The by-products are mainly obtained from olive cake, olive stones, and OMW by means of pyrolysis, combustion, and gasification processes, among others (Paraskeva & Diamadopoulos, 2006).

The olive cake can be an energy source: one can substitute 1100 tons of exhausted olive cake for 420 tons of No. 2 heavy fuel containing 4% sulfur (Masghouni & Hassairi, 2000). The use of this cake as an energy source is environmently friendly and avoids the emissions of sulfur. The cake can also be used for animal feeding or returned to the olive trees as mulch. Furthemore, the olive husk is used in molded products, plastics, in furfural manufacturing, and in activated carbon production, as well as fuel. OMW is also used as a growth medium for microbial production of enzymes (D'Annibale et al., 2006), algal biomass rich in polyunsaturated fatty acids (Anon., 2004), and for clay brick manufacturing (Mekki et al., 2006); its flocculated solid content is used as soil organic fertilizer (García-Gómez et al., 2003). Furthermore, various compounds are extracted from OMW and other by-products to be used as food additives, cosmetics, and medicines; OMW is, for example, a source of antioxidants as it contains approximately 53% of the olive phenols. Research efforts were recently made to develop cost-effective approaches to efficiently recover phytosterols from OMW and leaves, and to obtain pure crystals from the resulting sterol concentrate slurries.

# Acyl Lipids and Fatty Acids Composition

## Lipid Biosynthesis in Olives

Fatty acids account for up to 85% of olive oil total composition in the form of triacylglycerols (TAG). Thus, in discussing olive oil composition, one should first consider the biosynthesis of fatty acids. Labeling experiments demonstrated the existence in the pulp (pericarp) of developing olives of two carboxylating mechanisms —both in the light and dark—that involve the activity of three enzymes: ribulose-1,5-bisphosphate carboxylase, phosphoenolpyruvirate carboxylase, and NAD-malate dehydrogenase (Sánchez & Salas, 1997). Studies under autotrophic and heterotrophic conditions indicated that oil formation in olives requires the contribution of the two sources of reduced carbon, leaf, and fruit photosynthesis. Furthermore, experiments using either radio labeled acetate or pyruvate showed that both precursors are efficiently incorporated into acyl lipids by tissue slices from developing olives, suggesting that two operative pathways are present in the olive pulp for the formation of acetyl-CoA, the precursor for fatty acid biosynthesis. One pathway is based on the breakdown of six carbon sugars via glycolysis in the plastid, while the other involves collaboration with mitochondria (Harwood & Sánchez, 2000).

De novo fatty acid biosynthesis in olive fruit, as in most oil crops, involves the concerted activity of two enzymes, acetyl-CoA carboxylase and fatty acid synthase, that catalyze multiple reactions. In olive fruit, two isoforms of acetyl-CoA carboxylase exist, the plastid-localized isoform being a multienzyme complex. The fact that dicotyledons retain a multifunctional protein in the epithelial cytosol is important in providing malonyl-CoA for fatty acid elongation. Once malonyl-CoA is generated, it is used to form malonyl-ACP for fatty acid synthesis whose reactions comprise seven steps. Sánchez & Harwood (1992) studied the overall fatty acid synthase activity in soluble fractions of olive pulp by using radiolabeled malonyl-CoA as the precursor. They concluded that the activity was stimulated by ACP and inhibited by cerulenin, and it showed a strong dependence on NADH, NADPH, and thiol reagents. Chain termination of fatty acid synthesis is achieved through the activity of acyl-ACP thioesterase that shows better activity with C18 (oleoyl)-ACP. Further desaturations to produce polyunsaturated fatty acids occur either within the plastid or on the endoplasmic reticulum. Those fatty acids produced in the plastid are channeled out of the organelle to form the acyl-CoA pool in the cytosol. Further desaturation of oleate to linoleate and then to α-linoleate take place via phosphatidylcholine substrates on the endoplasmic reticulum.

The incorporation of fatty acids into complex lipids occurs via the well-known Kennedy pathway, which has a series of four reactions that yield triacyglycerols as the end product. In the case of olives, the incorporation of acyl-CoAs into TAGs by microsomes is mediated by the initial incorporation into phosphatidylcholine; oleoyl-CoA was found to be a better substrate for glycerolipid synthesis than palmitoyl-CoA (Sánchez et al., 1992). Furthermore, stereospecific analyses of olive oil showed that linoleonyl-CoA is also an effective substrate for olive G3PAT, while in vivo labeling experiments, using $14^{C}$-acetate and tissue slices, showed TAG formation is strongly reduced at temperatures above 40°C with a commensurate rise in the relative labeling of diacylglycerols (Rutter et al., 1997) that might be indicative of the limiting activity of DAGAT at high temperature.

Fleshy pericarp accumulates the vast majority of lipids of the olive, but TAGs are also stored within the seeds in oil bodies. Ultrastructural studies of olive pulp showed that no such oil bodies are observed in mature fruits, as they fuse on contact to form oil droplets during the fruit maturation. In fact, new oil bodies appear to form then, and they fuse with oil droplets to result in the formation of large oil droplets.

## Olive Oil Chemical Composition

Olive oil chemical composition is serarated, by a very simple classification, into major and minor compounds. The set of major compounds is made up primarily of triacylglycerols, and the set of glyceridic compounds is made up of free fatty acids and mono- and diglycerols (MAG and DAG). Other fatty acid derivatives such as phospholipids, waxes, and esters of sterols are traditionally classified into the set of minor compounds. The set of minor compounds forms an ample and heterogeneous group of compounds such as pigments, phenols, carotenoids, and fatty acid derivatives as phospholipids, waxes, and esters of sterols among others. Most of them are included in the unsaponifiable matter and have the common characteristic of being obtained by the saponification procedure (Morales and León, 2000).

## Acyl Lipids and Fatty Acids Composition

Olive oil, as most edible oils, mainly consists of triacylglycerols (TGs) that are glycerol esters of fatty acids. Fatty acids contribute 94–96% of the total weight of the TGs, and they follow a pattern whereby those in the 2-position are unsaturated, with linolenic acid being favored more than oleic and linoleic acids. Twenty TGs are identified and independently quantified, but only five are present in significant proportions: OOO (27.53–59.34%), POO+SOL (12.42–30.57%), OOL+LnPP (4.14–17.46%), POL+SLL (2.69–12.31%), and SOO+OLA (3.17–8.39%). TGs are comprised of most of the twelve identified fatty acids (C14–C24) though only six are major compounds: palmitic (6.30–20.93%), palmitoleic (0.32–3.52%), stearic (0.32–5.33%), oleic (55.23–86.64%), linoleic (2.7–20.24%), and linolenic (0.11–1.52%). The wide ranges of TGs and FAs show the ample variability of the olive oil chemical composition that is largely dependent on the cultivar (Table 1.4) and the geographical origin, and, to a certain extent, on olive ripeness and the extraction system. Furthermore, the determination of the equivalent carbon numbers (ECN) was shown to be decisive in authenticating olive oil (Table 1.1). Olive oil is characterized by ECNs from 44 to 50. Thus, additions to olive oil of as little as 2.5% of edible oils rich in linoleic acid (corn, sunflower, and soybean) are detected as lower ECN values.

Olive oil also has partial glycerides as a result of incomplete TG biosynthesis and hydrolytic reactions. Diacylglycerols (DAGs) are major compounds of the polar fraction in VOOs (1–3%), and they are found as 1,2- and 1,3-isomers. 1,2-DAGs are attributed to TGs' incomplete biosynthesis (Kennedy pathway), whereas 1,3-DAGs are attributed to TGs enzymatic or chemical hydrolysis; in fact, the amount of 1,3-TAGs is higher in cloudy VOOs. Monoacylglycerols (MAGs) are present in lesser amounts (≤0.25%), and their major constituents are glycerol oleate, linoleate, and palmitate. The amount of 1,2-DAGs decreases, while the amount of 1,3-DAGs increases during VOO storage. Based on this fact, the ratio between these isomers indicates VOO freshness, and acts as a hypothetical marker of the presence of deodorized olive oil

**Table 1.4.** Contents (mean ± SD, as %) of the Most Abundant Fatty Acids, Triacylglycerols and Their Equivalent Carbon Numbers (ECN) Determined in the Most Marketed Spanish VOOs

| Fatty acids/ triglycerides/ECNs | Variety | | | |
|---|---|---|---|---|
| | Cornicabra | Picual | Hojiblanca | Arbequina |
| C16:0 | 9.22±0.17 | 10.6±0.78 | 9.68±1.00 | 13.7±0.99 |
| C16:1 | 0.77±0.11 | 0.91±0.13 | 0.73±0.15 | 1.42±0.24 |
| C17:1 | 0.10±0.01 | 0.11±0.03 | 0.23±0.04 | 0.26±0.02 |
| C18:0 | 3.36±0.29 | 3.49±0.47 | 3.48±0.21 | 2.03±0.19 |
| C18:1 | 80.42±0.96 | 78.93±1.62 | 76.61±1.54 | 70.62±1.70 |
| C18:2 | 4.46±0.57 | 4.53±1.14 | 7.51±1.13 | 10.33±0.87 |
| C18:3 | 0.62±0.08 | 0.67±0.05 | 0.75±0.04 | 0.61±0.05 |
| OLL+PoOL | 0.75±0.16 | 0.68±0.23 | 1.74±0.44 | 2.42±0.38 |
| OOLn+PLL+PoPoO | 1.38±0.15 | 1.44±0.12 | 1.76±0.15 | 2.22±0.31 |
| OLO+LnPP | 7.79±0.91 | 7.16±1.10 | 11.72±1.60 | 13.93±0.78 |
| PoOO | 1.15±0.20 | 1.38±0.26 | 0.77±0.32 | 1.49±0.31 |
| POL+SLL | 2.69±0.32 | 2.86±0.43 | 4.26±0.66 | 7.72±0.78 |
| SPoL+SOLn | 0.20±0.04 | 0.26±0.04 | 0.37±0.10 | 1.01±0.21 |
| PPL | 0.21±0.04 | 0.24±0.05 | 0.41±0.06 | 0.40±0.04 |
| OOO | 51.72±1.84 | 48.51±1.63 | 45.42±2.94 | 35.53±2.61 |
| SOL+POO | 20.81±1.33 | 22.83±1.37 | 20.12±1.63 | 23.22±0.89 |
| PSL+PPO | 2.21±0.29 | 2.96±0.38 | 2.54±0.66 | 4.07±0.61 |
| OLA+SOO | 6.76±0.58 | 6.87±0.70 | 6.29±0.53 | 3.42±0.38 |
| OOA | 0.93±0.06 | 0.73±0.06 | 0.74±0.06 | 0.56±0.08 |
| ECN42 | 0.18±0.04 | 0.29±0.07 | 0.37±0.09 | 0.49±0.14 |
| ECN44 | 2.57±0.30 | 2.65±0.32 | 4.03±0.60 | 5.36±0.72 |
| ECN46 | 12.52±1.08 | 12.52±1.33 | 17.92±1.98 | 25.41±1.45 |
| ECN48 | 74.73±1.62 | 74.21±2.09 | 68.03±2.26 | 62.73±2.06 |
| ECN50 | 8.68±0.70 | 8.92±0.83 | 8.26±0.78 | 4.95±0.51 |
| ECN52 | 1.31±0.11 | 1.20±0.13 | 1.10±0.10 | 0.81±0.14 |

in VOO. Free fatty acids (FFAs) are fatty acids that are produced from TGs by hydrolytic reactions in any of the steps of the process. They are markers of the substandard processing of olives (e.g., infestation of olive fly) or poor handling during olive processing, and acidity is the basic criterion for olive-oil quality classification (Table 1.1).

## Minor Compounds

As stated, this ample and heterogeneous set of minor compounds includes several series of chemical compounds. These compounds originate from the unsaponifiable matter (e.g., sterols, alcohols, hydrocarbons) and are also derived from lipids (e.g., phospholipids, waxes); other compounds exist that are not related with lipids from a chemical viewpoint (e.g., pigments). By unsaponifiable matter, we mean the set of natural or accidental constituents that fail to react with NaOH and KOH to produce soaps while remaining soluble in classic fat solvents (hexane, ether) after saponification. Innumerable procedures can carry out the separation of the unsaponifiable constituents. Analytical methods are described in Morales and León-Camacho (2000).

## Sterols

Sterols make up an extensive series of compounds that are grouped into three classes: 4-demethylsterols, commonly called phytosterols; 4,4-dimethylsterols, called triterpenic alcohols; and 4-monomethylsterols, named methylsterols. The 4-demethylsterols identified in olive oil are: cholesterol, 24-methylenecholesterol, campesterol, campestanol, stigmasterol, $\Delta^7$-campesterol, $\Delta^{5,23}$-stigmastadienol, chlerosterol, β-sitosterol, sitostanol, $\Delta^5$-avenasterol, $\Delta^{5,24}$-stigmastadienol, $\Delta^7$-stigmastenol, and $\Delta^7$-avenasterol. However, some of them are not quantified in VOOs but appear in ROOs due to the refining process. This is the case of $\Delta^{5,23}$-stigmastadienol. Furthermore, $\Delta^{5,24}$-stigmastadienol increases its concentration in the refining process while the concentration of $\Delta^5$-avenasterol decreases. These compounds are used to distinguish ROO from VOO. The concentration of 4-demethylsterols also varies with cultivars, allowing their characterization (Aparicio, 2000). Table 1.5 shows the range of concentration of 4-demethylsterols in some European VOOs.

4,4-Dimethylsterols have a much more complex composition and, in fact, there is a large number of partially identified compounds (Azadmard-Damirchi & Dutta, 2006). Thirteen compounds (taraxerol, dammaradienol, β-amyrin, 28-nor-5α-olean-17-en-3β-ol, butirospermol, α-amyrin, 24-methylene-lanostenol, a $\Delta^8$-Sterol, cycloartenol, a $\Delta^7$-Sterol, a $\Delta^7$-Sterol, 24-methylene-cycloartanol, and a $\Delta^7$-Sterol) were used to characterize European VOOs by their geographical origin and cultivars (Aparicio, 2000). The presence of two penta cyclic triterpenic alcohols, erythrodiol and uvaol, has been also observed in the unsaponifiable matter of both olive–pomace oil and olive oil. These triterpene dialcohol compounds are used to detect the presence of olive–pomace oil in olive oil. With respect to 4-monomethylsterols, they are intermediate products in the biosynthesis of phytosterols and seven of them (gramisterol, ethyllophenol, fridelanol, and obtusifoliol) have been used in the characterization of European VOOs (Aparicio, 2000).

## Fatty Alcohols

Fatty alcohols are an important class of olive oil minor constituents because they are used as a criterion to differentiate various olive oil designations (Table 1.1). The main linear alcohols present in olive oil are docosanol, tetracosanol, hexacosanol, and oc-

**Table 1.5.** Ranges of Concentration (mg/kg) of Some 4-Demethylsterols, 4,4-Dimethyl Sterols and 4-Monomethyl Sterols in VOOs from Different European Geographical Origins

| 4-Demethyl sterols | | 4,4-Dimethyl sterols | | 4-Monometyl sterols | |
|---|---|---|---|---|---|
| Cholesterol | 5-8 | Taraxerol | 3-25 | Cycloeucalenol | 30-110 |
| Campesterol | 25-114 | β-amyrin | 8-120 | Citrostadienol | 17-228 |
| Campestanol | tr-5 | Butirospermol | 6-123 | 24-dimethyl-24-cholestadienol | 6-22 |
| Stigmasterol | 5-67 | 24-methylene-lanostenol | 2-35 | Gramisterol | 3-26 |
| β-Sitosterol | 683-2610 | α-amyrin | 10-100 | 24-Ethyllophenol | 12-53 |
| $\Delta^5$-Avenasterol | 34-266 | Cycloartenol | 36-600 | Fridelanol | nd-115 |
| $\Delta^{5,24}$-Stigmastadienol | 1-4 | 24-methylene-cycloartanol | 203-2190 | Obtusifoliol | 1-59 |
| $\Delta^7$-Stigmastenol | tr-2 | Cyclobranol | 1-200 | | |
| $\Delta^7$-Avenasterol | tr-2 | Erythrodiol | 7-92 | | |

Note: tr, trace; nr, not reported. Source: Aparicio and McIntyre, 1998; Aparicio, unpublished data.

tacosanol; the odd carbon atom alcohols (tricosanol, pentacosanol, heptacosanol) are quantified at trace levels. Total aliphatic alcohol content does not usually exceed 350 mg/kg in olive oil (Table 1.6) while it is ten times higher in olive-extracted oil.

## Hydrocarbons

Small quantities of normal, saturated, terpenic, and even aromatic hydrocarbons are always present in VOOs. Squalene is the most important hydrocarbon, and it accounts for around 50% of the unsaponifiable matter. During the refining processes (decoloration and deodoration), sterols undergo dehydration reactions that result in the corresponding unsaturated hydrocarbons possessing a steroidal nucleus. Thus, β-sitosterol produces 3,5-stigmastadiene, campesterol produces 3,5-campestadiene, and stigmasterol produces 3,5,22-stigmastatriene. The quantification of these hydrocarbons allows the detection of the presence of any refined edible oils in VOO. Hydrocarbons other than squalene (e.g., pentacosane, hexacosane, heptacosane, octacosane) are used for characterizing varietal VOO and VOO geographical traceability (Aparicio, 2000).

## Tocopherols

Tocopherols are heteroacid compounds with high molecular weight that are designated as α-, β-, γ-, δ-tocopherols. Only the first three are quantified in VOO in percentages that vary from 52–87% for α-tocopherol to 15–20% and 7–23% of β- and γ-tocopherols, respectively. The series of unsaturated analogs, tocotrienols, is not present in olive oil. Tocopherols contribute to the antioxidant properties of olive oil, and their profile is often used as a criteria of purity. The concentrations of these compounds are higher in olive-pomace oil (OPO). Thus, the range of their concentration (mg/kg) in European oils is: α-tocopherol (VOO: 38–412 versus OPO: 200–600),

**Table 1.6. Content of Aliphatic Alcohols (mean ± SD in mg/kg) of VOOs Harvested in Several Producer Regions of Spain and Italy**

| | Docosano | Tetracosano | Hexacosano | Octacosano |
|---|---|---|---|---|
| Andalusia (SP) | 33.4±0.9 | 61.7±1.7 | 88.8±2.2 | 27.9±0.7 |
| Extremadura (SP) | 48.8±5.1 | 72.0±8.5 | 75.5±7.1 | 33.6±6.4 |
| La Mancha (SP) | 39.9±4.8 | 57.1±7.0 | 48.5±2.7 | 20.3±1.7 |
| Valencia (SP) | 48.8±4.4 | 93.0±9.5 | 85.1±10.2 | 46.9±6.5 |
| Catalonia (SP) | 62.2±8.3 | 104.8±10.9 | 130.6±16.4 | 51.5±7.9 |
| Aragón (SP) | 29.0±3.6 | 48.8±4.6 | 53.9±4.5 | 25.5±1.9 |
| Basilicata (IT) | 23.7±2.6 | 34.7±3.6 | 56.1±5.1 | 28.3±2.4 |
| Liguria (IT) | 39.0±3.1 | 42.8±5.6 | 39.5±4.0 | 15.0±1.7 |
| Tuscany (IT) | 37.7±5.1 | 62.9±9.7 | 69.5±9.0 | 34.7±4.8 |

Source: Aparicio, unpublished data.

β-tocopherol (VOO: 0-17 versus. OPO: 1–6), and γ-tocopherol (VOO: 0–31 versus. OPO: 0–15).

## Phenols

Phenolic components have a great impact on the sensory and nutritional qualities of olive oil (Morales & Tsimidou, 2000), and they are responsible for some of the well known health benefits of olive oil (Covas, 2007). Furthermore, their presence in olive oil is related to its oxidative stability (Luna et al., 2006a). The content of phenols in VOOs varies from a few mg/kg to more than 500 mg/kg oil, expressed as caffeic acid equivalents; VOOs with concentrations that exceed 300 mg/kg have an intense bitter taste. Phenols are also found in olive leaves, and supplementing refined olive oils with phenolic extracts could increase their shelf life and could give them more color and bitter taste. The list of phenols identified in VOO includes: 4-acetoxy-ethyl-1,2-dihydroxybenzene, 1-acetoxy-pinoresinol, apigenin, caffeic acid, cinnamic acid (not a phenol), *o*- and *p*-coumaric acids, elenolic acid (not a phenol), ferulic acid, gallic acid, homovanillic acid, *p*-hydroxybenzoic acid, hydroxytyrosol, 1-(3'-methoxy-4'-hydroxy-phenyl)-6,7-dihydroxy-isochroman, 1-phenyl-6,7-dihydroxy-isochroman and derivatives, luteolin, oleuropein, pinoresinol, protocatechuic acid, sinapic acid, syringic acid, tyrosol, and derivatives (Boskou, 2006; Young et al., 2007). Table 1.7 shows the concentration of some phenols in four Spanish varietal VOOs.

## Pigments

Chlorophyll and carotenoid pigments are responsible for VOO color, ranging from yellow-green to greenish gold. The chlorophyll pigments are responsible for VOO greenish hues, with pheophytin (phenophytina) being the major compound. The major "yellow" pigments are the carotenoids, lutein, and β-carotene. But the amount of pigments greatly depends on the cultivar, olive ripeness, and olive oil processing system and storage conditions. Olive oils from cv. Arbequina, for instance, are characterized by high concentrations of xantophylls, and this feature can be used for their chemotaxonomic differentiation. These compounds also play an important role in health, food traceability, authenticity, and oxidation stability, acting as antioxidants in the dark and pro-oxidants in the light.

Carotenoids are pigments with the structure of tetraterpenoids, and their concentration does not usually exceed 10 mg/kg (Boskou, 2002). Lutein is the major carotenoid followed by β-carotene and other xantophylls (neoxanthin, violaxanthin, anteraxanthin, luteoxanthin) (Table 1.8). The presence of xantophylls esterified with fatty acids helps the stability of carotenoids. Carotenoids stabilize the color of the oils and ensure a sufficient level of antioxidant activity. The ripeness stage of the olives is determined by the ratio of minor carotenoids to lutein. Furthermore, the lutein/β-carotene ratio distinguishes Spanish VOOs from Italian and Greek ones that usually have values lower than 1.3.

**Table 1.7. Concentrations (mg/kg) of Phenols Quantified in Spanish Varietal VOOs**

| | Variety | | | |
|---|---|---|---|---|
| Compound | Picual | Arbequina | Hojiblanca | Cornicabra |
| Hydroxtyryosol | 110.1 | 40.8 | 88.5 | 57.3 |
| hydroxytyrosol acetate | 38.0 | 63.0 | 56.0 | 3.5 |
| Hy-EDA | 133.7 | 235.2 | 274.6 | 227.3 |
| Hy-EA | 264.7 | 23.4 | 196.2 | 129.1 |
| Tyrosol | 97.0 | 32.8 | 51.9 | 93.6 |
| Ty-EDA | 326.1 | 414.6 | 272.4 | 795.6 |
| Ty-EA | 199.5 | 28.0 | 76.0 | 167.9 |
| 1-Acetoxypinoresinol | 13.9 | 119.9 | 71.2 | 18.2 |
| Pinoresinol | 103.8 | 106.0 | 51.7 | 110.1 |
| Others | 9.8 | 15.4 | 15.0 | 5.6 |
| Total polyphenols | 1299.9 | 1083.4 | 1159.3 | 1610.5 |

Source: García et al., 2003. Note: Hy-EDA dialdehydic form of decarboxymethyl oleuropein aglycon, Hy-EA oleuropein aglycon, Ty-EDA dialdehydic form of decarboxymethyl ligstroside aglycon, Ty-EA ligstroside aglycon.

**Table 1.8. Main Concentration (μg/kg) of Individual Carotenoid and Chlorophyll Pigments Quantified in Several Spanish and Italian VOOs**

| Carotenoids/chlorophyll | Spain | Italy |
|---|---|---|
| Neoxanthin | 70-400 | 310-790 |
| Violaxanthin | 10-150 | 260-770 |
| Anteraxanthin | nd-475 | 470-640 |
| Mutatoxanthin[a] | 30-110 | nr |
| Luteoxanthin[a] | 90-250 | 510-800 |
| β-Cryptoxanthin | nd-60 | 380-620 |
| Lutein | 1200-9800 | 2280-4490 |
| β-Carotene | 110-2400 | 8060-16270 |
| Violaxanthin monoesterified | nd-20 | nr |
| Neoxanthin esterified | nd-30 | nr |
| Esterified xanthophylls[a] | nd-190 | nr |
| Chlorophyll a | nd-800 | nd-200 |
| Chlorophyll b | nd-1350 | 1030-1550 |
| Pheophytin a | 980-21700 | 19720-25040 |
| Pheophytin b | nd-840 | 150-2920 |
| Pheophorbide a | nd-565 | nr |
| Total chlorophylls | 400-25100 | 26100-31940 |

Note: nd, not detected; nr, not reported. Sources: Aparicio, unpublished data; Criado et al. 2007; Gandul-Rojas & Mínguez-Mosquera 1996; Giuffrida et al. 2007.

Chlorophyll is a chlorin with a magnesium ion at the center of the ring that has a side chain of propionic acid modified with a β-ceto cyclic ester and another with the propionic acid esterified with the phytol alcohol in C-17 that gives fat-soluble property to the molecule. Chlorophylls are transformed into their magnesium-free derivatives. They are quantified in olive oil at concentrations between 10 mg/kg and 30 mg/kg (Boskou, 2002). The quantitative studies of the chlorophylls (types a and b), and their derivatives (pheophytin a and b and pheophorbides) in ripe olives and in extracted VOO, has increased our understanding of the role of these compounds in the processes of fruit ripening and olive oil extraction, and also in shelf life and authenticity (Gandul-Rojas et al. 2000).

### Waxes

The content of waxes depends on the olive oil designations (Table 1.1). The profiles of these esters of fatty acids and fatty alcohols are of interest as indicators of both quality and purity. VOO is distinguished from refined olive oil and olive–pomace oils because VOO has a high content of C36 and C38 waxes and a low content of C40, C42, C44, and C46 while the other oils have an inverse relationship.

### Phospholipids

Studies of the phospholipids composition in olive oil are rather scarce. The total phospholipid content range is 40–135 mg/kg, and they are mainly phosphatidylglycerol, phosphatidylcholine, phosphatidylethanolamine phosphatidylserine, and phosphatidylinositol. These compounds can affect the physicochemical state of unfiltered olive oils and are considered as secondary antioxidants (Koidis & Boskou, 2006). The phospholipid composition is distinct compared to other edible oils; olive oil contains higher amounts of phosphatidylglycerol, while sunflower and almond oils are low in these phospholipids (Boukhchina et al., 2004).

## Flavor and Aroma Compounds

The demand of high-quality VOO can be attributed not only to its health benefits but also to reports of its fragrant flavor. Volatiles are responsible for its aroma while phenolic compounds are related to its taste. The presence of these compounds gives rise to the particular VOO flavor characterized by a unique balance of green, fruity, bitter, and pungent attributes that makes it a distinctive edible oil. The volatiles responsible for these positive attributes come from the olive, and they are direct metabolites produced in plant organs by intracellular biogenic pathways and oxidative processes. On the contrary, low-quality VOOs have complex profiles composed of numerous volatiles that are responsible for off-flavors such as rancid, mustiness-humidity, fusty, and muddy sediment.

Besides the volatiles present in the intact olive fruit as secondary products, other volatiles are formed very quickly during disruption of the cell structure due to enzymatic reactions in the presence of oxygen, and they are mainly responsible for VOO "green" aroma. The main precursors for volatile formation are fatty acids (linoleic and α-linolenic), mainly by means of lipoxygenase cascade (LOX) (Fig. 1.3), and amino acids (leucine, isoleucine, and valina). The profile of volatiles, however, depends on the maturity and olive variety (Aparicio & Morales, 1998). Table 1.9 summarizes the different pathways that lead to the formation of volatile compounds.

The volatiles responsible for VOO aroma are composed of compounds with diverse molecular weight and chemical nature, and they are present at a very low concentration. Therefore, the analytical techniques used for analyzing these compounds should include a preconcentration step or should be sensitive enough to detect all the compounds in a single analysis. Table 1.10 describes several of the methods used. Those methods that do not include preconcentration are generally characterized by low sensitivity, and they are rarely used in the analysis of olive-oil aroma today. The methods that include a preconcentration step increase the concentration of volatiles by means of distillation–extraction, supercritical fluid extraction, or by means of an adsorbent material or a cryogenic trap. Several kinds of traps have been used, from tubes filled with activated charcoal or Tenax to fibers of solid-phase microextraction (SPME) coated with absorbent material. Tenax traps have higher absorbent power for most of the volatiles, in comparison with SPME fibers and hence offer better results

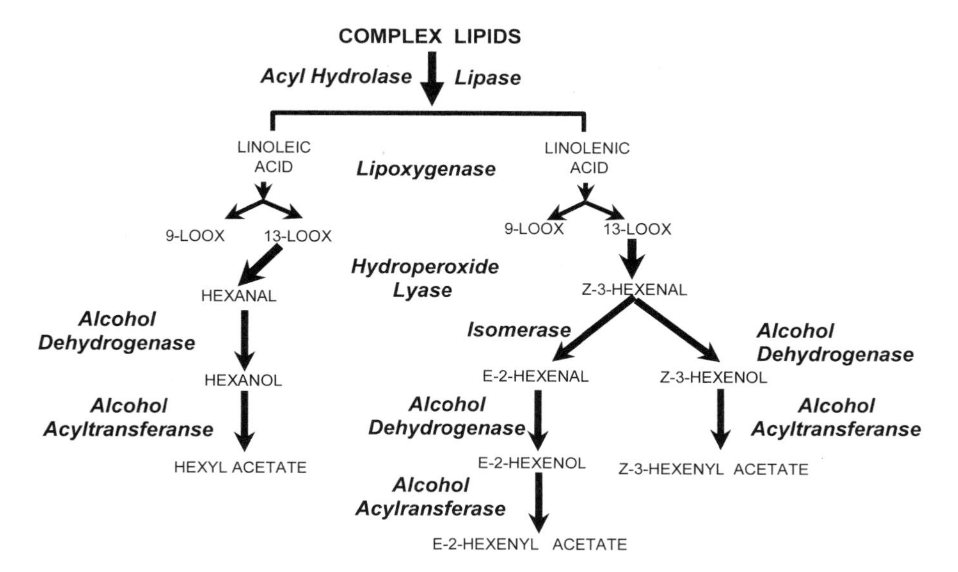


**Fig. 1.3.** Lipoxygenase cascade that is responsible for some of the most remarkable VOO volatiles.

**Table 1.9. Description of the Transformation of Precursors into Volatiles Identified in VOO**

| Precursors | Transformation | Description | Volatiles |
|---|---|---|---|
| Lipid | Fatty acid metabolism | Enzymes produce volatiles from fatty acids during olive ripening | Ketones, esters and alcohols |
| | Lipoxygenase cascade | $C_6$ volatiles are formed from 13-hydroperoxides of fatty acids (linoleic and α-linolenic) | Fig. 1.2 |
| | Cleavage reaction of the 13-hydroperoxide of linolenic acid | Reaction mediated through LOX giving rise to penten dimmers. Further activation of alcohol dehydrogenase enzyme could be responsible for $C_5$ compounds. | 1-Penten-3-ol, 2-Penten-1-ol |
| Aminoacids | Biochemical transformation of Amino acid branched chain | Valine, leucine, and isoleucine are transformed into branched aldehydes. They are transformed to alcohols and esters by alcohol dehydrogenase and acyltransferasa | 2-Methylpropanal, 2-Methyl-butanal, 3-Methyl-butanal. Alcohols and ester derivatives |

Note: nd, not detected; nr, not reported. Sources: Aparicio, unpublished data; Criado et al. 2007; Gandul-Rojas & Mínguez-Mosquera 1996; Giuffrida et al. 2007.

in terms of sensitivity (Vas & Vékey, 2004). However, SPME fibers are very easy to apply and allow automation; hence, they are more often used for VOO volatiles today (Tena et al., 2007). Among the SPME fibers, those coated with polydimethylsiloxane (PDMS) are less sensitive to most of VOO volatiles, while divinylbenzene/carboxen/polydimethylsiloxane (DVB/CAR/PDMS) fiber seems to have the most balanced sensitivity to all the volatiles present in VOO. The great advances in instrumentation in the past decades allow the identification of a great number of VOO chemical compounds (Morales & Tsimidou, 2000) that were used to explain VOO sensory attributes by means of several multidisciplinary approaches.

The sensory attributes of each volatile depends on its concentration and odor threshold. These parameters determine the odor activity value (OAV) that is the ratio between the concentration of the volatile and its odor threshold; volatiles with OAV $<1.0$ do not contribute to VOO aroma. A complementary method is the GC-sniffing technique that has been widely used to study the independent sensory impact of the volatiles to VOO aroma attributes (Morales & Tsimidou, 2000). An interesting approach is the statistical sensory wheel (SSW) that represents the global flavor matrix of EVOO (Aparicio & Morales, 1995) develped after compiling the information

**Table 1.10. Advantages and Disadvantages of the Most Common Techniques for the Quantification of VOO Volatiles.**

| Technique | Description | Advantages | Disadvantages |
|---|---|---|---|
| Without pre-concentration | | | |
| Direct injection | The sample is placed in a tube fitted to GC injector. Volatiles are purged by carrier gas into the column. | Rapid.<br>Simple | Very low sensitivity.<br>Degradation products |
| Static headspace | The sample is deposited in a sealed vial. An aliquot of the vapor phase is injected in a GC | Fine with multiple extractions. Appropriate for volatiles with highmolecular weight. | Poor sensitivity and reproducibility. Artifacts. Inappropriate for trace analysis.Leaks during syringe filling. |
| With pre-concentration | | | |
| Distillation- extraction | The vapor from distillation is condensed on a refrigerant or trapped in differentcryogenic traps or adsorbent materials, and later injected in a GC. | Small amount of solvents.<br>Rapid concentration process.<br>Low thermal degradation of volatiles | Not appropriate for thermolabile volatiles.<br>Lacks of solute by co-evaporation. |
| Static headspace with SPME | A SPME fiber is exposed to sample vapor phase. Volatiles adsorbed on the fiber are desorbed in the GC injection port. | Automatic. Rapid. Cheap.<br>Easy to use. All the steps in a single process. Various kinds of fibers.<br>Good repeatability | Differences in quantification of low weight molecules. Less number volatiles at low concentrations.<br>Some of the disadvantages of the static headspace. |
| Supercritical fluid extraction | Sample is placed in an extraction cell. A supercritical fluid pass through the cell and extract the volatile compounds | Easy to apply. Detection of volatiles in oil and olives.<br>Adequate for off-flavors. All the steps in a single process | Leaks. Selective with some volatiles (oxygenate compounds & by molecular weight and polarity)<br>Need of a concentration step (e.g. Tenax). |
| Dynamic headspace. Tenax traps. Thermal desorption & cool trap injection | An inert gas (e.g., $N_2$) sweep the sample headspace that is stirred or bubbled. Volatiles are trapped in Tenax. Trap is desorbed by GC | High adsorption capacity. Useful for almost all kind of volatiles. Good recovery factors. No artifacts | Less sensitive to some acids.<br>Temperature & flow-rate must be controlled An analysis per sample.<br>Expensive. |

Source: Morales and Tsimidou, 2000

of dozens of sensory attributes evaluated by assessors from several European-trained VOO sensory panels. The resulting information is reported as a circle divided into seven sectors, each sector representing a particular sensory perception of olive oil: bitter-pungent-astringent, green, sweet, fruity, ripe fruit, ripe olives, and undesirable, plus four miscellaneous sectors. The mathematical procedure ordered them as if each one of them was the main sensory perception qualifying VOOs obtained from olives harvested at successive levels of ripeness. The miscellaneous sectors are small sectors of the circle (Fig. 1.4) where there are diverse sensory attributes that cannot be clustered under a common sensory designation. The miscellaneous sectors form a gradual transition between qualified sectors, and provide a way to account for intermediate or eclectic sensory perceptions (Aparicio & Morales, 1995). Then, the information on the concentration of 85 volatile compounds was projected on the plot of the sensory attributes, and the result is the statistical sensory wheel (SSW) (Aparicio et al., 1996).

Figure 1.4 shows the green (from fruity-green to bitter-green) sector of SSW. This sector includes green and olive fruity notes such as cut grass, green olives, banana skin, etcetera. Most of the compounds placed in this sector are qualified with green and/or fruity odor when they are evaluated by GC-sniffing. Green odor is mainly associated with C6 aldehydes, alcohols, and their corresponding esters. The SSW obtained with this procedure allowed researchers to acquire a detailed knowledge of the volatiles that are responsible for the desirable sensory attributes as well as those volatiles that are responsible for unpleasant perceptions. The latter volatiles are detected at trace levels or at low concentration in EVOO and VOO. Table 1.11 summarizes the various volatiles associated with some SSW sectors and individual sensory attributes.

The volatile fraction can also be used to characterize varietal VOOs. The differences in volatiles of different botanical varieties are explained by the variation of the enzymatic activity of several biochemical pathways, which are also influenced by the pedoclimatic conditions. To try to gain an understanding of the latter factor, Luna et al. (2006b) studied 39 monovarietal VOOs obtained from trees cultivated in the same orchard under the same pedoclimatic conditions. Significant differences were found among the studied varieties. The total concentration of volatiles ranged from 9.83 to 35.32 mg/kg. The C6 compounds—"green compounds"—varied from 2.52 mg/kg to 18.11 mg/kg, E-2-hexenal being the major contributor. Table 1.12 shows volatiles with OAV>1.0 and the variety where each compound was quantified at the maximum concentration. The concentration values of these compounds, in combination with the information of the sensory wheel, explain the differences in sensory properties between oils from different cultivars, which constitute a reliable procedure of varietal VOO differentiation.

If VOO is obtained from infested olives (e.g., attacked by olive fly) or olives collected from the ground, or if VOO is inadequately processed or stored, the volatile profile will include compounds responsible for off-flavors (Morales et al., 2005). Lipolysis and oxidation are the processes leading to the most serious olive oil deteriora-

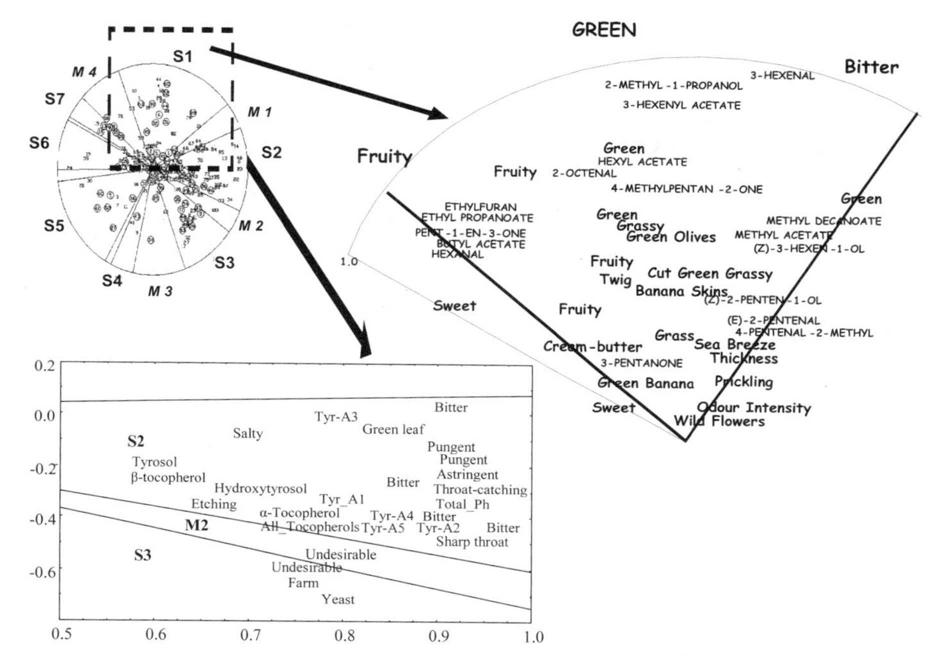


**Fig. 1.4.** Green and bitter-pungent sectors of Statistical Sensory Wheel (SSW). Note: S1: green sector, S2: bitter-pungent sector, S3: undesirable sector, S4: ripe-olives sector, S5: ripe-fruits sector, S6: fruity sector, S7: sweet sector, M1-M4: miscellaneous sectors; Tyr-A, Tyrosol aglycon; Total-Ph, Total phenols.

tion. Lipolysis usually starts in the olive fruit, while oxidation begins when VOO is stored. Rancid, fusty, winey–vinegary and mustiness-humidity are the most common VOO sensory defects, and the volatiles were studied to explain those unpleasant perceptions (Morales et al., 2005). Table 1.11 shows some of the volatiles identified in VOO and attempts to correlate them with defects at OAV>1.0.

Rancid oils are characterized by volatiles coming from the oxidation of unsaturated fatty acids, mainly aldehydes (e.g., pentanal, hexanal, heptanal, octanal, nonanal) and acids (e.g., acetic, butanoic, hexanoic, heptanoic). On the other hand, one of the main causes of VOO sensory defects is the storage of olive fruits in piles before extraction, which is associated with fermentative processes due to olive microbial contaminations. The winey–vinegary defect is produced by lactic acid (*Lactobacillus*) and acetic acid bacteria, among other microorganisms, that induce a fermentative process in the olives, thus, giving rise to the production of ethyl acetate and acetic acid that are responsible for this defective sensory attribute (Morales et al., 2005). The fusty defect is also a consequence of storing olives in piles. The enterobacteriaceae genera *Aerobacter* and *Escherichia* are found at the beginning of storage, while the genera

**Table 1.11. Main Volatiles That Contribute to Some VOO Sensory Attributes (OAV>1.0).**

| Sensory perception | Volatile compounds |
|---|---|
| Green fruity | Z-3-Hexen-1-ol, Z-3-Hexenal, Hexyl acetate, 3-Hexenyl acetate, Hexanal |
| Bitter, astringent. | E-2-hexenal |
| Green banana | 2-Penten-1-ol |
| Almond | E-2-hexenal[a] |
| Butter | 3-Methyl-2-butanal, 2-Methyl-butanal |
| Sweet tomato | 3-Pentanone, 1-Penten-3-one, Ethyl propanoate |
| Green tomato | 2-Heptanone, 2-Nonanone, 1-Hexanol, E-2-pentenal |
| Artichoke | E-3-hexenal |
| Rancid | Hexanal, Nonanal, E-2-Decenal, Nonanol, Hexanoic and Heptanoic acids, among others |
| Winey-vinegary | Butan-2-ol, 2-Methyl butan-1-ol, Acetic acid, among others |
| Fusty | Butanoic and Pentanoic acids, among others |
| Mustiness-humidity | 1-Octen-3-one, Heptan-2-ol, 1-Octen-3-ol, among others |

Note: [a], E-2-Hexenal also contributes to twig attribute.

**Table 1.12. Volatiles with OAV>1.0 Whose Concentrations Characterize Varietal VOOs.**

| Volatile | Variety (max. amount in µg/kg) | Volatile | Variety (max. amount in µg/kg) |
|---|---|---|---|
| 3-Methyl-Butanal | Tsounati (223) | E-2-Hexenal | Lechin (15459) |
| Ethyl propanoate | Cornicabra (249) | 3-Methyl butanol | Cañivano (5357) |
| 2-Methyl butanal | Picholine(510) | Z-3-Hexenyl acetate | Manzanilla (476) |
| 4-Methyl-pentan-2-one | Picudo (899) | Z-2-Penten-1-ol | Moraiolo (1313) |
| 1-Penten-3-one | Picudo (1689) | 6-Methyl-5-hepten-2-one | Picual (2337) |
| Butyl acetate | Picudo (455) | Hexan-1-ol | Empeltre (3297) |
| Hexanal | Cañivano (5353) | Z-3-Hexen-1-ol | Sourani (2764) |
| E-2-Methyl-2butenal | Zaity (404) | Nonan-2-one | Cañivano (728) |
| E-2-Pentenal | Moraiolo (3822) | E-2-Hexen-1-ol | Empeltre (4795) |
| 1-Penten-3-ol | Imperial (342) | | |

Source: Luna et al., 2006b.

*Pseudomonas*, *Clostridium,* and *Serratia* are the most significant after prolonged storage (Morales et al., 2005). The presence of these microorganisms results in an increase of esters and acids. Finally, the sensory perception of mustiness–humidity in VOO is due to the presence of various kinds of fungi on olives piled under humid conditions for several days. VOOs obtained from those olives are characterized by a low concentration of volatiles and the presence of volatiles with 7 and 8 atom carbons.

The current studies of virgin olive oil aroma are mainly focused on the development of new rapid methods (i.e., sensors) that allow detecting defects in VOO, which is demanded more and more by producers and sellers (García-González & Aparicio, 2007). Further research should contribute more information on chemical compounds, which, when present at very low concentration, influence VOO acceptability by consumers as well as predict VOO quality by analyzing the olive paste (García-González et al., 2007).

## Allergic and Toxic Compounds

Contact allergy to olive oil is considered rare. In fact, Kränke et al. (1997) tested 100 patients for an allergy to olive oil, and they had five positive reactions, only one of which they felt was authentically allergic in nature. They concluded that olive oil might be mildly irritating when used in occlusive dressings such as micropore tapes or bandages. However, 20 cases of contact allergy to olive oil were reported by pedicurists, masseurs, chefs, and three of these had occupational hand eczema as a result of olive oil. As long as they did not come into direct skin contact with the olive oil, they were without dermatitis and could also ingest the oil. No investigation has elucidated any special component in the olive oil as the cause of the patients' eczema. Furthermore, the majority of cases are, however, seen in patients treated for venous eczema. The reason that a few cases of olive oil allergy are described may be due to people who are exposed to olive oil through ingestion before handling it, thus, inducing tolerance.

## Health Benefits of the Oil and Oil Constituents

The traditional Mediterranean diet typically includes moderate energy intake, with low animal fat, high oleic oil, high cereals, high legumes, nuts and vegetables, and regular and moderate wine. Olive oil, one of the main ingredients of the Mediterranean diet, was always pointed out as mainly responsible for the health benefits of the nutritional pattern of the Mediterranean countries in numerous studies (Table 1.13). Its therapeutic properties, not only in topical applications, but also as nutraceutical, have been known since ancient times. Much has been reported about the protective effect of olive oil in cardiovascular morbidity. Today, the results of cohort studies that relate the low incidence of these diseases and the daily olive oil intake go beyond the experimental evidence. In addition, new healthy properties are extensively studied once the bioavailability and antioxidant properties of their micro-constituents are known. Intensive research studies are addressing the elucidation of the role of olive oil in diseases, such as cancer, diabetes, or cognitive decline, prompted by the fact that epidemiological studies on these diseases seem to confirm their relationship with olive oil consumption. In 2004 the Food and Drug Administration (FDA) evaluated the studies carried out on the healthy properties of olive oil up to that year and concluded that enough evidence is found to include a health claim on the labels of olive oil bottles to inform consumers about its health benefits (Covas, 2007).

**Table 1.13. Facts and Future Research Needs of a Mediterranean Diet Rich in Olive Oil with Respect to Its Health Benefits.**

| FACTS |
|---|
| Epidemiological studies relate olive oil and primary prevention of cardiovascular diseases. |
| It has been proved that it reduces plasma triglycerides and increases HDL-cholesterol levels. |
| It has been proved that improves of postprandial lipoprotein metabolism. |
| It has been demonstrated it causes an endothelial dependent vasodilatation and inflammatory response. |
| It has been proved that reduces of blood pressure and the risk of hypertension. |
| Anticancerogenic effect of olive oil has been proven in animal models and in human cell lines. |
| There are experimental evidences of the beneficial effect of olive oil in different steps of carcinogenesis. |
| The diet does not promote obesity while it increases the lipolytic activity in adipose tissue. |
| There is evidence that olive oil prevents age-related cognitive decline and dementia. |
| Evidence suggests an increase in survival and overall longevity with a diet rich in oleic acid. |
| Olive oil phenols are bioavailable in humans |
| FUTURE RESEARCH NEEDS |
| To know better the specific and differential healthy effects of olive oil micronutrients on atherogenesis. |
| To know in depth the molecular mechanisms by applying genomic and proteomic. |
| To end the controversy about olive oil consume and the lower risk of myocardial infarction. |
| To increase the current knowledge of the impact of Mediterranean diet on obesity and metabolic syndrome. |
| To know the effect of main olive oil components on cancer expression. |
| To have intervention studies in humans studying: (i) high risk populations, (ii) non-olive oil consumers, and (iii) cancer patients. |
| To elucidate the importance of associating the consumption of olive and fish oils. |
| To obtain more evidence about the antioxidant activity in humans. |
| To obtain more solid epidemiological evidences from intervention studies examining disease end-points. |

Source: Ordovás et al., 2005.

A direct relationship between the olive oil intake and a health benefit was first established by the Seven Countries Study (Keys et al., 1986). The objective of this research was to search for differences in the incidence of coronary heart diseases focusing on 16 areas in seven countries (USA, Finland, Holland, Italy, former Yugoslavia, Greece, and Japan). This study revealed that the populations of Mediterranean countries with a relatively high intake of fats (around 35% of energy) had a lower incidence of heart diseases compared to other European countries and the United States. This was correlated with a low intake of saturated fats. Since olive oil was the main fat source of this population, this cardioprotective property was attributed to the intake of this oil. The Seven Country Study was the first scientific observation of the health benefits of olive oil and was the beginning of an endless research to explain its mechanism effect. Table 1.14 shows the pharmacological effects reported for some olive oil compounds.

## Antioxidant and Anti-inflammatory Activities

The Seven Countries Study was followed by biochemical, clinical, and epidemiological research to identify the minor and major olive oil compounds responsible for its cardioprotective effect, and most recently, to determine the biological mechanisms that lead to this benefit. Thus, Renaud et al. (1995) carried out a study in 605 patients recovering from myocardial infarction, part of which was adapted to a Mediterranean diet rich in olive oil. The study corroborated the protective effect of olive oil, and ascribed this effect to an increase of n-3 fatty acids and oleic acid concentrations in serum, and a decrease in linoleic acid that resulted from higher intakes of oleic acid. Other studies proved that the monounsaturated fatty acids (MUFA), such as oleic acid, diminish the concentration of total cholesterol and LDL-cholesterol when these fatty acids replace saturated fats. In fact MUFA proved to be as effective in lowering

**Table 1.14. Pharmacological Effects of Some Olive Oil Compounds**

| Compound(s) | Pharmacological effects /biochemical evidences |
|---|---|
| (-)-Oleocanthal | Inhibits COX-1 and COX-2 |
| Oleuropein | Coronary vasodilator, anti-spasmodic, antihypertensive, anti-arrhythmic |
| Hydroxytyrosol<br>1-(3'-methoxy-4'-hydroxy-phenyl)-6,7-dihydroxy-isochroman<br>1-phenyl-6,7-dihydroxy-isochroman | Inhibits platelet aggregation |
| Caffeic acid | Exhilarates and scavenges ROS |
| p-coumaric acid | Anti-hyperlipidaemia |
| Pinoresinol | Inhibits cAMP phosphodiesterase |
| Verbascoside | Anti-inflammatory and inhibits 5-lipoxidase |
| Total phenols | Low lipid oxidation damage, anti-inflammatory effect, decrease in DNA oxidative damage, etc. |
| Tocopherols | Modulate the eicosanoid metabolism, interfere in the cyclooxygenase regulation |
| Sterols | Bile acid sequestrant, inhibit acyl-coenzyme A: cholesterol acyl transferase |
| Triterpenes (e.g. erythrodiol, uvaol, oleanolic and maslinic acids) | Anti-inflammatory, antioxidant, antidysrhythmic cardiotonic and vasodilatory activities |
| Oleic acid | Less oxidation of liposomes, prevention of endothelium activation by inhibition of the expression of adhesion of molecules, sensitivity Reduction of endothelial cell to oxidation, LDL-cholesterol lowering |

LDL cholesterol as n-6 PUFA. The MUFA content is not, however, enough to predict the lowering effect on cholesterol concentration (Harwood & Yaqoob, 2002).

The importance of oleic acid in cardiovascular diseases is supported by its bioavailability in humans. Oleic acid can account for up to 29% of the caloric intake in Mediterranean countries, and its concentration in plasma may 10 µmol/L–100 µmol/L (Perona et al., 2006). Several mechanisms have been proposed to explain the protective effect of oleic acid. Thus, it has been proposed that LDL rich in oleate is less susceptible to oxidation, in comparison to linoleate-rich LDL (Castro et al., 2000). This protective effect of oleic acid against oxidation can also be extend to the endothelial cells, which are involved in inflammation processes. Atherosclerosis is considered an inflammatory disease in which a series of mediators are released by the endothelium once it is activated (Covas et al., 2006). Oleic acid prevents the activation of endothelium by avoiding the expression of adhesion molecules or by improving nitrous oxide (NO) production (Christon, 2003). However, this effect is produced; it seems to be related to the reduced sensitivity of endothelial cells to oxidation in the presence of oleic acid as a consequence of the reduced production of reactive oxygen species (ROS).

Although MUFAs were thought to be the main components that cause the protective effect of olive oil, minor compounds with biological activity are present as well (Covas et al., 2006). That would explain why Truswell and Choudhury (1998) did not observe the same effect on plasma cholesterol attributed to oleic acid when they used various oils enriched with this compound. The susceptibility of LDL to oxidation also depends on other compounds with antioxidant activity bounded to them. For this reason, and as a consequence of the studies on the role of LDL oxidation in cardiovascular diseases, special attention was paid to compounds that showed antioxidant properties in in vitro or in vivo repeated studies; squalene, triterpenic, and phenolic compounds. The phenols showed antioxidant activity preventing the oxidation of LDL in vivo studies (Fitó et al., 2000). However, most of the studies carried out in vivo used quercetin as a model compound, and little is known about the phenolic compounds that are present in olive oil (Harwood & Yaqoob, 2002). The EUROLIVE project (Covas et al., 2006) studied the effect on humans of VOOs with different amounts of phenolic acids. VOOs with the highest amounts of phenols decreased the lipid oxidation damage to a very high degree, in comparison with other compounds. In particular, oleuropeina and hydroxytyrosol are scavengers of ROS near or within membranes (Saija et al., 1998). Also phenolic compounds take part in the processes that prevent endothelial activation and inflammation by inhibition of LB4 cytokine and eicosanoic production (de la Puerta et al., 1999). However, the effect of phenolic compounds does not seem to be limited to lipid oxidation since Salvini et al. (2006) pointed out that the intake of olive oils characterized by high phenol content reduces the DNA oxidative content. In recent times, a specific phenolic compound of olive oil awakened a high interest in the scientific community due to its high anti-inflammatory activity. Beauchampt et al. (2005) described the ibuprofen-like properties of the dialdehydic form of deacetoxy ligstroside aglycone, known as oleocanthal. This compound affects the eicosanoid pathway by inhibiting the activity of cyclooxygenase enzymes COX-1 and COX-2.

One should not overlook other minor compounds of olive oil protective cardiovascular effect of tocopherols, sterols terpenes, and squalene, which all showed antioxidant and anti-inflammatory properties. Perona et al. (2004), in a study carried out with incubated endothelial cells, concluded that sterols, tocopherols, and terpenoids may improve the balance between vasoprotective and prothrombotic factors. Tocopherols, in particular α-tocopherol, seems to prevent the adhesion of molecules and modulate eicosanoid metabolism by interfering in the cyclooxygenase regulation. Although the tocopherol concentration in olive oil may not be high enough to be considered a healthy factor in clinical studies, undoubtedly it contributes to the set of antioxidants present in olive oil (Covas et al., 2006). On the other hand, olive oil is rich in sterols, and these compounds are bile acid sequestrants and inhibitors of acyl-coenzyme A:cholesterol acyl transferase (Covas et al., 2006). The consumption of phytosterols in humans can lower cholesterol (Katan et al., 2003). Some studies, however, do not reveal anti-inflammatory properties (de Jongh et al., 2003), while others revealed that β-sitosterol can reduce the edema induced by TPA (2-o-tetradecanoylphorbol-13-acetate) as a consequence of its anti-inflammatory activity (de la Puerta et al., 2000). The contradictory results about the relationship between the levels of plasma sterols and the incidence of atherosclerosis are encouraging researchers to continue investigating to reach a conclusive theory. Very little is known about the potential benefits of olive oil triterpenes (oleanolic and maslinic acids, erythrodiol, and uvaol) that are much more concentrated in pomace oil, since they are originally located in the fruit skin. Some experimental studies with animals showed anti-inflammatory, antioxidant, cardiotonic, antidysrhythmic, and vasodilator activity (Covas et al., 2006). Thus, the oleanolic acid reduces the inflammation by means of its inhibition of LOX and COX-2, and also it may produce, together with erythrodiol, vasodilatation by the endothelial production of NO (Perona et al., 2006). Nevertheless, as in the case of sterols, more research is needed to determine the biological activity of triterpenes in humans. Likewise, one should include squalene in these studies since it proved to be a free radical scavenger (Aguilera et al., 2005).

## Blood Pressure

Blood pressure is truly related to cardiovascular diseases, and olive oil seems to have an important effect on blood pressure. The same compounds cited above are involved in blood pressure reduction observed in both men and women, when olive oil is regularly consumed due to the mechanism that leads to a vasodilatation due to the increase of NO. Thus, Ruiz-Gutiérrez et al. (1996) suggested that not only a diet rich in MUFA but also in minor compounds contributed to a lower blood pressure. Later, Fitó et al. (2005) associated a lower systolic blood pressure with a higher intake of phenols since phenols can increase the levels of NO, hence promoting vasodilatation. Triterpenoids (oleanolic acid and erythrodiol) also produce vasodilatation by the same mechanism (Rodríguez-Rodríguez et al., 2004). Table 1.15 summarizes the established and possible benefits, of a VOO-rich diet on cardiovascular and related diseases.

## Cancer

The relationship between the incidence of certain types of cancer and excessive lipid consumption is well-known. However, the recommendation of lowering the fat intake is not enough for a warranted prevention of cancer (Kushi & Giovannucci, 2002). Thus, for instance, some studies established the role of n-3 PUFA and n-6 PUFA in enhancing and inhibiting tumor growth, respectively (Harwood & Yaqoob, 2002; Rose, 1997). The consensus report of the international conference on the healthy effect of VOO (Ordovás et al., 2005) specifies in one of the conclusions that the incidence of some types of cancer is lower in Mediterranean countries where olive oil is the principal source of fat, such as Spain, Italy, and Greece. This conclusion is based on some epidemiological studies that compare the incidence rates of breast, ovarian, and colorectal cancers with those in Northern European countries (Serra-Majem et al., 1993). The epidemiological studies carried out to date show the major evidence in the case of breast cancer. Experimental animal research verifies the association of olive oil and cancer. Although MUFAs possibly are involved in the cancer protective effect of olive oil, due to their antioxidant and anti-aging properties, recent studies attribute this effect to minor compounds, such as flavonoids, Vitamin E, squalene, caffeic acid, and hydroxytyrosol, in the course of tumor growth (Owen et al., 2000). Thus, the cancer prevention effect of olive oil is explained by the action of bioactive compounds of olive oil altering tumor eicosanoid biosynthesis and cell-signaling pathways, modulating the gene expressions and preventing DNA damage induced by reactive oxygen

**Table 1.15. Benefits of a Virgin Olive Oil-rich Diet on Cardiovascular Risk Factors and Other Mechanisms Related to Atherogenesis.**

| ESTABLISHED BENEFITS |
|---|
| Reduces triglycerides and increases HDL cholesterol levels when it replaces carbohydrate-rich diets |
| Lowers LDL cholesterol levels when it replaces saturated fat-rich diets |
| Increases resistance of LDL to oxidation |
| Improves glucose metabolism in diabetes |
| POSSIBLE BENEFITS |
| Improves the endothelium dependent vasodilatation |
| Ameliorates the inflammation induced by intake of high saturated-fat diets |
| Reduces the activation of mononuclear cells |
| Reduces arterial blood pressure and the need for anti-hypertensives |
| Induces a less pro-thrombotic plasma environment |

Source: Pérez-Jiménez et al., 2007.

metabolites (Ordovás et al., 2005). Whatever the mechanism of olive oil reducing cancer incidence, the research in this field has demonstrated that the efficacy of this edible oil against tumor diseases may justify its use in nutritional–pharmacological combinations to reduce some types of cancers (Schwartz et al., 2004).

### Other Health Benefits

The healthy effect of olive oil intake on other pathologies was also studied although scarce knowledge exists about the mechanisms that lead to these effects. Thus, a notable improvement of carbohydrate metabolism control in diabetes patients was observed when the diet is supplemented with oleic acid (Soriguer et al., 2004).

Many pathologies in developed countries with the elderly population are related to the oxidation process and aging. Recently, the antioxidant property of some foods gained special interest among the population (Pala et al., 2006); olive oil is one of these anti-aging foods. Higher intakes of olive oil may prevent some cognitive disorders such as Alzheimer's and Parkinson's diseases. Scarmeas et al. (2006) concluded that a higher adherence to the Mediterranean diet is associated with a reduction in risk for Alzheimer's dysfunction. On the other hand, Parkinson's disease is related to a high fat diet, and some studies have related the intake of Vitamin E with a lower risk of this pathology (Etminan et al., 2005). However, more research is needed to explain the protection against these diseases before a general recommendation of olive oil as a protective agent is justified. In the meantime, current research is establishing the basis of known benefits. New healthy properties are being analyzed without neglecting olive oil's most ancient therapeutic utility in dermatological preparations.

## Purity, Authenticity, and Traceability

The higher price of olive oil and its reputation as a healthy and delectable oil make it a preferred target for defrauders. Several institutions (e.g., Antifraud Unit of EU—OLAF—and International Olive Council—IOC—among others) are, however, actively involved in the detection of any kind of fraud to avoid the image of a hypothetical uncontrolled distribution of adulterated olive oil into the market, and their standards contain those provisions needed to ensure fair trade and prevent fraud as well as safety and consumer protection. The result is a strictly regulated olive oil. Table 1.1 shows the purity and quality characteristics (criteria) of the olive designations described in the Introduction section. The limits established for each criterion include the precision values of the attendant recommended method. The methods are described in the IOC Web site (www.internationaloliveoil.org). The limits and designations approved by IOC are incorporated in the Norm of the Codex Alimentarius.

No rapid and universal method is officially recognized for all authenticity purposes, but the advances in knowledge and technology undoubtedly led to greater success in the fight against adulteration, though it is equally true that the same information is also used by defrauders to invalidate the usefulness of some standard methods. Such competition requires not only a considerable investment in perfecting techniques or developing new ones, but also a rapid pace of R&D for detection of malpractices (Aparicio et al., 2007). Table 1.16 shows the main authenticity issues and their paradigm; two of them are still challenges for analysts—detection of deodorized VOO in VOO and geographical traceability—despite the arsenal of analytical techniques at the analyst's disposal (Aparicio et al., 2007).

VOO with slight sensory defects is deodorized at a moderate temperature (≤100°C) to remove volatiles responsible for undesirable attributes. This deodorized oil cannot be marketed as VOO since it was subjected to a thermal process and can be fraudulently added to an irreproachable virgin olive oil (IOC, 2006). This is the so-called "deodorato." All the official methodologies are based on the analysis of saponifiable and unsaponifiable fractions that are not altered with this process that makes this adulteration potentially undetected. The ratio between a natural compound (pheophytin a) and pyropheophytin a may detect this adulteration since the latter is a thermal degradation compound though, unfortunately, it increases along VOO shelf life as well (Aparicio-Ruiz, 2008). Today, new strategies are being studied combining the information of chlorophylls and diacylglycerols with limited success.

Consumers' awareness of the importance of the geographical traceability in food safety and quality assurance increased the interest for new methods that can assess

**Table 1.16. Main Authenticity Issues and Sub-issues, and Their Current Examples.**

| Issue | Sub-issue | Paradigm | Technique |
|---|---|---|---|
| Adulteration | Addition of cheap oil to dear oils.<br>Addition of refined oils to VOO.<br>Addition of low to high olive oil designations. | Detection of refined hazelnut in ROO.<br>Detection of seed oils in VOO.<br>Detection of deodorized VOO in VOO. | LC<br>GC<br>GC+HPLC |
| Geographical origin | Inexact label.<br>Subsidy.<br>Traceability. | Detection of VOO from several origins.<br>Importations among countries.<br>Characterization of PDO. | GC<br>GC<br>GC+ICP-MS |
| Production system | Organic vs. conventional. | Detection of conventional in organic VOO. | GC;HPLC |
| Extraction system | Centrifugation & Percolation.<br>Cold-press vs. solvent. | Characterization of VOO obtained by two-phase centrifugation. | GC;HPLC |
| Type | Specie.<br>Variety. | Characterization of edible oils.<br>Characterization of varietal VOO. | GC,<br>DHS-GC; GC |

Note: DHS, Dynamic head space; GC, Gas chromatography; ICP, Inductive coupled plasma; LC, Liquid chromatography; MS, Mass spectrometry.

VOO geographical origin. In addition to consumers' interest, the high number of VOO designations of origin (PDO) also raised the concern of producers and consumers about the particular characteristics of the oils produced from PDOs. Inductively coupled plasma hyphenated to a mass spectrometer (ICP–MS) and elemental analyzer–pyrolysis–isotope ratio mass spectrometer ($\delta^2$H-EA-Py-IRMS) are being used to characterize the oils by their geographical origins (Aparicio et al., 2007). However, these methods need a large database to build a consistent classification model. The alternative is the quantification of major and minor compounds (e.g., fatty acids, hydrocarbons, alcohols, sterols) whose concentrations depend on the variety, the extraction system, and the pedoclimatic conditions. The information structured in a large database supervised by the knowledge base of an expert system, SEXIA™, allows the determination of the geographical origin of the most representative European VOO with a certainty factor (CF) (Aparicio, 2000).

The future work on geographical traceability will focus on building an Olive Oil Map where the most representative olive oils produced in each productive region are characterized by chromatographic, spectroscopic, and isotopic techniques. The resulting databases, in conjunction with new procedures of classification and visualization techniques, will allow the application of the best synergy between these techniques in VOO traceability.

## References

Aguilera, Y.; M.E. Dorado; F.A. Prada; J.J. Martínez; A. Quesada; V. Ruiz-Gutierrez. The protective role of squalene in alcohol damage in the chick embryo retina. *Exp. Eye Res.* **2005**, *80*, 535–543.

Angerosa, F. Influence of volatile compounds on virgin olive oil quality evaluated by analytical approaches and sensor panels. *Eur. J. Lipid Sci. Technol.* **2002**, *104*, **639–660.**

Anon. (2004) *By-Product Reusing from Olive and Olive Oil Production.* Project Report. Proj. FOOD-CT-2004-505524. http://www.tdcolive.net/

Aparicio, R. Characterization: Mathematical procedures for chemical analysis. *Handbook on Olive Oil: Analysis and Properties;* J. Harwood, R. Aparicio, R., Eds.; Aspen: Gaithersburg, MA, 2000; pp. 285–354.

Aparicio-Ruiz, R. Characterization of the reactions of thermodegredation of chlorophylls and carotenoid pigments in virgin olive oil. Ph.D. Thesis, University of Seville, Seville, Spain, 2008

Aparicio, R.; R. Aparicio-Ruiz; D.L. García-González. Rapid methods for testing of oil authenticity: The case of olive oil. *Rapid Methods;* A. van Amerongen,D. Barug, M. Lauwars, Eds.; Wageningen Academic: Wageningen, Holland, 2007; pp. 163–188.

Aparicio, R.; J.J. Calvente; M.V. Alonso; M.T. Morales. Good control practices underlined by an on-line fuzzy control database. *Grasas Aceites* **1994**, *45*, 75–81.

Aparicio, R.; P. McIntyre. Oils and fats. *Food Authenticity: Issues and Methodologies;* M. Lees, Ed.; Eurofins Scientific: Nantes, France, 1998; pp. 214–273.

Aparicio, R., M.T. Morales. Sensory wheels: A statistical technique for comparing QDA panels. Application to virgin olive oil. *J. Sci. Food Agric.* **1995**, *67*, 247–257.

Aparicio, R.; M.T. Morales. Characterization of olive ripeness by green aroma compounds of virgin olive oil. *J. Agric. Food Chem.* **1998**, *46*, 1116–1122.

Aparicio, R., M.T. Morales; M.V. Alonso. Relationship between volatile compounds and sensory attributes by statistical sensory wheel. *J. Am. Oil Chem. Soc.* **1996**, *73*, 1253–1264.

Aparicio-Ruiz, R. Characterization of the reactions of thermodegradation of chlorophylls and carotenoid pigments in virgin olive oil. Ph.D. Thesis, University of Seville, Seville, Spain, 2008.

Azadmard-Damirchi, S.; P.C. Dutta. Novel solid-phase extraction method to separate 4-desmethyl-, 4-monomethyl-, and 4,4′-dimethylsterols in vegetable oils. *J. Chromatogr. A* **2006**, *1108*, 183–187.

Bartolini, G.; G. Prevost; C. Messeri; G. Carignani; U.G. Menini. *Olive Germplasm: Cultivars and Word-wide Collections,* FAO, Rome, 1998.

Beauchamp, G.K.; R.S.J. Keast; D. Morel; J. Lin; J. Pika; Q. Han; C-H. Lee; A.B. Smith; P.A.S. Breslin. Phytochemistry: Ibuprofen-like activity in extra-virgin olive oil. *Nature* **2005**, *437*, 45–46.

Boskou, D. Olive oil. *Vegetable Oils in Food Technology: Composition, Properties, and Uses;* F. Gunstone, Ed.; Blackwell Publishing: Oxford, UK, 2002; pp. 244–277.

Boskou, D. Sources of natural phenolic antioxidants. *Trends Food Sci. Technol.* **2006**, *17*, 505–512.

Boukhchina, S.; K. Sebai.; A. Cherif; H. Kallel; P.M. Mayer. Identification of glycerophospholipids in rapeseed, olive, almond, and sunflower oils by LC–MS and LC-MS-MS. *Canadian J. Chem.* **2004**, *82*, 1210–1215.

Brenes, M.; A. Garcia; M.C. Dobarganes; J. Velasco; J. Romero. Influence of thermal treatments simulating cooking processes on the polyphenol content in virgin olive oil, *J. Agric. Food Chem.* **2002**, *50*, 5962–5967.

Castro, P., J. López Miranda; P. Gómez; D.M. Escalante; F. López Segura; A. Martín; F. Fuentes; A. Blanco.; J.M. Ordovás; F.P. Jiménez. Comparison of an Oleic Acid Enriched-Diet vs NCEP-I Diet on LDL Susceptibility to Oxidative Modifications. *Eur. J. Clin. Nutr.* **2000**, *54*, 61–67.

Christon, R.A. Mechanisms of action of dietary fatty acids in regulating the activation of vascular endothelial cells during atherogenesis. *Nutr. Rev.* **2003**, *61*, 272–279.

Covas, M.I. Olive oil and cardiovascular system. *Pharmacological Res.* **2007**, *55*, 175–186.

Covas, M.I.; V. Ruiz-Gutiérrez; R. de la Torre; A. Kafatos; R.M. Lamuela-Raventós; J. Osada; R.W. Owen; F. Visioli. Minor components of olive oil: Evidence to date of health benefits in humans. *Nutrition Rev.* **2006**, *64*, 20–30.

Criado, M.N.; M.J. Motilva; M. Goñi; M.P. Romero. Comparative study of the maturation process of the olive fruit on the chlorophylls and carotenoid fractions of drupes and virgin Oils from arbequina and farga cultivars. *Food Chem.* **2007**, *100*, 748–755.

D'Annibale, A.; G.G. Sermanni; F. Federici; M. Petruccioli. Olive-mill wastewaters: A promising substrate for microbial lipase production. *Biores. Technol.* **2006**, *97*, 1828–1833.

de Jongh, S.; M.N. Vissers; P. Rol; H.D. Bakker; J.J. Kastelein; E.S. Stroes. Plant sterols lower

LDL cholesterol without improving endothelial function in prepubertal children with familial hypercholesterolaemia. *J. Inherit. Metab. Dis.* **2003**, *26*, 343—351.

de la Puerta, R.; E. Martínez-Domínguez; V. Ruiz-Gutiérrez. Effect of minor components of virgin olive oil on topical antiinflammatory assays. *Z. Naturforsch*, **2000**, *55*, 814–819.

de la Puerta, R.; V. Ruiz Gutierrez; J.R. Hoult. Inhibition of leukocyte 5-lipoxygenase by phenolics from virgin olive oil. *Biochem. Pharmacol.* **1999**, *57*, 445–449.

Di Giovacchino, L. Technological aspects. *Handbook of Olive Oil: Analysis and Properties;* J. Harwood, R. Aparicio, Eds.; Aspen: Gaithersburg, MA, 2000; pp. 17–59.

Etminan, M.; S.S. Gill; A. Samii. Intake of vitamin E, vitamin C, and carotenoids and the risk of parkinson's disease: A meta-analysis. *Lancet Neurology*, **2005**, *4*, 362–365.

Fitó, M.; M. Cladellas; R. de la Torre; J. Martí; M. Alcántara; M. Pujadas-Bastardes; J. Marrugat; J. Bruguera; M.C. López-Sabater; J. Vila; M.I. Covas. Antioxidant effect of virgin olive oil in patients with stable coronary heart disease: A randomised, crossover, controlled, clinical trial. *Atherosclerosis* **2005**, *181*,149–158.

Fitó, M.; M.I. Covas; R.M. Lamuela-Raventós; J. Vila; J. Torrents; C. de la Torre-Boronat; J. Marrugat. Protective effect of olive oil and its phenolic compounds against low density lipoprotein oxidation. *Lipids*, **2000**, *35*, 633–638.

Gandul-Rojas, B.; M.R. López-Cepero; M.I. Mínguez-Mosquera. Use of chlorophylls and carotenoids pigments composition to determine authenticity of virgin olive oil. *J. Am. Oil Chem. Soc.* **2000**, *77*, 853–858.

Gandul-Rojas, B.; M.I. Mínguez-Mosquera. Chlorophylls and carotenoid: Composition in virgin olive oils from various spanish olive oil varieties. *J. Sci. Food Agric.* **1996**, *72*, 31–39.

García, A.; M. Brenes; P. García; C. Romero; A. Garrido. Phenolic content of commercial olive oils. *Eur. Food Res. Technol.* **2003**, *216*, 520–525.

García-Gómez, A.; A. Roig; M.P. Bernal. Composting of the solid fraction of olive mill wastewater with olive leaves: Organic matter degradation and biological activity. *Biores. Technol.* **2003**, *86*, 59–64.

García-González, D.L.; R. Aparicio. Olive oil research: Current challenges. *Inform* **2007**, *18*, 286–288.

García-González, D.L.; N. Tena; R. Aparicio. Characterization of olive paste volatiles to predict the sensory quality of virgin olive oil. *Eur. J. Lipid Sci. Technol.* **2007**, *109*, 663–672.

García-Martínez, M.; Z. Aragonés; N. Poole. A repositioning strategy for olive oil in the UK market. *Agribusiness* **2002**, *18*, 163–180.

Giuffrida, D.; F. Salvo; A. Salvo; L. La Pera; G. Dugo. Pigments composition in monovarietal virgin olive oils from various sicilian olive varieties. *Food Chem.* **2007**, *101*, 833–837.

Harwood, J.L.; J. Sánchez. Lipid biosynthesis in olives. *Handbook of Olive Oil: Analysis and Properties;* J. Harwood, R. Aparicio, Eds.; Aspen: Gaithersburg, MA, 2000; pp. 61–77.

Harwood, J.; P. Yaqoob. Nutritional and health aspects of olive oil. *Eur. J. Lipid Sci. Technol.* **2002**, *104*, 685–697.

International Olive Council (2006) *Trade Standard Applying to Olive Oil and Olive-Pomace Oil.* COI/T.15/NC n°3/Rev. 2, Madrid Spain.

Kalantzakis, G.; G. Blekas; K. Pegklidou; D. Boskou. Stability and radical-scavenging activity of heated olive oil and other vegetable oils. *Eur. J. Lipid Sci. Technol.* **2006**, *108*, 329–335.

Katan, M.B.; S.M. Grundy; P. Jones; M. Law; T. Miettinen; R. Paoletti. Efficacy and safety of plant stanols and sterols in the management of blood cholesterol levels. *Mayo Clin. Proc.* **2003**, *78*, 965–978.

Keys, A., A. Menotti; M.J. Karvonen; C. Aravanis; H. Blackburn; R. Buzina; B.S. Djordjevic; A.S. Dontas; F. Fidanza; M.H. Keys; et al. The diet and 15-year death rate in the seven countries study. *Am. J. Epidemiol.* **1986**, *124*, 903–915.

Koidis, A.; D. Boskou. The contents of proteins and phospholipids in cloudy (veiled) virgin olive oils. *Eur. J. Lipid Sci. Technol.* **2006**, *108*, 323–328.

Kränke B.; P. Komericki; W. Aberer. Olive oil—contact sensitizer or irritant? *Contact Dermatitis* **1997**, *36*, 5–10.

Kushi, L.; E. Giovannucci. Dietary fat and cancer. *Am. J. Med.* **2002**, *113*, 63S–70S.

Luna, G.; M.T. Morales; R. Aparicio. Changes induced by UV radiation during virgin olive oil storage. *J. Agric. Food Chem.* **2006a** , *54*, 4790–4794.

Luna, G.; M.T. Morales; R. Aparicio. Characterisation of 39 varietal virgin olive oils by their volatile composition. *Food Chem.* **2006b**, *98*, 243–252.

Masghouni, M.; M. Hassairi. Energy applications of olive oil industry by-products: I. The exhaust food cake. *Biomass Bioenergy* **2000**, *18*, 257–262.

Mataix, J.; F.J. Barbancho. Olive oil in mediterranean food. *Olive Oil and Health;* J.L. Quiles, M.C. Ramirez-Tortosa, P. Yaqoob, Eds.; CAB International: Wallingford, UK, 2006; pp. 1–44.

Mekki, H.; M. Anderson; E. Amar; G.R. Skerratt; M. BenZina. Olive oil mill waste water as a replacement for fresh water in the manufacture of fired clay bricks. *J. Chem. Technol. Biotechnol.* **2006**, *81*, 1419–1425.

Morales, M.T.; F. Angerosa; R. Aparicio. Effect of the extraction conditions of virgin olive oil on the lipoxygenase cascade: chemical and sensory implications. *Grasas Aceites*, **1999**, *50*, 114–121.

Morales, M.T.; M. León-Camacho. Gas and liquid chromatography: Methodology applied to olive oil. *Handbook on Olive Oil: Analysis and Properties;* J. Harwood, R. Aparicio, Eds.; Aspen: Gaithersburg, MA, 2000; pp. 159—208.

Morales, M.T.; G. Luna; R. Aparicio. Comparative study of virgin olive oil sensory defects. *Food Chem.* **2005**, *91*, 293–301.

Morales, M.T.; M. Tsimidou. The role of volatile compounds and polyphenols in olive oil sensory quality. *Handbook on Olive Oil: Analysis and Properties;* J. Harwood, R. Aparicio, Eds.; Aspen: Gaithersburg, MA, 2000; pp. 393–458.

Ordovás, J.M., et al. (2005) State of the art in olive oil, nutrition and health. IOC Scientific Seminar on Olive Oil and Health. 7–8 March, Madrid.

Owen, R.W.; W. Mier; A. Giacosa; W.E. Hule; B. Spiegelhalder; H. Bartsch. Phenolic compounds and squalene in olive oils: The concentration and antioxidant potential of total phenols, simple phenols, secoroids, lignans and squalene. *Food Chem. Toxicol.* **2000**, *38*, 647–659.

Pala, V., et al. Associations between dietary pattern and lifestyle, anthropometry and other health indicators in the elderly participants of the EPIC-Italy cohort. *Nutr. Metab. Cardiovasc. Dis.*

**2006**, *16*, 186–201

Paraskeva, P.; E. Diamadopoulos. Technologies for olive mill wastewater (OMW) treatment: A review. *J. Chem. Technol. Biotechnol.* **2006**, *81*, 1475–1485.

Peréz-Jiménez, F.; J. Ruano; P. Pérez-Martínez; F. López-Segura; J. López-Miranda. The influence of olive oil human health: not a question of fat alone. *Mol. Nutr. Food Res.* **2007**, *51*, 1199–1208.

Perona, J.S.; R. Cabello-Moruno; V. Ruiz-Gutierrez. The role of virgin olive oil components in the modulation of endothelial function. *J. Nutr. Biochem.* **2006**, *17*, 429–445.

Perona, J.S.; J. Martínez-González; J.M. Sánchez-Domínguez; L. Badimon; V. Ruiz-Gutiérrez. The unsaponifiable fraction of virgin olive oil in chylomicrons from men improves the balance between vasoprotective and prothrombotic factors released by endothelial cells. *J. Nutr.* **2004**, *134*, 3284–3289.

Renaud, S.; de M. Lorgeril; J. Delaye; J. Guidollet; F. Jacquard; N. Mamelle; J.-L. Martin; I. Monjaud; P. Salen; P. Toubol. Cretan mediterranean diet for prevention of coronary heart disease. *Am. J. Clin. Nutr.* **1995**, *61*, 1360S–1367S.

Rodríguez-Rodríguez, R.; M.D. Herrera; J.S. Perona; V. Ruíz-Gutiérrez. Potential vasorelaxant effects of oleanolic acid and erythrodiol, two triterpenoids container in "orujo" olive oil, on rat aorta. *Br. J. Nutr.* **2004**, *92*, 635–642.

Rose, D.P. Effects of dietary fatty acids on breast and prostate cancers: evidence from in vitro experiments and animal studies. *Am. J. Clin. Nutr.* **1997**, *66*, 1513S–1522S.

Ruíz-Gutiérrez, V.; F.J. Muriana; A. Guerrero; A.M. Cert; J. Villar. Plasma lipids, erythrocyte membrane lipids and blood pressure of hypertensive women after ingestion of dietary oleic acid from two different sources. *J. Hypertens.* **1996**, *14*, 1483–90.

Rutter, A.J.; J. Sánchez; J.L. Harwood. Glycerolipid synthesis by microsomal fractions from *Olea europaea* fruits and tissue cultures. *Phytochem.* **1997**, *46*, 265–272.

Saija, A.; D. Trombetta; A. Tomaino; R. Lo Cascio; P. Princi; N. Acella; F. Bonina; F. Castelli. In vitro evaluation of the antioxidant activity and biomembrane interaction of the plant phenols oleuropein and hydroxy-Tyrosol. *Int. J. Pharm.* **1998**, *166*, 123–133.

Salas, J.J.; D.L. García-González; R. Aparicio. Volatile compound biosynthesis by green leaves from an *Arabidopsis thaliana* hydroperoxide lyase knockout mutant. *J. Agric. Food Chem.* **2006**, *54*, 8199–8205.

Salvini, S., et al. Daily consumption of a high-phenol extra-virgin olive oil reduces oxidative DNA damage in postmenopausal women. *Br. J. Nutr.* **2006**, *95*, 742–751.

Sánchez, J.; M.T. Cuvillo; J.L. Harwood. Glycerolipid synthesis by microsomal fractions from olive fruits. *Phytochem.* **1992**, *31*, 129–134.

Sánchez, J.; J.L. Harwood. Fatty acid synthesis in soluble fractions from olive (*Olea europaea*) fruits. *J. Plant Physiol.* **1992,** 140, 402–408.

Sánchez, J.; J.J. Salas, Photosynthetic carbon metabolism of olives. *Physiology, Biochemistry and Molecular Biology of Plant Lipids;* J.P. Williams, M.U. Khan, N.W. Lenn, Eds; Kluwer Academic: Dordrecht, Holland, 1997; pp. 325–327.

Scarmeas, N.; Y. Stern; M.-X. Tang; R. Mayeux; J.A. Luchsinger. Mediterranean diet and risk for alzheimer's disease. *Annals Neurology*, **2006**, *59*, 912–921.

Schwartz, B.; Y. Birk; A. Raz; Z. Madar. Nutritional-pharmacological combinations. A novel approach to reducing colon cancer incidence. *Eur. J. Nutr.* **2004**, 43, 221–229.

Serra-Majem, L.; L. La Vecchia; F. Ribas-Barba; F. Prieto-Ramos; F. Lucchini; J.M. Ramon; L. Salleras. Changes in diet and mortality from selected cancers in southern mediterranean countries 1960–1989. *Eur. J. Clin. Nutr.* **1993**, *47*, S25–S34.

Servili, M.; R. Selvaggini; A. Taticchi; S. Esposto; G.F. Montedoro. Air exposure time of olive pastes during the extraction process and phenolic and volatile composition of virgin olive oil. *J. Am. Oil Chem. Soc.* **2003**, *80*, 685–695.

Soriguer, F., et al. Oleic acid from cooking oils is associated with lower insulin resistance in the general population (Pizarra Study). *Eur. J. Endocrinol.* **2004**, *150*, 33–39.

Tena, N.; A. Lazzez; R. Aparicio-Ruiz; D.L. García-González. Volatile compounds characterizing tunisian chemlali and chétoui virgin olive oils. *J. Agric. Food Chem.* **2007**, *55*, 7852–7858.

Truswell, A.S.; N. Choudhury. Monounsaturated oils do not all have the same effect on plasma cholesterol. *Eur. J. Clin. Nutr.* **1998**, *52*, 312–315.

Vas, G.; K. Vékey. Solid-phase microextraction: A powerful sample preparation tool to mass spectrometry analysis. *J. Mass Spectrom.* **2004**, *39*, 233–254.

Yangkyo, P.S.; M.J. Grove; H. Takamura; H.W. Gardner. Characterization of a C-5, 13-cleaving enzyme of 13(s)-hydroperoxide of linolenic acid by soybean seed. *Plant Physiol.* **1995**, *108*, 1211–1218.

Young, D.P.; D.X. Kong; H.Y. Zhang. Pharmacological effects of olive oil phenols. *Food Chem.* **2007**, *104*, 1269–1271.

# 2

# Avocado Oil

**Allan Woolf[1], Marie Wong[3], Laurence Eyres[2], Tony McGhie[1], Cynthia Lund[1], Shane Olsson[1], Yan Wang[3], Cherie Bulley[1], Mindy Wang[1], Ellen Friel[4], and Cecilia Requejo-Jackman[1]**

*[1]HortResearch, The Horticulture & Food Research Institute of New Zealand Limited, Auckland, New Zealand; [2]Oil and Fats Group, N.Z. Institute of Chemistry, Auckland, New Zealand; [3]Institute of Food Nutrition and Human Health, Massey University, Albany, Auckland, New Zealand. and [4] Diageo Baileys Global Supply, Nangor House, Nangor Road, Dublin 12, Ireland.*

## Introduction

Avocado fruit are well-known, with millennia of consumption in the Americas and an increasing popularity in the rest of the world. Throughout history, avocado oil was renowned for its healing and regenerating properties. Early writings from the sixteenth century reported the use of the oil obtained from the seed to treat rashes and scars (Argueta-Villamar et al., 1994). In skin care, the two major advantages of avocado oil are its marked softening and soothing nature and its notable absorption. For example, compared with almond, corn, olive, and soybean oils, avocado oil had the highest skin-penetration rate (Swisher, 1988).

Avocado oil obtained from the flesh (fruit), on the other hand, is a relatively new arrival in culinary circles. The predominant uses of avocado oil are in the cosmetic industry because of its stability and high level of vitamin E ($\alpha$-tocopherol). The volume of avocado oil produced (or traded) is relatively small compared with other oils, with $\cong$ 2000 tonnes/year.

Avocado (*Persea americana* Mil.) is a subtropical tree which is relatively frost-sensitive and grows to a height of 5–30 m (Scora et al., 2002). The fleshy fruit are borne yearly from the current season's wood, and the green fruit ripen only after being harvested. They are grown in most countries of the world that are frost-free, and are generally a "high input" crop requiring good horticultural management.

The bulk of avocado oil is extracted by relatively harsh methods (high temperature and solvent extraction), typically followed by standard refining steps (refining, bleaching, and deodorizing). The development in the twenty-first century was that of successful cold-pressing of avocado fruit by using technologies similar to those used to produce extra-virgin olive oil. This successful development was led by New Zealand, and its commercialization started in 2000. The demand for information by New Zealand commercial entities has driven much of the research on cold-pressed avocado oil, which we have published in various forms (e.g., Ashton et al., 2006; Wong et al., 2008), and we present additional unpublished material here.

Avocado oil is generally produced from fruit rejected from the fresh-fruit trade—either domestic or export-oriented— depending on the country. This is significantly different from olive or palm oils (grown and harvested solely for oil production) or from other oils produced from processing by-products (e.g., rice- bran oil). The ultimate market of the oil generally dictates the fruit quality required. For the production of good-quality cold-pressed oil, the fruit must be relatively "sound" with mainly cosmetic-quality issues or too small for sale. For cosmetic-oil production, one can use poor-quality rejected fruit (often with rots), although subsequent refining is required.

The cultivar Hass accounts for over 90% of total production in many large avocado-producing countries including Mexico, Chile and the United States, and this is also the case in other smaller producer countries (e.g., New Zealand, Spain, and Australia; Avocadosource, 2006). For this reason, Hass is the most likely cultivar to be used for avocado oil production. The other important cultivars grown worldwide in approximate order of priority include Fuerte, Ryan, Pinkerton, and Edranol. However, one must note that information is lacking on the cultivar diversity of large portions of avocado production, such as in Indonesia (the fourth-largest producer of avocados).

Overall, limited published information exists on avocado oil, and even less on cold-pressed avocado oil. Cold-pressed avocado oil (on which this chapter concentrates) is a new product with significant production, commercialization, and marketing only occurring in the twenty-first century. Thus, we include here information which is unpublished or not readily accessible. Some of these results are preliminary in nature, but provide important perspectives and point to research needs or commercial directions.

## Applications and Economics

### Cosmetic Applications

Although reliable data are not available for production and international trade of oil, the U.S. volume is said to be around 1000 tonnes. The main use for this trade is the cosmetics industry, where it is highly valued for its beneficial effects on the skin. For its use in cosmetics, crude avocado oil is further processed (refined, bleached. and deodorized: RBD). The resulting oil is pale yellow (instead of green) and has little remaining avocado odor or taste (Eyres et al., 2001). The refined oil is used in skin- care products since it is rapidly absorbed by the skin, and has sunscreen properties (Human, 1987; Swisher, 1988). Avocado oil is claimed to be good for tissue and massage creams, muscle oils, and other products where lubrication and penetration are essential, as it is one of the most penetrating oils available for cosmetics and soaps. It also forms finer emulsions because it reduces surface tension (Poucher, 1974). It is used in soaps to provide improved lathering, and it forms smoother creams (Human, 1987).

## Culinary Applications

Cold-pressed avocado oil has chemical properties similar to olive oil. At least 60% of the fatty acids are monounsaturated, and approximately 10% are polyunsaturated. In addition, cold-pressed avocado oil contains relatively high levels of pigments (chlorophylls and carotenoids) which act as antioxidants. The global trend of high-spending consumers is toward the consumption of fewer processed products because of controversy about the linkage of some chemicals with human diseases. This is reflected in the increase in the consumption of cold-pressed olive oil in the United Kingdom, where sales increased from 43% in 2005 to 51% of the total market (Mintel Report 2005 cited by Fletcher, 2006). Cold-pressed avocado oil appeals to the consumer looking for a delicate buttery flavor without the pungent notes of extra-virgin olive oil. In addition, avocado oil has a high smoke point (over 250°C) which makes it suitable for shallow pan frying. Currently, the oil retails at approximately US$5 per bottle (250 mL) on the New Zealand market and at a comparable price in Australia. Bulk avocado oil sales are at US$10 per liter and are increasing as U.K. and U.S. users become aware of the reliable supply (Eyres et al., 2001).

## Economics

The business model of the avocado oil industry in almost all countries relies on using "reject grade" fruit (low-price rated) from commercial packhouses, which have not met export or local-market quality standards. Generally, the grower does not receive a large economic benefit (in terms of price/kg of fruit). However, the removal of this "bottom" class of fruit from the local market does result in significant indirect benefits to growers by increasing the local-market price due to reduced fruit volumes.

Initial technical and market studies by New Zealand entrepreneurs showed that a premium avocado oil could achieve high market prices. Further research showed that avocado plantations in the main- producer countries are young and production volumes are increasing, and thus the volumes of undergrade fruit will consequently increase substantially (Requejo-Tapia, 1999). In addition, both a study of the market for edible oils and the changes in consumer preferences for healthier oils show an increase in consumption and a willingness to pay a premium price for alternatives to the standard oils (Mintel Report 2005 cited by Fletcher, 2006). In the United Kingdom alone, sales of all edible liquid oils are expected to rise in both value and volume, with the value of the market rising some 11% by 2010 to just under US$420 million. Avocado oil production is limited and specialized; therefore, it is considered a specialty oil. This means that avocado oil enjoys a high-value market.

## Plant Establishment and Oil-Yield Economics

The approximate cost of establishing an avocado oil processing plant is valued at US$1 million (Nathan, 2006). The processing equipment should be stainless-steel, state-of-the-art processing and bottling plants to achieve high food-grade quality standards. Avocado fruit need to be ripened to obtain maximal oil yield; yet they are highly perishable, with a significant propensity for rot development as fruit ripen. This means that a challenge exists in establishing the infrastructure and experience of handling, storing, and ripening avocados in a predictable and robust manner that minimizes fruit losses and rots. During processing, the skin and seeds are removed, and the oil is extracted from the flesh. Typical oil yields can vary from 10 to 15% of total fruit weight, depending on the time of harvest.

## Overview of the Cold-pressed Avocado Oil Industry

In the eight years since the first cold-pressed processing facility was set up in New Zealand, the avocado oil industry has grown relatively slowly. Numerous reasons may explain the slow expansion of the avocado oil industry. Firstly, because avocado oil is a side-industry of the fresh-fruit industry, the volume of fruit available for processing suffers significant annual fluctuations. This has occurred in Mexico, for instance (where high domestic consumption occurs), and extraction plants were mothballed during some years. In New Zealand annual oil production, figures varied significantly because of climate effects on fruit set and thus final production figures. Secondly, significant competition exists for fruit to be used for other processed products, such as guacamole produced by using either freezing or, more recently, high-pressure processing (San Martín et al., 2002), particularly in the main- producer countries.

If one considers the relative importance of countries by the volume of avocado traded (Wong et al., 2008), then countries such as Mexico, Chile, and the United States are the most important. However, also many other countries exist where large volumes of avocado are grown (e.g., Indonesia and Africa), yet these countries do not trade significant quantities. Countries such as these and small islands (e.g., the Pacific Islands or Haiti) might well contain significant under-utilized sources of fruit for oil processing. However, commercial feasibility is often a problem in these situations, and aid/development money may be required from donors such as the UN, FAO, and/or commercial partnerships.

Avocados are a high-input crop from a horticultural perspective, and because of this no orchards were dedicated solely to growing fruit for oil production. However, such a venture might be economically feasible if one can find a cultivar that produces very high oil yields and/or is of particularly high health-value because of higher levels of healthful phytochemicals (such as tocopherols, sterols, or carotenoids).

## Overview of Avocado: Fruit Physiology and Postharvest Ripening

### *Avocado Fruit Growth and Harvest*

Avocados are one of the few fruits, other than olives, that accumulate significant amounts of oil, and in some situations contain even more oil than olives. Avocado fruit consist of an exocarp or skin, the flesh (mesocarp), and the seed (stone or pit). Depending on the cultivar, the edible flesh makes up 50–80% of the total fruit weight, while the seed comprises 10–25% (Lewis, 1978). The seed is covered by two thin seed coats adhering to each other, and the woody seed is made up of two cotyledons made of starch-rich parenchyma tissue containing scattered oil droplets (Biale & Young, 1969).

Avocado-fruit development on the tree involves increased size and lipid content in the flesh tissue, while the moisture content decreases (Appleman, 1969; Hopkirk, 1989; Kikuta & Erickson, 1968; Lawes, 1980; Pearson, 1975). Avocados are unusual in that cell replication continues during development, whereas most other fruit-development patterns involve initial cell division and then, water-driven, increase in size approaching harvest (Schroeder, 1953). Lipids and water are the two major components of a mature avocado. Oil content in commercially mature avocados can range between 10 and 32% on a flesh fresh weight basis, depending on the cultivar and the time of harvest. For Hass avocados, this can be as high as 30% on a flesh fresh weight basis, depending on the growing region (Kaiser et al., 1992; Requejo-Tapia, 1999). The total oil content in the fruit increases during fruit development on the tree (Fig. 2.1).

Significantly, the level of oil and the fatty- acid composition do not change during ripening after the fruit are harvested (Luza et al., 1990). Similarly, no changes in

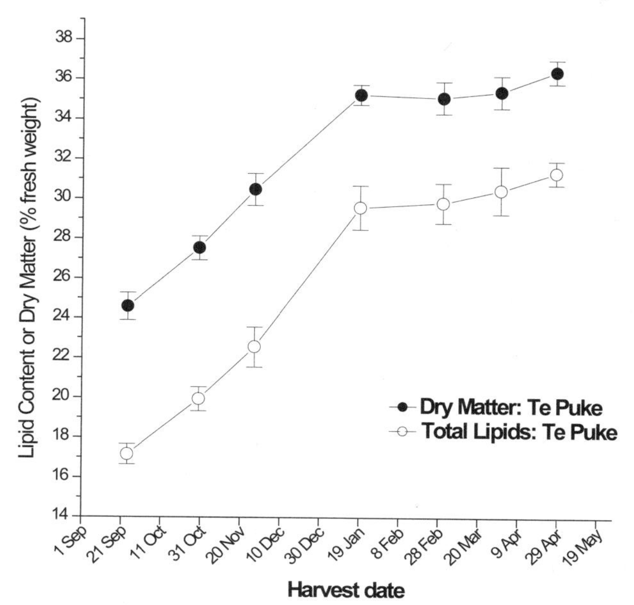


**Fig. 2.1. Mean lipid (oil) content and dry matter of Hass avocado fruit harvested over a commercial season from one New Zealand orchard in 1998–1999. Each point is the average of four replicates of five fruits. Vertical bars are the standard errors of the mean.**

fatty acid composition and concentration were observed at storage temperatures of 0–7°C for 2–6 weeks nor during subsequent ripening (Eaks, 1990; Kikuta & Erickson, 1968; Luza et al., 1990). These results were observed in the two key commercial cultivars Hass and Fuerte, and likely, similar results would be observed in other cultivars.

In the avocado flesh, the lipids are stored in two cell types: the parenchyma cells and in specialized oil cells or idioblasts. Approximately 85% of the total lipids present in an avocado are present as triglycerides, which occur as droplets or oil bodies scattered in the cytoplasm of the parenchyma cells (Platt-Aloia & Thomson, 1981). Ultrastructural studies showed that the idioblasts are larger than the parenchyma cells; however, these correspond only to approximately 2% of the cells in the mesocarp. The fate of the idioblasts in the cold-pressing process has not yet been determined. They may be broken down in the extraction process (and are therefore released into the oil), or remain in the solid phase (pomace), or liquid stream.

Much historical published information on avocado oil content was related to the use of oil content as a measure of the minimal maturity to harvest avocados (e.g., Lee et al., 1983). However, reliable and robust oil extraction is labor- intensive and expensive to implement on orchards. Because a strong relationship exists between oil content (as measured by solvent extraction) and dry matter, avocado industries worldwide base their regulations or recommendations for minimal harvest time on dry matter (Brown, 1984; Hopkirk, 1989; Lee et al., 1983).

### *Postharvest Ripening of Avocados*

A noteworthy characteristic of avocado fruit is that they do not ripen (soften to an edible state) while attached to the tree. Once removed from the tree, fruit ripening generally takes ≅6–10 days. The timing and variability of ripening are influenced by the cultivar, the stage of maturity, and other external factors such as storage time, temperature, and ethylene exposure. Therefore, ripening can be as short as 3–4 days or as long as 18–21 days. Ethylene treatment (a naturally occurring gaseous plant hormone) is used as a means of accelerating and synchronizing ripening, and is an important tool for the sale of "ripe tonight" fruit to consumers (as is carried out for bananas), but also for those processing avocados for cold-pressed oil.

During ripening, the flesh structure degrades as the pectin in the walls of the parenchyma cells solubilizes, causing the flesh to soften to a soft melting texture (Redgwell et al., 1997). For the main commercial cultivar Hass, changes in softening are accompanied by a distinct skin-color change from green to black/purple (Cox et al., 2004; Williams, 1977). For the majority of other cultivars ("green skins"), skin-color changes are more subtle, with a darkening of the green color from more emerald to darker green. Low-temperature storage of fruit for transport and sale is commonplace in the fresh-fruit industry. Cool storage at 4 to 8°C for up to about 4 weeks also reduces the ripening rate and fruit-to-fruit variability. However, longer storage periods can cause an internal-chilling injury (flesh discoloration) and general reduction of ripe-fruit quality (White et al., 2005).

### Tissue Types and Proportions of Hass Avocado Fruit

For Hass avocado, the fruit is made up of approximately 68% of flesh, 18% of seed, and 14% of skin (by fresh weight) (Wong et al., 2008). Fruit size has relatively little effect on the proportion of the fruit that is made up of flesh, and therefore, one can process all fruit sizes for oil extraction. This is a particularly useful attribute since at the end of the harvest season many growers "strip pick," and therefore these harvests have a wide range of fruit sizes.

## Avocado Oil Extraction

As has long been recognized, the release of the oil from the flesh of the avocado is not as easy as observed in other fruit tissue, such as olives. For example, one can observe oil release in mature olives by simply crushing the fruit (Kiritsakis, 1998); this is not the case for avocado flesh. A demonstration of the relative difficulty of extraction is shown in Fig. 2.2. Small disks of tissue (5-mm diameter by 1-mm thickness) were extracted by using chloroform/methanol (C/M) with three tissue-preparation techniques. The first was simply to agitate the slices for 24 hours on a tissue shaker in C/M ("Slices"), the second was to grind the tissue in liquid nitrogen in a mortar and pestle and then vortex the powder in C/M ("Liquid Nitrogen"), and finally, slices were ground using an overhead blender (Ultra-turrex (Jancke and Kundel, Munich, Germany) for approximately 30 seconds. Clearly, fruit ripeness has a major effect on oil extraction, and polytron grinding was the most effective technique.

Avocado oil was first extracted for use in the cosmetic industry. Since the flesh of the avocado has a relatively high water content, initial attempts to recover the oil from the flesh by using hydraulic pressures or organic solvents required the drying of the flesh prior to extraction (Human, 1987; Montano et al., 1962; Smith & Winter, 1970). Extraction by mechanical means leads to poor oil yields, while the use of solvents (petroleum ether, ethyl ether, or benzene) results in recoveries of between 60 and 90% of the total oil available. Boiling the flesh pulp and freezing it prior to solvent extraction were trialed to enhance extraction. Love (1944) added lime ($CaCO_3$) to the flesh pulp to form a dry cake from which the oil was extracted in one of three ways: by hydraulic press, by organic solvents, or by water flotation. They reported more than 80% of total oil recovery using these methods.

The extraction of the oil by centrifugation to obtain oil free of solvent impurities and suitable for food use was demonstrated in the 1980s (Buenrostro & López-Munguia, 1986; Swisher, 1988; Werman & Neeman, 1987). Using centrifugation techniques requires the fruit to be soft (i.e., ripe), but oil yields are significantly less than with solvent extraction, ranging from 30 to 80% of total oil (Buenrostro &

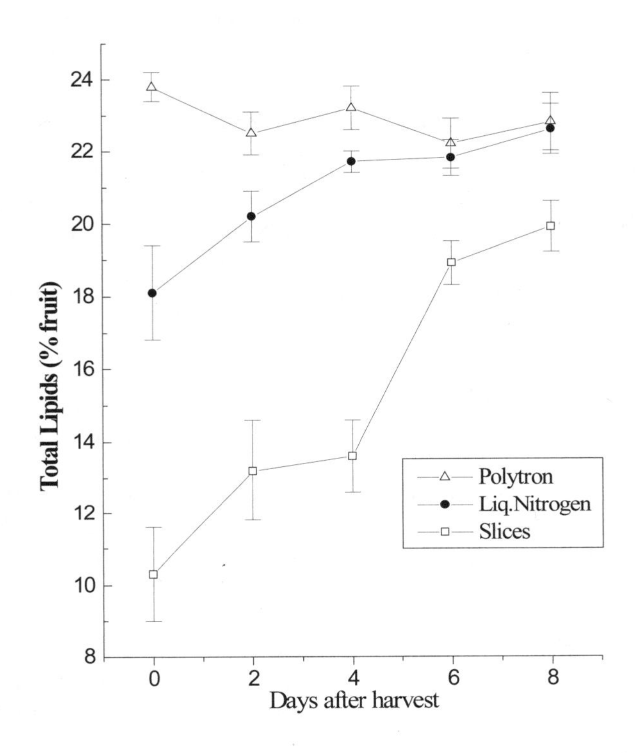


**Fig. 2.2. Mean percentage of lipids extracted from Hass avocado fruit immediately after harvest, and after 2, 4, 6, and 8 days (8 = soft ripe) after harvest. Fruit were ethylene-treated at 17°C for 48 hours. Each point is the average of six replicates of tissue taken from three fruits. Samples were extracted using three different methods. Vertical bars = standard errors of the mean (SEM).**

López-Munguia, 1986; Werman & Neeman, 1987). Wermen & Neeman (1987) compared solvent extraction and the centrifugation method. Prior to centrifugation, they also heated the pulp to temperatures of 40–75°C. They observed differences in the composition and quality of oil extracted by centrifugation or by solvents, including differences in fatty acid composition, unsaponifiables, chlorophyll content, acid values, and hydroxyl values. Differences in fatty acid profiles were attributed to the difference in the raw materials and to the extraction methods. Extraction by centrifugation resulted in higher chlorophyll concentrations in unrefined oils.

Southwell et al. (1990) extracted avocado oil from sun-dried avocado slices by using a screw-type expeller similar to that used in the extraction of seed oils. They compared peeled slices versus unpeeled, and evaluated the effect of heat conditioning (80°C for 30 minutes) prior to pressing by an expeller. A brown-colored oil was obtained from unpeeled or peeled slices, with extraction yields of clarified oil of 50–63% of total available oil. Ortiz et al. (2004) investigated the use of microwaves to heat the flesh to 95°C to disrupt cell structure. Even using simple manual squeezing of the pulp yielded 69% of total oil recovery.

Both Buenrostro and López-Munguia (1986) and Werman and Neeman (1987) extracted avocado oil by the centrifugation technique at 40°C. During the twentieth century, no commercial plants were set up for the cold-pressed extraction of avocado oil based on the centrifugation method. Historically, consumers have preferred refined oils for food use, but as the trend toward more "natural" foods has grown, unrefined cold-pressed oils have become more marketable. Soon after 2000, the commercial cold-pressed extraction of avocado oil by centrifugation successfully commenced in New Zealand. Two factories started producing natural-green, cold-pressed avocado oil for culinary food use (Board, 2007; Eyres et al., 2001). Both plants used an aqueous extraction similar to olive-oil processing, followed by centrifugation at temperatures below 50°C.

Botha and McCrindle (2003) reported on the use of supercritical fluid extraction for avocado oil recovery. The avocado flesh was first air-dried in an oven (80°C) prior to grinding; then the oil was extracted with supercritical carbon dioxide. They were able to extract >90% of the total available oil. Mostert et al. (2007) showed that the pre-treatment of the fruit and the grinding method influenced the degree of extraction achievable by supercritical carbon dioxide or by hexane solvent extraction.

Because of the relatively high water content of avocado flesh (compared with seed products), the flesh is dried to remove the excess water prior to any solvent extraction. Drying by hot air, freeze drying, or microwave was investigated to determine the effect on the cellular structure of the flesh (Human, 1987; Mostert et al., 2007; Ortiz et al., 2004). Numerous solvents were used, including petroleum ether, ethyl ether, benzene, hexane, ethanol/hexane mix, and supercritical carbon dioxide (Love, 1944; Mostert et al., 2007; Ortiz et al., 2004; Smith & Winter, 1970; Werman & Neeman, 1987). This method of extraction of avocado oil requires considerable capital investment, and if organic solvents are used, the recovery of these solvents can become costly. The removal of all traces of solvent from the oil is important, and one should minimize the loss of solvents into the environment.

## Refining

Avocado oil contains high levels of chlorophyll, and, depending on the fruit quality used for extraction, the oil can be highly colored or can contain high levels of free fatty acids (FFAs). The early production of avocado oil for cosmetic or food use required a light-colored, odorless oil; hence, the oil was bleached and refined (Human, 1987). In contrast, Poucher (1974) claimed that the green color of the oil provided a more natural appeal to cosmetic products.

Bleaching is carried out to remove any pigments in the oil, followed by deodorizing to remove any pungent odors. Other refining steps may include winterizing to settle out any high melting- point components or alkali refining if the FFA level is too high. Standard refining, bleaching, and deodorizing processes used currently for other edible oils (Anderson, 2005) can be used for these steps, and are not discussed further here.

## Cold-pressed Avocado Oil (Culinary)

### *Extraction*

The current aqueous extraction of avocado oil contradicts all earlier attempts to extract the oil from the avocado flesh. Instead of removing water prior to the recovery of the oil, water is added to the flesh pulp to help with mixing and the release of the oil from cells. The exact mechanism of oil release from cells is still unknown, but adequate grinding of the flesh is important to disrupt the cells, as is the malaxing (mixing) step where presumably endogenous enzymes assist with cell-wall breakdown and subsequent oil release.

The current aqueous-extraction procedures carried out in commercial plants in New Zealand were described by Wong et al. (2008). The process is based on a continuous process used commercially for olive-oil processing (Kiritsakis, 1998). Ripe, whole avocados are washed, followed by a seed/skin removal step. The flesh is then ground to a pulp by using a hammer mill or grinder. Water is sometimes added to this pulp to achieve a paste of lower viscosity which is then malaxed (mixed) in a temperature-controlled horizontal tank with a ribbon mixer. During malaxing, oil is released from cells. Mixing is slow, and emulsion formation was not a problem. The temperature during malaxing is maintained at 40–50°C. After malaxing, the paste is pumped to the horizontal decanter operating at 3000–4000 rpm. Often water is added during pumping, as the paste can be very viscous. In the decanter, the liquid phase (water and oil) is separated from the solids (pomace). The water and oil are then passed through polishing disc centrifuges to separate the water from the oil.

During the seed-and–skin-removal stage, approximately 90% of the skin is removed, although differences exist between producers, depending on the technique used. The inclusion of 40 or 100% of the available skin resulted in oils with higher levels of chlorophyll pigments (unpublished). The oils extracted with high skin levels also had different sensory characteristics. Near-complete seed removal is typically achieved, and seeds generally remain intact.

The quality of avocado oil extracted by using the aqueous-extraction method was very high compared with olive oil. Typical percentage of FFA levels and peroxide values (PVs) in fresh avocado oil are <0.5% w/w (as oleic acid) and <4 meq/kg of oil, respectively, while for fresh olive oil the values for FFA and PV are <0.5% w/w and <10 meq/kg of oil, respectively.

### *Typical Yields*

As discussed earlier, the amount of total oil available in the avocado increases with maturity. Even though the fruit are generally ripened to the same degree of firmness over the harvest season, an increase occurs in the oil yield from cold-pressed extraction as the season progresses. Early in the season (dry matter of ≅24%), the total oil content can be lower than 10% w/w, while it rises to ≅18% in the mid-season, and up to 25% in the late season.

To assist with oil extraction, numerous additives were investigated, including salts, enzymes, and pH adjustment. Werman and Neeman (1987) and Bizimana et al. (1993) found that, by adding water (3:1 or 5:1 water/avocado), adjusting the pH of the pulp to pH 5.5, and adding NaCl or $CaCO_3$ and $CaSO_4$, they could increase oil yield by 5–18%; both groups heated the pulp to 75–98°C. Buenrostro and López-Munguia (1986) added water at a ratio of 4:1 and a number of exogenous enzymes (cellulose, papain, α-amylase). They achieved 30–50% of increases in oil yield with enzyme addition, with α-amylase being the most effective.

In contrast to this earlier research, our research and the methodology used in New Zealand commercial plants use a water/avocado ratio of 1:3 or less, this being considerably less water than earlier attempts to extract avocado oil by centrifugation. With early-season fruit (dry matter of ≅24–28%), the addition of commercial enzymes containing various activities (pectinase, hemicellulase, cellulose) improved oil yields by 5–40%. No significant increase in oil yield with mid-season fruit was observed (unpublished). Overall, the addition of exogenous enzymes at the malaxing stage has not made the significant increases in oil yield hoped for.

### *Refining*

Avocado oil extracted by using the aqueous-extraction procedure is marketed as a "naturally" extracted (i.e., little to no chemical usage) culinary oil in the same market niche as extra-virgin cold-pressed olive oil. A key commercial goal is to retain the distinctive green color of the oil, as well as the natural flavors and aromas. Hence, to maintain as many natural attributes as possible, the oils are not refined prior to bottling. As occurs with other oils, oxidation can occur if the oil is not handled correctly or is "abused". Ideally, the oil is stored in stainless-steel or oxygen-impermeable tanks, and sparged with nitrogen to remove any free oxygen. The oils are typically stored to allow for the settling of waxes and high melting-point fatty acids or phospholipids. The degree of settling required depends on the original fruit used and the growing region and climate. Differences in the lipid composition in the oil were noticeable in oil from fruit from different regions and countries. For example, oil extracted from fruit grown in Queensland, Australia, contains more waxes and high melting-point fatty acids.

### *Waste Streams*

Three main waste streams are generated during the extraction of avocado oil: (i) seed and skin, and from the decanter; (ii) pomace (predominantly flesh tissue); and (iii) the water phase. The first stream is the skin and seeds of the avocado, which make up approximately 32% of the total fruit weight (Wong et al., 2008). In the past, this waste stream was dumped or used as landfill. However, because of growing environmental concerns, alternative uses for this material (e.g., mulch/fertilizer) are being investigated. This waste stream generally contains relatively little oil (≅3% by weight), even with the small amount of flesh tissue that adheres to the skin.

The seed generally is intact, except for late-season fruit where seed germination has commenced. In addition, the seed is viable, and thus one can use it for horticultural purposes if desired (e.g., seedling rootstocks). However, generally, the seeds are dumped as they contain minimal oil (<2%; Mazliak, 1965). One could investigate the seed as a food source, perhaps as stock feed, since it contains high amounts of carbohydrates, although one should consider possible toxicity (Werman et al., 1991).

Pomace exiting the horizontal decanter can be surprisingly high in water, depending on the water addition during malaxing. The pomace can contain as little as 20% of solids, depending on the operation of the decanter; it also contains 3–4% by weight of oil. The solid-phase pomace is generally used as animal feed. The oil loss in the pomace equates to a loss of 4–5% w/w of the total available oil. The liquid phase from the decanter is a mixture of oil and water, which is then separated in a series of two disc centrifuges.

In the finishing disc centrifuges, 98–99% of the oil leaving the decanter in the liquid phase is recovered. The final waste stream from the finishing centrifuge is primarily water. The amount of oil lost in this stream depends on the operation and separation efficiency of the centrifuge. Minimizing the loss of oil at this step is important for good recoveries, but also important is to ensure that no water is left in the oil. By ensuring that the oil contains ideally <0.1% of water prior to storage, inevitably a small loss of oil into the waste water stream will occur. Ideally, one should keep this oil loss to <5% of the total available oil in the liquid phase exiting the decanter. The recovery of oil from the water phase can be uneconomic because of the high volumes. The recovery of as much oil as possible from this wastewater stream is advisable, to reduce biological oxygen demand (BOD), and to reduce wastewater-disposal costs.

## Laboratory-based Extraction

A reliable and reproducible method for determining the oil content in avocado is needed: for determining the amount of oil available for commercial extraction, for determining fruit maturity, and for research purposes.

A range of techniques was employed to determine the total lipid content, many of them relatively time- consuming and expensive (Requejo-Tapia, 1999). The refractometric index (RI) method developed by Leslie and Christie (1929) in California, by using Halowax oil #1000 (monochloronaphthalene) as a solvent, was officially used for the measurement of the percentage of total lipid in avocado. However, because of the inconsistency of readings (very temperature-dependent) and equipment costs, this method was considered inconvenient for growers. In addition, Halowax is a suspected carcinogen and is no longer available (Lee et al., 1983).

The solvent-extraction technique using petroleum ether 40–60 (nonpolar solvent) is the standard method for extracting the nonpolar triglycerides and neutral fats in foods for analysis (AOCS Official Method Aa 4-38) (AOCS, 2004). This method, often referred to as the Soxhlet method, can take between 6 and 12 hours, and automated systems usually only run four to six samples at one time. The sample must be dried prior to extraction, and this technique is therefore considered too slow for the industry to be used as a routine test.

An adaptation of the Gerber method originally developed for the dairy industry showed accuracy in the determination of total lipids (polar and nonpolar) in avocados (Rosenthal et al., 1985). However, it not only uses a combination of flammable and dangerous solvents, but also equipment that is not always available in the horticultural industry (Requejo-Tapia, 1999). Several lipid-extraction methods originally developed for animal products were used for the determination of the total lipid in avocados with relative success. This is the case with the methods developed by Folch et al. (1957) and Bligh and Dyer (1959) using chlorofrom/methanol (C/M). However, a large sample size, large volumes of solvent, a relatively high level of difficulty, and a long time to achieve results are the main inconveniences of these methods. These methods are primarily used for "wet" products.

Lewis (1978) compared four methods of analyzing the lipid content of avocados including the Soxhlet method (using petroleum ether), homogenization with petroleum ether, the C/M technique, and the refractometric method (using Halowax oil). On average, results showed that C/M and the refractometric method gave 5–8% higher lipid yields than the first two methods. The researchers concluded that C/M is sufficiently polar to release some membrane-bound lipids, probably comprised of phospholipids and glycolipids, and that the similar results with Halowax oil may have been due to the prolonged ball milling rather than solvent polarity.

We found that solvent-extraction methods are relatively slow, and although they are generally reliable as a means of determining the oil content (AOCS, 2004), they result in the breakdown of other compounds of interest such as pigments like chlorophyll. We therefore sought to develop a protocol using the Accelerated Solvent Extraction system (ASE®300, Dionex Corporation, Sunnyvale, CA, USA) to extract a maximal proportion of the available oil reliably, with minimal effects on pigments and compounds such as sterols.

### *Development of Laboratory Oil Extraction by Using Accelerated Solvent Extractor (ASE®)*

ASE is a relatively new technology that uses pressurized solvent and heat to aid oil extraction. ASE uses stainless-steel cells, sealed closed with a screw cap and fitted with a cellulose filter. The cells are pressurized with oxygen-free nitrogen gas ($N_2$). Samples are heated to a designated temperature after pressurizing, and then held under this pressure and temperature— the "static time". The solvent is then flushed out of the cell (using $N_2$), and one can repeat this cycle. During our method development using the ASE machine, we used avocado tissue ground to a powder after freeze-drying. Various solvents and extraction combinations were tested. By using methanol, one could produce a white sediment in the final collection bottle. Methanol is a polar solvent that could also extract nonlipid compounds and protein-bound lipids such as phospholipids and sphingolipids and other waxy compounds. Hexane (nonpolar solvent) was more suitable for avocado oil extraction because we are primarily interested in the triglycerides. In addition, hexane is the "industry standard" solvent for oil extraction using the Soxhlet extraction method (AOCS, 2004). We sought to develop a method that would achieve oil recovery similar to that of the industry standard Soxhlet technique, while minimizing the deterioration of the compounds in the oil.

By using ASE, an increase in extraction temperature from 50 to 120°C improved oil recovery by 5.5% (Fig. 2.3). However, increasing the temperature also decreased the quality of the oil, destroying labile compounds such as pigments and antioxidants (data not shown). Destruction of chlorophyll pigments can occur at 60°C and above, but depends on time and temperature (Kidmose et al., 2002). Furthermore, temperatures lower than 60°C did not achieve oil recovery as well as the Soxhlet method. Thus, we examined increasing the static time from 15 to 40 minutes, and this increased oil, carotenoid, and chlorophyll recovery (data not shown). We noted that variable results were obtained if the ground tissue was tightly packed in the cell, presumably because of poor solvent contact and out-flow. Furthermore, the level of tissue grinding could also affect the oil recovery. We also examined a range of cycles and static times, and based on the percentage of oil recovered versus the industry standard (Soxhlet), and the recovery of pigments and other compounds, we decided on the following method, which was published by Ashton et al. (2006).

A weighed ground sample of ≅20 g is placed in a 100-mL stainless-steel closed cell. Extraction conditions are: a 5-minute pre-heating step, followed by a 100-minute total extraction time at 60°C and 10 MPa. The run is split into five cycles of 20 minutes with a $N_2$ gas purge cycle of 90 seconds. The oil dissolved in the solvent is collected in dark-glass bottles which are $N_2$-flushed during and after extraction by ASE®. Hexane is removed from the resulting solution over 2 hours at 30°C using a rapid solvent evaporator (RapidVap $N_2$ Evaporation Systems; Labconco® Corp., Kansas City, MO) under flowing $N_2$ and the oil yield is expressed as percentage of oil per dry weight of avocado tissue. The resulting oil samples are poured into dark-glass bottles, flushed with oxygen-free $N_2$, and stored at –80°C until analysis. All extrac-

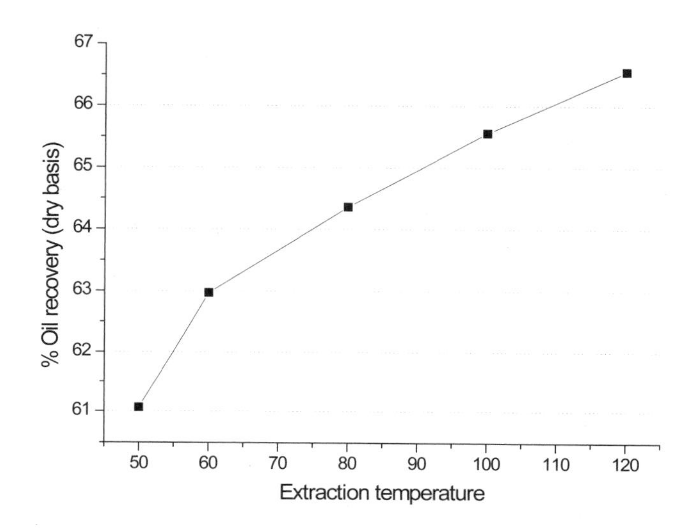


**Fig. 2.3. Effect of temperature (°C) on oil yield from ground, freeze-dried Hass avocado tissue by using hexane in ASE® with three cycles and a 15-minute static time.**

tions and handling are performed under minimal light (0.001 μmol $s^{-1}$ $m^{-2}$) by using 100% of hexane (liquid chromatography LiChrosolv®) and oxygen-free $N_2$ (99.99% of purity).

# Factors Affecting Oil Yield and Quality

## Factors Affecting Available Oil Yield

### *Preharvest Factors*

#### Maturity

The most important factor influencing the availability of avocado oil is the maturity of the fruit, or time in the harvest season. As noted previously, avocado fruit are unusual in that the fruit can remain on the tree for many months after the time at which the fruit are physiologically mature (i.e., they will ripen when removed from the tree). Over this period (from 7 to as long as 18 months), some important changes are occurring in the fruit. The one of main importance to oil extraction is an increase in dry matter, and the change in this is primarily an increase in oil content.

The most significant factor in examining maturity and oil content in avocado is the dry-matter content. This measure of maturity is routinely used by industry worldwide for the commercial harvesting of fresh fruit since it is simple, safe, and can be carried out by growers and packhouses. This simply involves removing a sample of flesh tissue and drying it to a constant weight, normally either by dehydrator or oven (≅65°C) or by using a domestic microwave.

A very high correlation is present between dry-matter accumulation and oil content as reported by Lee et al. (1983) and others in the past. We carried out two extensive projects examining dry matter and oil content in six main growing regions in New Zealand, and five in Australia. Fruit sampling was carried out over two seasons, from three growers from each region, three to six times through the commercial-harvest seasons. Dry matter was measured, and total oil content was determined by ASE® solvent extraction. The remarkably strong correlation ($r^2$ of 0.96) of dry matter with oil content is demonstrated in Fig. 2.4. Considering the very large differences in growing environments, the robustness of correlation shows the physiological significance of oil content in avocado maturity. A rough "rule of thumb" we found to hold true in nearly all situations was that one can estimate the total oil content as being 10% less than the dry-matter content.

## Dry Matter and Oil Accumulation for Hass

The very significant changes in dry matter and oil content of avocado flesh were introduced previously (Fig. 2.1). While the initial pattern of dry-matter accumulation is linear in some regions (Fig. 2.5A), dry matter (and oil content) reaches a plateau, while in others the changes continue in a relatively linear manner (Fig. 2.5B). The

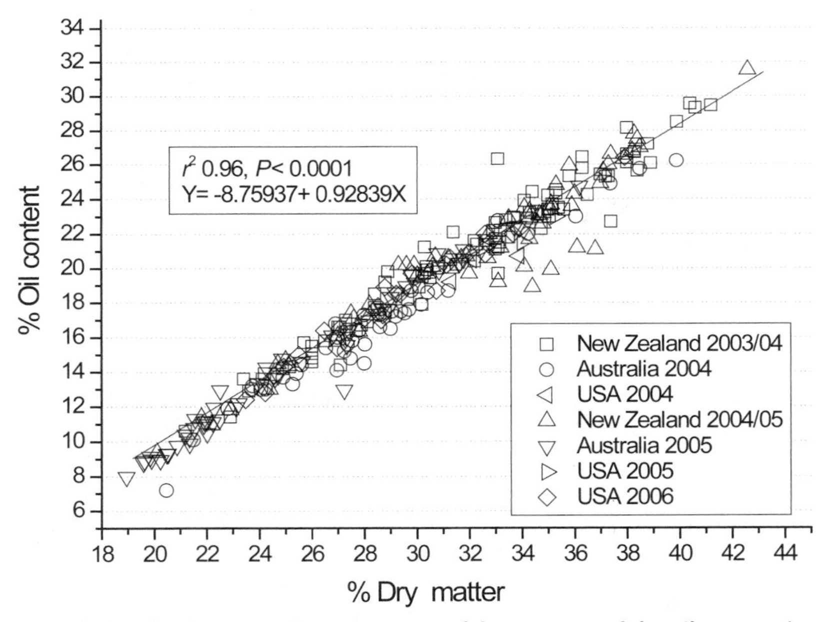


**Fig. 2.4. Relationship between the percentage of dry matter and the oil content (percentage of fresh-weigh basis) in the flesh of Hass avocado grown in New Zealand, Australia, and California from 2003 to 2006. Each data point is the average of the three replicate samples over a range of growers and times in the season.**

influence of climate in the plateau response appears significant, since this may occur in one season but not in another (data not shown).

## Orchard, Region, Country, and Seasonal Maturity Differences

Although we are not clear why, or what management practices influence maturity, undoubtedly, orchard differences occur, and these can occur over relatively small distances. Tree age can also influence fruit maturity since fruit on trees less than 5 years old tend to mature earlier (Jim Clark; personal communication, 2001). However, in this work, we sourced fruit from trees of at least 8–10 years of age.

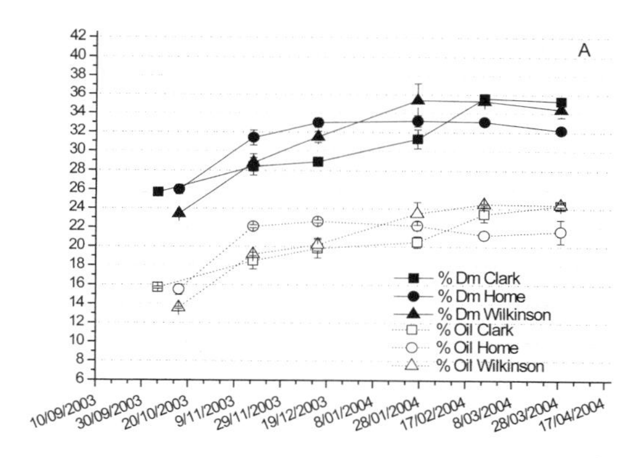


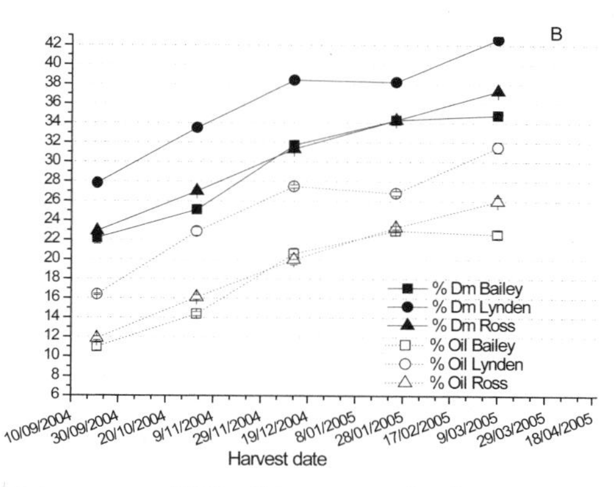


**Fig. 2.5. Percentage of dry matter (DM) of Hass avocados harvested from the Far North (A) and Te Puke (B) regions in New Zealand during the 2003–2004 and 2004–2005 production seasons, respectively. Each point is the mean of triplicate samples from 20 fruit. Vertical bars are standard errors of mean (SEM).**

As is well-known, (http://nzavocado.co.nz/ Monitoring Results; Fig. 2.6), the growing region has a significant impact on fruit maturity in terms of both timing and maximal dry matter attained.

A wider interpretation of growing region is, of course, the country in which it is grown. Clearly, large differences arise between Australia and New Zealand, and for example, California, where temperatures and rainfall differences can be extreme. These can affect tree growth, tree health, irrigation practices, and fruit quality (particularly the expression of ripe rot, which is higher in wetter growing environments). All of the above will affect maturity, and thus oil yield.

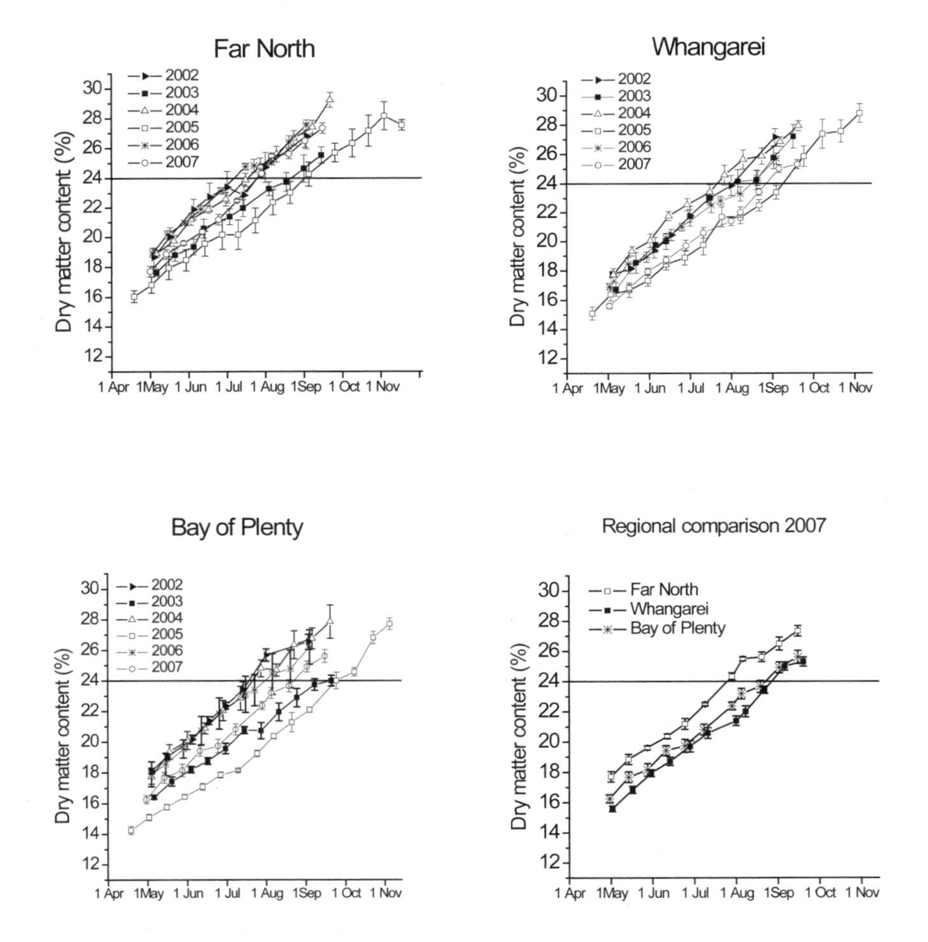


**Fig. 2.6. Dry-matter accumulation in Hass avocado for the years 2002–2007 for three New Zealand growing regions [Far North, Whangarei, and the Bay of Plenty (Katikati/Te Puke)], and a regional comparison for the 2007 season. Data and graphs from the New Zealand Avocado Industry Council (AIC) Website (http://nzavocado.co.nz/Monitoring Results).**

We have observed significant differences in the dry-matter and oil-accumulation patterns between seasons, being in some cases more significant than the grower or regional differences, both in terms of the timing of maturity changes.

**Dry matter and oil accumulation for other cultivars**

Other cultivars which are not as commonly grown as Hass—such as Reed, Fujikawa, Hayes, and Fuerte—are grown in smaller quantities. These cultivars are less well-studied in general, and very little information is available on their oils. In a small study conducted using fruit from Whangarei and the Far North regions of New Zealand, samples from these cultivars were evaluated for dry matter and oil content (by chemical extraction) during the season. As with Hass, a strong positive relationship existed between dry matter and oil content, and these characteristics increased as the fruit matured (Fig. 2.7).

These cultivars generally mature at different times than Hass. For instance, applying the minimum 24% of dry-matter commercial standard for the harvest of Hass, Reed, which is used as a pollinator in the field, and Hayes achieve commercial maturity in February in New Zealand, a time at which the Hass' harvest is coming to an end. Cold-press extraction of Reed avocados produces a very intense golden/yellow oil, resembling a refined oil and therefore making its marketing as an extra-virgin product difficult. In New Zealand, cold-pressed oil from Reed is usually refined and used to produce various flavor-infused avocado oils. Fuerte avocados dominated the world production for many years during the 1950s and 1960s mainly because of the fruit's excellent eating quality and relatively high yields (CAC, 1998). However, Fuerte also presents a number of problems such as alternate bearing and short postharvest life, and thus has become less dominant, in contrast to increased plantings of Hass. In our study, dry matter for the Fuerte cultivar was approximately 35% in August in New Zealand, meaning that this cultivar reached commercial maturity of 24% before Hass (usually September). Our informal observations of commercially extracted Fuerte oil flavor suggested significant differences from those of Hass, and research characterizing the flavor profiles of avocado oils from other cultivars than Hass would be useful.

### *Postharvest Factors*

**Ripeness**

Significantly, no changes occurred in either the oil content or levels of the various fatty acids during the ripening of Hass avocados, nor in storage at temperatures of 0°C (chilling injury- inducing), standard storage temperatures (≅6°C) nor up to 10°C (Eaks, 1990; Luza et al., 1990). The only significant change noted was in Fuerte avocados, where the monoglycerides and FFAs increased during ripening, probably because of the degradation of triglycerides (Kikuta & Erickson, 1968).

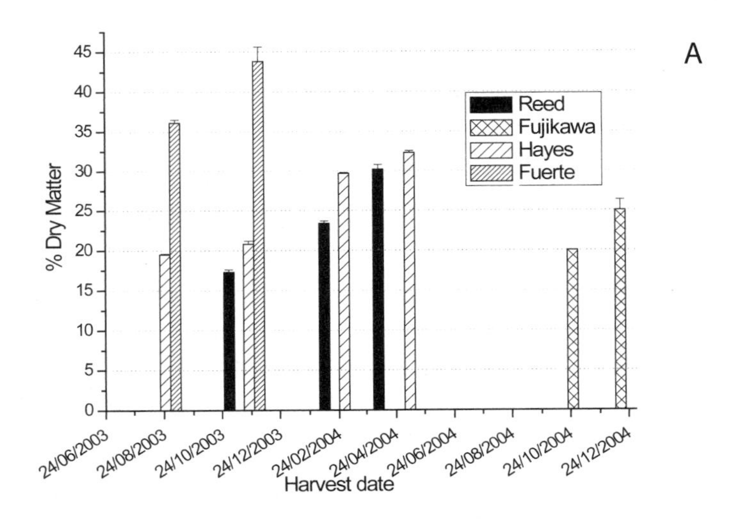


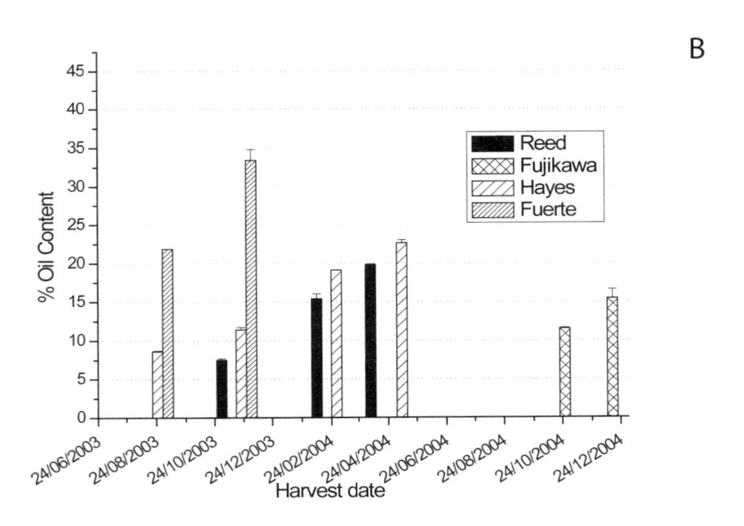


**Fig. 2.7.** Percentage of dry matter (a) and total oil content (b) (by chemical extraction) of avocado cultivars grown in the northern regions (Far North and Whangarei) of New Zealand during the 2003–2004 season. Error bars = standard error of mean.

## Factors Affecting Cold-pressed Oil Yield and Quality

The above pre- and postharvest factors clearly influence the total oil available, but a range of other factors influences the yield and quality of cold-pressed avocado oil.

### *Effect of Maturity (Time in Season) on Cold-pressed Yield*

This is perhaps the most problematic area for cold-pressed avocado oil manufacturers, since a significant difference exists between the available yield (i.e., total oil content as determined by laboratory-based chemical extraction) and the cold-pressed yield.

In the following figure, we demonstrated this by showing an approximation of the seasonal change in average dry matter (the key measure of the maturity of avocado fruit), the total oil content (as determined by hexane extraction), and the typical commercial cold-pressed oil yield (Fig. 2.8). Since commercial cold-pressed extraction is routinely carried out on fruit of the same firmness stage, the difference in yield is not due to ripeness differences. It is unclear why the yield from cold-pressed oils is so low in early-season fruit (less mature or lower dry matter), but this may possibly be due to the differences in the levels of endogenous cell-wall degrading enzymes, which may differ over the season for fruit of the same ripeness.

### *Ripening*

While information on oil yield with maturity (discussed above) is routinely collected commercially, other factors are difficult, or very expensive, to examine at a commercial level. Thus, for the data presented in this and the following section, we carried out laboratory-based trials by using a hammer mill to grind tissue, and 1- L malaxers stirred at 30 rpm for 75 minutes at 45°C, with Hass avocados of high maturity (late season with an average dry matter of 38%). Oil was separated by using a Heraeus centrifuge at 3000 x *g* at 40°C.

Hass avocados with three ripening stages were processed: minimal, fully ripe and overripe, where firmness hand ratings corresponded to 4, 5, and 6, respectively (White et al., 2005). Cold-pressed oil yield increased with fruit ripeness from 7, to 9, to 11% (g of oil/g of flesh fresh weight (FWt), respectively (Table 2.1). However, with increased fruit ripeness, FFAs also increased, from 0.03 to 0.12% w/w (as oleic acid).

### *Fruit Quality*

Because fruit must be ripened to maximize oil yield by cold-pressed extraction, a concomitant increase in fruit disorders affects oil quality. In avocado, one of the greatest challenges for fruit grown in wet environments is postharvest rots, which increase dramatically with ripening (Hopkirk et al., 1994). In addition, a long-term storage of

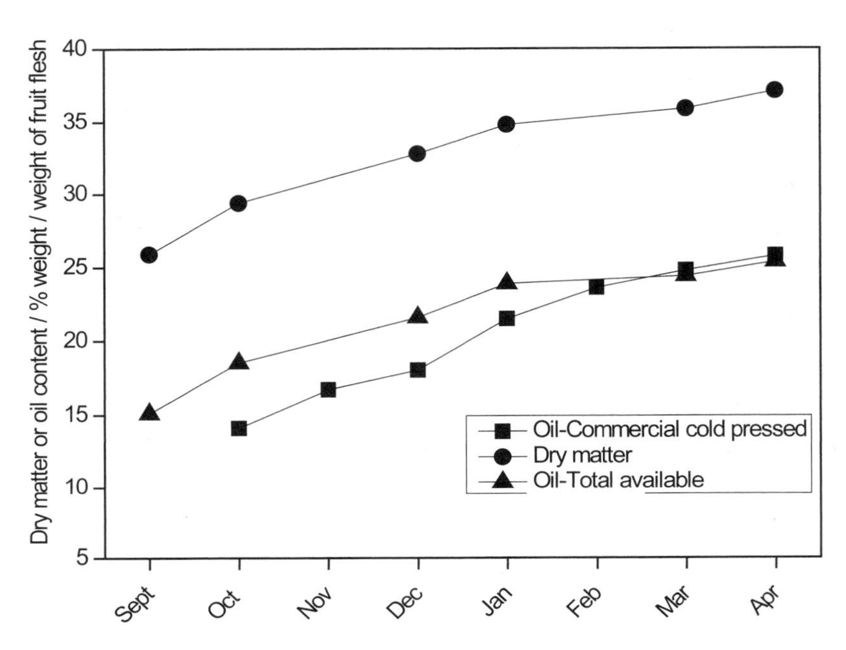


**Fig. 2.8. Schematic of typical changes in dry matter and oil content over a commercial harvest and oil- processing season in New Zealand Hass avocados. Fruit in July are physiologically mature (i.e., will ripen after harvest). Dry- matter content, total oil content (i.e., maximal available oil as determined by solvent extraction), and commercial cold-pressed yield are shown.**

**Table 2.1. Effect of Hass Avocado Fruit Ripeness on Oil Yield (Percentage of Flesh Tissue) and Percentage of Free Fatty Acids (FFAs). [Data Are Means of Three Extractions ± SEM (Standard Error of the Mean)]**

| Ripeness | 4 | 5 | 6 |
|---|---|---|---|
| Oil yield (g oil/g flesh FWt) | 7.0 ± 0.84 | 8.5 ± 0.22 | 11.04 ± 0.33 |
| % FFA (w/w) | 0.029 ± 0.017 | 0.093 ± 0.021 | 0.127 ± 0.037 |

fruit (over ≅4 weeks) results in physiological disorders, with one of the key expressions of chilling injury being "diffuse flesh discoloration" (or flesh greying) (White et al., 2005). Flesh bruising (due to physical damage) is also a common disorder.

In laboratory-based cold-pressed trials, we determined the effect of varying degrees of three disorders—body rots, bruising, and flesh greying—on oil quality as determined by FFA. Increased levels of rots, flesh bruising, or to some extent flesh greying all resulted in the reduction in oil quality as measured by FFA (Table 2.2).

These experiments demonstrate that increased fruit ripeness increases oil yield but that disorders also increase (thus decreasing oil quality), and therefore we demonstrated the need to balance oil yield with oil quality. As a best-practice recommendation, oil quality should be maximized by good postharvest handling of fruit to minimize rots and flesh disorders and thus maximize oil quality.

### *Fruit Storage*

In the normal operation of commercial avocado oil processing, fruit are often stored for 1–2 weeks prior to ripening and subsequent oil extraction. The nature of the avocado-production season means that during peak-harvest periods the capacity of the oil-processing facilities may be exceeded, and fruit must be stored for even longer periods. Fruit storage for longer than 3–4 weeks generally results in reduced fruit quality due to increases in physiological and pathological disorders (Hopkirk et al., 1994; Woolf et al., 2004). The commercial recommended storage temperature for avocados is generally 5–7°C, depending on the time in the season.

We examined the effect of storage duration on the quality of oil extracted under commercial conditions. Fruit were harvested in November (early/mid-season) at a dry matter of 26% (24% is the minimal commercial maturity in New Zealand). Fruit were randomized over a large number of commercial bins (250 kg) to eliminate any orchard effects. Fruit were placed immediately into cool storage at 6°C, and removed at weekly intervals for 7 weeks, ethylene-treated (100 μl $L^{-1}$ for 2 days), and ripened at 20°C to the same firmness level (≅0.4 kgf). Three replicate runs (3 × 250 kg of fruit) were processed under commercial conditions. Measures of fruit quality and firmness were taken on random samples of 20 fruit/250 kg bin, as described by White et al. (2005).

As the storage time increased, the amount of ripe rots (body and stem-end rots) and chilling disorders (flesh greying, vascular browning, and stringy vascular tissue) increased, while flesh browning (i.e., bruising) did not change significantly with storage time (Fig. 2.9). These disorders resulted in a significant increase in the proportion of unsound fruit (fruit with any significant disorders), from 15 to 80% following 1–7 weeks of cool storage.

**Table 2.2. Hass Avocado Oil Quality As Measured by Free Fatty Acids (FFAs w/w; as Oleic Acid) Immediately Following Extraction from Fruit with a Range of Fruit Disorders Including Body Rots, Bruising, and Graying. [Data Are Means of Three Extractions ± SEM (Standard Errors of the Mean)]**

| Rot level | Control | 5% | 10% | 15% | 30% |
|---|---|---|---|---|---|
| % FFA | 0.38 ± 0.02 | 0.36 ± 0.02 | 0.42 ± 0.03 | 0.80 ± 0.04 | 0.91 ± 0.14 |
| Bruising | Control | 20% | | | |
| % FFA | 0.77 ± 0.16 | 1.06 ± 0.48 | | | |
| Graying | Control | 20% | | | |
| % FFA | 0.77 ± 0.16 | 0.85 ± 0.09 | | | |

The percentage of FFAs in the oil, on the other hand, remained at approximately 0.5% for the duration of the study. Zauberman et al. (1985) showed that the peak of enzyme activity in avocados occurs during the climacteric rise and the lowest activity was always exhibited in the soft-ripe fruit stage at which the fruit is used for oil extraction.

The PV of the oil increased slowly from 0.75 to ≅1 mEq/kg as storage time increased up to 3 weeks, and then increased sharply reaching a maximum of 2.25 mEq/kg after 7 weeks of storage (Fig. 2.10). Hypothetically, fruit stored for over 3 weeks probably did not require long malaxing times, as they were already very ripe-soft after cool storage. The nature of malaxing operations can introduce oxygen into the paste, causing oxidation and therefore resulting in increased PVs. The increase in PV with storage time correlated strongly with the increase in fruit disorders. In addition, these higher PVs in the oil will also reduce its shelf life. On the basis of this research and our understanding of the effects of storage duration on fruit physiology and quality, we recommend that fruit are stored for no longer then 3–4 weeks after harvest prior to oil extraction.

### *Processing Conditions*

Factors that influence oil yield during processing are malaxer temperature, malaxing time, speed of decanter, and operation of the final polishing centrifuge. At too low a malaxer temperature, the oil does not easily release from the cells, so temperatures around 40 to 50°C give the best yields without a significant reduction in oil quality (FFA/PV). Oil release from the cells during malaxing requires adequate time, and this will usually be most strongly influenced by the time in the season (maturity). With early-season fruit, the release of oil is much slower; the flesh of the avocado does not break down as easily; hence, a longer malaxing time may be needed. Generally, malaxing times of 45 to 60 minutes are used. One must operate the decanter such that an efficient separation of solids and liquid phases occurs and that minimal oil is left occluded in the pomace. The degree of separation achieved in the final polishing centrifuges also will influence yield. A minimal loss of oil in the wastewater phase is desired for economic reasons.

Skin (peel) addition during oil processing does not affect oil quality (PV/FFA), but it will influence the composition of the oil. Avoid excessive aeration of the pulp and the oil. Oxidation can occur at any points along the process once the oil is released from the flesh. The oil in the malaxer is exposed to the air during this step; hence, one can cover and flush the tanks with a continual supply of nitrogen gas to minimize oxidation. With enclosed centrifuges, oxidation is minimized, but once the oil is pumped to the storage tanks, one should flush it with nitrogen immediately. Another precaution during processing is to minimize the exposure of the oil to light. Therefore, covered tanks and enclosed pipe work are recommended. Exposure to light will promote photooxidation and reduce the quality of the oil.

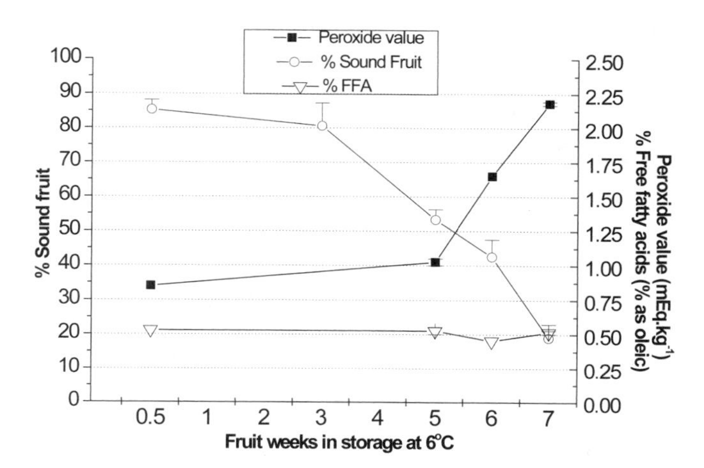


**Fig. 2.9.** Mean incidence of Hass avocado fruit rots and disorders as a percentage of total ripe fruit processed in the factory following up to 7 weeks in storage at 6°C. Disorders were rated on a scale of 0–3 (0 = none to 3 = severe). Error bars represent the standard error of means (SEM).

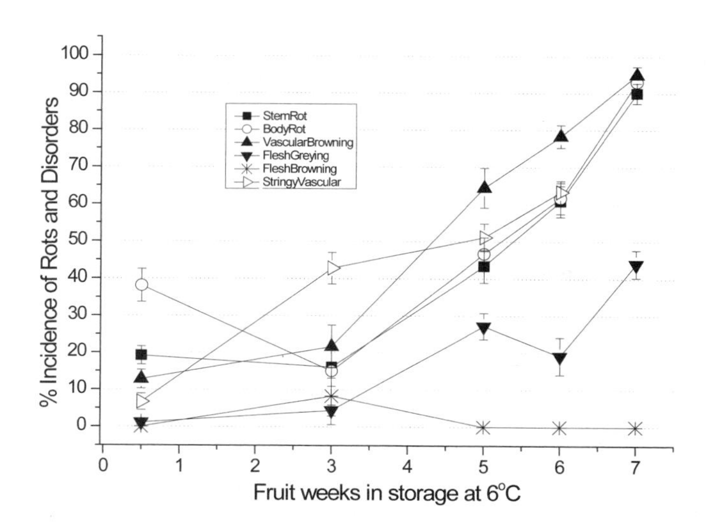


**Fig. 2.10.** Mean incidence of sound Hass avocado fruit as a percentage of total ripe fruit processed in the oil factory following up to 7 weeks in storage at 6°C. Percentage sound is shown as the proportion of fruit showing acceptable quality factors. Mean peroxide value (mEq/kg oil) and percentage of free fatty acid (w/w as oleic acid) in avocado oil obtained from the same fruit at each storage interval.

Also important is to ensure that the oil contains <0.1% of water. Any water in the oil will promote hydrolysis of the fatty acids, and will result in an oil with a higher than desired FFA content.

## Oxidative Stability of Avocado Oil

The presence of light has a strong influence on the oxidative stability of avocado oil, as was reported for other vegetable oils (Rahmani & Csallany, 1998). The high level of chlorophyll present in avocado oil acts as a photosensitizer promoting oxidation, while the tocopherols and carotenoids present are inhibitors of oxidation. Chlorophyll is excited by exposure to light, and reacts with lipids to form oxidation intermediates that can react further to form free radicals and oxidation-breakdown products; this consequently results in the breakdown of chlorophyll pigments in the oil (Hamilton, 1994). The presence of oxygen also has a significant influence on the oxidative stability of the avocado oil. Both photooxidation and autooxidation are promoted with the small increases in oxygen present. Increased temperature has the greatest influence on autooxidation reactions. No interaction was found between light and temperature (unpublished). Hence, in avocado oil under light conditions, photooxidation dominates, while at high temperatures autooxidation dominates, both reactions being independent, as also found by Rawls and Vansante (1970).

The presence of tocopherols will inhibit autooxidation by inhibiting the initiation and propagation of free radicals. Tocopherols donate their phenolic hydrogen to lipid-free radicals present. Carotenoids are considered to be antiphotooxidants by quenching singlet oxygen, an intermediate in the photooxidation reactions. The contribution of these antioxidants to the protection of avocado oil will depend on the storage conditions. Tocopherols are more protective at high temperatures and in dark situations, while carotenoids are protective in light conditions. Research shows that a reduction in both chlorophylls and carotenoids was found when the oil was stored under light or dark conditions at similar temperatures (unpublished).

Antioxidants are often added to reduce oxidation and to increase the shelf life of oils. Ascorbyl palmitate, citric acid, mixed tocopherols, and a combination of tocopherols and citric acid were all tested to see if they could slow the oxidation rate in avocado oil. The oils and the added antioxidants were stored in the dark at 60°C for over 20 days. The resulting PVs of the oils over this period are shown in Fig. 2.11. Ascorbyl palmitate provided significant protection against oxidation, citric acid provided some protection, but the mixed tocopherols added surprisingly little protection. High levels of tocopherols were reported to have prooxidant properties (Jung & Min, 1991).

To minimize the oxidation of avocado oil during storage, one should eliminate light and oxygen; hence, one should store the oil in clean, dark-colored-glass bottles or stainless-steel containers.

## Acyl Lipids

Mono-, di-, and triacylglycerides are all present in the lipid component of avocado flesh (Gaydou et al., 1987; Kaiser et al., 1992), but triacylglycerides by far represent the largest proportion of the lipids (88%; Kikuta & Erickson, 1968). Also present are low concentrations of glycol-, sulfo-, and galacto-lipids (Kaiser et al., 1992). Kikuta and Erickson (1968) reported the lipid breakdown in Fuerte avocados (Table 2.3). Lozano (1983) also determined the triglyceride composition of the mesocarp flesh of Fuerte avocados.

## Fatty Acid Composition During Development and Commercial Maturity

The main fatty acids in avocado oil are palmitic and stearic acids (saturates), palmitoleic and oleic acids (monounsaturates), and linolenic and linoleic acids (polyunsaturates). As the fruit grow and mature, the triglyceride content in the flesh increases. The main fatty acid (>50% of all lipids) found to increase during maturation in both Fuerte and Hass avocados was oleic acid, while linoleic acid decreased dramatically after the first 4 months (from ≅60 to ≅12%), and linolenic acid decreased from 15% at the start of the season to less than 2% at full maturity. All the other fatty acids

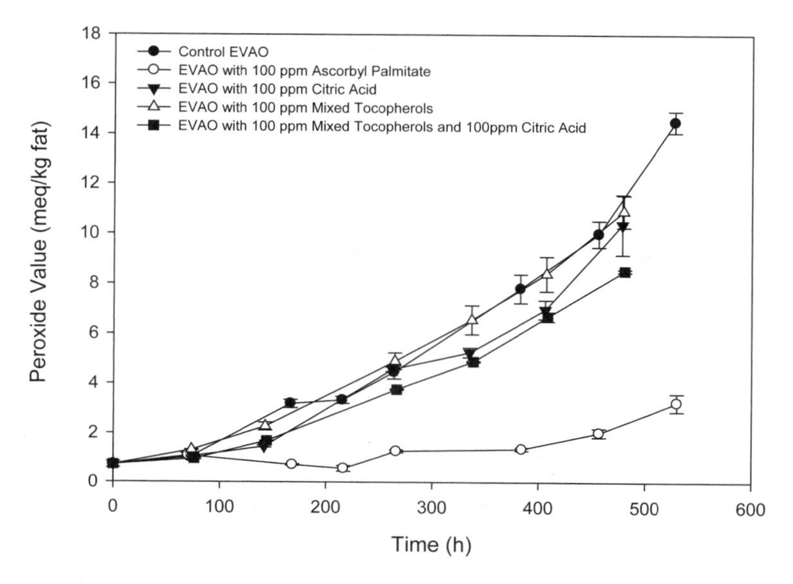


**Fig. 2.11. Effect of antioxidants on the peroxide value of extra virgin avocado oil (EVAO) stored at 60°C in dark conditions.**

**Table 2.3. Level of Acyl Lipids and Phospholipids in Fuerte Avocado Flesh (Percentage of Oil)**

| Class | % of Oil |
|---|---|
| Free fatty acids | 0.10 |
| Triglycerides | 19.96 |
| Diglycerides | 1.29 |
| Monoglycerides | 0.78 |
| Phospholipids | 0.39 |
| Others* | 0.28 |
| Total | 22.80 |

*Other substances extracted in chloroform/methanol (2:1).
Source: Kikuta and Erickson, 1968.

remained at relatively constant levels (palmitic, palmitoleic, and stearic; Eaks, 1990). Inoue and Tateishi (1995) followed the changes in fatty acid composition during part of the maturation of Fuerte avocado fruit. They found oleic acid increased from 37 to 50% of total lipids, palmitic acid remained constant at approximately 22%, linoleic acid decreased from 14 to 11%, linolenic acid decreased slightly from 0.3 to 0.1%, while palmitoleic acid remained fairly constant at about 10% of total lipids. Rotovohery et al. (1988) also reported changes in the fatty acid composition during fruit maturation (Table 2.4). However, one must note that the above results may not hold true if a significantly longer sampling period were included. We demonstrated that over a long harvest season the trends tended to follow an increase then a decrease in oleic acid, with palmitic and linoleic following the inverse pattern (Fig. 2.12). These trends were observed in three growing regions, although only one orchard was examined in each region.

### Fatty Acid Composition in Different Countries

A growing environment can have significant effects on the fatty acid profile of avocado oil. Table 2.5 shows typical values. However, consider differences cautiously, since a rigorous examination of fatty acids was not carried out in all countries, and these, of course, can also vary with season.

## Nonacyl (or Unsaponifiable) Components: Tocopherols, Sterols, Plant Pigments, and Phenolics

Oils extracted from plant sources also contain a number of phytochemicals, which may have important health benefits in terms of disease prevention. As well as being an oil considered to be healthy because of its high concentration of monounsaturated

**Table 2.4. Fatty acid Composition (Percentage by Weight) of Fuerte Avocado Mesocarp During Fruit Development (Grown Under Mediterranean Climate). Stages I, II, III, and IV = 20, 25, 31, and 36 Weeks After Full Bloom, Respectively**

| | Stage of development | | | |
|---|---|---|---|---|
| Fatty acid | Stage I (Mean) | Stage II (Mean) | Stage III (Mean) | Stage IV (Mean) |
| 16:0 | 12.8 | 10.2 | 10.9 | 9.3 |
| 16:1 | 2.6 | 1.9 | 2.2 | 1.5 |
| 18:0 | 0.9 | 0.6 | 0.5 | 0.6 |
| 18:1 | 60.8 | 68.4 | 68.6 | 76.8 |
| 18:2 | 12.3 | 10.2 | 10.8 | 8.3 |
| 18:3 | 1.1 | 1.1 | 0.9 | 0.6 |

From Ratovohery et al. (1988).

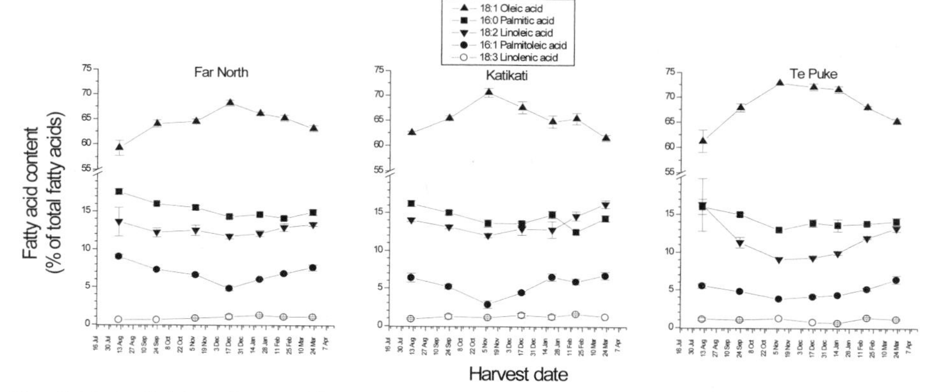


**Fig. 2.12. Individual fatty acid content as a percentage of total fatty acids for Hass avocado fruit harvested from August 1999 to March 2000 from one orchard of three growing regions in New Zealand. Each point is the average of four replicates of five fruit. Vertical bars = SEM.**

**Table 2.5. Fatty Acid Composition of Hass Avocado Oil from Five Countries (Wong et al., Unpublished Data)**

| Fatty acids (% of total) | New Zealand | | Australia | Chile | Mexico | California |
|---|---|---|---|---|---|---|
| | Range | Mean | Mean | Mean | Mean | Mean |
| Palmitic acid (16:0) | 9.7–15.2 | 12.3 | 21.7 | 13.1 | 14.8 | 14.5 |
| Palmitoleic acid (16:1) | 1.7–8.2 | 4.1 | 9.3 | 3.6 | 7.9 | 4.1 |
| Stearic acid (18:0) | 0.1–0.4 | 0.3 | 0.4 | 0.4 | 0.4 | 0.3 |
| Oleic acid (18:1) | 61.7–77.8 | 71.5 | 51.8 | 68.2 | 66.8 | 65.3 |
| Linoleic acid (18:2) | 7.7–18.9 | 11.6 | 16.0 | 13.2 | 9.5 | 15.0 |
| Linolenic acid (18:3) | 0.2–0.9 | 0.5 | 0.8 | 0.8 | 0.6 | 0.8 |

fatty acids (MUFAs), avocado oil contains relatively high concentrations of other non-acyl phytochemicals. These phytochemicals include pigments such as chlorophylls and carotenoids, which give the oil its distinctive color. Other phytochemicals present that are thought to contribute to health are the tocopherols, primarily α-tocopherol, and the high level of sterols (principally sitosterol). Phenolics so far were not found in high concentrations in avocado oils. A summary of the concentrations of various phytochemicals found in avocado oil is given in Table 2.6.

## Tocopherols and Tocotrienols

α-Tocopherol is the major form of vitamin E found in avocado oil (Table 2.6), with minimal or just detectable concentrations of β-tocopherol, δ-tocopherol and γ-tocopherol also present in oil from Hass avocados. Tocotrienols were not reported in avocado oil. Vitamin E is an essential vitamin which has good antioxidant properties. The concentration in avocado oil is similar to olive oil (0.1–0.14 mg $g^{-1}$; Boskou, 2006). Retaining vitamin E assists in extending the shelf life of avocado oil, since

**Table 2.6. Typical Concentrations of Phytochemicals Found in Avocado Oil**

| Pigments | Range | Approximate mean |
|---|---|---|
| Chlorophylls (μg g-1) | | |
| Total chlorophylls | 11.1-18.5 | 13.3 |
| Chlorophyll a | 2-9.5 | 4.9 |
| Chlorophyll b | 4.2-14.8 | 5.1 |
| Pheophytin a | | 1.1 |
| Pheophytin b | | 2.2 |
| Carotenoids (μg g-1) | | |
| Total carotenoids | 0.9-3.5 | 1.9 |
| Lutein | 0.5-3.3 | 1.6 |
| Neoxanthin | | 0.2 |
| Violaxanthin | | <0.5 |
| Antheraxanthin | | <0.5 |
| Tocopherols (mg g-1) | | |
| α-Tocopherol | 0.07-0.19 | 0.11 |
| β-Tocopherol | | <0.01 |
| γ- Tocopherol | | <0.01 |
| δ- Tocopherol | | <0.01 |
| Sterols (mg g-1) | | |
| Sitosterol | 2.23-4.48 | 3.28 |
| Δ-5-Avenasterol | | ≅0.3 |
| Campesterol | | ≅0.2 |
| Stigmasterol | | <0.1 |

α-tocopherol scavenges free radicals produced during oxidation reactions and also terminates the reaction chain. The reduction of oxidation reactions is important to reduce the formation of hydroperoxides and ultimately the formation of rancid off-flavor compounds (Coppen, 1994). As tocopherols are unstable and light-sensitive, tocopherol retention is maximized by carrying out oil extraction in low oxygen and light conditions.

## Sterols

Plant sterols are known to lower cholesterol adsorption in humans (Piironen et al., 2000). Sitosterol is the main plant sterol present in avocado oil. Other sterols also present but at much lower concentrations are Δ-5-avenasterol, campesterol, and stigmasterol. Sitosterol was stable in the oil during storage. The concentration of sitosterol in avocado oil (average of ≅3.3 mg $g^{-1}$ oil with up to 4.5 in some cases; Table 2.6) is significantly higher than that in olive oil, which was reported to be approximately 1.62–1.93 mg $g^{-1}$ of oil (Phillips et al., 2002; Verleyen et al., 2002).

## Pigments

The rich color of avocado oil is due to the extraction into the oil of various plant pigments such as substantial amounts of chlorophyll, carotenes, and xanthophylls.

### *Carotenoids*

Carotenoids have strong antioxidant activity, and help in reducing the incidence of various diseases (Lu et al., 2005) including AMD (age-related macular degeneration). The carotenoids α-carotene, β-carotene and β-cryptoxanthin have pro-vitamin A activity. Lu et al. (2005) reported that lutein alone from avocado did not contribute to the anticancer effect, but that this was due to the mix of bioactive components present in the avocado. Carotenoids are also thought to scavenge free radicals formed from oxidation reactions.

Ashton et al. (2006) reported the decline of individual carotenoids with fruit ripening. The level of total carotenoids fell to 30% of the original value at harvest in the pale-green flesh. Oil is extracted from ripe avocados, but the level is still relatively high, even though a loss in the fresh fruit occurs. Research shows that a high level of pigments is present in the skin of the avocado, and if more skin is included during extraction, then a higher level of carotenoids can be extracted into the oil. Carotenoid pigments levels will decrease during storage.

### *Chlorophylls*

Chlorophyll pigments are extracted from the skin and the mesocarp (Ashton et al., 2006). Chlorophyll pigments contribute significantly to the phytochemical components present in avocado oil. Their exact mechanism(s) is unknown, but the consumption of green vegetables was correlated with a reduced risk of cancer (Minguez-Mosquera et al., 2008). Our studies showed that increasing the amount of skin included in the avocado paste during malaxing can result in the extraction of more chlorophyll into the oil (data not shown). Chlorophyll is a strong photosensitizer, and if exposed to light will promote photooxidation reactions in the oil. Photooxidation leads to the oxidation of the lipids and the production of free radicals. Hence, during processing one should eliminate any light and oxygen, as the chlorophyll pigments are rapidly lost when exposed to both. Chlorophyll is thought to act as an antioxidant during the storage of oils in the dark, and therefore is lost during storage (Gutierrez-Rosales et al., 1992).

## Polyphenols

Compared with other oils, relatively little is known about the polyphenolic content of avocado oil. However, since phenolic compounds are present in avocado fruit, likely they would be present in avocado oil. The exact nature and concentration of the phenolic compounds present in avocado oil will depend on the avocados and the extraction conditions, since they are easily modified by oxidation. Phenolic compounds of avocado were studied as a component of enzymatic browning (Lelyveld et al., 1984), and are known to be present in mesocarp tissue (Torres et al., 1987; Van Lelyveld et al., 1984). The main phenolic acids reported following hydrolysis were *p*-coumaric, ferulic, and sinapic acids (Torres et al., 1987). Since the extracted compounds were subjected to hydrolysis, the identities of the parent or native phenolic compounds present in avocado are not known. The flavonoid epicatechin was also in avocado mesocarp tissue, and is thought to have a role in resistance to fungal pathogens.

In a preliminary study, we investigated the polyphenolic content of avocado oil extracted with different amounts of skin in the malaxer paste under fully commercial cold-pressed systems. The polyphenolics were extracted from avocado oil by using the method described by Kalua et al. (2005) for olive oil with some modifications. Briefly, hexane was added to an oil sample, and the polpyhenols were extracted with methanol/water/formic acid. The resulting extracts were analyzed for phenolic compounds by reversed-phase high-performance liquid chromatography (HPLC) by using a Zorbax SB column with a formic acid/acetronitrile solvent gradient. The results showed that avocado oils appear to contain four major and numerous minor phenolic components. The spectra of the two major components had absorption maxima at ≅280 nm, suggesting that these compounds are probably phenolic compounds, and the spectra of the other two major compounds had additional absorption maxima at ≅350 nm, indicating that they may be flavonoids. As yet, these compounds were not identified, and advanced spectroscopic techniques such as mass spectrometry and nuclear magnetic resonance (NMR) are required for positive identification. However, similar to the polyphenolics present in olive oil, the HPLC retention times of these avocado oil components suggest that they are nonglycosidic phenolic compounds, and are relatively lipophilic in nature, indicating they may be readily bioavailable when consumed. Further research is ongoing to identify the phenolic components of avocado oil.

## Flavor and Aroma Compounds

### Sensory Analysis

Although healthfulness (perceived or otherwise) of an oil plays a role in consumer-purchase decisions, ultimately the taste is a key factor. Although chemical analyses (e.g., PV and FFA) can tell us something of the quality of an oil, sensory analysis is an essential step in determining the overall oil or food quality. Humans can be viewed as highly sensitive "instruments" with the ability to perceive compounds such as 2-methoxy-3-isobutylpyrazine at parts per trillion (Lund et al., 2009). Not only is sensitivity important in the sensory experience, but so too is the way the sensory organs interpret the information holistically. Many chemical compounds are in food products that contribute to this sensory profile.

Nonvolatile compounds, such as polyphenols, can suppress or accentuate the perception of volatile compounds (Lund et al., 2009). Many nonvolatile and volatile compounds contribute to flavor attributes such as rancid, grassy, cherry, and citrus flavors. Sensory assessments are important in delineating between poor and high-quality oils in terms of negative "off-flavors" as well as positive flavors. Regulations of olive-oil quality are premised on the sensory assessment of flavor characteristics performed by trained panels.

In olive oil, many years of research and commercial application of sensory analysis were spent to develop oil- quality standards. In particular, the IOC (International Olive Council) accredits international sensory panels to verify the "extra-virgin" status of olive oils (COI/T.15/NC no./Rev), combined with basic chemical analyses. To date, no information is published on the description of cold-pressed avocado oil on which standards might be based. To meet this need, a trained sensory panel was established at HortResearch in 2003 to describe the sensory attributes of avocado oil. Although some similarities exist between avocado and olive oils, overall avocado oil is very different. In this section we present the use of descriptive analysis to define the flavor attributes of avocado oil.

#### *Use of Descriptive Analysis*

To develop a profile of avocado oil flavor, Generic Descriptive Analysis was used in sensory analyses. Ten panelists, with some experience of sensory descriptive analysis, generated descriptors for the avocado oil in a roundtable setting (Lawless & Heymann, 1999). Twenty-one descriptors were pared down to nine of the most salient reference terms for avocado oil. The panel leader and panelists developed reference standards for each descriptor. Once a reference standard was determined, the panel then established a value for each of the reference standards by smelling and/or tasting the reference standard while considering the attribute level in avocado oil. Upon panel consensus, each reference standard and value was set. A standard blue-glass cup with a watch-glass lid, as used for olive-oil assessment, was also used in avocado oil assessment to aid in warming the samples and masking possible color effects. The bowl-shaped glass tapers at the top to facilitate the concentration of the headspace volatiles and the watch-glass lids trap volatiles until the panelist is ready to evaluate the oil.

### *Oil Samples Examined*

A diverse range of avocado oils was examined over 5 years, mostly from oils extracted in commercial facilities (i.e., using a full commercial process rather than small-scale laboratory systems). Sensory studies were carried out on: freshly extracted oils; oils stored in bottles for 1 and 2 years; oils from previously opened bottles (i.e., oxidized); oils from fruit that had been stored for a range of time (up to 7 weeks); oils from fruit with a range of rot levels; oils from New Zealand, Mexico, and Australia; oils extracted from fruit which were ripened, stored, frozen, and then processed; and oils from different processing techniques (such as inclusion of more or less avocado-skin tissue in the malaxer). The majority of the oils were extracted from Hass avocados. Attributes were then determined to be positive and negative.

### *Attributes Identified*

Nine distinct attributes were identified. The attributes used to describe avocado oil by this panel were smoky, grassy, mushroom/butter-like, hoppy, aniseed-like, fishy, painty, glue-like, and greasy/oily. Fishy, painty, and glue-like are considered negative attributes, although when present in small amounts are not considered offensive. The panel was initially trained on eight of the nine attributes, as the panel debated what reference standard was used for oily/greasy. The term "oily/greasy" is a broad term, and narrowing the standard to specific anchor points can challenge the panelists. Table 2.7 shows the reference standards used for each flavor descriptor.

An example of sensorial differences that were measured is shown in Fig. 2.13. The trained panelists compared New Zealand- and Mexican-grown Hass avocado oils. Although the oils were both obtained from Hass avocados, significant differences were noted in the attributes, with the New Zealand oil being higher in grassy, hoppy, painty, and mushroom/butter-like characteristics, whereas the Mexican oil was significantly higher in the glue-like characteristic. The sensory assessment indicated that the quality of New Zealand avocado oil was superior to that of the Mexican oil. However, these oils were purchased through a retail outlet, and therefore assumptions about their representative qualities cannot be conclusively attributed to their region. To determine the effect of region, factors such as extraction methods, handling procedures,

**Table 2.7. Avocado Oil Sensory Reference Standards Used in Trained-Panel Evaluations**

| Flavor descriptor | Definition | Reference standard |
|---|---|---|
| Smoky | Odor of smoke off burning wood | Country Squire Smoke Flavor. Liquid smoke, 3 drops |
| Grassy | Odor of freshly cut grass | 25 μg/L cis-3 hexen-1-ol (Sigma) |
| Fishy | Odor of old, dead fish | Vita Pet Goldfish Granules Fish Food, 20 g |
| Glue-like | Sharp, acrid odor of PVA<Q> glue | Super Strong PVA, Holdfast NZ Ltd., 20 g |
| Hoppy | Hoppy, malted odor of beer | Cascade's Pale Ale Concentrated Beer Starter, 30 mL in 150 mL of warm water |
| Aniseed | Odor of aniseed, the spice, liquorice-like | Gregg's Aniseed (whole seeds), 20 g |
| Painty | Odor of very rancid linseed oil | Undiluted linseed oil, TMK Packers Ltd. 25 mL |
| Butter/Mushroom-like | Odor of freshly sautéed mushrooms in butter | Slices (5 mm) of white button mushrooms (200 g) sautéed in 25 g of butter for 5 minutes, medium-high heat |
| Oily/Greasy | Odor of oil from potato chips | Krispa, salted potato chips, newly opened bag |

and oil shelf life would need to be controlled. However, researchers have assessed the sensory characteristics of olive oils, and attributed specific sensory profiles to regional influences (Caporale et al., 2006). Future avocado oil sensory research could involve determining the regional effects on sensory attributes.

In another study, the sensory assessment of avocado oil processed from fruit with varying skin amounts showed that avocados pressed with 43% of skin content had higher intensities of negative attributes, specifically painty and fishy, than the oil derived from avocados with 11% of skin content at pressing. The 43% of skin-derived oil was also lowest in grassy characteristics, which is considered a positive attribute in avocado oil. These findings supported the attributes of painty and fishy, or grassy, as being definable markers for negative and positive characteristics, respectively, in the assessment of avocado oil.

## Aroma Compounds

Although sweetness and bitterness are perceived by the tongue, many other key flavor attributes are in fact perceived by the nose as aromas. This detection relies on a flavor compound being volatile— often characterized by a low molecular weight. Aroma volatiles are transported by air streams to the olfactory epithelium, where they stimulate sensory receptors, and hence in oils these are key to flavor/acceptability.

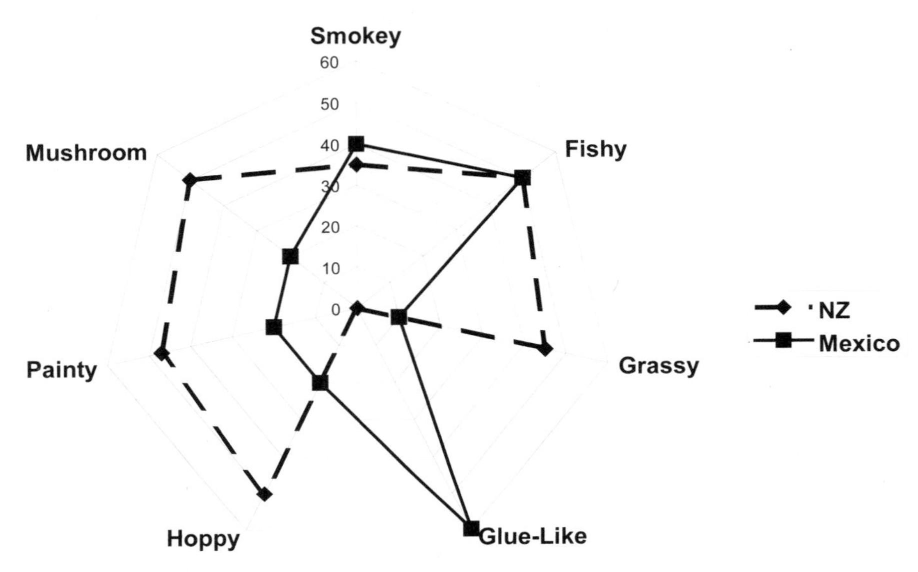


**Fig. 2.13. Sensory evaluation of the attributes of New Zealand and Mexican Hass avocado oils.**

Fruit's volatile components can differ depending on a number of variables, including the cultivar, fruit quality at extraction, various processing conditions, or according to how the oil is stored following extraction.

Although considerable work was done examining the volatile fractions from olive oils (Angerosa, 2002; Baccouri et al., 2008; Luna et al., 2006; Runcio et al., 2008; Vichi et al., 2003), only a paucity of published work regarding a similar analysis for avocado oil is available, especially in the relatively new area of cold-pressed avocado oil. Here we present some preliminary findings on a range of cold-pressed Hass avocado oils, including oils that have defects as determined by a trained sensory panel.

## *Methods*

Three avocado oils of differing quality examined in this preliminary work are described below. All oils examined were extracted, bottled, and stored in a commercial facility in the manner outlined above.

### High Quality

A "new season" oil, which was extracted and processed within the last 2 months, and the sample was obtained from an unopened bottle. One could view this as a "best case" oil.

### Moderate Quality

Oil from the current season was opened and closed over a 2-month period, but stored at ≅20°C in the dark. This oil would still be acceptable to the consumer.

### Poor Quality

An oil made from fruit with a high level of rots (resulting from a long storage of fruit), which was bottled and stored sealed at < –25°C for 2 years in the dark. This oil had a range of off-flavors, and was unacceptable from a sensory perspective.

Headspace volatiles were extracted separately from a 1-g sample of each oil, and analyzed using gas chromatography–mass spectrometry (GC–MS). Volatiles were extracted for 1 hour at 20°C into a sealed 10-mL vial by using 65 µm of polydimethylsiloxane-divinylbenzene solid-phase microextraction (SPME) fibers (Supelco).

The SPME fibers were thermally desorbed at 220°C at the injector, connected to a DB-Wax capillary column (J&W Scientific) into the GC (HP6890, Agilent). The GC oven-temperature program was 30°C for 3 minutes followed by an increase of 3°C/minute to 220°C. Peaks were detected and identified using MS (Leco Pegasus III).

## *Results and Discussion*

For the high-quality, new-season oil, high levels of the desirable hexanal and *E*-2-hexenal, *E*-2-hexenol, hexanol compounds—associated with fresh and green aromas (desirable volatiles)—were observed (Fig. 2.14a; Table 2.7). Conversely, the levels of the undesirable component, acetic acid, were low. This compound is probably associated with the "glue-like" descriptor (Table 2.7). Small amounts of terpenes were found in the headspace of the oil, including α- and β-pinene, β-myrcene, *cis*- and *trans*-β ocimene.

The moderate-quality oil (Fig. 2.14b) showed smaller amounts of desirable volatiles (hexanal, *E*-2-hexenal, and *E*-2-hexenol) than the high- quality oil. However, increased amounts of alcohols (propanol and hexanol) were present; these alcohols were derived from the reduction of aldehydes. These differences are consistent with the effect of the constant opening and closing of the bottle, which could have contributed to the loss of volatiles and also caused rancidity through exposure to oxygen.

Although the poor-quality oil showed some evidence of desirable compounds, the profile was dominated by acetic acid, reflecting high levels of oil oxidation (Fig. 2.14c).

Moreno et al. (2003) also recorded finding hexanal in oil extracted from microwaved avocado pulp, although they recorded other key compounds (octanal, nonanal, and β-caryophyllene) which we did not identify in our analyses. Such differences are likely to be due to differences in oil extraction or measurement techniques. They did not identify acetic acid in any of their oils, suggesting that these oils were not exposed to oxygen during storage following extraction.

We conclude that the quality of avocado fruit used for processing and the oil-storage conditions may be responsible for differences in the volatile levels of avocado oils. Our results, although preliminary, indicate that GC–MS analysis could prove to be a useful technique in the identification of positive and negative aroma attributes associated with avocado oil. Clearly, in-depth research is warranted, examining the effects of key quality factors, including the effects of fruit rots and time in the bottle (shelf life). This could then be used to compare and contrast the aroma volatile patterns from avocado oils and olive oils, on which more research was carried out.

## Allergic and Toxic Compounds

Three main areas of relevance pertain to this area: latex allergens, noted toxicities, and persins (dienes).

### Latex Allergens

The key allergen recorded in avocado fruit is the hevein-like allergy "latex-fruit response," so named primarily because of the urticaria and anaphylactic reactions to latex-containing rubber products that were recognized in the 1990s (Abeck et al.,

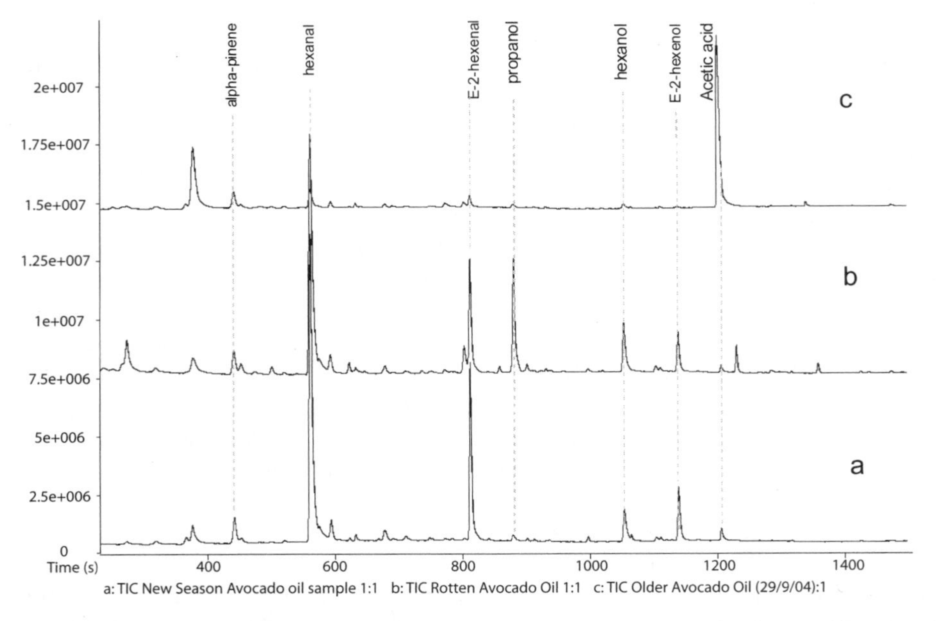


**Fig. 2.14. Gas chromatography–mass spectrometry (GC–MS) traces for three cold-pressed avocado oils: a) high-quality, new-season oil, b) medium-quality oil (opened for 2 months), and c) poor-quality oil (extracted from poor-quality, rotten fruit).**

1994). The avocado allergen (Prs a 1) was isolated and characterized as an IgE-binding peptide, which is a 32 kDa basic class I endochitinase (Sowka et al., 1998), homologous to PR-3 proteins (Breiteneder & Ebner, 2000). However, this was questioned somewhat by Karisola et al. (2005), who after clinical trials found that the isolated hevein-like domain (HLD) molecules alone, but not when linked to endochitinases, seemed to be responsible for IgE-mediated latex-fruit syndrome reactions. The latex-type avocado reaction is similar to that of other fruit, such as banana and chestnut (Breiteneder & Ebner, 2000).

Thus, although clearly, protein-based allergenic reactions to fresh avocado fruit exist, it is unlikely that these compounds cause allergies in avocado oil. This is because in the production of oil, particular attention is paid to eliminating proteins (present in solids' residue), since these proteins are likely to include enzymes that hydrolyze oils (e.g., lipoxygenases), and thus result in poor quality stored oils (Williams, 2005). Such proteins are removed at the decanter and oil "polishing" stages, and if not then, in the subsequent sparging or settling (or "racking") steps. Furthermore, these allergenic proteins are highly labile (Breiteneder & Ebner, 2000).

## Dienes and Related Compounds

Specialized idioblast cells are present in avocado-fruit tissue, and contain lipid and lipid-soluble compounds. One well-known lipid-soluble compound is persin [(+)-(*Z,Z*)-1-acetoxy-2-hydroxy-12,15-heneicosadie-4-one], originally isolated and identified from avocado leaves. Persin has insecticidal (Rodriguez-Saona et al., 1998) and antifungal activity (Prusky et al., 1991), and is believed to protect unripe avocado fruit from avocado anthracnose (*Colletotrichum gloeosporioides* Penz). Recently persin was found to have in vivo activity in the mammary gland, and induces Bim-dependent apoptosis in human-breast cancer cells (Butt et al., 2006). The concentration of persin is known to decrease during ripening, and likely, persin will not be present in avocado oil manufactured from ripe avocado fruit. Additional persin-related compounds were isolated from avocado fruit. These include: a monoene (1-acetoxy-2,4-dihydroxy-*n*-heptadeca-16-ene) (Prusky et al., 1991); a triene [(*E,Z,Z*)-1-acetoxy-2-hydroxy-4-oxo-heneicosa-5,12,15-triene] (Domergue et al., 2000); isopersin, a geometric isomer of persin with the same molecular formula (Rodriquez-Saona et al., 1998); and a series of triols (1,2,4-trihydroxynonadecane; 1,2,4-trihydroxyheptadec-16-ene; 1,2,4-trihydroxyheptadec-16-yne) with toxicity toward human tumor cells, including human-prostate carcinoma (PC-3) cells (Oberlies et al., 1998).

We sent three oils to the laboratory of Dr. Dov Prusky (Volcani Research Centre, Israel) for the determination of diene levels. These cold-pressed avocado oils were extracted commercially with the inclusion of 11, 43, and 100% of the available avocado skin. No dienes could be detected (<50 μg of $gFWt^{-1}$) in any of the oils. This suggests that dienes and presumably monoenes are unlikely to be significant factors in cold-pressed oils from a human health perspective.

### Avocado oil Toxicity and Allergies

We could find no formal studies into the allergies of avocado oil in humans. No allergies (or toxicity) were recorded to cold-pressed avocado oil at the New Zealand National Poisons Centre. However, reports were made of the effects of avocado fruit on animal and humans, as reported in the literature (Carman & Handley, 1999; Werman et al., 1989). Since cold-pressed avocado oil has been consumed widely by consumers in New Zealand for 6–8 years, the allergic compounds most likely are not extracted into the oil.

## Health Benefits of the Oil and Oil Constituents

### Avocado Oil

Because avocado oil was not consumed in significant quantities historically, and because the cold-pressed product is relatively new, no formal health tests of avocado oil were performed. However, here we will point to the key components of avocado oil that have strong scientific "healthfulness" support.

#### *Monounsaturated Fatty Acids*

The lipid content of avocados is made up of 15–20% of saturated fats, 60–70% of monounsaturates and ≅10% of polyunsaturates. A diet high in monounsaturated fatty acids (MUFAs) is recommended for a healthy Mediterranean diet (Birkbeck, 2002). Such a diet has favorable effects on lipoprotein levels, endothelium vasodilation, insulin resistance, metabolic syndrome, antioxidant capacity, and myocardial and cardiovascular mortality (Serra-Majem et al., 2006). The Mediterranean diet recommends abundant plant foods, and olive oil as the principal source of dietary lipids. Because avocado oil has a very similar lipid profile to olive oil, it clearly can be included as a healthy addition to the Mediterranean diet.

#### *α-Tocopherol*

Vitamin E is an essential vitamin and is widely recognized as a powerful antioxidant. These antioxidants prevent the formation of damaging free radicals from the body's normal oxidation processes, and are associated with the reduction in the incidence of cardiovascular diseases (Pryor, 2000). The favorable concentration of vitamin E in cold-pressed avocado oil (≅70–190 μg/g of oil) is comparable or greater to that in many olive oils (100–140 μg/g of oil; Boskou, 2006). Based on the Australia and New Zealand guidelines for nutrient intakes, one teaspoon of avocado oil would provide approximately one-ninth of the recommended dietary target of vitamin E for men (NHMRC, 2008). The "soft" processing conditions used in cold-pressed extraction methods help to retain higher levels of the relatively labile α-tocopherol than extraction and refining methods that use chemicals and/or heat.

### Sterols

Avocado fruit contain the highest levels of plant sterols of any fruit (Duester, 2001). Since plant sterols are soluble in fat, they are extracted with the oil from the avocado. The main plant sterol present in avocado oil is sitosterol, but other sterols were also identified in the oil, including Δ-5-avenasterol, campesterol, and stigmasterol. Plant sterols have a beneficial effect on lowering blood-cholesterol levels in humans (Piironen et al., 2000). Since the consumption of adequate amounts of plant sterols can reduce the risk of heart disease, they are included in food products such as margarine spreads (Weststrate & Meijer, 1998). These products have gained international acceptance, and have captured space in a competitive market despite their premium prices.

### Plant Pigments

Because of the nature of cold-pressed avocado oil extraction (i.e., the lack of post-extraction refinement), avocado oil contains significant levels of pigments such as carotenes, xanthophylls, and substantial amounts of chlorophyll, which are reflected in its rich green color. Since these pigments act as antioxidants, they are believed to provide protection from diseases (Lu et al., 2005). Of the pigments in avocado oil, the carotenoid lutein is probably the most important, and this has particular relevance to eye health, acting in the macular region of the retina. AMD is the leading cause of loss of vision in the elderly, and these pigments are believed to protect the cells of the macula from light-induced damage (Koh et al., 2004; Richer et al., 2004). Avocado oil contains approximately twice as much lutein as olive oil (Criado et al., 2007).

### Skin Health

Compared with almond, corn, olive, and soybean oils, avocado oil had the highest skin-penetration rate (Swisher, 1988). Due to its soothing and moisturizing effects, avocado oil was added to pharmaceutical creams for the treatment of common skin conditions such as psoriasis and dandruff with results superior to the traditional medicated therapy (Huang et al., 2004; Stücker et al., 2001).

## Avocado Fruit

Although no formal health trials were carried out on avocado oil itself, because most of the healthful compounds in avocado fruit are extracted into avocado oil (e.g., ≅90% of the carotenoids; Ashton et al., 2006), the findings relating to avocado fruit are very likely to apply to avocado oil, particularly cold-pressed avocado oil, which contains such important phytochemicals as pigments such as lutein. Examples of such benefits include those described by Lu et al. (2005), who found that an extract from Hass avocados inhibited the growth of cancer cells in vitro.

Other bioactivities of avocado extracts were also reported. For example, avocado extracts inhibited both nitric oxide and superoxide generation in cell-culture systems, suggesting that these compounds may have antioxidant activity and possibly chemopreventative activity against inflammation-associated carcinogenesis (Kim et al., 2000). This activity was associated with persin and the related triene. Avocado oil produced from high-temperature drying of whole fruit and solvent extraction appears to contain hepatoxic agents that modify hepatic lipid metabolism (Werman et al., 1989). A further study reported that avocado-fruit components have substantial acetyl-CoA carboxylase inhibitory activity, which may help to reduce fat accumulation and obesity. Inhibitory activity was isolated through bioassay-guided fractions, and was found to be associated with persin and the related triene [(*E,Z,Z*)-1-acetoxy-2-hydroxy-4-oxo-heneicosa-5,12,15-triene] (Hashimura et al., 2001). Additionally, two further persin-related compounds were found to have acetyl-CoA carboxylase inhibitory activity.

Avocado–soybean unsaponifiables (ASUs) were used widely in Europe as a complementary medicine for osteoarthritis (OA). ASU is claimed to improve the symptoms of OA, and is promoted as an alternative to the standard nonsteroidal anti-inflammatory drugs (NSAIDs) including paracetamol. A recent meta-analysis of randomized controlled studies concluded that ASUs have some efficacy against OA and that "ASUs are no worse and no better for treatment of OA than other medications". The combined evidence is stronger for knee OA than for hip OA (Chrostensen et al., 2008). ASU reduced the effects of the pro-inflammatory cytokine IL1$\beta$ by interfering with the induction of NF-$\kappa$B-activated elements in in vitro cell-based studies (Gabay et al., 2008). Further evidence shows that ASU suppresses TNF-$\alpha$, Cox-2, prostaglandin $E_2$, and inducible nitric oxide synthase expression, all consistent with strong anti-inflammatory effects (Au et al., 2007).

ASU is prepared by removing the acyl lipid components from oil by saponification, followed by the isolation of the lipid-soluble residue by solvent extraction. Avocado oil is rich in unsaponifiable components (2–7%), compared with other oils such as olive ($\cong$1%). Relatively little is known about the composition of avocado unsaponifiables (AUs), but they do contain sterols (sitosterol), vitamin E, and a series of aliphatic compounds containing a furyl nucleus conjugated with mono- or polyunsaturated chains of 13 to 17 carbon in length (Henrotin et al., 1998). More information is required about the composition of ASU, and studies are required to determine which of the components of AU are responsible for this bioactivity against OA.

## Authenticity and Adulteration

Avocado oil is a relatively new product in the culinary arena, and as such has no formal or even informal standards by which oils are certified, or for that matter were even described.

At this point, no anecdotal evidence points to the adulteration of cold-pressed avocado oil, although some product is labeled as "cold-pressed," yet has very low coloration, indicating a high likelihood of processing such as bleaching. Producers who are blending avocado oil with other oils do so to broaden the range of cold-pressed avocado oils available, and clearly state the oils used. This includes a wide range of flavored oils such as chilli, pepper, lemon, lime, garlic or basil-infused. In addition, oil blends with omega-3 and omega-6 fatty acids were formulated and are available commercially (e.g., OmegaPlus® Olivado Oil).

No organization exists for avocado oil to carry out the equivalent role of the International Olive Council (IOC—formerly the IOOC—International Olive Oil Council), which was a formal entity of the United Nations (IOC, 2006).

We believe that our experience over the last 8 years with cold-pressed avocado oil puts us in a credible position internationally to make the first detailed proposed standard for avocado oil. We have carried out many chemical analyses (of overall oil quality as well as individual phytochemicals), developed a sensory analysis system, and carried out many experiments examining a diverse range of factors that influence oil quality including avocado pre- and postharvest factors, extraction/processing conditions, and post-bottling conditions (i.e., oil storage). Much of this work was either published, is in preparation for publication, or was presented in this chapter.

Our standards proposed below (Table 2.8) are guided by the olive-oil standards of the IOC, although avocado oil has significant differences from olive oil.

## Summary

Avocado oil is a high-value oil with excellent qualities for both culinary and cosmetic uses. High-quality cold-pressed avocado oil for culinary consumption is a new product, and its production and sales are increasing consistently throughout the world. The supply of fruit with appropriate quality and price currently limits production, but the color, subtle flavor, and healthfulness of the oil should ensure its increasing growth.

## Acknowledgments

Most of the original work presented here was funded by the Foundation for Research Science and Technology (contract number C06X0203), and some work was funded by Horticulture Australia Ltd. (contract AVO3007) and the California Avocado Commission. The assistance of the Olivado staff (both in New Zealand and in Australia) is gratefully acknowledged. We also acknowledge the hard work of our under- and postgraduate students (Nimma Sherpa, Ofelia Ashton, Angela Yi, and Carlene Pulfer-Ridings). Thanks also to Dr. Dov Prusky (Volcani Research Institute, Israel) for the analysis of oil samples for diene concentrations. Thanks to Michelle Napier and Emma Clegg for assistance with literature searching. Finally, thanks to William Laing, Daryl Rowan, Sol Green, and Anne Gunson for comments on the manuscript.

**Table 2.8. Proposed International Quality Standards for Avocado Oil—Values Are Relevant to Oil Quality at Time of Bottling**

| | Extra virgin | Virgin | Pure | Blends |
|---|---|---|---|---|
| General | Oil extracted from high-quality fruit (minimal levels of rots and physiological disorders). Extraction to be carried out using only mechanical extraction methods including presses, decanters, and screw presses at low temperatures (<50 °C). Addition of water and processing aids (e.g., enzymes and talcum powder) is acceptable, but no chemical solvents can be used. | Oil extracted from sound fruit with some rots or physiological disorders. Extraction to be carried out using only mechanical extraction methods including presses, decanters, and screw presses at low temperatures (<50 °C ). Addition of water and processing aids (e.g., enzymes and talcum powder) is acceptable, but no chemical solvents can be used. | Fruit quality not important. Decolorized and deodorized oil with low acidity, low color, and bland flavor. Oil produced from good-quality virgin avocado oil; may be just avocado oil or infused with natural herb or fruit flavors. | Avocado oil is an excellent oil for blending, and complements extra-virgin olive, flaxseed, macadamia and pumpkin-seed oils. The specification and composition should match what is claimed on the label. |
| Organoleptic characteristics* | | | | |
| Odor and taste | Characteristic avocado flavor and sensory assessment show at least moderate (above 40 on a 100-point scale) levels of grassy and mushroom/butter with some smoky | Characteristic avocado flavor and sensory assessment show at some (above 20 on a 100-point scale) levels of grassy and mushroom/butter with some smoky | Bland or matches description of infused flavor: e.g., lemon, chilli, rosemary, etc. | Dependent on the blend |

**Table 2.8., cont. Proposed International Quality Standards for Avocado Oil—Values Are Relevant to Oil Quality at Time of Bottling**

| | Extra virgin | Virgin | Pure | Blends |
|---|---|---|---|---|
| Defects | Minimal to no defect such as painty, and fishy notes below 20 and glue-like below 35 as a sensory panel average on a 100- point scale | Low levels only of defects such as painty and fishy notes below 50 as a sensory panel average on a 100- point scale | Low defects such as painty and fishy notes below 50 as a sensory panel average on a 100-point scale | Low defects such as painty and fishy notes below 50 as a sensory panel average on a 100-point scale |
| Color | Intense and attractive green | Green with potential yellow hue | Pale yellow | Dependent on the blend |
| Free fatty acid (% as oleic acid) | ≤0.5% | 0.8–1.0% | ≤0.1% | As specified |
| Acid value | ≤1% | ≤2.0% | ≤0.2% | |
| Peroxide value (mEq/kg oil) | ≤4.0 | <8.0 | <0.5 | |
| Stability | 2 years at ambient temperature when stored under nitrogen and out of the light | 18 months at ambient temperature when stored under nitrogen and out of the light | >2 years at ambient temperature when stored under nitrogen and out of light | |
| Smoke point | ≥250 °C | ≥200 °C | ≥250 °C | |
| Moisture | ≤0.1% | ≤0.1% | ≤ 0.1% | |
| Fatty acid composition % (typical values) | | | | |
| Palmitic acid (16:0) | 10–25 | | | |

**Table 2.8., cont. Proposed International Quality Standards for Avocado Oil—Values Are Relevant to Oil Quality at Time of Bottling**

| | Extra virgin | Virgin | Pure | Blends |
|---|---|---|---|---|
| Palmitoleic acid (16:1) | 2–8 | | | |
| Stearic acid (18:0) | 0.1–0.4 | | | |
| Oleic acid (18:1) | 60–80 | | | |
| Linoleic acid (18:2) | 7–20 | | | |
| Linolenic acid (18:3) | 0.2–1 | | | |
| Antioxidants (mg/kg) | | | | |
| Vitamin E | 70–190 | | | |
| Trace metals (mg/kg) | | | | |
| Copper | ≤0.05 | ≤0.05 | ≤0.05 | ≤0.05 |

* These characteristics are measured with a trained sensory panel with a minimum of 15 hours of experience of tasting avocado oil.

## References

Abeck, D.; M. Borries; C. Kuwert; V. Steinkraus; D. Vieluf; J. Ring. Anaphylactic Reactions to Food Items with Latex Allergy. *Hautarzt.* **1994,** *45,* 364–367.

Anderson, D. A primer on oils processing technology. *Bailey's Industrial Oil and Fat Products;* F. Shahidi, Ed.; John Wiley & Sons: New Jersey, 2005; Volume 5 Chapter 1, pp. 1–56.

Angerosa, F. Influence of volatile compounds on virgin olive oil quality evaluated by analytical approaches and sensor panels. *Eur. J. Lipid Sci. Technol.* **2002,** *104,* 639–660.

AOCS. Official methods and recommended practices of the AOCS. American Oil Chemists' Society, Champaign, IL. 1994.

Appleman, D. Personal communication with Biale and Young 1969. *The Biochemistry of Fruits and Their Products;* A.C. Hulme, Ed.; Academic Press: London, 1969; pp. 16–19.

Argueta-Villamar, A.; L. Cano; M. Rodarte. *Atlas de las Plantas de la Medicina Tradicional Mexicana, México*; 1994..

Ashton, O.B.O.; M. Wong; T.K. Mcghie; R. Vather; Y. Wang; C. Requejo-Jackman; P. Ramankutty; A.B. Woolf. Pigments in avocado tissue and oil. *J. Agric. Food Chem.* **2006,** *54,* 10151–10158.

Au, R.Y.; T.K. Al-Talib; A.Y. Au; P.V. Phan; C.G. Frondoza. Avocado soybean unsaponifiables (ASU) suppress TNF-alpha, IL-1beta, COX-2, iNOS gene expression, and prostaglandin E2 and nitric oxide production in articular chonfrocytes and monocyte/macrophages. *Osteoarthritis Cartilage* **2007,** *15,* 1249–1255.

Avocadosource (2006). http://www.avocadosource.com/

Baccouri, O.; A. Bendini; L. Cerretani; M. Guerfel; B. Baccouri; G. Lercker; M. Zarrouk; D.D. Ben Miled. Comparative study on volatile compounds from Tunisian and Sicilian monovarietal virgin olive oils. *Food Chem.* **2008,** *111,* 322–328.

Biale, J.B.; R.E. Young. The avocado pear. *Biochemistry of Fruits and Their Products;* A.C. Hulme, Ed.; Academic Press: London, 1969; pp. 68–83.

Birkbeck, J. Health benefits of avocado oil. *Food New Zealand,* **2002,** *2,* 40–42.

Bizimana, V.; W.M. Breene; A.S. Csallany. Avocado oil extraction with appropiate technology for developing countries. *J. Am. Oil Chem. Soc.* **1993,** *70,* 821– 822.

Bligh, E.G.; W.J. Dyer. A rapid method of total lipid extraction and purification. *Can. J. Biochem. Physiol.* **1959,** *37,* 911–917.

Board, L. Fluctuations in production a headache for avocado oil producer. *The Orchadist.* **2007**, 80, 42–44.

Boskou, D. *Olive Oil: Chemistry and Technology;* AOCS Press: Champaign, Illinois, 2006; 268 pp.

Botha, B.M.; R.I. McCrindle. Supercritical fluid extraction of avocado oil. *South African Avocado Growers' Assoc. Yearbook.* **2003,** *26,* 11–13.

Breiteneder, H.; C. Ebner. Molecular and biochemical classification of plant-derived food allergens. *J. Allergy Clin. Immunol.* **2000,** *106,* 27–36.

Brown, B.I. Market maturity indices and sensory properties of avocados grown in Queensland.

*Food Technol. Australia.* **1984,** *36,* 474–476.

Buenrostro, M.; A.C. Lopez-Munguia. Enzymatic extraction of avocado oil. *Biotechnol. Lett.* **1986,** *8,* 505–506.

Butt, A.J.; C.G. Roberts; A.A. Seawright; P.B. Oelrichs; J.K. Macleod; T.Y.E. Liaw; M. Kavallaris; T.J. Somers-Edgar; G.M. Lehrbach; C.K. Watts; R.L. Sutherland. A novel plant toxin, persin, with in vivo activity in the mammary gland, induces Bim-dependent apoptosis in human breast cancer cells. *Mol. Cancer Ther.* **2006,** *5,* 2300–2309.

Caporale, G.; S. Policastro; A. Carlucci; E. Monteleone. Consumer expectations for sensory properties in virgin olive oils. *Food Quality and Preference.* **2006,** 116–125.

Carman, R.M.; P.N. Handley. Antifungal diene in leaves of various avocado cultivars. *Phytochemistry* **1999,** *50,* 1329–1331.

Chrostensen, R.; E.M. Bartels; A. Astrup; H. Bliddal. Symptomatic efficacy of avocado-soybean unsaponifiables (ASU) in osteoarthritis (OA) patients: a meta-analysis of randomized controlled trials. *Osteoarthritis Cartilage* **2008,** *16,* 399–408.

Coppen, P. P. The use of antioxidants. *Rancidity in Foods;* J.C. Allen; R.J. Hamilton, Eds.; Blackie Academic & Professional: London, 1994; pp. 84–103.

Cox, K.A.; T.K. Mcghie; A. White; A.B. Woolf.. Skin colour and pigment changes during ripening of "Haas" avocado fruit. *Postharvest Biol. Technol.* **2004,** *31,* 287–294.

Criado, M.N.; M.J. Motilva; M. Goni; M.P. Romero. Comparative study of the effect of the maturation process of the olive fruit on the chlorophyll and carotenoid fractions of drupes and virgin oils from Arbequina and Farga cultivars. *Food Chem.* **2007,** *100,* 748–755.

Domergue, F.; G.L. Helms; D. Prusky; J. Browse. Antifungal compounds from idioblast cells isolated from avocado fruits. *Phytochemistry* **2000,** *54,* 183–189.

Duester, K.C. Avocado fruit is a rich source of beta-sitosterol. *J. Am. Diet. Assoc.* **2001,** *101,* 404–405.

Eaks, I. Change in the fatty acid composition of avocado fruit during ontogeny, cold storage and ripening. *Acta Hortic.* **1990,** *269,* 141–151.

Fletcher, A. (January 2006). http://www.foodnavigator.com/Financial-Industry/Olive-oil-established-as-leading-UK-edible-oil

Folch, J.; M. Lees; G.H.S. Stanley. A simple method for the isolation and purification of total lipides from animal tissues. *J. Biol. Chem.* **1957,** *226,* 497–509.

Gabay, O.; M. Gosset; A. Levy; C. Salvat; C. Sanchez; A. Pigenet; A. Sautet; F. Berenbaum. Stress-induced signaling pathways in hyalin chondrocytes: inhibition by avocado-soybean unsaponifiables (ASU). *Osteoarthritis Cartilage* **2008,** *16,* 373–384.

Gaydou, E.M.; Y. Lozano; J. Ratovohery. Triglyceride and fatty acid compositions in the mesocarp of Persea americana during fruit-development. *Phytochemistry* **1987,** *26,* 1595–1597.

Gutierrezrosales, F.; J. Garridofernandez; L. Gallardoguerrero; B. Gandulrojas; M.I. Minguezmosquera. Action of chlorophylls on the stability of virgin olive oil. *J. Am. Oil Chem. Soc.* **1992,** *69,* 866–871.

Hamilton, R.J. The chemistry of rancidity in foods. *Rancidity in Foods;* J.C. Allen, R.J. Hamilton, Eds.; Blackie Academic & Professional: London; 1994, Chapter 1, pp. 1–21.

Hashimura, H.; C. Ueda; J. Kawabata; T. Kasai. Acetyl-CoA carboxylase inhibitors from avocado (Persea americana Mill) fruits. *Biosci. Biotechnol. Biochem.* **2001,** *65,* 1656–1658.

Henrotin, Y.E.; A.H. Labasse; J.M. Jaspar; D.D. De Groote; S.X. Zheng; G.B. Guillou; J.Y. Reginster. Effects of three avocado/soybean unsaponifiable mixtures on metalloproteinases, cytokines and prostaglandin E2 production by human articular chondrocytes. *Clin. Rheumatol.* **1998**, *17,* 31–39.

Hopkirk, G. Avocado Maturity Assessment. Auckland: New Zealand, 1989.

Hopkirk, G.; A. White; D.J. Beever; S.K. Forbes. Influence of postharvest temperatures and the rate of fruit ripening on internal postharvest rots and disorders of New Zealand "Hass" avocado fruit. *New Zealand J. Crop Hortic. Sci.* **1994,** *22,* 305– 311.

Huang, C.; A Huang; M. Chen. Preliminary study on application of extract from Baccae of Persea americana L. in cosmetic industry. *Chem. Industry of Forest Products* **2004,** *24,* 87–90.

Human, T.P. Oil as a byproduct of avocado. *South African Avocado Growers' Assoc. Yearbook.* **1987,** *10,* 159–162.

Inoue, H.; A. Tateishi. Ripening and fatty acid composition of avocado fruit in Japan. *Proceedings of the World Avocado Congress III.* Tel Aviv, Israel. 1998. pp. 366–369.

IOC. Trade standard applying to olive oils and olive-pomace oils. COI/T.15/NC no. 3/Rev. 2. 2006. Madrid, International Olive Council: 20p.

Jung, M.Y.; D.B. Min. Effects of quenching mechanisms of carotenoids on the photosensitized oxidation of soybean oil. *J. Am. Oil Chem. Soc.* **1991,** *68,* 653–658.

Kaiser, C.; M.T. Smith; B.N. Wolstenholme. Overview of lipids in the avocado fruit, with particular reference to the Natal Midlands. *South African Avocado Growers' Assoc. Yearbook,* **1992,** *15,* 78–82.

Kalua, C.M.; M.S. Allen; D.R. Bedgood; A.G. Bishop; P.D. Prenzler. Discrimination of olive oils and fruits into cultivars and maturity stages based on phenolic and volatile compounds. *J. Agric. Food Chem.* **2005,** *53,* 8054–8062.

Karisola, P.; A. Kotovuori; S. Poikonen; E. Niskanen; N. Kalkkinen; K. Turjanmaa; T. Palosuo; T. Reunala; H. Alenius; M.S. Kulomaa. Isolated hevein-like domains, but not 31-kd endochitinases, are responsible for 19E-mediated in vitro and in vivo reactions in latex-fruit syndrome. *J. Allergy Clin. Immunol.* **2005,** *115,* 598–605.

Kidmose, U.; M. Edelenbos; Norbaek; L.P. Christensen. Colour stability of vegetables. *Colour in Foods;* D.B. MacDougall, Ed.; Woodhead Publishing: Cambridge, 2002; pp. 179–232.

Kikuta, Y.; L.C. Erickson. Seasonal changes of avocado lipids during fruit development and storage. *California Avocado Soc. Yearbook* **1968,** *52,* 102–108.

Kim, O.K.; A. Murakami; Y. Nakamura; N. Takeda; H. Yoshizumi; H. Ohigashi. Novel nitric oxide and superoxide generation inhibitors, persenone A and B, from avocado fruit. *J. Agric. Food Chem.* **2000,** *48,* 1557–1563.

Kiritsakis, A.P. Olive Oil: From the Tree to the Table. Second Edition, Food and Nutrition Press, 348 pp., 1998.

Koh, H.H.; I.J. Murray; D. Nolan; D. Carden; J. Feather; S. Beatty. Plasma and macular response to lutein supplement in subjects with and without age-related maculopathy: a pilot study. *Exp.*

*Eye Res.* **2004,** *79,* 21–27.

Lawes, G.S. Maturity and quality in avocados. *The Orchardist* **1980,** *53,* 63–64.

Lawless, H.T.; H.G. Heymann. *Sensory Evaluation of Food —Principles and Practices,* 2nd ed.; Aspen Publishers: Gaithersburg, MD, 1999.

Lee, S.K.; R.E. Young; P.M. Schiffman; C.W. Coggins. Maturity studies of avocado fruit based on picking dates and dry-weight. *J. Am. Soc. Hortic. Sci.* **1983,** *108,* 390–394.

Lelyveld, L.J.V.; C. Gerrish; R.A. Dixon. Enzyme activities and polyphenols related to mesocarp discolouration of avocado fruit. *Phytochemistry* **1984,** *23,* 1531–1534.

Lesley, B.E. and A.W. Christie. Use of refractometric method in determination of oil in avocados. *J Ind Eng Chem Anal Ed* **1929**, *1,* 1–24.

Lewis, C.E. Maturity of avocados—general review. *J. Sci. Food Agric.* **1978,** *29,* 857–866.

Love, H.T. Avocado oil. *Tropical Agric.* **1944,** *21,* 7.

Lozano, Y. Analysis of triglycerides in the avocado oil by reversed-phase high-performance liquid-chromatography—scientific report. *Revue Francaise Des Corps Gras* **1983,** *30,* 333–346.

Lu, Q.Y.; J.R. Arteaga; Q.F. Zhang; S. Huerta; V.L.W. Go; D. Heber. Inhibition of prostate cancer cell growth by an avocado extract: role of lipid-soluble bioactive substances. *J. Nutr. Biochem.* **2005,** *16,* 23–30.

Luna, G.; M.T. Morales; R. Aparicio. Characterisation of 39 varietal virgin olive oils by their volatile compositions. *Food Chem.* **2006,** *98,* 243–252.

Lund, C.M.; M.K. Thompson; F. Benkwitz; M.W. Wohler; C.M. Triggs; R. Gardner; H.G. Heymann; L. Nicolau. New Zealand Sauvignon blanc distinct flavor characteristics: Sensory, chemical, and consumer aspects. *Am. J. Enol. Viticulture* **2009, in press.**

Luza, J.G.; L.A. Lizana; L. Masson. Comparative lipids evolution during cold storage of three avocado cultivars. *Acta Hortic.* **1990,** *269,* 153–160.

Mazliak, P. Les lipides de l'avocat. II Variation de la composition en acides gras des lipides du pericarpe selon la composition de l'atmosphere autor des fruits en maturation. *Fruits* **1965,** *20,* 117–120.

Minguez-Mosquera, M.I.; B. Gandul-Rojas; L. Gallardo-Guerrero; M. Roca; M. Jaren-Galan. Chlorophylls. *Methods of Analysis of Functional Foods and Nutraceuticals,* 2nd ed.; W.J. Hurst, Ed.; CRC Press: Boca Raton, 2008; pp. 337–400.

Montano, G.H.; B.S. Luh; L.M. Smith. Extracting and refining avocado oil. *Food Technol.* **1962,** *16,* 96–98.

Moreno, A.O.; L. Dorantes; J.Galindez; R.I. Guzman. Effect of different extraction methods on fatty acids, volatile compounds, and physical and chemical properties of avocado (Persea americana mill) oil. *J. Agric. Food Chem.* **2003,** *51,* 2216–2221.

Mostert, M.E.; B.M. Botha; L.M. Du Plessis; K.G. Duodu. Effect of fruit ripeness and method of fruit drying on the extractability of avocado oil with hexane and supercritical carbon dioxide. *J. Sci. Food Agric.* **2007,** *87,* 2880–2885.

Nathan, C. Olivado NZ, Keri Keri, New Zealand. Personal communication. 2006.

NHMRC (National Health and Medical Research Council). (2008). Nutrient reference values for

Australia and New Zealand. http://www.nhmrc.gov.au/PUBLICATIONS/synopses/_files/n35.pdf

Oberlies, N.H.; L.L. Rogers; J.M. Martin; J.L. McLaughlin. Cytotoxic and insecticidal constituents of the unripe fruit of Persea americana. *J. Nat. Prod.* **1998,** *61,* 781–785.

Ortiz, M.A.; A.L. Dorantes; M.J. Gallndez; S.E. Cardenas. Effect of a novel oil extraction method on avocado (Persea americana Mill) pulp microstructure. *Plant Foods for Human Nutr.* **2004,** *59,* 11–14.

Pearson, D. Seasonal English market variations in composition of South-African and Israeli avocados. *J. Sci. Food Agric.* **1975,** *26,* 207–213.

Phillips, K.M.; D.M. Ruggio; J.I. Toivo; M.A. Swank; A.H. Simpkins. Free and esterified sterol composition of edible oils and fats. *J. Food Composition and Analysis* **2002,** *15,* 123–142.

Piironen, V.; D.G. Lindsay; T.A. Miettinen; J. Toivo; A.M. Lampi. Plant sterols: biosynthesis, biological function and their importance to human nutrition. *J. Sci. Food Agric.* **2000,** *80,* 939–966.

Platt-Aloia, K.A.; W.W. Thomson. Ultrastructure of the mesocarp of mature avocado fruit and changes associated with ripening. *Annals of Botany* **1981,** *4,* 451–465.

Poucher, W.A. The materials of perfumery. *Perfumes, Cosmetics and Soaps;* Chapman Hall: London, 1974; pp. 44–47.

Prusky, D.; I. Kobiler; Y. Fishman; J.J. Sims; S.L. Midland; N.T. Keen. Identification of an antifungal compound in unripe avocado fruits and its possible involvement in the quiescent infections of Colletotrichum-Gloeosporioides. *J. Phytopathol.-Phytopathologische Zeitschrift* **1991,** *132,* 319–327.

Pryor, W.A. Vitamin E and heart disease: Basic science to clinical intervention trials. *Free Rad. Biol. Med.* **2000,** *28,* 141–164.

Rahmani, M.; A.S. Csallany. Role of minor constituents in the photooxidation of virgin olive oil. *J. Am. Oil Chem. Soc.* **1998,** *75,* 837–843.

Ratovohery, J.V.; Y.F. Lozano; E.M. Gaydou. Fruit-development effect on fatty acid composition of Persea-americana fruit mesocarp. *J. Agric. Food Chem.* **1988,** *36,* 287–293.

Rawls, H.R.; P.J. Vansante. A possible role for singlet oxygen in initiation of fatty acid autoxidation. *J. Am.Oil Chem. Soc.* **1970,** *47,* 121–125.

Redgwell, R.J.; E. Macrae; I. Hallett; M. Fischer; J. Perry; R. Harker. In vivo and in vitro swelling of cell walls during fruit ripening. *Planta* **1997,** *203,* 162–173.

Requejo-Tapia, C. International Trends in Fresh Avocado and Avocado Oil Production and Seasonal Variation of Fatty Acids in New Zealand-grown cv. Hass. Massey University, Palmerston North, New Zealand, 1999.

Richer, S.; W. Stiles; L. Statkute; J. Pulido; J. Frankowski; D. Rudy; K. Pei; M. Tsipursky; J. Nyland. Double-masked, placebo-controlled, randomized trial of lutein and antioxidant supplementation in the intervention of atrophic age-related macular degeneration: the Veterans LAST study (Lutein Antioxidant Supplementation Trial). *Optometry* **2004,** *75,* 216–229.

Rodriguez-Saona, C.; J.G. Millar; J.T. Trumble. Isolation, identification, and biological activity of isopersin, a new compound from avocado idioblast oil cells. *J. Nat. Prod.* **1998,** *61,* 1168–1170.

Rosenthal, I.; U. Merin; G. Popel; S. Bernstein. Determination of fat in vegetable foods. *J. Assoc. Offic. Analyt. Chem.* **1985,** *68,* 1226–1228.

Runcio, A.; L. Sorgona; A. Mincione; S. Santacaterina; M. Poiana. Volatile compounds of virgin olive oil obtained from Italian cultivars grown in Calabria. Effect of processing methods, cultivar, stone removal, and antracnose attack. *Food Chem.* **2008,** *106,* 735–740.

San Martin, M.F.; G.V. Barbosa-Canovas; B.G. Swanson. Food processing by high hydrostatic pressure. *Crit. Rev. Food Sci. Nutr.* **2002,** *42,* 627–645.

Schroeder, C.A. Growth and development of the "Fuerte" avocado fruit. *Proc. Am. Soc. Hortic. Sci.* **1953,** *6,* 103–109.

Scora, R.W and B.O. Bergh. Origin and taxonomic relationships within the genus *Persea*. In: Lovatt, C.; P.A. Holthe; and M.L. Arpaia (Eds.) Proceedings of the second world avocado congress Vol 2, University of California, Riverside, California, 1992. pp. 505–514.

Serra-Majem, L.; B. Roama; R. Estruch. Scientific evidence of interventions using the Mediterranean det: A systematic review. *Nutr. Rev.* **2006,** *64,* 27–47.

Smith, L.M.; F.H. Winter. Research on avocado processing at the University of California Davis. *California Avocado Soc. Yearbook* **1970,** *54,* 79–84.

Southwell, K.H.; R.V. Harris; A.A. Swetman. Extraction and refining of oil obtained from dried avocado fruit using a small expeller. *Tropical Sci.* **1990,** *30,* 121–131.

Sowka, S.; L.S. Hsieh; M. Krebitz; A. Akasawa; B.M. Martin; D. Starrett; C.K. Peterbauer; O. Scheiner; H. Breiteneder. Identification and cloning of Prs a 1, a 32-kDa endochitinase and major allergen of avocado, and its expression in the yeast Pichia pastoris. *J. Biol. Chem.* **1998,** *273,* 28091–28097.

Stücker, M.; U. Memmel; M. Hoffmann; J. Hartung; P. Altmeyer. Vitamin $B_{12}$ cream containing avocado oil in the therapy of plaque psoriasis. *Dermatology* **2001,** *203,* **141–147**.

Swisher, H.E. Avocado oil— from food use to skin care. *J. Am. Oil Chem. Soc.* **1988,** *65,* 1704–1706.

Torres, A.M.; T. Maulastovicka; R. Rezaaiyan. Total phenolics and high-performance liquid-chromatography of phenolic-acids of avocado. *J. Agric. Food Chem.,* **1987,** *35,* 921–925.

Verleyen, T.; M. Forcades; R. Verhe; K. Dewettinck; A. Huyghebaert; W. De Greyt. Analysis of free and esterified sterols in vegetable oils. *J. Am. Oil Chem. Soc.* **2002,** *79,* 117–122.

Vichi, S.; L. Pizzale; L.S. Conte; S. Buxaderas; E. Lopez-Tamames. Solid-phase microextraction in the analysis of virgin olive oil volatile fraction: Characterization of virgin olive oils from two distinct geographical areas of northern Italy. *J. Agric. Food Chem.* **2003,** *51,* 6572–6577.

Werman, M.J.; S. Mokady; I. Neeman; L. Auslaender; A. Zeidler. The effect of avocado oils on some liver characteristics in growing-rats. *Food Chem. Toxicol..* **1989,** *27,* 279–282.

Werman, M.J.; I. Neeman. Avocado oil production and chemical characteristics. *J. Am. Oil Chem. Soc.* **1987,** *64,* 229–232.

Werman, M.J.; I. Neeman; and S. Mokady. Avocado Oils and Hepatic Lipid Metabolism in Growing Rats. *Fd Chem. Toxic.* **1991**, *Vol. 29, No. 2,* 93–99

Weststrate, J.A.; G.W. Meijer. Plant sterol-enriched margarines and reduction of plasma total- and LDL-cholesterol concentrations in normocholesterolaemic and mildly hypercholesterolaemic subjects. *Eur. J.Clin.Nutr.* **1998,** *52,* 334–343.

White, A.; A.B. Woolf; P.J. Hofman; M.L. Arpaia. The International Avocado Quality Manual; HortResearch Ltd., 2005; p. 73.

Williams, L.O. The avocados, a synopsis of the genus Persea, subgenus Persea. *Economic Botany* **1977,** *31,* 315–320.

Williams, M. Recovery of oils and fats from oil seeds and fatty materials. Chapter 5.3. in Bailey's Industrial Oil and Fat Products 6th Edn. Ed. Shahidi, F. John Wiley & Sons, New Jersey. 2005.

Wong, M.; O.B.O. Ashton; C. Requejo-Jackman; T. Mcghie; A. White; L. Eyres; N. Sherpa; A.B. Woolf. Avocado oil—the colour of quality; C.A. Culver, R.E. Wrolstad, Eds.; Color Quality of Fresh and Processed Foods. ACS Symposium Series 983 American Chemical Society: Washington, DC, Copyright 2008; pp. 328–349.

Woolf, A.B.; A. White; M. Arpaia; K.C. Gross. Avocado. Agriculture Handbook 66: The Storage of Fruits, Vegetables and Florist and Nursery Stocks; K.C. Gross, C.Y. Wang, M. Salveit, Eds.;http://www.ba.ars.usda.gov/hb66/034avocado.pdf Accessed 22 September 2008.

Zauberman, G.; Y. Fuchus; M. Akerman. Peroxidase activity in avocado fruit stored at chilling temperatures. *Sci. Hortic.* **1985,** *26,* 261–265.

# 3

# Tree Nut Oils

**Afaf Kamal-Eldin[1] and Robert A. Moreau[2]**
[1] Department of Food Science, Swedish University of Agricultural Sciences, Box 7051, 750 07 Uppsala, Sweden; [2] Eastern Regional Research Center, Agricultural Research Service, United States Department of Agriculture, 600 East Mermaid Lane, Wyndmoor, Pennsylvania 19038, USA

## Introduction

The major tree nuts include almonds, Brazil nuts, cashew nuts, hazelnuts, macadamia nuts, pecans, pine nuts, pistachio nuts, and walnuts. Tree nut oils are appreciated in food applications because of their flavors, and are generally more expensive than other gourmet oils. Recently, some of the nut oils were suggested to have health-promoting effects, although all the compounds involved and their mechanisms of action are not yet identified. Tree nuts belong to several plant families (Table 3.1), but their common features are that they are rich in oil and the nuts (seeds) are larger in size than the seeds of oilseed species. The international production of tree nuts and peanuts is compared in Table 3.2. An excellent review on tree nut oils was published by Shahidi and Miraliakbari (2005). This chapter strives to cover the literature on this topic that has occurred in the last several years.

The almond tree (*Prunus dulcis*, Rosaceae) is distributed in the warm and temperate regions of West Asia (especially Iran, Syria, and Turkey), North Africa, and, more recently, North America (mostly California). It belongs to the family *Rosaceae* together with fruits such as plums, cherries, and peaches. Two main varieties of almonds are grown: bitter almond (*P. dulcis variety amara*) and sweet almond (*P. dulcis variety dulcis*). Bitter almonds contain amygdalin, a poisonous cyanogenic diglycoside (CAS# 29883-15-6) that breaks down to prussic acid, thus preventing its use in food. Sweet almond contains about 55% of fat, 12% of protein, 22% of carbohydrates, 5% of crude fiber, and 7% of ash. The fixed oil (nonvolatile oil) of almonds is extracted from both bitter and sweet varieties, but oils intended for human consumption are only prepared from the sweet varieties. Almond oil is clear, pale yellow, odorless, has a bland nutty taste, and is dominated by triolein (similar to olive oil), but is devoid of chlorophyll.

The Brazil-nut tree (*Bertholletia excelsa*, Lecythidaceae) is native to the Amazonian regions of South America, especially Brazil, Bolivia, Colombia, Venezuela, Peru, and Ecuador. The tree, 40–50 meters high, can live up to 800 years. The fruit, a large

**Table 3.1. Major Tree Nuts and Their Latin Names (Genus and Species) Classified According to Plant Family**

| Family | Common name | Latin name, genus and species |
|---|---|---|
| Anacardiaceae | Cashew | *Anacardium occidentale* |
| | Pistachio | *Pistacia vera* |
| Betulaceae | Hazelnut | *Corylus avellana* |
| Juglandaceae | Walnut | *Juglans regia* |
| | Pecan nut | *Carya illinoinensis* |
| Lecythidaceae | Brazil nut | *Bertholletia excelsa* |
| Pinaceae | Pine nut | *Pinus pinea* |
| Proteaceae | Macadamia nut | *Macadamia intergrifolia* |
| Rosaceae | Almond | *Prunus dulcis* |

**Table 3.2. Approximate Production Volumes and Main Producing Countries of Some Shelled Tree Nuts As Compared with Peanuts**

| Nut | Volume (t) | Major producing countries |
|---|---|---|
| Peanuts | 730,000 | United States, Argentina, India |
| Hazelnuts | 136,000 | Turkey, Italy |
| Almonds | 123,000 | United States, Spain |
| Walnut | 61,000 | United States, China |
| Cashews | 59,000 | India, Brazil |
| Brazil nuts | 20,000 | Brazil, Bolivia, Peru |

Source: Collinson et al. (2000).

woody capsule about the size of a grapefruit (≈2 kg), ripens between January and June, and falls with a characteristic cracking sound. Inside the shell, 12–25 Brazil-nut kernels are organized like orange segments. Brazil-nut trees reach maturity after 10–30 years, and then produce 300 or more fruit shells per year (Taylor, 2005). The Brazil-nut kernel, a staple food for many people in the producing countries, contains about 70% of fat, 18% of protein, and 12% of carbohydrates, and is an important source of selenium (Thomson et al., 2008). A list of literature reports about Brazil nuts is available at http://www.newcrops.uq.edu.au/listing/species_pages_B/Bertholletia_excelsa.htm.

The cashew-nut tree (*Anacardium occidentale*, Anacardiaceae) is another indigenous tree of Brazil and other Amazonian countries, but is also grown in West Africa, East Africa, Malaysia, and India. The cashew-nut tree, which grows up to 15 meters in Amazonia, is a multi-purpose tree with medicinal, food, and industrial applications of the nuts, fruits, leaves, and bark (Taylor, 2005). The fruit, a kidney-shaped hard shell, contains a cashew-nut kernel that is surrounded by an inner shell. Between the inner and outer shells (i.e., between the seed coat and the *fruit coat*) is a nonedible brown-reddish viscous cashew-nut-shell liquid (CNSL, CAS 108-95-2) mainly composed of cardols, which evoke immune responses and have industrial applications (Diogenes et al., 1996). The cashew-nut kernel contains 46% of oil, 21% of protein, and 25% of carbohydrates. The kernel is a rich source of minerals and vitamins especially vitamin C, which is about five times more than in an orange fruit.

The hazelnut tree (*Corylus avellana*, Betulaceae), a 3–8-meter high tree from the birch family, is an important tree in the Mediterranean countries and in the United States, with Turkey as the dominant country. The tree received its name from the Italian town of Avella when Linnaeus described it in 1542 as "*Avellana nux sylvestris,*" meaning the wild nut of Avella. The tree, which lives 75–100 years, produces 8–10 kg of nuts per year at maturity. Clusters of 1–6 nuts, each covered by a shell and a husk, grow together and ripen around October when the husks dry and release the kernels to fall on the ground. The hazel-nut kernels contain approximately 60% of oil, 18% of protein, and 15% of carbohydrates. A related species, *Corylus maxima,* commonly called fibert or filbert nut, is sometimes confused with hazelnuts. On the international scale, many more hazelnuts are produced than filbert nuts.

The macadamia-nut tree (*Macadamia* spp., Proteaceae), 2–13 meters high, is named after John MacAdam (1827–1865), who was a botanist and a colleague of Sir Ferdinand Jacob Heinrich von Mueller, who first described the genus (Mueller, 1857). Australia is the world's largest producer of the nuts, which are also produced in New Zealand, the United States (mostly Hawaii), Brazil, Costa Rica, Bolivia, South Africa, Malawi, and Kenya. Macadamia represents about 9 species, of which only *M. integrifolia* and *M. tetraphylla* and their hybrids are commercially important since the other species are inedible or toxic because they contain cyanogenic compounds. One spherical or two hemispherical nuts are enclosed in a very hard seed coat, which is surrounded by a 1-inch leathery two-valved green case that splits open when the nut matures and releases them to the ground. The creamy white macadamia nuts have a smooth hard shell that encases a white kernel. The kernels contain 65–75% of oil and about 9% of protein, and 9% of carbohydrates.

The pecan-nut tree (*Carya illinoinensis* (Wangenheim) K. Koch, Juglandaceae) is a large deciduous tree that can grow to 20–40 meters with a trunk of about 2 meters in diameter and a branch spread of 12–23 meters. At maturity (5–6 years), the trees produce nuts that ripen between mid-September to December and fall to the ground. The oval-to-oblong fruits of variable size (2.6–6 cm long × 1.5–3 cm broad, 30–90 nuts per pound) are dark brown with rough husks that split into four sections at maturity to release thin-shelled nuts. The production is dominated by the United States, which produces 80–95% of the world's supply of this nut, with the rest coming from Brazil, Mexico, Peru, South Africa, China, and Australia. The pecan nuts contain 72% of fat, 9% of protein, and 13% of carbohydrates. The nut is a rich source of several vitamins and antioxidants including tocopherols and ellagic acid.

About 20 species of the large pine-nut trees (*Pinus* spp., Pinaceae, 40–50) meters high and with a 1.5–2-meter trunk diameter—native to eastern parts of Russia and Asian countries and spread to the United States, Mexico, and other countries—

produce edible nuts. The Asian pine trees, *P. koraiensis* (Korean pine) and *P. Sibirica,* are used commercially for the production of cooking oil. Other North American pine species (e.g., *P. edulis, P. cembroides, P. sabiniana,* and *P. Pinea)* also produce oils. Pine nuts from different species vary widely in their proximate composition with 50–75% of fat, 19–30% of protein, and 12–23% of carbohydrates (FAO, 1995).

The pistachio-nut tree (*Pistacia vera*, Anacardiaceae), used as a food source since 7,000 B.C, is native to the Asian countries extending from Syria to the Caucasus region and Afghanistan. The tree was introduced to Italy from Syria, and subsequently spread to other Mediterranean countries. Today, the major producing countries are Iran, Turkey, Syria, Greece, India, Pakistan, and the United States. The fruits, botanically drupes, are borne in heavy clusters similar to grapes starting when the tree is 5–8 years of age. The shells split longitudinally along their sutures when mature, giving accessibility to the kernels, which vary in color from yellowish to greenish. Pistachio nuts contain about 55–60% of fat, 15–21% of protein, and 15–18% of carbohydrates.

The walnut tree (*Juglans* spp., Juglandaceae) is a deciduous tree growing 10–40 meters high, and is found throughout the world. The Persian walnut (*J. regia*) is the type of walnut most commonly used for edible purposes. The Black Walnut (*J. nigra*), a common wild species in North America and elsewhere, produces edible nuts having smaller kernels and extremely tough shells (much harder than the shells of *J. Regia*), and is not largely cultivated. *J. regia* is grown in the temperate regions of the world with its main production in Iran, Afghanistan and the northwestern Himalayas, Armenia, and Greece. Walnuts contain about 60–65% of fat, 15–24% of protein, and 10–14% of carbohydrates.

Besides the abovementioned nuts, other nuts—including apricot, peach, plum, and cherry, all *Prunus* species from the almond family Roseaceae, hickory and other *Carya* species from the pecan/walnut family *Juglandaceae*, beechnuts and chestnuts from the family *Fagaceae*, and kukui (*Aleurites moluccana, Euphorbiaceae*)—are also used to produce oil but to a much lesser extent. These minor tree nuts are not discussed in this chapter. Also, some evidence shows that peanuts (*Arachis hypogaea,* also called "ground nuts" in many countries) may also share some of the health-promoting properties of tree nuts (Griel et al., 2004; Pearson et al., 1998), but peanuts and peanut oil also are not covered in this chapter.

## Extraction and Processing of Nut Oils

The oil contents in tree nut kernels are shown in Table 3.3. Before oil extraction, the nuts first need to be removed from the shell (the fruit coat or pericarp) and then from the husk (the seed coat or epicarp). The various nut species differ in the hardness of their shells, and cashew nuts also contain the toxic CNSL between the shell and the seed coat. The nut oils also differ in their oxidative stabilities (see later section), which dictates the processing and storage conditions that are applied to each nut.

**Table 3.3. Oil Content of Nuts**

| Nut kernel | Oil (%) |
|---|---|
| Almonds | 41–60 |
| Brazil nuts | 61–69 |
| Cashew nuts | 40–49 |
| Hazelnuts | 49–67 |
| Macadamia nuts | 59–78 |
| Pecan nuts | 58–74 |
| Pine nuts | 59–71 |
| Pistachio nuts | 45–59 |
| Walnuts | 51–65 |

Sources: USDA, 2001; Maguire et al., 2004; Ryan et al., 2006; Kornsteiner et al., 2006.

Relative to other nuts, almond and Brazil nuts have a thin shell which one can remove by soaking the nuts in water for about 24 hours, followed by brief boiling (≈5 minutes) and hand cracking of the shell. Pistachio nuts are also easily removed from the partly open shells. Once removed from their shells, the nuts are dried and kept in a cool and dark environment to prevent rancidity. Bitter-almond oil—yellow in color with characteristic odor and flavor caused by benzaldehyde and a bitter taste attributed to hydrogen cyanide (or prusic acid)—is used in certain medicinal applications, and is toxic unless the cyanide is removed by maceration with water or by refining. When the hydrogen cyanide is removed, the oil is not toxic and is used as a flavoring agent. Do not confuse the fixed oil (nonvolatile oil) from bitter almonds with the essential oil (aromatic oil) extracted or distilled from the meal after the removal of the fixed oil.

Before oil extraction from cashew nuts, the shell is removed and is used for the extraction of CSNL. Generally, the nuts are roasted in rotating perforated cylinders inclined at an angle above a heat source to make the shells brittle and to collect the CNSL. The cracking of the shell and the removal of the nut kernels are usually performed manually, and may pose allergy problems to the hands of workers. After the removal of the shell and the CNSL, the nuts are sprayed with water and allowed to cool and dry before the removal of the thin skin that covers them. One can store the cashew nuts for up to one year in air-tight containers without problems (Axtell & Fairman, 1992). The shells of hazelnuts, macadamia nuts, pecans, pine nuts, and walnuts are hard. Before oil extraction, the nuts are lightly toasted for 15–20 minutes, and then ground. By using stone wheels, the dehulled nuts are usually ground into a paste, which is then pressed to produce the oil. Solvent or supercritical fluid extraction (SFE) also can extract the oil from the nut paste.

Cold-pressing is conducted by applying gentle pressure (so that the temperature does not exceed 30°C), producing a low yield of oil which is usually expensive. Cold-pressed oils have superior nutty flavor and fresh taste, and are often used in salad dressings and in the cooking of specialty dishes. The clear first-pressed oil is bottled as "extra-virgin nut oil," and the remaining meal is sold for applications such as specialty baking, energy bars, and animal feeds. Cold-pressing gives a low yield of oils with limited shelf lives that often require refrigeration. The oils are generally supplemented with antioxidants, packed under nitrogen to eliminate oxygen, and prevent rancidity, and then stored in refrigerators. Expeller-pressing at higher temperatures may also extract the paste to produce expeller-pressed oil. Expeller-pressed oils are of lower quality than cold-pressed oils, but sometimes these oils are mixed with certain flavors, and the resulting oil is still sold as cold-pressed. After filtration, cold- and expeller-pressed oils are bleached mostly by exposure to light. These oils usually contain water, resins, color and flavor compounds, and residual proteins and fiber that may make them dark and opaque. The oils need to be processed to remove these materials by a process called clarification: (i) The oils are allowed to stand undisturbed for a few days to yield a clear upper layer, (ii) a gentle heating follows to remove residual water and to destroy enzymes and any bacteria before filtration, and (iii) another brief heating occurs at ca. 100°C to remove the remaining traces of moisture. The rules for labelling pressed oils differ among various countries, with the temperature limit being a very strict quality parameter in the European Union (EU) but not so in the United States. Expeller-pressed oils, which are produced at very high temperatures, cannot be marketed as cold-pressed oils.

Solvents can extract whole kernels of low quality and the pastes obtained after pressing. Solvent extraction gives a higher yield than pressing, but the oils obtained are of lower quality compared to pressed oils and are generally refined by steps including neutralization, bleaching, degumming, and deodorization. The refined oils lose their nutty flavor, but are better suited for high-temperature uses.

Nut oils are also extracted from the nuts by supercritical fluid extraction (SFE) using compressed carbon dioxide (sometimes modified by the addition of ethanol) in wide temperature and pressure ranges. The oil yields obtained by SFE are generally lower than that extracted with *n*-hexane, but no significant differences are noted between the two oils regarding the composition of acyl lipids and sterols. Carbon dioxide-extracted oils are generally clearer, and may contain more tocopherols and be slightly more stable than oils extracted with *n*-hexane (Bernardo-Gil et al., 2002; Özkal et al., 2005; Rodrigues et al., 2005; Salgin & Salgin, 2006; Silva et al., 2008). Despite the high content of tocopherols in carbon dioxide-extracted oils, their oxidative stability seems to be lower than those extracted with solvents (Martinez et al, 2008).

## Edible and Nonedible Applications

Nut-kernel oils are used as: salad oils, in cooking and other food applications, in massages and as lubricants, as emollients in pharmaceuticals, and in cosmetics, soaps, shampoos and hair conditioning/repair products, skin lotions, and other cosmetic products. Some tree nut oils, such as hazelnut oil, may serve as cooking oils in specialty dishes because of their unique flavors.

## Composition of Fatty Acids and Acyl Lipids

As with other vegetable oils, the major components of tree nut oils are triacylglycerols (96–98%), with smaller amounts of diacylglycerols, monoacylglycerols, free fatty acids, and minor unsaponifiable components. Tree nut oils differ considerably in their oil contents (Table 3.3). In addition to triacylglycerols, tree nut oils also contain small amounts of phospholipids, sphingolipids, sterols, and sterol esters (Table 3.4). Considerable differences in the relative composition of fatty acids (Table 3.5) and triacylglycerol molecular species (Table 3.6) are reported for tree nut oils. Moreover, both the fatty-acid and triacylglycerol composition are affected by variety and geographical origin (Parcerisa et al., 1994). Similar to other Pinaceae plants, pine-nut oil contains considerable amounts of a nonmethylene-interrupted polyunsaturated fatty acid, pinolenic acid (18:3 Δ5,9,12) (Imbs et al., 1998; Wolff et al., 2001). Macadamia-nut oil is characterized by a high level of palmitoleic acid (16:1, 17–34%) (Ako et al., 1995; Kaijser et al., 2000).

Nut oils differ markedly in their degree of unsaturation (Table 3.5). The oils studied most for within-species variability in fatty-acid composition include almond oils (Kodad et al., 2008; Martín-Carratalá et al., 1999; Turan et al., 2007), hazelnut oils (Alasalvar et al., 2003, 2006; Amaral et al., 2006; Crews et al., 2005a; Özdemir et al., 2001; Savage & McNeil, 1998), and walnut oils (Amaral et al., 2003; Crews et al., 2005b; Martínez, 2006; Zwarts et al., 1999).

The main phospholipids found in nut oils include phosphatidylcholine (18–50%), phosphatidylinositol (18–45%), phosphatidylserine (20–45%), and phosphatidylethanolamine (8–16%), while phosphatidic acid, phosphatidylglycerol, lysophosphatidylcholine, and lysophosphatidylethanolamine were found to be minor phospholipid components (Table 3.4) (Miraliakbari & Shahidi, 2008; Zlatanov et al, 1999).

**Table 3.4. Lipid-Class Distribution (Relative %) of Selected Nuts**

| Compound | Almonds | Brazil nuts | Hazelnuts | Pecan nuts | Pine nuts | Pistachio nuts | Walnuts |
|---|---|---|---|---|---|---|---|
| Triacylglycerols | 98.2 | 96.7 | 98.0 | 96.4 | 97.6 | 96.2 | 97.2 |
| Sterols (free) | 0.22 | 0.18 | 0.21 | 0.26 | 0.13 | 0.19 | 0.26 |
| Sterol esters | 0.05 | 0.05 | 0.04 | 0.07 | 0.06 | 0.03 | 0.09 |
| Phosphatidylserine | 0.21 | 0.26 | 0.27 | 0.39 | 0.23 | 0.47 | 0.37 |
| Phosphatidylinositol | 0.11 | 0.09 | 0.06 | 0.15 | 0.14 | 0.21 | 0.25 |
| Phosphatidylcholine | 0.21 | 0.34 | 0.24 | 0.21 | 0.19 | 0.52 | 0.34 |
| Phosphatidic acid | — | — | 0.02 | — | — | — | — |
| Sphingolipids | 0.53 | 0.83 | 0.26 | 0.48 | 0.45 | 0.73 | 0.54 |

Source: Miraliakbari & Shahidi (2008a).

**Table 3.5. Composition (Relative %) of Major Fatty Acids in Nuts Oils**

| Fatty acid | Almonds[a] | Brazil nuts[b] | Cashew nuts[b] | Hazelnuts[a] | Macadamia nuts[a] | Pine nuts[c] | Pecans[c] | Pistachio nuts[b] | Walnuts[a] |
|---|---|---|---|---|---|---|---|---|---|
| 14:0 (M) | 0.1 | 0.1 | 0.1 | 0.1 | 1.0 | – | 0.09 | 0.1 | 0.1 |
| 16:0 (P) | 6.8 | 13.5 | 9.9 | 5.8 | 8.4 | 5.0 | 7.64 | 7.4 | 6.7 |
| 16:1 (Po) | 0.6 | 0.3 | 0.4 | 0.3 | 17.3 | 0.1 | 0.11 | 0.7 | 0.2 |
| 18:0 (S) | 1.3 | 11.8 | 8.7 | 2.7 | 3.2 | 2.7 | 2.52 | 0.9 | 2.3 |
| 18:1 | 69.2 | 29.1 | 57.2 | 79.3 | 65.1 | 28.6 | 49.60 | 58.2 | 21.0 |
| 18:2n-6 (L) | 21.5 | 42.8 | 20.8 | 10.4 | 2.3 | 44.1 | 37.71 | 30.3 | 57.5 |
| 18:2Δ5,9 (L*) | – | – | – | – | – | 2.2 | – | – | – |
| 18:3n-3 (Ln) | 0.2 | 0.2 | 0.2 | 0.5 | 0.1 | 0.1 | 1.47 | 0.4 | 11.6 |
| 18:3Δ5,9,12 (Ln*) | – | – | – | – | – | 13.9 | – | – | – |
| 20:0 | 0.2 | 0.5 | 1.0 | 0.2 | 2.3 | 0.5 | 0.34 | 0.6 | 0.1 |
| 20:1 | – | 0.2 | 0.2 | – | – | 1.3 | 0.52 | 0.6 | – |
| 22:0 | 0.1 | 0.1 | 0.4 | – | 0.2 | – | - | 0.3 | 0.1 |
| 22:1 | – | 0.3 | 0.3 | – | – | – | - | 0.6 | – |
| Calculated iodine valued | 97 | 100 | 86 | 87 | 60 | 141 | 115 | 105 | 148 |

Sources: [a]Maguire et al. (2004); [b]Ryan et al. (2006); [c]Imbs et al. (1998).
[d]Iodine values (IVs) were calculated as (% monounsaturated fatty acids x 0.860 + and 18:2 x 1.732 + %18:3 x 2.616) (Miraliakbari and Shahidi, 2008a).

**Table 3.6. Triacylglycerol (TAG) Molecular Species (Relative %) in Tree Nut Oils**

| TAG | Almonds[a] | Brazil nuts[a] | Cashew nuts[b] | Hazelnuts[a] | Macadamia nuts[a] | Pine nuts[de] | Pistachio nuts[a] | Walnuts[b] |
|---|---|---|---|---|---|---|---|---|
| LLnLn | — | — | — | — | — | —[e] | — | 2.1 |
| LLLn | — | 0.1 | — | 0.1 | — | — | 0.6 | 20.0 |
| LLL | 8.7 | 14.8 | 2.4 | 3.7 | — | 5 | 11.7 | 34.6 |
| PoLPo | — | — | — | — | 1.3 | — | — | — |
| PoPoPo | — | — | — | — | 2.6 | — | — | — |
| OLLn | 0.1 | 0.2 | — | 0.5 | — | — | 1 | 4.6 |
| PoPoM | — | — | — | — | 1.2 | — | — | — |
| OLPo | — | — | — | — | 3.9 | — | — | — |
| PoOPo | — | — | — | — | 8.2 | — | — | — |
| OPoM | — | — | — | — | 1.7 | — | — | — |
| PPoPo | — | — | — | — | 2.6 | — | — | — |
| PPoM | — | — | — | — | 1 | — | — | — |
| PMoPo | — | — | — | — | 0.4 | — | — | — |
| OPoMo | — | — | — | — | 0.7 | — | — | — |
| POPo | — | — | — | — | 6.1 | — | — | — |
| OOPo | — | — | — | — | 16.1 | — | — | — |
| OLO | — | — | 13.9 | — | 6.4 | — | — | — |
| LnLP | — | 0.1 | — | — | — | — | 0.3 | 3.1 |
| LnOP | — | — | 0.1 | — | — | — | — | — |
| OLL | 27.6 | 16.7 | 7.6 | 12.3 | — | 14.7 | 24.8 | 17.2 |
| PoLL | — | — | 0.1 | — | — | — | — | — |
| OLnO | — | — | 0.1 | 0.7 | — | 5.4 | 0.9 | — |
| LLP | 4.8 | 13 | 4.3 | 1.6 | — | 3.8 | 7.1 | 10.6 |
| OLO | 28 | 13.1 | — | 28.2 | — | 7.4 | 25.2 | 3.1 |
| LOP | 11.3 | 16.7 | 5.4 | 5.2 | 2.7 | 8.3 | 11.8 | 4.2 |
| LOPo | — | — | — | — | — | - | - | - |
| PLP | 0.5 | 2.6 | 1.8 | 0.2 | — | 0.6 | 1.1 | - |
| OOO | 13.3 | 4.6 | 15.7 | 36.5 | 19.4 | 7.6 | 8.9 | - |
| SLO | 1.8 | 10 | 6.4 | 1.4 | — | 3.5 | 1.3 | — |
| OOP | 2.7 | | 11.8 | 6.1 | 9.9 | 2.6 | 3.4 | — |
| SOO | 0.6 | 2.3 | 11.2 | 2.8 | — | 1.3 | 0.4 | - |
| SLL | — | — | 3.6 | — | — | 6.7 | - | - |
| SLP | — | — | 1.7 | — | — | 0.5 | — | - |
| POP | — | — | 4.3 | — | — | 0.1 | — | - |
| SOP | — | — | 5.1 | — | — | 0.1 | — | — |
| SOS | — | — | 2.6 | — | — | —[e] | — | — |

Sources: [a]Holcapek et al. (2003); [b]Amaral et al. (2004); [c]Holcapek et al. (2005); [d]Imbs et al. (1998).

[e]In addition, also contains PLLn* (5.5%), OOLn (5.4%), OLL* (3.4%) LLL* (3.1%), OLLn (6.5%), LLLn (10.7%).

## Nonacylglycerol Constituents

The content of unsaponifiable materials in tree nut oils was studied by Kornsteiner et al., 2006. In general, the level of unsaponifiable matter (%) was lower for these oils compared to seed oils: that is, almonds (0.35–0.53), Brazil nuts (0.44–0.66), cashew nuts (0.25–0.53), hazelnuts (0.20–0.30), macadamia nuts (0.30–0.33), pecan nuts (0.3–0.45), pine nuts (0.4–0.42), pistachio nuts (0.29–0.45), and walnuts (0.25–0.4). Similar to other vegetable oils, the sterols represent the major class in the unsaponifiable fractions of nut oils (Table 3.7). The major sterol is typically sitosterol followed by lower levels of campesterol, stigmasterol, and other desmethylsterols. Cholesterol levels in vegetable oils are generally insignificant, but a small amount of cholesterol was present in Brazil-nut oil (≈2% of total sterols). According to Kornsteiner et al. (2006), only pistachio-nut oil, among tree nut oils, contained carotenoids (in mg/100 g of oil): β-carotene (0–1) and lutein (1.5–9.6). Tree nut oils are generally light in color and devoid of chlorophyll and other pigments. As expected, the levels of the unsaponifiable components are influenced by cultivar and cultivation environment. Crews et al. (2005a,b) reported considerable variations in the levels of sterols, tocopherols, and fatty acids in oils obtained from different countries: walnut oils from China, India, France, Hungary, Italy, Spain and the United States and hazelnut oils from Croatia, France, Italy, Spain, and Turkey.

## Oxidative Stability of Tree Nut Oils

Published information on the relationship between the composition and the stability of tree nut oils is limited. In general, the oxidative stability of vegetable oils is determined by their fatty-acid composition and their antioxidant-compound content (Kamal-Eldin, 2006). As shown in Table 3.8, a study of the effects of tocopherols and polyphenolic compound content on diphenylpicrylhydrazyl (DPPH) radical scavenging and the Rancimat induction period confirmed that the oxidative stability of the tree nut oils is related to an interplay between their fatty-acid composition and their tocopherol content (Arranz et al., 2008). According to one study, the oils of pecans and pistachios were the most stable, whereas oils of pine nuts and walnuts were the least stable (Miraliakbari & Shahidi, 2008). Perhaps lipid classes other than triacylglycerols are involved in determining oil stability. For example, the oxidation of the fatty acids in the phospholipid fraction preceded the oxidation of triacylglycerols in pecans stored at room temperature for 8 months. Pentanal predominated in the oil headspace during the early stages, while hexanal predominated at later stages of storage. The strong negative correlation ($R = -0.98$) found between hexanal and its precursor linoleic acid in phospholipids suggested membrane lipids as the primary target for oxidation in the early stages (Eriksson, 1993). It is now known whether the phospholipids solubilized in the extracted oils would play a significant role in the oxidation of these oils.

**Table 3.7. Unsaponifiables of Nuts Oils (mg/kg oil)**

| Compound | Almonds[a] | Brazil nuts[b] | Cashew nuts[b] | Hazelnuts[a] | Macadamia nuts[a] | Pecan nuts[b] | Pine nuts[b] | Pistachio nuts[b] | Walnuts[a] |
|---|---|---|---|---|---|---|---|---|---|
| Sitosterol | 2071 | 1325 | 1768 | 991 | 1507 | 1572 | 1841 | 4586 | 1129 |
| Campesterol | 55 | 27 | 105 | 67 | 73 | 52 | 215 | 237 | 51 |
| Stigmasterol | 52 | 577 | 117 | 38 | 38 | 340 | 680 | 663 | 55 |
| Squalane | 95 | 1378 | 89 | 186 | 185 | 152 | 40 | 91 | 9 |
| α-Tocopherol | 439 | 83 | 4 | 310 | 122 | 12 | 124 | 16 | 21 |
| γ-Tocopherol | 12 | 116 | 57 | 61 | tr. | 168 | 105 | 275 | 300 |

Sources: [a]Maguire et al. (2004); [b]Ryan et al. (2006).

**Table 3.8. Oxidative Stability and the Characteristics of Antioxidants of Selected Nut Oils**

| Oil | Induction time (h) | Total tocopherols (ppm) | Total polyphenols (μg GAE/g of oil) | EC50 (g oil/g of DPPH) |
|---|---|---|---|---|
| Hazelnut oil | 52.7 | 455 | 80 | 478 |
| Pistachio-nut oil | 44.4 | 530 | 700 | 378 |
| Almond oil | 21.8 | 250 | 270 | 712 |
| Peanut oil | 14.6 | 48 | 80 | 1396 |
| Walnut oil | 4.7 | 249 | 320 | 1514 |

Source: Arranz et al. (2008).

Turkish Tombul hazelnuts were more stable in bulk than as an oil-in-water emulsion (Rodrigues, 2003). Oil inside shelled and broken Brazil nuts oxidizes more rapidly than when extracted (Vieira-Thais et al., 1999). The reason for the destabilizing effect of the shell is not known and may be enzymatic. Of particular interest is the oxidative instability of walnut oil and how it is modulated by fatty acids, tocopherols, and other components (Amaral et al., 2003; Li et al., 2007).

## Flavor and Taste

The flavors of nut oils depend mainly on the extraction and processing methods used for their preparation. For example, while sweet-almond oil (or *Oleum Amygdala Expressum*) is colorless-to-pale yellow and odorless, bitter-almond oil (from which hydrogen cyanide is not removed) is toxic and can be lethal at a low dose (10 mL). No literature reports were found on the molecules responsible for the flavor in nut oils, but a few reports are available on the volatile compounds in certain nuts and how they are affected by, for example, roasting (Alasalvar et al., 2006; Clark & Nursten, 1976; Fallico et al., 2003).

## Health Benefits of Nuts and Nut Oils

Research during the last decade provides much evidence that the consumption of nuts has beneficial effects on CHD risk. This evidence came from large prospective cohort studies: for example, the Adventist Health Study (Fraser et al., 1992), the Nurses' Health Study (Hu et al., 1998), the Cholesterol and Recurrent Events Study (Brown et al., 1999), the Iowa Women's Health Study (Ellsworth et al., 2001), and the Physicians' Health Study (Albert et al., 2002). On the basis of these studies, the conclusion is that the consumption of nuts and nut lipids, despite their tremendous variability with respect to fatty acid composition, is beneficial for health, with a special impact on cardiovascular-disease risk (Hu & Stampfer, 1999; Kris-Etherton et al., 1999). Accordingly, also another conclusion is that the consumption of nuts at least five times a week is associated with a lower risk of death from coronary heart disease and nonfatal myocardial infarction compared to consumption less than once a week (Fraser et al., 1992; Kris-Etherton et al., 2001; Kushi et al., 1996).

Besides epidemiological studies, controlled clinical intervention trials show significant reductions in serum total and low-density lipoprotein (LDL)-cholesterol concentrations in normo- and hyper- cholesterolemic people by the consumption of nuts including: almonds (Hyson et al., 2002; Jenkins et al., 2002; Sabate et al., 2003; Spiller et al., 1998), hazelnuts (Durak et al., 1999), macadamia nuts (Curb et al., 2000; Garg et al., 2003; Griel et al., 2008), pecans (Morgan & Clayshulte, 2000), pistachios (Edwards et al., 1999; Sheridan at al., 2007), and walnuts (Abbey et al., 1994; Almario et al., 2001; Chisholm et al., 1998; Morgan et al., 2002; Sabate et al., 1993; Zambón et al., 2000). As a result, the following health claim— "Scientific evidence suggests but does not prove that eating 1.5 ounces per day of most nuts as part of a diet low in saturated fat and cholesterol may reduce the risk of heart disease."—is allowed by the FDA (2003). The effects of nuts may not only be attributed to the fatty-acid composition, but also may also include proteins and other bioactive compounds, such as fiber, phytosterols, antioxidants, folate, copper, magnesium, potassium, boron, and some amino acids, such as arginine.

Conflicting results were reported about the relationship between nut consumption and body weight (Natoli & McCoy, 2007; Sabaté, 2003). The consumption of nuts and their oils suppresses appetite, enhances satiety, and limits the consumption of foods. For example, PinnoThin™, the oil pressed from pine nuts, stimulated the release of the endogenous satiety hormones cholecystokinin (CCK) and glucagon-like peptide (GLP-1) and reduced food intake by overweight women (Hughes et al,. 2008; Pasman et al., 2008; Scott et al., 2007). In addition, the consumption of almonds appears to reduce the glycemic impact of carbohydrate-rich foods (Josse et al., 2007). Results from the Nurses' Health Study suggest that nut consumption may also be inversely associated with the risk of type 2 diabetes mellitus in women (Jiang et al., 2002). In addition, nut consumption is associated negatively with stroke (Yochum et al., 2000), dementia (Zhang et al., 2002), advanced macular degeneration (Seddon et al., 2003), and gallstones (Tsai et al., 2004).

Although many more publications have been written about the health benefits of nuts than about nut oils, well-designed studies demonstrated the health benefits of several nut oils, including hazelnut oil, walnut oil, pine-nut oil, and almond oil. Three studies examined the effect of hazelnut oil on several cardiovascular parameters in rabbits and quail. Balkan et al. (2003) reported the influence of hazelnut oil on the peroxidation status of erythrocytes and on lipoproteins in rabbits fed a high-cholesterol diet. Although hazelnut oil did not significantly affect the levels of any of the cholesterol in any of the serum lipoproteins, it reduced lipid-peroxide levels in the plasma and in apolipoprotein B 100-containing lipoproteins. Hazelnut oil also reduced cholesterol-induced hemolytic anemia. Similar results were also reported by Hatipolgu et al. (2004), who also measured the effect of hazelnut oil on rabbits fed a high-cholesterol diet. They confirmed that hazelnut oil caused a reduction of lipid peroxidation, and had no effect on the levels of cholesterol in any of the serum lipoproteins. In addition, they reported that hazelnut oil ameliorated atherosclerotic lesions in the aortas of the rabbits. Finally, Guclu et al. (2008) reported that feeding hazelnut oil to laying quail significantly reduced the levels of serum triglycerides, but had no effect on the levels of serum LDL-cholesterol or on egg-quality parameters.

Three studies examined the effect of walnut oil on cardiovascular health. Lavedrine et al. (1999) reported that walnut-oil consumers in France had significantly higher levels of high-density lipoprotein (HDL)-cholesterol and apo A1, but walnut oil consumption caused no significant effect on the levels of LDL-cholesterol. Zibaeenezhad et al. (2003) conducted a human-feeding study, and reported that walnut oil (3 g/day) decreased plasma-triglyceride levels by 19–33%, but had no statistically significant effect on other serum lipids. Davis et al. (2005) reported results for a hamster study which revealed that diets rich in walnuts, α-tocopherol, and walnut oil reduced the levels of aortic cholesterol esters, which are considered to be associated with atherosclerotic plaque. Hyson et al. (2002) conducted a human-feeding study, and reported that both whole almonds and almond oil decreased the levels of plasma triglycerides, total cholesterol, and LDL-cholesterol (reduced by 14, 4, and 6%, respectively), and caused a 6% increase in the levels of HDL-cholesterol.

Two studies reported that pine-nut oil reduced blood pressure and reduced the levels of serum very low density lipoprotein (VLDL)-triglycerides and LDL-cholesterol in rats (Asset et al., 1999; Sugano et al., 1994). Lee et al. (2004) studied the effect of pinolenic acid (all-*cis*-5,9,12-18:3, a major fatty acid in pine oil) on LDL-receptor activity in human hepatoma HpeG2 cells. They concluded that the LDL-lowering properties of pine oil may be due to the ability of pinolenic acid to enhance hepatic LDL uptake.

## Allergenicity of Tree nut Oils

Unrefined oils, especially when obtained by pressing, usually contain variable levels of peptides and proteins (Hidalgo & Zamora, 2006; Madhaven, 2001; Teuber et al., 1997). Because of possible allergic reactions, products of nuts (including almonds, hazelnuts, walnuts, cashews, pecans, Brazil nuts, pistachios, and macadamia nuts) are listed in annex IIIa of the EU directive on the labelling of foods. Allergies to peanuts and tree nuts (almonds, cashews, hazelnuts, pecans, pistachios, and walnuts) are among the most common food allergies (Crespo et al., 2006; Roux et al., 2003; Sicherer et al., 1999). Although all food allergies have the potential to induce severe reactions, peanut and tree nuts are among the foods most commonly associated with anaphylaxis, a life-threatening allergic reaction ( Al-Muhsen et al., 2003; Flinterman et al., 2008, Malanin et al., 1995; Rosen & Fordice, 1994). Table 3.9 shows the main identified tree nut allergens (Barre et al., 2008). Cross-reactivity was found in tree nut allergies (Knott et al., 2008; Wallowitz et al., 2006).

**Table 3.9. Nut Allergens**

| Nut | Major allergens |
|---|---|
| Almonds | Pru du 2S albumin (2S albumin, seed-allergic protein 1, accession # P82944)<br>Pru du conglutin (conglutin γ, seed-allergic protein 2, accession # P82952)<br>Pru du 4 (profilin, accession # Q8GSL5)<br>Pru du amandin (11S legumin, seed-allergic protein) |
| Brazil nuts | Ber e 1 (2S albumin, accession # P04403),<br>Ber e 2 (11S globulin-like protein, accession # Q84ND2) |
| Cashew nuts | Ana o 1 (7S globulin, vicilin-like protein, accession # Q8L5L5)<br>Ana o 2 (11S globulin-like protein, accession # Q8GZP6)<br>Ana o 3 (2S albumin, accession # Q8H2B8) |
| Hazelnuts | Cor a 1 (PR-10 (Bet v 1 homologous), accession # Q9FPK2/3/4)<br>Cor a 2 (profilin)<br>Cor a 8 ( PR-14, lipid-transfer protein)<br>Cor a 9 (11S globulin-like protein, accession # Q8W1C2) |
| Macadamia nuts | Allergy established but allergen(s) not characterized. |
| Pecan nuts | Car i 1 (putative allergen 11, accession # AA032314) |
| Pine nuts | (17 kDa vicilin-like protein) |
| Pistachio nuts | Pis v 3 (vicilin-like protein, accession # AB036677) |
| Walnuts | Jug r 1 (2S albumin, accession # P93198)<br>Jug r 2 (vicilin-like protein, accession # Q9SEW4)<br>Jug r 3 (PR-14, lipid-transfer protein)<br>Jug r 4 (11S globulin-like protein) |

Sources: Robotham et al. (2005); Jin et al. (2008).

## Other Issues (Authenticity and Adulteration)

Due to its high cost, sweet-almond oil is often adulterated, particularly by the cheaper peach- and apricot-seed oils, which have very similar fatty-acid characteristics. In one survey in Brazil, 77% of almond-oil samples were adulterated (Salvo et al., 1980). Hazelnut oil is used to adulterate virgin olive oil, and the detection of this adulteration is difficult (Aparicio et al., 2007).

## References

Abbey, M.; M. Noakes; G.B. Belling; P.J. Nestel. Partial replacement of saturated fatty acids with almonds or walnuts lowers total plasma cholesterol and low-density-lipoprotein cholesterol. *Am. J. Clin. Nutr.* **1994,** *59,* 995–999.

Ako, H.; D. Okuda; D. Gray. Healthful new oil from macadamia nuts. *Nutrition* **1995**, *11*, 286–288.

Alasalvar, C.; J.S. Amaral; F. Shahidi. Functional lipid characteristics of Turkish Tombul hazelnut (*Corylus avellana* L.). *J. Agric. Food Chem.* **2006**, *54*, 10177–10183,

Alasalvar, C.; A.Z. Odabasi; N. Demir; M.Ö. Balaban; F. Shahidi; K. R. Cadwallader. Volatiles and flavor of five turkish hazelnut varieties as evaluated by descriptive sensory analysis, electronic nose, and dynamic headspace analysis/gas chromatography-mass spectrometry. *J. Food Sci.* **2006**, *69*, SNQ 99–106.

Alasalvar, C; F. Shahidi; C.M. Liyanapathirana; T. Ohshima. Turkish Tombul hazelnut (*Corylus avellana* L.). 1. Compositional characteristics. *J. Agric. Food Chem.* **2003**, *51*, 3790–3796.

Albert, C.M.; J.M. Gaziano; W.C. Willett; J.E. Manson. Nut consumption and decreased risk of sudden cardiac death in the Physicians Health Study. *Arch. Intern. Med.* **2002**, *162*, 1382–1387.

Almario, R.U.; V. Vonghavaravat; R. Wong; S.E. Kasim-Karakas. Effects of walnut consumption on plasma fatty acids and lipoproteins in combined hyperlipidemia. *Am. J. Clin. Nutr.* **2001,** *74*, 72–79.

Al-Muhsen, S.; A.E. Clarke; R.S. Kagan. Peanut allergy: an overview. *Can. Med. Assoc. J.* **2003**, *168*, 1279–1285.

Amaral, J.S.; S. Casal; J.A. Pereira; R.M. Seabra; B.P.P. Oliveira. Determination of sterol and fatty acid compositions, oxidative stability, and nutritional value of six walnut (*Juglans regia* L.) cultivars grown in Portugal. *J. Agric. Food Chem.* **2003**, *51*, 7698–7702.

Amaral, J.S.; S. Casal; R.M. Seabra; B.P. Oliveira. Effects of roasting on hazelnut lipids. *J. Agric. Food Chem.* **2006**, *54*, 1315–1321.

Aparicio, R.; R. Aparicio-Ruiz; D.L. García-González. Rapid methods for testing of oil authenticity: The case of olive oil. *Rapid Methods;* A. van Amerongen, D. Barug, M. Lauwars, Eds.; Wageningen Academic: Wageningen, Holland, **2007**; pp. 163–188.

Arranz, S.; R. Cert; J. Perez-Jimenez; A. Cert; F. Saura-Calixto. Comparison between free radical scavenging capacity and oxidative stability of nut oils. *Food Chem.* **2008,** *110*, **985–990.**

Asset, G.; B. Staels; R.L. Wolff; E. Bauge; Z. Madj; J.C. Fruchart; J. Dallongeville. Effects of *Pinus pinaster* and *Pinus koraiensis* seed oil supplementation on lipoprotein metabolism in the rat. *Lipids* **1999,** *34***,** 39–44.

Axtell, B.L.; R.M. Fairman. Minor oil crops. *FAO Agricultural Services Bulletin No. 94;* 1992, FAO: Rome, Italy, http://www.fao.org/docrep/x5043E/x5043E00.htm#Contents, accessed 23 Aug. 2008.

Balkan, J.; A. Hatipoglu; G. Aykac-Toker; M. Uysal. Influence of hazelnut oil administration on peroxidation status of erythrocytes and apolipoprotein B 100-containing lipoproteins in rabbits fed a high cholesterol diet. *J. Agric. Food Chem.* **2003**, *51,* 3905–3090.

Barre, A.; C. Sordet; R. Culerrier; F. Rancé; A. Didier; P. Rougé. Vicilin allergens of peanut and tree nuts (walnut, hazelnut and cashew nut) share structurally related IgE-binding epitopes. *Mol. Immunol.* **2008**, *45*, 1231–1240.

Bernardo-Gil, M.G.; J. Grenha; J. Santos; P. Cardoso. Supercritical fluid extraction and characterisation of oil from hazelnut, *Eur. J. Lipid Sci. Technol.* **2002,** *104***, 402–409.**

Brown, L., B.A. Rosner; W.C. Willett; F.M. Sacks. Nut consumption and risk of recurrent coronary heart disease. *FASEB J.* **1999**, *13*, A 538.

Chisholm, A.; J. Mann; M. Skeaff; C. Frampton; W. Sutherland; A. Duncan; S. Tiszavari. A diet rich in walnuts favourably influences plasma fatty acid profile in moderately hyperlipidaemic subjects. *Eur. J. Clin. Nutr.* **1998**, *52*, 12–16.

Clark, R.G.; H.E. Nursten. Volatile flavour components of Brazil nuts *Bertholletia excelsa* (Humb. and Bonpl.). *J. Sci. Food Agric.* **1976,** *27*, 713–720.

Collinson, C.; D. Burnett; V. Agreda. *Economic Viability of Brazil Nut Trading in Peru.* Report 2520, Natural Resources and Ethical Trade Programme, Natural Resources Institute: University of Greenwich, Kent, UK, 2000.

Crespo, J.F.; J.M. James; C. Fernandez-Rodriguez; J. Rodriguez. Food allergy: nuts and tree nuts. *Br. J. Nutr.* **2006**, *96*, S95–S102.

Crews, C.; P. Hough; J. Godward; P. Brereton; M. Lees; S. Guiet; W. Winkelmann. Study of the main constituents of some authentic hazelnut oils. *J. Agric. Food Chem.* **2005a**, *53*, 4843–4852,

Crews, C.; P. Hough; J. Godward; P. Brereton; M. Lees; S. Guiet; W. Winkelmann. Study of the main constituents of some authentic walnut oils. *J. Agric. Food Chem.* **2005b**, *53*, 4853–4860.

Curb, J.D.; G. Wergowske; J.C. Dobbs; R.D. Abbott; B. Huang. Serum lipid effects of a high-monounsaturated fat diet based on macadamia nuts. *Arch. Intern. Med.* **2000**, *160,* 1154–1158.

Davis, P.; G. Valacchi; E. Pagnin; Q. Shao; H.B. Gross; L. Calo; W. Yokoyama. Walnuts reduce aortic ET-1 mRNA levels in hamsters fed a high-fat, atherogenic diet. *J. Nutr.* **2006,** *136*, 428–432.

Diogenes, M.J.; F.M. DeMorais; F. F. Carvalho. Contact dermatitis among cashew nut workers. *Contact Dermatitis* **1996,** *35*, 114–115.

Durak, I.; J. Koksal; M. Kacmaz; S. Buyukkocak; B.M. Cimen; H.S. Ozturk. Hazelnut supplementation enhances plasma antioxidant potential and lowers plasma cholesterol levels. *Clin. Chim. Acta* **1999**, *284*, 113–115.

Edwards, K.; I. Kwaw; J. Matud; I. Kurtz. Effect of pistachio nuts on serum lipid levels in patients with moderate hypercholesterolemia. *J. Am. Coll. Nutr.* **1999**, *18*, 229–232.

Ellsworth, J.L.; L.H. Kushi; A.R. Folsom. Frequent nut intake and risk of death from coronary heart disease and all causes in postmenopausal women; the Iowa Women's Health Study. *Nutr. Metab. Cardiovasc. Dis.* **2001**, *11*, 372–377.

Eriksson, M.C. Contribution of phospholipids to headspace volatiles during storage of pecans. *J. Food Qual.* **1993,** *16*, 13–24.

Fallico, B.; E. Arena; M. Zappalà. Roasting of hazelnuts. Role of oil in colour development and hydroxymethylfurfural formation. *Food Chem.* **2003**, *81*, 569–573.

FAO. Non-wood forest products from conifers. *Non-Wood Forest Products 12;* Food and Agriculture Organization of the United Nations: Rome, Italy, 1995.

FDA, U.S. Food and Drug Administration, Center for Food Safety and Nutrition. *Summary of Qualified Health Claims Permitted;* 2003. Available at: http://www.cfsan.fda.gov/≈dms/qhc-sum.html#nuts), accessed 13 Sept. 2008.

Flinterman, A.E.; J.H. Akkerdaas; A.C. Knulst; R. van Ree; S.G. Pasmans. Hazelnut allergy: from pollen-associated mild allergy to severe anaphylactic reactions. *Curr. Opin. Allergy Clin. Immunol.* **2008**, *8*, 261–265.

Fraser, G.E.; J. Sabate; W.L. Beeson; T.M. Strahan. A possible protective effect of nut consumption on risk of coronary heart disease: The Adventist Health Study. *Arch. Intern. Med.* **1992**, *152*, 1416–1424.

Garg, M.L.; R.J. Blake; R.B. Wills. Macadamia nut consumption lowers plasma total and LDL cholesterol levels in hypercholesterolemic men. *J. Nutr.* **2003**, *133*, 1060–1063.

Griel, A.E.; Y. Cao; D.D. Bagshaw; A.M. Cifelli; B. Holub; P.M. Kris-Etherton. A macadamia nut-rich diet reduces total and LDL-cholesterol in mildly hypercholesterolemic men and women. *J. Nutr.* **2008**, *138*, 761–767.

Griel, A.E.; B. Eissenstat; V. Juturu; G. Hsieh; P.M. Kris-Etherton. Improved diet quality with peanut consumption. *J. Am. Coll. Nutr.* **2004**, *23*, 660–668.

Guclu, B.K.; F. Uyanik; K.M. Iscan, Effects of dietary oil sources on egg quality, fatty acid composition of eggs and blood lipids in laying quail. *S. African J. Animal Sci.* **2008**, *38,* 91–100.

Hatipoglu, A.; O. Kanbagli; J. Balkan; M. Kucuk; U. Cevikbas; G. Aykak-Toker; H. Berkkan; M. Uysal. Hazelnut oil administration reduces aortic cholesterol accumulation and lipid peroxides in the plasma, liver and aorta of rabbits fed a high-cholesterol diet. *Biosci. Biotechnol. Biochem.* **2004,** *68*, 2050–2057.

Hidalgo, F.J.; R. Zamora. Peptides and proteins in edible oils: Stability, allergenicity, and new processing trends. *Trends Food Sci. Technol.* **2006**, *17*, 56–63.

Holcapek, M.; P. Jandera; P. Zderadicka; L. Hrubá. Characterization of triacylglycerol and diacylglycerol composition of plant oils using high-performance liquid chromatography-atmospheric pressure chemical ionization mass spectrometry. *J. Chromatogr. A.* **2003**, *1010*, 195–215.

Holcapek, M.; M. Lísa; P. Jandera; N. Kabátová. Quantitation of triacylglycerols in plant oils using HPLC with APCI-MS, evaporative light-scattering, and UV detection. *J. Sep. Sci.* **2005**, *28*, 1315–1333.

Hu, F.B.; M.J. Stampfer. Nut consumption and risk of coronary heart disease: a review of epidemiologic evidence. *Curr. Atheroscler. Rep.* **1999**, *1*, 204–209.

Hu, F.B., M.J. Stampfer; J.E. Manson; E.B. Rimm; G.A. Colditz; B. A. Rosner; F.E. Speizer; C.H. Hennekens; W.C. Willett. Frequent nut consumption and risk of coronary heart disease in women: prospective cohort study. *Br. Med. J.* **1998**, *317*, 1341–1345.

Hughes, G.M.; E.J. Boyland; N.J. Williams; L. Mennen; C. Scott; T. C. Kirkham; J.A. Harrold; H.G. Keizer; J.C. Halford. The effect of Korean pine nut oil (PinnoThin) on food intake, feeding behaviour and appetite: a double-blind placebo-controlled trial. *Lipids Health Dis.* **2008,** *7*, 6.

Hyson, D.A.; B.O. Schneeman; P.A. Davis. Almonds and almond oil have similar effects on plasma lipids and LDL oxidation in healthy men and women. *J. Nutr.* 2002, *132*, 703–707.

Imbs, A.B.; N.V. Nevshupova; L.Q. Pham. Triacylglycerol composition of *Pinus koraiensis* seed oil, *J. Am. Oil Chem. Soc.* **1998**, *75,* 865–870.

Jenkins, D.J.; C.W. Kendall; A. Marchie; T.L. Parker; P.W. Connelly; W. Qian; J.S. Haight; D. Faulkner; E. Vidgen; K.G. Lapsley; G.A. Spiller. Dose response of almonds on coronary heart disease risk factors: blood lipids, oxidized low-density lipoproteins, lipoprotein(a), homocysteine, and pulmonary nitric oxide: a randomized, controlled, crossover trial. *Circulation* **2002**, *106*, 1327–1332.

Jiang, R.; J.E. Manson; M.J. Stampfer; S. Liu; W.C. Willett; F.B. Hu. Nut and peanut butter consumption and risk of type 2 diabetes in women. *JAMA* **2002**, *288*, 2554–2560.

Jin, T.; S.M. Albillos; Y-W. Chen; M.H. Kothary; T-J. Fu; Y-Z. Zhang. Purification and Characterization of the 7S Vicilin from Korean Pine (*Pinus koraiensis*), *J. Agric. Food Chem.*, **2008**, *56,* 8159–8165.

Josse, A.R.; C.W.C. Kendall; L.S.A. Augustin; P.R. Ellis; D.J.A. Jenkins. Almonds and postprandial glycemia: a dose-response study. *Metabolism Clin. Exp.* **2007**, *56,* 400–404.

Kaijser, A.; P. Dutta; G. Savage. Oxidative stability and lipid composition of macadamia nuts grown in New Zealand. *Food Chemistry* **2000**, *71,* 67–70.

Kamal-Eldin, A. Effect of fatty acid composition and tocopherol content on the oxidative stability of vegetable oils. *Eur. J. Lipid Sci. Technol.* **2006,** *58,* 1051–1061.

Knott, E.; C. Köse Gürer; J. Ellwanger; J. Ring; U. Darsow. Macadamia nut allergy. *J. Eur. Acad. Dermatol. Venereol.* **2008**, *22,* 1394–1395.

Kodad, O.; I. Socias; R. Company. Variability of oil content and of major fatty acid composition in almond (*Prunus amygdalus* Batsch) and its relationship with kernel quality. *J. Agric. Food Chem.* **2008**, *56,* 4096–4101.

Kornsteiner, M.; K-H. Wagner, I. Elmadfa. Tocopherols and total phenolics in 10 different nut types. *Food Chem.* **2006**, *98,* 381–387.

Kris-Etherton, P. Monounsaturated fatty acids and risk of cardiovascular disease. *Circulation* **1999**, *100,* 1253–1258.

Kris-Etherton, P.M; S. Yu-Poth; J. Sabate; H.E. Ratcliffe; G. Zhao; T. D. Etherton. Nuts and their bioactive constituents: effects on serum lipids and other factors that affect disease risk. *Am. J. Clin. Nutr.* **1999**, *70,* 504S–511S.

Kris-Etherton, P.M.; G. Zhao; A.E. Binkoski; S.M. Coval; T.D. Etherton. The effects of nuts on coronary heart disease risk. *Nutr. Rev.* **2001**, *59,* 103–111.

Kushi, L.H.; A.R. Folsom; R.J. Prineas; P.J. Mink; Y. Wu; R.M. Bostick. Dietary antioxidant vitamins and death from coronary heart disease in postmenopausal women. *N. Engl. J. Med.* **1996**, *334,* 1156–1162.

Lavedrine, F.; D. Zmirou; A. Ravel; F. Balducci; J. Alary. Blood cholesterol and walnut consumption: A cross sectional survey in France. *Preventive Med.* **1999**, *28,* 333–339.

Lee, J.W.; K.W. Lee; S.W. Lee; I.H. Kim; C. Rhee. Selective increase in pinolenic acid (all cis-5,9,12-18:3) in Korean pine nut oil by crystallization and its effect on LDL-receptor activity. *Lipids* **2004**, *39,* 383–387.

Li, L.; R. Tsao; R. Yang; J.K. Kramer; M. Hernandez. Fatty acid profiles, tocopherol contents, and antioxidant activities of heartnut (*Juglans ailanthifolia* Var. cordiformis) and Persian walnut (*Juglans regia* L.). *J. Agric. Food Chem.* **2007**, *55,* 1164–1169.

Madhaven, B.N. Final report on the safety assessment of *Corylus avellana* (Hazel) seed oil, *Corylus americana* (Hazel) seed oil, *Corylus avellana* (Hazel) seed extract, *Corylus americana* (Hazel) seed extract, *Corylus rostrata* (Hazel) seed extract, *Corylus vvellana* (Hazel) leaf extract, *Corylus americana* (Hazel) leaf extract, and *Corylus rostrata* (Hazel) leaf extract. *Int. J. Toxicol.* **2001,** *20,* 15–20.

Maguire, L.S.; S.M. O'Sullivan; K. Galvin; T.P. O'Connor; N.M. O'Brien. Fatty acid profile, tocopherol, squalene and phytosterol content of walnuts, almonds, peanuts, hazelnuts and the macadamia nut. *Int. J. Food Sci. Nutr.* **2004,** *55,* 171–178.

Malanin, K.; M. Lundberg; S.G.O. Johansson. Anaphylactic reaction caused by neoallergens in heated pecan nut. *Allergy* **1995**, *50*, 988–991.

Martín-Carratalá, M.L.; C. Llorens-Jordá; V. Berenguer-Navarro; N. Grané-Teruel. Comparative study on the triglyceride composition of almond kernel oil. A new basis for cultivar chemometric characterization. *J. Agric. Food Chem.* **1999**, *47*, 3688–3692.

Martínez, M.L. Varietal and crop year effects on lipid composition of walnut (*Juglans regia*) genotypes. *J. Am. Oil Chem. Soc.* **2006**, *83*, 791–796.

Martinez, M.L.; M.A. Mattea; D.M. Maestri. Pressing and supercritical carbon dioxide extraction of walnut oil. *J. Food Eng.* **2008,** *88,* **399–404.**

Miraliakbari, H.; F. Shahidi. Lipid class compositions, tocopherols, and sterols of tree nut oils extracted with different solvents. *J. Food Lipids* **2008a**, *15*, 81–96.

Miraliakbari, H.; F. Shahidi. Oxidative stability of tree nut oils. *J. Agric. Food Chem***. 2008b**, *56,* 4751–4759.

Morgan, W.A.; B.J. Clayshulte. Pecans lower low-density lipoprotein cholesterol in people with normal lipid levels. *J. Am. Diet. Assoc.* **2000**, *100*, 312–318.

Morgan, J.M.; K. Horton; D. Reese; C. Carey; K. Walker; D.M. Capuzzi. Effects of walnut consumption as part of a low-fat, low-cholesterol diet on serum cardiovascular risk factors. *Int. J. Vitam. Nutr. Res.* **2002**, *72*, 341–347.

Mueller, von F.J.H. Account of some New Australian Plants. *Transactions of the Philosophical Institute of Victoria 2: 72 Type: Macadamia ternifolia F. Muell,* 1857.

Natoli S.; P. McCoy. A review of the evidence: nuts and body weight. *Asia Pac. J. Clin. Nutr.* **2007,** *16,* 588–597.

Özdemir, M.; F. Açkurt; M. Kaplan; M. Yıldız; M. Löker; T. Gürcan; G. Biringen; A. Okay; F.G. Seyhan. Evaluation of new Turkish hybrid hazelnut (*Corylus avellana* L.) varieties: fatty acid composition, α-tocopherol content, mineral composition and stability. *Food Chem.* **2001,** *73,* 411–415.

Özkal, S.G.; U. Salgın; M.E. Yener. Supercritical carbon dioxide extraction of hazelnut oil. *J. Food Eng.* **2005,** *69,* 217–223.

Parcerisa J.; M. Rafecas; A.I. Castellote; R. Codony; R. Farran; J.Garcia; A. Lopez; A. Romero; J. Boatella. Influence of variety and geographical origin on the lipid fraction of hazelnut (*Coryllus avellana* L.) from Spain: II. Triglyceride Composition. *Food Chem.* **1994,** *50,* 245–249.

Parcerisa, J.; D.G. Richardson; M. Rafecus; R. Codony; J. Boatella. Fatty acid distribution in polar and nonpolar lipid classes of hazelnut oil (*Corylus avellana* L.). *J. Agric. Food Chem.* **1997**, *45*, 3887–3890.

Pasman, W.J.; J. Heimerikx; C.M. Rubingh; R. van den Berg; M. O'Shea; L. Gambelli; H.F.J. Hendriks; A.W.C. Einerhand; L.I. Mennen. The Effect of Korean pine nut oil on in vitro CCK release and on appetite sensations and gut hormones in post-menopausal overweight women. *Lipids Health Dis.* **2008**, *7, 10.*

Pearson, T.A.; T.D. Etherton; K. Moriarty; R. Reed; P.M. Kris-Etherton. High-MUFA diets with peanuts-peanut butter (P/PB) or peanut oil (PO) lower total cholesterol (TC) and LDL-C identically to a step 2 diet but eliminate the triglyceride (TG) increase. *FASEB J.* **1998**, *12,* A506.

Robotham, J.M.; F. Wang; V. Seamon; S.S. Teuber; S.K. Sathe; H.A. Sampson; K. Beyer; M. Seavy; K.H. Roux. Ana o 3, an important cashew nut (*Anacardium occidentale* L.) allergen of the 2S albumin family. *J. Allergy Clin. Immunol.* **2005**, *115*, 1284–1290.

Rodrigues, F.B. Turkish Tombul hazelnut (*Corylus avellana* L.). 2. Lipid characteristics and oxidative stability. *J. Agric. Food Chem.* **2003**, *51,* 3797–3805.

Rodrigues, J.E.; M.E. Araújo; F.F.M. Azevedo; N.T. Machado. Phase equilibrium measurements of Brazil nut (*Bertholletia excelsa*) oil in supercritical carbon dioxide. *J. Supercrit. Fluids* **2005,** *34,* 223–229.

Rosen, T.; D.B. Fordice. Cashew nut dermatitis. *South Med. J.* **1994**, *87,* 543–546.

Roux, K.H.; S.S. Teuber; S.K. Sathe. Tree nut allergens. *Int. Arch. Allergy Immunol.* **2003,** *131,* 234–244.

Ryan, E.; K. Galvin; T.P. O'Connor; A.R. Maguire; N.M. O'Brien. Fatty acid profile, tocopherol, squalene and phytosterol content of brazil, pecan, pine, pistachio and cashew nuts. *Int. J. Food Sci. Nutr*. **2006,** *57,* 219–228.

Sabaté, J. Nut consumption and body weight. *Am. J. Clin. Nutr*. **2003,** *78,* 647S–650S.

Sabaté, J.; G.E. Fraser; K. Burke; S.F. Knutsen; H. Bennett; K.D. Lindsted. Effects of walnuts on serum lipid concentrations and blood pressure in normal men. *N. Engl. J. Med*. **1993,** *328,* 603–607.

Sabaté, J.; E. Haddad; J.S. Tanzman; P. Jambazian; S. Rajaram. Serum lipid response to the graduated enrichment of a Step I diet with almonds: a randomized feeding trial. *Am. J. Clin. Nutr.* **2003**, *77,* 1379–1384.

Salgin, S.; U. Salgin. Supercritical fluid extraction of walnut kernel oil. *Eur. J. Lipid Sci. Technol.* **2006,** *108,* 577–582.

Salvo, F.; G. Dugo; I. Stagno D'Alcontres; A. Cotroneo. Composition of almond oil. II. Distinction of sweet almond oil from blends with peach and apricot seed oil. *Rivista Italiana delle Sostanze Grasse* **1980,** *57,* 24–26.

Savage, G.P.; D.L. McNeil. Chemical composition of hazelnuts (*Corylus avellana* L.) grown in New Zealand. *Int. J. Food Sci. Nutr.* **1998**, *49,* 199–203.

Scott, C.; W. Pasma; J. Hiemerikx; C. Rubingh; R. Van Den Berg; M. O'Shea; L. Gambelli; H. Hendricks; L. Mennen; A. Einerhand. Pinnothin™ suppresses appetite in overweight women. *Appetite* **2007**, *49,* 330.

Seddon, J.M.; J. Cote; B. Rosner. Progression of age-related macular degeneration: association with dietary fat, transunsaturated fat, nuts and fish intake. *Archives of Ophthalmol.* **2003**, *121,* 1728–1737.

Shahidi, F.; H. Miraliakbari. Tree nut oils. *Bailey's Industrial Oil & Fat Products*, 6th ed.; F. Shahidi, Ed.; Wiley-Interscience: Hoboken, NJ, 2005; pp. 175–193.

Sheridan, M.J.; J.N. Cooper; M. Erario; C.E. Cheifetz. Pistachio nut consumption and serum lipid levels. *J. Am. Coll. Nutr.* **2007**, *26,* 141– 148.

Sicherer, S.H.; A. Munoz-Furlong; A.W. Burks; H.A. Sampson. Prevalence of peanut and tree nut allergy in the US determined by a random digit dial telephone survey. *J. Allergy Clin. Immunol.*

**1999,** *103,* 559–562.

Silva, C.F.; M.F. Mendes; F.L.P. Pessoa; E.M. Queiroz. Supercritical carbon dioxide extraction of macadamia (*Macadamia integrifolia*) nut oil: experiments and modeling. *Braz. J. Chem. Eng.* **2008,** *25,* 175–181.

Spiller, G.A.; D.A. Jenkins; O. Bosello; J.E. Gates; L.N. Cragen; B. Bruce. Nuts and plasma lipids: an almond-based diet lowers LDL-C while preserving HDL-C. *J. Am. Coll. Nutr.* **1998,** *17,* 285–290.

Sugano, M.; I. Ikeda; K. Wakamatsu; T. Oka. Influence of Korean pine (*Pinus koraiensis*)-seed oil containing cis-5,cis-9,cis-12-octadecatrienoic acid on polyunsaturaed fatty acid metabolsim, eicosanoid production and blood preussure in rats. *Br. J. Nutr.* **1994**, *72*, 775–783.

Taylor, L. Brazil nut: Herbal properties and actions. *The Tropical Plant Database*, http://www.rain-tree.com/brazilnu.htm, accessed 23 Aug. 2008

Teuber, S.S.; R.L. Brown; L.A. Haapanen. Allergenicity of gourmet nut oils processed by different methods. *J. Allergy Clin. Immunol.* **1997**, *99*, 502–507.

Thomson, C.D.; A. Chisholm; S.K. McLachlan; J.M. Campbell. Brazil nuts: an effective way to improve selenium status. *Am. J. Clin. Nutr.* **2008,** *87,* 379–384.

Tsai, C.J.; M.F. Leitzmann; F.B. Hu; W.C. Willett; E.L. Giovannucci. Frequent nut consumption and decreased risk of cholecystectomy in women. *Am. J. Clin. Nutr.* **2004**, *80*, 76–81.

Turan, S.; A. Topcu; I. Karabulut; H. Vural; A.A. Hayaloglu. Fatty acid, triacylglycerol, phytosterol, and tocopherol variations in kernel oil of Malatya apricots from Turkey. *J. Agric. Food Chem.* **2007**, *55*, 10787–10794.

USDA, Agricultural Research Service. USDA, Nutrient Database for Standard Reference, 2001. http://www.nal.usda.gov/fnic, accessed 22 August 2008.

Vieira-Thais, M.F.; A.B. Regitano-d'Arce, Marisa. Antioxidant concentration effect on stability of Brazil nut (*Bertholletia excelsa*) crude oil. *Arch. Latinoam. Nutr.* **1999**, *49*, 271–274.

Wallowitz, M.; W.R. Peterson; S. Uratsu; S.S. Comstock; A.M. Dandekar; S.S. Teuber. Jug r 4, a Legumin Group Food Allergen from Walnut (*Juglans regia* Cv. Chandler). *J. Agric. Food Chem.* **2006,** *54,* 8369–8375.

Wolff, R.L.; O. Lavialle; F. Pédrono; E. Pasquier; L.G. Deluc; A.M. Marpeau; K. Aitzeműller. Fatty acid composition of Pinacaeae as taxonomic markers. *Lipids* **2001**, *36,* 439–451.

Yochum, L.A.; A.R. Folsom; L.H. Kushi. Intake of antioxidant vitamins and risk of death from stroke in post-menopausal women. *Am. J. Clin. Nutr.* **2000**, *72*, 476–483.

Zambon, D.; J. Sabate; S. Munoz; B. Campero; E. Casals; M. Merlos; J.C. Laguna; E. Ros. Substituting walnuts for monounsaturated fat improves the serum lipid profile of hypercholesterolic men and women. *Ann. Intern. Med.* **2000,** *7,* 538–546.

Zhang, S.M.; M.A. Hernan; H. Chen; D. Spiegelman; W.C. Willett; A. Ascherio. Intakes of vitamins E and C, carotenoids, vitamin supplements, and PD risk. *Neurology* **2002**, *59*, 1161–1169.

Zibaeenezhad, M.J.; M. Rezaiezadeh; A. Mowla; S.M.T. Ayatollahi; M.R. Panjehshahin. Antihypertriglyceridemic effect of walnut oil. *Angiology* **2003,** *54,* 411–414.

Zlatanov, M; S. Ivanov; K. Aitzetmüller. Phospholipid and fatty acid composition of Bulgarian nut oils. *Fett–Lipid* **1999**, *101*, 437–439.

Zwarts, L.; G.P. Savage; D.L. McNeil. Fatty acid content of New Zealand-grown walnuts (*Juglans regia* L.). *Int. J. Food Sci. Nutr.* **1999,** *50*, 189–194.

# 4

# Flax, Perilla, and Camelina Seed Oils: α-Linolenic Acid-rich Oils

**Clifford Hall III[1], Kelley C. Fitzpatrick[2], and Afaf Kamal-Eldin[3]**
*[1]Department of Cereal and Food Sciences, North Dakota State University, 210 Harris Hall, Fargo, North Dakota 58105, USA; [2]FLAX CANADA 2015, 465-167 Lombard Ave., Winnipeg, Manitoba R3B 0T6, Canada; [3]Department of Food Science, Swedish University of Agricultural Sciences, Box 7051, 750 07 Uppsala, Sweden*

## Introduction

Oils rich in α-linolenic acid (ALA) are gaining increased attention because of the anticipated health benefits related to the cardioprotective effects of this fatty acid. Flax and perilla seed oils are the richest sources of ALA, followed by the seed oils of camelina, hempseed, red and black currants, sea buckthorn, lingonberry, blueberry, cranberry, cloudberry, raspberry, and walnut. Hempseed oil, berry seed oils, nut oils, and other ALA-rich oils are discussed in separate chapters of this book. In this chapter, we present a review of the relevant literature on flaxseed oil and the limited data on perilla and camelina seed oils.

Mankind has used flaxseed or linseed (*Linum usitatissimum,* L., subspecies usitatissimum, Linaceae) for food and fiber since ancient times (Vaisey-Genser & Morris, 2003). Flaxseed cultivation dates back to around 9,000–8,000 B.C. in the Middle East: Turkey (van Zeiste, 1972), Iran (Helbaek, 1969), Jordon (Hopf, 1983; Rollefson et al., 1985), and Syria (Hillman, 1975; Hillman et al., 1989). The domestication of flaxseed dates back to 7000–4500 B.C. (Vaisey-Genser & Morris, 2003; Zohary & Hopf, 2000). About two million metric tons of flaxseed are produced annually, with Canada being the main producer (ca. 33%), followed by China (20%), the United States (16%), and India (11%) (Berglund, 2002).

Flaxseed contains lipids (40%), protein (21%), dietary fiber (28%), ash (4%), and other soluble components such as sugars, phenolic acids, and lignans (ca. 6%). The oil content in flaxseed falls between 29 and 45%, depending on the cultivar, location, and agroclimatic conditions (Daun et al., 2003; Oomah & Mazza, 1997; Wakjira et al., 2004). The main nutritional advantage of flaxseed oil is related to the high level of ALA in the oil (50–60%). About 20% of the flaxseed is a mucilaginous hull. Flaxseed mucilage is comprised of gum-like polysaccharides containing acidic (54.5% rhamnose and 23.4% galactose) and neutral arabinoxylan (62.8% xylose) (Cui et al., 1994a; Warrand et al., 2005). Flaxseed contains about 1–2% of total phenolic compounds (Hall & Shultz, 2001; Oomah et al., 1995), of which the lignan secoisolariciresinol diglucoside (SDG) is a major component. SDG is present in the seed as a mixture of oligomers, with hydroxymethylglutaric acid having an average molecular weight of 4000 Da (Kamal-Eldin et al., 2001). Numerous bioactivities are claimed for SDG, including antioxidant and estrogenic/oestrogenic effects (Adlercreutz et al., 1992; Hutchins & Slavin, 2003; Water & Knowler , 1982), leading to health benefits with respect to cardiovascular diseases (CVDs) and diabetes. Flaxseed also contains cyanogenic glycosides, namely linamarin, linustatin, lotasutralin, and neolinustatin, which release toxic hydrogen cyanide upon hydrolysis. Considered as anti-nutritional compounds, these limit the edible dose of flaxseeds. The consumption of 1–2 tablespoons of flaxseeds provides about 5–10 mg of hydrogen cyanide, which is below the estimated toxic dose of 50–60 mg/day (Rosling, 1994). Besides the cyanogenic glycosides, flaxseeds contain trypsin inhibitor, linatine, and phytic acid as anti-nutrients (Bhatty, 1993; Oomah et al., 1996).

Perilla (*Perilla frutescens* L., Britton, Lamiaceae), also known as beefsteak plant, Chinese basil, or purple mint, is a perennial herb that belongs to the mint family and grows mainly in India and Southeast Asia. The plant is well-known for an essential oil extracted from its leaves that contains perillaldehyde (*p*-mentha-1,8(9)-dien-7-al), limonene, linalool, β-caryophyllene, l-menthol, limonene, α-pinene, perillene (2-methyl-5-(3-oxolanyl)-2-pentene), and elemicin as main components (Yuba et al., 1995). The leaves also contain rosmarinic acid that might be responsible for their anti-allergic effect (Makino et al., 2003). Although the plant is not a botanical relative to sesame (*Sesamum indicum* L., Pedaliaceae, called goma in Japanese), perilla seeds are called *egoma* [荏胡麻,えごま] in Japanese meaning "sesame bean," and tul-kkae [들깨] in Korean meaning "wild sesame". Perilla seeds contain 35–45% oil that is comprised of approximately 55–60% of ALA (Lee et al., 1993).

Camelina (*Camelina sativa*, L., Crantz, Brassicaceae), known also as false flax or gold of pleasure, is a very old oil plant with cultivation history extending back to the Bronze Age (Schultze-Motel, 1979). The seed contains 30–40% of oil that is comprised of approximately 35–40% of ALA (Budin et al., 1995; Marquard & Kuhlmann, 1986; Plessers et al., 1962; Seehuber, 1984; Zubr, 1997), enabling its utilization as a drying oil in painting and coating applications (Luehs & Friedt , 1993). The camelina seed contains much lower levels of glucosinolates compared to other Brassicaceae species (Lange et al., 1995), which encourages the utilization of meals in feed applications.

The high content of ALA in the flax, perilla, and camelina seed oils has attracted the attention of the health industry. However, high ALA also renders these oils vulnerable to oxidation. Thus, one must take special measures during their processing, storage, and use, which are discussed in this chapter.

## Extraction and Processing of ALA-Rich Oils

Limited research is available on the extraction and processing of flaxseed oil and other ALA-rich oils. However, oil destined for the paints and coating industry is typically refined by processes similar to those used for commercial edible oils. In contrast, edible flaxseed oil is expeller-pressed at temperatures lower than traditional solvent-extraction processes. The term "cold-pressed" is used as a descriptor for expeller-pressed oil. However, this term does not have a legal definition in many countries, and the extracted-oil temperature of 50°C is a legal limit for cold-pressed oils in the United Kingdom, while 70°C is considered acceptable elsewhere (Panfilis et al., 1998; Singh & Bargale, 2000). The overall quality of flaxseed oil likely depends on original seed quality.

The information on flaxseed-oil pressing is limited. However, numerous studies reported a link between poor seed quality and poor milled-flaxseed quality. Pizzey & Luba (2002, 2004) established a relationship between the percentage of dark seed in a sample and milled-flaxseed lipid stability. They showed that number 1 grade flaxseed in the United States and Canada can contain high levels of dark seeds and that dark seed was detrimental to milled-flaxseed stability. They investigated lipid oxidation in milled-flaxseed samples containing up to 25% of dark flaxseed. A milled-flaxseed sample with 25% of dark seed produced a high free fatty acid (FFA) and a high peroxide value (Pizzey & Luba, 2002, 2004). White & Jayas (1991) found that a twofold increase in FFA corresponded to a discolored or charred appearance of the seeds.

Immature seed (green seed) is a flaxseed-grading factor in the United States and Canada. Malcolmson et al. (2000) speculated that the reason for observed high FFA content in a milled-flaxseed sample was due to the presence of immature seed in the sample. Wanasundara et al. (1999) reported that flaxseed lipase had an activity of 160 and 354 units/gram prior to germination and after germination, respectively. In addition to lipase, lipoxygenase (LOX) activity is generally higher in young and developing plant tissues than in mature tissues (Siedow, 1991; Zhuang et al., 1992). LOX prefers FFA substrates more than triacylglycerols (Hamilton, 1994). Thus, the combination of lipase and LOX may promote the oxidation of the oil in flaxseed.

Chlorophyll is a green photosynthetic pigment that is present in immature flaxseed (Preťová & Vojteková, 1985). Using a single flaxseed cultivar, these authors showed that the seed's chlorophyll content peaked at 16 days after fertilization, and plummeted to zero just prior to full maturity (24 days). Chlorophyll is potentially the most abundant photosensitizer in flaxseed. Number 1 and number 2 Canadian grade flaxseeds contain 0.7 to 0.9 mg/kg of chlorophyll, respectively (Daun, 1993). The presence of dark and immature seed is a source of FFA and chlorophyll that could directly or indirectly promote lipid oxidation in expeller-pressed flaxseed oil. Thus, to produce a high-quality flaxseed oil, one must use a good-quality flaxseed.

Mechanical pressing is the preferred method for producing edible flaxseed oil due to the healthful perception of the oil (Wiesenborn et al., 2005; Zheng et al., 2003). Numerous factors can affect the quality and yield of oil obtained from the expeller-pressing operation: The feed rate of the seed into the press, the speed of screw rotation, and the choke size influence oil extraction. In addition to press parameters, moisture content and seed fraction being pressed also can influence oil extraction and oil temperature.

Zheng et al. (2003, 2005) reported that the mean recovery of oil from whole flaxseed was not significantly different from oil recovery from a flaxseed that had 20% of the hull removed (FHR). Increasing the hull removal to 62 and 85% caused a reduction in oil recovery during pressing. The higher fiber in whole flaxseed might be the reason for the higher oil yields from whole flaxseed. The fiber causes friction, resulting in increased specific mechanical energy and heat during pressing. Zheng et al. (2005) reported that the specific mechanical energy decreased from approximately 84 kJ/kg to 40 kJ/kg during the expelling of whole flaxseed and flaxseed with 85% of the hull removed, respectively. The increased specific mechanical energy also enhanced oil and meal temperatures. Oil temperatures of 53, 46, 34, and 35°C were observed in expeller-pressed oil from whole flaxseed and 20, 62, and 85% dehulled flaxseeds, respectively. The higher oil temperatures also coincided with higher meal temperatures. Meal temperatures of 69, 60, 49, and 46°C were observed in expeller-pressed oil from whole flaxseed and 20, 62, and 85% dehulled flaxseeds, respectively. The choke size of the expeller press also influenced oil recovery. The 6-mm choke size generally enhanced oil recovery compared to the 8-mm choke size (Zheng et al., 2003). The smaller choke size probably was responsible for changes in specific mechanical energy and friction that contributed to increased oil yields.

The higher oil yield obtained from whole flaxseed versus dehulled flaxseed is likely due to a number of factors that include friction and temperature. The friction created during pressing likely caused a disruption in the flaxseed matrix, resulting in the disruption of the oil-containing spherosomes and thus more oil recovery. The increased temperature likely increased oil fluidity, thus causing more oil recovery from whole flaxseed. The moisture content of the flaxseed also contributes significantly to oil recovery. A reduction in the moisture content of flaxseed caused an increase in friction among the seeds and heat, resulting in increased oil recovery. Oil recovery from flaxseed and dehulled flaxseed with moisture contents of 11 – 12% was always lower than in seeds with moisture contents of 6–7% (Zheng et al., 2003, 2005). However, low- moisture contents (6–7%) and a small-choke size (6 mm) promoted plugging during pressing (Zheng et al., 2003). As a result, less oil production (kg/hour) was achieved.

The oil productivity or capacity (kg/hour) was inversely correlated with oil recovery or yield. The correlation between capacity and recovery was more evident in experiments involving whole and dehulled flaxseeds. For example, twice the oil output was achieved during the pressing of dehulled flaxseed compared to whole flaxseed (Zheng et al., 2003, 2005). These authors proposed that more dehulled material could be fed through the expeller press, resulting in more oil per unit of time. These authors also found that a larger choke size (8 mm) and moisture content near 10.5% produced the greatest amount of oil from dehulled flaxseed. The greater oil capacity from dehulled flaxseed and the lower temperature of the pressed oil illustrate an alternative to using whole flaxseed as the source of flaxseed oil. Furthermore, the oil quality of the pressed flaxseed oil was better under conditions that favored lower temperatures (Zheng et al., 2003). Abuzaytoun and Shahidi (2006) reported that the removal of natural antioxidants such as tocopherols greatly diminished the stability of cold-pressed flaxseed oil. Thus, cold-pressing without traditional refining is essential to producing good-quality flaxseed oil.

One could use supercritical carbon dioxide ($CO_2$) fluid extraction as an alternative to cold-pressing. Bozan and Temelli (2002) observed higher ALA content in oil obtained from supercritical $CO_2$ extraction compared to Soxhlet extraction. In contrast, tocopherol content was lower, suggesting the potential for lower oil stability. Temperatures (50 and 70°C) and pressures (35 and 55 MPa) had little effect on fatty-acid profiles. The stability of oil was not reported; thus, further research is required regarding shelf life and comparisons to the cold-pressing operation. When processed correctly, cold-pressed oil has low peroxide values and low FFA contents. Limited data are available on the effects of storage conditions on oil stability. However, the oxidation of camelina seed oil was affected by the temperature of storage, with greater effects by light, especially at the initial stage of storage (Abramovic & Abram, 2005). The same observations were reported for flaxseed oil. Flaxseed oil has a nutty flavor that can diminish over time and become painty and bitter if stored improperly. Wiesenborn et al. (2005) demonstrated that flaxseed oil stored 15 weeks at 4°C had statistically higher peroxide values than 7-day-old flaxseed stored at room temperature. However, the difference between 2.2 (cold-stored oil) and 1.6 mEq/kg may not be of practical significance. Thus, flaxseed oil should be stored under refrigeration Research (Hall et al., unpublished data) indicated that oil stored at room temperature in the dark was far more stable than flaxseed oil stored under light. The peroxide values of oils stored for 14 days were significantly affected by light. The peroxide value of oil stored at room temperature in the dark was approximately 12 mEq/kg, while the peroxide values in the sample stored under the light and room temperature reached 128 mEq/kg after 14 days of storage. A similar trend was observed in our laboratory (Tulbek et al., 2008) in milled flaxseed. Samples stored in sealed metalized pouches had lower peroxide values (1.3 mEq/kg) than samples stored in sealed clear plastic bags (5 mEq/kg) after 3 weeks of storage at 60°C under light in an environmental chamber. Secondary oxidation products were also higher in the sample stored in clear plastic bags. Collectively, these observations support the packaging of flaxseed oil in light-impermeable packaging. Manufacturers generally recommend storing flaxseed oil in unopened dark containers at 4–7°C for up to 12 months. When the container is opened, one should store the oil in the refrigerator for up to 12 weeks (Choo et al., 2007; Rudnik et al., 2001).

## Edible and Nonedible Applications

Traditionally, drying oils, particularly flax (usually called linseed oil when it is used for nonedible applications) and perilla seed oils were used in oil-based paints, varnishes, linoleum, printing inks, lacquers, waterproof coatings on cloth, and in the treatment of wood. As human food, flax seeds are traditionally used as ingredients in breakfast cereals and breads. However, since the recognition of their health potential, their use in foods has increased and includes products such as milk, yogurt, and bread topping. Currently, flax and perilla seed oils are important ingredients in health-promoting oils and soft-gel capsules. Flaxseed oil is also used in lipid infusions for patients with certain types of disorders, as well as in cosmetics and skin creams. Flaxseed oil was suggested as a valuable ingredient for ice cream and other frozen desserts, with improved fatty-acid composition (Hall & Schwarz, 2002).

## Composition of Fatty Acids and Acyl Lipids

The fatty-acid composition of flaxseed, which is cultivar-dependent, is dominated by ALA that constitutes about 50–60% of the total fatty acids (Fremont et al., 1998; Oomah & Mazza, 1997). The relative fatty-acid composition of the major fatty acids is palmitic (5.5–6.5%), stearic (2.2–4.1%), oleic (13.4–22.2%), linoleic (15.2–17.4%), and ALA (51.8–60.4%) (Choo et al., 2007; Wakjira et al., 2004). Linola, a low-ALA cultivar developed for the commercial vegetable-oil market, contains only 3–4% of ALA and 75% of linoleic acid (LA) (Lukaszewicz et al., 2004). The oil content and fatty-acid composition of flaxseed are both affected by agroclimatic conditions (Daun et al., 2003; Taylor & Morrice, 1991). Generally, the neutral lipids (NLs, acylglycerols and fatty acids) constitute 90–96% of the total lipids in flaxseed, while the remaining lipids are polar lipids [glycolipids (GLs) and phospholipids (PLs)] and unsaponifiable constituents (Stenberg et al., 2005; Wanasundra et al., 1999). In a detailed study, flaxseed oil was fractionated by silicic acid column chromatography into NLs (87.8–89.6%), GLs (5.8–6.6%), and PLs (3.8–5.8%). The NLs consist of triacylglycerols (93.5%), FFAs (3.0%), free sterols (1.1%), and small amounts of mono- and diacylglycerols (DAGs) (Khotpal et al., 1997). The PLs have higher percentages of saturated fatty acids and oleic acid and a lower percentage of ALA than the triacylglycerols and the FFAs of the NL fraction (El-Shattory, 1976). As a natural consequence of the fatty-acid composition and the dominance of triacylglycerols, flaxseed oil is dominated by tri-ALA (LnLnLn; 22–35%), which co-exists with 16 other triacylglycerols, mainly PLnLn (ca. 6%), SLnLn (ca. 10%), OLnLn (ca. 15%), LLnLn (ca. 14%), PLLn (ca. 5%), POLn (ca. 5%), SLLn (ca. 10%), and LOO (ca. 3%), while the main DAG in flaxseed oil is naturally LnLn (Ayorinde , 2000; Holcapek et al., 2003; Krist et al., 2006).

The oil content in perilla seed is 38.6–47.8%, with its major fatty acids being palmitic (4.0–8.9%), stearic (1.0–3.8%), oleic (12.9–19.9%), linoleic (14.3–20.0%), and ALA (56.8–64.0%). The oil consists of 91.2–93.9% of NLs, 3.9–5.8% of GLs, and 2.0–3.0% of PLs. The NLs are mostly triacylglycerols (88.1–91.0%), with smaller amounts of steryl esters, hydrocarbons, FFAs, free sterols, and partial glycerides. The GL fraction contained esterified sterylglycosides (48.9–53.2%), sterylglycosides

(22.1–25.4%), and smaller amounts of mono- and digalactosyldiacylglycerol. The PLs were comprised of phosphatidylethanolamine (50.4–57.1%), phosphatidylcholine (17.6–20.6%), and small amounts of phosphatidic acid, lysophosphatidylcholine, phosphatidylserine, and phosphatidylinositol (Shin & Kim, 1994).

Camelina seed oil is composed of palmitic acid (5.3–6.4%), stearic acid (1.4–3%), oleic acid (14.0–18.7%), LA (13.5–19.0%), ALA (27.1–38.9%), arachidic acid (20:0, 0.4–1.5%), gondoic acid (20:1, 11.6–16.2%), eicosadienoic acid (20:2, 1.7–2.1%), eicosatrienoic acid (20:3, 1.3–1.7%), and erucic acid (22:1, 1.6–4.0%) (Abramovic & Abram, 2005; Eidhin et al., 2003; Shukla et al., 2002).

## Nonacylglycerol Constituents

Flaxseed oil contains 0.7–1.3% of unsaponifiable compounds (Painter & Nesbitt, 1943) although Choo et al. (2007) reported unsaponifiable matter content as low as 0.4%. As with most other vegetable oils, the major unsaponifiable constituents are plant sterols. The total sterol content in flaxseed oil is about 700 mg/100 g, with the main sterols being sitosterol (30%), cycloartenol (29%), campesterol (15%), 24-methylene cycloartanol (9%), Δ5-avenasterol (8.5%), and stigmasterol (5%) (Schwartz et al., 2008). Other sterols, including cholesterol, brassicasterol, campestanol, sitostanol, stigmata-5,24-dienol, obtusifoliol, gramisterol, citrostadienol, and α-amyrin, are found as minor components (Capella, 1961; Kamm et al., 2001; Schwartz et al., 2008). Flaxseed oil contains about 4 mg/100 g of squalene, a level that is significantly lower than in olive, corn, and rice-bran oils (Dessi et al., 2002).

The major tocopherol in flaxseed oil is γ-tocopherol (ca. 130–575 mg/kg), which is usually accompanied by very small levels of α- and δ-tocopherols (tr.–1 mg/kg). Oomah et al. (1997) observed a correlation between seed-oil content and total tocopherols ($r = 0.42$) and γ-tocopherol ($r = 0.41$). Kamm et al. (2001) reported the distribution of tocopherols and tocotrienols in high- and low-ALA flaxseed, and Oomah et al. (1997) reported that γ-tocopherol content was greater (430–575 mg/kg oil) in high-ALA compared to low-ALA flaxseed oils (170 mg/kg of oil) in agreement with previous observations (Kamal-Eldin & Andersson, 1997). Flaxseed oil extracted with supercritical $CO_2$ fluid contained lower tocopherol levels compared to that obtained by Soxhlet extraction (764 mg/kg of oil) (Bozan & Temelli, 2002). In addition to tocopherols, flaxseed oil contained small amounts of geranyl geraniol (Fedeli et al., 1966). Moreover, flaxseed oil is characterized by relatively high levels (43–72 mg/kg) of plastochromanol-8 (Olejnik et al., 1997; Sondergaard & Dam, 1967; Vealsco & Goffman, 2000). Recently, plastochromanol-8 (Fig. 4.1) was shown to be present in many seed oils, including flax (17–30 mg/100 g of oil), rape (≈9 mg/100 g of oil), camelina (≈4 mg/100 g of oil), peanut (≈2 mg/100 g of oil), corn (≈2 mg/100 g of oil), and grape (≈1 mg/100 g of oil) seed oils (Gruszka & Kruk, 2007). The carotenoids, β-carotene, lutein, and violaxanthin, were detected in flaxseed oil (Pretova & Vojtekova, 1985). Flaxseed oils also exhibited a chlorophyll content in the range of

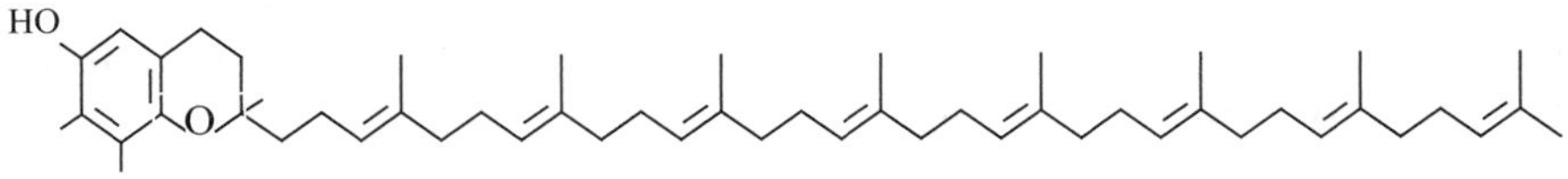

Fig. 4.1. Structure of plastochromanol-8, a unique tocopherol analog in flaxseed oil.

8–58 mg/kg, with the level being dependent on seed maturity (Choo et al., 2007). In addition, flaxseed oils contained 127–256 mg/kg of total flavonoids (luteolin equivalents) and 768–3073 mg/kg of total phenolic acids (Choo et al., 2007).

Tsuyuki et al. (1978) reported total sterols (0.72–0.89%) and total pigments (3.06–4.18%) in perilla seed oil. Perilla seed oil contains about 70 mg/100 g of waxy material, the major components of which are policosanols (25.5–34.8%), wax esters, steryl esters, aldehydes (49.8–53.0%), hydrocarbons (10.5–18.8%), acids (1.7–2.1%), and triacylglycerols (1.0–2.9%). The policosanols of perilla seeds, which exist in unesterified forms, were composed of octacosanol (C28:0, 67%), tetracosanol (C24:0, 1–2%), hexacosanol (C26:0, 17%), heptacosanol (C27:0, 3%), nonacosanol (C29:0, 3%), and triacontanol (C30:0, 8%) (Adhikari et al., 2006). The policosanols were claimed to reduce platelet aggregation, endothelial damage, and foam-cell formation and to have cardioprotective effects. The major sterols in perilla seed oil are sitosterol (78.0–81.7%), Δ5-avenasterol (12.6–14.8%), campesterol (4.4–6.5%), and stigmasterol (0.3–0.8%) (Shin, 1997). Perilla oil contains 490–676 ppm of tocopherols: mainly γ-tocopherol (92%), with very small amounts of α- and δ- tocopherols (Shin & Kim, 1994).

According to one report, camelina seed oil contained 0.54% of unsaponifiable materials, mainly composed of the following desmethyl sterols (0.36%): cholesterol (188 ppm), brassicasterol (133 ppm), campesterol (893 ppm), stigmasterol (103 ppm), sitosterol (1884 ppm), and $\Delta^5$-avenasterol (393 ppm) (Shukla et al., 2002).

## Oxidative Stability of ALA-Rich Oils

The fatty-acid composition is the main factor responsible for the stability of vegetable oils (Kamal-Eldin, 2006; Kamal-Eldin & Yanishlieva, 2002). Because of their high content of ALA, the oils of flax, perilla, and camelina are highly susceptible to oxidative deterioration, with the latter being the most stable (Eidhin et al., 2003). Oxidation is evident by the rancidity and the changes in viscosity, but until recently, it was not evaluated using appropriate chemical methods. The reason is that common oxidative markers, such as peroxide value and conjugated dienes, do not adequately describe the differences in oxidation between ALA and LA. For example, Lukaszewicz et al. (2004) reported that Linola (the high-LA flaxseed cultivar with the lowest content of ALA) exhibited the highest conjugated diene content compared to the cultivar Abby (high ALA) when heated for 40 min at 140°C. Possibly, factors other than the fatty-acid profile (e.g., the content of pro- and antioxidants) are responsible for the observed oxidation behavior of the different cultivars. A relationship between headspace volatile analysis and the sensory properties of cold-pressed flaxseed oil was established (Wiesenborn et al., 2005). Significant differences between samples, especially with respect to the painty–bitter flavors, were obtained, suggesting the solid-phase microextraction analysis as a method of choice for determining flaxseed-oil quality.

The high susceptibility to oxidation limits the use of flaxseed oil to food applications. Upon storage, flaxseed oil undergoes hydrolysis and oxidation reactions, leading to oxygen-rich volatile and polymeric products. The oxidation of high-ALA oils is accompanied by a noticeable color change (Kumarathasan et al., 1992). Three red pigment-forming substances derived from autoxidized ALA in linseed oil were identified as the stereoisomers of 3-(2-ethyl-5-hydroxy-3-oxo) cyclopentanyl-2-propenal (Nakamura, 1985).

Various antioxidants—including ascorbyl palmitate, citric acid, ascorbic acid, ethoxylated glycol, and α-tocopherol and their blends—can be applied in flaxseed oil (Rudnik et al., 2001). The endogenous antioxidant of flaxseed, secoisolariciresinol diglucoside, was found to significantly ($P < 0.05$) decrease the oxidation of corn oil and to synergize the antioxidant effect of added tocopherol (220 ppm) (Hall & Shultz, 2001). Van Ruth et al . (2001) used soybean extracts as an antioxidant in flaxseed oil, and obtained about a 30% reduction in the formation of primary oxidation products and up to a 99% reduction in the formation of secondary oxidation products.

When flaxseed oil was used for frying (177–191°C), its oxidation was very rapid and was accompanied by a fishy flavor, possibly caused by the formation of 1-penten-3-one (Hadley, 1996). Molecular changes during oxidation at high temperatures may differ from those occurring at low temperatures. For example, diunsaturated 5- and 6-membered ring cyclic fatty-acid monomers are formed from ALA during the heating of flaxseed oil (Mossoba et al., 1994; Waterman et al., 1949). Heating at high temperatures is usually also accompanied by the *cis,trans* isomerization of double bonds (Gourlay, 1951). Toxic furan fatty acids (e.g., 2-pentyl furan) are also formed in foods rich in ALA when subjected to high temperatures (Jacini, 1986; Krishnamurthy et al., 1967).

## Taste and Flavor

Flaxseed contains a number of cyclic peptides, cyclolinopeptides (Fig. 4.2), composed exclusively of the hydrophobic amino acids: Phe, Leu, Ile, Val, Met, Pro, and Trp. These cyclolinopeptides include CLA [cyclo(-Pro-Pro-Phe-Phe-Leu-Ile-Ile-Leu-Val-)], CLB [cyclo(-Pro-Pro-Phe-Phe-Val-Ile-Met-Ile-Leu-)], CLD [cyclo (-Pro-Phe-Phe-Trp-Ile-Met-Leu-Leu-)], CLE [cyclo (-Pro-Leu-Phe-Ile-Met-Leu-Val-Phe-)], CLF [cyclo (-Pro-Phe-Phe-Trp-Val-Met-Leu-Met-)], and CLG [cyclo (-Pro-Phe-Phe-Trp-Ile-Met-Leu-Met-)]. The cyclolinopeptides containing methionine undergo oxidation to form the cyclolinopeptides CLB*, CLD*, CLE*, CLF*, and CLG*, which contain methionine oxide (Stefanowicz, 2004). Some of these peptides are co-extracted with cold-pressed oil, which usually has a delicate nutty flavor when fresh. However, a lingering bitter off-taste develops after storing the oil at room temperature with the key bitter compound (threshold concentration of 12.3 μmol/L of water) being the methionine sulfoxide-containing CLE* originating from CLE present in the flaxseed oil at 485–925 mg/kg (Brühl et al., 2007, 2008). Although the cyclolinopeptides have this negative contribution to flavor, they might be beneficial for health. For example, cyclolinopeptide A (CLA), isolated in 1959 from linseed oil, possesses immunosuppressive activity. Similar to cyclosporin A, CLA inhibited the action of interleukin-1-α and interleukin-2, influenced human lymphocyte proliferation in vitro, and tempered the post-adjuvant polyarthritis in rats and hemolytic anemia of New Zealand Black mice (Wieczorek et al., 1991).

About 54 different volatile compounds were detected in flaxseed oil (Krist et al., 2006). The main volatile compounds identified in cold-pressed flaxseeds included hexanol (herbaceous, green, woody, sweet, 6.5–20.3%), acetic acid (strong, sour, pungent, 3.7%), trans-2-pentenal (pungent, green, fruity, 0.9%), trans-2-hexenal (sweet, fruity, vegetable, 0.8%), 2-heptanone (fruity, cinnamon, 0.3%), 1-octen-3-ol (2 %), trans, trans-2,4-hexadienal (moldy, 0.7%).

## Metabolism of ALA

ALA is found in plants, animals, plankton, and marine species, but flax and perilla seed oils seem to be the richest sources (Morris, 2007). ALA is an essential fatty acid

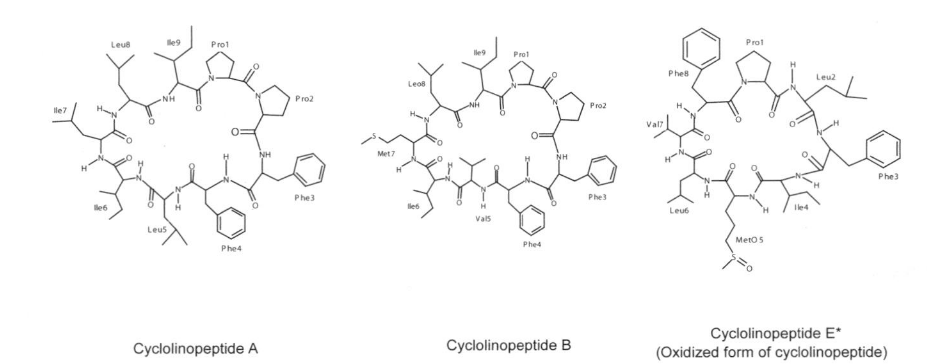


Cyclolinopeptide A

Cyclolinopeptide B

Cyclolinopeptide E*
(Oxidized form of cyclolinopeptide)

**Fig. 4.2.** Structures of selected cyclolinopeptides in flaxseed oil.

(EFA) that plays an important role in: growth and development, reproduction, vision, maintaining healthy skin, maintaining cell structure, the metabolism of cholesterol, and gene regulation. ALA, which is the most commonly consumed omega-3 fatty acid in the typical Western diet (Lanzmann-Petithory, 2001), is also linked to the prevention and/or amelioration of several chronic conditions, including cardiovascular disease (CVD), certain cancers, rheumatoid arthritis, and autoimmune disorders.

EFAs are required in the diet since humans can not synthesize them. The two established EFAs are the omega-6 fatty acid LA (C18:2n-6) and the omega-3 fatty acid ALA. LA and ALA are components of cellular membranes, and act to increase membrane fluidity. These fatty acids are necessary for cell-membrane function, as well as for the proper functioning of the brain and nervous system (Davis & Kris-Etherton, 2003; Harper & Jacobson, 2001). ALA is converted by a series of alternating desaturations and elongations (Fig. 4.3) to the long-chain omega-3 fatty acids, eicosapentaenoic acid (EPA), and docosahexaenoic acid (DHA), found in fish oil. Fish consumption and oil supplementation were studied extensively for their health benefits, and undoubtedly, EPA and DHA have very significant and well- supported positive effects in numerous chronic conditions and diseases. Similarly, LA is converted to long-chain omega-6 fatty acids, particularly arachidonic acid (AA), also by the same series of desaturations and elongations that metabolize ALA. EPA and AA serve as the starting points for the production of a number of important, very active, hormone-like compounds called "eicosanoids" with different activities (Fig. 4.4). The desaturation and elongation of LA and ALA, as well as the subsequent production of eicosanoids, occur competitively using the same group of enzymes. Thus, an excess of one family of fatty acids can interfere with the metabolism of the other, reducing its incorporation into tissue lipids and altering its biological effects (Harper & Jacobson, 2001). About 96% of dietary ALA appears to be absorbed in the gut (Burdge, 2006). After absorption, ALA has several metabolic fates, including desaturation and elongation to the longer-chain omega-3 fatty acids (Fig. 4.3) or shortening by β-oxidation.

Estimates of the amount of ALA converted to EPA range from 0.2% to 8% (Burdge et al., 2002; Burdge & Calder, 2005), with young women showing a conversion rate as high as 21% (Burdge & Wootton, 2002). The conversion of ALA to DPA is estimated to range from 0.13% to 6% (Burdge, 2006). The conversion rate for young women is on the higher end (6%) (Burdge & Wootton, 2002). The conversion of ALA to DHA appears to be limited in humans, with most studies showing a conversion rate of about 0.05% (Burdge, 2006; Pawlosky et al., 2001), although one study reported a value of 4% (Emken et al., 1994). A conversion rate of 9% was reported in young women (Burdge & Wootton, 2002). The large differences in the rates of ALA conversion may be due to major differences in study methodologies.

ALA conversion to its longer-chain metabolites is significantly affected by dietary intakes of LA. A diet rich in LA can reduce ALA conversion by as much as 40% (Emken, 1995), and a high intake of LA by pregnant women lowers EPA and DHA levels

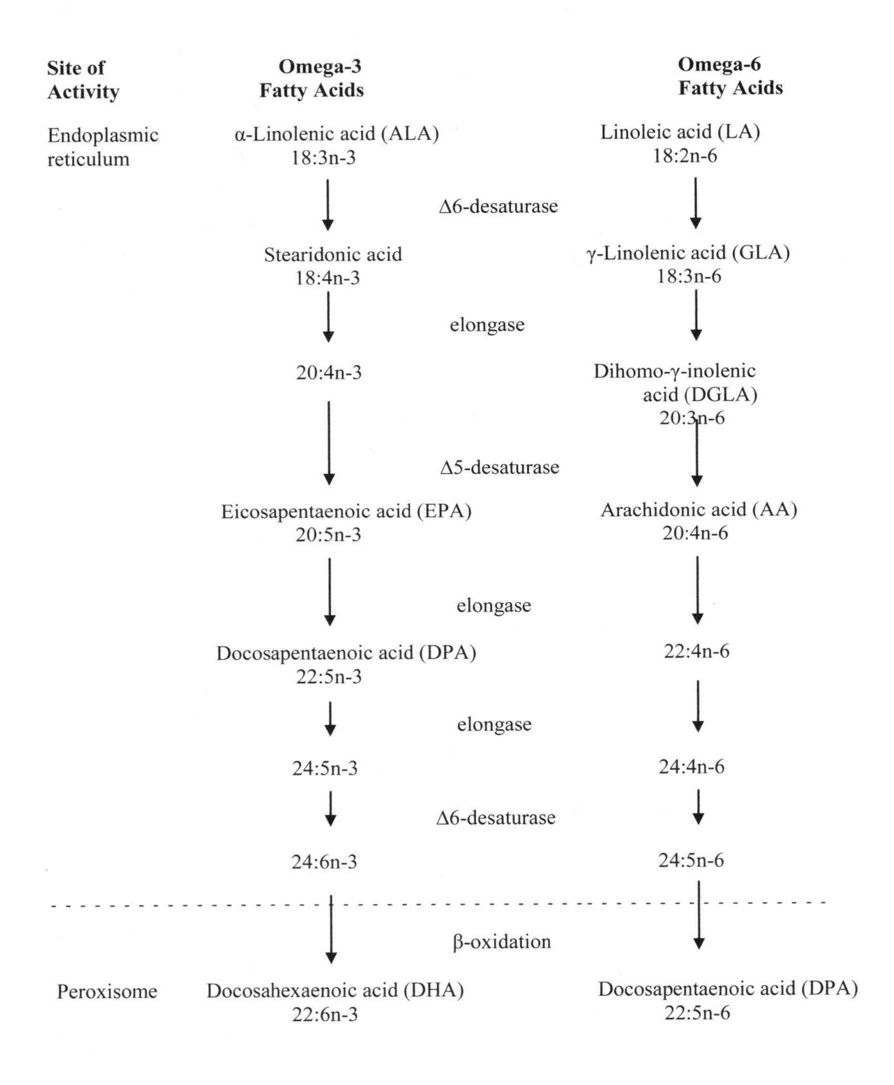


**Fig. 4.3.** Metabolic pathways of omega-3 and omega-6 fatty acids. The conversion pathway shown is the "Sprecher pathway," which is believed to be the major route (Burge, 2006).

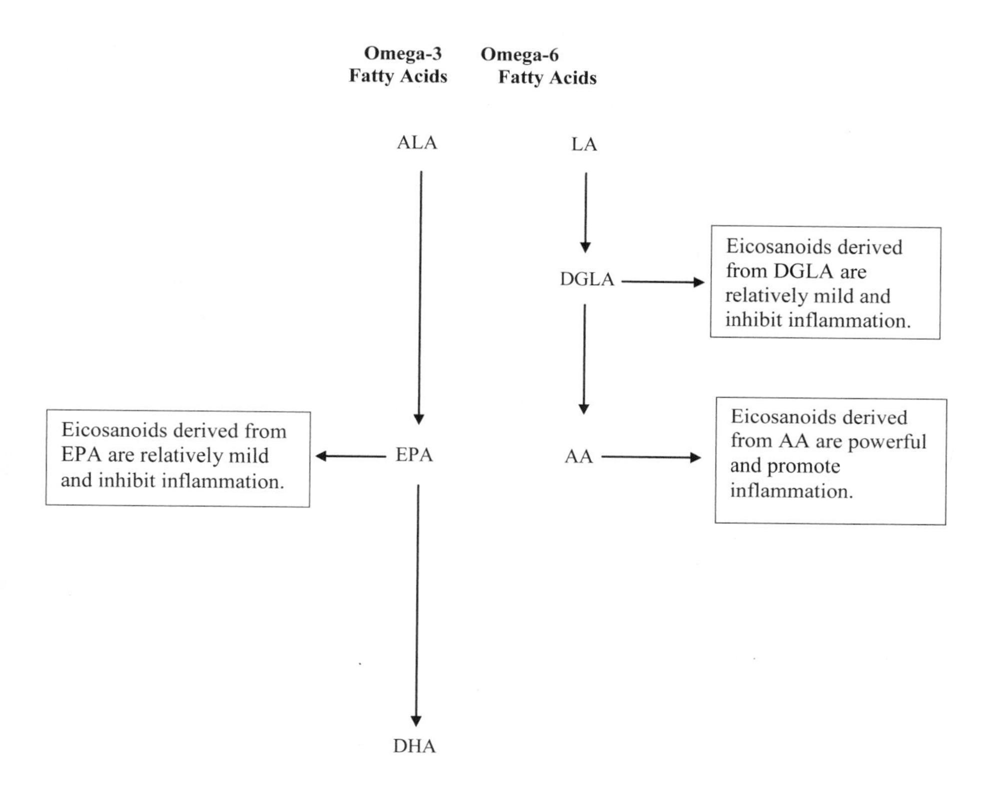


**Fig. 4.4.** Sources and actions of eicosanoids.

in umbilical plasma, suggesting a reduced ALA conversion and the availability of omega-3 fatty acids for the developing fetus (Al et al., 1996). In a study of 22 healthy men, an LA-rich diet (10.5% of energy) reduced the EPA content of plasma PLs significantly after four weeks compared with a low-LA diet (3.8% of energy), even though both diets contained the same amount of ALA (1.1% of energy) (Liou et al., 2007). The absolute amounts of ALA and LA in the diet also affect ALA conversion—decreasing the intake of LA increased the proportion of dietary ALA converted into EPA, while increasing the ALA intake increased the absolute amount of DHA synthesized (Goyens et al., 2006).

High intakes of EPA and DHA can also block ALA conversion, possibly by signaling that tissue levels of omega-3 fats are adequate. A diet containing more than 12 g of ALA each day can reduce ALA conversion (Cunnane et al., 1993). Other factors that influence the rates of ALA conversion include the intake of high levels of dietary cholesterol (Garg et al., 1988; Leikin & Brenner, 1987), saturated fat, oleic acid (Berger et al., 1992; Li et al., 1999), trans fatty acids (Houwelingen & Hornstra, 1994), and the ratio of polyunsaturated to saturated fats in the diet (Layne et al., 1996).

In men, 24–33% of an ingested dose of ALA undergoes β-oxidation (Bretillon et al., 2001; DeLany et al., 2000) compared to 19–22% in women (Burdge & Wootton, 2002; McCloy et al., 2004). The greater β-oxidation of ALA in men probably reflects a larger mass of active tissues such as muscles, heart, liver, and kidney compared with women. Furthermore, the figures may underestimate by ≈30% the actual amount of dietary ALA that undergoes β-oxidation due to the trapping of labeled CO2 in bicarbonate pools (Burdge, 2006). The amount of ingested ALA shunted into the β-oxidation pathway appears to be stable and is not affected by dietary intake. In a study of 14 healthy men ages 40–64 years, the proportion of ALA undergoing β-oxidation did not differ between the men who consumed a diet rich in ALA (10 g/day) versus those who consumed a diet rich in EPA + DHA (1.5 g/day) for 8 weeks (Burdge et al., 2003). Like other fatty acids, ALA can be stored in adipose tissue. In a typical 75-kg man with a fat mass of 15%, adipose tissue is calculated to contain 79 g of ALA. In a typical 65-kg woman with a fat mass of 23%, adipose tissue is calculated to contain 105 g of ALA (Burdge & Calder, 2005). The greater capacity for ALA storage by women probably reflects their greater fat mass compared with men.

## Health Effects of ALA

### ALA and the Prevention of CVD

Cardiovascular diseases (CVD) include all diseases of the blood vessels and circulatory system such as coronary heart disease (CHD), ischemic heart disease (IHD), myocardial infarction (MI), and stroke. CVD is the leading cause of death in the United States and Canada (Heart and Stroke Foundation of Canada, 2003; Rosamond, 2007). ALA affects cardiovascular health through its effects on blood lipid levels, blood pressure, endothelial function, and inflammation.

Most clinical studies reported no effect of flax-oil consumption on blood total cholesterol and low-density lipoprotein (LDL)-cholesterol levels (Clandinin et al., 1997; Goh et al., 1997; Kestin et al., 1990; Mantzioris et al., 1994; Nestel et al., 1997; Paschos et al., 2005; Rallidis et al., 2004; Sanders & Roshanai, 1983; Schwab et al., 2006; Singer et al., 1986, 1990). One study reported a significant decrease in blood total cholesterol among men who consumed 2 tablespoons of flax oil daily for 12 weeks (Wilkinson et al., 2005). High-density lipoprotein (HDL)-cholesterol decreased significantly between 4% and 10% in 4 of the 13 studies (Nestel et al., 1997; Paschos et al., 2005; Rallidis et al., 2004; Wilkinson et al., 2005). Triacylglycerols decreased significantly between 9% and 25% in 3 studies (Schwab et al., 2006; Singer et al., 1986, 1990).

Two population studies reported a benefit of ALA in reducing stroke risk. In the Edinburgh Artery Study, significantly lower levels of ALA were found in the red-blood-cell PLs of men and women who had experienced a stroke compared with participants who had no evidence of the disease (Leng et al., 1999). In the Multiple Risk Factor Intervention Trial (MRFIT), 96 men who had experienced a stroke were compared with 96 men without stroke who were matched for age. In the multivariate model, each increase of 0.13% in serum ALA level was associated with a 37% decrease in the risk of stroke (Simon et al., 1995). After controlling for risk factors of stroke such as smoking and blood pressure, ALA emerged as an independent predictor of stroke risk. In one clinical trial, supplementing the diet with flax oil (≈1 tablespoon providing 8 g of ALA each day) for 12 weeks lowered systolic and diastolic blood pressure significantly in middle-aged hypercholesterolemic men compared with a safflower-oil group. The magnitude of the effect (≈5 mmHg) was clinically relevant (Paschos et al., 2007).

Endothelial dysfunction is the earliest detectable stage in the development of atherosclerosis. An increase in systemic arterial compliance (SAC) denoting an improvement in endothelial function, combined with a reduction in mean arterial pressure, was reported among 15 obese adults who daily ate a diet enriched with flax oil (providing 20 g of ALA) for 4 weeks (Sanders & Roshanai, 1983). The increase in SAC with flax oil was similar to that achieved through exercise training. West and coworkers (West et al., 2005) measured endothelial function by the method of flow-mediated vasodilation (FMD) in 18 healthy adults with type 2 diabetes. FMD was measured before and 4 hours after 3 test meals, each providing 50 g of a specific type of fat—monounsaturated fatty acid (MUFA) obtained from high-oleic safflower and canola oils; the MUFA diet plus EPA and DHA from sardine oil; or the MUFA diet plus ALA from canola oil. In volunteers with high-fasting triacylglycerols, meals containing omega-3 fatty acids increased FMD by 50–80%. Sardine and plant omega-3 fats were equally effective in improving endothelial function as measured by FMD.

Endothelial dysfunction is also characterized by a tendency for leukocytes to adhere to the endothelium in a process controlled by cell-adhesion molecules, including E-selectin, vascular cell-adhesion molecule type 1 (VCAM-1), and intercellular-adhesion molecule type 1 (ICAM-1) (Hwang et al., 1997). A diet rich in ALA significantly decreased VCAM-1, ICAM-1, and E-selectin compared to an average American diet in 23 hypercholesterolemics (Zhao et al., 2004). Consuming 1 tablespoon of flax oil daily for 12 weeks reduced VCAM-1 levels by 18.7% in a group of male hypercholesterolemic subjects (Rallidis et al., 2004). These findings suggest that an ALA-rich diet containing flax oil has a beneficial effect on the endothelium.

An increasing amount of research suggests that the consumption of ALA may provide protection against heart disease and other inflammatory diseases by reducing inflammatory eicosanoids and cytokines. Pro-inflammatory eicosanoids such as thromboxane A2 ($TXA_2$) and leukotriene B4 ($LTB_4$) are derived from AA. $TXA_2$ is one of the most potent promoters of platelet aggregation known (Reiss & Edelman, 2006; Ross, 1999). LTB4 increases the release of reactive oxygen species and cytokines like tumor necrosis factor α (TNF-α), interleukin 1β (IL-1β), IL-6, and IL-8 (Calder, 2006). In a clinical study of healthy men, the consumption of 1¾ tablespoons of flax oil daily for four weeks led to a 30% reduction in the immune cells' concentration of TXB2, which is an inactive metabolite of TXA2 (Caughey et al., 1996). Concentrations of the pro-inflammatory cytokines TNF-α and IL-1β in immune cells decreased 26% and 28%, respectively. In 64 patients with chronic obstructive pulmonary disease (COPD), serum and sputum $LTB_4$ levels decreased 32% and 41%, respectively, in those patients who received an ALA-rich nutritional support (1.4% of ALA) daily for 24 months compared to those who received a low-ALA nutritional support (0.18% of ALA) (Matsuyama et al., 2005). Serum levels of IL-6 decreased 25% in men who consumed 1 tablespoon of flax oil daily for 12 weeks (Paschos et al., 2005). The serum levels of TNF-α decreased by 43%, and the production by immune cells of TNF-α, IL-6, and IL-1β decreased between 18% and 22% in hypercholesterolemics who consumed a diet rich in ALA compared with the average American diet (Zhao et al., 2007).

During inflammation, the liver releases acute-phase proteins such as C-reactive protein (CRP) and serum amyloid A (SAA) in response to acute injury, infection, malignancy, hypersensitivity reactions, and trauma. CRP and SAA are markers of systemic inflammation, and are present in the lesions of atherosclerosis. CRP is an independent risk factor for CVD (Getz, 2005). Consuming flax oil reduced CRP by 48% and serum SAA by 32% in 50 hypercholesterolemic men who consumed 1 tablespoon of flax oil daily for 12 weeks (Paschos et al., 2005). In a U.S. study of 23 adults with high blood-cholesterol levels, consuming a high-ALA diet based on walnuts, walnut oil, and flax oil resulted in a ≈75% decrease in CRP levels after 6 weeks (Zhao et al., 2004).

Four case-control studies (Baylin et al., 2003, 2007; Guallar et al., 1999; Lemaitre et al., 2003; Rastogi et al., 2004), one cross-sectional study (Manav et al., 2004), three prevention trials (de Lorgeril et al., 1994, 1999; Dolecek, 1992; Pietinen et al., 1997), and three cohort studies (Albert et al., 2005; Ascherio et al., 1996; Djoussé et al., 2001, 2003a,b, 2005a; Hu et al., 1999; Mozaffarian et al., 2005 ) found a benefit of ALA-rich diets in lowering the risk of CHD, IHD, nonfatal MI, and stroke. One prevention trial found no change in the estimated 10-year IHD risk, but reported a significant decrease in fibrinogen and CRP levels on ALA-rich diets (Bemelmans et al., 2002, 2004). The number of participants in these studies ranged from 233 to 76,283.

Modest intakes of ALA appear to have a significant effect on reducing nonfatal MI (Campos et al., 2008). A nonlinear inverse relationship between 0.7% of adipose tissue ALA and a dietary ALA intake of about 1.8 g/day (1/2 teaspoon of flax oil) and the risk of nonfatal MI was observed in a recent study of 1819 patients who survived an MI and 1817 matching controls. These results confirm several earlier studies in which the ALA content of adipose tissue was inversely related to the risk of MI in one case-control study conducted in Europe and Israel (Guallar et al., 1999) and inversely related to nonfatal acute MI in another study in Costa Rica (Baylin et al., 2003). In the Nurse's Health Study, which involved a 10-year follow-up of 76,283 women with no previously diagnosed CVD, a higher intake of ALA was associated with a lower relative risk of fatal and nonfatal MI (Albert et al., 2005; Hu et al., 1999). The most recent evaluation of food- balance sheets and CHD outcomes for eleven Eastern European countries revealed that populations that experienced the greatest increase in ALA consumption since 1990 also experienced a substantial decline in CHD mortality. These results were consistent in men and women (Zantonski et al., 2008).

In the Health Professional Follow-up Study, which began in 1986 with a cohort of 45,772 health professionals, the strongest evidence linking ALA intake and CHD risk reduction was observed in participants that consumed very little seafood. In men who consumed <100 mg of EPA + DHA each, 1 g/day of an ALA intake was associated with a 58% lower risk of nonfatal heart attack and a 47% lower risk of CHD. These data strongly support a direct role of ALA consumption in decreasing CHD risk, and further indicate that ALA may be particularly important in sectors of the population that do not eat fatty fish (Mozaffarian et al., 2005). The Lyon Diet Heart Study included participants who had previously survived an MI compared to an experimental group who consumed a typical Mediterranean-style diet rich in ALA. The control group consumed a typical Western-type diet low in ALA. The results were impressive: a 75% reduction in nonfatal MIs and a 70% reduction in total deaths noted among the ALA group compared to the control group (de Lorgeril et al., 1994, 1999).

One mechanism by which ALA may lower the risk of a fatal or a nonfatal MI appears to involve an effect on cardiac rhythm. In the Family Heart Study, Djoussé et al. (2005b) found that the higher the dietary ALA intake, the lower the risk of abnormally prolonged repolarization of the heart muscle, an indicator of cardiac arrhythmia. In a clinical study among women referred for elective coronary angiography, the ALA content of adipose tissue was positively correlated with a 24-hour heart-rate variability (Christensen et al., 2005). ALA can reduce ventricular fibrillation (Ander, 2004), and its cardioprotective effects were also attributed to improvements in arrhythmia and to reductions in platelet aggregation (Vos & Cunnane, 2003).

## ALA and the Prevention of Cancers

In contrast to the numerous studies focusing on the effect of ALA on cardiovascular risk, only a few studies reported on the relationship between ALA and breast cancer. Cohort and case-control studies showed negative associations between high dietary intakes of ALA and a reduced risk of breast cancer (Franceschi et al., 1996; Klein et al., 2000; Maillard et al., 2002; Voorrips et al., 2002). In a meta-analysis, Saadatian-Elahi et al. (2004) analyzed results from three cohorts and seven case-control studies, and reported a significant protective effect of n-3 fatty acids against breast cancer. Other studies showed low/no association between breast-cancer risk and subcutaneous ALA (London et al., 1993; Simonsen et al., 1998) and even an increased risk of breast cancer (De Stéfani et al., 1998). The epidemiological data on the association between ALA and prostate cancer are inconclusive (Bougnoux & Chajès, 2003). While Giovannacci et al. (1993) found an association between ALA intake and the incidence of prostate cancer, Leitzmann et al. (2004) reported a positive association between dietary ALA and the risk of advanced prostate cancer. Mannisto et al. (2003) suggested that factors other than dietary ALA, for example, smoking, need to be included in the models evaluating the ALA intake and prostate-cancer risk.

## The Dietary Omega-6 to Omega-3 Fatty-acid Ratio

The dietary n-6/n-3 ratio affects inflammation and gene expression, thus influencing the development of chronic disease. The n-6/n-3 ratio may be as high as 17:1 in some Western diets (Simpouolos, 2006), and is estimated to be 10:1 in the U.S. diet (Kris-Etherton et al., 2002). In the Women's Health Study, participants had an average dietary ratio of ≈8:1, although some women ate diets with a low ratio of about 1:1, while others ate diets with a ratio as high as 33:1 (Miljanovic et al., 2005). The n-6/n-3 ratio recommended by international agencies and some European countries ranges from 4:1 to 10:1 (Gebauer et al., 2006). The Institute of Medicine supports a ratio of 5:1 for the U.S. and Canadian populations (Institute of Medicine, 2002). ALA constitutes about 57% of the total fatty acids in flax, whereas the omega-6 fatty acids constitute about 16%. Thus, flax contains more than three times as many omega-3 as omega-6 fatty acids, giving an n-6/n-3 ratio of 0.3:1 (Morris, 2007), compared to the n-6/n-3 ratio of 58:1 in corn oil, 7:1 in soybean oil, and 2:1 in canola/rapeseed oil.

A dietary imbalance of omega-6 and omega-3 fats leads to a high ratio of omega-6 to omega-3 fatty acids in cell membranes (Harris et al., 2006). An imbalance in the n-6/n-3 ratio in tissues and blood can have adverse effects, including the overproduction of pro-inflammatory eicosanoids, many of which are derived from AA, which, in turn, stimulate the release of inflammatory cytokines and acute-phase proteins. The end result is low-grade chronic inflammation that contributes to health problems such as atherosclerosis, Alzheimer's disease, cancer, CVD, metabolic syndrome, obesity, osteoporosis, type 2 diabetes, and periodontitis (Calder, 2006; Cordain et al., 2005; Hotamisligil, 2006; Kornman, 2006). Table 4.1 outlines some consequences of eating a diet rich in omega-6 fats versus the benefits of eating a diet rich in omega-3 fats.

ALA appears to protect against CVDs by altering the omega-3 fat content of cell membranes (Harper et al., 2006), by improving blood lipids and endothelial function, and by exerting significant anti-inflammatory and anti-thrombotic effects (Bloedon & Szapary, 2004). These beneficial effects were reported with ranges of ≈3–20 g of ALA each day (equivalent to 1/2 teaspoon to 2½ tablespoons of flax oil). In epidemiologic studies, ALA intakes associated with reduced CVD risk averaged about 2 g/day (range = 0.7–6.3 g). A meta-analysis of prospective studies suggested

**Table 4.1. Comparison of Health Consequences of Diets Rich in Omega-6 Versus Omega-3 Fats**

| Consequences of eating a diet rich in omega-6 fats | Benefits of eating a diet rich in omega-3 fats |
|---|---|
| ↑ n-6/n-3 in cell-membrane phospholipids | ↓ n-6 fatty acids in cell membranes |
| ↑ production of arachidonic acid | ↓ n-6/n-3 in cell-membrane phospholipids |
| ↑ release of pro-inflammatory eicosanoids derived from arachidonic acid | ↓ levels of pro-inflammatory compounds like eicosanoids and cytokines |
| ↑ production of pro-inflammatory cytokines | ↓ clumping (aggregation) of blood platelets |
| ↑ expression (activation) of pro-inflammatory genes | ↓ expression (activation) of pro-inflammatory genes |
| ↑ biomarkers of inflammation such as C-reactive protein | ↓ biomarkers of inflammation such as C-reactive protein |
| ↑ blood viscosity | ↑ production of interleukin-10, an anti-inflammatory cytokine |
| ↑ constriction of blood vessels | |
| ↑ oxidative modification of low-density-lipoprotein (LDL) cholesterol | |
| ↓ | ↓ |
| Increased risk of chronic diseases | Decreased risk of chronic diseases |

Sources: Gebauer et al. (2006); Simopoulos (2006).

that increasing the intake of ALA by 1.2 g/day decreases the risk of fatal CHD by at least 20% (Brouwer et al., 2004). The Institute of Medicine has set an Adequate Intake for ALA, based on the median daily intake of healthy Americans who are not likely to be deficient in this nutrient as shown in Table 4.2 (Institute of Medicine, 2002). The Adequate Intake is 1.6 g of ALA each day for men and 1.1 g of ALA each day for women. EPA and DHA can provide up to 10% of the Adequate Intake for ALA.

New research suggests an effect of omega-3 fatty acid on gene expression (Deckelbaum & Worgall, 2006; Low et al., 2005). More studies are needed to clarify the role of ALA in reducing the CVD risk. In particular, the need is urgent for randomized, controlled clinical trials with good study designs, clearly defined outcomes, appropriate control groups, realistic dietary interventions, and thorough statistical analyses (Stark et al., 2008).

**Table 4.2. Estimated Dietary Intakes of Major Omega-6 and Omega-3 Fatty Acids in Canada and the United States (g/day)[a]**

| Country | Omega-6 fatty- acid intake (Linoleic acid) | Omega-3 fatty-acid intake | |
|---|---|---|---|
| | | ALA | Long-chain fatty acids (EPA, DHA, and/or DPA) |
| Europe[b] | 15 | 1-2 | 0.1-0.5 |
| Canada[c] | 8-11 | 1.3-1.6 | 0.14-0.24 |
| United States[d] | | | |
| Men | 18 | 1.7 | 0.13 |
| Women | 14 | 1.3 | 0.10 |

[a]Abbreviations: ALA, α-linolenic acid; DPA, docosapentaenoic acid; DHA, docosahexaenoic acid; EPA, eicosapentaenoic acid.
[b]Values for the general European population (Sanders, 2000).
[c]Values are for pregnant women living in British Columbia and Ontario (Denomme et al., 2005; Innis & Elias, 2003).
[d]Values are for men and women ages 20-39 years (Gebauer et al., 2006).

## Allergic and Toxic Compounds

Case reports of IgE-mediated anaphylaxis to flaxseed were reported (Alonso et al., 1996; Lezaun et al., 1998), and the allergic protein was described as a 28-kDa dimeric protein (León et al., 2002). Some individuals were reported to be allergic to perilla seed (Jeong et al., 2006). Cold-pressed oils might contain traces of the allergic proteins.

## References

Abramovic, H.; V. Abram. Physico-chemical properties, composition and oxidative stability of camelina sativa oil. *Food Technol. Biotechnol.* **2005**, *43*, 63–70.

Abuzaytoun, R.; F. Shahidi. Oxidative stability of flax and hemp seed oils. J. Am. Oil Chem. Soc. **2006**, 83, 855–861.

Adhikari, P.; K.T. Hwang; J.N. Park; C.K. Kim. Policosanol content and composition in perilla seeds. *J. Agric. Food Chem.* **2006**, *54*, 5359–5362.

Adlercreutz, H.; Y. Mousavi; J. Clark; K. Hockerstedt; E. Hämälainen; K. Wähälä; T. Mäkelä; T. Hase. Dietary phytoestrogens and cancer: *in vitro* and *in vivo* studies. *J. Steroid Biochem. Mol. Biol.* **1992**, *41*, 331–337.

Al, M.D.M.; A. Badart-Smook; A. Houwelingen; T. Hasaart; G. Hornstra. Fat intake of women during normal pregnancy: Relationship with maternal and neonatal essential fatty acid status. *J. Am. Coll. Nutr.* **1996**, *15*, 49–55.

Albert, C.M.; K. Oh; W. Whang; J.E. Manson; C.E. Chau; M. Stampfer; W. Willett; F.B. Hu. Dietary alpha-linolenic acid intake & risk of sudden cardiac death & coronary heart disease. *Circulation* **2005**, *112*, 3232–3238.

Alonso, L.; M. Marcos; J. Blanco; J. Navarro; S. Juste; M. del Mar Garcés; R. Perez; P. Carretero. Anaphylaxis caused by linseed (flaxseed) intake. *J. Allergy Clin. Immunol.* **1996**, *98*, 469–470.

Ander, B.P.; A.R. Weber; P.P. Rampersad; J. Gilchrist; G. Pierce; A. Lukas. Dietary flaxseed protects against ventricular fibrillation induced by ischemia-reperfusion in normal and hypercholesterolemic rabbits. *J. Nutr.* **2004**, *134*, 3250–3256.

Ascherio, A.; E.B. Rimm; E.L. Giovannucci; D. Spiegelman; M. Stampfer; W. Willett. Dietary fat and risk of coronary heart disease in men: Cohort follow-up study in the United States. *Br. Med. J.* **1996**, *313*, 84–90.

Baylin, A.; E.K. Kabagambe; A. Ascherio; D. Spiegelman; H. Campos. Adipose tissue α-linolenic acid and nonfatal acute myocardial infarction in Costa Rica. *Circulation* **2003**, *107*, 1586–1591.

Baylin, A.; E. Ruiz-Narvaez; P. Kraft; H. Campos. α-Linoleic acid, $\Delta^6$-desaturase gene polymorphism, and the risk of nonfatal myocardial infarction. *Am. J. Clin. Nutr.* **2007**, *85*, 554–560.

Bemelmans, W.J.E.; J. Broer; E.J.M. Feskens; A. Smit; F. Muskiet; J. Lefrandt; V. Bom; J. May; B. Meyboom-de Jong. Effect of an increased intake of α-linolenic acid and group nutritional education on cardiovascular risk factors: the Mediterranean Alpha-linolenic Enriched Groningen Dietary Intervention (MARGARIN) study. *Am. J. Clin. Nutr.* **2002**, *75*, 221–227.

Bemelmans, W.J.E.; J.D. Lefrandt; E.M.J. Feskens; P.L. van Haelst; J. Broer; B. Meyboom-de Jong; J.F. May; T. Cohen; A. Smit. Increased alpha-linolenic acid intake lowers C-reactive protein, but has no effect on markers of atherosclerosis. *Eur. J. Clin. Nutr.* **2004**, *58*, 1083–1089.

Berger, A.; M.E. Gershwin; J.B. German. Effects of various dietary fats on cardiolipin acyl composition during ontogeny of mice. *Lipids* **1992**, *27*, 605–612.

Berglund, D. Flax: new uses and demands. *Trends in New Crops and New Uses*; J. Janick, A. Whipkey, Eds.; ASHS Press: Alexandria, VA, 2002; pp. 258–360.

Bhatty, R.S. Further compositional analyses of flax: mucilage, trypsin inhibitors and hydrocyanic acid. *J. Am. Oil Chem. Soc.* **1993**, *70*, 899–904.

Bloedon, L.T.; P.O. Szapary. Flaxseed and cardiovascular risk. *Nutr. Rev.* **2004**, *62*, 18–27.

Bougnoux, P.; V. Chajes. α-Linolenic acid and heart disease. *Flaxseed in Human Nutrition,* 2nd ed.; S.C. Cunnane, L.U. Thompson, Eds.; AOCS Press: Champaign, IL, **2003,** pp. 232–241.

Bozan, B.; F. Temelli. Supercritical extraction of flaxseed oil. *J. Am. Oil Chem. Soc.* **2002**, *79*, 231–235.

Bretillon, L.; J.M. Chardigny; J.L. Sébédio; J. Noel; C. Scrimgeour; C. Fernie; O. Loreau; P. Gachon; B. Beaufrere. Isomerization increases the postprandial oxidation of linoleic acid but not α-linolenic acid in men. *J. Lipid Res.* **2001**, *42*, 995–997.

Brouwer, I.A.; M.B. Katan; P.L. Zock. Dietary α-linolenic acid is associated with reduced risk of fatal coronary heart disease, but increased prostate cancer risk: a meta-analysis. *J. Nutr.* **2004**, *134*, 919–922.

Brühl, L.; B. Matthäus; E. Fehling; B. Wiege; B. Lehmann; H. Luftmann; K. Bergander; K. Quiroga; A. Scheipers; O. Frank; T. Hofmann. Identification of bitter off-taste compounds in the stored cold pressed linseed oil. *J. Agric. Food Chem.* **2007**, *55*, 7864–7868.

Brühl, L.; B. Matthäus; A. Scheipers; T. Hofmann. Bitter off-taste in stored cold-pressed linseed oil obtained from different varieties. *Eur. J. Lipid Sci. Technol.* **2008**, *110*, 625–631.

Budin, J.T.; W.M. Breene; D.H. Putnam. Some compositional properties of camelina (*Camelina sativa* L. Crantz) seeds and oils. *J. Am. Oil Chem. Soc.* **1995**, *72*, 309–315.

Burdge, G.C. Metabolism of α-linolenic acid in humans. *Prostaglandins, Leukotrienes Essent. Fatty Acids* **2006**, *75*, 161–168.

Burdge, G.C.; P.C. Calder. Conversion of α-linolenic acid to longer-chain polyunsaturated fatty acids in human adults. *Reprod. Nutr. Dev.* **2005**, *45*, 581–597.

Burdge, G.C.; Y.E. Finnegan; A.M. Minihane; C.M. Williams; S.A. Wooton. Effect of altered dietary n-3 fatty acid intake upon plasma lipid fatty acid composition, conversion of [$^{13}$C] α-linolenic acid to longer-chain fatty acids and partitioning towards β-oxidation in older men. *Br. J. Nutr.* **2003**, *90*, 311–321.

Burdge, G.C.; A.E. Jones; S.A. Wootton. Eicosapentaenoic and docosapentaenoic acids are the principal products of α-linolenic acid metabolism in young men. *Br. J. Nutr.* **2002**, *88*, 355–363.

Burdge, G.C.; S.A. Wootton. Conversion of α-linolenic to eicosapentaenoic, docosapentaenoic and docosahexaenoic acids in young women. *Br. J. Nutr.* **2002**, *88*, 411–420.

Calder, P.C. n-3 Polyunsaturated fatty acids, inflammation, and inflammatory diseases. *Am. J. Clin. Nutr.* **2006**, *83(Suppl.)*, 1505S–1519S.

Campos, H.; A. Baylin; W.C. Willett. α-Linolenic acid and risk of nonfatal acute myocardial infarction, *Circulation* **2008**, *118*, 339–345.

Capella P. Cyclo-arthenol: a miner constituent of linseed oil. *Nature* **1961**, *190*, 167–168.

Caughey, G.E.; E. Mantzioris; R.A. Gibson; L.G. Cleland; M.J. James. The effect on human tumor necrosis factor α and interleukin 1β production of diets enriched in n-3 fatty acids from vegetable oil or fish oil. *Am. J. Clin. Nutr.* **1996**, *63*, 116–122.

Choo, W.; J. Birch; J.P. Dufour. Physicochemical and quality characteristics of cold-pressed flaxseed oils. *J. Food Comp. Anal.* **2007**, *20*, 202–211.

Christensen, J.H.; E.B. Schmidt; D. Mølenberg; E. Toft. Alpha-linolenic acid and heart rate variability in women examined for coronary artery disease. *Nutr. Metab. Cardiovasc. Dis.* **2005**, *15*, 345–351.

Clandinin, M.T.; A. Foxwell; Y.K. Goh; K. Layne; J. Jumpsen. Omega-3 fatty acid intake results in a relationship between the fatty acid composition of LDL cholesterol ester and LDL cholesterol content in humans. *Biochim. Biophys. Acta* **1997**, *1346*, 247–252.

Cordain, L.; S.B. Eaton; A. Sebastian; N. Mann; S. Lindeberg; B. Watkins; J. O'Keefe; J. Brand-Miller. Origins and evolution of the Western diet: health implications for the 21$^{st}$ century. *Am. J. Clin. Nutr.* **2005**, *81*, 341–354.

Cui, W.; G. Mazza; C. Biliaderis. Chemical structure, molecular size distributions and rheological properties of flaxseed gum. *J. Agric. Food Chem.* **1994**, *42*, 1891–1895.

Cunnane, S.C.; S. Ganguli; C. Menard; A.C. Liede; M.J. Hamadeh; Z.Y. Chen; T. Wolever; D. Jenkins. High α-linolenic acid flaxseed (Linum usitatissimum): Some nutritional properties in humans. *Br. J. Nutr.* **1993**, *69*, 443–453.

Daun, J.K. Oilseeds processing. *Grain Processing and Technology;* Canadian International Grains Institute: Manitoba, Canada, 1993; Vol, 2, pp. 883–936.

Daun, J.K.; V.J. Barthet; T.L. Chornick; S. Duguid. Structure, composition and variety development of flaxseed. *Flaxseed in Human Nutrition,* 2nd ed.; L.U. Thompson, S.C. Cunnane, Eds.; AOCS Press: Champaign, IL, 2003; pp. 1–40.

Davis, B.C.; P.M. Kris-Etherton. Achieving optimal essential fatty acid status in vegetarians: current knowledge and practical implications. *Am. J. Clin. Nutr.* **2003**, *78(Suppl.),* 640S–660S.

Deckelbaum, R.J.; T.S. Worgall; T. Seo. n-3 Fatty acids and gene expression. *Am. J. Clin. Nutr.* **2006**, *83*, 1520S–1525S.

DeLany, J.P.; M.M. Windhauser; C.M. Champagne; G.A. Bray. Differential oxidation of individual dietary fatty acids in humans. *Am. J. Clin. Nutr.* **2000**, *72*, 905–911.

De Lorgeril, M.; S. Renaud; N. Mamelle; P. Salen; J. Martin; I. Monjaud; J. Guidollet; P. Touboul; J. Delaye. Mediterranean alpha-linolenic acid-rich diet in secondary prevention of coronary heart disease. *Lancet* **1994**, *343,* 1454–1459.

De Lorgeril, M.; P. Salen; J-L., Martin; I. Monjaud; J. Delaye; N. Mamelle. Mediterranean diet, traditional risk factors, and the rate of cardiovascular complications after myocardial infarction: Final report of the Lyon Diet Heart Study. *Circulation* **1999**, *99*, 779–785.

De Stefani, E.H.; M. Deneo-Pellegrini; M. Mendilaharsu; A. Ronco. Essential fatty acids and breast cancer: a case-control study in Uruguay. *Int. J. Cancer* **1998**, *76*, 491–494.

Denomme, J.; K.D. Stark; B.J. Holub. Directly quantitated dietary (n-3) fatty acid intakes of pregnant Canadian women are lower than current dietary recommendations. *J. Nutr.* **2005**, *135*, 206–211.

Dessi, M.A.; M. Deiana; B. Dat; A. Rosa; S. Banni; F. Corongiu. Oxidative stability of polyunsaturated fatty acids: effect of squalene. *Eur. J. Lipid. Sci. Technol.* **2002**, *104*, 506–512.

Djoussé, L.; D.K. Arnett; J. Carr; J.H. Eckfeldt; P.N. Hopkins; M.A. Province; R.C. Ellison. Dietary linolenic acid is inversely associated with calcified atherosclerotic plaque in the coronary arteries: the National Heart, Lung, and Blood Institute Family Heart Study. *Circulation* **2005a**, *111*, 2921–2926.

Djoussé, L.; A.R. Folsom; M.A. Province; S.C. Hunt; R.C. Ellison. Dietary linolenic acid and carotid artherosclerosis: the National Heart, Lung, and Blood Institute Family Heart Study. *Am. J. Clin. Nutr.* **2003a**, *77*, 819–825.

Djoussé, L.; S.C. Hunt; D.K. Arnett; M.A. Province; J.H. Eckfeldt; R.C. Ellison. Dietary linolenic acid is inversely associated with plasma triacylglycerol: the National Heart, Lung, and Blood Institute Family Heart Study. *Am. J. Clin. Nutr.* **2003b**, *78*, 1098–1102.

Djoussé, L.; J.S. Pankow; J.H. Eckfeldt; A.R. Folsom; P.N. Hopkins; M.A. Province; Y. Hong; R.C. Ellison. Relationship between dietary linolenic acid and coronary artery disease in the National Heart, Lung, and Blood Institute Family Heart Study. *Am. J. Clin. Nutr.* **2001**, *74*, 612–619.

Djoussé, L.; P.M. Rautaharju; P.N. Hopkins; E.A. Whitsel; D.K. Arnett; J.H. Eckfeldt; M.A. Province; R.C. Ellison. Dietary linolenic acid and adjusted QT and JT intervals in the National Heart, Lung, and Blood Institute Family Heart Study. *J. Am. Coll. Cardiol.* **2005b**, *45*, 1716–1722.

Dolecek, T.A. Epidemiological evidence of relationships between dietary polyunsaturated fatty acids and mortality in the Multiple Risk Factor Intervention Trial. *Pro. Soc. Exp. Biol. Med.* **1992**, *200*, 177–182.

Eidhin, D.N.; J. Burke; D. O'Breirne. Oxidative stability of ω3-rich camelina oil and camelina oil-based spread compared with plant and fish oils and sunflower spread, *J. Food Sci.* **2003**, *68*, 345–353.

El-Shattory, Y. Chromatographic column fractionation and fatty acid composition of different lipid classes of linseed oil. *Nahrung* **1976**, *20*, 307–311.

Emken, EA. Influence of linoleic acid on conversion of linolenic acid to omega-3 fatty acids in humans. *Proceedings from the Scientific Conference on Omega-3 Fatty Acids in Nutrition, Vascular Biology, and Medicine;* American Heart Association: Dallas, TX, 1995; pp. 9–18.

Emken, E.A.; R.O. Adlof; R.M. Gulley. Dietary linoleic acid influences desaturation and acylation of deuterium-labeled linoleic and linolenic acids in young adult males. *Biochim. Biophys. Acta* **1994**, *1213*, 277–288.

Fedeli, E.; P. Capella; M. Cirimele; G. Jacini. Isolation of geranyl geraniol from the unsaponifiable fraction of linseed oil. *J. Lipid Res.* **1966**, *7*, 437–441.

Franceschi, S.; A. Favero; A. Decarli; E. Negri, C. La Vecchia; M. Ferraroni; A. Russo; S. Salvini; D. Amadori; E. Conti; M. Montella; A. Giacosa. Intake of macronutrients and risk of breast cancer. *Lancet* **1996,** *347*, 1351–1356.

Froment, M.A.; J. Smith; K. Freeman. Influence of environmental and agronomic factors contributing to increased levels of phospholipids in oil from UK linseed *Linum usitatissimum. Ind. Crops Prod.* **1999**, *10,* 201–207.

Froment, M.A.; J.M. Smith; D. Turley; E.J. Booth; S.P.J. Kightley. Fatty acid profiles in the seed oil of linseed and fibre flax cultivars (*Linum usitatissimum*) grown in England and Scotland. *Tests Agrochem. Cult.* **1998**, *19*, 60–61.

Garg, M.L.; A.A. Wierzbicki; A.B.R. Thomson; M.T. Clandinin. Dietary cholesterol and/or n-3 fatty acid modulate delta 9–desaturase activity in rat liver microsomes. *Biochim. Biophys. Acta* **1988**, *962*, 330–336.

Gebauer, S.K.; T.L. Psota; W.S. Harris; P.M. Kris-Etherton. n-3 Fatty acid dietary recommendations and food sources to achieve essentiality and cardiovascular benefits. *Am. J. Clin. Nutr.* **2006**, *83*, 1526S–1535S.

Getz, G.S. Immune function in atherogenesis. *J. Lipid Res.* **2005**, *46*, 1–10.

Giovannacci, E.; E.B. Rimm; G.A. Colditz; M.J. Stampfer; A. Ascherio; C.C. Chute; W.C. Willett. A prospective study of dietary fat and risk of prostate cancer. *J. Natl. Cancer Inst.* **1993**, *85*, 1571–1579.

Goh, Y.K.; J.A. Jumpsen; E.A. Ryan; M.T. Clandinin. Effect of ω-3 fatty acid on plasma lipids, cholesterol and lipoprotein fatty acid content in NIDDM patients. *Diabetologia* **1997**, *40*,

45–52.

Gourlay, J.S. Cis-trans isomerization of linseed oil. *Research* **1951**, *4*, 40–42.

Goyens, P.L.L.; M.E. Spilker; P.L. Zock; M. Katan; R. Mensink. Conversion of α-linolenic acid in humans is influenced by the absolute amounts of α-linolenic acid and linoleic acid in the diet and not by their ratio. *Am. J. Clin. Nutr.* **2006**, *84*, 44–53.

Gruszka, J.; J. Kruk. RP-LC for determination of plastochromanol, tocotrienols and tocopherols in plant oils. *Chromatographia* **2007**, *66*, 909–913.

Guallar, E.; A. Aro; F.J. Jiménez; J.M. Martin-Moreno; I. Salminen; P. vant Veer; A.F.M. Kardinaal; J. Gomez-Aracena; B. Martin; L. Kohlmeier; et al. Omega-3 fatty acids in adipose tissue and risk of myocardial infarction. The EURAMIC Study. *Arterioscler. Thromb. Vasc. Biol.* **1999**, *19*, 1111–1118.

Hadley, M. Stability of flaxseed oil used in cooking/stir frying. *Proceedings of the 56th Flax Institute of the United States*; Fargo, ND, 1996; pp. 55–61.

Hall, C. III; J. Schwarz. Functionality of flaxseeds in frozen desserts-preliminary report. *Proceedings of the 59th Flax Institute of the United States*; Fargo, ND, 2002; pp. 21–24.

Hall, C. III; K. Shultz. Phenolic antioxidant interactions. *Abstracts of the 92nd American Oil Chemists' Society Annual Meeting and Expo*; AOCS Press: Champaign IL, 2001; pp. S88.

Hamilton, R.J. The chemistry of rancidity in foods. *Rancidity in Foods*; J.C. Allen, R. J. Hamilton, Eds.; Blackie Academic and Professional: London, UK, 1994; pp. 1–21.

Harper, C.R.; M.J. Edwards; A.P. DeFilipis; T.A. Jacobson. Flaxseed oil increases the plasma concentrations of cardioprotective (n-3) fatty acids in humans. *J. Nutr.* **2006**, *136*, 83–87.

Harper, C.R.; T.A. Jacobson. The fats of life. *Arch. Intern. Med.* **2001**, *161*, 2185–2192.

Harris, W.S.; B. Assaad; W.C. Poston. Tissue omega-6/omega-3 fatty acid ratio and risk for coronary artery disease. *Am. J. Cardiol.* **2006**, *98(Suppl.)*, 19i–26i.

Heart and Stroke Foundation of Canada. The growing burden of heart disease and stroke in Canada; 2003. http://www.heartandstroke.ca. 2003. Accessed July 17, 2008.

Hillman, G. The plant remains from Tell Abu Hureyra: a preliminary report. *Proc. Prehist. Soc.* **1975**, *41*, 70–73.

Hillman, G.C.; S.M. Colledge; D.R. Harris. Plant-food economy during the Epipaleolithic period at Tell Abu Hureyra, Syria: dietary diversity, seasonality, and modes of exploitation. Foraging and Farming: the Evolution of Plant Exploitation; D.R. Harris, G.H. Hillman, Eds.; Unwin & Hyman: London, 1989; pp. 240–268.

Holcapek, M.; P. Jandera; P. Zderadicka; L. Hruba. Characterization of triacylglycerol and diacylglycerol composition of plant oils using high-performance liquid chromatography-atmospheric pressure chemical ionization mass spectrometry. *J. Chromatogr.* **2003**, *1010*, 195–215.

Hopf, M. Jericho plant remains. *Excavations at Jericho Vol. 5;* K.M. Kenyon, T.A. Holland, Eds.; British School of Archaeology: Jerusalem, London, 1983; pp. 576–621.

Hotamisligil, G.S. Inflammation and metabolic disorders. *Nature* **2006**, *444*, 860–867.

Houwelingen, A.C.; G. Hornstra. Trans fatty acids in early human development. *World Rev. Nutr. Diet.* **1994**, *75*, 175–178.

Hu, F.B.; M.J. Stampfer; J.E. Manson; E.B. Rimm; A. Wolk; G.A. Colditz; C.H. Hennekens; W.A. Willett. Dietary intake of α-linolenic acid and risk of fatal ischemic heart disease among women. *Am. J. Clin. Nutr.* **1999**, *69*, 890–897.

Hutchins, A.M.; J.L. Slavin. Effects of flaxseed on sex hormone metabolism. *Flaxseed in Human Nutrition,* 2nd ed.; L.U. Thompson, S.C. Cunnane, Eds.; AOCS Press: Champaign, IL, 2003; pp. 126–149.

Hwang, S.J.; C.M. Ballantyne; A.R. Sharrett; L. Smith; C. Davis; A. Gotto, Jr.; E. Boerwinkle. Circulating adhesion molecules VCAM-1, ICAM-1, and E-selectin in carotid atherosclerosis and incident coronary heart disease cases: the Atherosclerosis Risk in Communities (ARIC) study. *Circulation* **1997**, *96*, 4219–4225.

Innis, S.M.; S.L. Elias. Intakes of essential n-6 and n-3 polyunsaturated fatty acids among pregnant Canadian women. *Am. J. Clin. Nutr.* **2003,** *77*, 473–478.

Institute of Medicine. *Dietary Reference Intakes for Energy, Carbohydrate, Fiber, Fat, Fatty Acids, Cholesterol, Protein, and Amino Acids;* National Academies Press: Washington, DC, 2002; pp. 7-1— 7-69 (dietary fiber), 8-1— 8-97 *(fat and fatty acids).*

Jacini, G. Furanoid fatty acids and lipids. Fette Seifen Anstrich. **1986**, *88*, 290–292.

Jeong, Y.Y.; H.S. Park; J.H. Choi; S.H. Kim; K.U. Min. Two cases of anaphylaxis caused by perilla seed. *J. Allergy Clin. Immunol.* **2006**, *117*, 1505–1506.

Kamal-Eldin, A. Effect of fatty acid composition and tocopherol content on the oxidative stability of vegetable oils. *Eur. J. Lipid Sci. Technol.* **2006**, *58,* 1051–1061.

Kamal-Eldin, A.; R. Andersson. A multivariate study of the correlation between tocopherol content and fatty acid composition in different vegetable oils. *J.Am. Oil Chem. Soc.* **1997**, *74*, 375–380.

Kamal-Eldin, A.; N. Peerlkamp; P. Johnsson; R. Andersson; R.E. Andersson; L. Lundgren; P. Åman. An oligomer from flaxseed composed of secoisolariciresinoldiglucoside and 3-hydroxy-3-methyl glutaric acid residues. *Phytochemistry* **2001**, *58*, 587–590.

Kamal-Eldin, A.; N.V. Yanishlieva. n-3 Fatty acids for human nutrition: stability considerations. *Eur. J. Lipid Sci. Technol.* **2002**, *104*, 825–836.

Kamm, W.; F. Dionisi; C. Hischenhuber; K.H. Engel. Authenticity assessment of fats and oils. *Food Rev. Int.* **2001**, *17*, 249–290.

Kestin, M.; P. Clifton; G.B. Belling; P.J. Nestel. n-3 Fatty acids of marine origin lower systolic blood pressure and triglycerides but raise LDL cholesterol compared with n-3 and n-6 fatty acids from plants. *Am. J. Clin. Nutr.* **1990**, *51*, 1028–1034.

Khotpal, R.R.; A.S. Kulkarni; H.A. Bhakare. Studies on lipids on some varieties of linseed (*Linum usitatisimum*) of Ridarbha region. *Indian J. Pharm. Sci.* **1997**, *59*, 157–158.

Klein, V.; V. Chajès; E. Germain; G. Schulgen; M. Pinault; D. Malvy; T. Lefrancq; A. Fignon; O. Le Floch; C. Lhuillery; P. Bougnoux. Low alpha-linolenic acid content of adipose breast tissue is associated with an increased risk of breast cancer. *Eur. J. Cancer* **2000,** *36*, 335–340.

Kornman, K.S. Interleukin 1 genetics, inflammatory mechanisms, and nutrigenetic opportunities to modulate diseases of aging. *Am. J. Clin. Nutr.* **2006**, *83*, 475S–483S.

Kris-Etherton, P.M.; W.S. Harris; L.J. Appel for the Nutrition Committee. AHA Scientific

Statement–Fish consumption, fish oil, omega-3 fatty acids, and cardiovascular disease. *Circulation* **2002**, *106*, 2747–2757.

Krishnamurthy, R.G.; T.H. Smouse; B.D. Mookherjee; B.R. Reddy; S.S. Chang. Identification of 2-pentyl furan in fats and oils and its relationship to the reversion flavor of soybean oil. *J. Food Sci.* **1967**, *32*, 372–374.

Krist, S.; G. Stuebiger; S. Bail; H. Unterweger. Analysis of volatile compounds and triacylglycerol composition of fatty seed oil gained from flax and false flax. *Eur. J. Lipid Sci. Technol.* **2006**, *108*, 48–60.

Kumarathasan, R.; A.B. Rajkumar; N.R. Hunter; H.D. Gesser. Autoxidation and yellowing of methyl linolenate. *Prog. Lipid Res.* **1992**, *31*, 109–126.

Lange, R.; W. Schumann; M. Petrzika; H. Busch; R. Marquard. Glucosinolates in linseed dodder. *Fat Sci. Technol.* **1995**, *97*, 146–152.

Lanzmann-Petithory, D. Alpha-linolenic acid and cardiovascular diseases. *J. Nutr. Health Aging* **2001**, *5*, 79–183.

Layne, K.S.; Y.K. Goh; J.A. Jumpsen; E.A. Ryan; P. Chow; M.T. Clandinin. Normal subjects consuming physiological levels of 18:3(n-3) and 20:5(n-3) from flaxseed or fish oils have characteristic differences in plasma lipid and lipoprotein fatty acid levels. *J. Nutr.* **1996**, *126*, 2130–2140.

Lee, B.H.; J.I. Lee; C.B. Park; S.W. Lee; Y.H. Kim. Fatty acid composition and improvement of seed oil in perilla. *Crop Prod. Improvement Technol. in Asia,* **1993**, 471–479.

Leiken, A.I.; R.R. Brenner. Cholesterol-induced microsomal changes modulate desaturase activities. *Biochim. Biophys. Acta* **1987,** *922*, 294–303.

Leitzmann, M.F.; M.J. Stampfer; D.S. Michaud. Dietary intake of n-3 and n-6 fatty acids and the risk of prostate cancer. *Am. J. Clin. Nutr.* **2004,** *80*, 204–216.

Lemaitre, R.; I. King; D. Mozaffarian; L. Kuller; R. Tracy; D. Siscovick. N-3 polyunsaturated fatty acids, fatal ischemic heart disease, and nonfatal myocardial infarction in older adults: the cardiovascular study. *Am. J. Clin. Nutr.* **2003**, *77*, 319–325.

Leng, G.C.; G.S. Taylor; A.J. Lee; F.G. Fowkes; D. Horrobin. Essential fatty acids and cardiovascular disease: the Edinburgh Artery Study. *Vasc. Med.* **1999**, *4*, 219–226.

León, F.; M. Rodríguez; M. Cuevas. The major allergen of linseed. *Allergy* **2002**, *57*, 968.

León, F.; M. Rodríguez; M. Cuevas. Anaphylaxis to Linum. *Allergol. Immunopathol. (Madr).* **2003**, *31*, 47–49.

Lezaun, A.; J. Fraj; C. Colás; F. Duce; M.A. Domínguez; M. Cuevas; P. Eiras. Anaphylaxis from linseed. *Allergy* **1998,** *53,* 105–106.

Li, D.; N.J. Mann; A.J. Sinclair. Comparison of n-3 polyunsaturated fatty acids from vegetable oils, meat and fish in raising platelet eicosapentaenoic acid levels in humans. *Lipids* **1999**, *34,* S309.

Liou, Y.A.; D.J. King; D. Zibrik; S.M. Innis. Decreasing linoleic acid with constant α-linolenic acid in dietary fats increases (n-3) eicosapentaenoic acid in plasma phospholipids in healthy men. *J. Nutr.* **2007**, *137*, 945–952.

London, S.J.; F. Sacks; M.J. Stampfer; I.C. Henderson; H. Maclure; A. Tomita; W. Wood; S. Remine; N. Robert; J. Dmochowski. Fatty acid composition of the subcutaneous adipose tissue and risk of proliferative benign breast disease and breast cancer. *J. Natl. Cancer Inst.* **1993,** *85,* 785–793.

Low, Y.-L.; J.I. Taylor; P.B. Grace; M. Dowsett; E. Folkerd; D. Doody; A. Dunning; S. Scollen; A. Mulligan; A. Welch; et al. Polymorphisms in the CYP19 gene may affect the positive correlations between serum and urine phytoestrogen metabolites and plasma androgen concentrations in men. *J. Nutr.* **2005**, *135*, 2680–2686.

Lukaszewicz, M.; J. Szopa; A. Krasowska. Susceptibility of lipids from different flax cultivars to peroxidation and its lowering by added antioxidants. *Food Chem.* **2004**, *88*, 225–231.

Maillard, V.; P. Bougnoux; P. Ferrari. N-3 and n-6 fatty acids in breast adipose tissue and relative risk of breast cancer in a case-control study in Tours, France. *Int. J. Cancer* **2002,** *98*, 78–83.

Makino, T.; Y. Furuta; H. Wakushima; H. Fujii; K. Saito; Y. Kano. Anti-allergic effect of *Perilla frutescens* and its active constituents. *Phytotherapy Res.* **2003**, *17*, 240–243.

Malcolmson, L.J.; R. Przybylski; J. Daun. Storage stability of milled flaxseed. *J. Am. Oil. Chem. Soc.* **2000**, *77*, 235–238.

Manav, M.; J. Su; K. Hughes; H.P. Lee; C.N. Ong. ω-3 Fatty acids and selenium as coronary heart disease risk modifying factors in Asian Indian and Chinese males. *Nutrition* **2004**, *20*, 967–973.

Mannisto, S.; P. Pietinen; M. Virtanen. Fatty acids and risk of prostate cancer in a nested case-control study in male smokers. *Cancer Epidemiol. Biomarkers Prev.* **2003,** *12*, 1422–1428.

Mantzioris, E.; M.J. James; R.A. Gibson; L.G. Cleland. Dietary substitution with an α-linolenic acid rich vegetable oil increases eicosapentaenoic acid concentrations in tissues. *Am. J. Clin. Nutr.* **1994**, *59*, 1304–1309.

Marquard, R.; H. Kuhlmann. Investigations of productive capacity and seed quality of linseed dodder (*Camelina sativa* Crtz.). *Fette Seifen Anstrich.* **1986**, *88*, 245–249.

Matsuyama, W.; H. Mitsuyama; M. Watanabe; K. Oonakahara; I. Higashimoto; M. Osame; K. Arimura. Effects of omega-3 polyunsaturated fatty acids on inflammatory markers in COPD. *Chest* **2005**, *128*, 3817–3827.

McCloy, U.; M.A. Ryan; P.B. Pencharz; R. Ross; S. Cunnane. A comparison of the metabolism of eighteen-carbon $^{13}$C-unsaturated fatty acids in healthy women. *J. Lipid Res.* **2004**, *45,* 474–485.

Miljanović, B.; K.A. Trivedi; M.R. Dana; J. Gilbard; J. Buring; D. Schaumberg. Relation between dietary n-3 and n-6 fatty acids and clinically diagnosed dry eye syndrome in women. *Am. J. Clin. Nutr.* **2005**, *82*, 887–893.

Morris, D. *Flax Nutrition Primer*. Flax Council of Canada. www.flaxcouncil.ca. 2007.

Mossoba, M.M., M. Yurawecz; J. Roach; H. Lin; R. McDonald; B. Flickinger; E. Perkins. Rapid determination of double bond configuration and position along the hydrocarbon chain in cyclic fatty acid monomers. *Lipids* **1994**, *29*, 893–896.

Mozaffarian, D.; A. Ascherio; F.B. Hu; M.J. Stampfer; W.C. Willett; D.S. Siscovick; E.B. Rimm. Interplay between different polyunsaturated fatty acids and risk of coronary heart disease in men. *Circulation* **2005**, *111*, 157–164.

Nakamura, T. Red pigment-forming substances from autoxidized linolenate: identification of prostaglandin-like substances. *Lipids* **1985**, *20*, 180–186.

Nestel, P.J.; S.E. Pomeroy; T. Sasahara; T. Yamashita; Y. Liang; A. Dart; G. Jennings; M. Abbey; J. Cameron. Arterial compliance in obese subjects is improved with dietary plant n-3 fatty acid from flaxseed oil despite increased LDL oxidizability. *Arterioscler. Thromb. Vasc. Biol.* **1997**, *17*, 1163–1170.

Olejnik, D.; M. Gogolewski; M. Nogala-Kalucka. Isolation and some properties of plastochromanol-8. *Nahrung* **1997**, *41*, 101–104.

Oomah, D. Flaxseed as a functional food source. *J. Sci. Food. Agric.* **2001**, *81*, 889–904.

Oomah, B.D.; E. Kenaschuk; W. Cui; G. Mazza. Variation in the composition of water-soluble polysaccharides in flaxseed. *J. Agric. Food Chem.* **1995**, *43*, 1484–1488.

Oomah, B.D.; E. Kenaschuk; G. Mazza. Tocopherols in flaxseed. *J. Agric. Food Chem.* **1997**, *45*, 2076–2080.

Oomah, B.D.; G. Mazza. Effect of dehulling on chemical composition and physical properties of flaxseed. *Lebensm. Wiss. U. Technol.* **1997**, *30*, 135–140.

Oomah, B.D.; G. Mazza; R. Przybylski. Comparison of flaxseed meal lipids extracted with different solvents. *Leb. Wiss. Techol.* **1996**, *29*, 654–658.

Painter, E.P.; L.L. Nesbitt. Fat acid composition of linseed oil from different varieties of flaxseed. *Oil and Soap* **1943**, *20*, 208–211.

Panfilis, F.; T. Toschi; G. Lercker. Quality control for cold-pressed oils. *Inform* **1998**, *9*, 212–221.

Paschos, G.K.; F. Magkos; D.B. Panagiotakos; V. Votteas; A. Zampelas. Dietary supplementation with flaxseed oil lowers blood pressure in dyslipidaemic patients. *Eur. J. Clin. Nutr.* **2007**, *31*, 1–6.

Paschos, G.K.; N. Yiannakouris; L.S. Rallidis; I. Davies; B.A. Griffin; D Panagiotakos. Apolipoprotein E genotype in dyslipidemic patients and response of blood lipids and inflammatory markers to alpha-linolenic acid. *Angiology* **2005**, *56*, 56–59.

Pawlosky, R.J.; J.R. Hibbeln; J.A. Novotny; N. Salem, Jr. Physiological compartmental analysis of α-linolenic acid metabolism in adult humans. *J. Lipid Res.* **2001**, *42*, 1257–1265.

Pietinen, P.; A. Ascherio; P. Korhonen; A. Hartman; W. Willett; D. Albanes; J. Virtamo. Intake of fatty acids and risk of coronary heart disease in a cohort of Finnish men: the Alpha-Tocopherol, Beta-Carotene Cancer Prevention Study. *Am. J. Epidemiol.* **1997**, *145*, 876–887.

Pizzey, G.; T. Luba. Effect of seed selection and processing on stability of milled flaxseed. *Book of Abstracts of the 93rd AOCS Annual Meeting and Expo*; American Oil Chemists' Society: Champaign, IL, 2002; S143.

Pizzey, G.R.; T. Luba. Effect of seed selection and processing on stability of milled flaxseed. *Pizzey Milling Report* **2004**.

Plessers, A.G.; W.G. McGregor; R.B. Carson; W. Nadoneshny. Species trials with oilseed crops, Camelina. *Can. J. Plant Sci.* **1962**, *42*, 452–459.

Preťová, A.; M. Vojteková. Chlorophylls and carotenoids in flax embryos during embryogenesis. *Photosynthetica* **1985**, *19*, 194–197.

Rallidis, L.S.; G. Paschos; M.L. Papaioannou; G. Liakos; D. Panagiotakos; G. Anastasiadis; A. Zampelas. The effect of diet enriched with α-linolenic acid on soluble cellular adhesion molecules in dyslipidaemic patients. *Atherosclerosis* **2004**, *174*, 127–132.

Rastogi, T.; K.S. Reddy; M. Vaz; D. Spiegelman; D. Prabhakaran; W. Willett; M. Stampfer; A. Ascherio. Diet and risk of ischemic heart disease in India. *Am. J. Clin. Nutr.* **2004**, *79*, 582–592.

Reiss, A.B.; S.D. Edelman. Recent insights into the role of prostanoids in atherosclerotic vascular disease. *Curr. Vasc. Pharmacol.* **2006**, *4*, 395–408.

Rollefson, G.O.; A.H. Simmons; M.L. Donaldson; W. Gillespie; Z. Kafafi; I.U. Kohler-Rollefson; E. McAdam; S. Ralston; M. Tubb. Excavation at the Pre-Pottery Neolithic B village of 'Ain Ghazal (Jordan). *Mitteilungen der Deuschen Orient-Gesellschaft zu Berlin* **1985**, *117*, 69–116.

Rosamond, W. for the Writing Group Members. Heart disease and stroke statistics—2007 update. A report from the American Heart Association Statistics Committee and Stroke Statistics Subcommittee. *Circulation* **2007**, *115*, e69–e171.

Rosling, H. Measuring effects in humans of dietary cyanide exposure to sublethal cyanogens from Cassava in Africa. *Acta Hortic.* **1994**, *375*, 271–283.

Ross, R. Atherosclerosis—an inflammatory disease. *N. Engl. J. Med.* **1999**, *340*, 115–126.

Rudnik, A.; H. Szczucinska; H. Gwardiak; A. Szulc; A. Winiarska. Comparative studies of oxidative stability of linseed oil. *Thermochim. Acta* **2001**, *370*, 135–140.

Saadatian-Elahi, M.; T. Norat; J. Goudable. Biomarkers of dietary fatty acid intake and the risk of breast cancer: a meta-analysis. *Int. J. Cancer* **2004,** *111*, 584–591.

Sanders, T.A.B. Polyunsaturated fatty acids in the food chain in Europe. *Am. J. Clin. Nutr.* **2000,** *71,* 176S–178S.

Sanders, T.A.B.; F. Roshanai. The influence of different types of ω3 polyunsaturated fatty acids on blood lipids and platelet function in healthy volunteers. *Clin. Sci.* **1983**, *64*, 91–99.

Schultze-Motel, J. Die Anbaugeschichte des Leindotters, *Camelina sativa* (L.) Crantz. *Archaeo-Physika* **1979**, *8*, 267–281.

Schwab, U.S.; J.C. Callaway; A.T. Erkkilä; J. Gynther; M. Uusitupa; T. Jarvinen. Effects of hempseed and flaxseed oils on the profile of serum lipids, serum total and lipoprotein lipid concentrations and haemostatic factors. *Eur. J. Nutr.* **2006**, *45*, 470–477.

Schwartz, H.; V. Ollilainen; V. Piironen; A.M. Lampi. Tocopherol, tocotrienol and plant sterol contents of vegetable oils and industrial fats. *J. Food Comp. Anal.* **2008**, *21*, 152–161.

Seehuber, R. Genotypic variation for yield- and quality-traits in poppy and false flax. *Fette-Seifen-Anstrich.l* **1984**, *86*, 177–180.

Shin, H.S. Lipid composition and nutritional and physiological roles of perilla seed and its oil. *Perilla: The Genus Perilla;* H. Yu, K. Kosuna, M. Haga, Eds.; CRC Press: **1997**, pp. 93–107.

Shin, H.S.; S.-W Kim. Lipid composition of perilla seed. *J. Am. Oil Chem. Soc.* **1994**, *71*, 619–622.

Shukla, V.K.S.; P.C. Dutta; W. Artz. Camelina oil and its unusual cholesterol content. *J. Am. Oil Chem. Soc.* **2002**, *79*, 965–969.

Siedow, J.N. Plant lipoxygenase: Structure and function. *Ann. Rev. Plant Physiol. Plant Mol. Biol.* **1991**, *42*, 145-–149.

Simon, J.A.; J. Fong; J.T. Bernert; W.S. Browner. Serum fatty acids and the risk of stroke. *Stroke* **1995**, *26*, 778–782.

Simonsen, N.; N. Van't Veer; J. Strain; J. Martin-Moreno; J. Huttunen; J. Navajas; B. Martin; M. Thamm; A. Kardinaal; F. Kok; L. Kohlmeier. Adipose tissue omega-3 and omega-6 fatty acid content and breast cancer in the EURAMIC study. *Am. J. Epidemiol.* **1998,** *147*, 342–352.

Simopoulos, A.P. Evolutionary aspects of diet, the omega-6/omega-3 ratio and genetic variation: nutritional implications for chronic diseases. *Biomed. Pharmacother.* **2006**, *60*, 502–507.

Singer, P.; I. Berger; M. Wirth; W. Goedicke; W. Jaeger; S. Voigt. Slow desaturation and elongation of linoleic and α-linolenic acids as a rationale of eicosapentaenoic acid-rich diet to lower blood pressure and serum lipids in normal, hypertensive and hyperlipemic subjects. *Prostaglandins Leukotrienes Med.* **1986**, *24*, 173–193.

Singer, P.; M. Wirth; I. Berger. A possible contribution of decrease in free fatty acids to low serum triglyceride levels after diets supplemented with n-6 and n-3 polyunsaturated fatty acids. *Atherosclerosis* **1990**, *83*,167–175.

Singh, J.; P. Bargale. Development of a small capacity double stage compression screw press for oil expression, *J. Food Eng.* **2000,** *43*, 75–82.

Sondergaard, E.; H. Dam. Occurrence of a plastochromanol in linseed oil. *Acta Chem. Scand.* **1967**, *21*, 2582.

Stark, A..H.; M. Crawford; R. Reifen. Update on alpha-linolenic acid. *Nutr. Rev.* **2008**, *66*, 326–332.

Stefanowicz, P. Electrospray mass spectrometry and tandem mass spectrometry of the natural mixture of cyclic peptides from linseed. *Eur. J. Mass Spectrom.* **2004**, *10*, 665–671.

Stenberg, C.; M. Svensson; M. Johansson. A study of the drying of linseed oils with different fatty acid patterns using RTIR-spectroscopy and chemiluminescence (CL). *Ind. Crops Prod.* **2005**, *21*, 263–272.

Taylor B.R.; L.A.F. Morrice. Effects of husbandry practices on the seed yield and oil content of linseed in Northern Scotland. *J. Sci. Food Agric.* **1991**, *57*, 189–198.

Tsuyuki, H.; S. Itoh; Y. Nakatsukasa. Lipids in perilla seeds. *Nihon Daigaku Nojuigakubu Gijutsu Kenkyu Hokoku* **1978**, *35*, 224–230.

Tulbek, M.C.; C.A. Hall III; B. Zhao. Evaluation of the impact of packaging methods on the oxidative stability of milled flaxseed. *Abstracts of the 99th American Oil Chemists' Society Annual Meeting and Expo;* American Oil Chemists' Society: Champaign IL, 2008.

Vaisey-Genser, M.; D. Morris. Introduction: history of the cultivation and uses of flaxseed. *Flax: The Genus Linum*; A. Muir, N. Westcott, Eds.; Taylor and Francis, Ltd.: London, England, 2003; pp. 1–21.

Van Ruth, S.M.; E. Shaker. Influence of methanolic extracts of soybean seeds and soybean oil on lipid oxidation in linseed oil. *Food Chem.* **2001**, *75*, 177–184.

Van Zeiste, W. Palaeobotanical results in the 1970 seasons at Cayonu, Turkey. *Helinium* **1972**, *12*, 3–19.

Vealsco, L.; F.D. Goffman. Tocopherol, plastochromanol and fatty acid patterns in the genus *Linum. Plant Systematics Evolution* **2000**, *221*, 77–88.

Voorrips, L.; H. Brants; A. Kardinaal; G. Hiddink; P. van den Brandt. Intake of conjugated linoleic acid, fat, and other fatty acids in relation to postmenopausal breast cancer: the Netherlands cohort study on diet and cancer. *Am. J. Clin. Nutr.* **2002**, *76*, 873–882.

Vos, E.; S.C. Cunnane. α-Linolenic acid, linoleic acid, coronary artery disease, and overall mortality (letter). *Am. J. Clin. Nutr.* **2003**, *77*, 521–522.

Wakjira, A.; M.T. Labuschagne; A. Hugo. Variability in oil content and fatty acid composition of Ethiopian and introduced cultivars of linseed. *J. Sci. Food Agric.* **2004**, *84*, 601–607.

Wanasundara, P.K.J.P.D.; U.N. Wanasundara; F. Shahidi. Changes in flax (*Linum usitatissimum* L.) seed lipids during germination. *J. Am. Oil Chem. Soc.* **1999**, *76*, 41–48.

Warrand, J.; P. Michaud; L. Picton; G. Muller; B. Courtois; R. Ralainirina; J. Courtois. Structural investigations of the neutral polysaccharide of Linum usitatissimum L. seeds mucilage. *Int. J. Biol. Macromol.* **2005**, *35*, 121–125.

Waterman, H.I.; J. Cordia; B. Pennekamp. Formation of cyclic compounds in polymerization of methylesters of fatty acids from linseed oil and other drying oils. *Research* **1949**, *2*, 483–485.

West, S.G.; K.D. Hecker; V.A. Mustad; S. Nicholson; S. Schoemer; P. Wagner; A. Hinderliter; J. Ulbrecht; P. Ruey; P. Kris-Etherton. Acute effects of monounsaturated fatty acids with and without omega-3 fatty acids on vascular reactivity in individuals with type 2 diabetes. *Diabetologia* **2005**, *48*, 113–122.

White, N.D.G.; D.S. Jayas. Factors affecting the deterioration of stored flaxseed including the potential insect infestation. *Can. J. Plant Sci.* **1991**, *71*, 327–335.

Wieczorek Z.; B. Bengtsson; J. Trojnar; I. Siemion. Immunosuppressive activity of cyclolinopeptide A. *Peptide Res.* **1991**, *4*, 275–283.

Wiesenborn, D.; N. Kangas; K. Tostenson; C. Hall III; K. Chang. Sensory and oxidative quality of screw-pressed flaxseed oil. *J. Am. Oil Chem. Soc.* **2005**, *82*, 887–892.

Wilkinson, P.; C. Leach; E.E. Ah-Sing; E. Eric; N. Hussain; G. Miller; D. Millward; B. Griffin. Influence of α-linolenic acid and fish-oil on markers of cardiovascular risk in subjects with an atherogenic lipoprotein phenotype. *Atherosclerosis* **2005**, *181*, 115–124.

Yuba, A.; G. Honda; Y. Koezuka; M. Tabata. Genetic analysis of essential oil variants in Perilla frutescens. *Biochem. Genet.* **1995**, *33*, 341–348.

Zatonski, W.; H. Campos; W. Willett. Rapid declines in coronary heart disease mortality in Eastern Europe are associated with increased consumption of oils rich in alpha-linolenic acid. *Eur. J. Epidemiol.* **2008,** *23*, 3–10.

Zhao, G.; T.D. Etherton; K.R. Martin; P. Gillies; S. West; P. Kris-Etherton. Dietary alpha-linolenic acid inhibits proinflammatory cytokine production by peripheral blood mononuclear cells in hypercholesterolemic subjects. *Am. J. Clin. Nutr.* **2007,** *85*, 385–391.

Zhao, G.; T.D. Etherton; K.R. Martin; J. Vanden Heuvel; P. Gillies; S. West; P. Kris-Etherton. Dietary α-linolenic acid reduces inflammatory and lipid cardiovascular risk factors in hypercholesterolemic men and women. *J. Nutr.* **2004,** *134*, 2991–2997.

Zheng, Y.; D. Wiesenborn; K. Tostenson; N. Kangas. Screw pressing of whole and dehulled flaxseed for organic oil. *J. Am. Oil Chem. Soc.* **2003,** *80*, 1039–1045.

Zheng, Y., D.P. Wiesenborn; K. Tostenson; N. Kangas. Energy analysis in the screw pressing of whole and dehulled flaxseed. *J. Food Eng.* **2005,** *66*, 193–202.

Zhuang, H.; T. Hamilton-Kemp; R. Andersen; D. Hildebrand. Developmental change in C6-

aldehyde formation by soybean leaves. *Plant Physiol.* **1992,** *100*, 80–87.

Zohary, D.; M. Hopf. Oil and fibre crops. "Domestication of Plants in the Old World, 3rd ed.; D. Zohary, M. Hopf, Eds.; Oxford University Press: Oxford, 2000; pp. 125–132.

Zubr, J. Oil-seed crop: Camelina sativa. *Ind. Crops Prod.* **1997,** *6*, 113–119.

# 5

# Hempseed Oil

**J.C. Callaway[1,2] and David W. Pate[3]**
*[1]Finola ky, PL 236, Kuopio, FI-70101 Finland, www.finola.com; [2]Departments of Pharmaceutical Chemistry and Neurobiology, University of Kuopio, FI-70211 Kuopio, Finland; [3]Centre for Phytochemistry and Pharmacology, Southern Cross University, Lismore, NSW 2480, Australia*

## Introduction

Non-drug varieties of *Cannabis sativa* L., collectively known as "hemp," provide an important source of industrial fiber. Hemp fiber is used in the production of specialty paper (e.g., cigarette papers, bank notes, and tea bags), in addition to ropes, woven and nonwoven fabrics, automotive and building insulation, construction materials, and many other durable goods. Fiber varieties of hemp can be over four meters tall, and require over 150 days for seed maturation. In contrast, oilseed varieties are usually less than two meters tall at the time of harvest (110 to 150 days after sowing), which allows them to be efficiently harvested by conventional grain combines. Hemp contains very low amounts of Δ-9-tetrahydrocannabinol (THC), the main psychoactive component in drug varieties of *Cannabis*. The amount of THC in the mature hemp plant is typically less than 0.5% of the plant's dry weight, which is not sufficient for drug purposes. In addition, THC is not a toxic compound in humans, even at high dosages and over long periods of time.

Hempseed oil is a highly unsaturated product that is pressed or extracted from the achenes of *Cannabis*, which are also a source of highly digestible protein (Table 5.1). Thus, the tiny nut is an exceptionally good source of nutrition (Callaway, 2004; Deferne & Pate, 1996). This fruit of *Cannabis,* which has a relatively hard shell when mature, varies in shape—from almost spherical to somewhat oblong. Its overall size can vary considerably, but most often it approximates that of a match-head.

Not only does food-quality hempseed oil taste and smell delicious, it is extremely rich in lipid nutrients. The chemical analysis of hemp and other major seed oils was well underway in the late nineteenth century (Von Hazura, 1887) when linoleic acid (LA) was first identified (as "sativinsäure" or "sativic acid") as the main component of hempseed oil. The quantitative analysis of the major fatty-acid components in hempseed oil became of scientific interest in the early twentieth century (Kaufmann & Juschkewitsch, 1930). The essential fatty acids (EFAs) are well represented in hempseed oil. The "omega-6" LA (18:2n-6) component is present at about 55%, and "ome-

**Table 5.1. Typical Nutritional Composition (%) of Whole Hempseed, Dehulled Seed, and Seed Meal* (modified from Callaway, 2004)**

| | Whole seed | Dehulled seed | Seed meal |
|---|---|---|---|
| Oil | 36 | 44 | 11 |
| Protein | 25 | 33 | 34 |
| Carbohydrates | 28 | 12 | 43 |
| Moisture | 6 | 5 | 5 |
| Ash | 5 | 6 | 7 |
| Energy (kJ/100g) | 2200 | 2093 | 1700 |
| Total dietary fiber | 28 | 7 | 43 |
| Digestible fiber | 6 | 6 | 16 |
| Nondigestible fiber | 22 | 1 | 27 |

*Finola cultivar

ga-3" α-linolenic acid (ALA, 18:3n-3) occurs at about 20%. In addition, significant amounts of their respective metabolic products are found: the presence of γ-linolenic acid (GLA, 18:3n-6) ranges from 1 – 4%, and stearidonic acid (SDA, 18:4n-3) occurs at about 0.5 to 2%. While most vegetable oils have at least some EFAs, it is unusual to have this high amount of both in this proportion, in addition to GLA and SDA (Table 5.2). In fact, no other industrial crop can make this claim.

No great differences exist in the amounts or proportions of EFAs between an oilseed hemp cultivar from Northern Europe and a typical fiber hemp cultivar from Central Europe (Table 5.2). However, a considerable difference in the native abundance of both GLA and SDA was noted between northern and southern varieties of hempseed within the first report on the presence of SDA in hempseed (Callaway et al., 1997). Subsequent investigations confirmed this observation (Anwar et al., 2006; Bagci et al., 2003; Blade et al., 2005; Callaway, 2002, 2004; Matthäus et al., 2005; Mölleken & Theimer, 1997 ). The highest concentrations of GLA and SDA are found in seed of hemp varieties that are derived from extreme northern climes (Blade et al.. 2005; Callaway et al., 1997). Perhaps this higher content of super-unsaturated fatty acids protects the seeds from freezing solid during the harsh winter months, conferring the evolutionary advantage of an expanded range. Hempseed oil, for example, does not begin to thicken until it is stored for at least several days below –20°C. In an early study (Ross et al., 1996) on the fatty-acid profile of seed oils from drug-*Cannabis*, a remarkable uniformity was found in profiles of confiscated samples from Mexico, Columbia, Jamaica, and Thailand. These tropical seed samples were missing ALA, which complements the aforementioned observations of greater seed-oil unsaturation being correlated with more extreme latitudes.

If the nuisance of shell fragments between the teeth is disregarded, one can consume whole hempseed directly from the plant, because it lacks the anti-nutritive prop-

**Table 5.2. Typical Fatty-Acid Profiles (%) of Hemp and Other Seed Oils (Callaway, 2004)**

| Seed | Palmitic acid | Stearic acid | Oleic acid | LA | ALA | GLA | SDA | %PUFA | n-6/n-3 ratio |
|---|---|---|---|---|---|---|---|---|---|
| Oil hempseed* | 5 | 2 | 9 | 56 | 22 | 4 | 2 | 84 | 2.5 |
| Fiber hempseed | 8 | 3 | 11 | 55 | 21 | 1 | <1 | 77 | 2.7 |
| Black currant | 7 | 1 | 11 | 48 | 13 | 17 | 3 | 81 | 4.1 |
| Flax (linseed) | 6 | 3 | 15 | 15 | 61 | 0 | 0 | 76 | 0.2 |
| Evening primrose | 6 | 1 | 8 | 76 | 0 | 9 | 0 | 85 | >100.0 |
| Sunflower | 5 | 11 | 22 | 63 | <1 | 0 | 0 | 63 | >100.0 |
| Wheat germ | 3 | 17 | 24 | 46 | 5 | 5 | <1 | 56 | 10.2 |
| Rapeseed | 4 | <1 | 60 | 23 | 13 | 0 | 0 | 36 | 1.8 |
| Soy | 10 | 4 | 23 | 55 | 8 | 0 | 0 | 63 | 6.9 |
| Borage | 12 | 5 | 17 | 42 | 0 | 24 | 0 | 66 | >100.0 |
| Corn | 12 | 2 | 25 | 60 | 1 | 0 | 0 | 60 | 60.0 |
| Olive | 15 | 0 | 76 | 8 | <1 | 0 | 0 | 8 | >100.0 |

* Finola cultivar. LA = Linoleic Acid (18:2n-6), ALA = α-Linolenic Acid (18:3n-3), GLA = γ-Linolenic Acid (18:4n-6), SDA = Stearidonic Acid (18:4n-3), PUFA = Polyunsaturated Fatty Acid, n-6/n-3 Ratio = Percentages of ω-6 Fatty Acids Divided by ω-3 Fatty Acids
Reprinted with permission from Springer Science and Business Media.

erties that are commonly found in so many other raw foods and oilseeds (Mattäus, 1997). One can also feed the seed-press meal to pets and livestock. Hempseed does not contain gluten, which makes it an important source of vegetable protein for people who suffer from coeliac disease (also written as "celiac" disease), an autoimmune-based gluten intolerance disorder of the small bowel that affects approximately 1% of Indo–European populations, and a disease which is significantly underdiagnosed (Collin, 1999). In addition to its value for oil and protein, hempseed and the seed meal by-product of oil pressing contain respectable amounts of vitamins and minerals (Table 5.3).

The use of *Cannabis* as a source of food, fiber, and medicine is widespread in the Old World, and the whole seed continues to be used as a food and condiment by people in Asia (Xiaozhai & Clarke, 1995). After considering the historical accounts that demonstrate an intimate human relationship with this plant, imagining that this seed was overlooked by ancient humans in their transition from food gathering to the development of agriculture is difficult (Weiss et al., 2004). The oldest existing documents that describe the use of hempseed as both food and medicine are from China (de Padua et al., 1999), where *Cannabis* stalks, leaves, and seeds were found in tombs that are over 4500 years old (Jiang et al., 2006). Good evidence suggests that *Canna-*

**Table 5.3. Typical Nutritional Values for Vitamins and Minerals in Hempseed* (Callaway, 2004)**

| Vitamins and minerals | Nutritional values (mg/100 g) |
|---|---|
| Vitamin E (total) | 90 |
| α-tocopherol | 5 |
| γ-tocopherol | 85 |
| Thiamine (B1) | 0.4 |
| Riboflavin (B2) | 0.1 |
| Phophorus (P) | 1,160 |
| Potassium (K) | 859 |
| Magnesium (Mg) | 483 |
| Calcium (Ca) | 145 |
| Iron (Fe) | 14 |
| Sodium (Na) | 12 |
| Manganese (Mn) | 7 |
| Zinc (Zn) | 7 |
| Copper (Cu) | 2 |

* Finola cultivar

*bis* was used as a source of fiber and as a medicament in Ancient Egypt (Russo, 2007). Other tangible evidence suggests that *Cannabis* was used as a source of fiber for at least 6000 years (Schultes, 1970), and perhaps up to 12,000 years (Abel, 1980). Patterns of woven material, possibly *Cannabis* fibers in the form of nets for trapping small animals, were preserved as fossilized remains that are over 20,000 years old (Pringle, 1997). By providing a nutritious food from its seed, a durable fiber from its stalk, and an efficacious medicine from its flower and leaves, *Cannabis* has assisted in human development like no other plant species. A more intimate relationship for the co-evolution of *Cannabis* with humans has been advanced (McPartland & Guy, 2004).

## Hempseed Oil Processing

Hempseed oil that is used for human consumption is ideally produced from fresh, well-cleaned seeds that were air-dried at low temperatures (<25°C) over several days or weeks. At the time of harvest, the hempseed moisture content is typically 15–20%. The final moisture content of hempseed for storage and pressing should be just below 10%, and one must take special care to insure that the seed does not support mold growth between the time of harvest and the time of drying. If special care is not immediately taken at harvest, the aesthetic qualities of the product will suffer greatly, and the resulting oil will have a relatively short shelf life of only a few months, at best. The preferred containers for bulk oil storage, after pressing, are made of glazed-metal, ceramic, or glass. From these vessels, the oil is either filtered for immediate bottling into glass or allowed to settle before it is bottled for retail distribution. The fine sediment from freshly pressed hempseed oil has a high nutritive value, and one can use it directly as a nut-butter spread or in other human-food products, or in high-end pet foods.

As oxidation was noted as the main problem for the long-term storage of any polyunsaturated oil, one should take special care to insure that the seed is already under an inert atmosphere before it reaches the press head, and one must maintain this inert atmosphere throughout the processing until the oil is bottled and capped. After bottling, one should protect the product from light and store it at the coldest temperature possible. No worry exists for oil expansion and subsequent container damage when it has solidified at low temperatures (<–20°C), in contrast to frozen aqueous products.

In reality, the ideal scenario is seldom the case. Most contemporary producers of hempseed oil are either individual operators with a small press or small start-up enterprises with one or two presses. Unfortunately, hempseed is rarely pressed under an inert atmosphere, but most reputable processors and distributors will at least state a "best before" date on their products. However, these dates are often quite arbitrary and typically stated to meet some economic objective. Most distributors, and especially retailers, are very reluctant to accept products with short "best before" dates, and too many processors are willing to provide a date that is acceptable to the retailer, with little or no regard for the actual quality of the oil that eventually reaches the consumer. Most retailers that carry hempseed oil are small shops that specialize in "biological/ecological/health/organic" foods, and often have little experience with, or appreciation for, the storage requirements of such a highly unsaturated oil. Moreover, many retailers have little or no spare refrigerated space for "yet another" product and so, unfortunately, often leave highly unsaturated oils to more rapidly age on room-temperature shelves. In the past few years, hempseed oil has begun to appear on the shelves of main-stream food stores in the United Kingdom.

Currently, food-grade hempseed oil for human consumption is cold-pressed from hempseed (i.e., small-scale screw presses operating at 40–50°C). The oil is allowed to settle for at least one to two weeks, and is then decanted directly into smaller containers for retail sales. Larger screw presses are used for "industrial" production, and the more successful operations filter the fine sediment directly into bulk 1000 L containers, rather than wait for gravity sedimentation.

Bulk hydraulic pressing offers a viable economic alternative to cold-pressing for food grade hempseed oil, providing that subsequent processing is under inert atmosphere, but such facilities are usually not set up for the production of high-quality hempseed oil. More importantly, the current market for food grade hempseed oil is not nearly at the level required to take advantage of this processing method. In addition, industrial refining or bleaching of hempseed oil to remove chlorophyll and other components will remove the characteristic taste, antioxidants and other useful components from the oil.

Supercritical carbon dioxide can also be utilized for the extraction of food oils under low temperature and inert atmosphere. However, the main drawback of this technology is cost. Solvent extraction is used for the inexpensive industrial processing of many vegetable oils, although the defatted meal is not suitable for human food because residual solvents (typically hexanes) contaminate the final product.

## Hempseed Oil Composition

Good-quality, cold-pressed hempseed oil has a clear green to olive color, and ideally possesses a fresh nutty taste and smell (Table 5.4). It is an exceptionally rich source of polyunsaturated fatty acids (PUFAs), ranging from 75 to 85% (Table 5.2) of the total oil content (Blade et al., 2005; Callaway et al., 1997; Kriese et al., 2004; Matthäus et al., 2005). The absolute amounts of GLA and SDA in hempseed oil seem to be genetically determined, and their relative ratio is highly consistent, showing a range from 0.67 – 4.08% and 0.4 – 1.6%, respectively, of the total seed oil (Matthäus et al., 2005).

The ratio of n-6/n-3 EFAs in hempseed oil (i.e., the percentages of LA divided by ALA) is typically near an ideal value for their efficient and simultaneous metabolic conversion. Due to the metabolic competition between the two EFAs for access to the rate-limiting enzyme Δ-6 desaturase (Gerster, 1988), the significance of an ap-

**Table 5.4. Technical Characteristics of Cold-pressed Hempseed Oil**

| | |
|---|---|
| Solidification point | –20 °C |
| Flash point | 141 °C |
| Smoke point | 165 °C |
| Specific gravity | 0.9295 g/mL at 20 °C |
| Saponification value | 193 |
| Iodine value | 160 |
| Chlorophyll content | 5–80 ppm |
| Color: | |
| fresh cold-pressed oil | clear bright to dark green; fresh, nutty smell |
| old cold-pressed oil | clear olive green to yellow; fishy, paint smell |
| refined oil | clear colorless to light yellow; odorless to paint smell |
| Peroxide value: | |
| food-grade | < 2 meq/kg |
| cosmetic-grade | <10 meq/kg |
| industrial-grade | >10 meq/kg |
| Free fatty acid value: | |
| food-grade | <0.1% |
| other grades | >1.0% |

propriate dietary ratio of these fatty acids is important to consider in any discussion of general health (Okuyama et al., 1997) and within the interpretation of results from clinical studies that contain significant amounts of these oils (Simopoulos, 1999), especially in chronic-disease states. Only a decade ago, an optimal LA to ALA (n-6/n-3) ratio was considered to be somewhere between 5:1 and 10:1 (WHO & FAO, 1995). Soybean oil was popular and promoted as a healthy oil by the food industry and the agricultural community at the time, apparently because its ratio (about 7:1) is within that range. More recent considerations suggest an optimal n-6/n-3 ratio to be somewhere between 2:1 and 3:1 (Simopoulos et al., 2000), which reflects the ratio found in the traditional Japanese and Mediterranean diets, where the incidence of coronary heart disease was historically low. Fortuitously, the n-6/n-3 ratio in most commercial hempseed oils is typically near 2.5:1 (Table 5.2; Callaway et al., 1997; Kriese et al., 2004).

Flaxseed (*Linum usitatissimum* L.) oil typically contains about 60% of ALA, but lacks both GLA and SDA, while hempseed oils tend to have just over 20% of ALA

(Table 5.2). An excess of ALA can disturb the human metabolic balance of fatty-acid metabolism by leaving a net deficit in "omega-6" metabolites, which are derived from dietary LA. In fact, the daily use of only two tablespoons of flaxseed oil per day can significantly decrease the competitive metabolic production of GLA from LA (Schwab et al., 2006). The presence of both GLA and SDA in hempseed oil allows this competition for Δ-6-desaturase to be efficiently bypassed (Callaway et al., 2005; Okuyama et al., 1997; Schwab et al., 2006), while the favorable n-6/n-3 ratio in hempseed oil allows for the efficient metabolism of both EFAs to proceed in concert.

## Protein By-products of Hempseed

Hempseed and hempseed meal are excellent sources of digestible protein. Figure 5.1 compares the amino acid profile for the total protein in hempseed, soy bean and egg white. Protein concentrations vary between whole hempseed (ca. 25%), de-hulled hempseed (ca. 45%), soy bean (ca. 32%) and egg white (ca. 11%). Figure 5.1 illustrates individual amino acid values per 100 g of protein to provide a direct comparison between these products. Another important fact to keep in mind is that hempseed and egg white lack the anti-nutritional trypsin-inhibiting factors that are found in soy and many other vegetable products. This means that, like egg white, a greater proportion of the protein found in hempseed is digested and available for absorption. Recent interest in hempseed protein has increased due to its exceptional content of sulfur-containing amino acids (Callaway 2004, Tang et al. 2006), i.e., methionine and cystine (Odani & Odani 1998), and its surprisingly high amount of arginine (Fig. 5.1). As with most vegetable proteins, hempseed is considered to be low in the essential amino acid lysine, and is therefore not sufficient as the sole source of dietary protein for children under 10 years of age, according to FAO/WHO essential amino acid requirements (WHO & FAO 1995).

The major protein found in hempseed is edestin, which accounts for about 60-80% of the total protein content, with albumin making up the balance (Odani & Odani 1998, Tang et al. 2006). Edestin is a well-characterized protein, with a rich and detailed past (Osborn 1892) that has been nearly forgotten today. As with the soy protein glycinin, edestin is a hexamer, being composed of six identical AB protein subunits with molecular weights of about 33.0 and 20.0 kDa (Patel et al. 1994). An interesting non-food application for hempseed protein isolate derives from its ability to form cast films, which can be used in the production of biodegradable and even edible food packaging (Yin et al. 2007). In this study, the physical properties of cast films from hempseed protein isolate were investigated and compared to those of soy protein isolate. Their results suggest that hempseed protein isolates had good potential for the preparation of protein films and demonstrated some superior characteristics over soy isolates, such as low aqueous solubility and high surface hydrophobicity. Both of these properties are extremely important characteristics for durable food packaging of products having high moisture content.

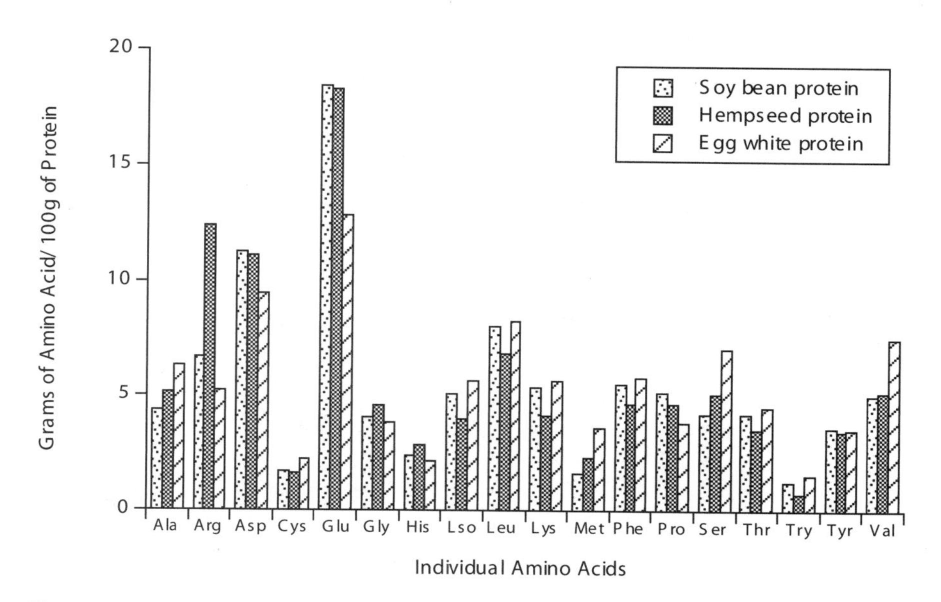


**Fig. 5.1.** Amino acid profiles of soy bean, hempseed, and egg white, as represented by their IUPAC abbreviations (Callaway 2004). Reprinted with the kind permission of Springer Science and Business Media.

## Economy of Hempseed and Hempseed Oil

Neither hempseed nor hempseed oil are among the 200 primary commodities that are tracked by the FAO. However, some limited amount of information is available for both. Hempseed oil production data from the FAO is important to consider (FAOSTAT 2007), and clearly shows China to be in the lead (Fig. 5.2). Figure 5.2 shows worldwide reported data for hempseed oil production from 1994 to 2004, with individual contributions from the major participants, Western Europe and China. It is important to note that the hempseed oil production reported to FAO is probably almost entirely for non-food, industrial purposes; e.g., in the production of paints, varnishes, resins and some bio-plastics. The main reason for presenting this information in this context is to give some idea of the recent scale of reported industrial hempseed oil production worldwide. Unfortunately, the FAO does not have more extensive information on hempseed oil production, and nothing at all on products made from hempseed oil at this time.

The FAO began to report worldwide values for hempseed production in 1961. From 1961 to 1975, Turkey led all countries in reported world exports of hempseed until Lebanon dominated the reported market from 1977 to 1985 (data not shown). France, Germany and Chile were also significant hempseed producers from 1961 to 1985, with the Netherlands, Spain, Italy and Yugoslavia making considerable contri-

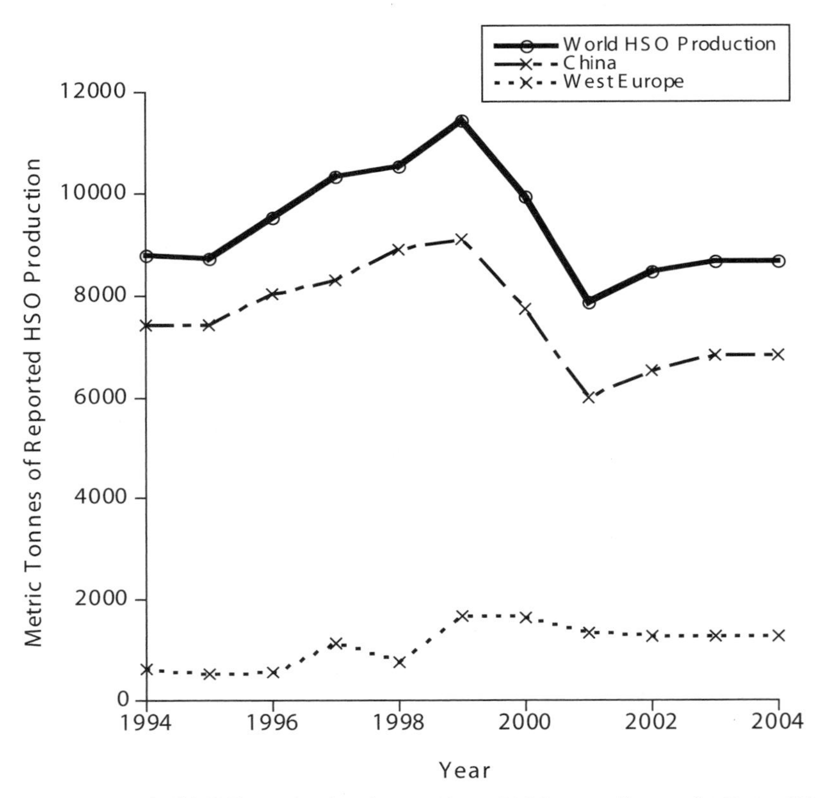


**Fig. 5.2.** Hempseed oil (HSO) production from 1994 to 2004, according to the United Nations Food and Agriculture Organization (http://faostat.fao.org/).

butions from Europe during this time. China's reported hempseed exports accounted for almost 77% of the total world exports reported in 1986, with 12,200 metric tons (MT). FAO data for China's hempseed exports (primarily for birdfeed) are only available from 1986 to 1991 (maximum reported value of 17,777 MT in 1991), and then again from 1998 to 2005 (average value of just over 10,300 MT/year). Unfortunately, there are no FAO data from the Soviet Union or Russia for hempseed or hempseed oil. This omission is unfortunate because of the long history of hemp production in Russia. Although production of hemp in this region of the world was primarily for bast fiber, which was used to produce durable fabrics, the production of dietary hempseed oil was also an important side product for "peasants" who could not afford butter (Grigoryev 2005). In Russia, hempseed oil is still referred to as 'black' oil because of its dark green color and, from 1925 to 1929, hempseed production was reported to be just over 500,000 MT/year (Kaufmann & Juschkewitsch 1930).

As a commodity, hempseed and hempseed oil occupy markets similar to flaxseed and flaxseed oil, with the latter production being about 70 times more than that of hempseed from 1999 to 2004, according to FAO data. Like hempseed, only a relatively small amount of flaxseed is cultivated for the production of human foods, while most flaxseed oil goes for industrial purposes, such as the production of paints, with the seed meal being used as a vegetable protein additive in animal feed.

Bulk quantities of food-grade hempseed and hempseed oil from Canada and Europe are still relatively expensive, as the cultivated area has not yet reached the levels of other grain commodities, and production costs are still high. Most hempseed for human food is produced in temperate climes throughout the world, although one early-maturing oilseed variety (i.e., Finola) grows well in the arable regions of subarctic Canada and Northern Europe (Callaway, 2002). So far, Canada seems to be the largest producer of food-grade hempseed and hempseed oil. From Canada, food-grade hempseed in lots of 1 MT or more costs about € 0.90/kg for conventional grain and about € 1.50/kg for certified organic grain. Bulk hempseed oil from Canada, in 1000-L bladders, may cost about € 5.00/L for conventional and € 9.50/L for certified organic. Retail unit prices vary considerably, along with quality; however, a direct correlation may not necessarily exist between price and quality, in most cases. Hempseed oil is typically sold in retail units of 250 or 500 mL, with conventional oil currently retailing at about € 6–7 for the smaller size and € 10–11 for the larger size. One can find certified organic hempseed oil for about € 9–10/250 mL and € 14–15/500 mL. European production of dietary hempseed and hempseed oil has lagged behind that of Canada, plus the costs of food production are higher in Europe, with the areas of cultivation being considerably smaller. For these reasons, retail prices for hempseed oil in Europe are currently about 20–30% higher than prices in Canada.

In Canada, 256 MTs of food-grade hempseed were exported in 2006, and 306 MTs were reported to have been exported by May of 2007, according to Agriculture and Agri-Food Canada. Assuming that a total of 600 MTs were exported from Canada by the end of 2007, this would amount to about 3% of the world's exported hempseed. Over 90% of Canadian hempseed export is sterilized and sent to the United States for the production of human foods, where one is still forbidden to cultivate industrial hemp crops for any purpose, and viable hempseed in the United States is indiscriminately lumped into the same legal category as drug-*Cannabis*.

According to the FAO, most varieties of hemp are cultivated for fiber. However, a harvest must produce sufficient seed to allow for cultivation in subsequent years, and some of this is inevitably utilized as food. Only a few varieties of hemp (e.g., Finola, Craig, and USO-31) are designated specifically as oilseed varieties.

## Current Applications of Hempseed Components

As mentioned in the Introduction, and presented in Tables 6.1, 6.2 and 6.3, hempseed is an incredibly rich source of beneficial dietary components. The demonstrated health benefits of hempseed and hempseed meal, primarily as animal food, is discussed in a later section. It is worth mentioning that a protein powder from sieved hempseed meal is currently the major human food product made from hempseed in Canada today. This is sold as a protein supplement that is added to baked goods or beverages. Pure edestin, a white protein powder isolated from hempseed meal, is commercially available from China, but the price of this material is currently more than twice that of whey protein. The health benefits of hempseed oil are also discussed in a later section of this chapter.

## Nonedible Applications

The primary non-food industrial use for hempseed oil originates from its high level (approximately 80%) of PUFAs (Table 5.2), which readily polymerize upon exposure to atmospheric oxygen. Such 'drying oils' are useful for the production of paints, varnishes, sealants and such durable goods as floor coverings e.g., linoleum (a floor covering made from flaxseed oil, that was invented in 1860) and other bio-plastics. Highly unsaturated vegetable oils from hemp and flax seeds could serve as hydrocarbon feed stocks for the production of plastics, glues and resins, in much the same way as plastics that are presently derived from petroleum. However, it is unreasonable to expect that seed oils can replace the enormous world requirements for petroleum fuels in the future. Because polymerization occurs more thoroughly with an increasing number of double bonds, the greater amount of unsaturated fatty acids in flaxseed oil results in a harder, more brittle quality in the dried product. This is also why flaxseed based paints, coatings and sealants tend to eventually crack after a few years, while similar products from hempseed oil remain pliable for longer periods of time.

Hempseed oil has also found some limited use in body care products, particularly as an additive to soaps and shampoos. Unfortunately, most products seek to gain value by simply stating the presence of hemp oil on the label, rather than having much of it present as a physical component. The idea of adding hempseed oil to a soap or shampoo does have appeal. However, without adequate precautions taken during formulation and packaging, these highly unsaturated hydrocarbons will oxidize faster than vegetable soaps made from palm or olive oils. Such an oxidized product, containing oil polymers, may leave a residual greasy feeling on the skin, which can be difficult to completely rinse away. On the other hand, EFAs and especially GLA and SDA, do find a demonstrated utility in skin creams and moisturizers, provided that these PUFAs are not oxidized before they are applied. Although triglycerides do not penetrate healthy skin to any significant extent, they do promote the healing of dry and damaged skin, where blood vessels are closer to the contact surface and PUFAs can act directly on locally damaged epidermal tissues.

## Applications of Hempseed and Hempseed Meal

Most of the world's hempseed is consumed by small birds, primarily as commercial birdseed, which is the major use of exported hempseed from China. Wild birds also take their share of seed in the field and after storage. The flocking of migratory birds in a field of mature hempseed is a very good indication for the time of harvest. Birds fare well on a diet of hempseed, either as whole seed for migratory songbirds, or as feed made from hempseed or hempseed meal for domestic poultry. Hemp is also an excellent game crop that provides cover for a wide variety of birds and other animals, with the short oilseed hemp varieties providing a safer environment for hunters to more easily see each other. It is not uncommon to observe geese trampling hemp crops at high northern latitudes, to gain access to the nutrient-rich seed (unpublished observations). Moreover, flocking migratory birds and small mammals that inhabit oilseed hemp fields throughout summer, and especially near harvest time in the late autumn, attract raptors such as hawks and falcons, offering dynamic displays of predator versus prey action.

The main advantage of hempseed meal over rapeseed and flaxseed meals in animal feed is the lack of anti-nutritional components and toxic glycosides (Matthäus 1997). Linamarin, lotaustralin, and other cyanogenic glycosides are present in flaxseed and flaxseed meal at about 0.2% each (Palmer et al. 1980). Raw, whole flaxseed and flaxseed meal that has been heated only to 'cold-pressing' temperatures (e.g., 40–50°C) can be toxic to animals, especially if the seed or the cake is wetted before processing into feed. Under moist conditions, the seed enzyme linease will release prussic acid (i.e., hydrogen cyanide gas, HCN) from the glycoside. Some naïve flaxseed oil producers even unwittingly advertise the 'delightful almond flavor' of HCN in their product! Bioactive, and especially toxic, levels of HCN limit the amount of flaxseed meal that may safely be fed to poultry and other animals (Wanasundara & Shahidi 1998). Under high-temperature treatment, such as boiling for 10 minutes or steaming under pressure, linease is destroyed and the immediate release of HCN from flaxseed meal can be significantly reduced, although these cyanogenic compounds also spontaneously release HCN over time (Frehner et al. 1990). Extraction with chloroform, dichloromethane, trichloroethylene or carbon tetrachloride removes the glucoside, but subsequent residues of these halogenated solvents may remain in the meal. Using hempseed, which lacks these toxic factors, avoids the additional costs of high-temperature treatment or solvent extraction to remove or reduce the risk of HCN in flaxseed products.

According to the FAO, the primary use of hempseed meal is for fattening cattle (supplements to 3 kg per day) and sheep (supplements to 0.5 kg per day). The utility of hempseed meal in feed for ruminants and laying hens has more recently been investigated in Canada. For example, Mustafa et al. (1999) described hempseed meal as an excellent source of rumen undegraded protein. Silversides et al. (2005) found that increasing the amount of hempseed meal to 20% of the feed in the diet of laying hens led to significant increases in levels of EFAs and decreased levels of palmitic acid in their eggs.

## Flavor and Aroma Components of Hempseed Oil

Fresh, cold-pressed hempseed oil from good quality seed typically offers a delicious combination of citrus, mint and pepper flavors from the oil. These organoleptic components can vary according to seed variety and growing conditions, but are particularly dependent upon the way the seed is dried and stored. The delicate flavors in hempseed oil result primarily from volatile terpenes (Mediavilla & Steinemann 1997, 1998, ElSohly 2002). Thus, it is extremely important to dry hempseed slowly, and at low temperatures (<25°C), to its target moisture content of just under 10%. Fast drying, especially at elevated temperatures, not only makes control of seed moisture difficult, but also results in a subsequent loss of the delicate flavors. The seed also acquires unwanted tastes and smells that are reminiscent of jute rope or burlap sacks (Fig. 5.3) due to oxidation. For these reasons, hempseed oil should ideally be pressed from seed that is not over one year old.

The taste and smell of paint in hempseed oil is the inevitable result of liberated free fatty acids from the triglyceride, and fatty acid oxidation. Fungal infection of seed that is not properly dried can leave sharp odors of fish or ammonia. A fruity smell can eventually develop in hempseed oil through the gradual production of aliphatic esters from oxidized fatty acids (deMan 2000).

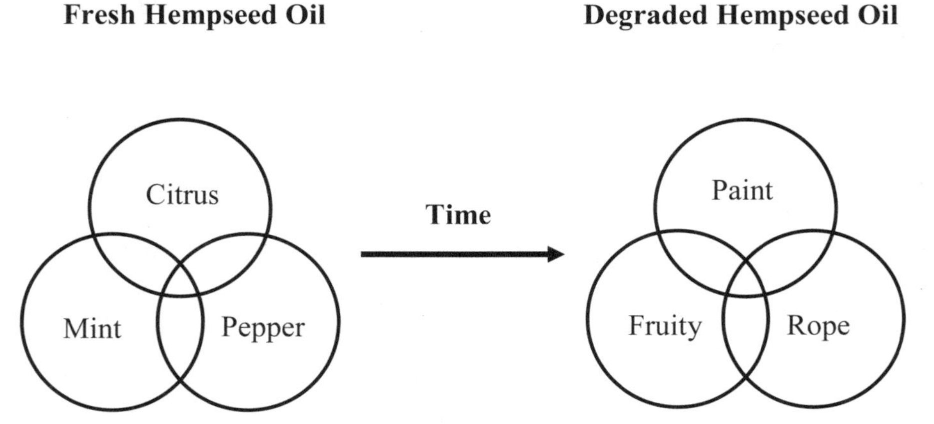


Fig. 5.3. Major organoleptic properties in fresh (left) and degraded (right) cold-pressed hempseed oil.

## Possible Allergic Reactions

To date, no reports exist of allergic reactions to hempseed oil. However, it is important to note that hempseed is technically a nut, and one of the proteins in hempseed is albumin, so those with nut and/or egg allergies should certainly be careful when trying hempseed foods, as with any new food. However, in light of this information and the wider availability of hempseed foods over the last decade, it is surprising that only one published medical report exists for an allergy that a hempseed food product could have reasonably precipitated. In 2003, this allergic reaction occured after an individual consumed part of a restaurant meal that contained dehulled hempseed (Stadtmauer, 2003).

Before hempseed foods became widely available, allergy reports were limited to medical studies on respiratory problems in workers who processed hemp fiber (also common with other natural fibers) under the very dusty industrial conditions which existed until the early decades of the 20th century. Similarly, one other report on a possible allergic reaction to hempseed, as birdseed, merits a mention. The reported context was vague, yet it seems as if the allergy was precipitated by dust from unclean birdseed (i.e., hempseed), which the person possibly inhaled while feeding a pet bird (Vidal et al. 1991).

From the evidence thus far, it seems safe to say that hempseed foods, and particularly hempseed oil, are no more allergenic that other common foods, and probably much less so.

## Health Benefits of Hempseed Oil

Although there are numerous articles on the health benefits of polyunsaturated oils (PUFAs), there are very few published studies, to date, on the nutritional effects of dietary hempseed oil, and fewer still that can be considered to be well-controlled clinical investigations into its putative health benefits. Aside from the inherent merits of the fatty acid profile for explaining the beneficial effects of hempseed oil (Table 5.2), the molecular distribution of individual fatty acids on the glycerol molecule of dietary triglycerides offers another important piece of evidence to consider (Table 5.5).

In dietary oils, the fatty acids on the outer positions of the triacylglycerol (TAG) backbone (i.e., the sn-1 and sn-3 positions) are readily hydrolyzed during digestion by pancreatic and lipoprotein lipases, primarily in the duodenum (for a recent review, see Karupaiah & Sundram 2007). The remaining fatty acid at the sn-2 (center) position of the newly formed monoacylglyceride is substantially preserved in chylomicrons, where a new triglyceride may then be reconstructed with other fatty acids in the body. Alternatively, this monoacylglyceride may serve as a precursor in either gut or liver phospholipid syntheses. Thus, preservation of the native fatty acid in the sn-2 position has a prolonging effect in subsequent metabolic processes, such as construction of the phospholipid bilayers of all living cells. Clearly, there is considerable benefit in having the EFAs (i.e., LA and ALA) at the center of the TAG structure, the sn-2 position, for long-term health effects from dietary oils. From the information in Table 5.5, it seems that hempseed oil is the only commercial food oil that can deliver both EFAs at the sn-2 position in significant amounts, as the triacylglycerols LnLLn, LLL and LLnLn make up about 45% of the total found in hempseed oil.

The motivation to investigate possible improvements in health from dietary intake of hempseed oil has come from folklore, traditional medicine, and modern anecdotes derived from the introduction of hempseed foods to modern consumers in Europe and North America. Such contemporary stories began to circulate during the 1990s, particularly concerning the ability of dietary hempseed oil to improve skin quality and certain skin conditions, within a couple of weeks, when taken in modest daily amounts as a dietary supplement. Due to the lipophilic character of the oil, and exceptionally high content of PUFAs, it is readily absorbed for metabolism and immediate use, or stored in adipose tissues for later use. For example, hempseed oil is not an effective laxative, even in relatively large amounts (100-200 ml), due to its rapid absorption and low content of saturated fats. Another consistent anecdotal report from consumers over the last decade is that dietary hempseed oil results in thicker hair and harder nails. These later effects are noticed only after several months, due to the slower growth of these dermal tissues, when compared with the rate of squamous cell turnover in normal skin. Not surprisingly, skin, hair, and nails develop from the same dermal stem cell line, and it seems that the PUFAs in hempseed oil are critical to the appropriate construction and subsequent function of these similar tissues.

Brittle hair and nails are certainly not life-threatening maladies, and it is quite possible to go through life with skin that does not retain water very well. However, these superficial imperfections are symptomatic of other cellular processes that are less easily observed, as all cells in the body require a sufficient amount of EFAs in a balance that supports the optimal metabolism of omega-6 and omega-3 fatty acids.

**Table 5.5. Stereospecificity (*sn*-1, *sn*-2, *sn*-3) of the Three Predominant Triacylglycerol (TAG) Species Found in HSO, Compared to Other Vegetable Oils**

| | | | |
|---|---|---|---|
| [1]Hempseed oil | LnLLn | LLL | LLnLn |
| [2]Flaxseed oil | LnLnLn | LnLnL | LnLnO |
| [3]Rapeseed oil | OOO | LOO | OOLn |
| [3]Olive oil | OOO | OOP | OLO |
| [3]Corn oil | LLL | LOL | LLP |
| [3]Soybean oil | LLL | LLO | LLP |
| [3]Sunflower oil | LLL | OLL | LOO |

Ln = α-linolenate (18:3n-3), L = linoleate (18:2n-6), O = oleate (18:1n-9), P = palmitate (16:0). Sources: [1]Larson & Graham, 2007; [2]Krist et al., 2006; [3]Karupaiah & Sundram, 2007.

Immunity and autoimmune disease are also linked to diet and EFA deficiencies or imbalances (Harbige 1998). So far, only two well-controlled, human clinical trials have been performed with hempseed oil, and both studies resulted in beneficial results (Callaway et al., 2005; Schwab et al., 2006).

In the first of these published studies (Callaway et al. 2005), hempseed oil versus olive oil were compared in a 20-week randomized crossover study of 20 patients with eczema, which is also known as atopic dermatitis. Eczema is a condition of chronic dry skin that is often itchy and painful, and has been linked (e.g., Thijs et al., 2000; Horrobin, 2000) to metabolic disturbances in EFA metabolism. Scratching, typically during sleep, results in abrasions that are slow to heal and easily infected. In the Callaway et al. study, patients were instructed to take two tablespoons of the study oil each day (30 ml/day), and suggestions were given to easily include this amount of oil within their normal diets. Each patient took one of the two study oils for eight weeks, followed by a washout period of four weeks. After the washout period, the patient crossed-over to the other study oil for the remaining eight weeks. Fatty acid profiles were determined from blood triglycerides, cholesteryl esters and the phospholipid fractions. Levels of both EFAs and GLA increased significantly in all fractions after administration of the hempseed oil, with no significant increase in arachidonic acid after the eight-week intervention. No adverse effects were reported by the patients or observed by the clinicians. Skin dryness and itchiness were also measured, and these symptoms improved significantly after hempseed oil administration, with no corresponding improvements after the olive oil treatment. Patient use of topical dermal medications decreased significantly only after hempseed oil consumption. This study concluded that the improvements in atopic symptoms resulted from the balanced and abundant supply of the PUFAs in hempseed oil (Callaway et al. 2005).

In the second published study (Schwab et al. 2006), dietary hempseed oil was compared with flaxseed oil in a group of 14 healthy volunteers. Hempseed oil and flaxseed oil each contain high amounts of both EFAs, but in approximately inverse proportions (Table 5.2). However, flaxseed oil has neither GLA nor SDA. While an excessive intake of one EFA over the other was thought to interfere with their respective metabolisms, through competition for Δ-6-desaturase, a controlled clinical study had never been conducted to prove this hypothesis in humans. Again, a randomized crossover design was used, wherein the volunteers consumed 30 ml/day of the study oil, but only for four weeks, with a four week wash-out period between each intervention. The results were striking. Increased levels of both EFAs were again found in blood cholesteryl esters and triglycerides after intervention with both study oils, and considerably more ALA was sequestered after the flaxseed oil intervention. GLA levels increased after ingesting the hempseed oil, but there was a dramatic decrease in GLA after taking flaxseed oil. Thus, it was demonstrated that excessive ALA (the omega-3 EFA) from flaxseed oil does compete for access to Δ-6 desaturase and inhibits the production of GLA from LA (the omega-6 EFA). It was not surprising that

GLA increased after ingesting hempseed oil, as this fatty acid is a natural component of that oil. The other remarkable finding in this study was a trend towards a lower "total-cholesterol" to "HDL-cholesterol" ratio after hempseed oil, which was not seen after the intervention with flaxseed oil. This effectively means an improvement in the "good cholesterol" over the "bad cholesterol" with dietary hempseed oil, but not with flaxseed oil. Neither any significant differences nor adverse effects were found between the experimental periods for either oil in measured values of fasting serum total lipids or lipoprotein lipids, plasma glucose, insulin or hemostatic factors. It is especially remarkable that these statistically significant findings were obtained from such a small group of volunteers.

SDA, a rare omega-3 biological metabolite of ALA that is present in hempseed oil, was detectable at low levels, but not quantified in either of these two clinical studies, due to its rapid metabolism to longer and more unsaturated omega-3 metabolites. It is a pity that more clinical information is not available for SDA, as it does not compete for delta-6-desaturase (as does its immediate precursor, ALA), and serves as an unrestricted dietary precursor to the biologically active EPA (20:5n3) (James et al., 2003). Moreover, SDA can provide similar metabolic benefits as fish oils, but does not suffer from the organic and inorganic contaminants that are found in fish oils. A recent study with dogs (Harris et al 2007) clearly indicated that SDA supplementation increased levels of EPA in the heart and red blood cells, and concluded that SDA may have utility as a safe, plant-based source of omega-3 fatty acid.

An observational study, with intention to treat, reported on the utility of hempseed oil topically applied to healing mucosal wounds after ear, nose, mouth and throat surgery, and concluded that hempseed oil provided rapid and complete support for wound healing (Grigoriev, 2002). Although triglycerides do not normally pass through intact skin tissues, the oil apparently affected wound healing by direct contact with blood capillaries and deeper tissues of the damaged mucosa. This finding is consistent with numerous other clinical studies that have demonstrated the utility of EFAs and other PUFAs in healing and immune response (e.g., Manku et al., 1982, 1984; Bordoni et al., 1988; Oliwiecki et al., 1991; Sakai et al., 1994; Yu & Björkstén, 1998; Derek & Meckling-Gill, 1999; Harbige et al., 2000; Harbige & Fisher, 2001; Horrobin, 2000; Simopoulos, 2002a & b; Simopoulos, 2006).

The fatty acid profile of hempseed oil (Table 5.2) is remarkably similar to that of black currant seed oil (Laakso & Voutilainen 1996), which also seems to have a beneficial impact on immunologic vigor (Wu et al., 1999; Barre, 2001). Borage, which is rich in GLA, but almost totally lacking in omega-3 PUFAs (Laakso & Voutilainen, 1996), is fairly well tolerated as a source of this fatty acid (Takwale et al., 2003), but perhaps more than just GLA is required in some disease states, such as atopy (Whitaker et al., 1996; van Gool et al., 2003; Callaway et al., 2005).

A porridge made from crushed oats and hempseed is a traditional food in the Czech Republic and other areas of Eastern Europe, much like the Chinese hempseed porridge *hou ma you*. One early published report from the former Czechoslovakia described the use of such porridge in treating children who had tuberculosis (Sirek, 1955). In this interesting report, improvements in the children's health were evaluated by medical diagnosis and confirmed by chest X-rays. Considering what is now known, it seems that a rapid improvement of nutritional input, due to hempseed proteins and PUFAs, was probably responsible for the dramatic results stated in this report. Recent investigations into the role of PUFAs for treating tuberculosis support this suggestion (Anes et al., 2003; Russel, 2003).

In rats, dietary hempseed oil reduced platelet aggregation upon the addition of 5% and 10% hempseed meal, containing residual hempseed oil, to their chow (Richard et al., 2007). These amounts were comparable to relative levels of human consumption. Platelet aggregation and rate of aggregation were significantly inhibited by the presence of either 5% and 10% hempseed meal in the rat chow. These effects were not seen with the control diet that contained a corresponding amount of palm oil. It was concluded that the observed effect of improved platelet function was due to the oil of the added hempseed meal.

Also in rats, an increase in plasma EFA profiles was observed (Al-Khalifa et al., 2007) after feeding them a diet that contained hempseed meal. The results were in line with those already seen in humans with hempseed oil (Callaway et al., 2005; Schwab et al., 2006). Moreover, improved heart function and significant cardio-protective effects were observed during post-ischemic reperfusion after the rats consumed 5 or 10% of their food as hempseed meal (Al-Khalifa et al., 2007). In another study (Karimi & Hayatghaib, 2006), rats were fed whole hempseed for only 20 days, which significantly decreased their mean fasting serum LDL level and significantly increased the mean fasting serum HDL and total protein levels. In that study, the authors concluded that short-term hempseed feeding improved rat blood lipid and protein profiles, and recommended that individuals with high cholesterol and high LDL levels, or those affected with coronary artery and liver diseases, incorporate hempseed into their food preparation. A similar improvement in LDL and HDL levels was observed as a statistical trend in the aforementioned human clinical trial with hempseed oil (Schwab et al., 2006).

## Hempseed Oil Stability

An unfortunate paradox of hempseed oil resides within its unsaturation, as this structural feature makes it both highly nutritious and chemically unstable (deMan, 2000). This vulnerability is not unique to hempseed oil, but is characteristic of any unsaturated oil, and is determined primarily by the degree of unsaturation of each of its component fatty acids, and the percentages of such compounds within the oil. This chemical difference explains why relatively saturated palm and coconut oils are more stable (and also solid at room temperature) than highly unsaturated hemp and flaxseed oils (which are still liquid at very cold temperatures).

Degradative processes in unsaturated oils were originally thought to begin with oxidation of the chemical double bond. Although reactions can occur at this site, the predominant mechanism is now recognized to be a free-radical reaction at the allylic carbon atoms adjacent to the double bond. These methylene carbons are more susceptible to reaction with oxygen due to the low dissociation energy of their hydrogen atoms (Ohloff, 1973). Thus, a predominantly monounsaturated mixture, such as olive oil, has one minor site (i.e., the double bond) and two major sites (i.e., the allylic carbons) of lability on its oleic acid constituent, which makes up over 70% of the fatty acids present.

In the case of LA, a *bis*-allylic methylene carbon helps to render this molecule 12-fold more oxidatively labile than oleic acid (Gunstone and Hilditch 1945). Multiple *bis*-allylic carbons are found on fatty acids of greater unsaturation, which makes them proportionately more labile. For example, the additional double-bond of ALA or GLA endows these compounds with twice the instability of LA (Cosgrove et al., 1987). It should be remembered that an unsaturated oil is only as stable as its most unstable component molecule, because once oxidative degradation begins, these reactions rapidly proliferate to include previously unaffected molecules. Excellent and detailed reviews of vegetable oil oxidation are available (e.g., deMan, 2000; SRI, 2005).

The potential instability of unsaturated oils is provoked to actually begin the process of oxidation by both intrinsic factors and assorted environmental influences (Sherwin, 1978). The first problem begins when enzymes are released from their proper sequestration within cellular structures of the oil-storage endosperm during the process of extraction. Lipase and lipoxygenase are consequently freed to disrupt triglyceride and fatty acid structures, respectively, immediately upon oil manufacture unless the oil "drip temperature" is hot enough (circa 60°C) to denature these proteins. However, even the relatively low temperatures (> 40°C) of "cold-pressed" oil manufacture promote oxidation when the press screw thoroughly adds atmospheric oxygen, so care must be taken to extract under an inert atmosphere (e.g., nitrogen). In addition, the use of transition metals such as iron, or particularly copper, to manufacture oil-exposed parts of the press is problematic. Trace amounts of these metals (i.e., <1 ppm) can catalyze degradation (Sherwin, 1978), especially if the oil contains significant amounts of the more reactive free fatty acids. Ideally, the working surfaces of press heads should be made from completely inert materials, such as high strength ceramic. Failing that, chelating agents (e.g., citric acid) should be promptly added to the oil, especially if the receiving or bulk-storage vessels are metallic.

Light also accelerates oxidation, and the chlorophyll found in raw, unprocessed oils will promote this process through its ability to capture light energy. Therefore, unnecessary exposure to light must be avoided during manufacture, and consumer bottles should be made of dark glass. Opaque plastic is sometimes used, but leaching of plasticizers and other chemical components into the oil may be a health hazard, unless the bottle interior is coated with an acceptable shielding resin.

Differences of stability between any two highly unsaturated oils, such as flaxseed and hempseed, can be observed and are at least partially attributable to the proportions of ancillary components within their respective seeds (Abuzaytoun & Shahidi, 2006), which are co-expressed with the oil. These components include phenolic pigments that act as anti-oxidants, or specific anti-oxidants such as the tocopherols. Tocopherols act as in situ preservatives during natural over-wintering of the seed, or during storage of its unadulterated oil. The various tocopherols differ in their respective anti-oxidant characteristics and biological activities. The alpha isomer of tocopherol (Vitamin E) has a high value in human nutrition, yet is relatively poor as an anti-oxidant (Helm, 2006). In contrast, the beta-, gamma- and delta-tocopherols are not nearly as effective as nutrients, but are 130, 200, or 500 times more powerful, respectively, in their anti-oxidant capacities (Helm, 2006). The primary tocopherol in flaxseed oil is the gamma isomer, with minor amounts of alpha-tocopherol (Abuzaytoun & Shahidi, 2006). Hempseed oil has somewhat more gamma-tocopherol as its major anti-oxidant, with collectively significant amounts of the alpha, beta, and delta isomers (Kriese et al., 2004; Blade et al., 2005; Abuzaytoun & Shahidi, 2006). The addition of ascorbyl palmitate is recommended to synergize the anti-oxidant action of these native tocopherols (Helm, 2006). Minor amounts of other native anti-oxidants, such as plastochromanol-8 (Kriese et al., 2004), cannabidiol (Hampson et al. 1998), or terpenes (Radonic & Milos, 2003), may also contribute to the natural stability of hempseed oil. Overall, differences in anti-oxidant quantity and composition, combined with the much greater amounts of unstable ALA in flaxseed oil, make hempseed oil the more stable of the two (Ramadana & Moersel, 2006). The previously mentioned review of vegetable oil oxidation (SRI, 2005) also contains a comprehensive review of anti-oxidants and oxidation test methods.

Because of the aforementioned oxidative properties, it should be obvious that hempseed oil and other polyunsaturated oils should not be used for frying, and moreover, that frying foods is an inherently unhealthy practice, as even the most saturated natural fats and oils contain some polyunsaturates. In general, the use of hempseed oil in any type of cooking should be limited to the temperature of boiling water. Interestingly, the internal temperature of baking bread does not surpass this threshold (Seiz, 2004). At most, the temperature of hempseed oil should not exceed about 120°C (Oomah et al., 2002), which is approximately the temperature found in pressure-cooking, and then only for relatively short periods of time. Not only are PUFAs vulnerable to high temperatures, but also tocopherols begin to degrade above 50°C, a process which accelerates above 100°C (Kerschbaum & Schweiger, 2007). Observing these use parameters to prevent oil oxidation will easily prevent the formation of the *trans* isomers (Mjos & Solvang, 2006) formed at higher temperatures (Wolff, 1993, Koletzko & Decsi, 1997), which have been linked to coronary heart disease and other chronic health problems (Woodside & Kromhout, 2005). In general, the best way to use hempseed oil is simply as an ingredient in salad dressings, dips and sandwich spreads, or as a substitute for butter on cooked foods such as bread, pasta, and vegetables.

## Whole and De-hulled Hempseed

The tough, fibrous shell of whole hempseed can limit the digestibility of the hempseed meal that is produced as a by-product of oil pressing or extraction. This limitation is dependent on the physical size of the shell particle present in the feed, especially in pig and poultry feeds, according to the FAO. However, whole hemp seed can be fed to poultry and other birds if access to sandy gravel is provided for the avian crop. The hempseed hull is not without value, as it contains considerable amounts of phytosterols such as sitosterol and other nutriceutical components (Jeong et al., 1974; Leizer et al., 2000).

De-hulled hempseed is a new and useful product that was not commercially available before the late 1990s. With the hull removed from hempseed, both oil and protein values increase dramatically (Table 5.1). De-hulled hempseed can be eaten directly or added as an ingredient in the preparation of foods, especially "smoothies" (Pate, 2008). Grinding de-hulled hempseed with water produces a tasty vegetable drink which looks like milk and tastes like walnuts or sunflower seeds. This taste is easily modified by a variety of flavors, ranging from sweet to salty. At a ratio of 1 cup de-hulled hempseed to one cup of water, the resulting "milk" can be used as a substitute for eggs and dairy products in the production of baked goods. For example, cakes and muffins can be made without milk or eggs, as hempseed proteins will denature in the same manner as egg white to produce the desired texture.

## Other Issues

The possibility always exists for the adulteration of vegetable oils, particularly when supplies are limited and unscrupulous individuals try to stretch a limited resource by diluting the more valuable component with something cheaper. Dilution by the addition of another vegetable oil will normally maintain the appearance of cold-pressed hempseed oil, due to the high level of chlorophyll in the latter. From our analysis of hundreds of hempseed oil samples, only very few samples of such adulteration of hempseed oil adulteration, presumably with cheaper rapeseed oil, were identified in the mid-1990s by gas chromatography. This was noticed primarily as an unusually high amount of oleic acid (>15%) in these particular samples. Possibly, these samples could have been adulterated with olive oil, but its higher price and higher level of oleic acid make this choice less likely.

Through carelessness or neglect, dirt and other potentially harmful contaminants can become part of hempseed oil during the crushing process. Detectable amounts of THC and other cannabinoids can also adhere to the hull of poorly cleaned hempseed. In the past, such poorly crafted oils resulted in a positive urinalysis for *Cannabis* metabolites in some individuals who were subjected to drug-testing procedures that are unable to differentiate between the consumption of illegally smoked drug-*Cannabis* and the legal consumption of hempseed oil (Callaway et al., 1997; Lehmann et al., 1997). A similar dilemma exists with poppy seeds (which contain small amounts of morphine and other opioids), which are popular in many baked goods. In the latter case, drug-testing officials quickly responded by raising the cutoff level for urinary opioid metabolites to a reasonably high level (2000 ng/mL).

Unfortunately, a special form of U.S. hysteria that can only be described as "cannabiphobia" has not allowed for a similarly lucid approach to be applied to hemp foods, apparently in the fear that a few occasional users of drug-*Cannabis* might slip by the detection process. Therefore, the cutoff level for a positive determination of cannabinoid metabolites in the urine remains at the exceptionally low value of 50 ng/mL, which is also the lower limit of quantitation for radioimmunoassay. The nascent hemp food industry in North America was quick to react to the potential negative implications for their customers under the so-called "zero tolerance" policies of the United States. Consequently, industry procedures were instituted to press hempseed oil only from well-cleaned hemp seed of varieties that produce only very low levels of THC (Leson et al., 2000). Such low levels of THC in hemp foods are no longer a cause for regulatory concern in workplace drug testing (Lachenmeier & Walch, 2005).

These onerous conditions originate within a U.S. political landscape that forbids the cultivation of industrial hemp, and intentionally seeks to confuse it with drug-*Cannabis*, even though hemp is otherwise universally acknowledged as having absolutely no value as a drug substance. Moreover, hemp varieties have been shown to produce more cannabidiol than THC, while drug varieties of *Cannabis* tend toward the inverse ratio (Hillig and Mahlberg, 2004; Mechtler et al. 2004). Aside from being a potent oil-soluble antioxidant and a useful phytochemical marker, cannabidiol can effectively attenuate the putative psychoactive effects of low THC levels by binding to cannabinoid (CB1) receptors in the brain (Pertwee, 2008). Unfortunately, the reasons for this intransigent position towards hemp in the U.S. are essentially doctrinal rather than rational, so it may take a bit more time and public education for hempseed products to become more widely recognized there as uniquely useful functional foods.

## References

Abel, E.L. *Marijuana: The First Twelve Thousand Years;* Plenum Press: New York, 1980.

Abuzaytoun, R.; F. Shahidi. Oxidative stability of flax and hemp oils. *J. Am. Oil Chem. Soc.* **2006,** *83(10),* 855–861.

Al-Khalifa, A.; T.G. Maddaford; M.N. Chahine; J.A. Austria; A.L. Edel; M.N. Richard; B.P. Ander; N. Gavel; M. Kopilas; R. Ganguly; P .K. Ganguly; G.N. Pierce. Effect of dietary hempseed intake on cardiac ischemia-reperfusion injury. *Am. J. Physiol. Regul. Integr. Comp. Physiol.* **2007,** *292(3),* R1198–1203.

Anes, E.; M.P. Kuhnel; E. Bos; J. Moniz-Pereira; A. Habermann; G. Griffiths. Selected lipids activate phagosome actin assembly and maturation resulting in killing of pathogenic mycobacteria. *Nat. Cell Biol.* **2003,** *5(9),* 793–803.

Anwar, F.; S. Latif; M. Ashraf. Analytical characterization of hemp (*Cannabis sativa*) seed oil from different agro-ecological zones of Pakistan. *J. Am. Oil Chem. Soc.* **2006,** *83(4),* 323–329.

Bagci, E.; L. Bruehl; K. Aitzetmuller; Y. Altan. A chemotaxonomic approach to the fatty acid and tocochromanol content of *Cannabis sativa* L. (Cannabaceae). *Turk. J. Bot.* **2003,** *27,* 141–147.

Barre, D.E. Potential of evening primrose, borage, black currant, and fungal oils in human health. *Ann. Nutr. Metab.* **2001,** *45,* 47–57.

Blade, S.F.; K. Ampong-Nyarko; R. Przybylski. Fatty acid and tocopherol profiles of industrial hemp cultivars grown in the high latitude prairie region of Canada. *J. Ind. Hemp* **2005,** *10(2),* 33–43.

Bordoni, A.; P.L. Biagi; M. Masi; G. Ricci; C. Fanelli; A. Patrizi; E. Ceccolini. Evening primrose oil (Efamol) in the treatment of children with atopic dermatitis. *Drugs. Exp. Clin. Res.* **1988,** *14(4),* 291–297.

Callaway, J.C. Hempseed as a nutritional resource: an overview. *Euphytica* **2004,** *140,* 65–72.

Callaway, J.C. Hemp as food at high latitudes. *J. Ind. Hemp* **2002,** *7(1),* 105–117.

Callaway, J.; U. Schwab; I. Harvima; P. Halonen; O. Mykkänen; P. Hyvönen; T. Järvinen. Efficacy of dietary hempseed oil in patients with atopic dermatitis. *J. Derm. Treat.* **2005,** *16,* 87–94.

Callaway, J.C.; T. Tennilä; D.W. Pate. Occurrence of "*omega*-3" stearidonic acid (*cis*-6,9,12,15-octadecatetraenoic acid) in hemp (*Cannabis sativa* L.) seed. *J. Int. Hemp Assoc.* **1997,** *3,* 61–63.

Callaway, J.C.; R.A. Weeks; L.P. Raymon; H.C. Walls; W.L. Hearn. A positive urinalysis from hemp (*Cannabis*) seed oil. *J. Anal. Toxicol.* **1997,** *21(4),* 319–320.

Chuan-He, T.; T. Zi; W. Xian-Sheng; Y. Xiao-Quan. Physicochemical and functional properties of hemp (*Cannabis sativa* L.) protein isolate. *J. Agric. Food Chem.* **2006,** *54(23),* 8945–8950.

Collin, P. New diagnostic findings in coeliac disease. *Ann. Med.* **1999,** *31(6),* 399–405.

Cosgrove, J.P.; D.F. Church; W.A. Pryor. The kinetics of the autoxidation of polyunsaturated fatty acids. *Lipids* **1987,** *22(5),* 299–304.

Deferne, J.L.; D.W. Pate. Hemp seed oil: a source of valuable essential fatty acids. *J. Int. Hemp Assoc.* **1996,** *3(1),* 1–7.

deMan, J.M. Chemical and physical properties of fatty acids. *Fatty acids in Foods and Their Health Implications,* 2nd ed.; Ching Kuang Chow, Ed.; Marcel Dekker, Inc.: Basel, 2000; pp. 17–46.

de Padua, L.S.; N. Bunyaprafatsara; R.H.M.J. Lemmens, Eds.; *Plant Resources of South-East Asia: Medicinal and Poisonous Plants,* Backhuys Publishers: Leiden, 1999; Vol. 1(12), pp. 167–175.

Derek, J.R.; K.A. Meckling-Gill. Both (n-3) and (n-6) fatty acids stimulate wound healing in the rat intestine epithelial cell line, IEC-6. *Am. Soc. Nutr. Sci.* **1999,** *129,* 1791–1798.

ElSohly, M. *Chemical Constituents of Cannabis;* F. Grotenhermen, E.B. Russo, Eds.; Haworth Press: Binghamton, NY, 2002; pp. 27–36.

FAOSTAT. (2007) http://faostat.fao.org/

Frehner, M.; M. Scalet; E.E. Conn. Pattern of the cyanide-potential in developing fruits: Implications for plants accumulating cyanogenic monoglucosides (*Phaseolus lunatus*) or cyanogenic diglucosides in their seeds (*Linum usitatissimum, Prunus amygdalus*). *Plant Physiol.* **1990,** *94(1),* 28–34.

Gerster, H. Can adults adequately convert α-linolenic acid (18:3n-3) to eicosapentanoic acid (20:5n-3) and docosahexanoic acid (22:6n-3)? *Int. J. Vit. Nutr. Res.* **1988,** *68,* 159–173.

Grigoriev, O.V. Application of hempseed (*Cannabis sativa* L.) oil in the treatment of the ear, nose and throat (ENT) disorders. *J. Ind. Hemp* **2002,** *7(2),* 5–15.

Grigoryev, S. Hemp of Russian Northern Regions and a source of spinning fiber. *J. Ind. Hemp* **2005,** *10(2),* 105–114.

Gunstone, F.D.; T.P. Hilditch. The union of gaseous oxygen with methyl oleate, linoleate, and linolenate. *J. Chem. Soc.* **1945,** 836–840.

Hampson, A.J.; M. Grimaldi; J. Axelrod; D. Wink. Cannabidiol and (-) Δ-9- tetrahydrocannabinol are neuroprotective antioxidants. *Proc. Natl. Acad. Sci. U.S.A.* **1998,** *95(14),* 8268–8273.

Harbige, L.S. Dietary n-6 and n-3 fatty acids in immunity and autoimmune disease. *Proc. Nutr. Soc.* **1998,** *57(4),* 555–562.

Harbige, L.S.; B.A.C. Fisher. Dietary fatty acid modulation of mucosally-induced tolerogenic immune response. *Proc. Nutr. Soc.* **2001,** *60,* 449–456.

Harbige, L.S.; L. Layward; M.M. Morris-Downes. The protective effects of *omega*-6 fatty acids in experimental autoimmune encephalomyelitis (EAE) in relation to transforming growth factor-beta 1 (TGF-β1) up-regulation and increased prostaglandin E2 (PGE2) production. *Clin. Exp. Immunol.* **2000,***122,* 445–452.

Harris, W.S.; M.A. DiRienzo; S.A. Sands; C. George; P.G. Jones; A.K. Eapen. Stearidonic acid increases the red blood cell and heart eicosapentaenoic acid content in dogs. *Lipids* **2007,** *42(4),* 325–333.

Helm New York, Inc. (2006) Difference between antioxidant and vitamin activity: Natural tocopherol isomers; http://www.helmnewyork.com/en/products /specification/ tocopherols-natural-mixed.html.

Horrobin, D.F. Essential fatty acid metabolism and its modification in atopic eczema. *Am. J. Clin. Nutr.* **2000,** *71(1),* 367–372S.

James, M.J.; V.M. Ursin; L.G. Cleland. Metabolism of stearidonic acid in human subjects: comparison with the metabolism of other n-3 fatty acids. *Am. J. Clin. Nutr.* **2003,** *77,* 1140–1145.

Jeong, T.M.; T. Itoh; T. Tamura; T. Matsumoto. Analysis of sterol fractions from twenty vegetable oils. *Lipids* **1974,** *9(11),* 921–927.

Jiang, H.E.; X. Li; Y.X. Zhao; D.K. Ferguson; F. Hueber; S. Bera; Y.F. Wang; L.C. Zhao; C.J. Liu; C.S. Li. A new insight into *Cannabis sativa* (Cannabaceae) utilization from 2500-year-old Yanghai Tombs, Xinjiang, China. *J. Ethnopharmacol.* **2006,** *108(3),* 414–422.

Karimi, I.;H. Hayatghaib. Effect of *Cannabis sativa* L. seed (hempseed) on serum lipid and protein profiles of rat, Pakistan. *J. Nutr.* **2006,** *5(6),* 585–588.

Karupaiah, T.; K. Sundram. Effects of stereospecific positioning of fatty acids in triacylglycerol structures in native and randomized fats: a review of their nutritional implications. *Nutr. Metab.* **2007,** *4(16),* [E-pub ahead of print].

Kaufmann, H.P.; S. Juschkewitsch. Quantitative analyse des hanföles. *Zeitschr. Für angew. Chemie* **1930,** *43,* 90–91.

Kerschbaum, S.; P. Schweiger. The influence of temperature and storage on the composition of hemp-, borage- and wheatgerm oils. *SOFW J.* **2007,** *127(1–2),* 2–7.

Koletzko, B.; T. Decsi. Metabolic aspects of *trans* fatty acids. *Clin. Nutr.* **1997,** *16(5),* 229–237.

Kriese, U.; E. Schumann; W.E. Weber; M. Beyer; L. Bruhl; B. Matthaus. Oil content, tocopherol composition and fatty acid patterns of the seeds of 51 *Cannabis sativa* L. genotypes. *Euphytica* **2004,** *137,* 339–351.

Krist, S.; G. Stuebiger; S. Bail; H. Unterweger. Analysis of volatile compounds and triacylglycerol composition of fatty seed oil gained from flax and false flax. *Eur. J. Lipid Sci Technol.* **2006,** *108,* 48–60.

Laakso, P.; P. Voutilainen. Analysis of triacylglycerols by silver-ion high-performance liquid chromatography- atmospheric pressure chemical ionization mass spectrometry. *Lipids* **1996,** *31(12),* 1311–1322.

Lachenmeier, D.W.; S.G. Walch. Current status of THC in German hemp food products. *J. Ind. Hemp* **2005,** *10(2),* 5–17.

Larson, T.; I.A. Graham. (2007) Personal communication.

Lehmann, T.; F. Sager; R. Brenneisen. Excretion of cannabinoids in urine after ingestion of cannabis seed oil. *J. Anal. Toxicol.* **1997,** *21(5),* 373–375.

Leizer, C.; D. Ribnicky; A. Poulev; S. Dushenkov; I. Raskin. The composition of hemp seed oil and its potential as an important source of nutrition. *J. Nutraceu., Function. Med. Foods* **2000,** *2(4),* 35–54.

Leson, G.; P. Pless; F. Grotenhermen; H. Kalant; M.A. ElSohly. Evaluating the impact of hemp food consumption on workplace drug tests. *J. Anal. Toxicol.* **2000,** *25(8),* 691–698.

Manku, M.S.; D.F. Horribin; N. Morse; V. Kyte; K. Jenkins; S. Wright; J.L. Burton. Reduced levels of prostaglandin precursors in the blood of atopic patients; defective *delta*-6-desaturase function as a biochemical basis for atopy. *Prostaglandins, Leukotrienes Med.* **1982,** *9(6),* 615–628.

Manku, M.S.; D.F. Horribin; N. Morse; S. Wright; J.L. Burton. Essential fatty acids in the plasma phospholipids of patients with atopic eczema. *Br. J. Dermatol.* **1984,** *110(6),* 643–648.

Matthäus, B. Antinutritive compounds in different oilseeds. *Fett/Lipids* **1997,** *99,* 170–174.

McPartland, J.M.; G.W. Guy. The evolution of *Cannabis* and coevolution with the cannabinoid receptor—a hypothesis. *The Medicinal Use of Cannabis and Cannabinoids;* G.W. Guy, B.A. Whittle, P. Robson, Eds.; Pharmaceutical Press: London, 2004; pp. 71–102.

Mediavilla, V.; S. Steinemann. Essential oil of *Cannabis sativa* L. strains. **1997,** *J. Int. Hemp Assoc. 4(2),* 82–84.

Mediavilla, V.; S. Steinemann. Factors influencing the yield and the quality of hemp (*Cannabis sativa* L.) essential oil. *J. Int. Hemp Assoc.* **1998,** *5(1),* 16–20.

Mjos, S.A.; M. Solvang. Geometric isomerization of eicosapentaenoic and docosahexaenoic acids at high temperatures. *Eur. J. Lipid Sci. Technol.* **2006,** *108,* 589–597.

Mölleken, H.; R.R. Theimer. Survey of minor fatty acids in *Cannabis sativa* L. fruits of various origins. *J. Ind. Hemp Assoc.* **1997,** *4(1),* 13–17.

Mustafa, A.F.; J.J. McKinnon; D.A. Christensen. The nutritive value of hemp meal for ruminants. *Can. J. Anim. Sci.* **1999,** *79(1),* 91–95.

Odani, S.; S. Odani. Isolation and primary structure of a methionine and cystine-rich seed protein of *Cannabis sativa* L. *Biosci. Biotechnol. Biochem.* **1998,** *62,* 650–654.

Ohloff, G. Fats and precursors. *Functional Properties of Fats in Foods;* J. Solms, Ed.; Forster Publishing: Zurich, 1973.

Okuyama, H.; T. Kobayashi; S. Watanabe. Dietary fatty acids–the n-6/n-3 balance and chronic elderly diseases. Excess linoleic acid and relative n-3 deficiency syndrome seen in Japan. *Prog. Lipid. Res.* **1997,** *3,* 409–457.

Oliwiecki, S.; J.L. Burton; K. Elles; D.F. Horrobin. Levels of essential and other fatty acids in plasma red cell phospholipids from normal controls and patients with atopic eczema. *Acta Derm. Venereol.* **1991,** *71(3),* 224–228.

Oomah, B.D.; M. Busson; D.V. Godfrey; J.C.G. Drover. Characteristics of hemp (*Cannabis sativa* L.) seed oil. *Food Chem.* **2002,** *76,* 33–43.

Osborn, T.B. Crystallised vegetable proteins. *Am. Chem. J.* **1892,** *14,* 662–689.

Palmer, I.S.; O.E. Olson; A.W. Halverson; R. Miller; C. Smith. Isolation of factors in linseed oil meal protective against chronic selenosis in rats. *J. Nutr.* **1980,** *110(1),* 145–150.

Pate, D.W. Hemp and Flax: The smoothie. *J. Ind. Hemp* **2008,** *13*(1), 93–95.

Patel, S.; R. Cudney; A. McPherson. Crystallographic characterization and molecular symmetry of edestin, a legumin from hemp. *J. Mol. Biol.* **1994,** *235,* 361–363.

Pringle, H. Ice age community may be earliest known net hunters. *Science* **1997,** *277,* 1203–1204.

Radonic, A.; M. Milos. Chemical composition and *in vitro* evaluation of antioxidant effect of free volatile compounds from *Satureja montana* L. *Free Radical Res.* **2003,** *37(6),* 673–679.

Ramadana, M.F.; J.-T. Moersel. Screening of the antiradical action of vegetable oils. *J. Food Comp. Anal.* **2006,** *19(8),* 838–842.

Richard, M.N.; R. Ganguly; S.N. Steigerwald; A. Al-Khalifa; G.N. Pierce. Dietary hempseed reduces platelet aggregation. *J. Thromb. Haemostasis* **2007,** *5,* 424–425.

Ross, S.A.; H.N. ElSohly; E.A. ElKashoury; M.A. ElSohly. Fatty acids of *Cannabis* seeds. *Phytochem. Analysis* **1996,** *7,* 279–283.

Russel, D.G. Phagosomes, fatty acids and tuberculosis. *Nat. Cell Biol.* **2003,** *5(9),* 776–778.

Russo, E. Review: History of *Cannabis* and its preparations in saga, science, and sorbriquet, *Chemistry and Biodiversity;* 2007; Vol. 4, pp. 1614–1648.

Sakai, K.; H. Okuyama; H. Shimazaki; M. Katagiri; S. Torii; T. Matsushita; S. Baba. Fatty acid compositions of plasma lipids in atopic dermatitis/asthma patients. *Arerugi* **1994,** *43(1),* 37–43.

Schultes, R.E. Random thoughts and queries on the botany of *Cannabis. The Botany and Chemistry of Cannabis;* R.B. Joyce, S.H. Curry, Eds.; J. & A. Churchill: London, 1970; pp. 11–38.

Schwab, U.S.; J.C. Callaway; A.T. Erkkilä; J. Gynther; M.I.J. Uusitupa; T. Järvinen. Effects of hempseed and flaxseed oil on the profile of serum lipids, serum total and liporpotein lipid concentrations and haemostatic factors. *Eur. J. Nutr.* **2006,** *45(8),* 470–477.

Seiz, K. (2004) Profiting from enzyme technology and process controls. Baking Management; http://bakingmanagement.bakery-net.com/article/6160.

Sherwin, E.R. Oxidation and antioxidants in fat and oil processing. *J. Am. Oil Chem. Soc.* **1978,** *55 (11),* 809–814.

Shou-Wei, Y.; T. Chuan-He; W. Qi-Biao; Y. Xiao-Quan.Properties of cast films from hemp (*Cannabis sativa* L.) and soy protein isolates. A comparative study. *J. Agric Food Chem.* **2007,** *55(18),* 7399–7404.

Silversides, F.G.; K.L. Budgell; M.R. Lefrançois. Effect of feeding hemp seed meal to laying hens. *Br. Poult. Sci.* **2002,** *46(2),* 231–235.

Simopoulos, A.P. Essential fatty acids in health and chronic disease. *Am. J. Clin. Nutr.* **1999,** *70,* 560–569S.

Simopoulos, A.P. The importance of the omega-6/omega-3 essential fatty acids. *Biomed. Pharmacother.* **2002a,** *56(8),* 365–379.

Simopoulos, A.P. ω-3 Fatty acids in inflammation and autoimmune disease. *J. Am. Coll. Nutr.* **2002b,** *21(6),* 495–505.

Simopoulos, A.P. Evolutionary aspects of diet, the ω-6/ω-3 ratio and genetic variation: nutritional implications for chronic diseases. *Biomed. Pharmacother.* **2006,** *60(9),* 502–507.

Simopoulos, A.P.; A. Leaf; N. Salem. Workshop statement on the essentiality of and recommended dietary intakes from omega-6 and omega-3 fatty acids. *Prostaglandins, Leukotrienes Essent. Fatty Acids* **2000,** *63(3),* 119–121.

SRI (Southwest Research Institute). (2005) Characterization of biodiesel oxidation and oxidation products (CRC Project No. AVFL-2b); http://www.nrel.gov/vehiclesandfuels/npbf/pdfs/39096.pdf.

Stadtmauer, G. Anaphylaxis to ingestion of hempseed. *J. Allergy Clin. Immunol.* **2003,** *112(1),* 216–217.

Takwale, A.; E. Tan; S. Agarwall; G. Barclay; I. Ahmed; K. Hotchkiss; J.R. Thompson; T. Chapman; J. Berth-Jones. Efficacy and tolerability of borage oil in adults and children with atopic eczema: randomised, double blind, placebo controlled, parallel group trial. *Br. Med. J.* **2003,** *327(7428),* 1358–1359.

Thijs, C.; A. Houwelingen; I. Poorterman; A. Mordant; P. van den Brandt. Essential fatty acids in breast milk of atopic mothers: comparison with non-atopic mothers, and effect of borage oil supplementation. *Eur. J. Clin. Nutr.* **2000,** *54(3),* 234–238.

van Gool, C.J.; C. Thijs; C.J. Henquet; A.D. van Houwelingen;P.C. Dagnelie; J. Schrander; P.P. Menheere; P.A. van den Brandt. γ-Linolenic acid supplementation for prophylaxis of atopic dermatitis—a randomized controlled trial in infants at high familial risk. *Am. J. Clin. Nutr.* **2003,** *77,* 943–951.

Vidal, C.; R. Fuente; A. Iglesias; A. Saez. Bronchial asthma due to the *Cannabis sativa* seed. *Allergy* **1991,** *46,* 647–649.

Von Hazura, K. Untersuchungen über die Hanfölsäure. *Monatsh* **1887,** *8,* 147–155.

Wanasundara, P.K.; F. Shahidi. Process-induced compositional changes of flaxseed. *Adv. Exp. Med.*

*Biol.* **1998,** *434,* 307–325.

Weiss, E.; W. Wetterstrom; D. Nadel; O. Bar-Yosef. The broad spectrum revisited: evidence from plant remains. *Proc. Natl. Acad. Sci. U.S.A.* **2004,** *101(26),* 9551–9555.

Whitaker, D.K. ; J. Chilliers ; C. de Beer. Evening primrose (Epogam) in the treatment of chronic hand dermatitis; disappointing therapeutic results. *Dermatotogy* **1996,** *193(2),* 115–120.

WHO & FAO Joint Expert Consultation Report. Fats and oils in human nutrition. *Nutr. Rev.* **1995,** *53(7),* 202–205.

Wolff, R.L. Heat-induced geometrical isomerization of alpha-linolenic acid: effect of temperature and heating time on the appearance of individual isomers. *J. Am. Oil Chem. Soc.* **1993,** *70(4),* 425–430.

Woodside, J.V.; D. Kromhout. Fatty acids and CHD. *Proc. Nutr. Soc.* **2005,** *64(4),* 554–564.

Wu, D.; M. Meydani; L.S. Leka; Z. Nightingale; G.J. Handelman; J.B. Blumberg; S.N. Meydani. Effect of dietary supplementation with black currant seed oil on the immune response of healthy elderly subjects. *Am. J. Clin. Nutr.* **1999,** *70,* 536–543.

Xiaozhai, L.; R.C. Clarke. The cultivation and use of hemp (*Cannabis sativa* L.) in ancient China. *J. Intl. Hemp Assoc.* **1995,** *2(1),* 26–33.

Yu, G.; B. Björkstén. Polyunsaturated fatty acids in school children in relation to allergy and serum IgE levels. *Pediatr. Allergy Immunol.* **1998,** *9,* 133–138.

# Berry Seed and Grapeseed Oils

**Anna-Maija Lampi and Marina Heinonen**
*Department of Applied Chemistry and Microbiology, P.O. Box 27, Latokartanonkaari 11, FI-00014 University of Helsinki, Finland*

## Introduction

Berry seed and grapeseed oils are major by-products of the berry and grape processing industry. Black currant, blueberry, boysenberry, cherries, cloudberry, cranberry, goldenberry, marionberry, and sea buckthorn, as well as some other berry seed oils and grapeseed oil, are produced and are commercially available (Parker et al., 2003; Yu et al., 2005). Among these oils, grapeseed oil is the one most commonly consumed, and because of its unique flavor, many consider it to be a gourmet oil. Grapeseeds were already listed as a minor oil crop in the early 1990s (FAO, 1992). Berry seed and grapeseed oils are generally used for food and cosmetic purposes because they contain fatty acids and other bioactive components, including vitamins and antioxidants that add to their technological properties and possible health benefits. A great variation occurs in the amount of seeds in the berries, ranging from less than 1% to over 10% among species (Table 6.1). Moreover, considerable variation occurs in the oil content of the seeds. Even within one species, the variation is explained by the purity and size of the seeds and the efficiency of the extraction procedures. In terms of production quantities, grapes are by far the most important commodity. The yearly production of grapes in the world in 2005 was 66.9 million t, followed by raspberries and other berries by 1.2, currants and gooseberries by 0.87, and cranberries and blueberries by 0.623 million t (http://faostat.fao.org/site/340/Desktopdefault.aspx?pageID=340, access Nov. 1, 2007). In developing value-added products from berry seeds and grapeseeds, such as seed oils, a practical approach is to screen for the value-adding factors in the seeds. Moreover, creating new uses for seeds promotes sustainable production of berries and grapes and reduces waste and resources (The Oregon Raspberry & Blackberry Commission, 2005).

## Processing

Berry seed oils, such as black currant, cranberry, grape, and raspberry, are commonly extracted from the berry press cake, a juice or wine production by-product. For grape-

**Table 6.1. Seed and Seed Oil Contents of Berries and Grapes**

| Plant species | Source of seeds | Seed content of berries | Oil content in seed (mean or range) | Reference |
|---|---|---|---|---|
| Red currant | | | | |
| *Ribes spicatum* | Wild in Finland | 9.5% | 14.0% | Johansson et al., 1997a |
| | Cultivated in Finland | | 12.5% | Johansson et al., 2000 |
| *Ribes rubrum* | Cultivated in Europe and USA | | 11.2–23.6% | Goffman and Galletti, 2001 |
| Black currant | | | | |
| *Ribes nigrum* | Wild in Finland | 4.9% | 15.9%. | Johansson et al., 1997a |
| | Cultivated in Europe | | 17.2–22.3% | Goffman and Galletti, 2001 |
| | Cultivated in UK | | 7.6–22.7% | Ruiz del Castillo et al., 2002, 2004 |
| | Cultivated in Southern Bulgaria | | 22% | Zlatanov, 1999 |
| Gooseberries | | | | |
| *Ribes grossularia* | Cultivated in Europe and Canada | | 16.2–26.4% | Goffman and Galletti, 2001 |
| Sea buckthron | | | | |
| *Hippophaë rhamnoides* ssp. *rhamnoides* | Wild and cultivated in Finland | 2.9–8.1% | 5.7–14.2% | Johansson et al., 1997a, Yang and Kallio, 2001, 2002 |
| ssp. *sinensis* | Wild in China | 3.6–10.5% | 5.8–9.8% | Yang and Kallio, 2001, 2002; Kallio et al., 2002 |
| spp. *monogolica* | Cultivated in Russia | 4% | 12.6% | Kallio et al., 2002 |
| cv. *Indian-summer* | Cultivated in Canada | | 11.2–12.1% | Gutiérrez et al., 2007 |
| Lingonberry | | | | |
| *Vaccinium vitsideaea* L. | Wild in Finland | 1.4% | 15–30.6% | Johansson et al., 1997a; Yang et al., 2003 |
| Blueberry | | | | |
| *Vaccinium myrtillus* L. | Wild in Finland | 2.9% | 20–30.5% | Johansson et al., 1997a; Yang et al., 2003 |
| Cranberry | | | | |
| *Oxycoccus quadripetalus* L. | Wild in Finland, | 0.7% | 30.6% | Johansson et al., 1997a |
| Cloudberry | | | | |
| *Rubus chamaemorus* L. | Wild in Finland | 12.2% | 9.1–12.4% | Johansson et al., 1997a, b |

**Table 6.1., cont. Seed and Seed Oil Contents of Berries and Grapes**

| Plant species | Source of seeds | Seed content of berries | Oil content in seed (mean or range) | Reference |
|---|---|---|---|---|
| Raspberry | | | | |
| *Rubus idaeus* L. | Wild in Finland | 10.1% | 23.2% | Johansson et al., 1997a |
| | Cultivated in Canada | | 10.7% | Oomah et al., 2000 |
| | Commercial by-product from USA | | 18.7% | Bushman et al., 2004 |
| Black raspberry | | | | |
| *Rubus occidentalis* L. | Commercial cold-pressed oil from USA | | | Parry and Yu, 2004 |
| Grape | | | | |
| *Vitis* spp. | Commercial by-product from USA and China | | 5.8–13.6% | Beveridge et al., 2005, Cao and Ito, 2003 |

seed and other berry oil, the available sources include by-products of juice, jam, and jelly manufacturing (Yu et al., 2006). Because the oils are usually used as ingredients in foods, as dietary supplements, or in the cosmetics industry (Harris, 2004), good standards of cleanliness are usually maintained when the seeds are handled, transported, stored, and the seed oils are obtained. Several techniques, including cold-pressing at temperatures not exceeding 45°C, pressing at higher temperatures, solvent extraction, or supercritical fluid extraction with carbon dioxide, are used to extract the berry seed oil. A further possibility is to use enzymes to break down cell walls followed by extraction (e.g., Soto et al., 2007). However, the choice of solvent is usually limited to those approved for food use. Depending on the abundance of the seeds in the berry and the extraction technique used, the yield of the seed oil varies. For example, 1 kg of seed oil is obtained from 100 kg of cloudberries by using supercritical fluid extraction. Supercritical fluid extraction with carbon dioxide is an acceptable possibility, but only a few manufacturers use this particular technique, partly due to the high cost of the plant and the oil obtained by this technology (e.g., www.aromtech.fi, www.essentialoil.in). The commercially available berry seed oils produced by supercritical fluid extraction include black currant, wild blueberry (bilberry), cloudberry, lingonberry, and sea buckthorn. Optimizing the extraction parameters of supercritical fluid extraction with carbon dioxide could produce grapeseed oil enriched with α-tocopherol (Bravi et al., 2007). Cold-pressing is a common processing procedure commercially used to produce edible oils. One of the advantages of this technique compared to conventionally processed oils is that cold-pressed oils may retain a greater level of antioxidants also resulting in greater oxidative stability of the seed oil (Yu et al., 2006). Berry seed oil from cherries is typically obtained by cold-pressing followed by filtering. Cold-pressing may be used to obtain both red and black raspberry seed oil. In this technique, the berry seeds are hydraulically pressed to extract cold-pressed oil. Use of hexane to extract oil from milled seeds such as red raspberry was described by Oomah et al. (1996). Hexane at room temperature may also be used to extract crude grapeseed oil. Red raspberry seed oils prepared either by cold-pressing or extracted with hexane exhibited similar fatty acid composition (Oomah et al., 2000; Parry & Yu, 2004a). The significantly greater tocopherol content of hexane-extracted raspberry seed oil (970 µg/g oil) compared to cold-pressed raspberry seed oil (610 µg/g oil) is explained by the presence of nonlipid contaminants from the cold-pressing procedure that dilute the tocopherol content of the cold-pressed seed oil (Oomah et al., 2000). In addition, the lipophilic tocopherols are extracted better by hexane than by cold-pressing. One may use some berry seed oils as such after filtering, but in many cases some refining is necessary to obtain good quality oils. When solvent extraction is used, desolventization is also necessary. Oils obtained by either hexane extraction or expeller pressing are then processed to reduce phosphatides (gums) in a degumming process similar to that employed in the vegetable oil industry (Kapoor & Nair, 2005). A bleaching process may further purify the berry seed oil resulting in oil with a lighter color. To retain the characteristic flavor in the berry seed oil, avoid deodorization at high temperatures. For supercritical extraction, the biggest advantage is that it eliminates the need for further processing of oil such as desolventization and degumming (Kapoor & Nair, 2005).

## Factors Affecting Seed Oil Quality

Oxidative stability is a critical factor that determines the potential uses of the berry seed oil in food and other products such as cosmetic preparations. The oxidative stability of oil mainly depends on the fatty acid composition and the presence of antioxidative compounds. Although berry seed oils are highly unsaturated oils, their oxidative stability may be comparable to refined seed oils. Pigments such as carotenoids and chlorophylls present in unrefined oils may decrease the stability of the oils although they may also be positive attributes, providing attractive color to the oils.

*p*-Anisidine values of 14.3 and 10.5 for red raspberry seed oil and grapeseed oil, respectively, were reported (Oomah et al., 2000) that indicate the presence of secondary lipid oxidation products (aldehydes), having impact on the sensory quality of the berry seed oils. The peroxide value of red raspberry seed oil was between 4.41 meq O-OH/kg (Parry & Yu, 2004a) and 8.25 meq O-OH/kg (Oomah et al., 2000). Freshly extracted sea buckthorn seed oil had a low peroxide value of 1.8 meq $O_2$/kg (Gutiérrez et al., 2007). For other berry seed oils, the following peroxide values were reported: black raspberry seed oil 88–91 mg/kg (Parry & Yu, 2004b), grapeseed oil from 0.96 to 5.6 meq O-OH/kg oil (Oomah et al., 1998, 2000; Parry & Yu, 2004a), blueberry seed oil 41.4 meq O-OH/kg, red raspberry seed oil 46.5 meq O-OH/kg, marionberry seed oil 85.2 meq O-OH/kg, and boysenberry seed oil 41.3 meq O-OH/kg (Parry et al., 2005). These values indicate a wide range of oxidation levels that are in every case very high. Refined highly unsaturated vegetable oils should have peroxide values close to 0 meq meq $O_2$/kg, and in general the acceptable value is below 10 meq $O_2$/kg. Virgin olive oil and refined olive oil should have peroxide values of less than 20 and 5 meq $O_2$/kg (European Commission, 1991). Moreover, the sum of *p*-anisidine value and peroxide values should not be greater than 15. Peroxide values and other indicators of oxidation status are highly dependent on the extent of processing, which was also shown in one study involving the drying method of grapeseeds. Oil extracted from microwave-dried seeds had a much higher peroxide and *p*-anisidine values, 5.6 meq $O_2$/kg and 7.2, than those of cold-pressed oil (1.9 meq $O_2$/kg and 3.2) (Oomah et al., 1998).

Using the oxidative stability index test at 80°C, cold-pressed extra virgin cranberry seed oil (68% PUFA) showed similar oxidative stability values than refined soybean and corn oils (Parker et al., 2003; Parry & Yu, 2004a; Parry et al., 2005). The oxidative stability index of 14–16 hours of black raspberry seed oil (92% PUFA) was much less than those reported by others. Parry et al. (2005) measured the oxidative stability index of different cold-pressed oils showing values of 20.1, 20.3, and 44.8 hours for marionberry, red raspberry, and boysenberry seed oils, respectively, showing that cranberry and boysenberry seed oils were the most stable ones among berry seed oils. Refined grapeseed oil had comparable oxidative stability at 110–140°C to that of soybean, safflower and sunflower oils, all having high iodine values, but lower oxidative stability than oils with less unsaturation (Tan et al., 2002), which might indicate that at high temperatures it is the fatty acid composition that dominates in oxidative stability.

The antioxidative capacity of berry seed oils was measured by using various in vitro radical scavenging and oxygen radical absorbance capacity (ORAC) assays that are generally performed on methanol or acetone extracts of the oils. After using the ORAC test with trolox as the standard, the antioxidative capacity of cranberry seed oil ranged from 3.5 to 3.9 μmol/trolox equivalents per gram of oil (TE/g oil), red raspberry seed oil was 48.8 TE/g of oil, and that of blueberry seed oil was 36.0 TE/g of oil (Oomah et al., 2000; Parry & Yu, 2004a). Although the ORAC test demonstrated that cranberry seed oil had low antioxidant properties, a methanol extract of cranberry seed oil at 11.3 mg oil/mL showed a very high DPPH• radical scavenging activity value of 95% (Yu et al., 2005). In this assay, black raspberry seed oil had lower scavenging activities with values of 11% and 49% (Parry & Yu, 2004). Cranberry seed oil extract was also much more effective than black raspberry seed oil extract in scavenging ABTS$^{+\bullet}$ cation radicals with 22.5 and $<1$ μg of trolox equivalents per gram oil (Parry & Yu, 2004; Yu et al., 2005). These values correlated positively with the amounts of total phenolics in the extracts, but the levels of phenolics in the oils were very low. Moreover, cold-pressed cranberry seed oil extract also inhibited $Cu^{2+}$ mediated human LDL oxidation significantly (Yu et al., 2005).

Although extracts of seed oils may possess antioxidant properties, greater activities were measured in the corresponding defatted seed meals. For example, the values for scavenging activity of acetone extracts of the oil and the meal of black raspberry were <1 and 361 TE/g, respectively (Parry & Yu, 2004). Furthermore, a comparable distribution of total phenolic compounds was also present in the extracts. This indicates that the antioxidant profiles of the two oily and defatted fractions are different and that most phenolic compounds are retained in the defatted meals. Thus, cold-pressed berry seed oils may not be good sources of water-soluble phenolic antioxidants, but they may have high contents of tocols (tocopherols and tocotrienols) and other lipid-soluble antioxidants.

Color is a quality attribute that greatly influences consumers'perception of oils. Green pigments, particularly chlorophyll content, impart undesirable color to berry seed oils and may also be able to promote oxidation. In raspberry seed oil, a negligible amount of chlorophyll was found (Oomah et al., 2000). Sea buckthorn seed oil has a bright yellow color (Gutiérrez et al., 2007; St George & Cenkowski, 2007). Grapeseed oil was dark green when 10% of ethanol was used as a modifier in supercritical fluid extraction while the color was light yellow when extracted with supercritical $CO_2$ without a modifier (Cao & Ito, 2003). Yellow color that is caused by carotenoids is claimed beneficial since they simulate the appearance of butter. One commercial cold-pressed cranberry oil had a small Hunter Lab lightness value of 1.8 and low values also in yellowness and redness that were comparable to commercial soybean and corn oils (Parker et al., 2003). In comparison, two cold-pressed black raspberry oils were relatively dark with Hunter Lab lightness values of 4–10, and high intensities in yellowness and redness that were comparable to those of carrot oil (Parry & Yu, 2004). The authors pointed out the great difference between the colors of the two batches of the same oil, and they concluded that variation in processing, growing, and harvesting conditions may have significant effects on the contents of minor compounds, like the colorants, in the oils. This also may lead into great differences in the stability of the oils.

## Edible and Nonedible Applications

### Food and Dietary Supplements

Grapeseed oil is primarily used as a gourmet oil; the berry seed oils are primarily used as dietary supplements. The oils are produced as by-products from the wine and berry industries, and the volumes, especially for berry seed oils, are very small. Grapeseed oil has a history of being classified as a minor vegetable oil (FAO, 1992), and is used in salad dressings, marinades, baking, and frying. It is advertised to health-conscious consumers due to its high contents of essential fatty acids, natural antioxidants, and phytochemicals, and to chefs who value its light and nutty flavor. Moreover, it supposedly has a high smoke point, which makes it suitable for frying. Berry seed oils are shown to have various positive health effects (e.g., Tahvonen et al., 2005; Wu et al., 1999), which opened up the marketing of these oils as dietary supplements. In general, oils from several berries are used as mixtures in supplements, while capsules made of only sea buckthorn and black currant are also commercially available. Although berry seed oils are known to possess many good attributes, their consumption is rather limited as food. As concluded in a final report of a commission promoting the uses of caneberries, the marketing of raspberry seed oil may be very difficult because its price was estimated to be approximately three-fold that of grapeseed oil (The Oregon Raspberry & Blackberry Commission, 2005). Due to the high price, manufacturers may mainly apply berry seed oils in dietary supplements and nonedible applications in the near future.

## Cosmetic Preparations

Berry seed oils from a number of different berry sources, including black currant, cloudberry, cranberry, grape, and sea buckthorn, are widely used in a broad range of cosmetic preparations. Berry seed oils are rich in linoleic acid, plant sterols, tocopherols (vitamin E), and provitamin A carotenoids. Various aroma compounds provide unique "berry flavors" and add to the value of facial creams, bath oils, lipsticks, mascara, moisturizers, shower gels, shampoos, and other hair products. Fatty acids such as both n-3 and n-6 fatty acids, especially gamma linolenic acid (GLA), are claimed essential for normal cell structure; thus, according to cosmetic manufacturers, they actively promote healthier skin appearance, reduce atopic dermatitis (eczema) problems, and improve circulation. For these purposes, black currant and cranberry seed oils are among the most favored ingredients of different skin cream products all over the world. The antioxidant compounds in the berry seed oils are also claimed to protect the skin from free radical damage. In addition, attributes linked to beneficial skin effects of sea buckthorn oil include improving skin's surface circulation. The cosmetic applications of by-products of arctic berry processing include the use of berry seed oils from cloudberries, wild blueberries, lingonberries, and sea buckthorn (www.lumene.com). Among the recent product innovations are the application of blueberry seed oil both in facial cream and in mascara for the suggested purpose of nourishing eye lashes. Berry oils, such as those from black currants and cloudberries, carry the flavor of the berry, adding to the pleasing quality of the cosmetic product. Also (because of its unique flavor and/or aroma), cranberry seed oil is used as one of the popular ingredients of lipsticks.

The claimed effects due to using berry seed oil as an ingredient in cosmetic products may be based on cosmeto-clinical research, resulting in patent applications or published journal articles. However, for many of the claimed effects, the scientific evidence relies on general textbook knowledge on the physiological effects of the berry seed oil ingredients. One example concerning patented effects of berry seed oils concerns raspberry seed oil. The incorporation of raspberry seed oil in cosmetic and pharmaceutical products, possibly based on its anti-inflammatory activity and notably for preventing gingivitis, rash, eczema, and other skin lesions, was patented (Pourrat & Pourrat, 1973). According to the patent, raspberry seed oil uses are as a sun screen and in toothpaste, crèmes for prevention of skin irritations, bath oil, aftershave cream, antiperspirants, shampoos, and lipsticks. The anti-inflammatory activity of raspberry seed oil was reported by Pourrat and Carnat (1981) to be superior compared to anti-inflammatory properties of avocado oil, grapeseed oil, hazelnut oil, and wheat germ oil.

## Acyl Lipids and Fatty Acid Composition

Berry and grapeseed oils consist mainly of neutral lipids, especially triacylglycerols (e.g., Yu et al., 2006). Depending on the extraction procedure and the extent of refining, they may also contain some phospholipids and free fatty acids. Most of the chemical and physical properties of oils are controlled by the fatty acid composition that also has a major impact on the nutritional quality of the oil. Thus, oils are commonly characterized by their fatty acid composition. Since acyl lipids other than triacylglycerols are considered harmful to oil quality, they are generally removed by refining. In addition to acyl lipids, seed oils also contain some other lipids that are presented as minor lipids.

### Other Lipids than Triacylglycerols

Berry and grapeseed oils contain other acyl lipids to some extent, because they are co-extracted during extraction. Phospholipids are natural components of seed membranes, and free fatty acids are mainly hydrolysis products or metabolic intermediates of glycerolipids. Laboratory-extracted black currant seed oil contained 1.3 mg of phospholipids/grams of oil (Zlatanov, 1999), crude raspberry seed oil 2.7% of phospholipids (Oomah et al., 2000), and commercial crude grapeseed oil 18.6 μg/g of phosphorus (Martinello et al., 2007), indicating the presence of phospholipids. Seeds of sea buckthorn contained 0.4–1.3% of glycerophospholipids after chloroform–methanol extraction, which means that they contributed to approximately 10% of total lipids (Kallio et al., 2002; Yang & Kallio, 2002), but the levels in hexane-extracted oil were 0.8–1.2% (Gutiérrez et al., 2007). These values highlight the great variation in the phospholipid contents in the oils, which reflects the efficacy of the extraction and the extent of refining. The phospholipid composition of black currant seed oil showed a typical profile with phosphatidylcholine, inositol, and ethanolamine being the major ones and contributing to 34.7%, 20.6%, and 11.2%, respectively, of the total amount of phospholipids (Zlatanov, 1999).

Free fatty acids are typically considered to be deterioration products of triacylglycerols. The acceptable free fatty acid content for extra virgin and virgin olive oils is ≤0.8% and ≤2%, respectively, and if the free fatty acid content is >2%, the oil is classified as lampante olive oil (European Commission, 1991). Crude raspberry seed oil contained 3.5% of free fatty acids (Oomah et al., 2000), crude grapeseed oil 2.75% (Martinello et al., 2007), and freshly extracted sea buckthorn seed oil 2.7–2.9% (Gutiérrez et al., 2007). One study showed that physical refining was effective enough to remove phospholipids totally from hexane-extracted grapeseed oil while molecular distillation was used to obtain oil with 0.1% of free fatty acids (Martinello et al., 2007).

## Fatty Acid Composition of Berry Seed Oils

Recently, a classification of vegetable oils based on the fatty acid profiles was developed (Dubois et al., 2007). The classification was based on different nutritional impacts of fatty acids and especially on their effects on the metabolism of lipoproteins (Dubois et al., 2007). In this classification, all berry and grapeseed oils were included in the polyunsaturated nutritional profile class, but they were divided into several subclasses based on the contents of linoleic, α-linolenic, and monounsaturated fatty acids. A summary of the fatty acid profiles of berry and grapeseed oils is presented in Table 6.2.

Grapeseed and black currant seed oils could be classified in the linoleic acid or the (n-6) polyunsaturated fatty acid class, because both of them contained more than 60% of linoleic acid or γ-linolenic acid (Dubois et al., 2007). The fatty acid profile of grapeseed oils from several grape varieties grown in different countries was dominated by 61–75% of linoleic acid (Beveridge et al., 2005; Crews et al., 2006). Thus grapeseed oil has more linoleic acid than other vegetable oils such as safflower and sunflower oils. Since only less than 2% of α-linolenic acid was in grapeseed oil, the (n-6) to (n-3) ratio of polyunsaturated fatty acids was one of the lowest among vegetable oils.

The linoleic acid content of black currant seed oil ranged from 43% to 58% and that of γ-linolenic acid from 12% to 25% (Goffman & Galletti, 2001; Ruiz del Castillo et al., 2002, 2004). In addition, black currant seed oil also contained a significant amount of α-linolenic acid, 10–19%, and a small amount of stearidonic acid [18:4(n-3)], 2.4–4.3%, resulting in a much higher ratio of (n-6) to (n-3) polyunsaturated fatty acids than other oils in this class. Black currant seed oil contains a relatively high amount of γ-linolenic acid, and it supposedly has positive effects on cardiovascular health and immune responses (e.g., Kapoor & Nair, 2005). Seed oils of wild and cultivated red currants had comparable fatty acid profiles as that of black currant except for the content of γ-linolenic acid that was in general ≤7% (Goffman & Galletti, 2001; Johansson et al., 1997, 2000).

**Table 6.2. Fatty Acid Composition of Seed Oils from Berries and Grapes**

| Plant species | Fatty acid content; g/100 g oil or mol% in oil [1)] | | | | | | | Reference |
|---|---|---|---|---|---|---|---|---|
| | 16:0 | 18:0 | 18:1 (n-9) | 18:2 (n-6) | 18:3 (n-3) | 18:3 (n-6) | Others | |
| Red currant | 3.8 [1)]–4.6 | 1.2 [1)]–2.0 | 13.5 [1)] –17.1 | 38.5 [1)] –41.3 | 19.1 [1)] –25.1 | 5.9–15.1 [1)] | 4.0–7.9 [1)] | Johansson et al., 1997; 2000 |
| Black currant | 5.2 [1)]–6.4 | 1.3–1.8 [1)] | 8.9–16. | 42.7–57.8 | 10.0–11.5 | 11.3[1)] –24.6 | ≤6.1 | Johansson et al., 1997; Zlatanov, 1999; Ruiz del Castillo et al., 2002; 2004 |
| Sea buckthorn | 6.8 [1)]–13 [1)] | 1.8 [1)] – 4 [1)] | 12.7[1)] –22.4[1)] | 34.7[1)] - 44.5[1)] | 26.5[1)] –34.8[1)] | - | 2.1 [1)]–5.6[1)] | Johansson et al., 1997; Yang and Kallio, 2001; 2002; Kallio et al., 2002; St. George and Cenkowski, 2007 |
| Lingonberry | 1.2 [1)] | 0.2 [1)] | 12.5 [1)] | 34.4 [1)] | 49.7 [1)] | - | 2.1 [1)] | Johansson et al., 1997a |
| Blueberry | 4.6 [1)]–5.7 | 1.1[1)] –2.8 | 21.6 [1)] – 22.9 | 34.4 [1)] – 43.6 | 25.1–36.5 [1)] | - | 1.9 [1)] | Johansson et al., 1997a; Parry et al., 2005 |
| Cranberry | 3.8 [1)]–7.8 | 0.8 [1)]–1.9 | 18.7 [1)]–22.7 | 38.8 [1)] -44.3 | 22.3 – 35.1 [1)] | - | 1.0 – 2.9 [1)] | Johansson et al., 1997a; Parker et al., 2003 |
| Cloudberry | 2.0–2.7 [1)] | 1.1–1.8 [1)] | 12.7–18.0 [1)] | 41.1[1)]–50.7[1)] | 26.2–36.6 [1)] | - | ≤5.5[1)] | Johansson et al., 1997a, b |
| Raspberry | 1.3–2.7 | 0.9–1.0 | 10.6[1)]–12.1 | 49.5[1)]–54.2 | 29.7–33.5 [1)] | ≤0.2 | 1.6–2.9[1)] | Johansson et al., 1997a; Oomah et al, 2000; Bushman et al., 2004; Parker and Yu, 2004; Parry et al., 2005 |
| Black raspberry | 1.2–1.6% | | 6.2–7.7% | 55.9–58.6% | 35.2–35.3% | - | | Parry and Yu, 2004b |
| Grapeseed | 6.3–11.6 | 3.6–5.4 | 12.7–20.9 | 61.3–74.6 | 0.3–1.8% | - | | Beveridge et al, 2005; Crews et al., 2006 |

Because these oils have more monounsaturated fatty acids than the others in this group, and they could thus be considered as a subgroup, keep this in mind as a separate section. Cranberry seed oil is a linoleic acid-rich oil with large amounts of monounsaturated fatty acids and α-linolenic acid (Dubois et al., 2007). In wild and cultivated cranberry seed oils, the content of linoleic acid was the greatest being approximately 40%, with 20% oleic acid and 22–35% of γ-linolenic acid (Johansson et al., 1997a; Parker et al., 2003). Blueberry seed oil could also be classified in the same group, because its fatty acid profile resembles that of cranberry seed oil (Johansson et al., 1997a; Parry et al., 2005). Due to the relatively high amount of α-linolenic acid, the ratio of (n-6) to (n-3) polyunsaturated fatty acids in these oils ranged from 1 to 2.

The last subgroup in the polyunsaturated nutritional profile is that of oils high in α-linolenic and linoleic acids. Two berry seed oils, namely raspberry and sea buckthorn, were classified in this group (Dubois et al., 2007), although sea buckthorn berry seed oil also could have been included in the previous class due to its relatively high content of oleic acid. Linoleic and α-linolenic acids were the two major fatty acids contributing to approximately 35–41% and 30%, respectively, of the profile and yielding in a relatively balanced ratio of (n-6) and (n-3) of 1.3 (e.g., St. Geroge & Cenkowski, 2007; Yang & Kallio, 2001). Moreover, sea buckthorn berry seed oil contained 2–4% of vaccenic acid [18:1 (n-7)]. The fatty acid composition of the seeds remained relatively constant during the harvesting period, but the origin of berries had a significant effect on the fatty acid composition and the level (St George & Cenkowski, 2007; Yang & Kallio, 2002). The fatty acid composition of phospholipids in sea buckthorn seed oil differed from that of triacylglycerols. The content of α-linolenic acid was significantly smaller, while that of linoleic acid and most other fatty acids was greater (Yang & Kallio, 2001a, 2002). Bear in mind that the fatty acid compositions of buckthorn berry and seed oils are different and that the seed oil contained more polyunsaturated fatty acids than the whole berry oil (Yang & Kallio, 2001).

Raspberry seed oil contains approximately 30% of α-linolenic and 50–55% of linoleic acid, and the ratio of n-6 to n-3 fatty acids is between 1.5 and 1.9 (Buschman et al., 2004; Johansson et al., 1997a; Oomah et al., 2004; Parry et al., 2005). In addition, the oil has an exceptionally low amount of saturated fatty acids (<5%). The fatty acid profiles of oils of wild and cultivated varieties and of berries from different origin were comparable. Studies of black raspberry seeds show that they have a fatty acid profile similar to red raspberries, and thus could be considered an excellent source of both essential fatty acids. (Buschman et al., 2004; Parry & Yu, 2004b). Oil from cloudberry seeds that were grown wild in Finland was reported to have a fatty acid composition rich in linoleic and α-linolenic acids, like raspberry seed oil (Johansson et al., 1997ab).

## Minor Components in Grape and Berry Seed Oils

In addition to acyl lipids, berry and grapeseed oils also contain minor lipid components such as tocopherols and tocotrienols (i.e., tocols), carotenoids, phytosterols, and other fat-soluble compounds. Tocols are the most important fat-soluble antioxidants in the diet (Institute of Medicine, 2000). In addition, α-tocopherol possesses vitamin E activity, and together with other tocols it also has other biological activities (Institute of Medicine, 2000). Carotenoids give strong yellow color to edible oils, and they may also contribute to the antioxidant activity of oils (Institute of Medicine, 2000; Krinsky, 2001). In addition, some carotenoids, namely β-carotene, α-carotene, and β-cryptoxanthin, have provitamin A activity. Phytosterols are another important class of nonsaponifiable lipids. The consensus is that they lower serum cholesterol values when consumed at levels of at least 2 g per day (e.g., Katan et al., 2003), which has led to a vast variety of phytosterol-enriched foods (e.g., Moreau, 2005). Strong evidence supports the role of polyphenols in human health, including the onset of many diseases such as cardiovascular diseases and cancer (Kroon & Williamson, 2005; Williamson & Manach, 2005). In the past few years, major advances occurred regarding the knowledge of polyphenol absorption, metabolism, and bioefficacy (Day & Williamson, 2001; Manach et al., 2005; Scalbert & Williamson, 2000; Williamson & Manach, 2005).

### Tocopherols and Tocotrienols

Berry and grapeseed oils are rich sources of tocols, mainly tocopherols. A few studies exist in which the variation of tocopherol compositions within and between species was made (Table 6.3), making it possible to compare the oils with each other. When seeds from different *Ribes* species (e.g., currants and gooseberry) were extracted with isooctane for the determination of tocols, total tocol contents decreased from 1.7 mg/g oil in black currant seeds to 1.4 and 0.8 mg/g in red currant and gooseberry seeds, respectively (Goffman & Galletti, 2001). The ranges within the same species were as great as 1.42–2.46, 0.86–2.48, and 0.56–1.19 mg/g oil, respectively. In another study, the content in a laboratory-extracted black currant seed oil was much smaller (0.25 mg/g) (Zlatanov, 1999). In both studies, γ-tocopherol was the major tocol followed by α- tocopherol and a small amount of δ-tocopherol. The same tocopherols dominated also the tocol profiles in red and black raspberry and boysenberry seed oils (Bushman et al., 2004). Hexane-extracted and cold-pressed raspberry seed oils contained 3.60 and 1.98 mg/g of total tocols, respectively (Oomah et al., 2000), indicating that the extraction method had an important effect on the tocol yield. Commercial cold-pressed berry seed oils contained much lower levels of tocols than hexane-extracted oils. Total tocol contents of blueberry, red raspberry, marionberry, and boysenberry seed oils ranged from 0.11 to 0.94 mg/g of total tocols (Parry et al., 2005). Total tocol amount was the lowest in blueberry seed oil. Its major tocol was

**Table 6.3. Tocopherol and Tocotrienol Contents in Berry and Grapeseed Oils from Different Sources**

| Plant species | Number of samples | Tocopherols, μg/g oil α | β | γ | δ | Tocotrienols, μg/g oil α | γ | Reference |
|---|---|---|---|---|---|---|---|---|
| Red currant | 15 | 147-888 | 51-107 | 487-1104 | 215-377 | | - | Goffman and Galletti, 2001 |
| Black currant | 10 | 358-1074 | - | 799-1317 | 66-144 | - | - | Goffman and Galletti, 2001 |
| Gooseberries | 15 | 91-237 | - | 444-773 | 24-183 | - | - | Goffman and Galletti, 2001 |
| Grapeseed | 8 | 36-309 | 22-153 | 21-141 | - | 102-228 | 217-383 | Beveridge et al., 2005 [1)] |
| | 30 | <10-229 | <10-133 | <10-168 | <10-69 | <10-352 | <10-785 | Crews et al., 2006 |

Results from two extraction methods combined (extractions with supercritical carbon dioxide and petroleum ether).

α-tocopherol, while in other berry seed oils it was γ-tocopherol. Among the commercial berry seed oils, red raspberry had the highest level of α-tocopherol, 0.15 mg/g, and also a high total tocol content, 0.89 mg/g. Red raspberries and boysenberries had the greatest amounts of γ-tocopherol.

The most promising potential health benefits of sea buckthorn oil are probably attributable to its high level of vitamin E compounds (i.e., tocols) (Beveridge et al., 1999). Total tocol contents were reported to range from 0.6 to 3.18 mg/g, with α- and γ-tocopherols being the major tocols. A large variation existed in the tocol composition of sea buckthorn berry seed oils in two subspecies, namely spp. *mongolica* grown in Russia and ssp. *sinensis* grown in China. Total tocol contents ranged from 0.872 mg/g to 2.910 mg/g in laboratory-extracted seed oil, with significantly higher levels in spp. *mongolica* than in ssp. *sinensis*. Moreover, γ-tocopherol was the major tocol in the first subspecies and α-tocopherol in the second one, respectively (Kallio et al., 2002). In another study, the total tocol contents of sea buckthorn seed oils (spp. *saneness*) were reported to range from 2.17 to 2.62 mg/g, with α-tocopherol as the major tocol (St George & Cenkowski, 2007). Although the tocol contents and those of vitamin E active α-tocopherol are very high compared to other vegetable oils, note that the oil of the fruit fraction of sea buckthorn has even higher levels of tocols than the seed oils (Beveridge et al., 1999; Kallio et al., 2002; St George & Cenkowski, 2007).

Grapeseed oil has a clearly different tocol profile than berry seed oil, because it is dominated by tocotrienols. For example, in commercial grapeseed oil containing 0.55 mg/g of total tocols, the major tocols were γ- and α-tocotrienols, with only minor amounts of tocopherols (Oomah et al., 2000). γ-Tocotrienol was the major tocol in other studies followed by smaller amounts of α- tocotrienol and α-, β-, and γ-tocotrienols (Beveridge et al., 2005; Crews et al., 2006). Thus, grapeseed oil is one of the few examples of a tocotrienol-rich oil; the others are palm oil and oils from cereal grains (Bramley et al., 2000). Total tocol contents of eight grapeseeds ranged from 0.6 to 1.0 mg tocols/g of oil (Beveridge et al., 2005), and those of ten grapeseeds from three countries ranged from 0.06 to 1.21 mg tocols/g of oil (Crews et al., 2006), indicating a very large variation among varieties and production origins. In addition, tocol contents are affected by processing. Crude grapeseed oil contained 1.01 mg/g of total tocopherols (Martinello et al., 2007), and the content decreased only by 10% after physical refining (degumming, dewaxing, and bleaching). By using supercritical fluid extraction under optimal conditions, the α-tocopherol content of grapeseed oil was increased to 265 μg/g, a level that was about six-fold higher than that reported after hexane extraction (Bravi et al., 2007).

## Carotenoids

Very little published data exist on the carotenoid contents of berry seed oils. Among four cold-pressed berry seed oils (i.e., blueberry, red raspberry, marionberry, and boysenberry), the latter contained the highest level of 16.8 μg/g of carotenoids in oil, and red raspberry seed oil contained the lowest level of 7.1 μg/g. Carotenoids consist mainly of zeaxanthin, cryptoxanthin, and β-carotene (Parry et al., 2005). Blueberry seed oil was the richest source of β-carotene with a content of 1.3 μg/g.

Sea buckthorn seed oils have a strong yellow color and thus a high content of carotenoids. The carotenoid contents of sea buckthorn seed oils vary greatly from a trace amount to 410–850 μg/g as reviewed by Beveridge and co-authors (1999). Total carotenoid contents of cultivated sea buckthorn seed oils (ssp. *sinensis*) were 244–276 μg/g (St George & Cenkowski, 2007) which is severalfold greater than those of other berry seed oils.

## Phytosterols

In general, phytosterols are the most abundant class of nonsaponifiable lipids in oils. They may occur as free alcohols or esters of fatty acids. In lingonberry seeds, the oil contained 5–6 mg/g of free and esterified sterols, and in blueberry seeds, the respective values were 6.7 mg/g and 2.6 mg/g (Yang et al., 2003). Black currant seed oil contained 1.4 mg/g of phytosterols (Zlatanov, 1999).

Sitosterol contributed to approximately 80% of total sterol in all of these seeds. Other sterols in blueberry and lingonberry seeds included campesterol and isofucosterol (Yang et al., 2003), and in black currant seeds campesterol and stigmasterol (Zlatanov, 1999). Esterified sterols in blueberry and lingonberry seeds contained approximately 25% of dimethyl sterols, namely cycloartenol and 24-methylenecycloartanol that usually occur only as minor components in oils (Yang et al., 2003).

The phytosterol composition of sea buckthorn seed oil was characterized in several studies. The phytosterol profile has been relatively consistent among studies, but the amount of phytosterols has varied considerably due to the method used to separate the oil. Phytosterol contents of sea buckthorn seed oils produced for nutraceutical use were 8.97, 13.26, and 16.40 mg/g after cold-pressing, hexane and supercritical fluid extractions, respectively (Li et al., 2007).

Laboratory-scale extraction of the seeds using chloroform–methanol extraction yielded oil containing 12.4–23.0 mg/g of total sterols (Yang et al., 2001) and 5.21–5.67 mg/g of sitosterol (St George & Cenkowski, 2007). Sitosterol was the major sterol in sea buckthorn seed oils contributing to 48–76% of total sterols and followed by Δ5-avenasterol and sitostanol. Other sterols found were, for example, campesterol, β-amyrin, citrostadienol, and cycloartenol (Li et al., 2007; St George & Cenkowski, 2007; Yang et al., 2001b). Phytosterol levels of hexane- and supercritical fluid-extracted sea buckthorn oils are among the highest found in vegetable oils.

A large variation exists in the contents of phytosterols in grapeseed oil from various varieties. The total amounts reported for eight grape varieties ranged from 3.16 to 18.61 mg/g oil, and among ten grape varieties grown in three countries from 2.58 to 11.25 mg/g oil (Beveridge et al., 2006; Crews et al. 2006). In both studies, sitosterol was found to be the major sterol contributing approximately 60–80% of total sterols, and followed by several other sterols including campesterol, stigmasterol, and Δ5-avenasterol. A similar phytosterol profile (74% sitosterol in a composite sample) was found in commercial grapeseed oils (Normén et al., 2007). The levels of sitosterol in grapeseeds themselves were, however, much lower (2.15 mg/g), which indicates that after industrial extraction and refining the levels may be remarkably reduced, compared to laboratory extraction. Grapeseeds also contained minor amounts of steradienes, compounds that are dehydration products of sterols and are formed during heating or under acidic conditions, such as during refining. Steradiene contents of industrially extracted grapeseed oils were from less than 0.05 to 6.7 µg/g (Crews et al., 2006). The range of values are rather high because the acceptable level for virgin olive oils is ≤0.15 µg/g (European Commission, 1991).

### Phenolic Compounds

Some berry seed oils were reported to contain trace amounts of phenolic compounds measured as total phenolic content using a spectrophotometric method based on reducing capacity (Folin-Ciocalteau) and expressed as milligrams of gallic acid equivalents per gram of oil (mg GAE/g). In cold-pressed black raspberry oil, the amount of total phenolics was 0.04–0.09 mg of GAE/g of oil, and that of cold-pressed cranberry seed oil was 1.6 mg of GAE/g of oil (Parry & Yu, 2004a). The highest total phenolic content of 2.0 mg of GAE/g of oil was observed in red raspberry seed oil compared to cold-pressed marionberry, boysenberry, and blueberry seed oils (Parry et al., 2005). As most of the phenolic compounds are rather hydrophilic in nature, they are most likely not present in significant amounts in berry seed oils rich in lipophilic constituents.

## Possible Allergic and Toxic Compounds in the Oil

As with any other food or cosmetic ingredient, berry seed oils' safe use must be assured. Allergic reactions due to berry seed oils are not common. The most likely cause would be residual amounts of berry proteins in the seed oil. The risk for allergic reactions due to berry seed oil residual proteins is further diminished because of the low amounts of berry seed oils used in food apart from food supplements or cosmetic products. A very limited risk exists of toxic compounds in berry seed oils, except for those caused by contamination with pesticides or other chemicals to kill harmful microorganisms or weeds during the growing season of berries. Toxicity problems may also be caused by various isolation techniques (e.g., use of extraction solvents) to obtain berry seed oil from the berries or berry by-products. As an example, some samples of grapeseed oil are reported to have higher levels of polyaromatic hydrocarbons (PAH) than desired (Gunstone, 2006). Moret et al. (2000) described the effect of processing on the PAH content of the oil.

## Health Benefits of the Oil and Oil Constituents

In berry seed oils, the fatty acids, especially omega-3 fatty acids, and antioxidant compounds are components with potential beneficial effects toward human health. Nutrition claims regarding omega-3 fatty acids derived from berry seed oils are already used in marketing commercial products such as fruit juices and some dairy products. Also dietary supplements containing black currant seed oil and sea buckthorn seed oil are marketed using claims related to their composition of bioactive substances and their postulated effects on human health. Limited data are available concerning health effects of berry seed oils. The berry seed oils subjected to human clinical studies include black currant, cranberry, and sea buckthorn. Black currant seed oil was investigated relative to its possible beneficial cardiovascular effects. In one study, 12 healthy subjects ingested 5 g of black currant seed oil for four weeks, resulting in minor effects in the blood lipids (Johansson et al., 1999). In another study a low-density lipoprotein (LDL)-cholesterol-lowering effect was reported with a daily supplement of 3 g of black currant seed oil for four weeks (Tahvonen et al., 2005). Wu et al. (1999) investigated the effect of dietary supplementation with black currant seed oil on the immune response of healthy elderly subjects. They concluded that the oil may moderately enhance the immune function by reducing the production of prostaglandin $E_2$. The oil of another type of currant, alpine currant, was investigated as a source of polyunsaturated fatty acids in the treatment of atopic eczema (Johansson et al., 1999). Parry and Yu (2004b) reported a reduction of human lLDL oxidation by cranberry seed oil, suggesting possible benefits of cranberry seed oil in the prevention of heart disease. Several health-related effects are claimed for sea buckthorn oil. Most of the physiological effects of sea buckthorn oil relate to the fatty acid content of the seed oil. Sea buckthorn seed oil is rich in oleic, linoleic, and linolenic acids, but the berry oil, which is not always kept separate from the seed oil, is rich in palmitoleic acid. Sea buckthorn seed oil reportedly alleviated atopic dermatitis (Yang et al., 1999; Yang, 2001). In these studies, volunteer subjects with diagnosed atopic dermatitis ingested 5 g of either sea buckthorn seed oil or sea buckthorn pulp oil daily for four months. Sea buckthorn oil ingested as a daily dose of 5 g for four weeks did not affect blood lipids but had an influence on platelet aggregation (Johansson et al., 2000). Microencapsulated sea buckthorn and black currant seed oil were reported to increase the glycemic response (Tuomasjukka et al., 2006). A recent innovation from the cosmetic industry (www.lumene.com) suggested a "beauty claim" for sea buckthorn seed oil containing facial cream with dietary supplements, thus boosting the claimed skin beneficial effect of sea buckthorn seed oil.

## Conclusion

The potential benefits of berry seed oils and grapeseed oil as ingredients in food and cosmetic products, including possible beneficial health uses, are being recognized. Applications of berry seed oils and grapeseed oil provide an environmentally friendly use of by-products of the berry and fruit-processing industry. The development of new economical processes may permit industries to better utilize and obtain more profit from these value-added products.

## References


Beveridge, T.H.J.; B. Girard; T. Kopp; J.C.G. Drover. (2005) Yield and composition of grape seed oils extracted by supercritical carbon dioxide and petroleum ether: varietal effect. *J. Agric. Food Chem.* **2005,** *53,* 1799–1804.

Beveridge, T.; T.S.C. Li; B.D. Oomah; A. Smith. Sea buckthorn products: manufacture and composition. *J. Agric. Food Chem.* **1999,** *47,* 3480–3488.

Bramley, P.M.; I. Elmadfa; A. Kafatos; F.J. Kelly; Y. Manios; H.E. Roxborough; W. Schuch; P.J.A. Sheehy; K.-H. Wagner. Review vitamin E. *J. Sci. Food Agric.* **2000,** *80,* 913–938.

Bravi, M.; F. Spinoglio; N. Verdone; M. Adami; A. Aliboni; A. D'Andrea; A. De Santis; D. Ferri. Improving the extraction of α-tocopherol-enriched oil from grape seeds by supercritical $CO_2$. Optimisation of the extraction conditions. *J. Food Eng.* **2007,** *78,* 488–493.

Bushman, B.S.; B. Phillips; T. Isbell; B. Ou; J.M. Crane;S.J. Knapp. Chemical composition of caneberry (*rubus* spp.) seeds and oils and their antioxidant potential. *J. Agric. Food Chem.* **2004,** *52,* 7982–7987.

Cao, X.; Y. Ito. Supercritical fluid extraction of grape seed oil and subsequent separation of free fatty acids by high-speed counter-current chromatography. *J. Chromatogr.* **2003,** *1021,* 117–124.

Crews, C.; P. Hough; J. Godward; P. Brereton; M. Lees;S. Guiet; W. Winkelmann. Quantitation of the main constituents of some authentic grape-seed oils of different origin. *J. Agric. Food Chem.* **2006,** *54,* 6261–6265.

Day, A.J.; G. Williamson. Biomarkers for exposure to dietary flavonoids. A review of the current evidence for identification of quercetin glycosides in plasma. *Br. J. Nutr.* **2001,** *86,* S105–110.

Dubois, V.; S. Breton; M. Linder; J. Fanni; M. Parmentier. Fatty acid profiles of 80 vegetable oils with regard to their nutritional potential. *Eur. J. Lipid Sci. Technol.* **2007,** *109,* 710–732.

FAO. Minor oil crops. *Agricultural Services Bulletin No. 94;* FAO: Rome, Italy, 2002; available online: http://www.fao.org/docrep/X5043E/X5043E00.htm (12. Nov. 2007)

Goffman, F.D.; S. Galletti. Gamma-linolenic acid and tocopherol contents in the seed oil of 47 accessions from several *Ribes* species. *J. Agric. Food Chem.* **2001,** *49,* 349–354.

Gunstone, F. Minor speciality oils. *Nutraceutical and Speciality Lipids and Their Co-Products;* F. Shahidi, Ed.; Taylor & Francis: Boca Raton, FL, 2006; pp. 91–125.

Harris, N. Putting specialty and novel oils to good use. *Oils Fats Int.*, **2004,** *20,* 20–21.

Johansson, A. Availability of Seed Oils from Finnish Berries with Special Reference to Compositional, Geographical and Nutritional Aspects. Dissertation. University of Turku, Finland; 1999.

Johansson, A.; E. Isolauri; S. Salminen; P. Laakso; K. Turjanmaa; J. Katajisto; H. Kallio. Alpine currant seed oil as a source of polyunsaturated fatty acids in the treatment of atopic eczema.

Functional Foods—*A New Challenge for the Food Chemists;* R. Lásztity, W. Pfannhauser, L. Simon-Sarkadi, S. Tömösközi, Eds.; Publishing Company of TUB, Hungary, 1999; pp. 530–536.

Johansson, A.; H. Korte; B. Yang; J.C. Stanley; H.P. Kallio. Sea buckthorn berry oil inhibits platelet aggregation. *J. Nutr. Biochem.* **2000,** *11,* 491–495.

Johansson, A.K.; P.H. Kuusisto; P.H. Laakso; K.K. Derome;P.J. Sepponen; J.K. Katajisto; H.P. Kallio. Geographical variations in seed oils from *rubus chamaemorus* and *empetrum nigrum. Phytochem.* **1997b,** *44,* 1421–1427.

Johansson, A.; P. Laakso; H. Kallio. Characterization of seed oils of wild, edible finnish berries. *Z. Lebensm. Unters. Forsch. A* **1997a,** *204,* 300–307.

Johansson, A.; T. Laine; M.-M. Linna; H. Kallio. Variability in oil content and fatty acid composition in wild nothern currants. *Eur. Food Res. Technol.* **2000,** *211,* 277–283.

Kallio, H.; B. Yang; P. Peippo; R. Tahvonen; R. Pan.Triacylglycerols, glycerophospholipids, tocophreols, and tocotrienols in berries and seeds of two subspecies (ssp. *sinensis* and *mongolica*) of sea buckthorn (*hippopheaë rhamnoides*). *J. Agric. Food Chem.* **2002,** *50,* 3004–3009.

Kapoor, R.;H. Nair. Gamma linolenic acid oils, *Bailey's Industrial Oil and Fat Products,* Sixth edition;F. Shahidi, Ed.; John Wiley & Sons: Hoboken, NJ, 2005.

Katan, M.B.; S.M. Grundy; P. Jones; M. Law; T. Miettinen; R. Paoletti. Efficacy and safety of plant stanols and sterols in the management of blood cholesterol levels. *Mayo Clin. Proc.* **2003,** *78,* 965–978.

Krinsky, N.I. Carotenoids as antioxidants. *Nutrition* **2001,** *17,* 815–817.

Kroon, P.; G. Williamson. Polyphenols. Dietary components with established benefits for health ? *J. Sci. Food Agric.* **2005,** *85,* 1239–1240.

Li, T.S.C.; T.H.J. Beveridge; J.C.G. Drover. Phytosterol content of sea buckthorn (*hippophae rhamnoides* l.) seed oil: extraction and identification. Food Chem. **2007,** *101,* 1633–1639.

Manach, C.; G. Williamson; C. Morand; A. Scalbert;C. **Rémésy.** Bioavailability and bioefficacy of polyphenols in humans. I. Review of 97 bioavailability studies. *Am. J. Clin. Nutr.* **2005,** *81,* 230S–42S.

Martinello, M.; G. Hecker; M. de Carmen Pramparo. Grape seed oil deacidification by molecular distillation: analysis of operative variables influence using the response surface methodology. *J. Food Engineering* **2007,** *81,* 60–64.

Moreau, R.A. Phytosterols and phytosterol esters. *Healthful Lipids;* C.A. Akoh, O.-M. Lai,, Eds.; AOCS Press: Champaign, IL, 2005; pp. 335–360..

Moret, S.; A. Dudine; L.S. Conte. Processing effect on the polyaromatic hydrocarbon content of grapeseed oil. *J. Am. Oil. Chem. Soc.* **2000,** *77,* 1289–1292.

Normén, L.; L. Ellegård; H. Brants; P. Dutta; H. Andersson. A phytosterol database: fatty foods consumed in sweden and the Netherlands. *J. Food Comp. Anal.* **2007,** *20,* 193–201.

Oomah, B.D.; S. Ladet; D.V. Godfrey; J. Liang; B. Girard. Characteristics of raspberry (*rubus idaeus* l.) seed oil. *Food Chem.* **2000,** *69,* 187–193.

Oomah, B.D.; G. Mazza; R. Przybylski. Comparison of flaxseed meal lipids extracted with different solvents. *Lebensm.-Wiss.i.-Technol.* **1996,** *29,* 654–658.

The Oregon Raspberry & Blackberry Commission (2005) Final Report: Assessing Market Opportunities for Raspberry and Blackberry Seeds. www.ams.usda.gov/tmd/FSMIP/FY2001/OR320.pdf. Access 12 Nov. 2007.

Parker, T.D.; D.A. Adams; K. Zhou; M. Harris; L. Yu. Fatty acid composition and oxidative stability of cold-pressed edible seed oils. *J. Food Sci.* **2003,** *68,* 1240–1243.

Parry, J.; L. Su; M. Luther; K. Zhou; M. Yurawecz;P. Whittaker; L. Yu. Fatty acid composition and antioxidant properties of cold-pressed marionberry, boysenberry, red raspberry, and blueberry seed oils. *J. Agric. Food Chem.* **2005,** *53,* 566–573.

Parry, J.W.; L. Yu (2004a). *Phytochemical Composition and Free Radical Scavenging Capacities of Selected Cold-Pressed Edible Seed Oils,* abstracts of papers, 228th National Meeting of the American Chemical Society, Philadelphia, PA, August 22–26, 2004.

Parry, J.; L. Yu. Fatty acid content and antioxidant properties of cold-pressed black raspberry seed oil and meal. *J. Food Sci.* **2004b,** *69,* 189–193.

Pourrat, H.; A.P. Carnat. Chemical composition of raspberry seed oil (*rubus idaeus* l. *rosaceae*). *Rev. Fr. Corps Gras* **1981,** *28,* 477–479.

Pourrat, H.; A. Pourrat. Compositions Cosmétiques et Pharmaceutiques. French Patent 7345501 (1973).

Ruiz del Castillo, M.L.; G. Dobson; R. Brennan; S. Gordon. Genotypic variation in fatty acid content of blackcurrant seeds. *J. Agric. Food Chem.* **2002,** *50,* 332–335.

Ruiz del Castillo, M.L.; G. Dobson; R. Brennan; S. Gordon. Fatty acid content and juice characteristics in black currant (*ribes nigrum* l.) genotypes. *J. Agric. Food Chem.* **2004,** *52,* 948–952.

Scalbert, A.; G. Williamson. Dietary intake and bioavailability of polyphenols. *J. Nutr.* **2000,** *130,* 2073S–2085S.

Soto, C.; R. Chamy; M.E. Zúñiga. Enzymatic hydrolysis and pressing conditions effect on borage oil extraction by cold pressing. *Food Chem.* **2007,** *102,* 834–840.

St George, S.; S. Cenkowski. Influence of harvest time on the quality of oil-based compounds in sea buckthorn (*hippophae rhamnoides* l. ssp. *sinensis*) seed and fruit. *J. Agric. Food Chem.* **2007,** *55* 8054–8061.

Tahvonen, R.L.; U.S. Schwab; K.M. Linderborg; H.M. Mykkänen; H.P. Kallio. Black currant seed oil and fish oil supplements differ in their effects on fatty acid profiles of plasma lipids, and concentrations of serum total and lipoprotein lipids, plasma glucose and insulin. *J. Nutr. Biochem.* **2005,** *16,* 353–359.

Tan, C.P.; Y.B. Che Man; J. Selamat; M.S.A. Yusoff. Comparative studies of oxidative stability of edible oils by differential scanning calorimetry and oxidative stability index methods. *Food Chem.* **2002,** *76,* 385–389.

Tuomasjukka, S.; H. Kallio; P. Forssell. Effect of microencapsulation of dietary oil on postprandial lipemia. *J. Food Sci.* **2006,** *71,* S225–S230.

Williamson, G.; C. Manach. Bioavailability and bioefficacy of polyphenols in humans. ii. Review of 93 intervention studies. *Am. J. Clin. Nutr.* **2005,** *81,* 243S–255S.

Wu, D.; M. Meydani; L.S. Leka; Z. Nightingale; G.J. Handelman; J.B. Blumberg; S.N. Meydani. Effect of dietary supplementation with black currant seed oil on the immune response of healthy elderly subjects. *Am. J. Clin. Nutr.* **1999,** *70,* 536–543.

Yang, B. Lipophilic Components of Sea Buckthorn (*Hippophaë rhamnoides*) Seeds and Berries and Physiological Effects of Sea Buckthorn Oils. Dissertation. University of Turku, Finland; 2001.

Yang, B.; K.O. Kalimo; L.M. Mattila; S.E. Kallio; J.K. Katajisto; J. Peltola; H.P. Kallio. Effects of dietary supplementation with sea buckthorn (*hippophaë rhamnoides*) seed and pulp oils on atopic dermatitis. *J. Nutr. Biochem.* **1999,** *10,* 622–630.

Yang, B.; H. Kallio. Fatty acid composition of lipids in sea buckthorn (*hippophaë rhamnoides* l.) berries of different origins. *J. Agric. Food Chem.* **2001,** *49,* 1939–1947.

Yang, B.; H. Kallio. Effects of harvesting time on triacylglycerols and glycerophospholipids of sea buckthorn (*hippophaë rhamnoides l.*) berries of different origin. *J. Food Comp. Anal.* **2002,** *15,* 143–157.

Yang, B.; R.M. Karlsson; P.H. Oksman; H.P. Kallio. Phytosterols in sea buckthorn (*hippophaë rhamnoides* l.) berries: identification and effects of different origins and harvesting times. *J. Agric. Food Chem.* **2001b,** *49,* 5620–5629.

Yang, B.; J. Koponen; R. Tahvonen; H. Kallio. Plant sterols in seeds of two species of vaccinium (v. myrtillus and v. vitis-idaea) naturally distributed in finland. *Eur. Food Res. Technol.* **2003,** *216,* 34–38.

Yu, L.; J. Perry; K. Zhou. Fruit seed oils. *Nutraceutical and Speciality Lipids and their Co-Products;* F. Shahidi, Ed.; Taylor & Francis: Boca Raton, 2006; pp. 73–90.

Yu, L.; K. Zhou; J. Parry. Antioxidant properties of cold-pressed black caraway, carrot, cranberry, and hemp seed oils. *Food Chem.* **2005,** *91,* 723–729.

Zlatanov, M.D. Lipid composition of bulgarian chokeberry, black currant and rose hip seed oils. *J. Sci. Food Agric.* **1999,** *79,* 1620–1624.

# 7

# Borage, Evening Primrose, Blackcurrant, and Fungal Oils: γ-Linolenic Acid-rich Oils

**D.E. Barre**
*Cape Breton University, P.O. Box 5300, Sydney, Nova Scotia, Canada B1P-6L2*

## Introduction

Evening primrose (EPO), borage (BO), blackcurrant (BCO), and fungal (FUO) oils are receiving increasing attention for their potential in human medicine. In part, this interest arises from the conversion of γ-linolenic acid [GLA, 18:3 (n-6)], contained in the oils (Table 7.1), to dihomo-γ-linolenic acid [DGLA, 20:3 (n-6)] and in the case of BCO and sometimes FUO, α-linolenic acid [ALA 18:3 (n-6)] to eicosapentaenoic acid [EPA, 20:5 (n-3)] in the body. The structures of these precursor fatty acids are shown in Fig. 7.1 while their metabolism including derivation to prostaglandins and leukotrienes is found in Fig. 7.2. DGLA is a precursor to a variety of the 1-series prostaglandins (PG; e.g., $PGE_1$) and three-series leukotrienes (LT; e.g., $LTC_3$). EPA is converted to the 3-series PG (e.g., $PGE_3$) and 5-series LT (e.g., $LTB_5$). As hypothesized, elevation of DGLA and/or EPA in body phospholipid (PL) will be at the expense of the sometimes disease-promoting 2-series PG (e.g., $PGE_2$) and 4-series LT (e.g., $LTB_4$); (Fig. 7.2) derived from arachidonic acid [AA, 20:4 (n-6)]. This chapter critically examines studies using these oils relative to their actual or potential use in selected human pathologies.

## Oil Content

The oil content of the borage seed is 30–40% by weight (Beaubaire & Simon 1987; Simpson, 1993; Piquette & Laflamme, 2000). Evening primrose's (Ghasemnezhad et al., 2006) and fungi (Certik & Horenitzky, 1999) oil content is about 25%, with blackcurrant oil occurring at about 16% (Johannson et al., 1997)

## Oil Cake

Among the GLA containing oilseeds, only evening primrose and borage seed oil cakes were studied regarding their potential effects on human health. The EPO oil cake was noted to have antioxidant properties (Birch et al., 2001). Shahidi et al. (1997) and

O

OH

Linoleic acid

OH

O

γ-linolenic acid

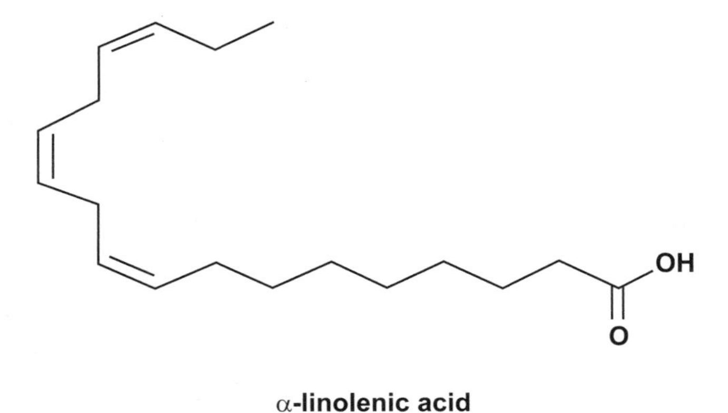


α-linolenic acid

Fig. 7.1. Structures of linoleic acid, γ-linolenic acid, and α-linolenic acid.

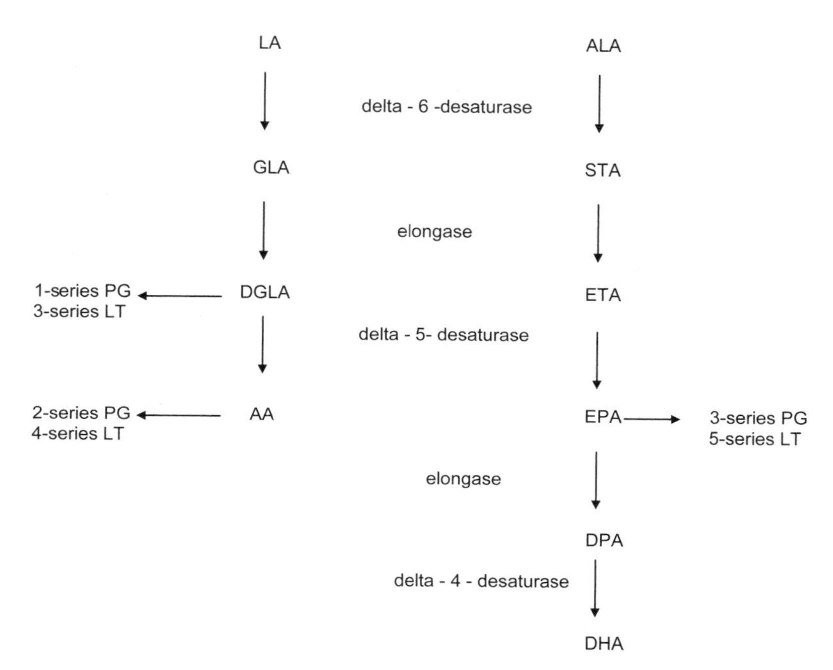


**Fig. 7.2.** Metabolic pathways of omega-3 and -6 fatty acids and the fatty acid-derived prostaglandins and leukotrienes. AA = arachidonic acid (20:4n-6), ALA = α-linolenic acid (18:3 n-3), DGLA = dihomo-γ-linolenic acid (20:3 n-6), DHA = docosahexaenoic acid (22:6 n-3), DPA = docosapentaenoic acid (22:5 n-3), EPA = eicosapentaenoic acid (20:5n-3) , ETA = eicosatetraenoic acid (20:4 n-3), GLA = γ-linolenic acid (18:3 n-6), LA = linoleic acid (18:2 n-6), LT = leukotrienes, PG = prostaglandins, STA = stearidonic acid (18:4 n-3).

Wattasinghe et al. (2002) noted the presence of the phenolics, catechin, epicatchin, and gallic acids, all known to have antioxidant properties. Shahidi (2000) noted the presence of phenolic antioxidants, rosamarinic, syringic, and sinapic acids in borage oil. Mechanisms that include oxidation are features of numerous diseases including diabetes, atherosclerosis, and cancer. A defatted evening primrose seed extract was shown to induce apoptosis in a cancer cell line. More specifically, the phenolic antioxidant gallic acid in EPO cake was shown to have anti-tumor potential via its promotion of apoptosis (Pellegrini et al., 2005). Apoptosis also occurs due to suppression of the tumor-promoting polyamines by this extract (Arimura et al., 2003; 2005). Apoptosis is critical to the control of cancer.

## Economy

Average retail prices for 30-count containers (30 capsules per bottle) run about 10 dollars Canadian for evening primrose oil, borage oil, and blackcurrant oil though prices can vary considerably. Fungal oil appears not to be available in the public retail market. Evening primrose oil, borage oil, and blackcurrant oilseed production contribute significantly to the world economy. Fungal oil appears to be more of an experimental entity at this time.

## Processing

The oils may be extracted by the four methods shown in Table 7.1. Enzymatic digestion boosts oil yield as shown below. Cold pressing and supercritical carbon dioxide extraction alleviate health concerns about the presence of residual hexane or other solvents. Reputable manufacturers take great care to evaporate solvents though it is up to the customer to self-satisfy regarding any safety concerns about residual solvents in the oil.

## Edible and Nonedible Applications

The only applications discussed herein are for health issues as this is the focus of this chapter. Evening primrose and borage, and blackcurrant oils are found in lotions and creams for skin care. Consumption of the extracted oil is the only way to obtain therapeutic doses of GLA. Greater detail on health applications are found in the section on health benefits of the oils and oil constituents.

## Acyl Lipids and Fatty Acid Content of Various Oils, Including GLA and ALA

In all the oils discussed, the fatty acids are almost entirely associated with the triglyceride fraction. EPO is derived from the seed of *Oenothera biennis*. It contains about 9% (9 g/100 g of total fatty acids) as GLA. This oil received the most attention of the four oils discussed here. BO is derived from the seed of *Borago officinalis*. It contains about 23% as GLA. The much higher GLA content compared to EPO might be more efficacious at a lower dose of oil. However, the groups of Engler (1993), Fan et

**Table 7.1. Oil Extraction Methods for Various GLA Containing Oils**

| Oil extraction method | Evening primrose oil | Borage oil | Blackcurrant oil | Fungal oil |
|---|---|---|---|---|
| Solvent extraction | yes | yes | yes | yes |
| Supercritical $CO_2$ | yes | yes | yes | yes |
| Cold pressing | yes | yes | yes | No published reports |
| Enzymatic | Yes-assist to cold pressing | Yes-assist to cold pressing | Yes-assist to cold pressing | No published reports |

al. (1996), and Dines et al. (1996) produced data that BO produces effects less than that predicted by its GLA content. This is perhaps due to the complexity of BO, with Dines et al. (1996) suggesting that the presence of one or more components of BO diminishes its efficacy. BCO is derived from the seed of *Ribes nigrum*. One of the attractive features of BCO is its content of both ALA (12%) and GLA (16%; g/100 g of total fatty acids). FUO is derived from a variety of fungi species, but *Mucor javanicus* is the one most frequently cited. The GLA content is about 25% with a small amount of ALA (2%). A summary of the fatty acid compositions of each of the four oils is found in Table 7.2.

## Minor Lipid Components

Evening primrose oil has 5% minor lipid components, mainly ergosterol (Takada et al., 1994) and triterpenes and caffeoyl esters of betulinic, morolic, and oleanolic acid, the latter four only in cold-pressed, nonraffinated (not exposed to solvent) oil (Puri, 2004; Hamburger et al., 2002; Zaugg et al., 2006). Such caffeoyl esters are progressively lost with increasing raffination (Knorr & Hamburger, 2004). It also has β-carotene, an antioxidant. Evening primrose seed oil has tocopherols, tocotrienols, phytosterols, and phenolic compounds. Borage oil has tocopherols and phytosterols. Blackcurrant oil has tocopherols (Yu et al., 2006). However, the focus of this chapter is on the health benefits of GLA contained in these oils and ALA contained in blackcurrant oil.

**Table 7.2. Typical Fatty Acid Composition (g/100 g total fatty acids) of Evening Primrose Oil (EPO), Borage Oil (BO), Blackcurrant Oil (BCO), and Fungal oil (FUO)**

| Fatty acids | EPO | BO | BCO | FUO |
|---|---|---|---|---|
| 16:0 | 6 | 11 | 7 | 15 |
| 18:0 | 2 | 4 | 1 | 2 |
| 18:1 (n-9) | 12 | 16 | 11 | 37 |
| 18:2 (n-6) | 72 | 37 | 47 | 17 |
| 18:3 (n-6) | 8 | 22 | 16 | 25 |
| 18:3 (n-3) | – | – | 12 | 2 |
| 20:1 | – | 5 | – | – |
| 22:1 (n-9) | – | 3 | – | – |
| 24:1 (n-9 | – | 2 | – | – |

Source: EPO data from Ishikawa et al (1989); BO data from Barre and Holub (1992a); BCO data from Lawson and Hughes (1988); FUO data from Takada et al. (1994).

## Flavor and Aroma Compounds

The flavors of evening primrose oil and borage oil are not documented. The aroma of evening primrose oil and borage oil is light and sweet. The aroma and flavors of blackcurrant oil and fungal oil are not published.

## Toxicity and Allergic and Toxic Compounds in the Oil

Increased risk of seizure in epileptics induced by EPO was suggested by Werneke et al. (2004) and Miller (1998), the latter author extending that caution also to borage oil. Thus, epileptics should avoid GLA containing oils until such time as dosing and frequency studies are carried out in epileptics to fully elucidate the restrictions that must be in place for that patient population. Miller suggested that such oils may very well increase the dosage requirement for anti-seizure medications. The comments of Werneke et al. (2004) were directed at epileptics taking the anti-seizure medication, sodium valproate. Halat and Denneby (2003) recommended that EPO not be taken with any antiplatelet or anticoagulant medication as EPO was shown to decrease platelet aggregation. Kast (2001) suggested increased miscarriage due to borage oil consumption in humans, but this was never borne out with human studies. EPO was reported to cause GI symptoms (soft stools, belching, and abdominal bloating) [reviewed by Fan & Chapkin (1998)] and headache [Boon & Wong (2004)]. The work of Wainright et al. (2003) in mice suggested no borage oil-induced increased loss of mice pups between birth and day 21, but they did not comment on miscarriage rates in the mice. Burns et al. (1999) found that a mix of algal and fungal oils was safe during pre-mating, mating, gestation, and lactation in rats. Similarly in pre-term infants, no fungal/fish oil combination-induced side effects occurred (Clandinin et al., 2005). However, pure fungal oil was not examined in this study nor were any such fungal oil safety studies done in human subjects, though they are clearly needed. No published reports exist that these oils have any other induced allergy or toxicity. However, none of these oils were tested for their safety regarding conception, pregnancy, or lactation in humans.

## Health Benefits of the Oil and Oil Constituents

### *Evidence For and Against Dietary GLA and ALA as Essential Fatty Acids*

Dietary GLA is seen by at least some as essential in a preventative or ameliorative role in various disease states where the levels of GLA, DGLA, and AA (and DGLA- and AA- derived eicosanoids) are abnormal in tissues central to the etiologies of these pathologies (Brush et al., 1984; Corrigan et al., 1988; Doris et al., 1998; Horrobin, 1993a; Manku et al., 1984, 1988; Strannegard et al., 1987). One can convert linoleic acid (LA) via $\Delta 6$-desaturase to GLA even though slowly and in a rate-limiting fashion. In that sense, GLA is not an essential dietary fatty acid. ALA is essential if it is the only source of the derivative n-3 fatty acids (EPA and DHA). One can obtain ALA from its

16 carbon precursors available in the human diet (Cunnane, 1996). Otherwise, one can obtain EPA and DHA by eating foods such as fish. EPA and DHA are essential in that they are associated with improved neural function, visual activity (Kinsella, 1991), and IQ (Lucas et al., 1992). In summary, GLA and ALA are essential only if they are not present in amounts sufficient to meet the requirement for amelioration of the pathology or status of function or the prevention of a particular disease.

### Evaluation of GLA- and ALA-containing Oils in Clinical Conditions

This discussion focuses on dermatitis, rheumatoid arthritis, platelet aggregation, plasma lipid profiles, and blood pressure control where most of the work using these oils was done. Other areas in which at least one of the oils was investigated are also presented.

***Dermatitis.*** Fiocchi et al. (1994) fed GLA (3 g/day) in the form of EPO for 4 weeks to children with infantile atopic dermatitis. A statistically significant decrease in sleep interruption and itching requiring antihistamine or corticosteroidal therapy was observed after 10 days. This study was uncontrolled, and its short duration failed to account for the infantile atopic dermatitis's cyclical nature. Biagi et al. (1994) fed children with atopic dermatitis a daily dose of 22.5 or 45 mg GLA/kg body weight in the form of a 50/50 EPO/olive oil mix or pure EPO, respectively, for 8 weeks. Clinical improvement was noted only with the 45-mg dose and was supported by statistically significant increase in RBC membrane DGLA before and after treatment. Yet, Johnston et al. (2004) indicated that only 13% of children given EPO by their parents reported improvement in their atopic dermatitis. However, no attempt was made apparently to determine the dose of GLA consumed by these children. Bahmer (1992) indicated a positive effect of feeding BO on atopic dermatitis using the atopic dermatitis area and severity index; however, because data from only one patient was reported, the observation is viewed as inconclusive. Tollesson and Frithz (1993a) studied 37 children (19 boys and 18 girls) suffering from infantile seborrhoeic dermatitis and 25 children (14 boys and 11 girls), but not individually matched as controls. These investigators (Tollesson & Frithz, 1993a) topically applied 0.05 mg/kg body weight GLA two times each day for 3–4 weeks to the nappy (diaper) area. The lack of a control group with the disease renders the data more difficult to interpret. In a study similar to their earlier investigation [Tollesson & Frithz (1993a)], Tollesson and Frithz (1993b) studied 48 children with infantile seborrhoeic dermatitis, but with no mention of any control group, and observed similar rates of improvement in terms of loss of lesions. Hederos and Berg (1996) observed a significant improvement in the clinical picture of pediatric atopic dermatitis patients relative to a placebo group. However, assessment of clinical parameters using an analog scale of up to 100 (worst even seen by the physician) and parental assessment of itching obviously raise very troubling questions about their accuracy. The virtual absence of data on the dermal proinflammatory mediators ($LTB_4$, $C_4$, and $D_4$) and anti-inflammatory $PGE_1$ (Ziboh, 1996) are areas rich for investigation of the clinical application of these oils to atopic dermatitis. Leonhardt et al. (1997) observed no difference in the plasma $PGE_1$ and $PGE_2$ profiles between the atopic dermatitis patients and healthy controls. However, plasma levels of these eicosanoids may not be an accurate reflection of their impact on atopic dermatitis. Other tissue levels of these or other eicosanoids might better characterize this pathology. Lehmann et al. (1995) fed EPO (540 mg GLA/day) for 12 weeks to six adult atopic dermatitis patients. Substantial decreases in pro-inflammatory $LTB_4$ by A23187 (a calcium ionophore) stimulation of isolated PMNs occurred after EPO consumption relative to the healthy controls. The substantial rise in DGLA content of the PMNs may well explain these data. The AA content did not change, and so one might conclude a competition between AA and DGLA for the eicosanoid pathway. Such a notion is supported by the decreased AA/DGLA ratio after 12 weeks of EPO administration. Whitaker et al. (1996) found no statistically significant difference in improvement of atopic dermatitis measures between patients receiving 600 mg GLA/day for 16 weeks in the form of EPO compared to patients in a placebo group. However, the impact, if any, of the use of unlimited quantities of standard emollient and limited amounts of semi-potent type-III topical steroid cream on the study results is unclear. Yet Morse and Clough (2006) provided convincing evidence of EPO's effectiveness in the presence of steroid medications though the impact of EPO becomes less in the face of increased use of potent steroids, consistent with other meta-analyses. Several other studies (Anstey et al., 1990; Bamford et al., 1985; Berth-Jones & Graham-Brown, 1993; Biaggi et al., 1988; Bordoni et al., 1987; Horrobin & Morse, 1995; Humphreys et al., 1994; Morse et al., 1989; Schalin-Karrila et al., 1987; Stewart et al., 1991; Wright & Burton, 1982) are generally more positive in outcome than the above EPO work but subject to many of its criticisms. Indeed, Williams (2003) has strongly questioned the meta-analyses by Morse et al. (1989) that indicated convincing success of EPO in atopic dermatitis. In contrast, Andreassi et al. (1997) fed 30 patients 548 mg of GLA in the form of BO twice per day. A significant drop occurred in patient-assessed itching and physician-assessed vesicle formation, erythema, and oozing after 12 weeks of BO compared to placebo. Patient assessment of itching is very subjective; however, the vesiculation, erythema, and oozing scores assigned by the patients after 12 weeks were not statistically significant from the physicians giving credence to the statistically significant drops after 12 weeks seen by the physicians. In another study by Henz et al. (1999), subjects were administered 690 mg of GLA (in the form of BO) per day for 24 weeks. Henz et al. (1999) observed a significant improvement in atopic dermatitis only in a subpopulation of BO consumers with increased erythrocyte DGLA levels, adherence to inclusion criteria, and study protocol. However, a sizeable number of the patients taking BO showed no increase in erythrocyte DGLA, suggesting noncompliance or malabsorption or defective metabolism and possibly suggesting that BO treatment may be good only in a select group of atopic dermatitis patients. The heavier use of corticosteroid creams by the BO recipients in some of the research centers also confounded the results. Takwale et al. (2003) found the use of BO in adults and children was without effect for atopic dermatitis; it is not clear why the data of Takwale et al. (2003) are so different from the majority of other BO studies in this arena particularly given the higher dose of GLA used in the study. However, it was only done for 12 weeks compared to the 24 weeks of Henz et al. (1999).

***Rheumatoid arthritis.*** Belch et al. (1988) fed 16 rheumatoid arthritis (RA) patients 540 mg GLA/day in the form of EPO for 12 months in a double-blind study. After the 12-month period, a significant improvement occurred in the patient's self-assessment of their conditions. However, this was not reflected in any of the conventional measurements of disease activity (erythrocyte sedimentation rate, rheumatoid factor, and c-reactive protein levels). While well-controlled, this study failed to provide convincing evidence of EPO efficacy. Jantti et al. (1989a) administered EPO (GLA dose not given) or olive oil in a double-blind randomized study to RA patients for 12 weeks and noted no difference in clinical improvement. Similarly, Brzeski et al. (1991) found no advantage using EPO (560 mg GLA/day for 6 months) vs. olive oil to treat RA patients. Pullman-Mooar et al. (1990) fed 1,100 mg of GLA/day in the form of BO for 12 weeks to 7 RA patients and 7 healthy controls. They observed a decrease in pro-inflammatory $PGE_2$ resulting from aggregated γ-globulin stimulation of peripheral blood monocytes and a decrease in the pro-inflammatory $LTB_4$ by similarly stimulated neutrophils. These findings were supported by the observation of clinical improvement observed in 6 of the 7 RA patients. Kast (2001) suggested that a borage oil mediated decrease in inflammation may also be due to increased c-AMP that suppresses tumor necrosis factor-alpha. In a later study, Pullman-Mooar et al. (1990) treated 7 RA patients and 7 healthy controls with 1,100 mg/day GLA in the form of BO for 12 weeks. Some of these patients were taking nonsteroidal anti-inflammatory drugs (dosage, frequency, and duration not noted), and no statistical assessment was made of the relative strength of the contributions of nonsteroidal anti-inflammatory drugs and BO therapies to the improvement of RA. Consequently, one cannot be sure of the impact of the BO therapy. Leventhal et al. (1993) fed 1.4 g GLA/day in the form of BO dietary supplements to 37 RA patients. Unlike Pullman-Mooar et al. (1990), a double-blind randomized controlled distribution was done of the BO and the placebo cotton oil capsules among the RA patients. Zurier et al. (1996) fed RA patients 2.8 g GLA/day for a minimum of 6 months in a double-blind randomized cross-over controlled study. This study (Zurier et al., 1996), while using subjective clinical measurements such as swollen joint count, was nonetheless double-blind controlled and showed substantial clinical improvement relative to the placebo-treated patients. Watson et al. (1993) studied 25 female and 5 male RA patients (mean age 40 years) taking nonsteroidal anti-inflammatory drugs and 20 healthy student volunteers (10 male, 10 females, mean age 20 years). Subjects were randomly assigned to receive capsules containing safflower oil or 525 mg/day GLA in the form of BCO capsules for 6 weeks followed immediately by a 6-week wash-out period. This study found a significant decrease in pro-inflammatory cytokines [interleukin-1β (IL-1β), interleukin-6 (IL-6)] and $PGE_2$ in patients after 6 weeks of BCO but not safflower oil consumption. Healthy controls taking BCO had statistically significant drops in all measured cytokines from lipopolysaccharide-stimulated cultured monocytes of patients except IL-1β over the same period. A reduction in IL-6 may not be desirable because of its protective role in the stimulation of acute phase proteins as well as the resulting absent negative control of (i) cellular adhesion molecules for endothelial cells, (ii) phospholipase $A_2$, and (iii) synovial collagenase (Dinarello, 1992). All effects were lost during the wash-out period in healthy controls and in the patient populations in both measures of clinical symptoms and cytokine production. However, the 6-week period over which this trial was conducted may very well be insufficient to determine true efficacy in such patients. Ziboh and Fletcher (1992) examined the effects of BCO and BO on PMN activity in terms of production of $LTB_4$. They supplemented the diets of 12 healthy subjects (6 male, 6 female) with olive oil for 2 weeks followed by BO or BCO delivering 0.48 g GLA/day for the next 6 weeks. Both BO and BCO lowered the $LTB_4$ production in $Ca^{2+}$ ionophore-stimulated cultured PMNs by an equal amount. Whether this reflects an equal decrease for BO and BCO in the in vivo stimulated production of the pro-inflammatory $LTB_4$ is open to question. Additionally Darlington and Stone (2001) reviewed the literature which indicates that $PGE_1$ derived from DGLA, in turn arising from GLA, results in decreased levels of the inflammatory IL-1β. As well, they indicate that increased GLA consumption promotes elevated Mn-superoxide dismutase activity, thus possibly reducing oxidative stress promotion of RA. Belch and Hill (2000) reviewed studies (non-meta-analysis) using evening primrose and borage oils in RA and concluded that they hold promise in the management of this disease. They, however, also note that more studies of greater than 4 months dosing duration are required to adequately address the use of GLA containing oils.

***Platelet aggregation***. A reduction in the ability of platelets to aggregate is important to decrease the risk of coronary atherosclerotic heart disease. Corbett et al. (1991) observed an increase in platelet aggregation in response to ADP and 5-hydroxytryptamine in subjects administered EPO (GLA dose 300 mg/day for 1 week) relative to a safflower oil placebo. Interestingly, Barre et al. (1993) feeding BO also observed an increase in platelet aggregation. Thus, apparently the heavy emphasis placed on the potential elevated production of DGLA-derived anti-aggregatory $PGE_1$ may be overcome by other factors facilitating platelet aggregation. In contrast, McGregor et al. (1989) studying multiple sclerosis and healthy individuals observed no difference before and after EPO supplementation (1.8 g GLA/day for 6 weeks) in platelet aggregation in response to collagen, thrombin, or ADP. Their data were supported by the absence of difference in fibrinogen binding to platelets before and after EPO supplementation. Vericel et al. (1987) supplemented the diets of elderly males with 984 mg/day of GLA in a single-blind randomized cross-over study for two months on EPO or safflower oil followed by another two months on the other oil. An EPO-attributable significant rise in the DGLA content of platelet choline-containing phospholipids (PC), but not phosphatidylinositol (PI) or ethanolamine-containing phospholipids (PE) occurred with no change in platelet reactivity. Boberg et al. (1986) fed 360 mg GLA/day in the form of EPO and olive oil to hypertriglyceridemic patients (8 weeks each oil) in a double-blind randomized cross-over study. No significant change was noted in platelet reactivity to ADP or collagen despite a significant rise in the $PGE_1$ precursor DGLA in the platelet phospholipids. Darcet et al. (1980), in a study involving elderly males treated with 200 mg/day GLA in the form of EPO for 5 weeks, noted a significant decrease in platelet aggregation in response to collagen and adrenaline in the hyper-aggregants but not those with normal platelet function. Interestingly, only the rise in platelet PL-DGLA in the hyper-aggregants was significant. Guiverneau et al. (1994), feeding 240 mg/day of GLA in the form of EPO, observed a significant decrease in ADP and epinephrine-induced platelet aggregation in hyperlipidemic men. Hyperlipidemia is known to enhance platelet reactivity, and so perhaps these individuals were like the hyper-aggregants in the study of Darcet et al. (1980). Mills et al. (1989a) had subjects (n = 10) take BO (supplying 1.3 g GLA/day for 4 weeks) to examine the effect of BO supplementation on psychological stress-induced platelet reactivity. The psychological stressors, given in random order, were mental arithmetic, a favorable impression interview, and a Stroop color word conflict test. The BO had no significant impact on platelet aggregation upon presentation of platelet-rich plasma to epinephrine and ADP. However, the degree of stress may have been too mild, and no plasma catecholamine was reported suggesting that the opportunity for significant platelet aggregation due to the stress-induced production of catecholamines that were perhaps never there in the first place. Finally the dose of BO may have been too low to adequately bring about changes in the platelet aggregation. Such an absence of change in platelet aggregation associated with a low GLA dose is supported by Bard et al. (1997) feeding 3 g BO/day to healthy male volunteers aged 18–45 years (approximately 0.66 g GLA/day assuming 22 wt% GLA content). In contrast, Barre et al. (1993), feeding six healthy human subjects an average of 5.23 g/day of GLA for 42 days, observed that platelet aggregation was significantly enhanced when platelets were exposed to collagen in platelet-rich plasma. This effect was lost upon withdrawal of BO supplementation. The significant rise in DGLA content of all the platelets' major phospholipid classes, except sphingomyelin (Barre & Holub, 1992a), was not manifested in elevated $PGE_1$ or decreased pro-aggregatory $PGE_2$ and thromboxane $A_2$ ($TXA_2$)(Barre et al., 1993). Indeed, no decrease occurred in the AA content of PE or PC (Barre & Holub, 1992a), the two major phospholipid classes contributing to the formation of eicosanoids by platelets (McKean et al., 1981). However, an increase did occur in the ratio of $PGE_1/PGE_2$, $PGE_1/TXA_2$ or $PGE_2/TXA_2$ in both resting platelet-rich plasma and collagen-stimulated platelets in platelet-rich plasma. One criticism of the work by Barre et al. (1993) is the lack of bleeding time data which is a more physiological measure of platelet activity in vivo (Rodgers & Levin, 1990). Nakahara et al. (1990) feeding FUO containing ALA found an increase in thrombus formation time in rats.

***Plasma lipid profiles.*** Alteration of plasma lipid profiles is also important in reducing the risk of coronary atherosclerotic heart disease. Chaintreuil et al. (1984) noted a drop in serum triglycerides and cholesterol in insulin-dependent diabetics taking 2 g/day but not 0.5 g GLA in the form of EPO for 6 weeks. This study was not placebo-controlled although the lipid response to the higher GLA dose is suggestive of an effect of GLA on lipid metabolism. Jantti et al. (1989b) performed a double-blind randomized study with RA patients (EPO for 12 weeks, no GLA mass dose given; control olive oil). This study noted a significant decrease only in the apo B content of the plasma but no change in the triglycerides, total or high-density lipoprotein cholesterol (HDLc) attributable to the EPO. Abraham et al. (1990) noted no impact of EPO (GLA mass dose not given) or safflower oil on plasma cholesterol levels in a cross-over double-blind randomized study over 4 months. Richard et al. (1990) observed a significant decrease in total cholesterol and low-density lipoprotein cholesterol (LDLc) and increase in HDLc in healthy normolipidemic males due to the administration of 1.1 followed by 1.6 g GLA/day for 1 month at each dosage. In contrast, Vikkari and Lehtonen (1986), studying hyperlipidemic men, found no change in serum lipids after 1 month of EPO supplementation (GLA mass dose not given). The GLA dose may have been insufficient to bring about a change in the plasma lipids. Ishikawa et al. (1989), using hypercholesterolemic individuals observed an EPO-induced significant decrease in plasma cholesterol only in patients without hypertriglyceridemia. A significant rise in HDLc occurred only in individuals with hypertriglyceridemia (Ishikawa et al., 1989). The study design was 300 mg/day GLA, 8 weeks) in a double-blind randomized cross-over design using a safflower oil placebo (Ishikawa et al., 1989). Horrobin and Manku (1983) in a double-blind randomized cross-over study fed 360 mg GLA/day for 12 weeks to individuals with plasma cholesterol levels of more or less than 5 mmol/L. Only individuals in the former category showed a decrease in cholesterol levels. This study is consistent with most others in the EPO field showing strong lipid-lowering in hyperlipidemic individuals. Also consistent in that regard is the study of Guiverneau et al. (1994), feeding 240 mg/day of GLA in the form of EPO. Guiverneau et al. (1994) observed a significant decrease in plasma triglycerides, total and LDLc and a rise in HDLc in hyperlipidemic men. Barre and Holub (1992b), feeding an average of 5.23 g GLA/day in the form of BO for 42 days to 6 healthy normolipidemic subjects, observed no impact on plasma lipid levels (triglycerides, total cholesterol, HDLc, and LDL). The individuals with the highest triglyceride levels showed the greatest triglyceride drop, but this was not statistically significant. The absence of an increase in HDLc reflects the nonsignificant drop in triglyceride levels. A number of studies, most notably those in Inuit populations, suggested that high levels of both DGLA (from GLA) and EPA (in part from ALA) combine to lower levels of plasma triglycerides, total cholesterol, and LDLc as well as to decrease platelet aggregation (Dyerberg et al., 1975). Interestingly, Seri et al. (1993) found that BCO or a mix of BCO and fish oil resulted in lower plasma triglyceride levels in hemodialysis patients compared to fish oil alone. This observation suggests the possibility that the dietary molar ratio of n-6 to n-3 fatty acids may be important. However, the mix of BCO and fish oil was no more effective than either oil alone in terms of the magnitude of decrease in systolic blood pressure compared to pre-supplementation levels. Spielmann et al. (1989) noted a significant drop in total cholesterol and LDLc as well as a rise in HDLc in type IIa and IIb hyperlipidemia with BCO providing 450 mg GLA/day for 12 weeks vs. grape seed oil. This is in contrast to much higher levels of GLA in BO offered to normolipidemics that were without effect, perhaps due to the normolipidemic status of the subjects or possibly differences in GLA doses or oil components (Barre and Holub, 1992b). Agostini et al. (1995) observed no statistically significant improvement in the plasma lipid profiles of pediatric phenylketonurics taking 23 mg GLA and 18 mg ALA/kg body weight/day for 6 months. However, the control group (healthy children) was not fed BCO, and so the absence of change relative to the controls is meaningless. Again it would have been better to feed a placebo to a group of pediatric phenylketonurics. One human study by Innis and Hansen (1996) using a mix of microalgal and FUOs improved the plasma lipid profiles. The contribution of the fungal component to this amelioration is unclear. However, some potential for success does exist in humans based on animal experiments. Dietary FUO (no ALA) resulted in decreased body fat content and plasma total cholesterol but not triglycerides in a dose-related manner (Takada et al., 1994) compared to control rats. Sugano et al. (1986) also noted a drop in cholesterol with mold oil supplementation in rats. Bergstrom (1995) reported a 50–60% lowering of cholesterol by the fungal metabolites, zaragozic acids, in monkeys. Zaragozic acids are squalene synthase inhibitors and may explain the cholesterol results of Sugano et al. (1986) and Bergstrom (1995). Another explanation for the results of Sugano et al. (1986) and Bergstrom (1995) may be the finding of Harada et al. (1998) who noted that a sesquiterpene ester found in fungi accelerates the LDL catabolism in HepG2 cells. LDL is a major carrier of cholesterol. FUO, rich in the lipoprotein antioxidant β-carotene (Jialal et al., 1991; Cerda-Olmeda & Avalos, 1994), might be useful, given the putative connection of oxidized lipoproteins to atherosclerotic risk (Parthasarathy, 1991). Additional support for this type of antioxidant mechanism was reported by Phylactos et al. (1994), who observed elevated activities of the antioxidant enzyme, manganese superoxide dismutase, in rats fed FUO compared to a stock diet that did not contain ALA.

***Blood pressure control.*** Blood pressure is yet another area in which EPO supplementation was studied. Fievet et al. (1985) observed a 5% decrease in the mean blood pressure (diastolic/systolic combined) before and after EPO supplementation (7 weeks, 270 mg GLA/day) in pregnant patients at high risk of toxemia. The well-controlled blood pressure study by Mills et al. (1990) observed that the plasma norepinephrine and the vasoconstrictor responses to –10 and –40 mm Hg lower body negative pressure were substantially augmented by BO consumption (91.04 g GLA/day for 4 weeks). Indeed, a small but significant rise appeared in diastolic blood pressure at both the –10 and –40 mm Hg lower body negative pressure stressor levels. Yet these benefits may not be due to improved endothelial function and vascular tone as no such changes occurred as the result of EPO administration, even though in healthy subjects (Khan et al., 2003). As well, the forearm vascular resistance increased, before and after BO consumption, in these normotensive subjects (Mills et al., 1990). Mills et al. (1989b) fed BO, delivering 1.3 g GLA/day for 28 days to 30 normotensive male university students. Before and after 28 days of supplementation the subjects did a Stroop color-word conflict test (Stroop, 1935) to induce psychological stress. A small but statistically significant decrease occurred (before and after supplementation) in the basal systolic and diastolic blood pressures. They further observed that BO significantly reduced the systolic blood pressure response relative to pre-supplementation. The work of Mills et al. (1989b, 1990) suggests that BO can correct perturbations in blood pressure in either direction. Deferne and Leeds (1996) were apparently the first to use BCO to control blood pressure. Deferne and Leeds (1996) administered 6 g of BCO (968 mg GLA) or safflower oil/day for 8 weeks to borderline hypertensive individuals. Resting blood pressure was unchanged in both groups, but the BCO group showed a 40% drop, relative to the safflower oil group in blood pressure reactivity to stress induced by mental arithmetic. This result is far more dramatic than the work of Mills et al. (1989b) using BO at a higher GLA dose. However, the stress of the conflict test used by Mills et al (1989b) may be greater than mental arithmetic. The presence of ALA in BCO may also play a role in the generation of additional blood pressure-lowering eicosanoids and other factors important in the regulation of hypertension. However, Engler (1993) observed decreased blood pressure in spontaneously hypertensive rats red FUO with no ALA.

## Other Health-promoting Properties of GLA-rich Oils

Immune function, diabetic neuropathy and microvascular complications, ulcerative colitis, migraine, pediatric respiratory infections, premenstrual syndrome, cancer, gastric ulceration, uremic pruritis, mental development, osteroporosis, asthma, hot flashes in menopause, inflammation, obesity management, Raynaud's or Sjorgen's syndrome, growth, visual acuity, information processing, general development, language, temperament, and epidermal hyperproliferation are other areas in which at least one EPO, BO, BCO, or pure GLA was studied in humans.

Fisher and Harbige (1997) examined the use of BO consumption by humans on cytokine production by isolated blood mononuclear cells upon stimulation with phytohemagglutinin ($TGF\beta_1$, IL-4) and α-CD3 (IL-10). IL-10 and IL-4 cytokines showed a significant drop and $TGF\beta_1$ a significant rise by week 12 of the borage oil administration. Both agonists failed to stimulate lymphocyte proliferation. $TGF\beta_1$ boosts immune function to streptococcus. The impact on IL-4 and 10 was of unclear significance. Wu et al. (1999) administered BCO (675 mg GLA/day) in healthy elderly subjects (65 years or older) in a randomized double-blind placebo-controlled (soybean oil) trial for 2 months. It was observed that the BCO group had increased lymphocyte proliferation, no difference in the response to dermally applied antigens except tetanus toxoid, and no significant change in agonist-stimulated blood mononuclear production of IL1β and IL-2, but a drop in $PGE_2$ formation. Immune enhancement was seen as the drop in BCO induced pro-inflammatory $PGE_2$ formation.

Horrobin (1997) reviewed two human EPO supplementation trials that indicated very promising effects on the reduction of diabetic neuropathy and hence improved nerve conduction velocity as measured by improvement in neurophysiological parameters, thermal thresholds, and clinical sensory evaluations. These findings are supported by multiple other human (Fedele & Guigliano, 1997; Horrobin, 1992; Jamal, 1994; Keen et al., 1993; Wolever, 1993) and animal studies (Ford et al., 2001; Coste et al., 1999; Head et al., 2000; Kuruvilla et al., 1998). Another successful approach, in animal studies only, in this pathology is the use of GLA supplementation combined with antioxidants (Cameron & Cotter, 1996a, 1996b; Cameron et al., 1998; Hounsom et al., 1998; Peth et al., 2000) such as ascorbate, lipoic acid, and others. Oxidative stress is causally associated with diabetic neuropathy (Reljanovic et al., 1999; Vinik, 1999). The combination of GLA and antioxidants produced synergistic improvements in measures of diabetic neuropathy. Interestingly, Shotton et al. (2003) showed that in streptozotocin induced diabetes in rats, lipoic acid is more effective than EPO in terms of preventing diabetes-induced changes in autonomic nerves to the heart and penis, but that neither treatment could affect diabetes-induced changes to the autonomic innervation of the ileum. Unfortunately, this study did not examine synergistic impacts of these two agents, and thus far no such studies were done in humans. Along with improvements in neural performance, evening primrose oil appears to increase endothelium-dependent hyperpolarizing activity in streptozotocin-induced diabetes rats, thus potentially decreasing microvascular complications of diabetes (Jack et al., 2002).

Regarding ulcerative colitis, EPO (200 mg GLA/day for 1 month followed by 100 mg/day for 6 months) improved stool consistency in ulcerative colitis compared to fish oil and olive oil (Greenfield et al., 1993). Yet no difference was observed in stool frequency, rectal bleeding relapse, appearance upon sigmoidoscopy or rectal histology. However, treatments were in addition to other medications, and no indication was given as to the similarity of the three groups in that regard. Consequently, possibly no meaningful conclusions are drawn about the efficacy of EPO relative to the fish oil or olive oil groups.

Wagner and Nootbaar-Wagner (1997) gave migraine patients 1,800 mg/day each of GLA and ALA (in unstated form) for 6 months. At the end of the period, a significant improvement occurred in the severity and frequency of the attacks. However, the authors admit that without a placebo group control one cannot be sure that the effects seen are due to the GLA/ALA combination. They raise the specter of altered eicosanoid formation but have made no attempt to include such data in their paper. Nonetheless, further study of migraine is of interest. Dietary supplementation with GLA and ALA combined was also found to reduce the incidence of migraine (Wagner & Nootbaar-Wagner, 1997). The combination of the GLA and ALA makes the impact of GLA unclear.

Budieri et al (1996) in a meta-analysis gave EPO little credence for premenstrual syndrome (PMS) treatment. They pointed out that only five of the seven existing trials are well-controlled, and the rest suffer from inconsistent scoring and response criteria, thus rendering a thorough meta-analysis impossible. Khoo et al. (1990), in a double-blind randomized cross-over study, administered 360 mg GLA/day in the form of EPO through three complete menstrual cycles and supported the ineffectiveness of EPO. However, Horrobin (1983; 1993b), in a review of a number of studies, noted EPO is highly successful for depression, irritability, breast pain and tenderness, and fluid retention associated with PMS, the findings with breast tenderness are supported by a literature review (Dickerson et al., 2003) Clearly this area is highly disputed.

van der Merwe et al. (1990) found EPO (1.44 g GLA/day) to be ineffective in terms of survival time or liver size in liver cancer patients relative to an olive oil placebo. The large size of the tumors and low doses of GLA may have accounted for this lack of significant difference. The large tumor size suggests advanced carcinoma, and earlier intervention may have been of benefit. Das et al. (1995) observed effective cancer suppression when GLA (1 mg/day for 10 days) was applied locally to malignant gliomas. Fearon et al. (1996) noted significantly enhanced survival times on pancreatic cancer patients given lithium GLA (2-4 g/day). However, neither study (van der Merwe et al., 1990; Fearon et al., 1996) was double-blind or controlled, thus rendering highly questionable results. Kollias et al. (2000) observed no impact of EPO in reducing breast fibroadenomas relative to a placebo. Thus it appears that more studies need to be done to confirm the effect of EPO and indeed the other oils featured in this chapter to determine what cancers and what individuals, if any, are susceptible to improvement using these approaches. Borage oil mixed with rat chow attenuated tumor growth in a rat model of prostatic adenocarcinoma (Pham et al., 2006). However no borage oil trials are held in humans with cancer.

Potential mechanisms for cancer suppression were studied. Cai et al. (1999) observed a dose-related GLA inhibition of angiogenesis in cultured endothelial cells.

Angiogenesis is an essential feature of malignant tumor development. Jiang et al. (1997) found that cultured cancer cells incubated with GLA produced more maspin (a tumor suppressor gene with antimotility activity), suggesting a mechanism by which GLA may inhibit cancer. These papers by Cai et al. (1999) and Jiang et al. (1997) suggest anticancerous mechanisms for GLA. However, as is the case of all cell culture studies, one is dealing with isolated cells and not the whole body, rendering such testing as a screening tool but not a guaranteed way to elucidate the mechanism of carcinogenesis in patients.

Prichard et al. (1988) observed in a double-blind randomized placebo-controlled trial that EPO (360 mg GLA/day for 2 weeks) consumption did not affect gastric ulceration due to aspirin use in humans. However, the GLA dose may have been too low relative to the aspirin challenge.

Yoshimoto-Furuie et al. (1999) in a double-blind randomized study observed that EPO (180 mg GLA/day for 6 weeks) consumption significantly improved pruritis, erythema, and skin dryness in hemodialysis patients compared to baseline before supplementation in the treatment group. Measures of uremic skin symptoms showed no difference between treatment and control groups at week 6, vastly diminishing the authors' conclusion that EPO improved such symptoms. Tamimi et al. (1999) found no significant change in any of the hematological, biochemical, or other indices of uremic pruritis in dialysis patients after 58 mg/day of GLA in the form of EPO for 6 months in a double-blind placebo-controlled study. This suggests insufficient doses or perhaps complete inefficacy of EPO directed at this complication of uremia.

Fewtrell et al. (2004) showed that a BO/fish oil-supplemented formula improved growth in pre-term infants (males and females) but only improved mental development index in the males compared to nonsupplemented formula. Unfortunately, the specific impact of BO was not measured.

Whelan et al. (2006) reviewed the literature and suggested that EPO has a benefit in osteoporosis, though it is not clear the mechanism by which such a benefit occurs. For example, the relationship between eicosanoids generated by evening primrose oil use and the pathophysiology of osteoporosis is not clear. Ziboh et al. (2004) noted that $LtB_4$ generation in ex vivo neutrophils from asthma patients is suppressed in borage oil consumers. However, this was not matched by clinical improvement in asthma, and they suggested that the dose of 2 grams per day of GLA was insufficient to produce such clinical improvement. Carroll (2006), Low Dog (2005), Haimon-Kochman and Hochner-Celinker (2005), Rock and Demichele (2003), Philp (2003), and Kronenberg and Fugh-Berman (2002) in reviews of the literature noted that EPO is without effect in reducing hot flashes in menopause.

Consistent with a number of studies using EPO in various disease states, de la Puerta Vaquez et al. (2004) showed that stimulated leukocytes from EPO fed rats produce less of the pro-inflammatory eicosanoid $PGE_2$ compared to the basal diet. However, EPO also produced higher levels of various pro-inflammatory reactive oxygen species compared to the basal diet. Clearly, the anti-inflammatory mechanisms of EPO in human clinical trials need to be more specifically delineated.

Schirmer and Phinney (2007) found that formerly obese humans given 890 mg/d GLA in the form of a dietary borage oil supplement did not regain weight. Takada et al. (1994) showed that EPO induces fat loss in rats via increased β-oxidation of fatty acids. However, the precise mechanism(s) explaining the lack of weight regain in humans were not elucidated and are worthy of investigation, given the current worldwide explosion of obesity.

Belch and Hill (2000) in a review of the literature found no conclusive benefit of GLA containing oils in the management of Raynaud's or Sjorgen's syndrome. However, no further studies were done since 2000 and it may well be a question of tailoring specific doses and durations of these oils to specific individuals to get a more accurate grasp on the efficacy and safety of such oils.

A mix of fungal and fish oils was without impact on growth, visual acuity, information processing, general development, language, and temperament (Auestad et al., 2001) in term infants. Unfortunately, the fish and fungal oils were not examined separately so it is impossible to know which, if not both, contributed to this negative finding. It is necessary to conduct such separate trials to come to any firm conclusions about the combination of oils or the oils separately regarding these parameters. In contrast, Clandinin et al. (2005) found that in pre-term infants that a blend of fish and fungal oils did improve psychomotor and mental development. Thus it appears that the fungal/fish oil combination is of benefit in pre-term infants. Regardless, more studies are required to draw any such definitive conclusion.

BO was found to be more effective compared to EPO in reversing epidermal hyperproliferation in guinea pigs (Chung et al., 2002). The sn-2 GLA in triglyceride is more bioavailable and, when combined with the higher amount of GLA in BO compared to EPO, provides the greater impact of BO in this study (Chung et al., 2002), though Redden et al. (1996) indicated that the sn-2 position bioavailability is also dictated by fatty acid composition of the sn-1 and -3 positions. Redden et al. (1996) feel that the GLA (mostly in the sn-3 position) of EPO is more bioavailable than that of BO, despite the higher level of GLA in BO compared to EPO. Regardless, no human trials were carried out to date regarding the use of GLA-containing oils in potential of reversing epidermal hyperproliferation.

## Conclusions

Based on alterations in eicosanoid profiles and some clinical outcomes, the use of GLA or GLA/ALA-containing oils appears to have some potential success in human medicine. For dermatitis such alterations include increases in DGLA-derived $PGE_1$ and a decline in the AA resulting in decreases in the pro-inflammatory $LTB_4$, $C_4$, and $D_4$. RA is expected to benefit due to reported decreases in AA and subsequent declines in AA-derived pro-inflammatory eicosanoids $PGE_2$ and $LTB_4$ and cytokines

IL-1β, and IL-6. Platelet aggregation is anticipated to decrease based on elevated DGLA-derived $PGE_1$ and decreases in pro-aggregatory AA-derived $TxB_2$ and $PGE_2$. The various oils may improve lipid levels via increases in polyunsaturation levels or perhaps alterations in eicosanoid profile though in the case of fungal oils the presence of zaragozic acids and sesquiterpene ester is believed to be a possible regulatory factor. Blood pressure is expected to be improved due to potential increases in DGLA-derived hypotensive PGs and decreases in AA-derived hypertensive PGs. Further, in these various disease states, a DGLA deficiency exists suggesting a possible remedy via delivery of the DGLA precursor, GLA.

However, various problems with some of these studies have arisen. First, studies using all these oils for a given pathology have yet to be completed to give a complete picture of the comparative efficacy and safety of these treatments. Further, an absence of control groups, subjective assessments of treatment efficacy, and simultaneous use of drugs directed at the amelioration of the pathology under consideration must be overcome in some dermatitis studies. Certain RA studies have the weaknesses of non-specification of GLA mass, simultaneous use of drugs confounding the findings and the short trial periods. A few of the platelet aggregation investigations suffer variously from potentially insufficient stressors, low doses of GLA, and inappropriate selection of outcome. In some instances, problems with plasma lipid profile studies include the absence of GLA mass specification and failure to feed the test oil to an appropriate control group. Some trials suffer from simultaneous use of result-confounding drugs (ulcerative colitis), lack of a control group (migraine), potential lack of double-blind control (respiratory infections), poorly-controlled studies (PMS), no double-blind control and late treatment onset (cancer), low GLA dose (gastric ulceration) and comparisons within group changes rather than to control groups, and too low GLA doses (uremic pruritus).

Hence, the overall evidence for success of each of these oils in treating or preventing human disease is at best troubled. Studies that are much longer in term with vastly increased patient numbers and better controls need to be done to arrive at definitive conclusions about the efficacy and safety of these oils in preventative health and established pathologies. While a lack of correlation exists between the biochemical and clinical outcomes in this area, generally speaking, it is the clinical data that matters. However, the conduct of the clinical trials is almost always lacking in meeting the regulatory standards for clinical trials. Thus, the use of GLA-containing oils cannot be conclusively recommended for any pathology until the issues outlined in this Conclusions section are addressed adequately. Even though testing of GLA, most notably in eczema, arthritis and diabetic neuropathy, produced clinical trial data comparable to many drugs, such results were derived from trials contrasting sharply with the rigorous standards employed by pharmaceutical company trials. However, the medicinal use of GLA is an area rich for investigation.

## Other Issues

No published issues regarding authenticity or adulteration of these oils exist.

## References

Abraham, R.D.; R.A. Riermersma; R.A. Elton; C. Macintyre; M.F. Oliver. Effects of safflower oil and evening primrose oil in men with a low dihomo-γ-linolenic acid. *Atherosclerosis* **1990,** *81,* 199–208.

Agostoni, C.; E. Riva; G. Biasucci; D. Luotti; M.G. Bruzzese; F. Marangoni; M. Giovannini. The effects of n-3 and n-6 polyunsaturated fatty acids on plasma lipids and fatty acids of treated phenylketonuric children. *Prostaglandins Leukot. Essent. Fatty Acids* **1995,** *53,* 401–404.

Andreassi, M.; P. Forleo; A. Di Lorio; S. Masci; G. Abate; P. Amerio. Efficacy of γ-linolenic acid in the treatment of patients with atopic dermatitis. *J. Int. Med. Res.* **1997,** *25,* 266–274.

Anstey, A.; M. Quigley; J.D. Wilkinson. Topical evening primrose oil as a treatment for atopic eczema. *J. Dermatol. Res.* **1990,** *1,* 199–201.

Arimura, T.; A. Kojima-Yuasa; M. Suzuki; D.O. Kennedy; I. Matsui-Yuasa. Caspase-independent apoptosis induced by evening primrose extract in Ehrlich ascites tumor cells. *Cancer Letters* **2003,** *201,* 9–16.

Arimura, T.; A. Kojima-Yuasa; Y. Tatsumi; D.O. Kennedy; I. Matsui-Yuasa. Involvment of polyamines in evening primrose extract-induced apoptosis in Erlich ascites tumor cells. *Amino Acids* **2005,** *28,* 21–27.

Auestad, N.; R. Halter; R.T. Hall; M. Blatter; M.L. Bogle; W. Burks; J.R. Erickson; K.M. Fitzgerald; V. Dobson; S.M. Innis; et al. Growth development in term infants fed long-chain polyunsaturated fatty acids: A double-masked, randomized, parallel, prospective, multivariate study. *Pediatrics.* **2001,** *108,* 372–381.

Bahmer, F.A. ADASI Score: Atopic dermatitis area and severity index. *Acta Derm. Venereol. Suppl. (Stockh.)* **1992,** *176,* 32–33.

Bamford, J.T.M.; R.W. Gibson; C.M. Renier. Atopic eczema unresponsive to evening primrose oil (linoleic and γ-linolenic). *J. Am. Acad. Dermatol.* **1985,** *13,* 959–965.

Bard, J.M.; G. Luc; B. Jude; J.C. Bordet; B. Lacroix; J.P. Bonte; H.J. Parra; P. Duriez. A therapeutic dosage (3 g/day) of borage oil supplementation has no effect on platelet aggregation in healthy volunteers. *Fundam. Clin. Pharmacol.* **1997,** *11,* 143–144.

Barre, D.E.; B.J. Holub. The effect of borage oil consumption on the composition of individual phospholipids in human platelets. *Lipids* **1992a,** *27,* 315–320.

Barre, D.E.; B.J. Holub. The effect of borage oil consumption on human plasma lipid levels and the phosphatidylcholine and cholesterol ester composition of high density lipoprotein. *Nutr. Res.* **1992b,** *12,* 1181–1194.

Barre, D.E.; B.J. Holub; R.S. Chapkin. The effect of borage oil supplementation on human platelet aggregation, thromboxane $B_2$, prostaglandin $E_1$ and $E_2$ formation. *Nutr. Res.* **1993,** *13,* 739–751.

Beaubaire, N.A.; J.E. Simon. Production potential of *Borago officinalis L. Acta Hort.* **1987,** *208,*

101–114.

Belch, J.J.F.; D. Ansell; R. Madhok; A. O'Dowd; R.D. Sturrock. Effects of altering dietary essential fatty acids on requirements for non-steroidal anti-inflammatory drugs in patients with rheumatoid arthritis: A double-blind placebo controlled study. *Ann. Rheum. Dis.* **1988,** *47,* 96–104.

Belch, J.J.F.; A. Hill. Evening primrose oil and borage oil in rheumatologic conditions. *Am. J. Clin. Nutr.* **2000,** *71(suppl.),* 352S–356S.

Bergstrom, J.D. Is squalene synthase a good target for cholesterol lowering therapy? *Studies with zaragozic acids (abstract). Drugs Affecting Lipid Metabolism Conference Houston,* **1995**, 6.

Berth-Jones, J.; R.A.C. Graham-Brown. Placebo controlled trial of essential fatty acid supplementation in atopic dermatitis. *Lancet* **1993,** *341,* 1557–1560.

Biagi, P.L.; A. Bordoni; S. Hrelia; M. Celadon; G.P. Ricci; V. Cannella; A. Patrizi; F. Specchia; M. Masi. The effect of gamma-linolenic acid on clinical status, red cell fatty acid composition and membrane microviscosity in infants with atopic dermatitis. *Drugs Exp. Clin. Res.* **1994,** *20,* 77–84.

Biagi, P.L.; A. Bordoni; M. Masi; G. Ricci; C. Fanelli; A. Patrizi; E. Ceccolini. A long-term study on the use of evening primrose oil (Efamol) in atopic children. *Drugs Exp. Clin. Res.* **1988,** *14,* 285–290.

Birch A.E; G.P. Fenner; R. Watkins; L.C. Boyd. Antioxidant properties of evening primrose seed extracts. *J. Agric Food Chem.* **2001,** *49,* 4502–4507.

Boberg, M.; B. Vessby; I. Selinius. Effects of dietary supplementation with n-6 and n-3 long chain polyunsaturated fatty acids on serum lipoproteins and platelet function in hypertriglyceridaemic patients. *Acta. Med. Scand.* **1986,** *220,* 153–160.

Boon, H.; J. Wong. Botanical medicine and cancer: A review of safety and efficacy. *Expert Opin. Pharmacother.* **2004,** *5,* 2485–2501.

Bordoni, A.; P.L. Biagi; M. Masi; G. Ricci; C. Fanelli; A. Patrizi; E. Ceccolini. Evening primrose oil (Efamol) in the treatment of children with atopic eczema. *Drugs Exp. Clin. Res.* **1987,** *14,* 291–297.

Brush, M.G.; S.J. Watson; D.F. Horrobin; M.S. Manku. Abnormal essential fatty acid levels in plasma of women with premenstrual syndrome. *Am. J. Obstet. Gynecol.* **1984,** *150,* 363–366.

Brzeski, M.; R. Madhok; H.A. Capell. Evening primrose oil in patients with rheumatoid arthritis and side-effects of non-steroidal anti-inflammatory drugs. *Br. J. Rheumatol .* **1991,** *30,* 370–372.

Budeiri, D.; A. Li Wan Po; J.C. Dornan. Is evening primrose oil of value in the treatment of premenstrual syndrome? *Control Clin. Trials* **1996,** *17,* 60–68.

Burns. R.A.; G.J. Wibert; D.A. Diersen-Schade; C.M. Kelly. Evaluation of single-cell sources of docosahexaenoic acid and arachidonic acid: 3-month rat oral safety study with an *in utero* phase. *Food and Chem. Toxicol.* **1999,** *37,* 23–36.

Cai, J.; W.G. Jiang; R.E. Mansel. Inhibition of angiogenic factor-and tumour-induced angiogenesis by gamma-linolenic acid. *Prostaglandins Leukot. Essent. Fatty Acids* **1999,** *60,* 21–29.

Cameron, N.E.; M.A. Cotter. Comparison of the effects of ascorbyl γ-linolenic acid and γ-linolenic acid in the correction of neurovascular deficits in diabetic rats. *Diabetologica* **1996a,** *39,* 1047–1054.

Cameron, N.E.; M.A. Cotter. Interaction between oxidative stress and γ-linolenic acid in impaired neurovascular function of diabetic rats. *Am. J. Physiol.* **1996b,** *271*, E471–E476.

Cameron, N.E.; M.A. Cotter; D.F. Horrobin; H.J. Tritschler. Effects of α-lipoic acid on neurovascular function in diabetic rats: Interaction with essential fatty acids. *Diabetologica* **1998,** *41,* 390–399.

Carroll, D.G. Nonhormonal therapies for hot flashes in menopause. *Am. Fam. Physician* **2006,** *73*, 457–464.

Cerda-Olmedo, E.; J. Avalos. Oleaginous fungi: Carotene-rich oil from phycomyces. *Prog Lipid Res.* **1994,** *33,* 185–192.

Certik, M.; R. Horenitzky. Supercritical $CO_2$ extraction of fungal oil containing γ-linolenic acid. *Biotech. Techniques.* **1999,** *13*, 11–15.

Chaintreuil, J.; L. Monnier; C. Colette; P. Crastes de Paulet; A. Orsetti; D. Spielmann; F. Mendy; A. Crastes de Paulet. Effects of dietary γ-linolenate supplementation on serum lipids and platelet function in insulin dependent diabetic patients. *Hum. Nutr. Clin. Nutr.* **1984,** *38*, 121–130.

Chung, S.; S. Kong; K. Seong; Y. Cho. γ-linolenic acid in borage oil reverses epidermal hyperproliferation in guinea pigs. *J. Nutr.* **2002,** *132*, 3090–3097.

Clandinin, M.T.; J.E. van Aerde; K.L. Merkel; C.L. Harris; M.A. Springer; J.W. Hansen; D.A. Diersen-Schade. Growth and development of preterm infants fed infant formulas containing docosahexaenoic acid and arachidonic acid. *J. Pediatr.* **2005,** *146*, 461–468.

Corbett, R.; J.B.A. Meagher; B.E. Leonard. The effect of acute alcohol intoxication on psychometric testing and 5-HT-induced platelet aggregation in normal subjects; modulatory role of evening primrose oil. *Hum. Psychopharm.* **1991,** *6*, 253–256.

Corrigan, F.M.; D.F. Horrobin; E.T. Skinner; J.A.O. Besson; M.B. Cooper. Abnormal content of n-6 and n-3 long-chain unsaturated fatty acids in the phosphoglycerides and cholesterol esters of parahippocampal cortex from Alzheimers' disease patients and its relationship to acetyl CoA content. *Int. J. Biochem. Cell Biol.* **1998,** *30*, 197–207.

Coste, T.; M. Pierlovisi; J. Leonardi; D. Dufayet; A. Gerbi; H. Lafont; P. Vague; D. Raccah. Beneficial effects of gamma linolenic acid supplementation on nerve conduction velocity, $Na^+$, $K^+$ ATPase activity and membrane fatty acid composition in sciatic nerve of diabetic rats. *J. Nutr. Biochem.* **1999,** *10,* 411–420.

Cunnane, S.C. The Canadian Society for Nutritional Sciences 1995 Young Scientist Award Lecture. Recent studies on the syntheis, β-oxidation, and deficiency on linoleate and alpha-linoleate: Are essential fatty acids more aptly named indispensable or conditionally dispensable fatty acids? *Can. J. Physiol. Pharmacol.* **1996,** *74,* 629–639.

Darcet, P.; F. Driss; F. Mendy; N. Delhaye. Exploration du metabolism des acides gras des lipids totaux plasmatiques et de l'aggregation plaquettaire d'une population d'hommes ages a l'aide d'une alimentation enrichie en acide y-linolenique. *Ann. Nutr. Aliment.* **1980,** *34*, 277–290.

Darlington, L.G.; T.W. Stone. Antioxidants and fatty acids in the amelioration of rheumatoid arthritis and related disorders. *Brit. J. Nutr.* **2001,** *85,* 251–269.

Das, U.N.; V.V.S.K. Prasad; D.R. Reddy. Local application of a γ linolenic acid in the treatment of human gliomas. *Cancer Lett.* **1995,** *94*, 147–155.

de la Puerta Vázquez. R.; E. Martínez-Domínguez; J.S. Perona; V. Ruiz-Gutiérrez. Effects of different dietary oils on inflammatory mediator generation and fatty acid composition in rat neutrophils. *Metabolism* **2004,** *53,* 59–65.

Deferne, J.-L.; A.R. Leeds. Resting blood pressure and cardiovascular reactivity to mental arithmetic in mild hypertensive males supplemented with black currant seed oil. *J. Hum. Hypertens.* **1996,** *10,* 531–537.

Dickerson, L.M.; P.J. Mazyck; M.H. Hunter. Premenstrual syndrome. *Am. Fam. Physician.* **2003,** *67,* 1743–1752.

Dinarello, C.A. Anti-cytokine strategies. *Eur. Cytokine Netw.* **1992,** *3,* 7–17.

Dines, K.C.; M.A. Cotter; N.E. Cameron. Effectiveness of natural oils as sources of γ-linolenic acid to correct peripheral nerve conduction velocity abnormalities in diabetic rats: Modulatin by thromboxane $A_2$ inhibition. *Prostaglandins Leukot. Essent. Fatty Acids* **1996,** *55,* 159–65.

Doris, A.B.; K. Wahle; A. MacDonald; S. Morris; I. Coffey; W. Muir; D. Blackwood. Red cell membrane fatty acids, cytosolic phospholipase-$A_2$ and schizophrenia. *Schizophrenia* **1998,** *31,* 185–196.

Dyerberg, J.; H.O. Bang; N. Hjome. Fatty acid composition of the plasma lipids of Greenland Eskimoes. *Am. J. Clin. Nutr.* **1975,** *28,* 958–966.

Engler, M.M. Comparative study of diets enriched with evening primrose, black currant, borage or fungal oils on blood pressure and pressor responses in spontaneously hypertensive rats. *Prostaglandins Leukot Essent Fatty Acids* **1993,** *49,* 809–814.

Fan, Y-Y.; R.S. Chapkin. Importance of dietary γ-linolenic acid in human health and nutrition. *J. Nutr.* **1998,** *128,* 1411–1414.

Fan Y-Y.; R.S. Chapkin; K.S. Ramos. Dietary lipid source alters murine macrophage/vascular smooth muscle cell interactions in vitro. *J. Nutr* **1996,** *126,* 2083–2088.

Fearon, K.C.H.; J.S. Falconer; J.A. Ross; D.C. Carter; J.O. Hunter; P.D. Reynolds; Q. Tuffnell. An open-label phase I/II dose escalation study of the treatment of pancreatic cancer using lithium gammalinolenate. *Anticancer Res.* **1996,** *16,* 867–874.

Fedele, D.; D. Giugliano. Peripheral diabetic neuropathy. Current recommendations and future prospects for its prevention and management. *Drugs* **1997,** *54,* 414–421.

Fewtrell, M.S.; R.A. Abbott; K. Kennedy; A. Singhal; R. Morley; E. Caine; C. Jamieson; F. Cockburn; A. Lucas. Randomized, double-blind trial of long-chain polyunsaturated fatty acid supplementation with fish oil and borage oil in pre-term infants. *J. Pediatr.* **2004,** *144,* 471–479.

Fievet, P.; B. Tribout; B. Castier; J. Dieval; J.C. Capoid; J. Delobel; B. Coevoet; A. Fournier. Effects of evening primrose oil (EPO) on platelet functions during pregnancy in patients with high risk of toxemia. *Kidney Int.* **1985,** *28,* 234.

Fiocchi, A.; M. Sala; P. Signoroni; G. Banderali; C. Agostoni; E. Riva. The efficacy and safety of γ-linolenic acid in the treatment of infantile atopic dermatitis. *J. Int.. Med. Res.* **1994,** *22,* 24–32.

Fisher, B.A.C.; L.S. Harbige. Effect of omega-6-lipid rich borage oil feeding on immune function in healthy volunteers. *Biochem. Soc. Trans.* **1997,** *25,* 343S.

Ford, I.; M.A. Cotter; N.E. Cameron; M. Greaves. The effects of treatment with α-lipoic acid or evening primrose oil on vascular hemostatic and lipid risk factors, blood flow, and peripheral nerve conduction in the streptozotocin-diabetic rat. *Metabolism.* **2001,** *50,* 868–875.

Ghasemnezhad, A.; S. Cergel; B. Honermeier. Evening primrose oil (EPO) quality changes dependent on storage temperature and storage time. *Planta Med.* **2006**, *72,* 207.

Greenfield, S.M.; A.T. Green; J.P. Teare; A.P. Jenkins; N.A. Punchard; C.C. Ainley; R.P.H. Thompson. A randomized controlled study of evening primrose oil and fish oil in ulcerative colitis. *Aliment. Pharmacol. Ther.* **1993,** *7,* 159–166.

Guivernau, M.; N. Meza; P. Barja; O. Roman. Clinical and experimental study on the long-term effect of dietary gamma-linolenic acid on plasma lipids, platelet aggregation, thromboxane formation and prostacyclin production. *Prostaglandins Leukot. Essent. Fatty Acids.* **1994,** *51,* 311–316.

Haimov-Kochman, R.; D. Hochner-Celnikier. Hot flashes revisited: Pharmacological and herbal options for hot flashes management. What does the evidence tell us? *Acta Obstet. Gynecol Scand.* **2005**, *84,* 972–979.

Halat, K.M.; C.E. Denneby. Botanicals and dietary supplements in diabetic peripheral neuropathy. *J. Am. Fam Pract.* **2003**, *16,* 47–57.

Hamburger, M.; U. Riese; H. Graf; M.F. Melzig; S. Ciesielski; D. Baumann; K. Dittmann; C. Wegner. Constituents in evening primrose oil with radical scavenging, cyclooxygenase and neutrophil elastase inhibitory activities. *J. Agric. Food Chem.* **2002**, *50,* 5533–5538.

Harada, T.; K. Hasumi; A. Endo. Induction of low density lipoprotein catabolism in Hep G2 cells by a fungal sesquiterpene ester, FR111142. *Biochem. Biophys. Res. Commun.* **1998,** *251,* 830–834.

Head, R.J.; P.L. McLennan; D. Raederstorff; R Muggli; S.L. Burnard; E.J. McMurchie. Prevention of nerve conduction deficit in diabetic rats by polyunsaturated fatty acids. Am. *J. Clin Nutr.* **2000,** *71 (suppl),* 386S–392S.

Hederos, C.-A.; A. Berg. Epogam evening primrose oil treatment in atopic dermatitis and asthma. *Arch. Dis. Child* **1996,** *75,* 494–497.

Henz, B.M.; S. Jablonska; P.C.M. van de Kerkhof; G. Stingl; M. Blaszczyk; P.G.M. Vandervalk; R. Veenhuizen; R. Muggli; D. Raederstorff. Double-blind, multicentre analysis of the efficacy of borage oil in patients with atopic eczema. *Br. J. Dermatol.* **1999,** *140,* 685–688.

Horrobin, D.F. The role of essential fatty acids and prostaglandins in the premenstrual syndrome. *J. Reprod Med.* **1983,** *28,* 465–468.

Horrobin, D.F. (1992) The use of gamma-linolenic acid in diabetic neuropathy. *Agents Actions Suppl.* **1992,** *37,* 120–144.

Horrobin, D.F. Fatty acid metabolism in health and disease: The role of Δ6-desaturase. *Am. J. Clin. Nutr.* **1993a,** *57(suppl),* 732S–737S.

Horrobin, D.F. The effects of gamma-linolenic acid on breast pain and diabetic neuropathy: Possible non-eicosanoid mechanisms. *Prostaglandins Leukot. Essent. Fatty Acids* **1993b,** *48,* 101–104.

Horrobin, D.F. Essential fatty acids in the management of impaired nerve function in diabetes. *Diabetes* **1997,** *46 (suppl. 2),* S90–S93.

Horrobin, D.F.; M.S. Manku. How do polyunsaturated acids lower plasma cholesterol levels?

*Lipids* **1983,** *18*, 558–562.

Horrobin, D.F.; P.F. Morse. Evening primrose and atopic eczema. *Lancet* **1995,** *345*, 260–261.

Hounsom, L.M.; D.F. Horrobin; H. Tritschler; R. Corder; D.R. Tomlinson. A lipoic acid-gamma linolenic acid conjugate is effective against multiple indices of experimental diabetic neuropathy. *Diabetologica* **1998,** *41,* 839–843.

Humphreys, F.; J.A. Symons; H.K. Brown; G.W. Duff; J.A.A. Hunter. The effects of gamolenic acid on adult atopic eczema and premenstrual exacerbation of eczema. *Eur. J. Dermatol.* **1994,** *4,* 598–603.

Innis, S.M.; J.W. Hansen. Plasma fatty acid responses, metabolic effects and safety of microalgal and fungal oils rich in arachidonic and docosahexaenoic acids in healthy adults. *Am. J. Clin. Nutr.* **1996,** *64*, 159–167.

Ishikawa, T.; Y. Fujiyama; O. Igarashi; M. Morino; N. Tada; A. Kagami; T. Sakamoto; M. Nagano; H. Nakamura. Effects of gammalinolenic acid on plasma lipoproteins and apolipoproteins. *Artherosclerosis* **1989,** *75*, 95–104.

Jack, A.M.; A. Keegan; M.A. Cotter; N.E. Cameron. Effects of diabetes and evening primrose oil treatments on responses of aorta, corpus cavernosum and mesenteric vasculature in rats. *Life Sci.* **2002,** *71*, 1863–1877.

Jamal, G.A. The use of gamma linolenic acid in the prevention and treatment of diabetic neuropathy. *Diabetic Med.* **1994,** *11*, 145–149.

Jantti, J.; E. Seppala; H. Vapaatalo; H. Isomaki. Evening primrose oil and olive oil in treatment of rheumatoid arthritis. *Clin. Rheumatol.* **1989a,** *8*, 238–244.

Jantti, J.; T. Nikkari; T. Solakivi; H. Vapaatalo; H. Isomaki. Evening primrose oil in rheumatoid arthritis: Changes in serum lipids and fatty acids. *Ann. Rheum. Dis.* **1989b,** *48*, 124–127.

Jialal, I.; E.P. Norkus; L. Cristol; S.M. Grundy. β-Carotene inhibits the oxidative modification of low-density lipoprotein. *Biochim. Biophys. Acta.* **1991,** *1086,* 134–138.

Jiang, W.G.; S. Hiscox; D.F. Horrobin; R.P. Bryce; R.E. Mansel. Gamma linolenic acid regulates expression of maspin and the motility of cancer cells. *Biochem. Biophys. Res. Commun.* **1997,** *237,* 639–644.

Johansson, A.; P. Laakso; H. Kallio. Characterization of seed oils of wild, edible Finnish berries. *Z. Lebensm Unters Forsch A.* **1997**, *204,* 300–307.

Johnston, G.A.; R.M. Bilbao; R.A.C. Graham-Brown. The use of dietary manipulation by parents of children with atopic dermatitis. *Brit. J. Dermatol.* **2004**, *150,* 1186–1189.

Kast, R.E. Borage oil reduction of rheumatoid arthritis activity may be mediated by increased cAMP that suppresses tumor necrosis factor-alpha. *Int. Immunopharm.* **2001**, *1,* 2197–2199.

Keen H.; J. Payan; J. Allawi; J. Walker; G.A. Jamal; A.I. Weir; L.M. Henderson; E.A. Bissessar; P.J. Watkins; M. Sampson; et al. Treatment of diabetic neuropathy with γ-linolenic acid. *Diabetes Care* **1993**, *16,* 8–15.

Khan, F.; K. Elherik; C. Bolton-Smith; R. Barr; A. Hill; I. Murrie; J.J.F. Belch. The effects of dietary fatty acid supplementation on endothelial function and vascular tone in healthy subjects. *Cardiovasc. Res.* **2003**, *59,* 955–962.

Khoo, S.K.; C. Munro; D. Battistutta. Evening primrose oil and the treatment of premenstrual syndrome. *Med. J. Aust.* **1990,** *153*, 189–192.

Kinsella, J.E. α-Linolenic acid: Functions and effects on linoleic acid metabolism and eicosanoid-mediated reactions. *Adv. Food Nut. Res.* **1991,** *35*, 1–184.

Knorr, R.; M. Hamburger. Quantitative analysis of anti-inflammatory and radical scavenging triterpenoid esters in evening primrose oil. *J. Agric. Food Chem.* **2004,** *52*, 3319–3324.

Kollias, J.; R.D. Macmillan; D.M. Sibbering; H. Burrell; J.F.R. Robertson. Effect of evening primrose oil on clinically diagnosed fibroadenomas. *The Breast.* **2000,** *9*, 35–36.

Kronenberg, F.; A. Fugh-Berman. Complementary and alternative medicine for menopausal symptoms: A review of randomized, controlled trials. *Ann. Intern. Med.* **2002**, *137*, 805–813.

Kuruvilla, R.; R.G. Peterson; J.C. Kincaid; J. Eichberg. Evening primrose oil treatment corrects reduced conduction velocity but not depletion of arachidonic acid in nerve from streptozotocin-induced diabetic rats. *Prostaglandins Leukot. Essent. Fatty Acids* **1998,** *59*, 195–202.

Lawson, L.D.; B.G. Hughes. Triacylglycerol structure of plant and fungal oils containing γ-linolenic acid. *Lipids* **1988,** *23*, 313–317.

Lehmann, B.; C. Hubner; H. Jacobi.; A. Kampf; G. Wozel. Effects of dietary γ-linolenic acid-enriched evening primrose seed oil on the 5-lipoxygenase pathway of neutrophil leukocytes in patients with atopic dermatitis. *J. Dermatol. Treat.* **1995,** *6*, 211–218.

Leonhardt, A.; M. Krauss; U. Gieler; H. Schweer; R. Happle; H.W. Seyberth. In vivo formation of prostaglandin $E_1$ and prostaglandin $E_2$ in atopic dermatitis. *Br. J. Dermatol.* **1997,** *136*, 337–340.

Leventhal, L.J.; E.G. Boyce; R.B. Zurier. Treatment of rheumatoid arthritis with gammalinolenic acid. *Ann. Intern. Med.* **1993,** *119*, 867–873.

Low Dog, T. Menopause: A review of botanical dietary supplements. *Am. J. Med.* **2005**, *118*, 98S–108S.

Lucas, A.; R. Morley; T.J. Cole; G. Lister; C. Leeson-Payne. Breast milk and subsequent intelligence quotient in children born preterm. *Lancet* **1992,** *339*, 261–264.

Manku, M.S.; D.F. Horrobin; N.L. Morse; S. Wright; J.L. Burton. Essential fatty acids in the plasma phospholipids of patients with atopic eczema. *Br. J. Dermatol.* **1984,** *110*, 643–648.

Manku, M.S.; N. Morse-Fisher; D.F. Horrobin. Changes in human plasma essential fatty acid levels as a result of administration of linoleic acid and gamma-linolenic acid. *Eur. J. Clin. Nutr.* **1988,** *42*, 55–60.

Mcgregor, L.; A.D. Smith; M. Sidey; J. Belin; K.J. Zilkha; J.L. McGregor. Effects of dietary linoleic acid and gamma linolenic acid on platelets of patients with multiple sclerosis. *Acta Neurol. Scand..* **1989,** *80*, 23–27.

McKean, M.L.; J.B. Smith; M.J. Silver. Formation of lysophosphatidylcholine by human platelets in response to thrombin. *J. Biol. Chem.* **1981,** *256*, 1522–1524.

Miller, L.G. Herbal medicinals. Selected clinical considerations focusing on known or potential drug-herb interactions. *Arch. Int. Med.* **1998**, *158*, 2200–2211.

Mills, D.E.; M. Mah; P.P. Ward; B.L. Morris; J.S. Floras. The effects of stress and dietary n-6 and n-3 rich oils on platelet aggregation in man. *Proc ISF/JAOCS World Congress* **1989a,** *1*, 281–287.

Mills, D.E.; K.M. Prkachin; K.A. Harvey; R.P. Ward. Dietary fatty acid supplementation alters stress reactivity and performance in man. *J. Hum. Hypertens.* **1989b,** *3,* 111–116.

Mills, D.E.; M. Mah; R.P. Ward; B.L. Morris; J.S. Floras. Alteration of baroreflex control of forearm vascular resistance by dietary fatty acids. *Am J Physiol.* **1990,** *259,* R1164–1171.

Morse, N.L.; P.M. Clough. A meta-analysis of randomized, placebo controlled clinical trials of Efamol® evening primrose oil in atopic eczema. Where do we go from here in light of more recent discoveries? *Curr. Pharmaceut. Biotech.* **2006,** *7,* 503–524.

Morse, P.F.; D.F. Horrobin; M.S. Manku; J.C.M. Stewart; R. Allen; S. Littlewood; S. Wright; J. Burton; D.J. Gould; P.J. Holt; et al. Meta-analysis of placebo-controlled studies of the efficacy of Epogam in the treatment of atopic eczema. Relationship between plasma essential fatty acid changes and clinical response. *Br. J. Dermatol.* **1989,** *121,* 75–90.

Nakahara, T.; T. Yokochi; Y. Kamisaka; M. Yamaoka; O. Suzuki; M. Sato; S. Okazaki; N. Ohshima. Inhibitory effect of mold oil including gamma-linolenate on platelet thrombus formation in mesenteric microvessels of the rat. *Thromb. Res.* **1990,** *57,* 371–381.

Parthasarathy, S. Novel atherogenic, oxidative modification of low-density lipoprotein. *Diabetes Metab. Rev.* **1991,** *7,* 163–171.

Pellegrina, C.D.; G. Padovani; F. Mainente; G. Zoccatelli; G. Bissoli; S. Mosconi; G. Veneri; A. Peruffo; G. Andrighetto; C. Rizzi; and R. Chignola. Anti-tumour potential of a gallic acid-containing phenolic fraction from *Oenethera biennis. Cancer Letters.* **2005,** *226,* 17–25.

Peth, J.A.; T.R. Kinnick; E.B. Youngblood; H.J. Tritschler; E.J. Henriksen. Effects of a unique conjugate of α-lipoic acid and γ-linolenic acid on insulin action in obese Zucker rats. *Am. J. Physiol. Regul. Integr. Comp. Physiol.* **2000,** *278,* R453–R459.

Pham, H.; K. Vang; V.A. Ziboh. Dietary γ-linolenate attenuates tumor growth in a rodent model of prostatic adenocarcinoma via suppression of elevated generation of $PGE_2$ and 5S-HETE. *Prost. Leukotr. Ess. Fatty Acids* **2006,** *74,* 271–282.

Philp, H.A. Hot flashes—a review of the literature on alternative and complementary treatment approaches. *Altern. Med. Rev.* **2003,** *8,* 284–302.

Phylactos, A.C.; L.S. Harbige; M.A. Crawford. Essential fatty acids alter the activity of manganese-superoxide dismutase in rat heart. *Lipids* **1994,** *29,* 111–115.

Piquette, K.; P. Laflamme. Borage (*Borago officinalis*). Alberta Agr., Food and Rural Dev. Library, Edmonton, Alberta. 2000.

Prichard, P.; G. Brown; N. Bhasker; C. Hawkey. The effect of dietary fatty acids on gastric production of prostaglandins and aspirin-induced injury. *Aliment. Pharmacol. Ther.* **1988,** *2,* 179–184.

Pullman-Mooar, S.; M. Laposata; D. Lem; R.T. Holman; L.J. Leventhal; D. DeMarco; R.B. Zurier. Alteration of the cellular fatty acid profile and the production of eicosanoids in human monocytes by gamma-linolenic acid. *Arthritis Rheum.* **1990,** *33,* 1526–1533.

Puri, B.K. The clinical advantages of cold pressed, non-raffinated evening primrose oil over refined preparations. *Med Hypoth.* **2004,** *62,* 116–118.

Redden, R.R.; X. Lin; D.F. Horrobin. Comparison of the Grignard deacylation TLC and HPLC methods and high resolution of $^{13}C$-NMR for the sn-2 positional analysis of triacylglycerols containing γ-linolenic acid. *Chem. Phys. Lipids.* **1996,** *79,* 9–19.

Reljanovic, M.; G. Reichel; K. Rett; M. Lobisch; K. Schuette; W. Moller; H-J. Tritschler; H. Mehnert. Treatment of diabetic polyneuropathy with the antioxidant thioctic acid (α-lipoic acid): A two year multicenter randomized double-blind placebo-controlled trial (Aladdin II). *Free Radic Res.* **1999,** *31*, 171–179.

Richard, J.-L.; C. Martin; M. Maille; F. Mendy; B. Delplanqe; B. Jacotot. Effects of dietary intake of gamma-linolenic acid on blood lipids and phospholipid fatty acids in healthy human subjects. *J. Clin. Biochem. Nutr.* **1990,** *8,* 75–84.

Rock, E.; A. Demichele. Nutritional approaches to late toxicities of adjuvant chemotherapy in breast cancer survivors. *J. Nutr.* **2003,** *133*, 3785S–3793S.

Rodgers, R.P.C.; J. Levin. A critical reappraisal of the bleeding time. *Semin Thromb. Hemost.* **1990,** *16,* 1–20.

Schalin-Karrila, M.; L. Mattila; C.T. Jansen; P. Uotila. Evening primrose oil in the treatment of atopic eczema: Effect on clinical status, plasma phospholipid fatty acids and circulating blood prostaglandins. *Br. J. Dermatol.* **1987,** *117*, 11–9.

Schirmer, M.A.; S.D. Phinney. γ-linolenate reduces weight gain in formerly obese humans. *J. Nutr.* **2007,** *137*, 1430–1435.

Seri, S.; A. D'Alessandro; S. Acitelli; U. Giammaria; M. Cocchi; R.C. Noble. Effect of dietary supplementation by alternative oils on blood lipid levels in hemodialysed patients. *Med. Sci. Res.* **1993,** *21*, 315–316.

Shahidi, F. Antioxidant factors in plant foods and selected oilseeds. *Biofactors.* **2000,** *13*, 179–185.

Shahidi, F.; R. Amarowicz; Y. He; M. Wettasinghe. Antioxidant activity of phenolic extracts of evening primrose (*Oenothera biennis*): A preliminary study. *J. Food Lipids.* **1997,** *4*, 75–86.

Shotton, H.R.; S. Clarke; J. Lincoln. The effectiveness of treatments of diabetic autonomic neuropathy is not the same in autonomic nerves supplying different organs. *Diabetes* **2003,** *52*, 157–164.

Simpson, M.J.A. Comparison of swathing and desiccation of borage (*Borago officinalis*) and estimation of optimum harvest stage. *Ann. Appl. Biol.* **1993,** *123*, 105–108.

Spielmann, D.; H. Traitler; G. Crozier; M. Fleith; U. Bracco; P.A. Finot; M. Berger; R.T. Holman. Biochemical and bioclinical aspects of blackcurrant seed oil ω-3 and ω-6 balanced oil. *Dietary ω3 and ω6 Fatty Acids: Biological Effects and Nutritional Essentiality;* C. Galli, A.P. Simopoulos, Eds.; Plenum Press: New York, 1989**;** pp. 309–221.

Stewart, J.C.M.; P.F. Morse; M. Moss; D.F. Horrobin; J.L. Burton; W.S. Douglas; D.J. Gould; C.E.H. Grattan; T.C. Hindson; J. Anderson; C.T. Jansen; C.T.C. Kennedy; R. Lindskov; A.M.M. Strong; S. Wright. Treatment of severe and moderately severe atopic dermatitis with evening primrose oil (Epogram): A multi-centre study. *J. Nutr. Med.* **1991,** *2,* 9–15.

Strannegard, I.L.; L. Svennerholm; O. Strannegard. Essential fatty acids in serum lecithin of children with atopic dermatitis and in umbilical cord serum of infants with high or low IgE levels. *Int .Arch. Allergy Applied Immunol.* **1987,** *82,* 422–423.

Stroop, J.R. Studies of interference in serial verbal reactions. *J. Exp. Psychol.* **1935,** *18*, 643–662.

Sugano, M.; T. Ishida; K. Yoshida; K. Tanaka; M. Niwa; M. Arima; A. Morita. Effects of mold oil containing y-linolenic acid on blood cholesterol and eicosanoid levels in rats. *Agric. Biol. Chem.*

**1986,** *50,* 2483–2491.

Takada, R.; M. Saitoh; T. Mori. Dietary γ-linolenic acid-enriched oil reduces body fat content and induces liver enzyme activities relating to fatty acid β-oxidation in rats. *J. Nutr.* **1994,** *124,* 469–474.

Takwale, A.; E. Tan; S. Agarwal; G. Barclay; I. Ahmed; K. Hotchkiss; J.R. Thompson; T. Chapman; J. Berth-Jones. Efficacy and tolerability of borage oil in adults and children with atopic eczema: randomized, double blind, placebo controlled parallel group trial. *BMJ.* **2003,** *327,* 1385–1388.

Tamimi, N.A.M.; A. Mikhail; P.E. Stevens. Role of γ-linolenic acid in uraemic patients. *Nephron* **1999,** *83,* 170–171.

Tollesson, A.; A. Frithz. Transepidermal water loss and water content in the stratum corneum in infantile seborrhoeic dermatitis. *Acta Derm. Venereol. (Stockh.)* **1993a,** *73,* 18–20.

Tollesson, A; A. Frithz. Borage oil, an effective new treatment for infantile sebborrhoeic dermatitis. *Br. J. Dermatol.* **1993b,** *129,* 95.

van der Merwe, C.F.; J. Booyens; H.F. Joubert; C.A. van der Merwe. The effect of gamma-linolenic acid, an in vitro cytostatic substance contained in evening primrose oil, on primary liver cancer. A double blind placebo-controlled trial. *Prostaglandins Leukot Essent Fatty Acids.* **1990,** *40,* 199–202.

Vericel, E.; M. Lagarde; F. Mendy; P. Courpron; M. Dechavanne. Comparative effects of linoleic acid and gamma-linolenic acid intake on plasma lipids and platelet phospholipids in elderly people. *Nutr. Res.* **1987,** *7,* 569–580.

Viikari, J.; A. Lehtonen. Effect of primrose oil on serum lipids and blood pressure in hyperlipidemic subjects. *Int. J. Clin. Pharmacol. Ther. Toxicol.* **1986,** *24,* 668–670.

Vinik, A.I. Diabetic neuropathy: Pathogenesis and therapy, *Am. J. Med.* **1999,** *107,* 17S–26S.

Wagner, W.; U. Nootbaar-Wagner. Prophylactic treatment of migraine with gamma-linolenic acid and alpha-linolenic acids. *Cephalagia* **1997,** *17,* 127–130.

Wainwright, P.E.; Y-S. Huang; S.J. DeMichele; H.C. Xing; J-W. Liu; L-T. Chuang; J. Biederman. Effects of high-γ-linolenic acid canola oil compared with borage oil on reproduction, growth, and brain and behavioral development in mice. *Lipids.* **2003,** *38,* 171–178.

Watson, J.; M.L. Byars; P. McGill; A.W. Kelman. Cytokine and prostaglandin production by monocytes of volunteers and rheumatoid arthritis patients treated with dietary supplements of black-currant seed oil. *Br. J. Rheumatol.* **1993,** *32,* 1055–1058.

Werneke, U.; J. Earl; C. Seydel; O. Horn; P. Crichton; D. Fannon. Potential health risks of complementary alternative medicines in cancer patients. *Brit. J. Cancer.* **2004,** *90,* 408–413.

Wettasinghe, M.; F. Shahidi; R. Amarowicz. Identification and quantification of low molecular weight phenolic antioxidants in the seeds of evening primrose (*Oenothera biennis* L.) *J. Agric. Food Chem.* **2002,** *50,* 1267–1271.

Whelan, A.M.; T.M. Jurgens; S.K. Bowles. Natural health products in the prevention and treatment of osteoporosis: systemic review of randomized controlled trials. *Ann. Pharmacother.* **2006,** *40,* 836–849.

Whitaker, D.K.; J. Cilliers; C. deBeer. Evening primrose oil (Epogam®) in the treatment of chronic hand dermatitis: Disappointing therapeutic results. *Dermatology* **1996,** *193,* 115–120.

Williams, H.C. Evening primrose oil for atopic dermatitis. *Brit. Med. J.* **2003,** *327,* 1358–1359.

Wolever, T.M.S. Comments on "treatment of diabetic neuropathy with γ-linolenic acid" by the γ-linolenic multicenter trial group. *Diabetes Care* **1993,** *16,* 1309–1310.

Wright, S.; J.L. Burton. Oral evening-primrose seed oil improves atopic eczema. *Lancet* **1982,** *ii,* 1120–1122.

Wu, D.; M. Meydani; L.S. Leka; Z. Nightingale; G.J. Handelman; J.B. Blumberg; S.N. Meydani. Effect of dietary supplementation with black currant seed oil on the immune response of healthy elderly subjects. *Am. J. Clin. Nutr.* **1999,** *70,* 536–543.

Yoshimoto-Furuie, K.; K. Yoshimoto; T. Tanaka; S. Saima; Y. Kikuchi; J. Shay; D.F. Horrobin; E. Hirotoshi. Effects of oral supplementation with evening primrose oil for six weeks on plasma essential fatty acids and uremic skin symptoms in hemodialysis patients. *Nephron* **1999,** *81,* 151–159.

Yu, K. L.; J.W. Parry; K. Zhou. Fruit seed oils. *Nutraceutical and Specialty Lipids and their Co-products*; F. Shahidi, Eds.; Nutraceutical Science and Technology; CRC, Taylor and Francis: New York, 2006; Vol. 5, p. 86.

Zaugg, J.; O. Potterat; A. Plescher; B. Honermeier; M. Hamburger. Quantitative analysis of anti-inflammatory and radical scavenging triterpenoid esters in evening primrose seeds. *J. Agric. Food Chem.* **2006,** *54,* 6623–6628.

Zhu, M.; P.P. Zhou; L.J. Yu.. Extraction of lipids from *Mortiella alpina* and enrichment of arachidonic acid from fungal lipids. *Biores. Tech.* ***2002,*** *84,* 93–95.

Ziboh, V.A. The significance of polyunsaturated fatty acids in cutaneous biology. *Lipids* **1996,** *31(suppl),* S249–S253.

Ziboh, V.A.; M.P. Fletcher. Dose-response effects of dietary γ-linolenic acid-enriched oils on human polymorphonuclear-neutrophil biosynthesis of leukotriene $B_4$. *Am. J. Clin. Nutr.* **1992,** *55,* 39–45.

Ziboh, V.A.; S. Naguwa; K. Vang; J. Wineinger; B.M. Morrissey; M. Watnik; M.E. Gershwin. Suppression of leukotriene $B_4$ generation by ex-vivo neutrophils isolated from asthma patients on dietary supplementation with gammalinolenic acid-containing borage oil: Possible implication in asthma. *Clin. Dev. Immunol.* **2004,** *11,* 13–21.

Zurier, R.B.; R.G. Rossetti; E.W. Jacobsen; D.M. DeMarco; N.Y. Liu; J.E. Temming; B.M. White; M. LaPosata. Gamma-linolenic acid treatment of rheumatoid arthritis. A randomized placebo-controlled study. *Arthritis Rheum.* **1996,** *39,* 1808–1817.

# Sesame Seed Oil

**Ali Moazzami and Afaf Kamal-Eldin**
*Department of Food Science, Swedish University of Agricultural Sciences, 750 07 Uppsala, Sweden*

## Introduction

The sesame plant *Sesamum* spp. (family: *Pedaliaceae*) is spread throughout the tropical and subtropical areas in Asia, Africa, and South America. The genus *Sesamum* is comprised of about 35 wild species, in addition to the only cultivated species, *Sesamum indicum*. The name "sesame" comes from the Arabic word "simsim." Linnaeus and then Bidigian and Harlan (1986) cited India as the origin of sesame, but a belief also exists that the actual origin was Africa, where many wild species are found (Namiki, 1995). According to FAOSTAT, about 60% of the world's sesame crop is grown in Asia (mainly India, China, Myanmar, Pakistan, Bangladesh) followed by ca. 15–20% in Africa (mainly Sudan, Uganda, Nigeria, Ethiopia) and ca. 5% in South America (mainly Mexico, Guatemala, Venezuela). Four countries, India, China, Myanmar, and Sudan, contribute >60% to the total world production of sesame seeds.

In old literature, the sesame seed was referred to as the "queen of oilseeds," and sesame oil is among the first oils known and consumed by man. The seed is flat, pear-shaped, ca. 2–3 mm long, and 2–3.5 mg in weight. Eight rows of seeds are encapsulated (50–100 seeds per capsule) in shattering capsules that open when dry and scatter the mature seeds. In the tale of Ali Baba and the 40 thieves in *The Thousand and One Nights*, the password "open sesame" was perhaps chosen to portray how the ripe seeds burst from their pods suddenly with a sharp pop like the springing open of a lock. This feature of shattering capsules is one obstacle that has slowed the expansion of cultivation of sesame. Currently, cultivars with nonshattering capsules are under development.

Sesame seeds are rich in oil (ca. 50%) and protein (ca. 20–25%), and the rest is composed of mineral ash (ca. 6%), crude fiber (ca. 4.5%), oxalates (ca. 2.5%), and soluble carbohydrates and phytate (12.5%) (Johnson et al., 1979). Sesame proteins are especially rich in two essential amino acids, methionine and tryptophan, but are deficient in lysine. Their digestibility is adversely affected by antinutritional factors in the seed including fiber and phytate. Until recently, the uses of sesame seeds and oil were mainly limited to their countries of production with a few countries, mainly Japan, being major importers.

In Japan, a strong belief exists that sesame seed and sesame oil are healthy, and much of the research on the beneficial physiological effect of sesame seed lignans was conducted there (see later section Physiological effects of sesame seed oil). In the Middle East, dehulled sesame seeds are ground into a smooth paste called tahini (>50% oil) or mixed with an equal amount of sugar to provide a confection called Halwah tahenieh. The cake or meal remaining after the pressing of virgin oil from unroasted seeds is very rich in protein (45–50%), but it is mostly used as animal feed. The cake of oil from roasted seeds is very rich in antioxidant compounds, and should be exploited as a source of potent antioxidants. Currently, the use of sesame seed products is expanding into Europe, the United States, and Canada, but unfortunately cases of sesame seed allergy are increasingly reported in these countries. This has prompted research aiming to identify the allergic proteins and their cross reactivity with other food allergens.

## Oil Extraction and Processing

Sesame oil is used in various edible applications, as a solvent for intramuscular injections and in the production of drugs, perfumes, cosmetics, creams, lubricants, insecticides, and fungicides (Pecquet et al., 1998). In the seeds, sesame oil is stored in oil bodies composed mainly of triacylglycerols surrounded by a layer of phospholipids embedded in encapsulating proteins, namely oleosin, caleosin, and steroleosin (Tzen et al., 1993, 2003). One can extract and process several types of sesame oils that vary tremendously in their properties and bioactivities. One can cold press sesame seeds to yield virgin sesame oil, which is regarded as superior to solvent-extracted oils. Crude (unrefined) sesame oils sometimes contain high levels of free fatty acids and have a strong distinct flavor that is appreciated in several African countries but considered too strong by Western consumers. A study by Ajibola et al. (1993), using a laboratory press, showed that mild heat treatment (40–85°C) to reduce the moisture content increases the oil extraction yield. The crude oil has the following characteristics: Lovibond color 5 ¼ (18–30 yellow, 2.5 red), refractive index at 20°C (1.473–1.477), melting point (–3 to –6°C), viscosity at 37.8°C (39.6 centistokes), and a dextro optical rotation. The crude and refined oils vary significantly regarding the composition of minor components, especially lignans (see later).

In Western countries, crude sesame oils are refined (neutralization, acid bleaching, and deodorization) to produce several grades of refined sesame oil, each with a light color and a bland flavor and taste. This oil is very different from the crude oil especially regarding its content and composition of lignans, the unique constituents of sesame oil (Moazzami et al., 2007). In countries of the Far East (e.g., China, Japan, and Korea), sesame seeds are roasted (180–220°C) before oil extraction. Sesame oils extracted from roasted seeds are red in color due to the formation of unidentified Maillard reaction products that contribute significantly to the extreme oxidative stability of these oils. Oils from roasted seeds also differ markedly in their flavor and taste compared to virgin and refined oils. Due to these differences, it is important to specifically describe what type of sesame oil is under consideration when we study or report their effects. Unfortunately, several literature reports do not specify this vital information.

## Composition of Acyl Lipids and Fatty Acids

Like most edible oils, sesame oil is mainly composed of triacylglycerols (ca. 90%) and lesser amounts of diacylglycerols, free fatty acids, and phospholipids and unsaponifiables. The ranges of fatty acid composition summarized by the FAO/WHO Codex Alimentarius Committee on Fats and Oils are palmitic acid (7–12%), stearic acid (3.5–6%), oleic acid (35–50%), and linoleic acid (35–50%) (Spencer et al., 1976). The predominant triacylglycerols species in sesame oil is 1,2-dilinoleoyl-3-oleoyl-*rac*-glycerol (LLO, ca. 25%) followed by trilinoleoylglycerol (LLL, ca. 20%), 1-linoleoyl-2,3-dioleoyl-*rac*-glycerol (LOO, 15%), 1-palmitoyl-2,3-dilinoleoyl-*rac*-glycerol (PLL, ca. 10%), and trioleoylglycerol (OOO, ca. 8%) together with less prevalent molecular species containing saturated fatty acids (Kamal-Eldin et al., 1992a). The composition of fatty acids and their molecular species affects the melting point of the oil and its crystallization properties.

The fatty acid composition differs among the various acyl lipid classes (Fig. 8.1) with the steryl esters, phospholipids, and free fatty acids containing higher levels of saturated fatty acids than the tri- and diacylglycerols (Kamal-Eldin et al., 1994a). No published detailed studies exist on the composition of the phospholipid fraction of sesame oil.

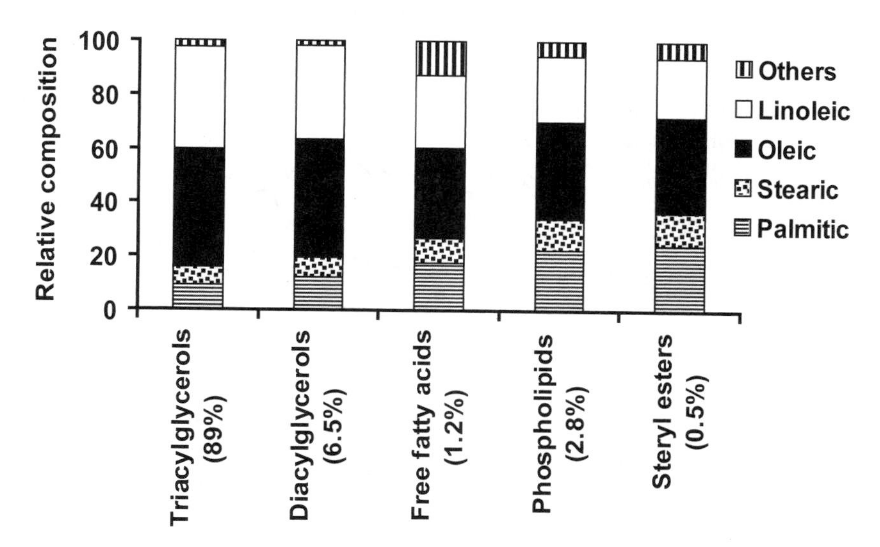


**Fig. 8.1.** Relative distribution of acyl lipids and their respective fatty acid compositions in sesame oil (Kamal-Eldin & Appelqvist, 1994a).

## Nonacylglycerol Components in Sesame Oil

Nonacyl minor components (also called the unsaponifiable fraction) comprise a total of about 1–2% of sesame oil. The actual size of the unsaponifiable fractions varies with the source of seed and processing conditions. In most vegetable oils, phytosterols represent the main constituents of this fraction. Sesame oil presents an exception, where two lignans sesamin and sesamolin (Fig. 8.2) are generally the major components (Namiki, 1995). However, small amounts of sesaminol, piperitol, sesamolinol, pinoresinol, (+)-episesaminone, hydroxymatairesinol, allohydroxymatairesinol and larisiresinol, which possess free phenolic groups and therefore antioxidant activity, were reported in sesame seeds (Fukuda et al., 1985, 1988; Marchand et al. 1997; Nagashima & Fukuda, 2004; Osawa et al., 1985). The concentrations (mg/100 g seeds) and some physical and chemical properties of the most important sesame seed lignans appear in Table 8.1. Unlike other vegetable oils that are levorotatory due to the optical activity of phytosterols (–34.4°), sesame oil is dextrorotatory due to the optical activities of sesamin (+68.6°) and sesamolin (+218°) (Budowski & Markley, 1951), the two major lignans in unrefined sesame oil.

Some important transformations in the structures of lignans occur during sesame oil refining, thus making the refined oils extremely different from the crude oils. During the bleaching step, in the presence of acidic clay and heat, sesamolin is first decomposed into sesamol and an oxonium ion by protonolysis, followed by rearrangement of these two to produce sesaminol. This intermolecular transformation was first suggested for conversion sesamolin to sesaminol during bleaching by Fukuda et al., (1986a). As a result, refined sesame oils were found to contain considerable amounts of sesaminol (6–60 mg/100 g oil) (Fukuda et al., 1986b). Epimerization of the sesame oil lignans also happens during bleaching, in which episesamin and episesaminol are formed (Fukuda et al., 1986b) (Fig. 8.3). Roasting was not reported to cause isomerization of sesamin to episesamin or of sesamolin to sesaminols, but it leads to a small release of sesamol from sesamolin.

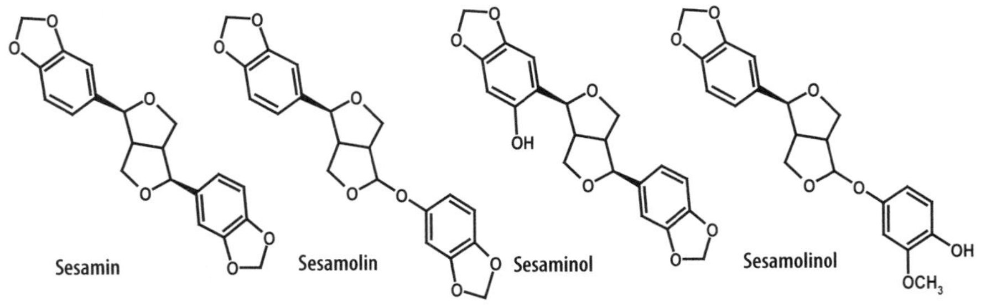

Sesamin Sesamolin Sesaminol Sesamolinol

**Fig. 8.2.** Structures of lignans present in sesame oil. Sesamin and sesamolin are major lignans while sesaminol and sesamolinol are minor lignans in the oil. Sesaminol and sesamolinol are also present in glucosilated forms in the cake.

**Table 8.1. Concentration, Molecular Formula, UV Absorption ($\lambda_{max}$, nm), and Extinction Coefficient ($\varepsilon_{max}$, for 1 cm cell) of Major Sesame Oil Lignans**

| Lignan | Concentration mg/100 g seed | Molecular formula (Mwt) | $\lambda_{max}$ (nm) | $\varepsilon_{max}$ ($g^{-1}L$) |
|---|---|---|---|---|
| Sesamin | 77–930[a] | $C_{20}H_{18}O_6$ (354) | 287<br>236 | 23.02[d]<br>26.01 |
| Sesamolin | 61–530[a] | $C_{20}H_{18}O_7$ (370) | 288<br>235 | 21.79[d]<br>24.85 |
| Sesaminol | 0.3–1.4[b] | $C_{20}H_{18}O_7$ (370) | 238<br>295 | 3.99[e]<br>3.85 |
| Sesamolinol | 6–28[c] | $C_{20}H_{20}O_7$ (372) | 231<br>287 | 3.95[e]<br>3.8 |

[a]The range was based on the data reported previously by Fukuda et al. (1988), Beroza and Kinman (1955), and Hemalatha and Ghafoorunissa (2004).
[b]The range was based on the data reported previously by Fukuda et al. (1988).
[c]The range was based on the data reported previously by Ryu et al. (1998).
[d]Measured in isooctane (Budowski & Markley, 1951).
[e]Measured in chloroform (Nagata et al., 1987; Osawa et al., 1985).

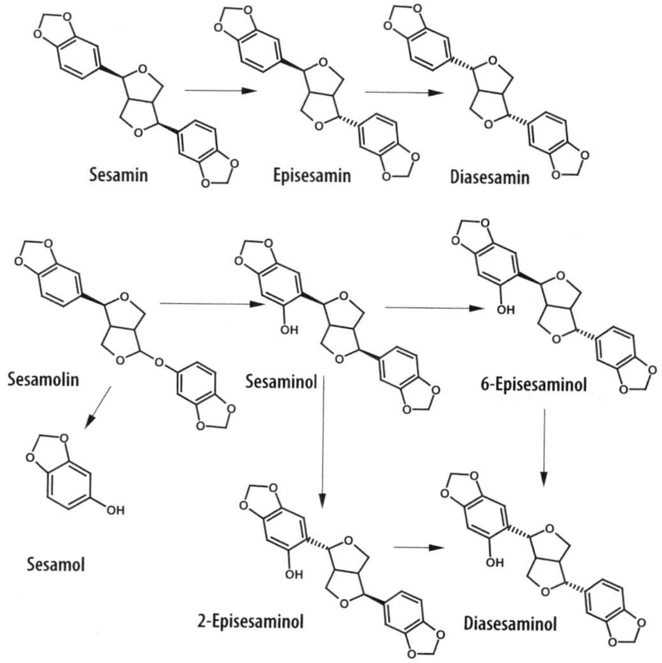

Sesamin Episesamin Diasesamin

Sesamolin Sesaminol 6-Episesaminol

Sesamol 2-Episesaminol Diasesaminol

**Fig. 8.3.** Transformation of natural sesame oil lignans (sesamin and sesamolin) during refining of sesame oil to isomeric forms (episesamin and diasesamin from sesamin and sesaminols from sesamolin). Small amounts of sesamol are also produced from sesamolin.

Phytosterols are common constituents of plant lipids. Sesame oil contains about 0.5% of phytosterols, of which ca. 65–80% exists in free form and ca. 20–35% in esterified forms (Kamal-Eldin & Appelqvist, 1994b) and much smaller amounts of sterylglycosides (≈1%) (Murui et al,. 1994). The main sterols in crude sesame oil are campesterol (12–17%), stigmasterol (6–9%), sitosterol (57–62%), and $\Delta^5$-avenasterol (8–11%) (Itoh et al., 1973a,b; Johansson & Croon, 1981; Kamal-Eldin et al., 1992b; Kamal-Eldin & Appelqvist, 1994b). Like other vegetable oils, the sterol content in sesame oils is variably reduced by refining.

Besides the lignans and sterols, sesame oil contains much lower amounts (µg/g) of a wide range of other minor lipid components. The oil contains intermediate levels (440–550 µg/g) of total tocopherols, predominantly γ-tocopherol (95.5–99%) with a small amount of δ-tocopherol and trace amounts of α-tocopherol and β-tocopherol (Kamal-Eldin & Appelqvist, 1994b). High levels (200–400 µg/g) of C18–C34 long-chain hydrocarbons, such as squalene and *n*-dotriacontane, were also reported in the oil (Bhatia et al., 1986). Colored unsaponifiable constituents (a total of about 1 mg/g oil) including chlorophylls A and B, pheophytins A and B, β-carotene, lutein, and zeaxanthin are the main natural pigments in crude oils (Vogel, 1977; Yen, 1990a). Also, as mentioned previously, red-brown colored Maillard reaction products dominate the color of roasted sesame oil (Yen, 1990b).

## Flavor and Aroma Compounds

Volatile components of sesame oil contribute strong, light, aromatic, mild, nutty, bitter, astringent, complicated, degraded, and aftertaste flavor attributes. The aroma of sesame oil is significantly influenced by roasting, with the main flavor components being alkyl pyrazines and sulfur-containing compounds and other minor components including hydrocarbons, alcohols, aldehydes, ketones, acids, furans, phenols, lactones, pyrroles, indoles, pyridines, and oxazoles (Elsawy et al., 1988; Endo, 1999; Manley et al., 1974; Nakamura et al., 1989; Ryu et al., 1999; Soliman et al., 1975, 1986).

## The Oxidative Stability of Sesame Seed Oil

Sesame seed oil is acclaimed for its superior oxidative stability, compared to other vegetable oils both at common storage temperatures (Beroza & Kinman, 1955; Budowski & Markley, 1951) and during frying (Yen, 1991). As discussed above, a large variation occurs in the oxidative stability among the various types of sesame oil (e.g., virgin, refined, or extracted from roasted seeds. Sesamol (3,4-methylene dioxyphenol), which is present in very small amounts in sesame oil (Budowski et al., 1950), and γ-tocopherol are the main antioxidants in virgin oils, and they may form a synergy with phospholipids in protecting these oils. The hull appears to contain high levels of antioxidant compounds since oils extracted from whole seeds were found to be more stable than those extracted from dehulled seeds, and extraction with polar solvents provided much more stable oils than cold-pressing or extraction with nonpolar solvents (Kamal-Eldin & Appelqvist, 1995). Chang et al. (2002) suggested that seed coat of sesame contains antioxidant factors including phenolic compounds, tannins, and homoterpenes. The presence of trace metal ions in sesame oil may also destabilize it (Kamal-Eldin & Appelqvist, 1995).

Han and Ahn (1993) reported that refining (degumming, alkali-refining, bleaching, and deodorization) decreases the oxidative stability of sesame oil. During refining, the level of γ-tocopherol is significantly lowered, and sesamolin is reduced and transformed into much lower levels of sesaminol. Reportedly, sesamolin also decomposes to sesamol during frying (Fukuda et al., 1986a), a reaction that is enhanced by elevated temperatures and acidic conditions (e.g., the presence of free fatty acids).

Sesame oil extracted from roasted seeds, by hexane or cold-pressing, is extremely stable, having a Rancimat stability at 110°C of >70 hours compared to 14 hours for virgin sesame oil (Kamal-Eldin, unpublished). The very high oxidative stability of roasted sesame seed oil is claimed to be due to synergistic interactions between uncharacterized Maillard reaction products, γ-tocopherol, sesamin, sesaminols, and other unknown factors (Fukuda et al., 1996; Kumazawa et al,. 2003; Yen, 1991; Yen & Shyu, 1989;). More research is needed to identify the role of various components in the oxidative stability of roasted sesame seed oils.

## Physiological Effects of Sesame Seed Oil

Sesame seed has been an important oil-seed since ancient times (Namiki, 1995). Ancient Chinese books claimed that consumption of sesame seeds (*Chih-Ma* in Chinese) provides increased energy and prevents aging. Sesame oil (*Tila* in ancient Indian language, Sanskrit) was used as a domestic Ayurvedic medicine (roughly translated to science of life). Recently, scientific studies attributed many of the health-promoting effects of sesame seed to its lignans, especially sesamin, which is the major oil-soluble lignan.

### Reduction of Serum Cholesterol

Coronary heart disease (CHD) is the number one cause of morbidity and mortality in industrialized Western countries. A positive relationship between high blood cholesterol concentrations, particularly in low-density lipoprotein cholesterol (LDL-C), and the incidence of atherosclerosis and CHD was clearly established. Lowering blood cholesterol by dietary intervention is one of the first measures in prevention of CHD (Assmann et al., 1999). Sesame oil and sesame lignans were investigated for their cholesterol-lowering effect. Sesame oil reduced serum cholesterol levels in rats compared to corn oil in spite of the comparable fatty acid composition of the two oils (Koh, 1987). Hirose et al. (1991) showed that serum and liver cholesterol were reduced (25% and >50%, respectively) in rats fed a hypercholesterolemic diet containing 0.5% sesamin. They demonstrated that the hypocholesterolemic activity of sesamin can, at least in part, be explained by the inhibition of the intestinal absorption of cholesterol as reflected by the significant reduction in cholesterol in the thoracic lymph and a significant reduction in the activity of liver microsomal 3-hydroxy-3-methylglutaryl coenzyme A reductase, which is a rate-limiting enzyme in the biosynthesis of cholesterol. Consumption of 32 mg sesamin capsules for 4 weeks followed by 65 mg sesamin capsules for 4 weeks reduced total cholesterol (9%), LDL-C (16.5%) and apoprotein B (10.5%) in 12 males with hypercholesterolemia (Hirata et al., 1996). In another study, 21 hypercholesterolemic subjects consumed 40 g of roasted sesame seeds for 4 weeks. The results showed that the sesame diet significantly decreased the levels of serum total cholesterol (6.4%) and LDL-C (9.5%). However, the effect of sesame seeds on serum cholesterol ceased when patients stopped the consumption of the sesame diet (Chen et al., 2005). Wu et al. (2006) also observed similar reductions in total cholesterol and LDL-C in 24 postmenopausal subjects following a 5-week intervention with 50 g of pulverized roasted sesame seed. High density lipoprotein cholesterol (HDL-C) levels were unchanged in all of these human studies that involved the intervention with sesamin or sesame seeds (Chen et al., 2005; Hirata et al., 1996; Wu et al., 2006). Since changes in the diet can be an effective way to lower blood levels of total and LDL cholesterol (Delahanty et al., 2001; Sikand et al., 2000), drug therapy may be reserved for patients who are at high risk for CHD (Expert Panel on Detection, Evaluation and Treatment of High Blood Cholesterol in Adults, 1993). Sesame and its lignans can reduce TC and LDL-C in hypercholesterolemic patients, including postmenopausal women who have a higher incidence of hypercholesterolemia (Chang et al., 2002).

## Enhancement of Vitamin E Level

The tocopherol and tocotrienol vitamers that comprise the vitamin E family are considered to be the most important lipophilic radical-quenching antioxidants in cell membranes. While their function is most often associated with the reduction of peroxyl radicals, novel vitamer-specific roles for tocopherols in signal transduction and in the quenching of other reactive chemical species such as nitrogen dioxide and peroxynitrite are now being investigated (Brigelius-Flohe & Traber, 1999). While α-tocopherol has attracted much attention, recent studies indicate that several of these important roles may be specific to γ-tocopherol. For example, γ-tocopherol and its major metabolite, 2,7,8-trimethyl-2-(beta-carboxyethyl)-6-hydroxychroman (γ-CEHC), inhibit cyclooxygenase activity in stimulated macrophages and epithelial cell, which reduces the synthesis of prostaglandin $E_2$ ($PGE_2$) (Jiang et al., 2000, 2001). Moreover, in carrageenan-induced inflammation in male Wistar rats, administration of γ-tocopherol (33 mg/kg) can reduce the levels of $PGE_2$, Leukotriene $B_4$, and tumor necrosis factor-alpha (TNF-α) (Jiang & Ames, 2003). Administration of sesame oil and it lignans increases blood and tissue concentrations of γ-tocopherol without altering those of α-tocopherol in rats (Ikeda et al., 2002; Kamal-Eldin et al., 1995, 2000; Yamashita et al., 1992). The inclusion of unrefined sesame oil in the daily diet of healthy women over a 4-week intervention period raised plasma level of γ-tocopherol by 42% without altering α-tocopherol (Lemcke-Norojärvi et al., 2001). In another study, the levels of plasma γ-tocopherol increased 19% in volunteers after only 3 days of daily consumption of muffins containing sesame seed (equal to a daily dose of 35 mg of sesamin and 13 mg of sesamolin) (Cooney et al., 2001).

Sontag and Parker (2002) suggested a cytochrome P450 4F2 mediated ω-hydroxylation pathway for γ-tocopherol catabolism, which was inhibited by sesamin. Also, the urinary excretion of γ-tocopherol metabolites was significantly lower in volunteers after the consumption of sesame oil muffins (Frank et al., 2004). These data suggest that sesamin can possibly inhibit the catabolism of γ-tocopherol, which results in its higher bioavailability observed in human and animal studies (Cooney et al., 2001; Ikeda et al., 2002; Kamal-Eldin et al., 1995, 2000; Lemcke-Norojärvi et al., 2001; Sontag et al., 2002; Yamashita et al., 1992 ). Recently, Abe et al. (2005) demonstrated that dietary sesame seeds elevate α-tocopherol concentration in rat brain. They showed that the concentration of α-tocopherol in the brain of rats (Cerebrum, cerebellum, brain stem, and hippocampus) fed 50 mg of α-tocopherol/kg with sesame seeds was higher than that of the rats fed 500 mg α-tocopherol/kg without sesame seed. These results suggest that the dietary sesame seeds are more useful than the intake of an excessive amount of α-tocopherol, for maintaining a high α-tocopherol concentration and inhibiting lipid peroxidation in the various regions of the rat brain (Abe et al., 2005).

## Effect On Lipid Metabolism and Inflammatory Responses

An experimental diet containing sesame seeds (200 g/kg) increased both the hepatic mitochondrial and peroxisomal fatty acid oxidation rate in rats. Noticeably, peroxisomal activity levels were increased >3 times in rats fed diets containing sesame seeds, compared to those fed a control diet without sesame seed. A sesame seed diet also significantly increased the activity of hepatic fatty acid oxidation enzymes including acyl-CoA oxidase, carnitine palmitoyltransferase, 3-hydroxyacyl-CoA dehydrogenase and 3-ketoacyl-CoA thiolase (Sirato-Yasumoto et al., 2001). In contrast, a sesame seed diet lowered the activity of enzymes involved in fatty acid synthesis including fatty acid synthase, glucose-6-phosphate dehydrogenase, ATP-citrate lyase, and pyruvate kinase (Sirato-Yasumoto et al., 2001). Similar effects on hepatic fatty acid oxidation and synthesis were also observed when rats were fed diets containing pure sesamin (1) or episesamin (0.2%) for 15 days, suggesting the possible involvement of sesame oil lignans in the induction of the observed changes in hepatic fatty acid metabolism (Kushiro et al., 2002). Considering the lower levels of serum triacylglycerol observed in rats fed sesame seed diets, Sirato-Yasumoto et al. (2001) suggested that sesame seeds can possibly exert their triglyceride-lowering-effect through increasing the rate of fatty acid oxidation and reducing the rate of fatty acid biosynthesis in the liver. However, neither a 5-week intervention with 50 g of pulverized roasted sesame seed in postmenopausal women (Wu et al., 2006) nor a 4-week intervention of 40 g of roasted sesame seed in hypercholesterolemic patients (Chen et al., 2005) altered the level of serum triacylglycerols.

Sesame seed lignans, such as sesamin, episesamin, sesamolin, sesaminol as well as sesamol, are known to inhibit Δ5-desaturase activity, which is an enzyme in the biosynthesis of arachidonic acid from dihomo-γ-linolenic acids in vitro (Chavali & Forse, 1999; Shimizu et al., 1991). Moreover, sesamin increased the levels of dihomo-γ-linolenic acid and increased the ratio of eicosapentaenoic acid/α-linolenic acid and decreased the ratio of arachidonic acid/dihomo-γ-linolenic acids in liver lipids of rats fed sesamin (0.5% w/w). Therefore, sesamin inhibits the Δ5-desaturase activity involved in the biosynthesis of n-6 fatty acids but has no effect on the biosynthesis of n-3 fatty acids in vivo (Fujiyama-Fujiwara et al., 1995; Mizukuchi et al., 2003).

Conflicting data exist concerning the arachidonic acid content of the liver lipids in rats fed sesamin, which can possibly be explained by different fatty acid profiles in the diet (Fujiyama-Fujiwara et al., 1995; Kamal-Eldin et al., 2000; Mizukuchi et al., 2003). Wu et al. (2006) showed that consumption of 50 g of roasted sesame seed for 5 weeks reduced the content of arachidonic acid in the serum lipids of postmenopausal women, with no effect on dihomo-γ-linolenic acid content. Arachidonic acid is the precursor of eicosanoids, which include prostaglandins, thromboxanes, and leukotrienes. The release of arachidonic acid from membrane phospholipids, which is the first stage in the biosynthesis of eicosanoids, can occur as a result of tissue-specific stimuli. The released arachidonic acid then undergoes oxygenation by cyclooxygenase to yield prostaglandin $H_2$ ($PGH_2$), which serves as precursor to other prostaglandins and thromboxanes (Mathews et al., 2000). Chavali et al. (1998) showed that the lipopolysaccharide-induced production of $PGE_2$ is lower and that of TNF-α is higher in mice fed a diet containing 0.25% sesamin. In addition, they showed that sesamol, a sesamolin hydrolysis product, can reduce lipopolysaccharide-induced production of $PGE_2$ and interleukin 6 (IL-6) without increasing TNF-α (Chavali & Forse, 1999). Proinflammatory mediators, such as $PGE_2$, can influence the production of cytokines, which mediate inflammatory responses during inflammation and infection (Pruimboom et al., 1994). However, the reduction in $PGE_2$ in mice, exerted by sesamin and sesamol, poses a different effect on interleukins and TNF-α possibly because of different mechanisms. Despite a lack of effect on the levels of arachidonic acid, the $PGE_2$ levels were reported to be significantly lower in mice fed sesamin- or sesamol-supplemented diets. Therefore, Chavali et al. (1998, 1999) suggested that sesamin, sesamol, or their metabolites decrease the activity of cyclooxygenase and decrease the biosynthesis of prostaglandins.

## Allergic and Toxic Compounds in Sesame Oil

Numerous studies reported allergic reactions to sesame seed oil (Chiu & Haydik, 1991; Gangur et al., 2005; Kanny et al., 1996; Robenstein, 1950). Symptoms of sesame seed allergy include cutaneous, gastrointestinal, and respiratory symptoms including anaphyactic shock. These symptoms, especially acute reactions and anaphylactic shock, suggest IgE-mediated or type 1 allergies. Sesame seed contains a number of allergic proteins, *viz. Ses i* 1, *Ses i* 2, *Ses i* 3, *Ses i* 4, *Ses i* 5, *Ses i* 6, and *Ses i* Profilin (Gangur et al., 2005). Residual proteins/peptides suspended in the oil can cause the allergenicity of sesame oil. (Fremont et al., 2002). Other studies claimed that sesamin and sesamolin can cause contact dermatitis in humans by elevating plasma IgG (Hayakawa et al., 1987; Kubo et al., 1986; Neering et al., 1975; Nonaka et al., 1997; van Dijk et al., 1972, 1973).

## Conclusion

Sesame oil is unique because it contains the furofuan lignan sesamin and its analog sesamolin, which together can comprise 0.5–1.5% to the oil. These ligans can affect the desaturation and β-oxidation of fatty acids as well as inhibit cholesterol biosynthesis and tocopherol metabolism and excretion. These effects may exert beneficial health effects.

## References


Abe, C.; S. Ikeda; K. Yamashina. Dietary sesame seeds eleva–te α-tocopherol concentration in rat brain. *J. Nutr. Sci. Vitaminol.* **2005,** *51,* 223–230.

Ajibola, O.O.; O.K. Owolarafe; O.O. Fasina; K.A. Adeeko. Expression of oil from sesame seeds. *Canadian Agricultural Engineering* **1993,** *5,* 83–88.

Assmann, G.; P. Cullen; F. Jossa; B. Lewis; M. Mancini. Coronary heart disease: Reducing the risk: The scientific background to primary and secondary prevention of coronary heart disease. A world wide view. International task force for the prevention of coronary heart disease. *Arterioscl. Thromb. Vascu. Biol.* **1999,** *19,* 1819–19824.

Bedigian, D.; J.R. Harlan Evidence for cultivation of sesame in the ancient world. *Economic Botany* **1986,** *40,* 137–154.

Beroza, M.; M.L. Kinman. Sesamin, sesamolin and sesamol content of the oil of sesame as affected by strain, location grown, ageing and frost damage. *J. Am. Oil Chem. Soc.* **1955,** *32,* 348–350.

Bhatia, A.; M. Benbouzid; R.J. Hamilton; P.A. Sewel. Separation of triglycerides and hydrocarbons from seed oils by high performance liquid chromatography with an infra-red detector. *Chemistry and Industry* **1986,** *20,* 70–71.

Brigelius-Flohe, R.; M.G. Traber. Vitamin E: Function and metabolism. *FASEB J.* **1999,** *13,* 1145–1155.

Budowski, P.; K.S. Markley. The chemical and physiological properties of sesame oil. *Chemistry*

*Rev.* **1951,** *48,* 125–151.

Budowski, P.; F.G.T. Menezes; F.G. Dollear. Sesame oil. V. The stability of sesame oil, *J. Am. Oil Chem. Soc.* **1950,** *27,* 377–380.

Chang, L.W.; W.J. Yen; S.C. Huang; P.D. Duh. Antioxidant activity of sesame coat, *Food Chem.* **2002,** *78,* 347–354.

Chavali, S.R.; R.A. Forse. Decreased production of interleukin-6 and prostaglandin $E_2$ associated with inhibition of $\Delta^5$ desaturation of ω6 fatty acids in mice fed safflower oil diets supplemented with sesamol. *Prostagl. Leukotri. Essential Fatty Acids,* **1999,** *58,* 185–191.

Chavali, S.R.; W.W. Zhong; R.A. Forse. Dietary α-linolenic acid increases TNF-α, and decreases IL-6, IL-10 in response to LPS: Effect of sesamin on the Δ-5 desaturation of ω6 and ω3 fatty acids in mice. *Prostagl. Leukotri. Essential Fatty Acids* **1998,** *58,* 185–191.

Chen, P.R.; K.L. Chien; T.C. Su; C.J. Chang; T. Liu; H. Cheng; C. Tsai. Dietary sesame reduces serum cholesterol and enhances antioxidant capacity in hypercholesterolemia. *Nutri. Res.* **2005,** *25,* 559–567.

Chiu, J.T.; I.B. Haydik. Sesame seed oil anaphylaxis. *J. Allergy Clinical Immunol.* **1991,** *88,* 414–415.

Cooney, R.V.; L.J. Custer; L. Okinaka; A.A. Franke. Effects of dietary sesame seeds on plasma tocopherol levels. *Nutr. Cancer* **2001,** *39,* 66–71.

Delahanty, L.M.; L.M. Sonnenberg; D. Hayden; D.M. Nathan. Clinical and cost outcomes of medical nutrition therapy for hypercholesterolemia: A controlled trial. *J. Am. Dietetics Assoc.* **2001,** *101,* 1012–1016.

Elsawy, A.A.; M.M. Soliman; H.M. Fadel.Identification of volatile flavour components of roasted red sesame seeds. *Grasasy Aceites* **1988,** *39,* 160–162.

Endo, Y. Flavour components in edible fats and oils. *J. Japanese Oil Chem. Soc.* **1999,** *48,* 1133–1143.

Expert Panel on Detection, Evaluation and Treatment of High Blood Cholesterol in Adults. Summary of the second report of the national cholesterol education program (NCEP) expert panel on education, evaluation and treatment of high blood cholesterol in adults (adult treatment Panel II). *JAMA* **1993,** *269,* 3015–3023.

Frank, J.; A. Kamal-Eldin; M. Traber. Consumption of sesame oil muffins decreases the urinary excretion of γ-tocopherol metabolites in humans. *Annals New York Academy of Sci.* **2004,** *1031,* 365–367.

Fremont, S.; Y. Errahali; M. Bignol; M. Metche; J.P. Nicolas. Allergenicity of some isoforms of white sesame proteins. *Allergy Immunol. (Paris)* **2002,** *34,* 91–94.

Fujiyama-Fujiwara, Y.; R. Umeda-Sawada; M. Kuzuyama; O. Igarashi. Effect of sesamin on the fatty acid composition of the liver of rats fed *n*-6 and *n*-3 fatty acid-rich diet. *J. Nutri. Sci. Vitaminol.* **1995,** *41,* 217–225.

Fukuda, Y.; T. Osawa; M. Namiki; T. Ozaki. Studies on antioxidative substances in sesame seed. *Agric. Biolog.Chemistry* **1985,** *49,* 301–306.

Fukuda, Y.; M. Isobe; M. Nagata; T. Osawa; M. Namiki. Acidic transformation of sesamolin, the sesame-oil constituent, into an antioxidant bisepoxylignan, sesaminol. *Heterocycles* **1986a,** *24,*

923–926.

Fukuda, Y.; M. Nagata; T. Osawa; M. Namiki. Contribution of lignan analogues to antioxidative activity of refined unroasted sesame seed oil. *J. Am. Oil Chem. Soc.* **1986b,** *63,* 1027–1031.

Fukuda, Y.; M. Nagata; T. Osawa; M. Namiki. Chemical aspects of the antioxidative activity of roasted sesame seed oil and the effect of using the oil for frying. *Agric. Biolog. Chemistry* **1986c,** *50,* 857–862.

Fukuda, Y.; T. Osawa; S. Kawagishi; M. Namiki. Comparison of contents of sesamolin and lignan antioxidants in sesame seeds cultivated in Japan. *Nippon Shokuhin Kogyo Gakkaishi* **1988,** *35,* 483–486.

Fukuda, Y.; Y. Koizumi; R. Ito; M. Namiki. [Synergistic action of the antioxidant components in roasted sesame seed oil]. *Nippon Shokuhin Kogaku Kaishi* **1996,** *43,* 1272–1277.

Gangur, V.; C. Kelly; L. Navuluri. Sesame allergy: A growing food allergy of global proportions? *Annals Allergy, Asthma & Immunol.* **2005,** *95,* 4–11.

Han, J.H.; S.Y. Ahn. Effects of oil refining processes on oil characteristics and oxidation stability of sesame oil. *J. Korean Chemical Soc.* **1993,** *36,* 284–289.

Hayakawa, R.; K. Matsunaga; M. Suzuki; K. Hosokawa; Y. Arima; C.S. Shin; M. Yoshida. Is sesamol present in sesame oil?, *Contact Dermatitis* **1987,** *17,* 133–135.

Hemalatha, S.; Ghafoorunissa. Lignans and tocopherols in indian sesame cultivars. *J. Am. Oil Chem. Soc.* **2004,** *81,* 467–470.

Hirata, F.; K. Fujita; Y. Ishikura; K. Hosoda; T. Ishikawa; H. Nakamura. Hypercholesterolemic effect of sesame lignan in human. *Atherosclerosis* **1996,** *122,* 135–136.

Hirose, N.; T. Inoue; K. Nishihara; M. Sugano; K. Akimoto; S. Shimizu; S. Yamada. Inhibition of cholesterol absorption and synthesis in rats by sesamin. *J. Lipid Res.* **1991,** *32,* 629–638.

Ikeda, S.; T. Tohyama; K. Yamashita. Dietary sesame seed and its lignans inhibit 2,7,8-trimethyl-2(2′-carboxyethyl)-6-hydroxychroman excretion into urine of rats fed γ-tocopherol. *J. Nutr.* **2002,** *132,* 961–966.

Itoh, T.; T. Tamura; T. Matsumoto. Sterol composition of 19 vegetable oils. *J. Am. Oil Chem. Soc.* **1973a,** *50,* 122–125.

Itoh, T.; T. Tamura; T. Matsumoto. Methyl sterol composition of 19 vegetable oils. *J. Am. Oil Chem. Soc.* **1973b,** *50,* 300–303.

Jiang, Q.; B.N. Ames. γ-Tocopherol, but not α-tocopherol, decreases proinflammatory ecosanoids and inflammation damage in rats. *FASEB J.* **2003,** *17,* 816–822.

Jiang, Q.; S. Christen; M.K. Shigenaga; B.N. Ames. γ-Tocopherol, the major form of vitamin E in the US diet, deserves more attention. *Am. J. Clinical Nutr.* **2001,** *74,* 714–722.

Jiang, Q.; I. Elson-Schwab; C. Courtemanche; B.N. Ames. γ-Tocopherol and its major metabolite, in contrast to α-tocopherol, inhibit cyclooxygenase activity in macrophages and epithelial cells. *Proceedings of the National Academy of Sciences of the USA* **2000,** *97,* 11494–11499.

Johansson, A.; L.B. Croon. 4-Desmethyl, 4-monomethyl-, and 4,4-dimethlysterols in some vegetable oils. *Lipids* **1981,** *16,* 285–291.

Johnson, L.A.; T.M. Suleiman; E.W. Lusas. Sesame protein: A review and prospectus. *J. Am. Oil*

*Chem. Soc.* **1979,** *56,* 463–468.

Kamal-Eldin, A.; G. Yousif; L.Å. Appelqvist; G.M. Iskander. Seed lipids of *Sesamum indicum,* L. and related wild species in Sudan. 1. Fatty acids and triacylglycerols. *J. Fat Sci. Technol.* **1992a,** *94,* 254–259.

Kamal-Eldin A.; L.Å. Appelqvist; G. Yousif; G.M. Iskander. Seed lipids of *Sesamum indicum,* L. and related wild species in Sudan. The sterols. *J. Sci. Food Agric.* **1992b,** *59,* 327–334.

Kamal-Eldin, A.; L.Å. Appelqvist. The effect of extraction methods on sesame oil stability. *J. Am. Oil Chem. Soc.* **1995,** *72,* 967–969.

Kamal-Eldin, A.; L.Å. Appelqvist. Variations in fatty acid composition of the different acyl lipids in seed oils from four *Sesamum* species. *J. Am. Oil Chem. Soc.* **1994a,** *71,* 135–139.

Kamal-Eldin, A.; L.Å. Appelqvist. Variations in the composition of sterols, tocopherols and lignans in seed oils from four *Sesamum* species. *J. Am. Oil Chem. Soc.* **1994b,** *71,* 149–156.

Kamal-Eldin, A.; D. Pettersson; L. Appelqvist. Sesamin (a compound from sesame oil) increases tocopherol levels in rats fed ad libitum. *Lipids* **1995,** *30,* 499–505.

Kamal-Eldin, A.; J. Frank; A. Razdan; S. Tengblad; S. Basu; B. Vessby. Effects of dietary phenolic compounds on tocopherol, cholesterol, and fatty acids in rats. *Lipids* **2000,** *35,* 427–435.

Kanny, G.; C. Hauteclocque; D.A. Moneret-Vautrin. Sesame seed and sesame seed oil contain masked allergens of growing importance. *Allergy* **1996,** *51,* 952–957.

Koh, E.T.; Comparison of hypolipemic effects of corn oil, sesame oil, and soybean oil in rats. *Nutr. Reports International* **1987,** *36,* 903–917.

Kubo, Y.; S. Nonaka; H. Yoshida. Contact sensitivity to unsaponifiable substances in sesame oil. *Contact Dermatitis* **1986,** *15,* 215–217.

Kumazawa, S.; M. Koike; Y. Usui; T. Nakayama; Y. Fukuda. Isolation of sesaminols as antioxidants from roasted sesame seed oil. *J. Oleo Sci.* **2003,** *52,* 303–307.

Kushiro, M.; T. Masaoka; S. Haeshita; Y. Takahashi; T. Ide; M. Sugano. Comparative effect of sesamin and episesamin on the activity and gene expression of enzymes in fatty acid Oxoidation and synthesis in rat liver. *J. Nutr. Biochem.* **2002,** *13,* 289–295.

Lemcke-Norojärvi, M.; A. Kamal-Eldin; L.Å. Appelqvist; L.H. Dimberg; M. Öhrvall; B. Vessby. Corn and sesame oils increase serum γ-tocopherol concentrations in healthy Swedish women. *J. Nutr.* **2001,** *131,* 1195–1201.

Manley, C.H.; P.P. Vallon; R.E. Ericksson. Some aroma components of roasted sesame seed (*Sesamum indicum,* L.) *J. Food Sci.* **1974,** *39,* 73–76.

Marchand, P.A.; M.J. Kato; N.G. Lewis (+)-Episesaminone, a *Sesamum indicum* furofuran lignan. Isolation and hemisynthesis. *J. Natural Products* **1997,** *60,* 1189–1192.

Mathews, C.K.; K.E. van Holde; K.G. Ahern. *Biochemistry;* Third Edition; Benjamin/Cummings, an imprint of Addison Wesley Longman 2000; pp. 700–704.

Mizukuchi, A.; R. Umeda-Sawada; O. Igarashi. Effect of dietary fat level and sesamin on the polyunsaturated fatty acid metabolism in rats. *J. Nutr. Sci. Vitaminol.* **2003,** *49,* 320–326.

Moazzami, A.A.; R.E. Andersson; A. Kamal-Eldin. Lignan content in sesame seeds and products. *European J. Lipid Sci. Technol.* **2007,** *109,* 1022–1027.

Murui, T.; K. Wanaka; H. Seki. Content of sterylglycosides and dibasic acids in commercial sesame oil. *J. Japan Oil Chem. Soc.* **1994,** *43,* 158–161.

Nagashima, M.; Y. Fukuda. Lignan-phenols of water-soluble fraction from 8 kinds of sesame seed coat according to producing district and their antioxidant activities. *Nippon Nogeikagaku Kaishi* **2004,** *38,* 45–53.

Nagata, M.; T. Osawa; M. Namiki; Y. Fukuda; T. Ozaki. Stereochemical sStructures of antioxidative bisepoxylignans, sesaminol and its isomers, transformation from sesamolin. *Agric. Biolog. Chemistry* **1987,** *51,* 1289–1987.

Nakamura, S.; O. Nishimura; H. Masuda; S. Mihara. Identification of volatile flavour components of the oil from roasted sesame seeds. *Agric. Biolog. Chem.* **1989,** *53,* 1891–1899.

Namiki, M. The chemistry and physiological function of sesame. *Food Rev. Intern.* **1995,** *11,* 281–329.

Neering, H.; B.E. Vitanyi; K.E. Malten; W.G. van Ketel; E. van Dijk. Allergens in sesame oil contact dermatitis. *Acta Dermato-Venereologica* **1975,** *55,* 31–34.

Nonaka, M.; K. Yamashita; Y. Iizuka; M. Namiki; M. Sugano. Effects of dietary sesaminol and sesamin on eicosanoid production and immunoglobulin level in rats given ethanol. *Biosci. Biotechnol. Biochem.* **1997,** *61,* 836–839.

Osawa, T.; M. Nagata; M. Namiki; Y. Fukuda. Sesamolinol, a novel antioxidant isolated from sesame seeds. *Agric. Biolog. Chem.* **1985,** *49,* 3351–3352.

Pecquet, C.; F. Leynadier; P. Saiag. Immediate hypersensitivity to sesame in foods and cosmetics, *Contact Dermatitis* **1998,** *39,* 313.

Pruimboom, W.M.; J.A.P.M. van Dijk; C.J.A.M.

Tak; I. Garrelds; I.L. Bonta; P.J.H. Wilson; F.J. Zijlstra. Interactions between cytokines and eicosanoids: A study using human peritoneal macrophages. *Immuno. Lett.* **1994,** *41,* 255–260.

Robenstein, L. Sensitivity to sesame seed and sesame oil. *NY State J. Medicine* **1950,** *50,* 343–344.

Ryu, S.; C.T. Ho; T. Osawa. High performance liquid chromatographic determination of antioxidant lignan glycosides in some variations of sesame. *J. Food Lipids* **1998,** *5,* 17–28.

Ryu, S.N.; S.M. Kim; J. Xi; C.T. Ho. Influence of Seed Roasting Process on the Changes in Volatile Compounds of the Sesame (*Sesamum indicum* L.) Oil, *Flavor Chemistry of Ethnic Foods*; F. Shahidi, C.T. Ho, Eds.; Kluwer Academic/Plenum Publishers: New York, NY, 1999; pp. 229–237.

Shimizu, S.; K. Akimoto; Y. Shinmen; H. Kawashima; M. Sugano; H. Yamada. Sesamin is potent and specific inhibitor of Δ5-desaturase in polyunsaturated fatty acid biosynthesis. *Lipids* **1991,** *26,* 512–516.

Sikand, G.; M.L. Kashyap; N.D. Wong; J. Hsu. Dietitian intervention improves lipid values and saves medication costs in men with combined hyperlipidemia and a history of niacin noncompliance. *J. Am. Dietetics Assoc.* **2000,** *100,* 218–224.

Sirato-Yasumoto, S.; M. Katsuta; Y. Okuyama; Y. Takahashi; T. Ide. Effect of sesame seeds rich in sesamin and sesamolin on fatty acid oxidation in rat liver. *J. Agric. Food Chem.* **2001,** *49,* 2647–2651.

Soliman, M.M.; A.A. Elsawy; H.M. Fadel; F. Osman. Identification of volatile flavour components of roasted white sesame seed. *Acta Alimentaria* **1986,** *15,* 251– 263.

Soliman, M.M.; S. Kinoshita; T. Yamanishi. Aroma of roasted sesame seeds. *Agric. Biolog. Chem.* **1975,** *39,* 973–977.

Sontag, T.J.; R.S. Parker. Cytochrome P450 ω-hydroxylase pathway of tocopherol catabolism. *Biol. Chem.* **2002,** *277,* 25290–25296.

Spencer, G.F.; S.F. Herb; P.J. Gormisky. Fatty acid composition as a basis for identification of commercial fats and oils. *J. Am. Oil Chem. Soc.* **1976,** *53,* 94–96.

Tzen, J.T.C.; Y.Z. Cao; P. Laurent; C. Ratnayake; A.H.C. Huang. Lipids, proteins, and structure of seed oil bodies from diverse species. *Plant Physiol.* **1993,** *101,* 267–276.

Tzen, J.T.C.; M.M.C. Wang; J.C.F. Chen; L.J. Lin; M.C.M. Chen. Seed oil body proteins: Oleosin, caleosin, and steroleosin. *Currant Topics in Biochem. Res.* **2003,** *5,* 133–139.

van Dijk, E.; E. Dijk; H. Neering; B.E. Vitanyi. Contact hypersensitivity to sesame oil in patients with leg ulcers and eczema. *Acta Dermato-Venereologica* **1973,** *53,* 133–135.

van Dijk, E.; H. Neering; B.E. Vitnay. Contact hypersensitivity to sesame oil in zinc oxide ointment. *Nederlands Tijdschrift voor Geneeskunde* **1972,** *116,* 2255–2259.

Vogel, P. Determination of xanthophyll content of vegetable oils. *Fette Seifen Anstrichmittel* **1977,** *79,* 97–103.

Wu, W.; Y. Kang; N. Wang; H. Jou; T. Wang. Sesame ingestion affects sex hormones, antioxidant status, and blood lipids in postmenopausal women. *J. Nutr.* **2006,** *136,* 1270–1275.

Yamashita, K.; Y. Nohara; K. Katayama; M. Namiki. Sesame seed lignans and gamma-tocopherol act synergistically to produce vitamin E activity in rats. *J. Nutr.* **1992,** *122,* 2440–2446.

Yen, G.C. Influence of seed roasting process on the changes in composition and quality of sesame seed with different roasting temperatures. *Food Chem.* **1990a,** *31,* 215–224.

Yen, G.C. Influence of seed roasting process on the changes in composition and quality of sesame (*Sesame indicum*) oil. *J. Sci. Food Agri.* **1990b,** *50,* 563–570.

Yen, G.C. Thermal stability of sesame/soybean oil blends. *Food Chem.* **1991,** *41,* 355–360.

Yen, G.C.; S.L. Shyu. Oxidative stability of sesame oil prepared from sesame seed with different roasting temperatures. *Food Chem.* **1989,** *31,* 215–224.

# Niger Seed Oil

**Mohamed Fawzy Ramadan**
*Biochemistry Department, Faculty of Agriculture, Zagazig University, Zagazig 44511, Egypt*

## Introduction

In many parts of the world a surplus of some conventional crops exists, and a search for new crops, including some that produce useful oils, is ongoing. Until now the agricultural revolution has produced raw material mainly for the food industry, but in the future more of its crops will be directed toward chemical applications. Unless a new vegetable oil has some specific property, it will have to compete with existing supplies of oils available at bulk prices (Gunstone, 1996). Nontraditional oilseeds are being considered because their constituents have unique chemical properties and may augment the supply of edible oils (Cherry & Kramer, 1989). No oil from any single source was found to be suitable for all edible and nonedible applications because oils from different sources generally differ in their composition. This necessitates the search for new sources of novel oils. So far, numerous plants were analyzed, and some of these were cultivated as new oil crops. Among these new sources of edible oils, niger (*Guizotia abyssinica* Cass.) seed oil is of interest and may play an important role in human nutrition and health, because of its fatty acid composition and its high levels of fat-soluble bioactive components (Dutta et al., 1994; Ramadan & Moersel, 2002a,b, 2004).

Niger is an oilseed that humans have cultivated for approximately 5000 years. *G. abyssinica* is widely grown in South India and Ethiopia, which are the two main producing countries. In other parts of the world, niger is mainly cultivated as an oil-seed crop. Niger seed belongs to the same botanical family as sunflower and safflower (Compositae). Six species of *Guizotia* exist with *G. abyssinica* as the only cultivated species (Baagøe, 1974). It is a dicotyledonous herb, moderately to well-branched, and grows up to 2 m high (Getinet & Teklewold, 1995). The crop grows best on poorly drained, heavy clay soils (Alemaw & Wold, 1995; Francis & Campbell, 2003). Niger is cultivated in both temperate and tropical climates, being considered a temperate-region plant that has adapted to a semi-tropical environment. It prefers moderate temperatures for growth, from about 19 to 30°C (Quinn & Myers, 2002).

## Oil Content

Niger seeds were reported to contain 483 calories/ 100 g seeds, 2.8–7.8% moisture, 17–30% protein, 34–39% total carbohydrate, 9–13% fiber, 1.8–9.9 g ash/ 100g seeds, 50–587 mg calcium /100 g seeds, 180–800 mg phosphorous/100 g seeds, 0.43 mg thiamine/100 g seeds, 0.22–0.55 mg riboflavin/100g seeds and 3.66 mg niacin/100 g seeds (Bhardwaj & Gupta, 1977; Kachapur et al., 1978; Rao, 1994). Reports in the literature showed the oil content in niger seed to be in the range of 30–50% (Dange & Jonsson, 1997; Getinet & Teklewold, 1995), and a large variation appeared in oil content (28.5–38.8%) of different samples collected from Ethiopia (Dutta et al., 1994). The oil has an attractive pale yellow color and a nutty taste. With high levels of linoleic acid, it is very similar to sunflower and safflower oils (Francis & Campbell, 2003). The fatty acid composition of niger seed oil is described in detail in a later section of this chapter.

## Protein and By-products

Niger meal, remaining after oil extraction, contains approximately 30% protein and 23% crude fiber. In general, Ethiopian niger meal contains less protein and more crude fiber than the niger meal from seeds grown in India (Seegeler, 1983). The oil, protein, and crude fiber contents of niger are affected by the hull thickness. Thick-hulled seeds tend to have less oil and protein and more crude fiber (Getinet & Teklewold, 1995). The meal was reported to be free from any toxic substance but contains more crude fiber than most oilseed meals. The utilization of niger seed proteins in human food is very limited due to the presence of a high fiber content and a dark color of the cake. An integrated method of processing was developed to dehull niger seeds with a dehulling efficiency of 96–98%. The oil extracted from dehulled seeds was of good quality, and the cake was high in protein and low in fiber (Bhagya & Sastry, 2003). Chemical and biological properties of protein fractions and lipoprotein concentrates from niger seed were studied (Eklund, 1971a,b; 1974). The amino acid composition of niger protein was deficient in tryptophan. The lipoprotein contained 4% moisture, 12% ash, 46% protein, 20% fat, 7% crude fiber, and 11% soluble carbohydrate.

Shahidi et al. (2003) examined the antioxidant potential of defatted niger seed and its extracts. The extracts were also evaluated for their antioxidant and radical scavenging activities. UV spectroscopy suggested that the major active component present was a chlorogenic acid-related compound. Furthermore, high performance liquid chromatography (HPLC) analysis established that chlorogenic acid was dominant in the free phenolic fraction (2.6 mg/g). Upon hydrolysis, a substantial amount of caffeic acid (42.8 mg/g) was released, presumably from esterified and glycosylated chlorogenic acid. Very small amounts of free sinapic and caffeic acids were present, but these were not detected by thin-layer chromatography (TLC). The conclusion was that defatted niger seed and its extracts can provide a natural source of antioxidants (Shahidi et al., 2003). The use of niger seed was recommended in applications where functionality of seed components, such as their fat- and moisture-binding ability and textural effects, are important in addition to their antioxidant activity.

## Economy

Today, niger seed is particularly important to the economy of Ethiopia, where it accounts for 50–60% of its edible oil supply. In India, niger is extensively grown on the hill slopes and in coastal plains of West Bengal but provides only 2% of India's total oilseed production (Mishra et al., 2007). Niger seed is also a minor oilseed crop in some other African countries (Ramadan & Moersel, 2002a; Riley & Belayneh, 1989; Shahidi et al., 2003). Yields of niger seed reported in literature varied from 0.25 to 0.50 tons per hectare (Riley & Belayneh, 1989), to over 1.25 tons per hectare (Weiss, 2000). Another report mentioned that the niger seed production in India and Ethiopia together is estimated to be 318–340 thousand tons (Quinn & Myers, 2002). Yields in excess of 1.5 tons per hectare are recorded, but generally yields are less than 0.8 tons per hectare (Francis & Campbell, 2003).

*G. abyssinica* seed is exported from Ethiopia, India, Myanmar, and Nepal, which account for over 90% of the niger seed in the global market. Depending on the country of import, the wholesale price averages 0.46–0.82 USD/kg that, after heat treatment and shipment, scales up to 1.1 USD/kg (compared to 0.26 USD/kg for sunflowers). The limited amount of the niger seed in the market coupled with its high price indicates that the market potential of niger seed is still promising (Quinn & Myers, 2002).

## Processing

In Ethiopia, the conventional procedure for oil extraction from niger is through a combination of warming, grinding, and mixing with hot water followed by centrifugation. After centrifugation on a smooth soft surface, the pale yellow oil settles over the meal. Niger is also crushed in small cottage expellers and large oil mills. Several models of small, electrically powered cottage expellers are manufactured in Ethiopia. The meal remaining after oil extraction using Ethiopian expellers contains 6–12% oil depending on the expeller. In India, the oil is extracted by mechanized expellers and hydraulic presses in large industrial areas (Getinet & Teklewold, 1995). Dehulling of niger seeds increased the oil and protein contents and decreased the crude fiber. The protein content of the dehulled flour increased from 44 to 63%. The calculated nutritional indices, essential amino acid index, biological value, and nutritional index were higher in dehulled flour compared to undehulled flour. On the other hand, good quality oil was extracted from dehulled seeds (Bhagya & Sastry, 2003).

## Edible and Nonedible Applications

Niger seeds are used fried, milled into flour, pressed with honey into cakes and for livestock feed after oil extraction, while the plants are used as green manure (a type of cover crop grown primarily to add nutrients and organic matter to the soil) (Bhagya & Sastry, 2003; Riley & Belayneh, 1989). In North America, Europe, and Japan, niger seed is used as a bird feed (Dutta et al., 1994; Marini et al., 2003; Shahidi et al., 2003).

Niger seed is used in human food mainly as a spice in the frying of vegetables (Bhagya & Sastry, 2003). Seeds pressed with honey are made into cakes in Ethiopia. The oil is used as a substitute for olive oil and a substitute for sesame oil for pharmaceutical purposes. In India, niger seeds are also fried or used as a condiment. Niger seed oil is also used in the manufacture of soap and as a lubricant or lighting fuel. The oil is also used to a limited extent in paints (being slow-drying), for which the Ethiopian seed is superior to the Indian seed as it has higher linoleic acid content. Niger seed or niger seed oil is also used in perfumes as a carrier of the scents and fragrances (Getinet & Teklewold, 1995).

## Acyl Lipids and Fatty Acid Composition

### *Fatty Acids*

Niger seed oil has a fatty acid composition typical for seed oils of the Compositae family (e.g., safflower and sunflower), with linoleic acid as the dominant fatty acid. The fatty acid composition of niger seed oil typically 7–8% palmitic and stearic acids, 5–25% oleic acid, and 55–80% linoleic acid (Getinet & Teklewold, 1995). This fatty acid composition is comparable to those of safflower (*Carthamus tinctorius*) and sunflower (*Helianthus annus*) oils, which are low in saturated acids, contains virtually no n-3 acids, and is rich in linoleic acid (up to 70%) (Gunstone, 1996). The Indian varieties of niger seed contain 25% oleic and 55% linoleic acids (Nasirullah et al., 1982), with the linoleic acid percentage being lower than in seeds grown in Ethiopia (75%) (Getinet & Teklewold, 1995; Nasirullah et al., 1982; Riley & Belayneh, 1989;Seegeler, 1983).

Dutta et al. (1994) studied the variation in the fatty acid profile of some uncertified niger seed varieties and of a few certified varieties collected from different regions in Ethiopia. The main fatty acids were palmitic, stearic, oleic, and linoleic acids (Table 9.1). The linoleic acid content was in the range of 71–79% irrespective of variety and/or location. Marini et al. (2003) analyzed the fatty acid composition of samples collected from different Ethiopian and Indian regions. Linoleic acid was the predominant fatty acid; its content lies in the range of 68.9–73.0%, irrespective of the region of origin (Table 9.1). The results agreed with those reported in the literature for Ethiopian and Indian niger oils (Ramadan & Moersel, 2003a; Seegeler, 1983; Seher & Gundlach, 1982). In addition, the oil can contain up to 1% arachidic acid, and

**Table 9.1. Variation in the Fatty Acid Levels (% of FAME) of Indian and Ethiopian Niger Seed Oil [a, b, c]**

| Acid | Ethiopian seed oil | Indian seed oil |
|---|---|---|
| Palmitic acid (C16:0) | 8.0–9.7 | 9.4–16.9 |
| Palmitoleic acid (C16:1) | 0.2 | 0.2 |
| Stearic acid (C18:0) | 5.6–8.1 | 6.8–6.5 |
| Oleic acid (C18:1) | 5.9–11.0 | 8.3–9.2 |
| Linoleic acid (C18:2) | 70.7–79.2 | 63.0–71.1 |
| Linolenic acid (C18:3) | 0.4 | 0.3 |
| Arachidic acid (C20:0) | 0.4 | 0.5 |
| Behenic acid (C22:0) | 0.3 | 0.4–0.5 |

[a] Ramadan & Moersel (2002 a,b, 2006).
[b] Dutta et al. (1994).
[c] Marini et al. (2003).

in special varieties up to 3% linolenic acid (Bockisch, 1998). Moreover, palmitoleic, linolenic, arachidic, eicosenoic, behenic, erucic, and lignoceric acids were detected in trace amounts in the chloroform/methanol extract of Indian niger seed (Ramadan & Moersel, 2002a).

## *Acyl Lipids*

Only a few papers reported the composition of various lipid classes in niger seeds (Dagne & Jonsson, 1997; Seegeler, 1983; Seher & Gundlach, 1982). Dutta et al. (1994) studied the lipid profile of three released and three local cultivars of Ethiopian niger oilseeds. Most of the total lipid (TL) was triacylglycerols (TAG), and polar lipids accounted for 0.7–0.8% of the TL content. The significant difference in the fatty acid compositions between TAG and polar lipids was that the TAG had lower 16:0 percentages, ca. 10% compared with ca. 16% in polar lipids. The higher percentages of 16:0 in PL were mainly compensated for by lower percentages of 18:2. Analyses by TLC plates demonstrated that TAG was the predominant lipid class in TL. Minor fatty acids [e.g., *cis*-vaccenic acid ($18{:}1^{\Delta 11}$)] were detected in trace amounts in both TAG and polar lipid fractions of all the samples.

Fractionation of neutral lipid (NL), glycolipid (GL), and phospholipid (PL) was carried out on a silica column with solvent of increasing polarity (Ramadan & Moersel, 2003b). The analysis of the lipid classes in the *n*-hexane extract of niger seed showed the NL to be 97%, GL to be 1.92%, and PL to be 0.28% of total lipids. The fatty acids present in the three fractions were generally similar, but the polar lipid fractions were characterized by higher palmitic and lower linoleic acid percentages than NL.

### Neutral Lipids

Major NL subclasses of *G. abyssinica* seed oil were separated using TLC (Ramadan & Moersel, 2002a). TAG accounted for the major subclass (91.9% of total NL), while monoacylglycerol (MAG), diacylglycerol (DAG), and free fatty acids (FFA) were presented in lower levels in the TL (Table 9.2). By using high- temperature gas liquid chromatography, the actual TAG molecular species of *G. abyssinica* seed oil was analyzed (Ramadan & Moersel, 2002a). Results showed that the major TAG peaks were C54:6 and C54:3, corresponding to trilinolein and triolein. The TAG molecular species distribution in niger seed oil from Ethiopian and Indian regions was recently reported (Marini et al., 2003). The results showed that LLL was the predominant TAG, as expected for oil rich in linoleic acid (Table 9.3). Also the percentage of the other TAG molecular species was in good agreement with the fatty acid distribution in the corresponding samples: besides LLL, PLL, POL, and OLL appear as the other relevant TAG. Apart from a slight but significant variation in the OLL and OOO content, the triglyceridic distribution was quite the same for all the oils, irrespective of the geographical origin.

### Glycolipids

Ramadan & Moersel (2003a) analyzed glycolipid (GL) subclasses in crude niger seed oil using normal phase high performance liquid chromatography (NP-HPLC). The proportion of each component was estimated by the lipid-carbohydrate determination. A relatively high level of GL was found in all niger seed oil. Among the total GL of niger seed oil, acylated steryl glucoside (ASG), steryl glucoside (SG), and cerebroside (CER) were the detected components (Fig. 9.1), where each fraction was comprised of about one-third of total GL. The average daily intake of GL in humans was reported to be 140 mg of ASG, 65 mg of SG, and 50 mg of CER (Sugawara &

**Table 9.2. The Major Nonpolar (Excluding Triacylglycerols) and Polar Lipid Subclasses and Their Fatty Acid Composition (% of Total FAME) in the Indian Crude Seed Oil**

| | Neutral lipids | | | | Glycolipids | | Phospholipids | | | |
|---|---|---|---|---|---|---|---|---|---|---|
| | MAG | DAG | STE | FFA | ASG | CER | PE | PI | PS | PC |
| | 0.49[a] | 0.73[a] | 0.29[a] | 4.82[a] | 38.5[b] | 31.0[b] | 22.5[c] | 14.6[c] | 8.74[c] | 48.7 |
| Palmitic acid (C16:0) | 28.1 | 24.5 | 23.2 | 21.8 | 21.4 | 25.7 | 22.6 | 19.6 | 18.9 | 20.2 |
| Stearic acid (C18:0) | 7.00 | 3.77 | 8.10 | 7.26 | 8.42 | 6.99 | 6.81 | 8.20 | 8.10 | 9.90 |
| Oleic acid (C18:1) | 17.4 | 19.2 | 20.3 | 18.9 | 16.4 | 14.3 | 13.1 | 11.10 | 16.9 | 11.7 |
| Linoleic acid (C18:2) | 47.2 | 49.2 | 44.0 | 46.2 | 51.2 | 50.5 | 55.3 | 60.2 | 55.5 | 57.6 |

[a]g/100 g of total neutral lipids, MAG, monoacylglycerols; DAG, diacylglycerols; FFA, free fatty acids STE, sterol esters, the rest of neutral lipids includes mainly triacylaglycerols.
[b]g/100 g of total glycolipids, ASG, acylated steryl glucoside; and CER, cerebrosides.
[c]g/100 g of total phospholipids, PS, phosphatidylserine; PI, phosphatidylinositol; PC, phosphatidyl choline; PE, phosphatidylethanolamine.
Source: Ramadan and Moersel (2002a; 2003a,b)

**Table 9.3.** TAG Molecular Species Profile (%) of Indian and Ethiopian Niger Seed Oil[a]

| | Ethiopian seed oil | Indian seed oil |
|---|---|---|
| LLL | 38.0 | 37.8 |
| OLL | 11.9 | 12.7 |
| PLL | 14.9 | 15.0 |
| OOL | 2.57 | 2.66 |
| POL | 14.9 | 14.8 |
| PPL | 1.67 | 1.59 |
| OOO | 4.15 | 3.66 |
| POO + SOL | 2.54 | 2.49 |
| SOO | 1.68 | 1.64 |
| Other TAG | 7.50 | 7.40 |

[a] Marini et al. (2003).

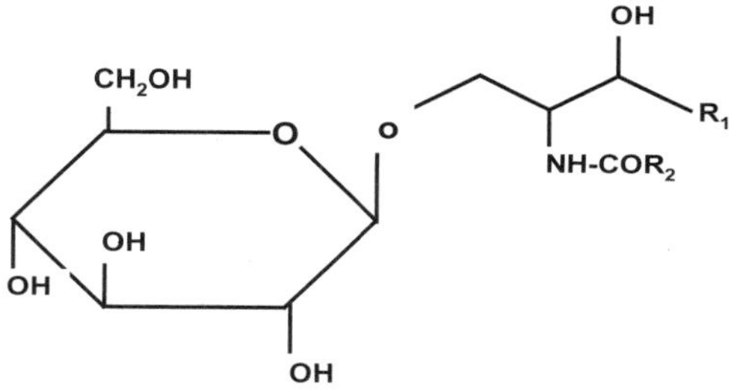

Cerebroside (CER)

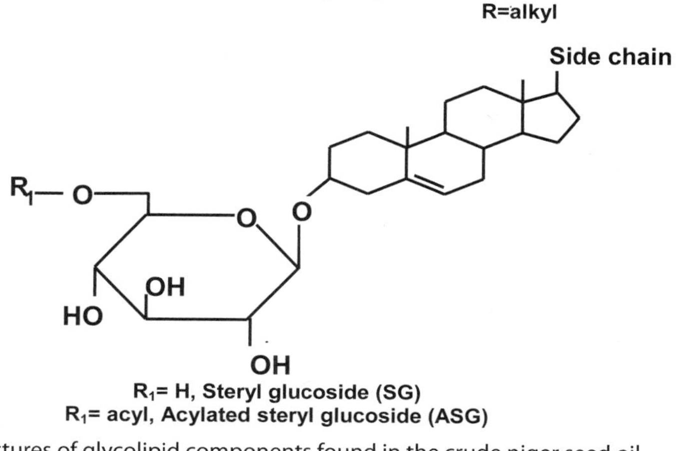

R=alkyl

$R_1$= H, Steryl glucoside (SG)
$R_1$= acyl, Acylated steryl glucoside (ASG)

**Fig. 9.1.** Structures of glycolipid components found in the crude niger seed oil.

Miyazawa, 1999). The fatty acid profile of the individual GL subclasses obtained from *G. abyssinica* seed oil is presented in Table 9.2. As expected among polar lipids, individual subclasses contained more saturated and less unsaturated fatty acids than the corresponding NL. The fatty acid profile of the GL lipid classes from niger seed oil was generally similar, wherein linoleic acid (>50% of FAME) was the most abundant among GL acyl residues followed by oleic acid as the second unsaturated fatty acid. Palmitic acid was the major saturated fatty acid in all individual GL subclasses, while the second major saturated acid was stearic acid which was detected in lower levels. The sterol components of SG and ASG subclasses obtained from niger seed oil were analyzed by GLC-FID after saponification. In order of decreasing prevalence, β-sitosterol> campesterol> stigmasterol were the major sterols found in SG and ASG fractions (Ramadan & Moersel, 2003a). As for the component sugars, glucose was the only sugar detected in the analyzed GL fractions.

### *Phospholipids*

Phospholipid (PL) subclasses from *G. abyssinica* crude seed oil were separated into seven fractions by NP-HPLC (Ramadan & Moersel, 2003b). Quantitative evaluation of the HPLC fractions by phosphorus analysis (Table 9.2) revealed that the predominant PL isolated from niger seed oil was phosphatidylcholine (PC) followed by phosphatidylethanolamine (PE), phosphatidylinositol (PI), and phosphatidylserine (PS), respectively. Phosphatidylglycerol (PG) and lysophosphatidylcholine (LPC), on the other hand, were present in smaller amounts, whereas lysophosphatidylethanolamine (LPE) was only detected in trace amounts. The PL fractions showed the highest amount of PC (ca. 49% of the total PL). PE and PI constituted approximately 22% and 14%, respectively, of total PL content, whereas, PS was present in approximately 8% of total PL content. The fatty acid profile of the PL major classes showed an overall similarity in composition (Table 9.2). Linoleic acid was the predominant fatty acid in all PL fractions followed by palmitic, oleic, and stearic acids, respectively. The ratio of saturated to unsaturated fatty acids was approximately 1:2 in all PL subclasses. This agrees with the general pattern of the fatty acids present in individual PL of different seed oils. The differences in fatty acid composition between neutral and polar lipids are of interest because the fatty acid pattern of the neutral lipids will predominate in refined, bleached, and deodorized oil, whereas that of the polar lipid fraction may be of interest if a "lecithin" fraction is recovered (Dutta et al., 1994).

## Minor Lipid Components

### *Tocopherols*

From different samples collected from Ethiopia, α-tocopherol was the major component of the total tocopherols in the samples investigated (Dutta et al., 1994). The levels varied from 680 – 850 μg/g oil, comprising more than 90% of the total tocopherol content (Table 9.4). γ-tocopherol was present at a low of about 1% of the total tocopherols and β-tocopherol was present at 3–5%. Niger seed oil contained much higher levels of tocopherol (660–850 μg tocopherol/g oil) than sunflower and safflower oils, belonging to the Compositae family (510 and 400 μg tocopherol/g oil), which may result in great stability of the oil toward oxidation in spite of higher linoleic acid content than sunflower and safflower oils. Recently, Ramadan & Moersel (2002b) analyzed tocopherols from Indian niger crude seed oil, and the data agreed with Dutta et al. (1994), wherein α-tocopherols were major components but comprised only 44% of total tocopherol. More recently, oil extracted from niger seed samples from different regions in India and in Ethiopia was analyzed for the tocopherol distribution (Marini et al., 2003), and the results also agreed with Dutta et al. (1994). The total tocopherols content was similar for all samples and is significantly greater than that of the seed oils from the same botanical family.

## *Sterols*

Analysis of sterols (ST) sometimes provides a powerful tool for quality control of vegetable oils, and for the detection of oil as well as blends not recognized by the fatty acids profile. The common phytosterols, β-sitosterol, campesterol, stigmasterol, and Δ5-avenasterol were among the major components in niger seed oil (Dutta et al., 1994). The predominant component was β-sitosterol, comprising 38–43% of the total sterols (Table 9.4), followed by campesterol and stigmasterol, both comprising ca. 14%. The Δ5-avenasterol was present at a level of ca. 5–7%, and Δ7-avenasterol was present at ca. 4%. A small amount (0.2–0.4%) of cholesterol, identified by retention time only, was present in all the samples analyzed (Dutta et al., 1994).

**Table 9.4. Levels of Sterols and Tocopherols (% Composition) of Indian and Ethiopian Niger Seed Oil**

| Sterols | Ethiopian seed oil | Indian seed oil | Tocopherols | Ethiopian seed oil | Indian seed oil |
|---|---|---|---|---|---|
| Cholesterol | 0.3 | 0.3 | α-Tocopherol | 94-96 | 44.2 |
| Campesterol | 11.8-13.8 | 7.1-10.4 | β-Tocopherol | 1.0 | 17.0 |
| Stigmasterol | 13.4-14.2 | 6.7-13.2 | γ-Tocopherol | 3-5 | 29.2 |
| β-Sitosterol | 38.4-43.6 | 20.3-45.2 | δ-Tocopherol | — | 9.5 |
| Δ7-Stigmastenol | 11.3 | 11.7 | | | |
| Δ5-Avenasterol | 5.1-6.6 | 5.3-6.0 | | | |
| Δ7-Avenasterol | 4.1-4.6 | 1.6-4.5 | | | |
| Lanosterol | — | 1.1 | | | |

Marini et al. (2003), Dutta et al. (1994), Ramadan and Moersel (2002a,b; 2006).

High levels of ST (ca. 0.42% of TL) were measured in the *n*-hexane extract of niger seeds (Ramadan & Moersel, 2002a). The ST profile was in line with the results reported by Dutta et al. (1994), whereas the major ST were, in order of decreasing prevalence, β-sitosterol> campesterol> stigmasterol> Δ5-avenasterol> Δ7-avenasterol> lanosterol. The major component was β-sitosterol, constituting about 48% of ST content. Campesterol and stigmasterol were detected at approximately equal amounts in both samples, and comprised of 17% of total ST. β-sitosterol, campesterol, stigmasterol, and Δ5-avenasterol comprised together *ca* 90% of ST content. Small amounts of Δ7-avenasterol (ca. 3.9%) and lanosterol (ca. 2.7%) were identified in the analyzed samples. Marini et al. (2003) examined the sterol distribution of the oil extracted from niger seed samples coming from different regions in India and in Ethiopia. The sterol distribution of the Ethiopian oils resembled those of the samples analyzed by Dutta et al. (1994). In the Indian samples, the campesterol content (and to a lesser extent the percentages of β-sitosterol and Δ5-avenasterol) was significantly different in niger seeds grown in various regions. Numerous published results exist on sterols of some other oils of the Compositae family (e.g., sunflower and safflower). These oils had higher sitosterol content, at a range of 50–75% of total ST, compared to niger seed oils (38–43%).

### Vitamin $K_1$ and β-carotene

Niger seed oil was characterized by an extremely high level of Vitamin $K_1$ (more than 0.2% of TL) and β-carotene (ca. 0.06% of TL) (Ramadan & Moersel, 2002b). The Vitamin $K_1$ (Fig. 9.2) level is very low in most foods (<10 mg/100 g), and the majority of the vitamin is obtained from a few green and leafy vegetables (e.g., spinach and broccoli). Many studies show that some vegetable oils (especially soybean, cottonseed, and rapeseed oils) are important dietary sources of phylloquinone (Booth & Suttie, 1998; Gao & Ackman, 1995; Jakob & Elmadfa, 1996; Koivu et al., 1999; Piironen et al., 1997). Among edible oils, the best sources of phylloquinone were rapeseed oil (ca. 1.5 µg/g) and soybean oil (ca. 1.30 µg/g). Sunflower oil was the poorest source (ca. 0.10 µg/g) of phylloquinone (Piironen et al., 1997). In the soft margarines with 80% fat, the phylloquinone levels were 0.89–1.1 µg/g. Blended and hard margarines contained less phylloquinone than the soft margarines with corresponding fat contents (Piironen et al., 1997). Its levels are also moderate in olive oil (Cook et al., 1999; Ferland & Sadowski, 1992; Jakob & Elmadfa, 1996; Piironen et al., 1997; Shearer et al., 1996). A recent study reported that in olive oil, the mean content of phylloquinone ranged from 12.7 to 18.9 µg/100 g while in human plasma, phylloquinone content varied between 0.22 and 0.56 ng/mL (Otles & Cagindi, 2007). The addition of phylloquinone-rich oils in the processing and cooking of foods that are otherwise poor sources of Vitamin K (for example, peanut and corn oils) makes them potentially important dietary sources of the vitamin.

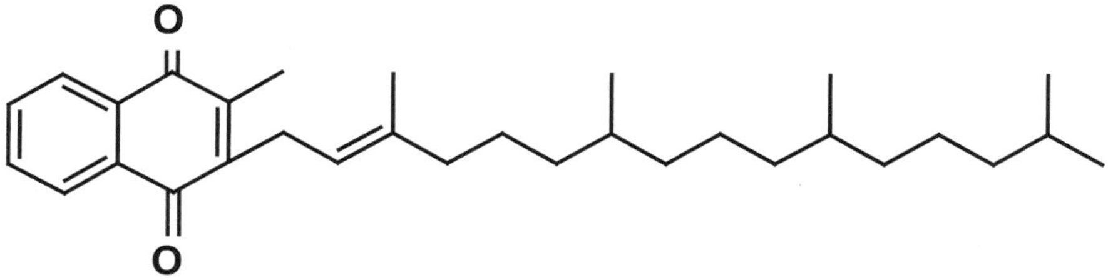

Fig. 9.2. Vitamin K1 (phylloquinone).

### *Phenolic Compounds*

Low levels of phenolic compounds (5 mg/kg oil) were determined in crude niger seed oil. To assist in characterizing phenolic compounds, absorption was scanned between 200 and 400 nm. The UV spectra of methanolic solutions of niger seeds displayed one maximum at 280 nm. The absorption maximum at this shorter wavelength (280 nm) may be due to the presence of *p*-hydroxybenzoic acid and flavone/flavonol derivatives (Ramadan et al., 2003).

## Health Benefits of the Oil and Oil Constituents

The high level of Vitamin $k_1$ may be the most unique health-promoting characteristic of niger seed oil. The perceived importance of dietary Vitamin K recently increased. Vitamin K is a fat-soluble vitamin that functions as a coenzyme and is involved in the synthesis of numerous proteins participating in blood clotting and bone metabolism (Damon et al., 2005). Vitamin K also plays a role as a co-factor for post-translational carboxylation of specific glutamate residues to gamma-carboxyglutamate residues in several blood coagulation factors and coagulation inhibitors in the liver; as well as a variety of extra hepatic proteins such as the bone protein osteocalcin (Shearer, 1992). Vitamin K's importance as a blood-clotting agent is well known. Moreover, Vitamin K may play a variety of health-promoting roles. Vitamin K reduces the risk of heart disease, kills cancer cells, and enhances skin health and may have antioxidant properties (Otles & Cagindi, 2007). A recent study concluded also that high phylloquinone intakes are markers of a dietary and lifestyle pattern that is associated with lower coronary heart disease (CHD) risk in men (Erkkilä et al., 2007).

Niger seed oil appears to be nutritionally valuable, as the high content of linoleic acid is thought to prevent cardiovascular diseases and is a precursor of structural components of plasma membranes and of some metabolic regulatory compounds (Vles & Gottenbos, 1989). Even less information is available on the associations of linoleic acid with CHD than is available for PUFA intake overall. Linoleic acid lowers LDL cholesterol with minimal effects on HDL cholesterol (Mensink & Katan, 1992). Based on experimental or in vitro studies, linoleic acid and other n-6 fatty acids are even purported to be proinflammatory or prothrombotic (Calder, 2001). Linoleic acid may also decrease arrhythmias (Charnock et al., 1991) and improve insulin sensitivity (Erkkilä et al., 2008; Laaksonen et al., 2002).

The levels of total tocopherols were significantly higher than that estimated in other oils rich in linoleic acid: this identifies the oil as nutritionally valuable. Furthermore, it is a good source of Vitamin E because almost all of the tocopherols are α-tocopherol (Dutta et al., 1994; Marini et al., 2003; Ramadan & Moersel, 2002 a,b).

Other health benefits were also attributed to niger seed oil. Rao (1994) reported that niger seed oil can be used to treat rheumatism. In addition, niger sprouts mixed with garlic are used to treat coughs (Getinet & Teklewold, 1995).

Recently, radical scavenging activity (RSA) tests were used to evaluate the health impact of many bioactive compounds found in foods. The radical scavenging properties of the crude niger seed oil were compared to those of other common as well as nonconventional vegetable oils toward two different stable free radicals (DPPH and galvinoxyl) (Ramadan et al., 2003; Ramadan & Moersel, 2006). Both ESR and spectrophotometric assays showed similar trends in the quenching of free radicals. The results showed that crude niger seed oil had the weakest antiradical activity. After a one-hour incubation, 35% of DPPH radicals was quenched by coriander seed oil, while niger seed oil was able to quench 14%. ESR measurements also showed a similar pattern, coriander and niger crude seed oils quenched 32.4 and 12.8% of galvinoxyl radical, respectively (Ramadan et al., 2003). Generally, the higher the degree of unsaturation of an oil, the more susceptible it is to oxidative deterioration. The low RSA of niger seed oil could be partly explained by the fact that niger seed oil has a high level of PUFA. Although, the amounts of tocopherols in niger seed oil were higher than other vegetable oils, no correlation was noted between RSA and the levels of tocopherols in oils. The Vitamin E scavenging effect is probably overwhelmed by the amount of radicals formed from PUFA, which is reflected in the highest initial peroxide value of niger seed oil.

RSA of niger seed oil fractions (NL, GL, and PL) was also compared. The results revealed that the PL fraction had the strongest antiradical action followed by GL and NL, respectively (Ramadan et al., 2003). The radical quenching property of GL was expected to be due to reducing sugars in all GL components and the sterol moiety in steryl glucoside. Moreover, less polar phenolic compounds that were extracted with GL may be responsible for the strong antiradical action. On the other hand, four postulates were proposed to explain the antioxidant activity of PL: (i) synergism between PL and tocopherols; (ii) chelation of pro-oxidant metals by phosphate groups; (iii) formation of Maillard-type products between PL and oxidation products, and (iv) action as an oxygen barrier between oil/air interfaces.

## Other Issues

### *Oxidative Stability*

Oxidative stability (OS) of stripped (refined) and crude niger oil was studied during 21 days under accelerated oxidative conditions at 60°C (Ramadan & Moersel, 2004). Peroxide value (PV), ansidine value (AV), and UV absorptivity were determined to monitor lipid oxidation during the experiment. The crude oil had a much lower PV than that of stripped oil over the entire storage period. PV in crude oil remained slightly increased over 21 days, whereas the peroxides accumulated in the stripped oils to high levels. Niger seed oil in comparison with coriander and black cumin seed oils had the highest AV and was oxidized rapidly. Absorptivity at 232 nm and 270 nm, due to the formation of primary and secondary compounds of oxidation, showed a pattern similar to that of the PV. The high content of conjugated oxidative products in niger oil was attributed to its high linoleic acid content which is readily decomposed to form conjugated hydroperoxides. Thus, the lower OS of niger seed oil could be partly explained by the fact that niger seed oil has a high proportion of PUFA. Aside from fatty acid profile, factors, such as oxygen concentration, metal contaminants, lipid hydroxy compounds, enzymes, and light may also influence the OS of the oil.

### *Authenticity*

Despite the promising features of niger seed oil, only a few articles reported comprehensive analyses of the lipid composition of niger seed oil. Dutta et al. (1994) extended their investigation to characterize the tocopherol and sterol distribution of niger oils, but they considered only a few samples from Ethiopia. Marini et al. (2003) examined the fatty acid, sterol, and triglyceride distribution and the total tocopherol content of the oil extracted from niger seed samples coming from different regions in India and in Ethiopia. Moreover, they studied the possibility of elaborating the analytical data by supervised pattern recognition methods; in particular Linear Discriminant Analysis (LDA) and Artificial Neural Network (ANN), in order to build a classification model that could discriminate the geographical origin of the samples. The results showed that 8 and 11 variables were necessary to achieve a complete discrimination of the country of origin and of the region of origin, respectively, of the oils under examination, when using LDA, whereas ANN required a smaller number of experimental variables (4 and 6), due to its nonlinearity.

## References

Alemaw, G.; A.T. Wold. An agronomic and seed-quality evaluation of noug (*Guizotia abyssinica* Cass.) germplasm in Ethiopia. *Plant Breed.* **1995**, *114*, 375–376.

Baagø, E.J. The genus *Guizotia* (Compositae). A taxonomic revision. *Bot. Tidsskr.* **1974**, *69*, 1–39.

Bhagya, S.; M.C.S. Sastry. Chemical, functional and nutritional properties of wet dehulled niger (*Guizotia abyssinica* Cass.) seed flour. *Lebensm.-Wiss. u.-Technol.* **2003**, *36*, 703–708.

Bhardwaj, S.P.; R.K. Gupta. Tilangi, a potential rich yielding oil seed crop. *Indian Farming* **1977**, *27*, 18–19.

Bockisch, M. Vegetable fats and oils. *Fats and Oils Handbook;* AOCS Press: Champaign, IL, 1998; pp. 301–303.

Booth, S.L.; J.W. Suttie. Dietary intake and adequacy of vitamin K. *J. Nutr.* **1998**, *128*, 785–788.

Calder, P.C. Polyunsaturated fatty acids, inflammation, and immunity. *Lipids* **2001**, *36*,1007–1024.

Charnock, J.S.; K. Sundram; M.Y. Abeywardena; P.L. McLennan; D.T. Tan. Dietary fats and oils in cardiac arrhythmia in rats. *Am. J. Clin. Nutr.* **1991**, *53*,1047S–9S.

Cherry, J.P.; W.H. Kramer. Plant sources of lecithin. *Lecithins: Sources, Manufacture and Uses;* F.B. Szuhaj, Ed.; AOCS Press: Champaign, IL,1989; pp. 16–33.

Cook, K.K.; G.V. Mitchell; E. Grundel; J.I. Rader. HPLC analysis for transvitamin $K_1$ and dihydro-vitamin $K_1$ in margarines and margarine-like products using C30 stationary phase. *Food Chem.* **1999**, *67*, 79–88.

Damon, M.; N.Z. Zhang; D.B. Haytowitz; S.L. Booth. Phylloquinone (vitamin $K_1$) content of vegetables. *J. Food Comp. Anal.* **2005**, *18*, 751–758.

Dange, K.; A. Jonsson. Oil content and fatty acid composition of seeds of *Guizotia* Cass (compositae). *J. Sci. Food Agric.* **1997**, *73*, 274–278.

Dutta, P.C.; S. Helmersson; E. Kebedu; A. Getinet; L. Appliqvist. Variation in lipid composition of niger seed (*Guizotia abyssinica* Cass) samples collected from different regions in Ethiopia. *J. Am. Oil Chem. Soc.* **1994**, *71*, 839–843.

Eklund, A. Biological evaluation of protein quality and safety of a lipoprotein concentrate from niger seed (*Guizotia abyssinica* Cass.). *Acta Physiol. Scand.* **1971a,** *82*, 229–235.

Eklund, A. Preparation and chemical analyses of a lipoprotein concentrate from niger seed (*Guizotia abyssinica* Cass.). *Acta Chem. Scand.* **1971b**, *25*, 2225–2231.

Eklund, A. Some chemical and biological properties of a protein fraction from nigerseed (*Guizotia abyssinica* Cass.) soluble in hot aqueous ethanol. *Acta Physiol. Scand.* **1974**, *90*, 602–608.

Erkkilä, A., D.F. de Mello V; U. Riserus; D.E. Laaksonen. Dietary fatty acids and cardiovascular disease: An epidemiological approach. *Progress Lipid Res.* **2008**, *47*, 172–187.

Erkkilä, A.T.; S.L. Booth; F.B. Hu; P.F. Jacques; A.H. Lichtenstein. Phylloquinone intake and risk of cardiovascular diseases in men. *Nutrition, Metab. Cardiovasc. Dis.* **2007**, *17*, 58–62.

Ferland, G.; J.A. Sadowski. Vitamin $K_1$ (phylloquinone) content of edible oils: effects of heating and light exposure. *J. Agric. Food Chem.* **1992**, *40*, 1869–1873.

Francis, C.M.; M.C. Campbell. *New High Quality Oil Seed Crops for Temperate and Tropical Australia.* A report for the Rural Industries Research and Development Corporation, RIRDC Publication No 03/045, ISBN 0642 586136; 2003.

Gao, Z.H.; R.G. Ackman. Determination of vitamin $K_1$ in canola oils by high performance liquid chromatography with menaquinone-4 as internal standard. *Food Res. Inter.* **1995**, *28*, 61–69.

Getinet, A.; A. Teklewold. An agronomic and seed-quality evaluation of niger (*Guizotia abyssinica* Cass.) germplasm grown in Ethiopia. *Plant Breed.* **1995**, *114*, 375–376.

Gunstone, F.D. The major source of oils, fats and other lipids. *Fatty Acid and Lipid Chemistry*; F.D. Gunstone, Ed.; Blackie Academic & Professional: Glasgow; 1996; pp. 61–86.

Jakob, E.; I. Elmadfa. Rapid and simple HPLC analysis of vitamin K in food, tissue and blood. *Food Chem.* **2000**, *68*, 219–221.

Kachapur, M.D.; A.S. Hasimani; K.S.K. Sastry. Influence of presowing seed hardening on early growth of niger (*Guizotia abyssinica* Cass.). *Curr. Res.* **1978**, *7*, 86–87.

Koivu, T.; V. Piironen; A.-M. Lampi; P. Mattila. Dihydrovitamin $K_1$ in oils and margarines. *Food Chem.* **1999**, *64*, 411–414.

Laaksonen, D.E.; T.A. Lakka; H.M. Lakka; K. Nyyssonen; T. Rissanen; L.K. Niskanen. Serum fatty acid composition predicts development of impaired fasting glycaemia and diabetes in middle-aged men. *Diabetic Med.* **2002**, *19*, 456–464.

Marini, F.; A.L. Magrì; D. Marini; F. Balestrieri. Characterization of the lipid fraction of niger seeds (*Guizotia abyssinica cass.*) from different regions of Ethiopia and India and chemometric authentication of their geographical origin. *Eur. J. Lipid Sci. Technol.* **2003**,*105*, 697–704.

Mensink, R.P.; M.B. Katan. Effect of dietary fatty acids on serum lipids and lipoproteins. A meta-analysis of 27 trials. *Arterioscler. Thromb.* **1992**, *12*, 911–919.

Mishra, J.S.; B.T.S. Moorthy; M. Bhan; N.T. Yaduraju. Relative tolerance of rainy season crops to field dodder (*Cuscuta campestris*) and its management in niger (*Guizotia abyssinica*). *Crop Protection* **2007**, *26*, 625–629.

Nasirullah, M.T.; S. Rajalakshmi; K.S. Pashupathi; K.N. Ankaiah; S. Vibhakar; M.N. Krishna Murthy; K.V. Nagaraja; O.P. Kapur. Studies on niger (*Guizotia abcyssinica*) seed oil. *J. Food Sci. Technol.* **1982**, *19*, 147–149.

Otles, S.; O. Cagindi. Determination of vitamin $K_1$ content in olive oil, chard and human plasma by RP-HPLC method with UV-Vis detection. *Food Chem.* **2007**, *100*, 1220–1222.

Piironen, V.; T. Koivu; O. Tammisalo; P. Mattila. Determination of phylloquinone in oils, margarines and butter by high-performance liquid chromatography with electrochemical detection. *Food Chem.* **1997**, *59*, 473–480

Quinn, J.; R.L. Myers. Nigerseed: Specialty grain opportunity for Midwestern US. *Trends in New Crops and New Uses;* J. Janick, A. Whipkey, Eds.;, ASHS Press: Alexandria, VA, 2002; pp. 174–182.

Ramadan, M.F.; L.W. Kroh; J.-T. Moersel. Radical scavenging activity of black cumin (*Nigella sativa* L.), coriander (*Coriandrum sativum* L.) and niger (*Guizotia abyssinica* Cass.) crude seed oils and oil fractions. *J. Agric. Food Chem.* **2003**, *51*, 6961–6969.

Ramadan, M.F.; J.-T. Moersel. Proximate neutral lipid composition of niger (*Guizotia abyssinica* Cass.) seed. *Czech J. Food Sci.* **2002a**, *20*, 98–104.

Ramadan, M.F.; J.-T. Moersel. Direct isocratic normal-phase assay of fat-soluble vitamins and beta-carotene in oilseeds. *Eur. Food Res. Technol.* **2002b**, *214*, 521–527.

Ramadan, M.F.; J.-T. Moersel. Analysis of glycolipids from black cumin (*Nigella sative* L.), coriander (*Coriandrum sativum* L.) and niger (*Guizotia abyssinica* Cass.) oilseeds. *Food Chem.* **2003a**,

*80*, 197–204.

Ramadan, M.F.; J.-T. Moersel. Determination of lipid classes and fatty acid profile of niger (*Guizotia abyssinica* Cass.) seed oil. *Phytochemical Anal.* **2003b**, *14*, 366–370.

Ramadan, M.F.; J.-T. Moersel. Oxidative stability of black cumin (*Nigella sativa* L.), coriander (*Coriandrum sativum* L.) and niger (*Guizotia abyssinica* Cass.) upon stripping. *Eur. J. Lipid Sci. Technol.* **2004**, *106*, 35–43.

Ramadan, M.F.; J.-T. Moersel. Screening of the antiradical action of Vegetable oils. *J. Food Comp. Anal.* **2006**, *19*, 838–842.

Rao, P.U. Nutrient composition of some less-familiar oil seeds. *Food Chem.* **1994**, *50*, 379–382

Riley, K.W.; H. Belayneh. Niger. *Oil Crops of the World*; G. Röbbelen, R.K. Downey, A. Ashri, Eds.; McGraw Hill: New York, 1989; pp. 394–403.

Seegeler, C.J.P. Oil plants. *Ethiopia: Their Taxonomy and Agricultural Significance.* Center for Agricultural Publishing and Documentation: Wageningen, The Netherlands, 1983; pp. 122–146.

Seher, A.; U. Gundlach. Isomeric monoenoic acids in vegetable oils. *Fette Seifen Anstrichm.* **1982**, *84*, 342–349.

Shahidi, F.; C. Desilva; R. Amarowicz. Antioxidant activity of extracts of defatted seeds of niger (*Guizotia abyssinica*). *J. Am. Oil Chem. Soc.* **2003**, *80*, 443–450.

Shearer, M.J. Vitamin K metabolism and nutriture. *Blood* **1992**, *6*, 92–104.

Shearer, M.J.; A. Bach; M. Kohlmeier. Chemistry, nutritional sources, tissue distribution and metabolism of vitamin K with special reference to bone health. *J. Nutr.* **1996**, *126*, 1181–1186.

Sugawara, T.; T. Miyazawa. Separation and determination of glycolipids from edible plant by high-performance liquid chromatography and evaporative light-scattering detections. *Lipids* **1999**, *34*, 1231–1237.

Vles, R.O.; J.J. Gottenbos. Nutritional characteristics and food uses of vegetable oils. *Oil Crops of the World;* G. Röbbelen, R.K. Downey, A. Ashri, Eds.; McGraw Hill: New York, 1989; pp. 63–86.

Weiss, E.A. *Oilseed Crops*, Second edn.; Blackwell Science: Malden, MA, 2000; pp. 259–273.

# 10

# Nigella (Black Cumin) Seed Oil

**Afaf Kamal-Eldin**
*Department of Food Science, Swedish University of Agricultural Sciences, 750 07 Uppsala, Sweden*

## Introduction

The annual herbaceous Nigella plants (family: Ranunculaceae) are spread throughout the Mediterranean and West Irano-Turanian countries. The seed of *Nigella sativa* (Linn.) is also known as black cumin or black caraway, although this species is not related to cumin or caraway, both belonging to the family Umbelliferaceae. This very aromatic seed is known in Arabian countries as Al-Haba-El-Sauda or Habat-Al-Barka. It is used as a spice in foods (bread, cheese, pickles, yogurt) and is highly appreciated for medicinal purposes (Riaz et al., 1996). The religious belief in the healthiness of this seed is very strong in the Islamic countries with reference to words attributed to prophet Mohammed that regard Al-Haba-El-Sauda as a panacea that cures every disease but death (Ali & Blunden, 2003). The seed has several bioactivities including antineoplastic (Salomi et al., 1992), emmenagogue and diuretic (Ballero & Fresu, 1993), oestrogenic (Agradi et al., 2002), analgestic (Bekemeier et al., 1967), as well as antibacterial, antifungal, and antiparasitic properties (Hanafy & Hatem, 1991; Hasan et al., 1989; Salomi et al., 1992).

*Nigella sativa* seeds contain approximately 3.8–7.0% moisture, 3.7–4.9% ash, 20.6–31.2% proteins, 23.5–34% carbohydrates, 34.5–42% fat, and 0.5–1.5% yellowish essential oil (Abdel-Aal & Attia, 1993; Atta, 2003; Babayan et al., 1978; Cheikh-Rouhou et al., 2007; Dandik & Aksoy, 1992; El-Dhaw & Abdel-Moneim, 1996; Gad et al., 1963; Nergiz & Ötles, 1993; Salem, 2001; Takruri & Dameh, 1998; Üstun et al., 1990 ). The seeds contain substantial amounts of a wide range of minerals including potassium, magnesium, calcium, phosphorus, sodium, iron, copper, zinc, and manganese (Takruri & Dameh, 1998; Cheikh-Rouhou et al., 2007). The seeds also contain reducing sugars, organic acids, mucilage, resins, tannins, melanin, glucosylated saponins, bitter components, alkaloids, melanthin resembling helleborin, melanthigenin and vitamins including ascorbic acid, thiamine, niacin, pyridoxine, and folic acid (Al-Gaby, 1998; Gilani et al., 2004; Takruri & Dameh, 1998). Most of the medicinal properties were attributed to alkaloidal quinines, but only thymoquinone has received research interest. Structures of main bioactive compounds that were identified in *N. sativa* are shown in Fig. 10.1.

## Oil Extraction and Processing

Commercially produced Nigella seed oil is generally obtained by cold pressing of raw or slightly roasted seeds. The oil production is minor and is limited to countries where it is cultivated mainly for its perceived in health benefits. One can also obtain the oil by solvent extraction, and this provides numerous advantages over cold pressing. The solvent-extracted and pressed oils differ in composition and properties (Table 10.1). In addition to a higher yield, solvent extraction provides oils with lower free fatty acids and peroxide values. The presence of an active nonspecific lipase in the seeds (Dandik & Aksoy, 1992, 1996; Mert et al., 1995) catalyzes the hydrolysis of 10–40% free fatty acids from the triacylglycerols during cold mechanical pressing while solvent extraction inactivates this lipase (Abdel-Aal & Attia, 1993; Atta, 2003; Babayan et al., 1978; Gad et al., 1963; Salem, 2001; Üstun et al., 1990). Cold pressing yields oils with lower free fatty acid content compared to hot pressing and pre-heating (50–80°C) of moist seeds prior to extraction may inactivate the lipolytic enzymes and consequently reduce the levels of free fatty acids in the oil (Üstun et al., 1990). The degree of enzymatic hydrolysis of free fatty acids is dependent on the grinding method, the storage time and temperature, the moisture content of the seeds, and the relative humidity during storage (Dauksas et al., 2002; Üstun et al., 1990). Indeed, the time between seed crushing and oil extraction should be minimal to obtain an oil with the lowest possible free fatty acids. On the other hand, solvent extraction provided brownish-yellow oil compared to golden-yellow oil in the case of cold pressing. Color measurements (L*, a*, b*) showed that the b* value for Nigella seed oil is higher than those of other vegetable oils indicating the presence of more yellow carotenoid pigments (Cheikh-Rouhou et al., 2007).

Extraction of the oil with pure supercritical carbon dioxide is inefficient as it only gives 10.6% oil, but the addition of 1% ethanol and optimization of pressure in-

**Table 10.1. Composition and Properties of Nigella sativa Seed Oils Obtained by Solvent Extraction and Pressing**

| Property | Cold pressing | Petroleum ether extraction |
|---|---|---|
| Oil yield (%) | 24.8 | 34.8 |
| Specific gravity (g/L) | 0.9110 | 0.9210 |
| Melting point (°C) | -1.7 | -3.3 |
| Lovibond color (R/Y/B) | 8/42/14 | 11/42/81 |
| Free fatty acids (%) | 11 | 6.7 |
| Peroxide value (meq.$O_2$/kg oil) | 13.5 | 10.7 |

Source: Atta (2003)

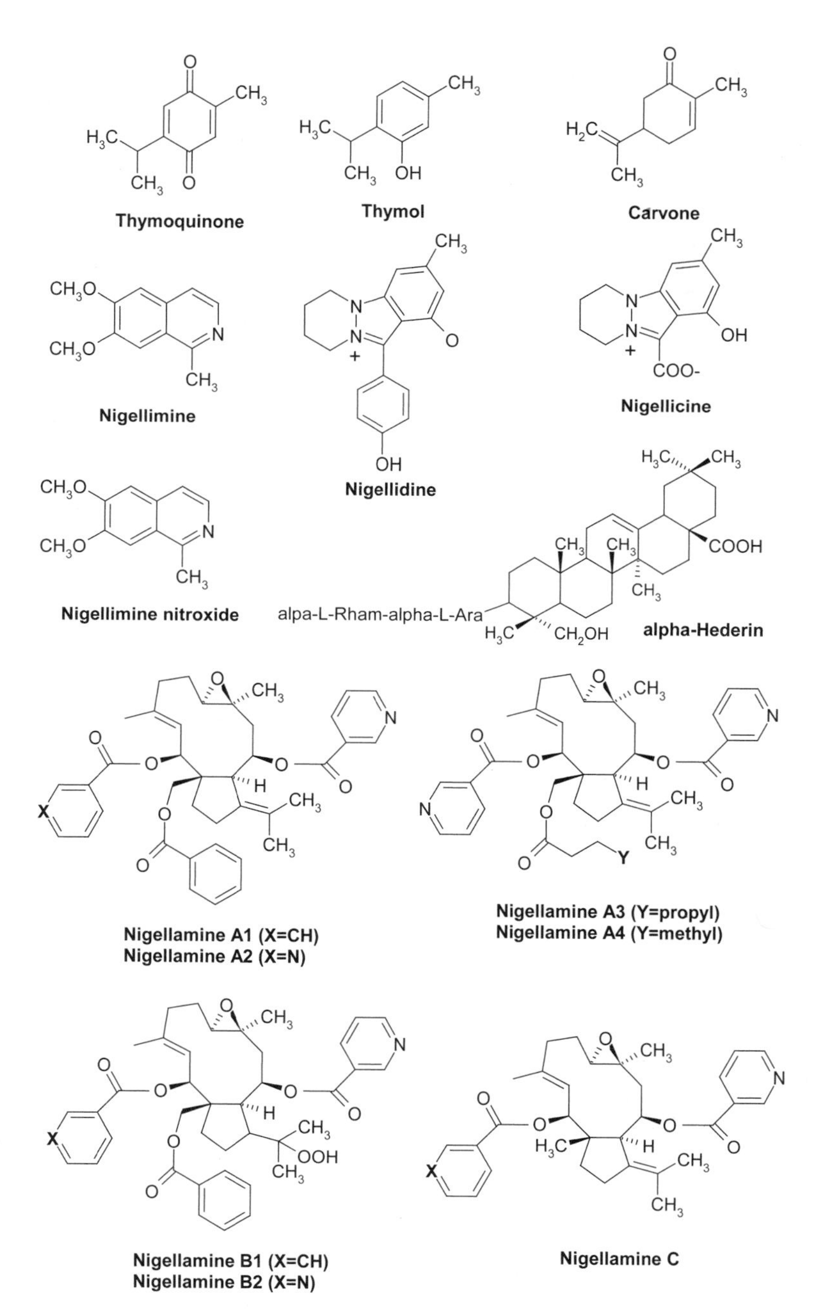


**Fig. 10.1.** Structures of some important bioactive compounds in the seed of *Nigella sativa.*

creased the yield to levels comparable to those obtained by solvent extraction (Dauksas et al., 2002). A few aroma compounds were characterized in this study, mainly α-, β-, γ-, and δ- elemenes and a major unknown volatile constituent. Since free fatty acids are unhealthy, they need to be removed if the oil is to be used for edible applications. This is generally achieved by chemical or physical refining for oils containing <10% free fatty acids (Türkay et al., 1996). Supercritical carbon dioxide extraction may provide an alternative processing method for the deacidification of Nigella seed oil (Dauksas et al., 2002; Türkay et al., 1996). Extraction with supercritical carbon dioxide at 15 MPa, and 60°C was found to reduce the acidity of the oil from 38 to 8% (Türkay et al., 1996).

Because it possesses an off-flavor, the Nigella seed cake (meal), remaining after oil extraction, has yet to be efficiently utilized into animal feeds. However, this meal might provide a wide range of bioactive compounds for medicinal uses. For example, Atta & Imaizumi (1998) showed that the ethanol extract of the seed has an antioxidant potential comparable to that of TBHQ. In fact, a wide range of antioxidant compounds was described in Nigella seeds (Ramadan, 2007).

## Composition of Acyl Lipids and Fatty Acids

The oil yield and fatty acid composition of the non-volatile oils of 10 different Nigella species (i.e., *N. arvensis* L., *N. damascene* L., *N. elata* Boiss, *N. lancifolia* Hub.-Mor.) *N. latisecta* P.H. Davis, *N. nigellastrum* (L) Willk., *N. orientalis* L., *N. oxypetala* L., *N. segetalis* Bieb., and *N. unguicularis* (Lam.) Spenner) from Turkey were studied (Kökdil & Yilmaz, 2005). The results (Table 10.2) showed a wide variation in the oil yield of the seeds (17.6–41.3%). Oils from all *Nigella* species contained the characteristic fatty acid, eicosadienoic acid (20:2 n-6), with considerable variation (1.9–9.4%). Among the ten oils, *N. nigellastrum* and *N. unguicularis* were unique because their oils contained considerable amounts of eicosenoic acid (22:1); 23.1% and 17.5%, respectively.

Ramadan and Moersel (2002a,b,c, 2003) studied in detail the lipid composition of *N. sativa* seed oil extracted with *n*-hexane or a mixture of chloroform and methanol (2:1, v/v). The composition of the hexane extracts is summarized in Table 10.3. The neutral lipids accounted for 97.2% of the crude extract followed by glycolipids (2.6%) and phospholipids (0.3%). The neutral lipids included triacylglycerols, diacylglycerols, sterol esters, and free fatty acids, with expected differences in fatty acid composition. For example, the percentage of palmitic acid (16:0) is much higher in the monoacylglycerol fraction, consistent with predominance of palmitic acid and linoleic acid (18:2) at the *sn*-1,3 and the *sn*-2 positions of the glycerol molecule, respectively. Except for the monoacylglycerol fraction, different lipid classes show comparable fatty acid profiles (Table 10.3).

The triacylglycerol species had the following composition: PPP (3.5%), SSS (3.8%), POP (1.5%), PLP + POO (17%), SLO + OOO (42.9%), and LLL (31.3%)

**Table 10.2. Oil Yield (via hexane extraction) and Fatty Acid Composition of Oils from Various Nigella Species**

| | | Relative fatty acid composition (%) | | | | | | |
|---|---|---|---|---|---|---|---|---|
| Species | Oil yield (%) | 14:0 | 16:0 | 18:0 | 18:1 | 18:2 | 20:2 | Others* |
| *N. arvensis* | 41.3 | 0.2 | 8.8 | 2.8 | 16.9 | 63.4 | 3.2 | 4.7 |
| *N. damascena* | 32.9 | 0.2 | 12.7 | 3.6 | 36.0 | 39.7 | 4.0 | 3.8 |
| *N. elata* | 36.3 | 0.2 | 11.2 | 3.3 | 22.9 | 47.4 | 8.4 | 6.6 |
| *N. lancifolia* | 24.3 | 0.2 | 7.2 | 2.9 | 16.0 | 67.6 | 3.5 | 2.6 |
| *N. latisecta* | 20.0 | 0.2 | 7.4 | 2.8 | 17.3 | 66.3 | 3.5 | 2.5 |
| *N. nigellastrum* | 32.1 | 0.1 | 5.9 | 3.4 | 23.6 | 31.2 | 9.4 | 26.4 |
| *N. orientalis* | 28.0 | 0.2 | 10.9 | 3.1 | 16.5 | 64.5 | 2.6 | 2.2 |
| *N. oxypetala* | 17.6 | 0.3 | 9.0 | 2.5 | 18.0 | 66.3 | 1.9 | 2.0 |
| *N. segetalis* | 37.5 | 0.1 | 6.7 | 2.7 | 15.8 | 69.5 | 3.2 | 2.0 |
| *N. unguicularis* | 36.5 | | 7.8 | 3.1 | 24.9 | 36.5 | 7.2 | 20.5 |
| *N. sativa* (Tunisia) | 28.5 | 0.4 | 18.4 | 2.8 | 25.0 | 50.3 | 0.3 | 2.8 |
| *N. sativa* (Iran) | 40.4 | 0.4 | 19.2 | 3.7 | 23.7 | 49.2 | 0.3 | 3.5 |

*Other fatty acids include 16:1, 17:1, 20:0, 18:3, 20:1, 22:0, 22:1 at levels of 0.1–0.5% in the different oils except 17:1 in *N. elata* (3.3%).

Source: All data from Kökdil and Yilmaz (2005) except *N. sativa* (Cheikh-Rouhou et al., 2007).

(Ramadan & Moersel, 2002a). Differential scanning calorimetric studies compared the melting thermograms of oils from Tunisian and Iranian Nigella seeds. The seed oils from both countries each exhibited two distinct melting peaks, suggesting polymorphism in Nigella seed oil (Cheikh-Rouhou et al., 2007). As mentioned above, the presence of the active lipase in Nigella seeds explains the high free fatty acid content in the oil (Dandik & Aksoy, 1992).

Six glycolipid classes were found in *N. sativa* hexane extract, dominated by digalactosyldiacylglycerol (DGDG) followed by glucocerebroside (CER) (Table 10.3). Linoleic acid was the major fatty acid in glycolipids followed by oleic acid, and glucose was the only sugar detected (Ramadan & Moersel, 2003). The phospholipid fraction contained a higher palmitic/linoleic acid ratio compared to the glycolipid and the neutral lipid fractions. The major phospholipid classes in this oil were phosphatidylcholine (46%), and phosphatidylethanolamine (25%) together with smaller amounts of phosphatidylserine (12.3%), phosphatidylinositol (9.6%), phosphatidylglycerol (1.5%), *lyso*-phosphatidylethanolamine (1.2%), and *lyso*-phosphatidylcholine (4.2%) (Ramadan & Moersel, 2002b).

## Nonacylglycerol Components in Nigella Oil

Various amounts of unsaponifiable materials, 0.7–5.4% range, were reported in different studies (Üstun et al., 1990 and references cited therein). The variation in this

**Table 10.3.** Fatty Acid Composition of Lipid Classes of *N. sativa* Seed Oil Extracted with *n*-Hexane

| Lipid class | Percent of total lipids | Relative distribution (%) | Relative fatty acid composition (%) | | | | |
|---|---|---|---|---|---|---|---|
| | | | 16:0 | 18:0 | 18:1 | 18:2 | 20:2 |
| **Total lipids** | **100.0** | | 13.0 | 3.2 | 24.1 | 57.3 | 2.4 |
| **Neutral lipids** | **97.2** | | 13.7 | 1.5 | 2.2 | 71.6 | 11.0 |
| Triacylglycerols | | 81.4 | 12.2 | 3.1 | 22.8 | 59.9 | 2.0 |
| Diacylglycerols | | 0.7 | 9.0 | - | 26.4 | 61.5 | 3.1 |
| Monoacylglycerols | | 0.4 | 32.5 | 0.6 | 7.0 | 53.4 | 6.5 |
| Free fatty acids | | 14.3 | 10.5 | 2.2 | 22.3 | 62.4 | 2.6 |
| Sterol esters | | 0.4 | 16.9 | 2.7 | 26.4 | 51.5 | 2.5 |
| **Glycolipids** | **2.6** | | 17.6 | 3.6 | 23.6 | 52.6 | 2.6 |
| Sterolglucosides | | 9.45 | 0 | 0 | 0 | 0 | 0 |
| Acylsterolglucosides | | 9.95 | 19.9 | 2.0 | 24.9 | 50.9 | 2.3 |
| Monogalactosyldiacyl-glycerol | | 7.88 | 20.1 | 1.6 | 22.9 | 52.7 | 2.0 |
| Digalactosyldiacyl-glycerol | | 55.6 | 16.9 | 2.3 | 26.1 | 51.7 | 2.3 |
| Cerebroside | | 11.9 | 20.1 | 2.0 | 27.2 | 48.5 | 2.2 |
| Sulfoquinovosydiacyl-glycerol | | 5.08 | 18.5 | 2.0 | 23.9 | 52.8 | 2.4 |
| **Phospholipids** | **0.3** | | 23.3 | 4.2 | 24.3 | 45.8 | 2.4 |
| Phosphatidylcholine | | 46.1 | 20.3 | 4.7 | 23.3 | 49.2 | 2.5 |
| Phosphatidyleth-anolamine | | 25.1 | 25.4 | 4.4 | 19.3 | 48.9 | 2.0 |
| Phosphatidylserine | | 12.3 | 35.6 | 3.0 | 18.2 | 41.2 | 2.0 |
| Phosphatidylinositol | | 9.56 | 29.9 | 3.7 | 23.5 | 40.7 | 2.2 |
| Phosphatidylglycerol | | 1.51 | 20.6 | 4.0 | 25.74 | 46.8 | 2.9 |
| *lyso*-phosphatidyleth-anolamine | | 1.2 | 24.9 | 2.5 | 22.3 | 47.6 | 2.7 |
| *lyso*-phosphatidylcho-line | | 4.23 | 20.9 | 4.7 | 25.1 | 46.5 | 2.8 |

Source: Ramadan and Moersel (2002a, b, 2003)

**Table 10.4.** Composition of Desmethyl Sterols Present as Free Sterols, Sterolglucosides and Acylated Sterolglucoside

| Sterol | Free sterols and sterol esters | Sterolglucosides | Acylatedsterolglucosides |
|---|---|---|---|
| Campesterol | 6.2 | - | - |
| Stigmasterol | 8.6 | 12.0 | 8.2 |
| Sitosterol | 34.1 | 35.3 | 20.8 |
| $\Delta^5$-Avenasterol | 27.9 | 9.0 | 5.0 |
| $\Delta^7$-Avenasterol | 22.0 | 43.7 | 66 |

Source: Ramadan and Moersel (2002a, 2003)

amount might relate to variety, agroclimatic condition, and method of oil extraction. *Nigella sativa* seed oil was found to contain 360 mg/100 g of total desmethyl sterols in which sitosterol (34.1%), $\Delta^5$-avenasterol (27.9%), and $\Delta^7$-avenasterol (22%) predominate while campesterol (6.2%), stigmasterol (8.6%), and lanosterol (1.2%) are present in relatively small amounts (Ramadan & Moersel, 2002a, 2003). This sterol composition is very different from that of most edible oils where sitosterol and campesterol are usually the predominant sterols. These sterols are mainly present in free form and to a smaller extent as sterolglucosides and acylated sterol glucosides (Table 10.4), mainly in the form of β-D-glucopyranoside. In acylsterolglucosides, the fatty acids are usually esterified at position 6 of glucose. Possibly, one can use the sterol pattern, e.g., the sitosterol/campesterol ratio, as a marker for the purity and authenticity of the oil (El-Hinnawy et al., 1983).

A sample of *N. sativa* seed oil extracted with *n*-hexane contained the following oil-soluble vitamins (in ppm): α-tocopherol (284), β-tocopherol (40), γ-tocopherol (225), δ-tocopherol (48), 2-methyl-3-phytyl-1,4-naphthochinon (vitaminK1) (1162), and β-carotene (593) (Ramadan & Moersel, 2002c). The UV spectra of methanolic solutions of black cumin phenolics exhibited one absorption maxima at 282 nm. Novel alkaloids (nigellicine), an isoquinoline alkaloid (nigellimine) and an indazole alkaloid (nigellidine), were isolated from the black cumin seeds (Atta-ur-Rahman et al., 1985, 1992, 1995).

## Flavor and Aroma Compounds

Solvent extraction of Nigella seeds gives a green oil extract with a strong aromatic flavor (Nickavar et al., 2003). Hydrodistillation of the extracted oil provides a volatile oil (also called essential oil) with 32 compounds constituting ca. 87% of this volatile fraction. The major constituents in this extract were phenyl propanoid compounds (46%, mainly *trans*-anethole (38%), monoterpene hydrocarbons (27%, mainly *p*-cymene 15%), while thymoquinone represented only 0.6% of this fraction.

Several volatile compounds were identified in Nigella oil including *n*-nonane, 3-methyl nonane, *n*-decane, *n*-tetradecane, *n*-pentadecane, *n*-hexadecane, *n*-octadecane, hexadec-1-ene, octadec-1-ene, tricos-9-ene, docos-1-ene, pentacos-5-ene, 2-methyltetracosane, 1-phenyldecan-2-one, 1-phenylhepta-2,4-dione, bis(3-chlorophenyl) ketone, henicosan-10-one, hexadecanoic acid, octadeca-9,12-dienoic acid, 2-methyl octadecanoic acid, methyl pentadecanoate, methyl hexadecanoate, methyl octadecanoate, ethyl octadecanoate, 12-methyltricosane, methyl hept-6-enoate, ethyl octadec-7-enoate, methyl octadec-15-enoate, 1,3,5-trimethylbenzene, 1-methyl-3-propylbenzene, 1-ethyl-2,3-dimethylbenzene, diethyl phthalate, and dibutyl phthalate (Nickavar et al., 2003). The active constituents of the seeds include the volatile oil constituents estragole, anisaldehyde, *trans*-anethole, myristician, dill apiole, apiole, β-bourbonene, carvacrol, carvone, dihydrocarvone, β-caryophyllene, *o*- and *p*-cymenes, α-, β-, γ-, and δ-elemenes, germacrene A, limonene, α-longipinene, longifolene, myrcene, myrtenol, α- and β-pinenes, sabinene, *trans*-sabinene hydrate, 7-epi-selinene, α-and γ-terpinene, thymoquinone, thujene, thujol, thymol, α-phellandrene, fenchone, terpinen-4-ol, p-cymene-8-ol (D'Autuono et al., 2002; Nickavar et al., 2003). Thymoquinone (45%), β-cymene (15.5%), carvacrol (7.2%), longifolene (7.2%), and 4-terpineol (3.1%) were identified as the predominant essential oil components in one study (Burits & Bucar, 2000), while *trans*-anethole (38.3%), β-cymene (14.8%), limonene (4.3%), and carvone (4%) were the major components reported in another study by Nickavar et al. (2003).

The levels of volatile components and their contributed aroma in Nigella seed oil depend on the seed variety and the extraction and processing conditions (Ghosheh et al., 1999). D'Antuono et al. (2002) reported significant within-species and between-species differences in the composition of the volatile components This might lead to significant variations in the medicinal and sensory quality of oils from seeds of various species and subspecies.

## The Oxidative Stability of Nigella Seed Oil

The oxidative stability of *N. sativa* seed oils varied tremendously. For example, the Rancimat Oxidative Index of two seed oil samples was found to vary between 12 and 66 hours (Cheikh-Rouhou et al., 2007). The oxidative stability of the two Nigella oils was higher than that of sunflower oil (7.7 h) and is more comparable to that of olive oil. The oxidative stability (Table 10.5) correlated positively, as expected, with total phenols, and negatively with iodine value, free fatty acids, viscosity, and chlorophyll in the oil (Cheikh-Rouhou et al., 2007). The radical scavenging capacity of crude black cumin seed oil and its fractions was confirmed using the stable galvinoxyl radical detected by electron spin resonance spectrometry (ESR) and 1,1-diphenyl-2-picrylhydrazyl (DPPH) radical, followed by color changes, measured spectrophotometrically (Ramadan et al., 2003). The essential oil components including thymoquinone but especially carvacrol may contribute to the antioxidant activity (Burits & Bucar, 2000). The influence of compositional factors on the stability of Nigella seed

**Table 10.5. Rancimat Oxidation Stability Index (OSI, 100°C) and Relevant Compositional Factors in Tunisian and Iranian Nigella Seed Oils Extracted with Hexane and Refined (RBD)**

| Property | Tunesian Nigella seed oil | Iranian Nigella seed oil |
|---|---|---|
| OSI (h) | 12 | 50.3 |
| Viscosity (mPa s) | 11 | 6 |
| Iodine value (g iodine/100 g oil) | 119 | 101 |
| Free fatty acids (%) | 22.7 | 18.6 |
| Chlorophyll (mg/kg oil) | 6 | 2.3 |
| Polyphenols (mg gallic acid/kg oil) | 245 | 309 |

Source: Cheikh-Rouhou et al. (2007).

oil needs further more elaborate studies, taking into account the prooxidant effects of free fatty acids, chlorophyll, and other components.

## Health Effects of the Nigella Seed Oil

The pressed oil of *N. sativa* is traditionally used topically to treat skin eruptions, paralysis, back pain, rheumatism, and other inflammatory diseases (Gilani et al., 2004). The oil is used in skin care as a lubricant, in moisturizers, and for treating conditions such as psoriasis and eczema or as a mixture with beeswax for the treatment of burns, infections, joint pains, and as an anti-wrinkle agent. The oil and derived thymoquinone were shown to illicit an anti-inflammatory effect by inhibiting cyclooxygenase, and 5-lipoxygenase induced eicosanoid generation as well as membrane lipid peroxidation in leukocytes (Houghton et al., 1995). The oil and its antioxidant thymoquinone were shown to inhibit lipid peroxidation in liposomes made using a mixture of brain phospholipids. An active constituent, 2-(2-methoxypropyl)-5-methyl-1,1,4-benenediol, as well as thymol and carvacrol showed inhibitory effects against arachidonic acid-induced platelet aggregation and blood coagulation (Enomoto et al., 2001).

Incorporation of the hexane extract of *N. sativa* seeds into a rat diet for 12 weeks was found to reduce their serum cholesterol, triglycerides, glucose, and the number of leukocytes and platelets and to increase hematocrit and hemoglobin levels (Zaoui et al., 2002a). In another study, *N. sativa* oil decreased glucose, prolactin, triacylglycerol, and cholesterol levels and increased hemoglobin level in healthy females (Ibraheim, 2002). Treatment of normal rats with *N. sativa* oil, equivalent to 2 g seed/kg body weight/day) for 4 weeks resulted in a transient initial weight loss as a result of sustained reduction in food but not water intake, a reduction in plasma triacylglycerol and plasma insulin levels, and an increase in HDL-cholesterol (Le et al., 2004). Thymoquinone was shown to reduce the blood levels of triacylglycerols, cholesterol, HDL, and LDL in rats (Bamosa et al., 2002), but whether any other active principles is fixed in the oil are responsible for the observed lipid-lowering effects is not known.

Two insulin-responsive kinases, namely MAPK p42/44erk and PKB/Akt, showed greater response to in vitro insulin stimulation in the hepatocytes of rats treated with Nigella oil. MAPK p44/42erk participates in cell proliferation and protein synthesis while PKB/Akt is important for glucose transport in the skeletal muscle (Le et al., 2004). Thus, *N. sativa* seed oil may help to increase insulin sensitivity and resistance to type 2 diabetes mellitus, rather than increase insulin release as suggested by Fararh et al. (2002). Despite the reducing effect on food intake observed in the last study, the authors claimed that the oil lacks toxicity, in agreement with others (El-Dakhakhny et al., 2002; Zaoui et al., 2002a,b).

Nigella seed oil was also shown to cause an increase in the activity of certain liver enzymes, namely alanine aminotransferase (ALT), aspartate aminotransferase (AST), alkaline phosphatase (ALP), and gamma glutamyltransferase and to stimulate the immune response to infections (Ibraheim, 2002; Mahmoud et al., 2002). The hepatoprotective effects of Nigella oil and thymoquinone were attributed to antioxidant mechanisms (Mansour et al., 2002). The hexane extract of *N. sativa* seeds caused a significant contraceptive activity in rats (Keshri et al., 1998).

### Allergic and Toxic Compounds in *N. sativa* Oil

In certain individuals, *N. sativa* oil was shown to induce irritant (Lee & Lam, 1989) and allergic contact dermatitis (Steinmann et al., 1997; Zeditz et al., 2002). The chemical components responsible for this effect are not yet identified. A need may exist for toxicology studies for Nigella oil, since it contains some unique components, whose safety is not known.

## Conclusion

The crude seed oil of *N. sativa* is a valuable source of essential fatty acids, phytosterols, and numerous bioactive compounds (Fig. 10.1) that might have important in nutritional applications. The potential of the oil is, however, underexplored and deserves further attention. Of special importance is the unique aromatic flavor, the nutritive value of the uncommon fatty acid, 20:2, and the identification of the different minor components in the oil and their health effects.

## References

Abdel-Aal, E.S.M.; R.S. Attia. Characterization of black cumin (*Nigella sativa*) seeds. 2-proteins. *Alex. Sci. Exch*. **1993**, *14*, 483–496.

Agradi, E.; G. Fico; F. Cillo; C. Franicisci; F. Tomè. Estrogenic activity of *Nigella damascene* extracts evaluated using a recombinant yeast screen. *Phytotherapy Res*. **2002**, *16*, 414416.

Ali, B.H.; G. Blunden. Pharmacological and toxicological properties of *Nigella sativa*. *Phytotherapy Res*. **2003**, *17*, 299–305.

Atta, M.B. Some characteristics of Nigella (*Nigella sativa* L.) seed cultivated in Egypt and its lipid profile. *Food Chem*. **2003**, *83*, 63–68.

Atta, M.B.; K. Imaizumi. Antioxidant activity of Nigella (*Nigella sativa* L.) seed extracts. *J. Jpn. Oil Chem. Soc*. **1998**, *47*, 475–480.

Atta-ur-Rahman; S. Malik; S.S. Hasan; M.I. Chaudhary; C.Z. Ni; J. Clardy. Nigellidine—A new indazole alkaloid from the seeds of *Nigella sativa*. *Tetrahedron Lett*. **1995**, *36*, 1993–1996.

Atta-ur-Rahman; S. Malik; C.H. He; J. Clardy. Isolation and structure determination of Nigellicine, a novel alkaloid from the seeds of *Nigella sativa*. *Tetrahedron Lett*. **1985**, *26*, 2759–2762.

Atta-ur-Rahman; S. Malik; K. Zaman. Nigellimine: A new isoquinoline alkaloid from the seeds of *Nigella sativa*. *J. Natural Products* **1992**, *55*, 676–678.

Babayan, V.K.; D. Koottungal; G.A. Halaby. Proximate analysis, fatty Acid, and amino acid composition of *Nigella sativa* L. seeds. *J. Food Sci*. **1978**, *43*, 1315–1319.

Ballero, M.; I. Fresu. Le Piante di uso Officinale nella Barbagia di Seui (*Sardegna Centrale*). *Fitoterapia* **1993**, *64*, 141–150.

Bamosa, A.O.; B. A. Ali; Z.A. Al-Hawsawi. The effect of thymoquinone on blood lipids in rats. *Indian J. Physiol. Pharmacol.* **2002**, *46*, 195–201.

Bekemeier, H.; G. Leuschner; W. Schmollack. Antipyretische, antiödematöse und analgetische wirkung von damascenin im vergleich mit acetylsalicylsäre und phenylbutazon. *Archives Int de Pharmacodynamie Thérapie* **1967**, *168*, 199211.

Burits, M.; F. Bucar. Antioxidant activity of *Nigella sativa* essential oil. *Phytotherapy Res.* **2000**, *14*, 323–328.

Cheikh-Rouhou, S.; S. Besbes; B. Hentati; C. Blecker; X. Deroanne; H. Attia. *Nigella sativa* L.: Chemical composition and physicochemical characteristics of lipid fraction. *Food Chem.* **2007**, *101*, 673–681.

D'Antuono, L.F.; A. Moretti; A.F.S. Lovato. Seed yield, yield components, oil content and essential oil content and composition of *Nigella sativa* L. and *Nigella damascene* L. *Ind. Crops Prod.* **2002**, *15*, 59–69.

Dandik, L.; H.A. Aksoy. The kinetics of *Nigella sativa* (black cumin) seed oil catalyzed by native lipase in ground seed. *J. Am. Oil Chem. Soc.* **1992**, *69*, 1239–1241.

Dandik, L.; H.A. Aksoy. Applications of *Nigella sativa* seed lipase in oleochemical reactions. *Enz. Microb. Technol.* **1996**, *19*, 277–281.

Dauksas, E.; P.R. Venskutonis; B. Sivik. Comparison of oil from *Nigella damascene* seed recovered by pressing, conventional solvent extraction and carbon dioxide extraction. *J. Food Sci.* **2002**, *67*, 1021–1024.

El-Dakhakhny, M.; N. Mady; N. Lembert; H.P. Ammon. The hypoglycemic effect of *Nigella sativa* oil is mediated by extrapancreatic actions. *Planta Med.* **2002**, *68*, 465–466.

El-Dhaw, Z.Y.; N.M. Abdel-Muneim. Chemical and biological values of black cumin seeds. *J. Agric. Sci. Mansoura Univ.* **1996**, *21*, 4149–4159.

El-Hinnawy, S.E.; M.A. Torki; Z.A. El-Hadidy; A.R. Khalil. Studies on the unsaponifiable matters in some vegetable oils. *Annals Agric. Sci. Fac. Agric. Ain Shams Univ.* **1983**, *28*, 395415.

Enomoto, S.; R. Asano; Y. Iwahori; T. Narui; Y. Okada; A.N. Singab; T. Okuyama. Hematologic studies on black cumin oil from the seeds of *Nigella sativa*. *Biol. Pharmaceut. Bull.* **2001**, *24*, 307–310.

Fararh, K.M.; Y. Atoji; Y. Shimizu; T. Takewaki. Insulinotropic properties of *Nigella sativa* in streptozotocin plus nicotinamide diabetic hamsters. *Res. Vet. Sci.* **2002**, *73*, 279–282.

Gad, A.M.; M. El-Dakhakhny; M.M. Hassan. Studies on the chemical composition of Egyptian *Nigella sativa* L. oil. *Planta Med.* **1963**, *11*, 134136.

Gilani, A.H.; Q. Jabeen; M.A.U. Khan. A review of medicinal uses and pharmacological activities of *Nigella sativa*. *Pakistan J. Biol. Sci.* **2004**, *7*, 441–451.

Hanafy, M.S.; M.E. Hatem. Studies on the antimicrobial activity of *Nigella sativa* seed (black cumin). *J. Ethnopharmacol.* **1991**, *34*, 275–278.

Hasan, C.M.; M. Ahsan; N. Islam. *In vitro* antibacterial screening of the oils of *Nigella sativa* seeds. *Bangladesh J. Bot.* **1989**, *18*, 171–174.

Houghton, P.J.; R. Zarka; B. De-las-Heras; J.R. Hoult. Fixed oil of *Nigella sativa* and derived thymoquinone inhibit eicosanoid generation in leukocytes and membrane lipid peroxidation. *Planta Med.* **1995**, *61*, 33–36.

Ibraheim, Z.Z. Effect of *Nigella sativa* seeds and total oil on some blood parameters in female doxorubicinvolunteers. *Saudi Pharmaceit. J.* **2002**, *10*, 54–59.

Keshri, G.; V. Lakshmi; M.M. Singhe. Postcoital contraceptive activity of some indigenous plants in rats. *Contraception* **1998**, *57*, 357–360.

Kökdil, G.; H. Yilmaz. Analysis of the fixed oils of the genus *Nigella* L. (Ranunculaceae) in turkey. *Biochem. Sys. Ecol.* **2005**, *33*, 1203–1209.

Le, P.M.; A. Benhaddou-Andaloussi; A. Elimadi; A. Settaf; Y. Cherrah; P.S. Haddad. The petroleum ether extract of *Nigella sativa* exerts lipid-lowering and insulin-sensitizing actions in the Rrt. *J. Ethnopharmacol.* **2004**, *94*, 251–259.

Lee, T.Y.; T.H. Lam. Irritant contact dermatitis due to the herbal oil, black man oil. *Contact Dermatitis* **1989**, *20*, 229–230.

Mahmoud, M.R.; H.S. El-Abhar; S. Saleh. The effect of *Nigella sativa* oil against the liver damage induced by *Schistosoma mansoni* infection in mice. *J. Ethnopharmacol.* **2002**, *79*, 1–11.

Mansour, M.A.; M.N. Nagi; A.S. El-Khatib; A.M. Al-Bekairi. Effects of thymoquinone on antioxidant enzyme activities, lipid peroxidation and DT-diaphorase in different tissues of mice: A possible mechanism of action. *Cell Biochem. Funct.* **2002**, *20*, 143151.

Mert, S.; L. Dandik; H.A. Aksoy. Production of glycerides from glycerol and fatty acids by native lipase of *Nigella sativa* seed. *Appl. Biochem. Biotechnol.* **1995**, *50*, 333–342.

Nergiz, C.; S. Ötles. Chemical composition of *Nigella sativa* L. seeds. *Food Chem.* **1993**, *48*, 259–261.

Nickavar, B.; M. Mojab; K. Javidnia; M.A.R. Amoli. Chemical composition of the fixed and volatile oils of *Nigella sativa* L. from Iran. *Naturforsch* **2003**, *58c*, 629–631.

Ramadan, M.F. Nutritional value, functional properties and nutraceutical applications of black cumin (*Nigella sativa* L.): An overview. *Intern. J. Food Sci. Technol.* **2007**, *42*, 1208–1218.

Ramadan, M.F.; L.W. Kroh; J.T. Moersel. Radical scavenging activity of black cumin (*Nigella sativa* L.), coriander (*Coriandrum sativum* L.) and niger (*Guizotia abyssinica* Cass.) crude seed oils and oil fractions. *J. Agric. Food Chem.* **2003**, *51*, 6961–6969.

Ramadan, M.F.; J.T. Moersel. Neutral lipid classes of black cumin (*Nigella sativa* L.) seed oils. *Eur. Food Res. Technol.* **2002a**, *214*, 202–206.

Ramadan, M.F.; J.T. Moersel. Characterization of phospholipid composition of black cumin (*Nigella sative* L.) seed oil. *Nahrung* **2002b**, *46*, 240–244.

Ramadan, M.F.; J.T. Moersel. Direct isocratic normal-phase assay of fat-soluble vitamins and beta-carotene in oilseeds. *Eur. Food Res. Technol.* **2002c**, *214*, 521–527.

Ramadan, M.F.; J.T. Moersel. Analysis of glycolipids from black cumin (*Nigella sative* L.), coriander (*Coriandrum sativum* L.) and niger (*Guizotia abyssinica* Cass.) oilseeds. *Food Chem.* **2003**, *80*, 197–204.

Riaz, M.; M. Syed; F.M. Chaudhary. Chemistry of the medicinal plants of the genus *Nigella*. *Hamdard Medicus* **1996**, *39*, 40–45.

Salem, M.L.; M.S. Hossain. *In vivo* acute depletion of CD8(+) T cells before murine cytomegalovirus infection upregulated innate antiviral activity of natural killer cells. *Int. J. Immunopharmacol.* **2000**, *22*, 707–718.

Salomi, N.J.; S.C. Nair; K.K. Jayawardhanan; C.D. Varghese; K.R. Panikkar. Antitumour principles from *Nigella sativa* seeds. *Cancer Lett.* **1992**, *63*, 41–46.

Steinman, A.; M. Schätzle; M. Agathos; R. Berit. Allergic contact dermatitis from black cumin (*Nigella sativa*) oil after topical use. *Contact Dermatitis* **1997**, *36*, 268–269.

Takruri, H.R.H.; M.A.F. Dameh. Study of the nutritional value of black cumin seeds (*Nigella sativa*). *J. Sci. Food Agric.* **1998**, *76*, 404–410.

Türkay, S.; M.D. Burford; M.K. Sangün; E. Ekinic; K.D. Bartle; A.A. Clifford. Deacidification of black cumin seed oil by selective supercritical carbon dioxide extraction. *J. Am. Oil Chem. Soc.* **1996**, *73*, 12651270.

Üstun, G.; L. Kent; N. Cekin; H. Clvelekoglu. Investigation of the technological properties of *Nigella sativa* (black cumin) seed oil. *J. Am. Oil Chem. Soc.* **1990**, *67*, 958–960.

Zaoui, A.; Y. Cherrah; K. Alaoui; N. Mahassine; H. Amarouch; M. Hassar. Effects of *Nigella sativa* fixed oil on blood homestasis in rat. *J. Ethnopharmacol.* **2002a**, *79*, 23–26.

Zaoui, A.; Y. Cherrah; N. Mahassine; K. Alaoui; H. Amarouch; M. Hassar. Acute and chronic toxicity of *Nigella sativa* fixed oil. *Phytomedicine* **2002b**, *9*, 69–74.

Zeditz, S.; R. Kaufmann; H. Boehncke. Allergic contact dermatitis from black cumin (*Nigella sativa*) oil-containing ointment. *Contact Dermatitis* **2002**, *46*, 188.

# 11

# Camellia Oil and Tea Oil

**Kevin Robards[1], Paul Prenzler[1], Danielle Ryan[1], and Haiyan Zhong[2]**
*[1]School of Agricultural and Wine Sciences, Charles Sturt University, Locked Bag 588, Wagga Wagga 2678, Australia; [2]Faculty of Food Science and Engineering, Central South University of Forestry and Technology, Changsha 410004, Hunan, P. R. China*

## Introduction

The camellia is valued not only for its aesthetic contribution as an ornamental tree or shrub but also for its economic importance as it provides the beverage, tea, and edible oil in some countries, notably China. Many countries issued postage stamps honoring the importance of the camellia. The first to do so was Portuguese India in 1898 followed by Japan, China, Albania, Belgium, Haiti, Korea, Vietnam, Poland, Rwanda, the United States, France, and New Zealand (Rolfe, 1992).

The genus *Camellia* was named in 1735 by the Swedish botanist and physician, Carolus Linnaeus, to honor a Jesuit apothecary and naturalist, Georg Josef Kamel. Genus *Camellia* belongs to the botanical tribe Gordonieae, within the family Theaceae. This tribe is characterized by the formation of seeds within a capsule. Linnaeus listed two species of the genus, *C. japonica* and *C. sinensis,* but the botanical hierarchy expanded considerably since the 1730s as the number of species named by botanists now totals more than 250 (Savige, 1993). Camellia are evergreen flowering trees and shrubs native to a number of areas throughout China, Southeast Asia, and Japan. The various species are characterized according to floral and leaf characteristics.

One species that is very familiar internationally is *C. sinensis,* the tea plant. Black, green, jasmine, and oolong teas are produced from the dried leaves of this species. The English word "tea" is derived from the Chinese character, pronounced as tê in the Min Nan dialect. Indeed, camellia first arrived in the West early in the seventeenth century in the form of *C. Sinensis,* thus enabling preparation of the beverage tea. The demand for this tea plant ultimately led to the interest in ornamental species such as *C. japonica*. Many of the *C. sinensis* plants that arrived in Europe at this time were, in fact, *C. japonica* (Rolfe, 1992). This substitution possibly was a result of language confusion or a deliberate swap. However, it led to the growth in popularity of the plant as an ornamental shrub and to its distribution throughout Europe, the United States, and Australia.

Apart from its uses for ornamental purposes and production of tea, camellia is one of the four main oil-bearing trees (palm, coconut, olive, and tea) in the world (Anon., 2007a). The oil from camellia seeds has long been important in China and Japan, to a lesser extent. Curiously, the introduction of camellia oil into regions outside China was delayed by some three centuries after introduction of the ornamental and tea-producing varieties. The growing recognition of the nutritional value of olive oil led to a dramatic increase in its consumption in nontraditional areas outside the Mediterranean. Similarly, camellia oil is now associated with a healthy image, and it is reasonable to expect a similar upturn in its use. Interest has emerged in the commercial production of camellia oil in Western nations, such as the United States and Australia.

All camellia species contain oil, but *Camellia oleifera* was probably the earliest species exploited for its edible oil, with its use in China dating back for more than 1000 years (Shanan & Ying, 1982). Extensive trials were conducted in various provinces of China in the 1960s and 1970s (Fang, 1994), and varietal selections were made for superior fruit production. A lack of funding halted Chinese research in about 1990 although it has now resumed, and research is once again underway (Niankang et al., 1996), with a number of varieties now being cultivated by selective breeding for oil production. Research is also being conducted in the United States (Ruter, 2002) and in Australia jointly with China (Zhong et al., 2006, 2007ab). The oil produced from camellia is variously known as camellia oil, oil-tea camellia, tea oil, tea oil camellia, and tea seed oil. It was used as an edible oil in China but also for cosmetic and medicinal purposes in both China and Japan. Don't confuse camellia or tea oil with tea tree oil, an essential oil extracted from the leaves of the paperbark, *Melaleuca alternifolia*.

China currently is the only country with a significant production of camellia oil. Oil-tea camellia (*Camellia oleifera* Abel) is a small tree occurring naturally from 18° to 34°N and growing in acidic soils with pH 4.5–6.0. The cultivation area in China covers about 3.5 million hectares in 17 provinces which accounts for 80% of woody oil plants cultivation acreage (Fig. 11.1) (Anon., 2007a). It yields 0.15 million tons oil annually (Tang et al., 1993) and costs 6 billion Yuan (about 1 billion AUD$ and about 1 billion USD$) according to current oil sale price. The cultivated species are *Camellia oleifera, C. meiocarpa, C. vietnamensis, C. yuhsienensis, C. chekiangoleosa, C. semiserrata, C. reticulata, C. gigantocarpa, C. octopetala, C. semiserrata* var. *abliflora* etcetera. *Camellia oleifera* ranks first in cultivated acreage, accounting for 98% of the total cultivated area (Yao et al., 2005).

Oil content (and quality) of the seed vary with species, harvesting time, and agronomic practices, but it is typically 40 to 50% (Shanan & Ying, 1982; Xia et al., 1993). Most producers of camellia oil in China are small companies with outdated technology (Anon., 2006a). Thus, the quality of the oil also varies greatly due to the various technologies, and it appears that incorrect labelling and the presence of *trans*-fatty acids could be a problem (Zhai et al., 2004). Analytical chemistry will play a key

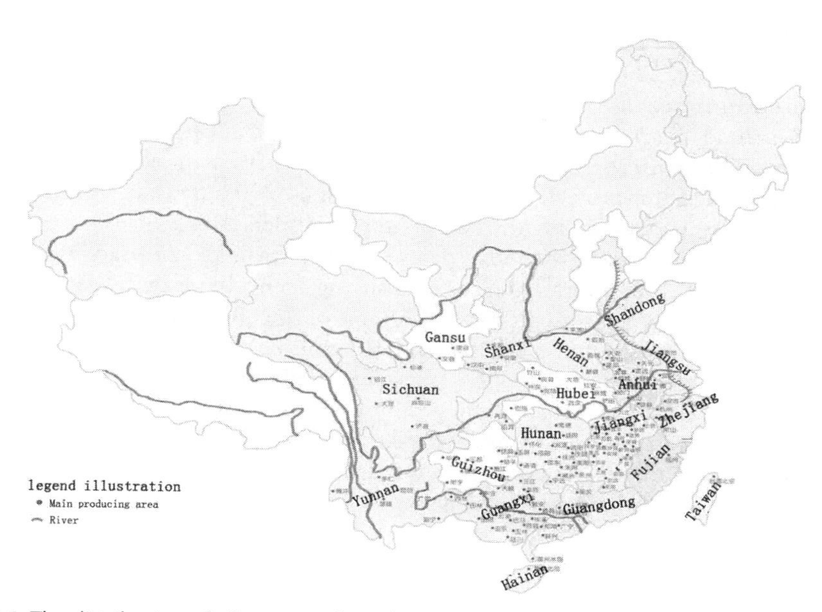


**Fig. 11.1.** The distribution of oil-tea camellia cultivation in China. Source: (Yao et al., 2005).

role in ensuring oil quality (Torto et al., 2007). Camellia oil is generally produced by cold-pressing, producing a pale amber-green oil with a sweet, herbal aroma. Typical gross features of the oil are listed in Table 11.1. If processed correctly, it is described as a very high quality oil with a long storage life. The latter can be attributed to its fatty acid profile and the high content of the antioxidant Vitamin E.

Camellia oil with its low saturated fatty acid and high monunsaturated oleic acid content is a natural competitor for olive and grape seed oils. It has a relatively high

**Table 11.1. Gross Properties of Camellia Oil**

| Property | Range | Test method |
|---|---|---|
| Physical state at 25°C | Liquid | Visual |
| Color | <40Y/4.0R Yellow/Red | Lovibond cell 5-1/4" |
| Density at 20°C | 0.905–0.925 g/cc | ASTM D1298-85 |
| Refractive index | 1.468–1.474 | ASTM 1248-92 |
| Acid value | <1.0 mg KOH/g | AOCS Cd3a-63 |
| Iodine value | 75–95 g $I_2$/100 g | AOCS Tg2a-64 |
| Saponification value | 185–205 mg KOH/g | ISO 3657-1988 |
| Peroxide value | <5.0 meq $0_2$/kg | Internal method |

Source: (Anon. 2007b).

smoke point (Table 11.2) and is the main cooking oil in the southern provinces of China, especially Hunan, where more than 50% of the vegetable cooking oil is from camellias (Ruter, 2002). It is also used in dips, salad dressings, sauces, and in production of margarine.

Nonfood uses of camellia oil and its by-products include production of paint, and fertilizer, while cosmetic uses include preparation of soaps, hair oil, lipstick, antiwrinkle creams, and sun-protection preparations (Ruter, 2002; Sabetay, 1972). Japanese camellia oil is used for setting the hair of Sumo wrestlers. Extraction of the seed hulls yields such compounds as tannins, pentosans, triterpenoids, and saponins (Shanan & Ying, 1982). The latter are used as emulsifying agents in pesticides and detergents and for foam-forming fire extinguishers. A number of pests were effectively controlled by tea oil residues including rice blast, sheath and culm blight of rice, wheat rust, rice hopper, cutworms, cotton aphids, certain scale insects, long-horned beetles, and leeches (Shanan & Ying, 1982). The possibility of developing new biologically based pesticides exists for this product as extracts of the seed cake are known to deter larval development in insects (Duke & Ayensu, 1985). Processing by-products were also used as livestock feed and to formulate fertilizers (Ruter, 2002).

## Quality Requirement

The final products of oil-tea and camellia oil are designated as either pressed (2 grades) or solvent extracted (4 grades) oils (He et al., 2004). Table 13.2 includes sensory and physicochemical indices.

## Processing

Processing involves several steps that begin with seed harvest.

### *Harvest*

Camellia fruit-harvesting time mainly depends on variety, but is generally from October to early November in south China. Fruit maturation is judged by seed appearance with the color change from green to red or yellow and with a small crack on the pericarp. After harvest the fruits are piled indoors for about one week. This storage period is claimed to cause an increase of oil content. The fruits are then dried in the sunshine until they crack. The seeds are separated after the pericarps are removed and the seeds are dried in the sunshine for a further 12 days (Chen & Zhang, 1999).

### *Dehulling*

The purpose of dehulling is to remove the major part of the fiber and a group of pigments which, if allowed to enter into the oil, would lower its value. The weight and volume of individual camellia seeds vary greatly between 0.461–1.463 g and 0.3–2.5 $cm^3$, respectively. However, the space between the kernel (or meat) and the wall of

**Table 11.2. Quality Criteria of Final Products of Oil-tea Camellia Oils**

| | | Pressed oil | | Solvent extracted oil | | | |
|---|---|---|---|---|---|---|---|
| | | Grade 1 | Grade 2 | Grade 1 | Grade 2 | Grade 3 | Grade 4 |
| Color | (Lovibond cuvette, 25.4 mm) ≤ | Yellow 35, Red 2.0 | Yellow 35, Red 3.0 | – | – | Yellow 35, Red 2.0 | Yellow 35, Red 5.0 |
| | (Lovibond cuvette, 133.4 mm) ≤ | – | – | Yellow 30, Red 3.0 | Yellow 35, Red 4.0 | – | – |
| Odour and flavour | | Have inherent odour and flavour of camellia oil, no off-flavour | Have inherent odour and flavour of camellia oil, no off-flavour | No odour, have a better taste | No odour, have a good taste | Have inherent odour and flavour of camellia oil, no off-flavour | Have inherent odour and flavour of camellia oil, no off-flavour |
| Transparency | | Clear and transparent | Clear and transparent | Clear and transparent | Clear and transparent | – | – |
| Moisture and volatile matter (%), ≤ | | 0.10 | 0.15 | 0.05 | 0.05 | 0.10 | 0.20 |
| Insoluble impurity (%), ≤ | | 0.05 | 0.05 | 0.05 | 0.05 | 0.05 | 0.05 |
| Acid value (mg KOH /g), ≤ | | 1.0 | 2.5 | 0.20 | 0.30 | 1.0 | 3.0 |
| Peroxide value, PV (mmol/kg), ≤ | | 6.0 | 7.5 | 5.0 | 5.0 | 6.0 | 6.0 |
| Residual solvent content in oil (mg/kg) | | Not be detected | Not be detected | Not be detected | Not be detected | ≤50 | ≤50 |
| Heating test (280 °C) | | No precipitate; Colour: Yellow≤35, Red≤2.4 | A little precipitate; Colour: Yellow≤35, Red≤7.0 | – | – | No precipitate; Colour: Yellow≤35, Red≤2.4 | A little precipitate; Colour: Yellow≤35, Red≤9.0 |
| Saponified matter content (%), ≤ | | – | – | – | – | 0.03 | 0.03 |
| Smoking point/ °C | | – | – | 215 | 205 | – | – |
| Refrigeration test (0 °C, 5.5 h) | | – | – | Clear and transparent | – | – | – |

–: not specified. Source: He et al, 2004

the sun-dried seed is 2–5 mm, and this gap facilitates dehulling (Huang et al., 2006). Because of the diversity of seed volume and weight, the currently used disc mill, in which the disc spacing and rotational speed can be varied, requires previous grading of seeds. Huang (Huang et al., 2006) invented a new machine for *Camellia oleifera* seed dehulling that doesn't require grading. The study showed that the moisture content and linear velocity of the roller were the most significant factors for dehulling and separating. Within the range of moisture contents and roller linear velocities of 5–20% and 8.90–10.00 m/s, respectively, the dehulling rate was more than 98.5% while the content of kernels in the hull fraction and the content of hulls in the kernel fraction were less than 4% and 1%, respectively (Table 11.3) (Huang et al., 2006).

### Cracking and Flaking

After dehulling, the kernels are cracked in a crusher. The disc mill is also used for cracking, but the strip worms of the disc need to be replaced with grid worms (Chen & Zhang, 1999). The cracked kernels are flattened into flakes of uniform thickness (ca. 3–5 mm) using a roller.

### Cooking

The flakes are cooked in a steam jacket cooker. The purpose of cooking is to cause heat denaturation and subsequent coagulation of proteins, in addition to causing the coalescence of oil droplets. A covered pan is now widely used in China for cooking in camellia oil processing. The flakes are kept at 110–115°C, and 20% of water is added first while the flakes are stirred, and then cooked in the covered pan for 30 min (Chen & Zhang, 1999).

### Extraction

Since the 1960s, hydrolytic presses have been used for camellia oil extraction in China. Hydraulic presses are still widely used in camellia seed producing areas because the scale of camellia cultivation is small, and for individual farmers the lower cost of the hydraulic press is attractive. The hydraulic press is a batch press procedure that requires hand labor. The cake is made from flakes with the addition of fibers such as rice

**Table 11.3. Effect of Dehulling Conditions for *Camellia oleifera* Seed**

| Moisture content (%) | Roller linear velocity (m/s) | Dehulling rate (%) | Content of kernels in hull (%) | Content of hulls in kernel (%) |
|---|---|---|---|---|
| 12.51 | 9.52 | 99.42 | 1.84 | 0.81 |
| 14.65 | 9.63 | 99.34 | 1.58 | 0.77 |
| 16.80 | 9.73 | 99.25 | 2.02 | 0.89 |
| 18.19 | 9.94 | 99.18 | 2.26 | 0.92 |

Source: (Huang et al., 2006)

straw prior to pressing. Also, a higher temperature (~ 40°C) is maintained in the press room to increase the oil fluidity. For most factories equipped with solvent extractors, the pressed cake with about 7% residual oil is the main feedstock. The hydraulic pressed crude oil is also the main material used to make refined camellia oil.

In the 1980s the screw press was adopted to produce camellia oil (Zhuang, 1988). Screw pressers are now used in a continuous extraction process in larger scale plants. The commonly used presses are divided into Type-95 and Type-200 by the throughput of presses. The main worm shaft, the drainage barrel, thrust bearing, and the motor transmission are the essential elements in the screw presses. The condition of the cooked flakes influences the oil content in the cake (Table 11.4) (Zhuang, 1988).

Solvent extraction is the most efficient method extensively used for removing oil from cakes after pressing in the camellia oil industry. The rate of extraction depends on the levels of oil in the cake, the solvent, and temperature. No. 6 solvent is a petroleum fraction used as the main solvent in the Chinese solvent extraction industry. This solvent is comprised predominantly of acyclic hexanes (74% vol/vol) and cyclohexane (16.5% vol/vol) with lesser amounts of pentanes (2.6% vol/vol)), heptanes (3.5% vol/vol), cyclopentane (1.6% vol/vol), and cycloheptane (0.2% vol/vol) with trace amounts of benzene and toluene (<0.1% vol/vol) (Zhuang, 1988).

Recent studies (Wu et al., 2007; Zhong et al., 1999, 2001) reported the application of supercritical $CO_2$ (SC-$CO_2$) extraction for camellia oil production. SC-$CO_2$ has a high extracting ability for oil-tea camellia seed oil. The oil yield was markedly affected by the pressure, temperature, and extracting time (Figs. 11.2–4). The oil yield was above 90% by SC-$CO_2$ at 40–45 MPa, 45–50°C and for 3–4 hours with the highest yield reaching 95.1% (Zhong et al., 1999). Quality parameters for SC-$CO_2$ extracted camellia oil were closely related to SC-$CO_2$ extracting conditions. The effect of pressure on the amount of extracted FFA was more than that of temperature, but the water and volatile materials extracting amount were more affected by temperature. The changes in phospholipid solubility in SC-$CO_2$ with different temperatures, pressures and extracting times showed the same trends as camellia oil solubility, but the solubility of phospholipids was very low. The color of SC-$CO_2$ extracted camellia oil was rather light. Without any purification, the quality of SC-$CO_2$ extracted camellia oil met the standard of pressed oil, Grade two in GB11765-2003 (Table 11.5).

**Table 11.4. The Relationship Between Conditions (Moisture and Temperature) of Cooked Flake and Oil Content in the Cake, After Screw Pressing**

| Moisture content in cooked flake (%) | Temperature of cooked flake before extraction (°C) | Oil content in cake (%) |
|---|---|---|
| 1.5 | 132 | 4.6 |
| 1.8 | 128 | 5.0 |
| 2.2 | 129 | 5.2 |
| 2.3 | 126 | 5.5 |

Source: (Zhuang, 1988)

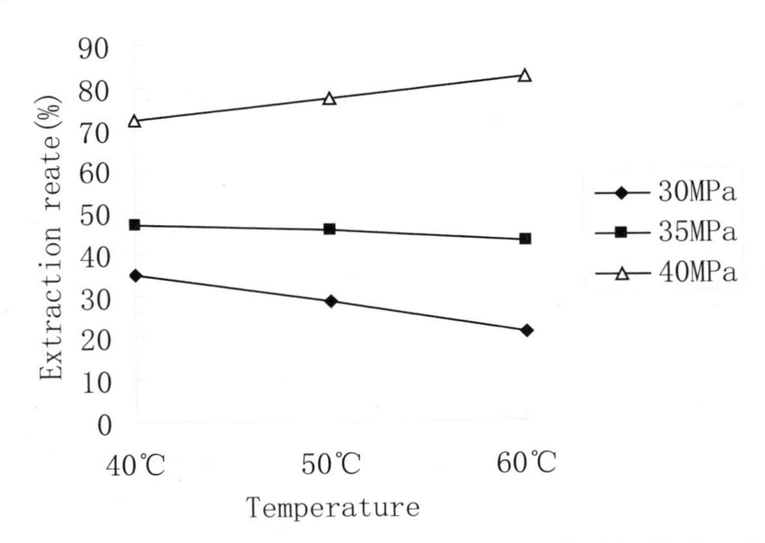


**Fig. 11.2.** Effect of temperature on extraction rate. Extraction for 3 h $CO_2$ flow of 3 L/min separation at 5 MPa 25°C. Source: (Zhong et al., 1999).

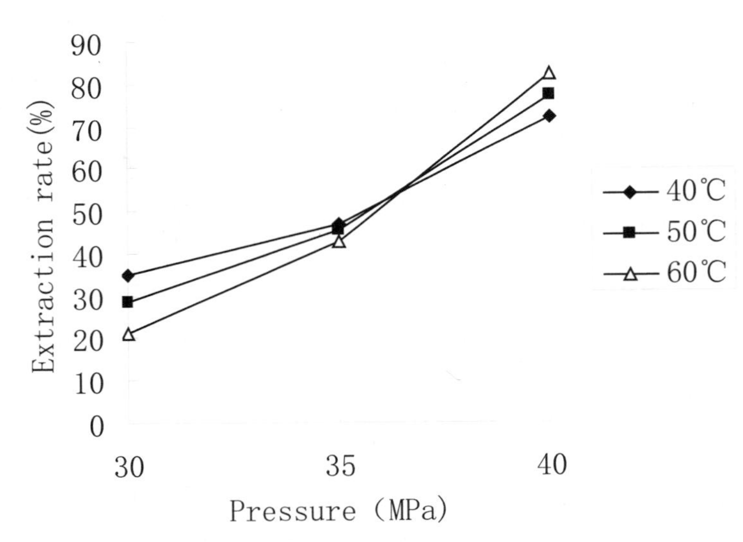


**Fig. 11.3.** Effect of pressure on extraction rate. Extraction for 3 h $CO_2$ flow of 3 L/min separation at 5 MPa 25°C. Source: (Zhong et al., 1999).

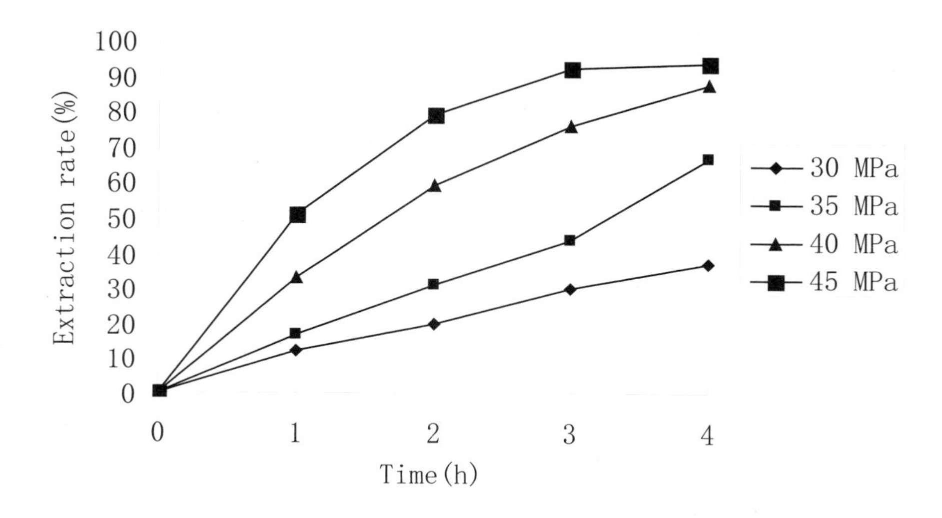


**Fig. 11.4.** Effect of time on extraction rate. Extraction at 50°C $CO_2$ flow of 3 L/min separation at 5 MPa 25°C Source: (Zhong et al., 1999).

**Table 11.5. The Quality of SC-$CO_2$ Extracted Camellia Seed Oil Compared with Standard Oils Obtained by Screw Pressing GB 11765-2003**

| | GB11765-2003 | | SC-$CO_2$ extracted |
|---|---|---|---|
| | Grade one | Grade two | |
| Odor and flavor | Have inherent odor and flavor of camellia oil, no off-flavor | Have inherent odor and flavor of camellia oil, no off-flavor | Have inherent odor and flavor of camellia oil |
| (Lovibond cuvette, 25.4 mm) ≤ | Yellow 35, Red 2.0 | Yellow 35, Red 3.0 | Y30，R0.6 |
| Acid value (mg KOH/g), ≤ | 1.0 | 2.5 | 1.018 |
| Insoluble impurity %, ≤ | 0.05 | 0.05 | 0.045 |
| Moisture and volatile matter %, ≤ | 0.10 | 0.15 | 0.145 |
| PV (mmol/kg), ≤ | 6.0 | 7.5 | 2.085 |

Source: (Wu et al., 2007)

The oxidative stability of SC-$CO_2$ extracted oil was poor although the addition of α-tocopherol to the SC-$CO_2$ extracted oil enhanced the stabilizing effect (Wu et al., 2007; Zhong et al., 2001).

## Refining

*Degumming*: Degumming is the removal of impurities, especially phospholipids, in the crude oil. The presence of phospholipids can darken the color of the oil and cause the oil to bubble if it is used for frying. Zhong et al. (2004) compared various degumming methods to show their effects for camellia oil (Table 11.6).

Membrane separation was applied in camellia oil processing to increase the relative content of oleic acid. This separating technique is conducted at a lower temperature, and the polymerization of unsaturated fatty acid under high temperature is avoided. Pan et al. (2006) reported the selection of membrane materials and their effects on degumming (Table 11.7).

**Table 11.6. The Results of Different Degumming Methods for Camellia Oil**

| Methods | Acid value (KOH, mg/g) | PV (meq/kg) | Phospholipids (ppm) | Heating test (280 °C) |
|---|---|---|---|---|
| 1 | 1.60 | 18.7 | 31.24 | No precipitate |
| 2 | 1.62 | 16.3 | 25.73 | No precipitate |
| 3 | 1.60 | 21.7 | 12.35 | No precipitate |
| 4 | 1.68 | 21.1 | 26.12 | No precipitate |

Method 1, conventional hydrated degumming: the crude oil was heated to 80 °C and mixed with 3% soft water that was heated to 80 °C, and then stirred for 30 min. Water and oil were separated by centrifugation at 3000 rpm for 20 min.
Method 2, dried degumming: the crude oil was heated to 65 °C and mixed with 0.05% of 85% $H_3PO_4$, and then stirred for 20 min. Centrifugation conditions as for Method 1.
Method 3, wet degumming: the crude oil was heated to 80 °C and mixed with 0.1% of 85% $H_3PO_4$, and 3% soft water that was heated to 80 °C, and then stirred for 30 min. Centrifugation conditions as for Method 1.
Method 4, Unilever super-degumming: the crude oil was heated to 75 °C and mixed with 0.3% of 85% $H_3PO_4$, and then stirred for 20 min. The mixture was cooled to 25 °C and 3% soft water was added, then it was stirred for 1 h. Centrifugation conditions as for Method 1.

**Table 11.7. Results of Degumming by Different Organic Membranes**

| Membrane materials | Membrane flux ($m^3/m^2 \cdot h$) | Degumming rate (%) (as phospholipids ) |
|---|---|---|
| Polypropylene | 1.85 | 99.6 |
| Polysulfone | 1.6 | 96.7 |
| Polyvinylidene fluoride | 1.2 | 97.6 |

Source: (Pan et al., 2006)

*Alkali refining and washing*: Sodium hydroxide is used in refining to remove free fatty acids (FFAs). The FFA content must be determined, and the mass of NaOH added is calculated according to the formula (Lu et al., 2003):

$$G_{NaOH} = 7.73 \times 10^{-4}\, G_o\, V_A$$

where:

$G_{NaOH}$ is the amount of added NaOH, kg;
$G_o$ is the oil weight, kg;
$V_A$ is the acid value, KOH mg/g.

The crude oil is stirred at 60–70 rpm, while NaOH of 12–16°Be′ (Degree Baum′e) is added for about 20 min. Then the temperature is increased at 1°C/min. The stirring rate is reduced to 30–40 rpm for 10–15 min until the temperature reaches 55–65°C. The oil is left to stand for 8 hours and is then heated to 78–80°C and mixed with 6–10% wt/wt soft water that is heated to 83–85°C. The mixture is cooled and left to stand for 1 hour to allow the separation of oil and water. The procedure is repeated three to six times until the separated water is clear (Pang & Mao, 2004).

*Bleaching*: The alkali refined oil is pumped into a bleaching kettle and heated to 90–100°C at 2°C/min. Activated clay (2–4%) is added, and the oil is sequentially heated to 105°C and left to stand at 105°C for 20–30 min (Pang & Mao, 2004). The activated earth bleaching conditions also affect the quality of refined oil. Zhong et al. (2000) used an orthogonal design [$L_{16}4^5$] to optimize the bleaching conditions of activated earth amount, temperature, and time. Table 11.8 shows that the acid value (FFA content) increased with the increase of activated earth amount, temperature, and time. Peroxide value and phospholipids exhibited an inverse trend, compared with that of acid value. The carbonyl value (CV) increased with the increase of time and temperature. CV also increased with 2–4% of activated earth, but decreased when the clay was over 4%.

*Deodorization and Winterization*: Deodorization is a high-temperature, high-vacuum steam-distillation process that is used for removal of volatiles. Tests showed that camellia oil was completely deodorized at 260 Pa and 200°C with 3–8% steam (Lu et al., 2003). Winterization is the process that allows the solid portion of the oil under chilling conditions to crystallize, which is necessary to meet the requirement of the cold test. Lu et al. (2003) reported that the oleic acid content of camellia oil was enhanced by 5–12% after the oil was chilled at –5 to 5°C. However, Zhang (2003) found no significant increase in the oleic acid content after camellia oil was cooled at 0, 3, 7°C for up to 72 hours, but the cold stability of the oil was more than 200 hours at 0°C.

**Table 11.8. The Bleaching Orthogonal Experiment Results**

| No. | clay (%) | Temp. (°C) | Time (min) | Absorbance ($A^{5cm}{}_{535nm}$) | AV (mg/kg) | PV (meq/ kg) | CV (meq/kg) | Phospho-lipids (ppm) |
|---|---|---|---|---|---|---|---|---|
| 1 | 2 | 80 | 5 | 2.250 | 0.63 | 4.62 | 2.10 | 1149 |
| 2 | 2 | 100 | 10 | 0.225 | 0.73 | 2.52 | 2.68 | 563 |
| 3 | 2 | 120 | 15 | 0.150 | 0.80 | 2.69 | 2.81 | 301 |
| 4 | 2 | 140 | 20 | 0.155 | 0.95 | 1.67 | 2.56 | 304 |
| 5 | 4 | 80 | 10 | 0.325 | 0.66 | 2.55 | 3.24 | 549 |
| 6 | 4 | 100 | 5 | 0.175 | 0.84 | 3.02 | 4.02 | 282 |
| 7 | 4 | 120 | 15 | 0.080 | 1.10 | 2.32 | 5.82 | 314 |
| 8 | 4 | 140 | 20 | 0.130 | 1.08 | 1.51 | 4.66 | 329 |
| 9 | 6 | 80 | 15 | 0.175 | 0.72 | 2.50 | 2.97 | 490 |
| 10 | 6 | 100 | 20 | 0.090 | 0.89 | 2.41 | 3.34 | 273 |
| 11 | 6 | 120 | 5 | 0.110 | 0.82 | 2.74 | 3.47 | 280 |
| 12 | 6 | 140 | 10 | 0.100 | 1.09 | 1.72 | 3.12 | 293 |
| 13 | 8 | 80 | 20 | 0.200 | 0.70 | 2.33 | 2.23 | 510 |
| 14 | 8 | 100 | 15 | 0.080 | 0.86 | 2.48 | 2.66 | 250 |
| 15 | 8 | 120 | 10 | 0.085 | 0.94 | 2.54 | 2.87 | 259 |
| 16 | 8 | 140 | 5 | 0.095 | 1.12 | 1.50 | 2.20 | 264 |

AV= anisidine value, PV = peroxide value, CV = carbonyl value.

## *Utilization of Defatted Cake*

*Thea saponins:* Thea saponins in *Theaceae* plants are triterpenoid saponins with 5–7 types of sapogenin (Fig. 11.5). The sugar portions of the saponins are composed of glucuronic acid, arabinose, xylose and galactose, and the acid portion is composed of *trans, cis*-1, 2-dimethylacrylic acid and acetic acid (Li & Xue, 1994). The saponins in oil-tea camellia (also referred to as saponin of thea sasanqua or *oleiferin*) exist in various organs of the oil-tea camellia tree. The concentrations of saponins are highest in the cake with 12.8–13.8% of content (Zhuang, 1988). Sokol'skii et al. (1976) reported that the saponins in the oil of *oleiferin* were comprised of pentacyclic polyhydroxytriterpene aglycones which included dihydropriverogenin A, barringtogenol C and theasapogenol A. The acid hydrolyzates of oleiferin contained D-glucuronic acid, D-glucose, D-galactose, and D-xylose. In the alkaline hydrolyzate, they found angelic acid, tiglic acid, and methylbutyric acids.

The surface activity of the saponins of thea sasanqua is mainly due to the presence of a hydrophobic fatty acid on one end of the molecule, and a hydrophilic sugar moiety on the other end. The saponins of thea sasanqua had a strong antimicrobial activity on *Escherichia coli*, *Penicillium citrinum*, *Aspergillus niger,* and *Candida uitilis* (Table 11.9) with an MIC (minimal inhibiting concentration) below 0.625%. However, they had no effect on the growth of *Staphylcoccus aureus* (Huang et al., 2002).

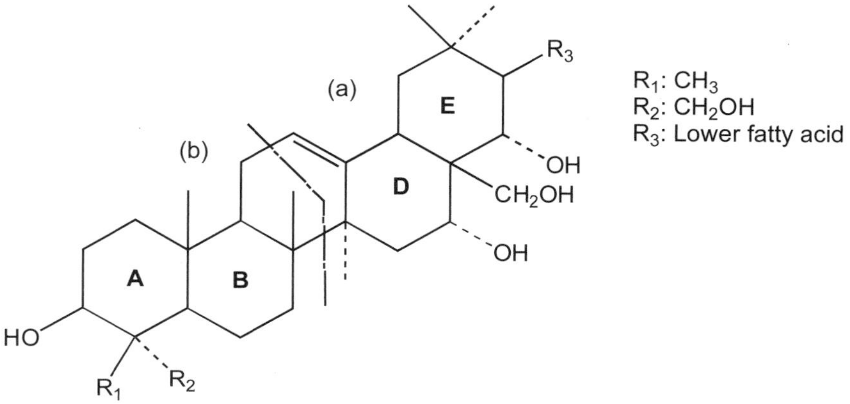

$R_1$: $CH_3$
$R_2$: $CH_2OH$
$R_3$: Lower fatty acid

**Fig. 11.5.** The structure of Thea Saponin.

**Table 11.9. The Antimicrobial Effects of Saponins of Thea Sasanqua Expressed As the Diameter of Anti-microorganism Circle (mm)**

| | Saponins of thea sasanqua concentration (%) | | | | | |
|---|---|---|---|---|---|---|
| Microorganisms | 10 | 5 | 2.5 | 1.25 | 0.625 | 0 |
| Aspergillus niger | 33.9 | 26.8 | 20.2 | 11.1 | 8.0 | 0 |
| Penicillum citrinum | 21.0 | 19.7 | 13.5 | 10.5 | 8.3 | 0 |
| E. Coli | 33.3 | 15.8 | 10.0 | 7.8 | 6.3 | 0 |
| 117 Candida uitilis | 18.8 | 15.5 | 10.2 | 9.3 | 8.2 | |
| Staphylococcus aureus | 0 | 0 | 0 | 0 | 0 | 0 |

*Protein in defatted cake:* Zhong et al. (2001) reported that ethanol removed saponins from oil-tea camellia defatted cake with very little nutritional loss. Microbes such as *Aspergillus niger, Mucor mucedo,* and *Cadida uitilis* could grow and ferment the cake after removal of saponins with a net protein increase of 66.7–99.7%. *Pleurotus ostreatus* and *Auricularia polytricha* also grew well in the cake after removal of saponins, but *Collybia velutipes* did not. *Pleurotus ostreatus, Auricularia polytricha,* and *Mucor mucedo* were used in the first solid fermentation for 25 days, then *Candida uitilis* was added and fermentation was allowed to continue for an additional 20 days. The protein content of final fermented products was close to 20%. Their amino acid composition was balanced (Table 11.10). This product should be useful as livestock feed.

## Edible and Nonedible Applications

Camellia oil has numerous applications that include both edible and nonedible uses. It is widely used in China but finds limited application elsewhere although this may change as its health benefits are recognized.

**Table 11.10. The Amino Acid Content in Oil-Tea Camellia Cake and Its Solid Fermented Products (Dry Basis, %)**

| Removal of saponins | Mean of fermentation | Amino acid content | | | | | | | | | | | | | | | |
|---|---|---|---|---|---|---|---|---|---|---|---|---|---|---|---|---|---|
| | | Asp | Thr | Ser | Glu | Gly | Met | Val | Ile | Tyr | Phe | Lys | His | Arg | Ala | Leu | Total |
| No | No | 1.08 | 0.68 | 0.74 | 1.54 | 0.68 | 0.13 | 0.58 | 0.57 | 0.28 | 0.34 | 0.36 | 0.19 | 0.83 | 0.65 | 0.78 | 9.43 |
| By hot water | No | 1.08 | 0.63 | 0.64 | 1.50 | 0.60 | 0.10 | 0.49 | 0.44 | 0.24 | 0.34 | 0.29 | 0.14 | 0.79 | 0.58 | 0.63 | 8.49 |
| | *Mucor mucedo* + *Candida uitilis* | 1.49 | 0.69 | 0.85 | 2.34 | 0.86 | 0.07 | 0.95 | 0.73 | 0.24 | 0.92 | 0.83 | 0.31 | 0.76 | 0.92 | 1.23 | 13.19 |
| By ethanol | No | 0.77 | 0.31 | 0.51 | 1.96 | 0.42 | 0.05 | 0.40 | 0.37 | 0.19 | 0.46 | 0.47 | 0.22 | 0.96 | 0.45 | 0.67 | 8.21 |
| | *Auricularia polytricha* | 1.32 | 0.58 | 0.80 | 2.53 | 0.80 | 0.18 | 0.81 | 0.64 | 0.26 | 0.78 | 0.87 | 0.37 | 1.01 | 0.75 | 1.04 | 12.74 |
| | *Auricularia polytricha* + *Candida uitilis* | 1.76 | 0.91 | 1.11 | 3.12 | 1.05 | 0.12 | 1.11 | 0.85 | 0.36 | 1.15 | 0.97 | 0.45 | 1.11 | 1.01 | 1.40 | 16.48 |

*Cosmetics and traditional medicines:* Camellia oil is used widely in the cosmetics industry, particularly in China and Japan, and can be found in shampoos, lotions, hair conditioners, soaps, eye creams, anti-wrinkle creams, and lipsticks. The Victani skincare company claims that camellia seed oil is highly moisturizing and protects skin from harsh environmental damage, preserves skin cells against scarring, lightens stretch marks and age spots, and prevents freckles and wrinkling (Anon. 2005). These properties are attributed to polyphenols in the oil which can protect the skin from free radical-induced cell death (apoptosis), and the subsequent fibrosis (scar formation) (Anon., 2005). Indeed, diphenolic lignan components in tea seed oil, for example, sesamin are known to possess exceptional antioxidant activity (Lee & Yen, 2006). Such antioxidant properties are highly regarded in the "cosmaceuticals" industry. The Boots Company PLC, an internationally renowned skincare company, recently patented skincare compositions containing *Camellia sinensis* extracts to fight free radical skin damage (Pykett et al., 2007).

Healthy skin requires an effective barrier functioning to protect the body from allergens and infections (Wang et al., 2004). Because of this, topically applied cosmetics contain an oil component to serve as a skin barrier. However, the oil component in cosmetics must be correctly formulated, since it is reported that free fatty acids (Morimoto et al., 1995, 1996) and terpenes (Kitahara et al., 1993) are skin-penetration enhancers. This results in reduced skin barrier function with respect to the permeation of drugs which is undesirable in cosmetic formulations. In light of this phenomenon, Wang et al. (2004) investigated the permeation of the model drugs flurbiprofen and diclofenac sodium through in vitro rat and pig skin pre-treated with either distilled camellia oil (DCO) or filtered camellia oil (FCO). Compared to DCO, FCO has a higher free fatty acid content (16.9% vs. 11.6%) and a lower triacylglycerol and free sterol content (40.6% vs. 42.6% and 2.1% vs. 5.2%, respectively), and presumably FCO, subjected to filtration without heating, may contain higher concentrations of minor components (Wang et al., 2004). The results demonstrated that permeation of flurbiprofen through the skins pretreated with FCO and DCO was enhanced, while that of diclofenac sodium was suppressed. Such results were surprising since flurbiprofen and diclofenac sodium have similar lipophilicity and pKa values. The authors hypothesized that the higher free fatty acid composition of FCO compared to DCO may explain the disparate results, particularly with respect to concentrations of oleic acid (FCO 83.6% and DCO 62.8%). However, rat skin pre-treated with oleic acid resulted in enhanced permeation of both flurbiprofen and diclofenac sodium, and similar instances of penetration enhancement effects of oleic acid were reported (Karande & Mitragotri, 2003; Yu et al., 2003). The results suggest that other minor components of FCO may be responsible for its permeation suppression effects. Although camellia oil appears to enhance penetration, more research is needed to objectively evaluate its superiority over other penetration enhancers currently used in the cosmetics industry.

Protection of the skin against harmful UV rays is a major need of the skincare industry. The protective effect of topically applied extra virgin olive oil and Camellia oil against photocarcinogenesis following UVB exposure of mice was compared (Budiyanto et al., 2000). While results showed that mice coated with olive oil after UVB exposure (post-UVB group) had significantly lower numbers of tumors per mouse than those in the UVB control group, camellia oil, applied using the same experimental protocol, did not have a suppressive effect. This was surprising, since triterpene alcohols from camellia oil were shown to have an anti-inflammatory effect on 12-*O*-tetradecanonylphorbol-13-acetate (TPA)-induced inflammation of mouse skin (Akihisa et al., 1997) and also to inhibit two-stage chemical carcinogenesis of mouse skin (Yasukawa et al., 1991). This further suggests that general nonspecific effects of plant or vegetable oils cannot be responsible for the suppressive effects of olive oil, but rather are attributed to components such as the antioxidants present in the oil.

Oil-tea seed oil (*Camellia oleifera* Abel.) was also used as a traditional medicine in China for treating stomach aches, burns (Lee & Yen, 2006), ringworm (Duke & Ayensu, 1985), and dandruff.

*Cooking:* The oil from oil-tea seed (*Camellia oleifera* Abel.) is used extensively in Chinese cooking (Yu et al., 1999). Chefs prefer it due to its high temperature resistance and smoke point, which ensure that the oil maintains its quality and consistency, even when cooked at high temperatures. Furthermore, it has almost no odors and taste of its own, enabling chefs to create tastes which are entirely free of oily undertones and aftertastes (Anon., 2000a). Camellia oil is believed to be the best oil for tempura since it makes an especially light and crisp coating due to the oil's low viscosity (Anon., 2007c). Camellia oil is used widely in Southern China and Southeast Asia, and because of its similar chemical composition to olive oil, is often referred to as "Eastern olive oil" (Wang et al., 2004).

### Acyl Lipids

Oil yields of camellia seeds extracted with *n*-hexane were approximately 30% w/w (Wang & Lin, 1990). Free fatty acids comprise a minor component of edible oils, but as the acyl moiety of triacylglycerols they comprise the major part of an edible oil. The levels of free acids in the oil increase during storage or heating, causing its deterioration. Little is reported in the refereed literature on the fatty acid profile of camellia oil, making it difficult to define acceptable ranges of individual fatty acids (Brady et al., 2006; Wang & Lin, 1990; Xu et al., 1995ab; Zhong et al., 2006). In the nonrefereed literature (e.g., company trade specifications), the oleic acid content of camellia oil is generally reported to be above 80% in agreement with the published data. Along with this high content of monounsaturated lipid, camellia oil has one of the lowest levels of saturated fats. Table 11.11 compares the fatty acid composition of several crude

**Table 11.11. Fatty Acid Profiles (%W/W) of Selected Oils**

| Oil | Refined avocado | Cold-pressed avocado oil | Refined camellia oil | Cold-pressed camellia oil | Refined pumpkin seed | Cold-pressed pumpkin oil | Expeller pressed soybean | Cold-pressed sesame oil |
|---|---|---|---|---|---|---|---|---|
| 16:0 Palmitic | 16.3 | 14.1 | 7.9 | 9.3 | 12.4 | 11.5 | 10.0 | 9.4 |
| 16:1 Palmitoleic | 7.7 | 5.7 | 0.1 | 0.2 | 0.1 | 0.1 | 0.2 | 0.1 |
| 18:0 Stearic | 0.6 | 0.4 | 1.8 | 1.9 | 4.9) | 5.6 | 3.9 | 5.6 |
| 18:1 ω9 Oleic | 62.7 | 69.1 | 82.0 | 79.1 | 37.7 | 37.7 | 24.4 | 41.8 |
| 18:2 ω6 Linoleic | 11.4 | 9.6 | 6.8 | 8.4 | 43.3 | 44.0 | 52.8 | 41.3 |
| 18:3 Linolenic | 0.8 | 0.6 | 0.4 | 0.3 | 0.6 | 0.2 | 7.3 | 0.7 |
| 20:0 Arachidic | 0.1 | 0.1 | <0.1 | <0.1 | 0.4 | 0.4 | 0.3 | 0.6 |
| 20:1 ω9 Eicosaenoic | 0.2 | 0.2 | 0.6 | 0.4 | 0.1 | 0.1 | 0.2 | 0.2 |
| 22:0 Behenic | <0.1 | <0.1 | <0.1 | <0.1 | <0.1 | 0.1 | 0.3 | 0.1 |

Source: (Zhong et al., 2007a)

and refined camellia oils. The oils exhibit wide variations in the contents of palmitic, palmitoleic, stearic, oleic, and linoleic acids, leading to differences in total saturated and unsaturated fatty acids, monounsaturated and polyunsaturated fatty acids. The high levels of oleic acid in camellia oil make it a direct competitor with canola oil and particularly olive oils.

Two oils having the same fatty acid profile can still differ greatly in physical and chemical properties. The triacylglycerols that constitute the bulk of an edible oil are not a single chemical entity; rather the oil contains a mixture of mixed triacylglycerols. The distribution of fatty acids both between and within the triacylglycerols is selective rather than random. Data for camellia oil are limited, but the data for other oils suggest that unsaturated fatty acids preferentially occupy the sn-2 position, and saturated fatty acids are usually located in the sn-1 and sn-3 positions (Yoshida et al., 2001ab). Xu Xuebing et al. (Xu, Hu, & Zhang, 1994) reported that triacylglycerols of *Camellia oleifera* oil included 58.8% of OOO, 19.2% of POO, 3.4% of StOO, and 18.6% of others that were made up of OLO, OOL, POL, PLO, and StOP (P = palmitate, St = stearate, O = oleate, L = linoleate).

## Minor Lipid Components

The minor lipid components reported for camellia oil include tocopherols (Wang et al., 1994), sterols (Wang et al., 1994), triterpenes (Akihisa et al., 1999), and phenolic compounds (Zhong et al., 2007a).

Wang et al. (1994) reported α-, β-, γ-, and δ-tocopherol levels in the oils of *C. Oleifera* and *C. Tenuifolia* (Table 11.12). The total tocopherol levels (45 mg/100 g oil) of the oil of *C. Oleifera* place it near the top of the list of tocopherol-rich oils, between sunflower oil (56 mg/100 g) and hazelnut oil (26 mg/100 g) according to Bauernfeind (1980). More recent (2000) charts (Anon., 2000b) of tocopherol content now list many oils with >50 mg/100 g total tocopherols, and this may reflect modern breeding and/or horticultural practices. Interestingly, camellia oil may have followed this trend. As can be seen from Table 11.12, reasonable proportions exist of each of the four tocopherols, in contrast to many oils that are dominated by one or two (e.g., olive oil contains almost solely α-tocopherol).

Wang et al. (1994) also reported the levels of the three common plant sterols, stigmasterol, sitosterol, and campesterol, in camellia oil.The levels were very low compared to many other vegetable oils (Phillips et al., 2002). However, Wang et al.

**Table 11.12. Tocopherol and Sterol Content in Tea Oils (*Camellia* spp.) (Wang et al., 1994)**

| | Tocopherol Content (mg/100 g) | | | | Sterol Content (mg/100 g) | | |
|---|---|---|---|---|---|---|---|
| Oil | α– | β– | γ– | δ– | Stigmasterol | Sitosterol | Campesterol |
| C. Oleifera | 21 | 8 | 13 | 3 | 10 | 8.5 | 5 |
| C. Tenuifolia | 10 | 6 | 9 | 2 | 12 | 11 | 6 |

(2004) reported higher values, so additional studies are needed to understand the actual levels of sterols in camellia oil. Yamaguchi and Kurata (2005) in a forensic study of vegetable oils (see below) report the presence of squalene, campesterol, sitosterol, stigmasterol, and spinasterol (Fig. 11.6). The latter is unique to camellia oil, but no quantitative data were given for the sterol components.

Li et al. (2006) reported that the levels of squalene in the seed oil of *Camellia oleifera* changed with extraction method and extraction solvent (Table 11.13). Very recently, the search for bioactive compounds in *C. oleifera* Abel oil led Lee and Yen

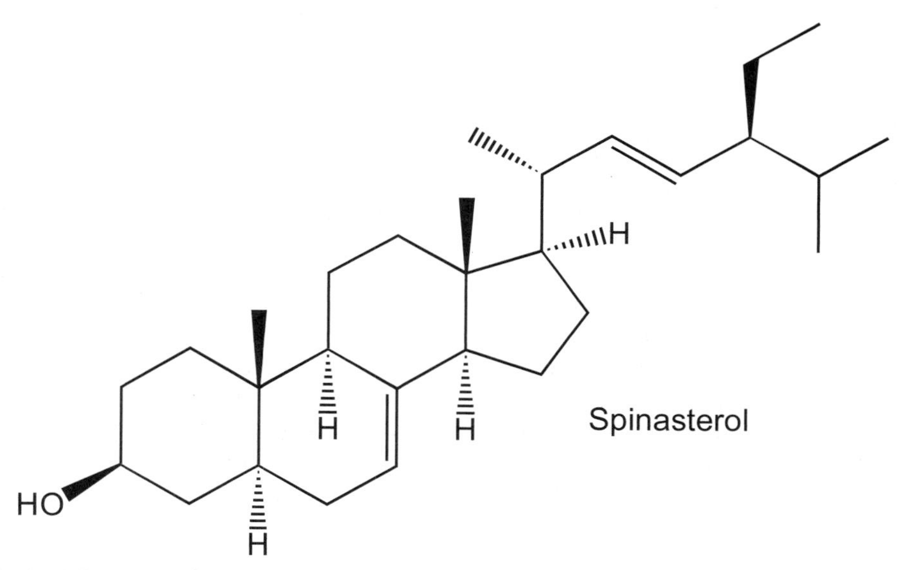


**Fig. 11.6.** Structure of spinasterol.

**Table 11.13. GC Analysis Results of Methylated Seed Oil of** *Camellia oleifera* **(%)**

| Extraction methods | Methyl palmitate | Methyl stearate | *Methyl oleate* | *Methyl linoleate* | *Squalene* |
|---|---|---|---|---|---|
| Direct expression | 8.61 | 1.60 | 77.52 | 7.17 | 2.98 |
| Extraction with hexane | 8.23 | 1.76 | 69.43 | 6.26 | 7.62 |
| Extraction with petroleum ether | 8.71 | 0.57 | 80.65 | 7.13 | 0.94 |
| Extraction with acetone | 8.91 | 0.37 | 81.51 | 7.15 | 0.29 |
| Extraction with chloroform | 8.64 | 0.46 | 79.55 | 7.77 | 0.54 |

Source: (Li et al. 2006)

(2006) to undertake activity- guided fractionation of the methanol extract of the oil. Two active compounds were detected—sesamin and "Compound B" (Fig.11.7)—and both exhibited potent in vitro antioxidant activity using a range of assays.

Liu and Zhao (2002) used TLC and phenolic-specific spray reagents to detect phenols in *Camellia sinensis* seed oil. They suggested that polar antioxidant compounds played a major role in the stability of the oil. Zhong et al. (2006,. 2007a) measured total phenols and reported the phenolic profile of a water/methanol extract of a cold-pressed camellia oil. Total phenols were similar to other cold-pressed oils (avocado and pumpkin), but the profiles differed considerably (Fig. 11.8). Interestingly, the sesamin and "Compound B" peaks from Lee and Yen's extract (2006) are not evident in Zhong et al.'s chromatogram. This could be a result of different processing, different oil ages, or may be related to the original phenolic profiles of the seeds. Further work is required to ascertain the variability in phenolic profiles of camellia oils as a function of these variables as was recently demonstrated for olive oil (Kalua et al. 2005, 2006ab).

The triterpene alcohol fraction (accounting for 57.9%) in unsaponifiable matter from *Camellia oleifera* seed oil was separated by HPLC (Fang et al., 1999). The triterpene alcohol acetate fraction was separated into β-amyrin acetate, germanicol acetate, taraxerol acetate, Ψ-taraxasterol acetate, tirucalla-7,24-dienol acetate, butyrospermol acetate, and dammaradienol acetate, using silver nitrate-impregnated silica gel. The contents of the triterpene alcohol acetates described above were determined by GC as 41.7%, 4.3%, 3.4%, 2.9%, 24.5%, 15.1%, and trace, respectively.

Triterpenoids, including cyclized and incompletely cyclized triterpene alcohols, were featured in the studies on the unsaponifiable fraction of camellia oil from *Camellia japonica* L. and *C. sasanqua* THUNB. (Akihisa et al., 2004) and *Camellia weiningen-*

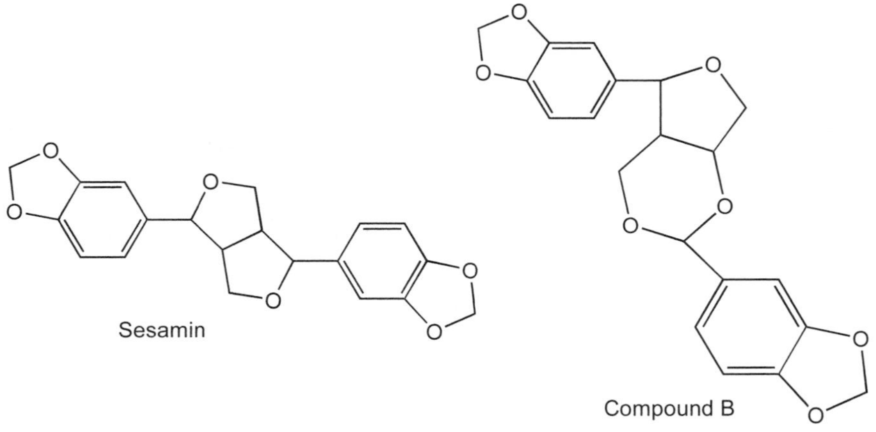

Sesamin

Compound B

**Fig. 11.7.** Structures of sesamin and compound B isolated from the methanol extract of camellia oil.

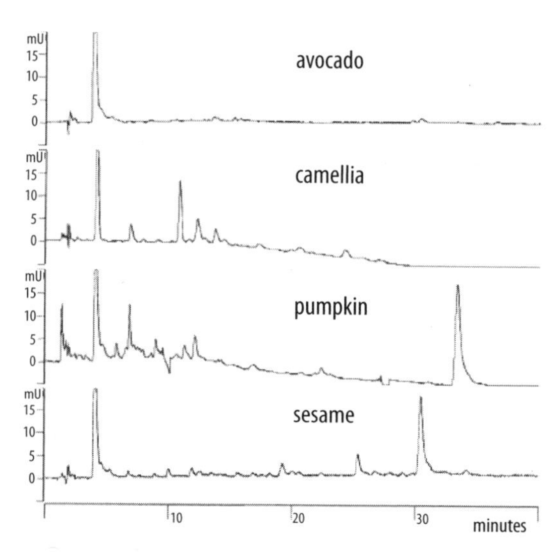


**Fig. 11.8.** HPLC phenolic profile of selected oils. Source: (Zhong et al., 2006).

*sis* L. (Li et al., 2000). Camelliols A(1), B(3), and C(5) were reported (Akihisa et al., 1999) as novel triterpene alcohols from *C. sasanqua* THUNB. featuring tri-, bi-, and mono-cyclic systems, respectively (Fig. 11.9).

## *Flavor and Aroma Compounds*

Various commercial suppliers describe camellia oil as having nutty or smoky flavours, which are mild in comparison to other more flavorsome oils such as extra virgin olive oil. The results of a detailed sensory analysis of camellia oil (*C. sinensis*) support the claims of the commercial literature (Chen et al., 2003) . In their study a 12-person sensory panel found that camellia oil was less strongly flavored than olive oil using five descriptors: toasty/nutty flavor, buttery flavor, fruity flavour, and overall flavor. However, on the "desirability" scale, camellia oil was ranked as more desirable. No flavor or aroma compounds were reported in this work.

Zhong et al. (2006, 2007a) reported nine major volatile compounds in the headspace of camellia oil as determined by SPME-GC-MS (Fig. 11.10). The major compounds were saturated aldehydes ranging from pentanal to nonanal. No attempt was made to correlate the volatiles to sensory properties. Despite the predominance of aldehydes, which are common to most oils, the pattern of headspace volatiles in camellia oil may be unique. A recent report by Hai and Wang (2006) showed that an electronic nose was able to discriminate between camellia, maize, and sesame oils. No individual compounds were reported.

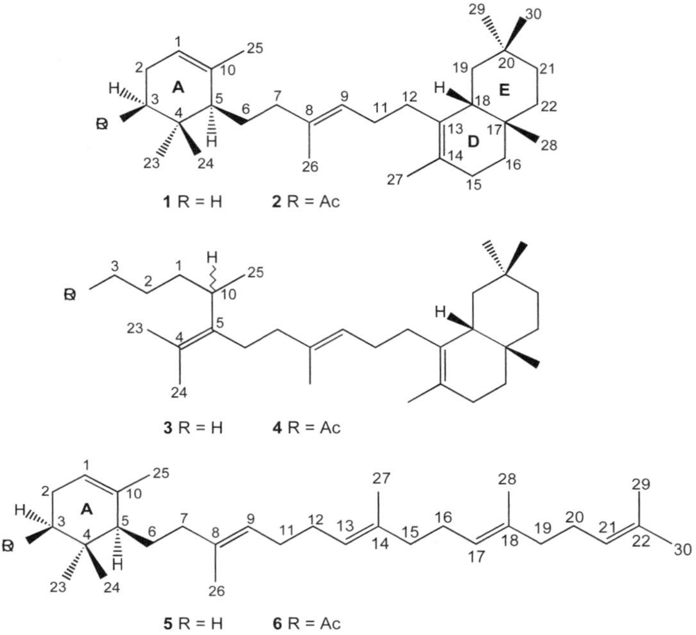

**1** R = H **2** R = Ac

**3** R = H **4** R = Ac

**5** R = H **6** R = Ac

**Fig. 11.9.** Structures of Camelliol A (1), B (3), C (5).

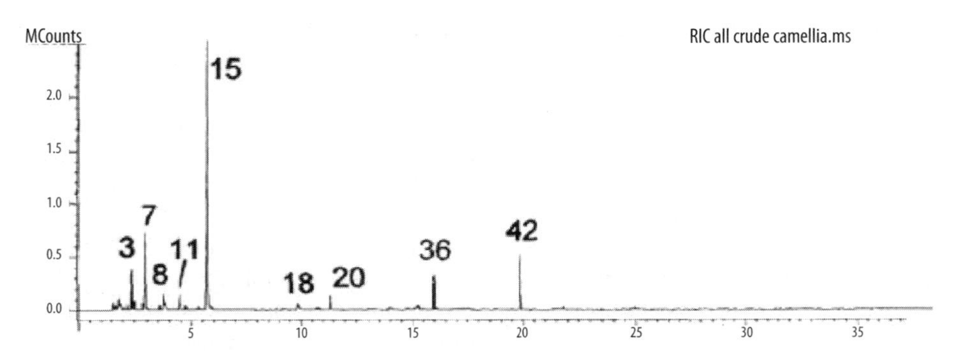


**Fig. 11.10.** Volatile profile of camellia oil using SPME and GC–MS. 3. 3-methylbutanal 7. pentanal 8. 3-methyl-1-butanol 11. toluene 15. hexanal 18. 1-hexanol 20. heptanal 36. octanal 42. nonanal. The numbering used in the source was retained. Source: (Zhong et al., 2006).

Thus, many opportunities are available for further research to understand the unique sensory properties of camellia oil and to explore its possible gourmet oil applications.

### *Health Benefits*

The high incidence of diet-related disorders has generated a worldwide interest in consumption of healthy foods. Camellia oil has a healthy image. For instance, camellia oil was launched in Canada as "a new ally in healthy cooking" (Anon., 2007c), and in Australia its promotion was under the banner "Move over, olive, camellia has the good oil" (Anon., 2006b). It was the excellent reputation of camelia oil in Chinese medicine that first attracted the Australian distributor to camellia oil. Not surprisingly, the virtues of camellia oil have long been recognized in China, where traditional medicines have a history spanning many thousands of years (Deng et al., 2007). Indeed, the distinction between medicinal and food use of commodities is relatively recent (Andlauer & Furst, 2002). As summarized by Hippocrates in the fifth century B.C.E., "Let food be your medicine and medicine be your food."

The health attributes of an edible oil are determined by its chemical composition. The primary determinant of chemical composition is the genetics of a plant. Nevertheless, the basic features of composition are the same in all edible oils. Thus, the major constituents (generally up to 98% by mass) in all oils are triacylglycerols, with numerous minor components comprising the remainder of the oil. The latter include free fatty acids, partial acylglycerols, biophenols, tocopherols, sterols, stanols, phospholipids, waxes, squalene, and other hydrocarbons. The precise molecular structure and the quantities of these minor constituents vary greatly between edible oils and are affected by both horticultural practices and processing. Hence, the composition of commercial oils from the same species may differ greatly, and a need exists to consider both the variety and geographic origin of an oil in deciding its potential health benefits. Processing-induced changes are of two fundamental types—simple reduction in the level of a component(s) or, alternatively, chemical modification of one or more components.

Data specifically relating to camellia oil composition are limited, but information is gleaned from work on other oils and their constituents. However, the particular composition and synergism may contribute a unique benefit to camellia oil. Claims for the health benefits of camellia oil are based primarily on a favorable fatty acid profile. Although defining the precise composition of the ideal healthy edible oil is not feasible, some general principles can be established (Beardsell et al., 2002). Thus, desirable features for maximal nutritional benefit are minimal levels of saturated fats, especially lauric and myristic acids, and minimal levels of *trans* fatty acids. Moreover, monounsaturated fatty acids typically contributing 80–90% should dominate the fatty acid profile. U.S. studies on canola oil recommend that healthy oils should contain four to ten times as much linoleic acid as linolenic acid (Dupont et al., 1989).

Camellia oil has a fatty acid profile approaching this ideal (Table 11.11) and is very similar to that of olive oil but with a slightly higher monounsaturated fatty acid content coupled with lower saturated fatty acid content.

Healthy oils should contain minimal amounts of *trans* fatty acids as most published work shows their detrimental effects on human health. The extent of processing modifications or losses will depend on processing conditions, particularly temperature. *Trans*-fatty acid isomers form when oils are heated, and each heating episode increases the levels of *trans* fatty acids (Christy & Egeberg, 2006; Griguol et al., 2007; Herzallah et al., 2005; Ledoux et al., 2000; Precht & Molkentin, 2000; Tasan & Demirci, 2003). Heat stability of oils, including the propensity to form *trans* fatty acids, depends on the fatty acid profile and antioxidant content. Numerous studies show that linolenic acid is 13–14 times more prone to isomerization than linoleic acid (Wolff, 1992). The type and level of antioxidants in an oil also determine the degree of *trans* fatty acid formation. Camellia oil presents an interesting case because it typically has a relatively high antioxidant content. In a study of an expressed refined camellia oil, unsaturated fatty acids were present as the *cis*-isomers with only traces of the corresponding *trans*-isomers (<0.02 mg $kg^{-1}$) (Zhong et al., 2006). Heating the camellia oils at temperatures up to 180°C for periods up to 24 hours did not produce significant changes in the fatty acid profile. Moreover, no evidence was present for *cis-trans* isomerization induced by heating the oil.

Human and animal feeding studies show (Kubow, 1996) effects of triacylglycerol stereospecific composition on fat absorption, plasma cholesterol, and plasma triacylglycerol concentrations and atherogenesis. Nevertheless, data considering the regiospecificity of acyl substitution and the fate of lipids beyond the bloodstream (absorption, chylomicron formation, and deposition in adipose tissue and in different liver lipids) are rather scarce, and animal model studies are needed. The nutritional impact of the regiospecific distribution of the acyl groups on the glycerol backbone was not addressed in camellia oil.

Secondary metabolites including phenols, proanthocyanidins, tocopherols, and carotenoids are an important fraction of the minor components of camellia oil. In studies of seed oils from other species, these compounds were shown to possess antioxidant properties (Lee & Yen, 2006), and they were implicated in both reducing certain cancers (Miura et al., 2007) and lowering harmful LDL cholesterol (Brufau et al., 2008). In a study involving young males whose diets were supplemented with camellia oil, serum HDL-cholesterol levels were raised from those supplemented with beef tallow (Hong, 1988). However, data on the biological effects of camellia oil are limited and often restricted to ex vivo and in vitro studies (Wu et al., 1998). In one study, topically applied camellia oil did not have a suppressive effect on UVB-induced murine skin tumors (Budiyanto et al., 2000). Various triterpenoids were isolated from Camellia and investigated for bioactivity (Akihisa et al.,1998). For example, a triterpenoid saponin from camellia oil exhibited both antioxidant and antimutagenic properties in humans and animals (Zhan, 1999). Triterpenoild saponins also improved immune function and enhanced antibacterial and antiviral activities (Francis et al., 2002). Numerous other triterpenoids also exhibited anti-viral activities (Akihisa et al., 2004).

## Other Issues

Adulteration of expensive vegetable oils with cheaper, inferior quality oils constitutes major economic fraud. Generally, adulteration does not pose a threat to public health, but rather the fundamental rights of consumers are violated by fraudulent malpractice (Ulberth & Buchgraber, 2000). Camellia oil commands a higher market price compared to other vegetable oils because of its nutritional and medicinal properties. However, unlike the more popular and similarly expensive olive oils, adulteration of camellia oil has received little attention (Wang et al., 2006). The occurrence of adulteration in camellia oil has been a long-term problem in Taiwan (Weng et al., 2006), and the most commonly used adulterants are sunflower oil, maize oil, and bean oil (Hai & Wang, 2006).

Methods for the determination of adulteration include spectroscopic- and chromatographic-based techniques. Spectroscopic techniques focused on the near- and mid-infrared regions of the electromagnetic spectrum. Wang et al. (2006) successfully applied total reflectance infrared spectroscopy (MIR-ATR) and fiber optic diffuse reflectance near infrared spectroscopy (FODR-NIR) for the identification and authentification of camellia oils adulterated with soybean oil. Adulteration was determined based on slight differences in the raw spectra in the MIR ranges of 1132–885 $cm^{-1}$ and NIR ranges of 6200–5400 $cm^{-1}$ between the pure camellia oil and those adulterated with soybean oil. Such differences reflect the compositional variation between the two oils, with oleic acid and linoleic acid representing the main constituent in camellia oil and soybean oil, respectively. Adulteration levels of 5–25% w/w could be successfully detected at the 95% confidence level; however, the authors caution that additional work, incorporating larger numbers of camellia oils and adulterants of increased complexity, is required to further validate this rapid and cost-effective methodology. Near Infrared Raman spectroscopy was also applied to the quantitative determination of adulteration in *Camellia oleifera* Abel oil (Weng et al., 2006). Results showed a direct relationship existed between the intensity ratio $I_{v(1656)}/I_{v(1439)}$ with the magnitude of double bonds in the binary mixtures of the camellia oil blended with other edible oil. A linear relationship with a high correlation coefficient ($R^2 = 0.9938$) between the aforementioned Raman intensity ratio and the percentage of camellia oil was obtained, which might determine the authenticity of the camellia oils collected from various markets.

Chromatographic methods were developed (Kurata et al., 2005; Yamaguchi & Kurata, 2005) for the discrimination of animal and vegetable fats including camellia oil, whereby discrimination was facilitated through fatty acid and sterol compositions, respectively. More specifically, adulteration of camellia seed oil and sesame oil with maize oil was investigated using an electronic nose (E-nose) (Hai & Wang, 2006). Results showed that signals corresponding to the three different oils were significantly different; however, while principal component analysis (PCA) could successfully discriminate adulteration in sesame oil, the same results were not achieved in the case of camellia seed oil. The use of an artificial neural network incorporating the E-nose results can be used in the quantitative determination of adulteration in sesame oil, but could not quantitatively predict the percentage adulteration in camellia seed oil. Such results therefore imply that further method development is necessary before E-nose can be routinely employed for adulteration determinations. These results may also indicate the apparent chemical complexity of camellia oil.

## References

Akihisa, T.; K. Arai; Y. Kimura; K. Koike; W. Kokke; T. Shibata; T. Nikaido. Camelliols a-C, three novel incompletely cyclized triterpene alcohols from sasanqua oil (*Camellia sasanqua*). *J. Nat. Prod.* **1999,** *62,* 265–268.

Akihisa, T.; H. Tokuda; M. Ukiya; T. Suzuki; F. Enjo; K. Koike; T. Nikaido; H. Nishino. 3-Epi-cabraleabydroxylactone and other triterpenoids from camellia oil and their inhibitory effects on epstein-barr virus activation. *Chem. Pharm. Bull.* **2004,** *52,* 153–156.

Akihisa, T.; K. Yasukawa; Y. Kimura; S. Takase; S. Yamanouchi; T. Tamura. Triterpene alcohols from camellia and sasanqua oils and their anti-inflammatory effects. *Chem. Pharm. Bull.* **1997,** *45,* 2016–2023.

Akihisa, T.; K. Yasukawa; Y. Kimura; S. Yamanouchi; T. Tamura. Sasanquol, a 3,4-seco-triterpene alcohol From sasanqua oil, and its anti-inflammatory effect. *Phytochemistry* **1998,** *48,* 301–305.

Andlauer, W.; P. P. Furst. Nutraceuticals: a Piece of History, Present Status and Outlook, Food Res. Int. **2002,** *35,* 171–176.

Anon. (2000a) Nature Products: Network. Wholesale list of TianZi's Medicinal, Agricultural and Horticultural Crops from China, http://www.natureproducts.net/Wholesale/Camellia.html. Last accessed 16 June 2007.

Anon. *Dietary Reference Intakes for Vitamin C, Vitamin E, Selenium and Carotenoids;* 2000b; p. 246.

Anon. (2005) Victani Products. http://www.victani.com/camellia.asp. Last accessed 3 July 2007.

Anon. (2006a) Ministry of Commerce of the People's Republic of China, Jiangxi Qinglong Oil Company: Pioneer of Oil Maker. http://bzb2.mofcom.gov.cn/aarticle/representativecases/200608/20060802961074.html. Last accessed 20 June 2007.

Anon. (2006b) Fairfax Digital, The Age, Move Over, Olive, Camellia Has the Good Oil. http://www.theage.com.au/articles/2006/11/30/1164777723114.html. Last accessed 23 August 2007.

Anon. (2007a) International Plant Nutrition Institute, Oil-Tea and Its Production in China. http://www.ipni.net/ppiweb/sechina.nsf/$webindex/CB8B90655EB7426648256F63002DFC7A?opendocument&navigator=special+crops. Last accessed 16 June 2007.

Anon. (2007b) Oils by Nature Inc. Camellia Oil (Expeller Pressed). http://www.oilsbynature.com/products/camellia-oil.htm. Last accessed 6 June 2007.

Anon. (2007c) PR Newswire: News & Information, A New Ally in Healthy Cooking. http://www.prnewswire.com/cgi-bin/stories.pl?ACCT=109&STORY=/www/story/02-23-2006/0004288384&EDATE=. Last accessed 30 July 2007.

Bauernfeind, J. *Vitamin E: A Comprehensive Treatise*; Marcel Dekker: New York, 1980; p. 99.

Beardsell, D.; J. Francis; D. Ridley; K. Robards. Health Promoting Constituents in Plant Derived Edible Oils.*J. Food Lipids* **2002,** *9,* 1–34.

Brady, A.; R. Loughlin; D. Gilpin; P. Kearney; M. Tunney. In vitro activity of tea-tree oil against clinical skin isolates of meticillin-resistant and -sensitive *Staphylococcus aureus* and coagulase-negative staphylococci growing planktonically and as biofilms. *J. Med. Microbiol.* **2006, *55,*** 1375–1380.

Brufau, G.; M.A. Canela; M. Rafecas. Phytosterols: physiologic and metabolic aspects related to cholesterol-lowering properties. *Nutr. Res.* **2008,** *28,* 217–225.

Budiyanto, A.; N.U. Ahmed; A. Wu; T. Bito; O. Nikaido;T. Osawa; M. Ueda; M. Ichihashi. Protective effect of topically applied olive oil against photocarcinogenesis following UVb exposure of mice. *Carcinogenesis* **2000,** *21,* 2085–2090.

Central South Forestry University. *The Utilization and Analysis of Non-wood Forest Products;* Chinese Forestry Publishing House: Beijing, China, 1986.

Chen, F.; X. Wang; J.C.S. Chen.(2003) Composition and Sensory Evaluation of Tea Seed Oil. http://greenteaseedoil.com/media/AOCS-2003-tea.seed1.ppt. Last accessed 30 August, 2007.

Chen, Q.; X.W. Zhang. The technique of camellia oil with type 95 screw presser. *Quarterly of Forest By-Product and Speciality in China* **1999,** *199,* 23–24.

Christy, A.A.; P.K. Egeberg. Quantitative determination of saturated and unsaturated fatty acids in edible oils by infrared spectroscopy and chemometrics. *Chemometrics and Intelligent Laboratory Systems* **2006,** *82,* 130–136.

Deng, C.H.; N. Liu; M.X. Gao; X.M. Zhang. Recent developments in sample reparation techniques for chromatography analysis of traditional chinese medicines. *J. Chromatogr. A* **2007,** *1153,* 90–96.

Duke, J.A.; E.S. Ayensu. *Medicinal Plants of China,* Reference Publications: Algonac, MI, 1985.

Dupont, J.; P.J. White; K.M. Johnston; H.A. Heggtveit; B.E. Mcdonald; S.M. Grundy; A. Bonanome. Food safety and health-effects of canola oil. *J. Am. Coll. Nutr.* **1989**, *8,* 360–375.

Fang, J. Advances in science and technology on tea oil tree and tung oil tree in china. *Forest Sci.* **1994,** *7,* 30–38.

Fang, Y.D.; L.Q. Liu; H. Li. The pentacyclic triterpene alcohol and tetracyclic triterpene alcohols in unsaponifiable matters from camellia seed oil. *J. Chinese Cereals Oils Assoc.* **1999,** *14,* 18–21, 26.

Francis, G.; Z. Kerem; H.P.S. Makkar; K. Becker. The biological action of saponins in animal systems: a review. *Br. J. Nutr.* **2002,** *88,* 587–605.

General Administration of Quality Supervision, Inspection and Quarantine of People's Republic of China. (2003) Chinese National Standard, Oil-tea camellia oil, Chinese Standard Publishing

House.

Griguol, V.; M. Leon-Camacho; I.M. Vicario. Review of the levels of *trans* fatty acids reported in different food products. *Grasas Y Aceites* **2007,** *58,* 87–98.

Hai, Z.; J. Wang. Detection of adulteration in camellia seed oil and sesame oil using an electronic nose. *Eur. J. Lipid Sci. Technol.* **2006,** *108,* 116–124.

He, F.; B. He; Z.H. Li; R.Q. Zhong; J.C. Dai; Y.F. Sun. Criterion constituion of quantity classification of oil tea camellia. *Nonwood Forest Research,* **2004,** *22*(4), 105–108.

Herzallah, S.M.; M.A. Humeid; K.M. Al-Ismail.Effect of heating and processing methods of milk and dairy products on conjugated linoleic acid and *trans* fatty acid isomer content. *J. Dairy Sci.* **2005,** *88,* 1301–1310.

Hong, Y.S. A study on metabolic effects of lipid supplemented diets. *Korea University Med. J.* **1988,** *25,* 829–842.

Huang, F.H.; W.L. Li; F.J. Xia; Y.X. Niu. Research and application of dehulling machine for *camellia oleifera* seed. *Transactions of the CSAE* **2006,** *22,* 147–151.

Huang, W.W.; C.W. Ao; H.Y. Zhong. The antimicrobial effect of saponin of thea sasanqua. *Nonwood Forest Res.* **2002,** *20,* 17–19.

Kalua, C.M.; M.S. Allen; D.R. Bedgood; A.G. Bishop; P.D. Prenzler. Discrimination of olive oils and fruits into cultivars and maturity stages based on phenolic and volatile compounds. *J. Agric. Food Chem.* **2005,** *53,* 8054–8062.

Kalua, C.M.; D.R. Bedgood; A.G. Bishop; P.D. Prenzler. Changes in volatile and phenolic compounds with malaxation time and temperature during virgin olive oil production. *J. Agric. Food Chem.* **2006a,** *54,* 7641–7651.

Kalua, C.M.; R.J. Mailer; J. Ayton; M.S. Allen; D.R. Bedgood; A.G. Bishop; P.D. Prenzler. Discrimination of olive oils and fruits into cultivars and maturity stages based on phenolic and volatile compounds (Vol. 53, pp. 8054, 2005). *J. Agric. Food Chem.* **2006b,** *54,* 8390.

Karande, P.; S. Mitragotri. Dependence of skin permeability on contact area. *Pharm. Res.* **2003,** *20,* 257–263.

Kitahara, M.; F. Ishiguro; K. Takayama; K. Isowa; T. Nagai.Evaluation of skin damage of cyclic monoterpenes, percutaneous-absorption enhancers, by using cultured human skin cells. *Biol. Pharm. Bull.* **1993,** *16,* 912–916.

Kubow, S. The influence of positional distribution of fatty acids in native, interesterified and structure-specific lipids on lipoprotein metabolism and atherogenesis. *J. Nutr. Biochem.* **1996,** *7,* 530–541.

Kurata, S.; K. Yamaguchi; M. Nagai. Rapid discrimination of fatty acid composition in fats and oils by electrospray ionization mass spectrometry. *Anal. Sci.* **2005,** *21,* 1457–1465.

Ledoux, M.; L. Laloux; D. Sauvant. *Trans* fatty acid isomers: origin and occurrence in food. *Sci. Aliments* **2000,** *20,* 393–411.

Lee, C.P.; G.C. Yen. Antioxidant activity and bioactive compounds of tea seed (*camellia oleifera* abel.) oil. *J. Agric. Food Chem.* **2006,** *54,* 779–784.

Li, D.M.; J. Wang; L.W. Bi; Z.D. Zhao. Influence of extraction method on the bioactive component squalene in seed oil of *camellia oleifera. Biomass Chem. Eng.* **2006,** *40,* 9–12.

Li, J.G.; C.T. Ho; H. Li; H.R. Tao; L.Q. Liu. Separation of sterols and triterpene alcohols from unsaponifiable fractions of three plant seed oils. *J. Food Lipids* **2000,** *7,* 11–20.

Li, Y.S.; H.B. Xue. Studies on the chemical and physical properties of saponin of thea sasanqua as well as development utilization. *Acta Botanica Boreali-Occidentalia Sinica* **1994,** *5,* 149–153.

Liu, L.; Z.H. Zhao. Study on polar antioxidative compounds in tea (*camellia sinensis*) seed oil. *J. Chin. Cereals and Oils Assoc.* **2002,** *17,* 4–9.

Lu, S.Z.; H. Zen; X.Q. Cai. The refining technology of camellia oil. *Guangxi Forestry Sci.* **2003,** *32,* 182–184.

Miura, D.; Y. Kida; H. Nojima. Camellia oil and its distillate fractions effectively inhibit the spontaneous metastasis of mouse melanoma bl6 cells. *FEBS Lett.* **2007,** *581,* 2541–2548.

Morimoto, K., T. Haruta; H. Tojima; V. Takeuchi.Enhancing mechanisms of saturated fatty-acids on the permeations of indomethacin and 6-carboxyfluorescein through rat skins. *Drug Dev. Ind. Pharm.* **1995,** *21,* 1999–2012.

Morimoto, K.; H. Tojima; T. Haruta; M. Suzuki; M. Kakemi. Enhancing effects of unsaturated fatty acids with various structures on the permeation of indomethacin through rat skin. *J. Pharm. Pharmacol.* **1996,** *48,* 1133–1137.

Niankang, X.; W. Kongxion; C. Xiangping; G. Xueqin; Z. Shiyou. The studies on the results of improving the low-yield stands of *camellia oleifera* by means of grafting clones and the determination of its clones. *Forest Sci.* **1996,** *9,* 184–188.

Pan, C.R.; J.Y. Lin; S.L. Qiu. Technology for increasing oleic acid content in camellia oil. *Transactions of the CSAE* **2006,** *22,* 163–165.

Pang, W.S.; X.R. Mao. The investigation of refining technology of camellia oil. *China Forestry Sci. Technol.* **2004,** *18,* 54–55.

Phillips, K.M.; D.M. Ruggio; J.I. Toivo; M.A. Swank; A.H. Simpkins. Free and esterified sterol composition of edible oils and fats. *J. Food Comp. Anal.* **2002,** *15,* 123–142.

Precht, D.; J. Molkentin. Recent trends in the fatty acid composition of german sunflower margarines, shortenings and cooking fats with emphasis on individual c16:1, c18:1, c18:2, c18:3 and c20:1 *trans* isomers. *Nahrung-Food* **2000,** *44,* 222–228.

Pykett, M.A.; A.H. Craig; E. Galley; C. Smith; S.P. Long. Skincare Composition Against Free Radicals. 27 March 2007. U.S. Patent 07195787. The Boots Company PLC.

Rolfe, J. *Gardening with Camellias: A Complete Guide;* Kangaroo Press: Kenthurst, NSW, 1992.

Ruter, J.M. Nursery production of tea oil camellia under different light levels. *Trends in New Crops and New Uses,* 222–224, ASHS Press: Alexandria, VA, 2002.

Sabetay, S. Camellia seed oil: the seed oil of *camellia japonica* l. and its uses in cosmetology and dermo-pharmacy. *Soap Perfumery Cosmetics* **1972,** *45,* 244, 252.

Savige, T.J. *The International Registrar of the Genus Camellia,* The International Camellia Society: Wirlinga, Australia, 1993.

Shanan, H.; G. Ying. The comprehensive utilization of camellia fruits. *Am. Camellia Yearbook* **1982,** *37,* 104–107.

Sokol'skii, I.N.; A.I. Ban'kovskii; E.P. Zinkevich.Triterpene glycosides from *camellia oleifera* and *camellia sasanqua*. *Chem. Nat. Compd.* **1976,** *11,* 116–117.

Tang, L.; E. Bayer; R. Zhuang. Obtain, properties and utilization of chinese teaseed oil. *Fett Wissenschaft Technologie-Fat Sci. Technol.* **1993,** *95,* 23–27.

Tasan, M.; M. Demirci. *Trans* fatty acids in sunflower oil at different steps of refining. *J. Am. Oil Chem. Soc.* **2003,** *80,* 825–828.

Torto, N.; L.C. Mmualefe; J.F. Mwatseteza; B. Nkoane; L. Chimuka; M.M. Nindi; A.O. Ogunfowokan. Sample preparation for chromatography: an african perspective. *J. Chromatogr. A* **2007,** *1153,* 1–13.

Ulberth, F.; M. Buchgraber. Authenticity of fats and oils. *Eur. J. Lipid Sci. Technol.* **2000,** *102,* 687–694.

Wang, A.P.; T. Seki; D. Yuan; Y. Saso; O. Hosoya; S. Chono; K. Morimoto. Effect of camellia oil on the permeation of flurbiprofen and diclofenac sodium through rat and pig skin. *Biol. Pharm. Bull.* **2004,** *27,* 1476–1479.

Wang, C.I.; H.W. Yin; W.Y. Liu. Stability examination of oiltea oil and analysis of oil's tocopherol and sterol components. *Bull. Taiwan Forestry Res. Inst. New Series* **1994,** *9,* 73–85.

Wang, C.L.; Y.H. Lin. The extraction and analysis of oils from selected species of oil tea camellia in taiwan. *Bull. Taiwan Forestry Res. Inst. New Series* **1990,** *5,* 11–16.

Wang, L.; F.S.C. Lee; X.R. Wang; Y. He. Feasibility study of quantifying and discriminating soybean oil adulteration in camellia oils by attenuated total reflectance MIR and fiber optic diffuse reflectance NIR. *Food Chem.* **2006,** *95,* 529–536.

Weng, R.H.; Y.M. Weng; W.L. Chen. Authentication of *camellia oleifera* abel oil by near infrared fourier transform raman spectroscopy. *J. Chin. Chem. Soc.* **2006,** *53,* 597–603.

Wolff, R.L. *Trans*-polyunsaturated fatty-acids in french edible rapeseed and soybean oils. *J. Am. Oil Chem. Soc.* **1992,** *69,* 106–110.

Wu, K.Y.; Y.X. Weng; X.Q. Fei; W.Q. Yang; X.Z. Sun. Comparison of antisenile effects of seed oil of camellia grijsii and certain other oil from woody crops on 2bs cell culture. *Forest Res.* **1998,** *11,* 355–360.

Wu, X.H.; B.G. Chen; Y.F. Wang; L. Chen; W. Zhou. Study on technology of supercritical $co_2$ extraction of the camellia oils. *Food Sci. Technol.* **2007,** 139–141.

Xia, L.; A. Zhang; T. Xiao. An introduction to the utilization of camellia oil in china. *Am. Camellia Yearbook* **1993,** *48,* 12–15.

Xu, J.S.; S. Meguro; S. Kawachi. Oil comparison of camellia species of japan and china. *Mokuzai Gakkaishi* **1995a,** *41,* 92–97.

Xu, J.S.; S. Meguro; S. Kawachi. Variations of fatty-acid compositions in growing and stored camellia seeds. *Mokuzai Gakkaishi* **1995b,** *41,* 98–102.

Xu, X.B.; X.Z. Hu; G.W. Zhang. CBE from teaseed oil by lipase catalysed modification. *Proceedings of International Symposium and Exhibition on New Approaches in the Production of Food Stuffs and Intermediate Products from Cereal Grains and Oilseeds;* Beijing, China, 1994.

Yamaguchi, K.; S. Kurata. Forensic discrimination of unsaponifiables of fats and oils using gas chromatography/mass spectrometry. *Bunseki Kagaku* **2005,** *54,* 1091–1100.

Yao, X.H.; K.L. Wang; X.F. Luo; H.D. Ren; B.C. Gong; X.Q. Fei. The present state and development of industrialization of oiltea camellia. *China Forestry Sci. Technol.* **2005,** *19,* 3–6.

Yasukawa, K.; M. Takido; T. Matsumoto; M. Takeuchi; S. Nakagawa. Sterol and triterpene derivatives from plants inhibit the effects of a tumor promoter, and sitosterol and betulinic acid inhibit tumor-formation in mouse skin 2-stage carcinogenesis. *Oncology* **1991,** *48,* 72–76.

Yoshida, H.; S. Abe; Y. Hirakawa; S. Takagi. Roasting effects on fatty acid distributions of triacylglycerols and phospholipids in sesame (sesamum indicum) seeds. *J. Sci. Food. Agric.* **2001a.** *81,* 620–626.

Yoshida, H.; Y. Hirakawa; S. Abe. Influence of microwave roasting on positional distribution of fatty acids of triacylglycerols and phospholipids in sunflower seeds (*Helianthus annuus* L.). *Eur. J. Lipid Sci. Technol.* **2001b,** *103,* 201–207.

Yu, B.; K.H. Kim; P.T.C. So; D. Blankschtein; R. Langer. Evaluation of fluorescent probe surface intensities as an indicator of transdermal permeant distributions using wide-area two-photon fluorescence microscopy. *J. Pharm. Sci.* **2003,** *92,* 2354–2365.

Yu, Y.S.; S.X. Ren; K.Y. Tan. Study on the climatic regionalization and layer and belt distribution of oiltea camellia quality in china. *J. Natural Res.* **1999,** *14,* 123–127.

Zhai, F.Y.; S.F. Du; B.M. Popkin; J.C. Wallingford; R. Yuhas. Fatty acids in chinese edible oils: value of direct analysis as a basis for labeling. *Food Nutr. Bull.* **2004,** *25,* 330–336.

Zhan, Y. Animal Feed Compositions and Uses of Triterpenoid Saponin Obtained From Camellia L. Plants. 1999. U.S. Patent 6,007,822.

Zhang, Y. Study on Refining Technics of Oil-Tea Camellia Oil for Cosmetics. Master's Dissertation, 2003. Central South University of Forestry and Technology.

Zhong, H.Y.; D.R. Bedgood; A.G. Bishop; P.D. Prenzler; K. Robards. Effect of added caffeic acid and tyrosol on the fatty acid and volatile profiles of camellia oil following heating. *J. Agric. Food Chem.* **2006,** *54,* 9551–9558.

Zhong, H.Y.; D.R. Bedgood; A.G. Bishop; P.D. Prenzler; K. Robards. Endogenous biophenol, fatty acid and volatile profiles of selected oils. *Food Chem.* **2007a,** *100,* 1544–1551.

Zhong, H.Y.; Y.H. Huang; Q.H. Long; Z.H. Li; P. Prenzler; K. Robards. Effect of SPME fibers on volatile extraction of camellia oil. *J. Chin. Cereals and Oils Assoc.*, **2007b.**

Zhong, H.Y.; C.N. Wan; B.X. Xie. Supercritical $co_2$ extraction of oiltea camellia seed oil. *Food and Machinery* **1999,** *2,* 13–14.

Zhong, H.Y.; C.N. Wan; B.X. Xie. The effect of supercritical $co_2$ extracting conditions on the quality of oil-tea camellia seed oil.*J. Chin. Cereals and Oils Assoc.* **2001,** *16,* 9–13.

Zhong, H.Y.; C.N. Wan; B.X. Xie; M.Y. Zhao. The bleaching technique and colour determination of oil-tea camellia seed oil. *J. Central South Forestry University* **2000,** *4,* 25–29.

Zhong, H.Y.; C.N. Wang; J.P. Huang; B.X. Xie; Y. Liu. The technique of solid-state fermentation on oil-tea camellia defatted cake. *J. Central South Forestry University* **2001,** *21,* 21–25.

Zhong, H.Y.; Y.Q. Zhang; H.Z. Sun; Z.H. Li.Hydrated degumming technique for camellia oil. *Nonwood Forest Res.* **2004,** *22,* 29–31.

Zhuang, R. *Chinese Oil-tea Camellia,* Chinese Forestry Publishing House: Beijing, China, 1988.

# 12

# Pumpkin Seed Oil

**Michael Murkovic**
*Graz University of Technology, Institute for Food Chemistry and Technology*
*Petersgasse 12/2, A-8010 Graz, Austria*

## Introduction

People have cultivated cucurbits (*Cucurbitaceae*) for over 10,000 years to provide food and other products for humankind. The main cultivated species today include cucumber, gherkins, melons, muskmelon, watermelons, pumpkins, and squashes. The cucurbits are a significant worldwide food crop, mostly on account of their fruits. The edible plant parts also include flowers, leaves, shoot tips, storage roots, and seeds.

Pumpkin seed oil, a local specialty produced mainly in southeastern Austria, is extracted by physical means from the seeds of a variety of *Cucurbita pepo* that do not fully develop seed coats, and is denominated as *C. pepo* var. *styriaca* or var. *oleifera*. Teppner (2004) gives a detailed description of the oil pumpkin from a botanical point of view. Styrian pumpkin seed oil, the product, is protected within the European Union as PGI (protected geographical indication): http://www.lebensmittelnet.at/article/articleview/49482/1/17831). Similar types of oil are produced in certain regions of eastern Slovenia, northwestern Croatia, and the adjacent regions of Hungary.

The production area in Austria is comprised of about 16,300 ha with a production of 7,700 tons of seeds (Rupprechter et al., 2006). Some of the seeds are used for snacks as well as for bread making, and some are used for oil production. Additionally, the seeds are used for many meat and vegetable dishes, as well as in soups and desserts, imparting these foods with a distinct aroma and taste. The production area in Austria changes from year to year, mainly depending on the demand for the oil and the expected price for the seeds.

## Development of New Varieties

The main objectives of current breeding programs in Austria are to increase the harvest, to optimize the oil productivity, and to improve the technological properties, especially the separation of the seeds from the placenta. One can achieve the optimization of oil productivity by increasing the seed yield, the seed size, or the seed number within the fruits. The seed size is not an important trait for oil productivity but for using the pumpkin seeds as a snack. In high-yielding varieties, the fruit weight is in the range of 1.0–1.5 kg, and the seed weight's range is 170–220 mg. Each fruit contains 275–425 seeds (Cui & Loy, 2002; Lelley et al., 2008).

An outbreak of zucchini yellow mosaic virus (ZYMV) in the Austrian pumpkin fields in 1997 showed that the varieties grown in Austria did not have a genetic resistance to this virus. This outbreak initiated a breeding program to introduce a resistance gene in the *C. pepo* breeds. As resistance was not described in *C. pepo,* the origin of all resistant cultivars is a Nigerian landrace of *C. moschata* called Nigerian Local. The crossing of the resistance into the oil pumpkin was done via zucchini. The introduction of the resistance gene was successful, and currently stable production lines are being developed (Lelley et al., 2000).

## Oil Extraction and Processing

For extraction of the oil, small batches of seeds (50–100 kg) are milled, and a small amount of salt and water is added prior to roasting. The seed meal is then roasted at about 115°C for 50–60 minutes. During the heating, the water evaporates, and the oil emerges from the subcellular structures, giving the ground meal an oily consistency. The oil dissolves the green color that is present in the thin hulls of the seeds. This color is attributed to the presence of mainly protochlorophyll (Teppner, 2004). Figure 12.1 shows a typical temperature profile during roasting.

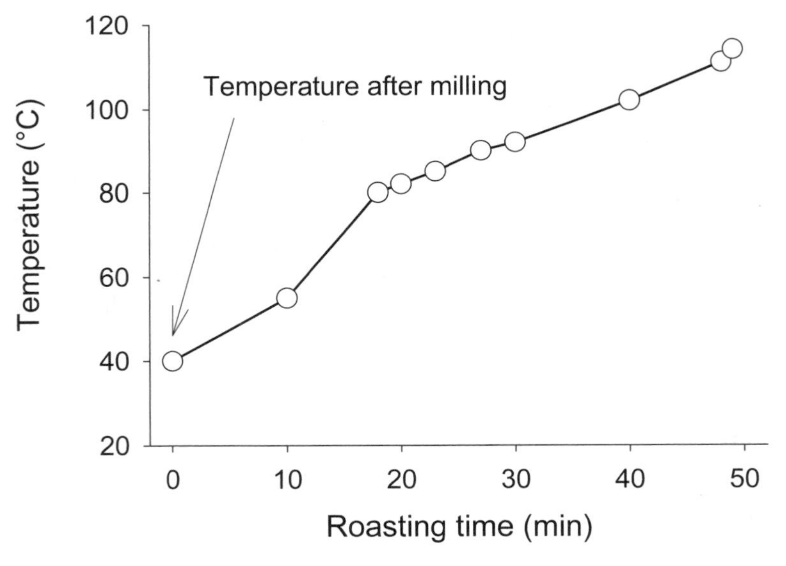


**Fig. 12.1.** Temperature increase of milled pumpkin seeds during the "roasting" process.

Many chemical changes take place during roasting, including the formation of the typical aroma that is associated with a nutty impression. The aroma is comprised of substances arising from: (i) the Maillard reaction (e.g., 2,5(6)-dimethyl-pyrazine, 2-ethyl-5(6)methyl-pyrazine, 2-methyl-pyrazine, 3(2)-ethyl-2,5(3,6)-dimethyl-pyrazine, and 2-acetyl-pyrrole), (ii) fatty-acid oxidation (e.g., hexanal, *E*-2-heptenal, 2-pentyl-furan, pentenal, and 2-butanone), and (iii) amino-acid degradation (e.g., dimethyl sulfide) (Siegmund & Murkovic, 2004). The Sontag group showed that the lignans are degraded during the first 20–30 minutes of the roasting process. The temperature reached at this time is around 95°C (Murkovic et al., 2004).

After the roasting process, the oily meal (still having a temperature of ca. 80°C) is extracted using a ram press. The freshly pressed oil is stored in small stainless-steel tanks to allow the particulate substances to settle. Without any further treatment, the oil is transferred to dark-glass bottles that one can store for several months. The press cake—which contains all the protein, carbohydrates, and dietary fiber as well as up to 14% of remaining fat—is mainly used for animal feed. The product is dark-green oil that shows a strong red fluorescence.

## Pumpkin Seed Oil Composition

### Fatty Acids

The dominant fatty acids occurring in different pumpkin- seed oils are palmitic acid (16:0), stearic acid (18:0), oleic acid (18:1), and linoleic acid (18:1) (Table 12.1; Stevenson et al., 2007). In Styrian pumpkin seed oil–which is produced from *C. pepo* seeds–the average content of linoleic acid, the most abundant fatty acid, is 54.2% (range: 35.6–60.8%), and oleic acid is 26.6% (21.0–46.9%) (Fig. 12.2). The saturated fatty acids occur at lower concentrations, with a palmitic-acid range of 9.5–15.9% and a stearic-acid range of 3.1–7.4%. These four fatty acids together represent 98.3% of the total fatty acids with a very narrow distribution. All other fatty acids are present at very low concentrations. Recent data from Bravi et al. (2006) were also in the abovementioned ranges. Applequist and co-workers (2006) compared different *Cucurbita* species and varieties there of and they also obtained similar results. In contrast to these results, Spangenberg and Ogric (2001) found a much greater variation in the fatty-acid composition.

Interestingly, good linear correlation of oleic acid to linoleic acid exists in pumpkin seeds (Fig. 12.3). This might reflect the direct desaturation of oleic acid to linoleic acid with no significant new synthesis of oleic acid at the stage of desaturation. The desaturation process might be temperature-induced similar to that in sunflower seeds that were investigated in detail (Garces et al., 1992). Lower temperatures induce desaturase activity, which is reflected in the composition of the pumpkin seeds harvested in different climatic conditions. This effect was shown when pumpkin seeds were investigated at either different years with different climates or when pumpkins

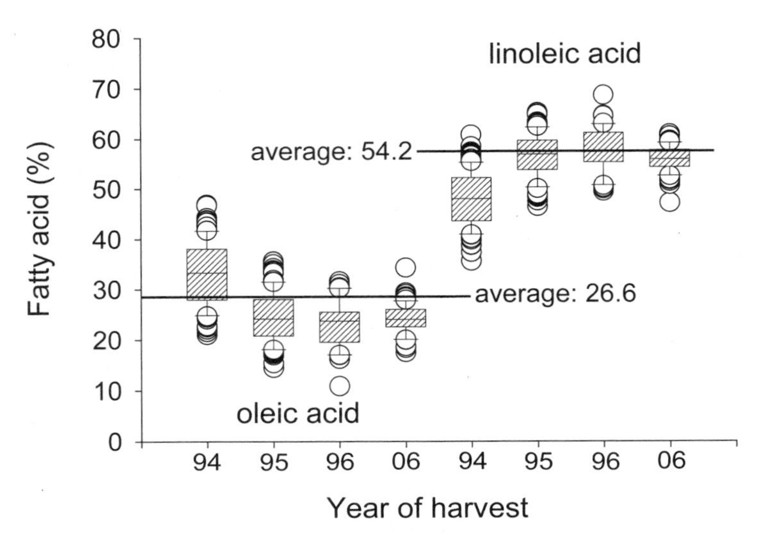


**Fig. 12.2.** Box-Plot of the fatty-acid distribution of the oil of breeding lines of *Cucurbita pepo* in the years 1994–1996 and 2006 (updated from Murkovic et al., 1996b).

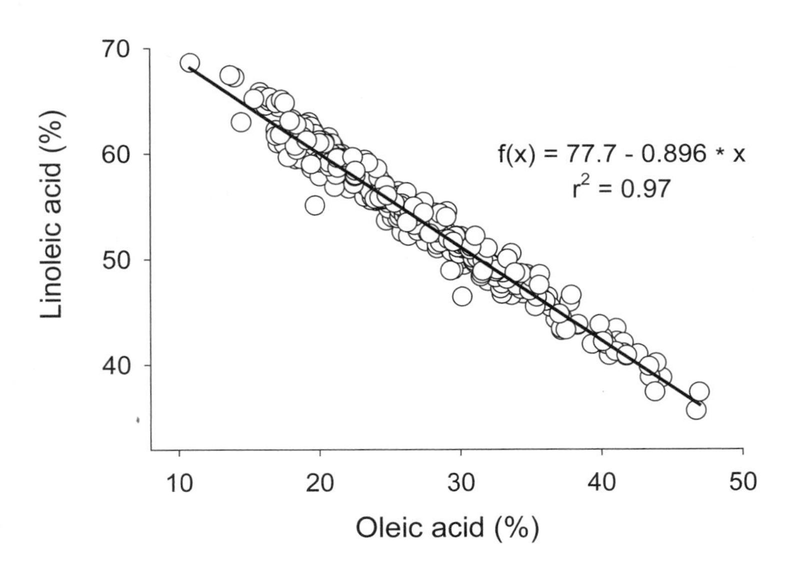


**Fig. 12.3.** Correlation of oleic acid to linoleic acid in the analyzed samples of the years 1994–1996 (Murkovic et al., 1996b).

harvested later in the year were exposed to colder temperatures, especially during the nights for a longer time (Murkovic et al., 1999).

For other species like *C. maxima* L., the fatty-acid distribution in the seed's triglycerides is very similar to *C. pepo* (Stevenson et al., 2007). Additionally, new varieties of *C. moschata* were identified by Lelley as also having a similar fatty-acid composition (Lelley et al., in press). The seeds of this variety have a similar morphology as the Styrian pumpkin seeds but without the green color. As the green color can act as a photosensitizer, likely, the oil from these seeds has a better light stability.

Stevenson and co-workers (2007) reviewed the existing literature on the oil content and fatty-acid composition of pumpkin seeds from different varieties and species. In all *Cucurbita, Langenaria, Luffa, Momordica, Telfairia,* and *Trichosanthes* species and varieties, the fatty-acid composition is dominated by the same four fatty acids: palmitic acid, stearic acid, oleic acid, and linoleic acid. Other fatty acids are occurring in very low concentrations, and ω-3 fatty acids are not found in any sample (Table 12.1).

## Vitamin E

The vitamin-E content in the pumpkin seed oil shows a great variability. All eight vitamers are found in the oil. However, α- as well as γ-tocopherol are the most prominent vitamers occurring in the oil. Besides the tocopherols, significant amounts of tocotrienols are also found in the oil. γ-Tocopherol occurs at concentrations in the range of

**Table 12.1. Average Content of Oil and Distribution of the Four Dominant Fatty Acids (Stearic Acid, Palmitic Acid, Oleic Acid, Linoleic Acid) and Total Unsaturated Fatty Acids (%) in Seeds of Different Pumpkin Species (adapted from Stevenson et al., 2007)**

| | | Relative fatty-acid composition (%) | | | | |
|---|---|---|---|---|---|---|
| Pumpkin species | Oil content (%) | C16:0 | C18:0 | C18:1 | C18:2 | Total fatty acids* |
| Cucurbita argyrosperma | 36.0 | 13.6 | 8.6 | 30.8 | 44.8 | 71.9 |
| C. ficifolia | 43.5 | 10.6 | 5.3 | 24.9 | 56.6 | 82.8 |
| C. foetidissima | 32.3 | 10.0 | 3.5 | 28.6 | 57.7 | 85.6 |
| C. maxima | 35.4 | 14.4 | 6.2 | 36.0 | 41.3 | 77.6 |
| C. mixta | 50.6 | 13.8 | 5.9 | 21.4 | 58.9 | 80.3 |
| C. moschata | 37.8 | 16.6 | 6.7 | 24.0 | 51.6 | 75.0 |
| C. pepo | 38.3 | 16.2 | 7.5 | 32.2 | 41.1 | 73.6 |
| Langenaria sicceraria | 21.9 | 13.0 | 5.1 | 11.4 | 71.0 | 80.3 |
| Luffa acutangula | 44.3 | 20.9 | 10.8 | 24.1 | 43.7 | 67.9 |
| Momordica charantia | unknown | 2.8 | 21.7 | 30.0 | — | 73.7 |
| M. cochinchinensis | unknown | 5.6 | 60.5 | 9.5 | 20.3 | 34.5 |
| Telfairia occidentalis | 45.4 | 16.3 | 13.5 | 29.8 | 39.6 | 69.3 |

*Total unsaturated fatty acids.

41–620 mg/kg, and α-tocopherol in the range of 0–140 mg/kg. This means that the concentration of γ-tocopherol is 5 to 10 times higher than that of α-tocopherol. The concentration of β- and δ-tocopherol is normally below that of α-tocopherol. Figure 12.4 shows the correlation of α- to γ-tocopherol (from Murkovic et al., 1996a).

## Sterols

The main steroids found in pumpkin seed oil are typically Δ7-sterols. Mandl and coworkers (1999) analyzed these sterols in pumpkin seed oil, and found Δ7,22,25-stigmastatrienol (326 µg/mL of oil), and Δ7-avenasterol (164 µg/mL of oil). They were not able to separate Δ7-stigmastenol and Δ7,25-stigmastadienol (sum: 310 µg/mL of oil). Additionally, they found the Δ5-sterols β-sitosterol (58 µg/mL of oil) and spinasterol (300 µg/mL of oil).

## Pigments

Seeds of *C. pepo* contain pigments—mainly protochlorophylls—in the innermost layer of the chlorenchyma (Teppner, 2004). The identified green pigments comprise four different esters of protochlorophyllide and protopheophorbide, respectively (Mukaida et al., 1993). Chlorophylls are not formed inside the fruits since the com-

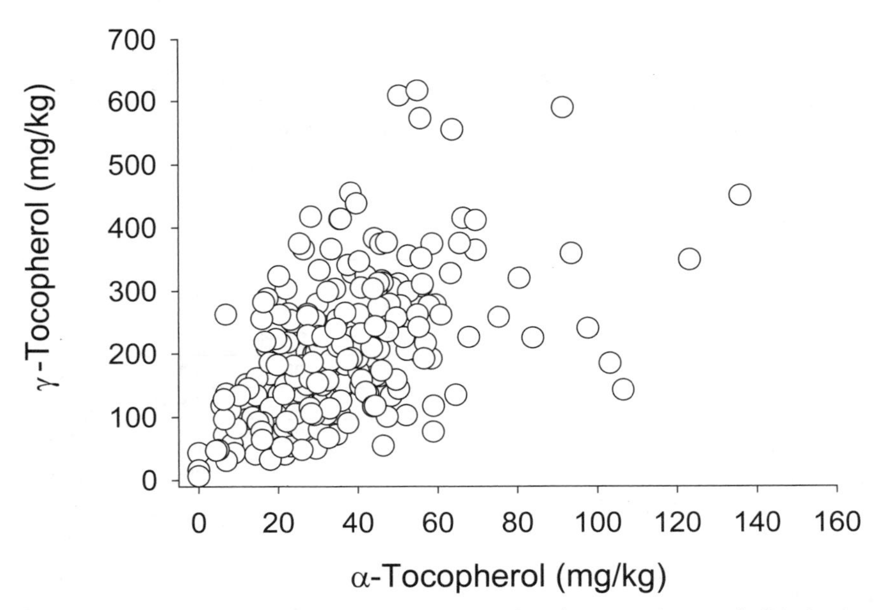


**Fig. 12.4.** Correlation of α-tocopherol and γ-tocopherol in pumpkin seeds (Murkovic et al., 1996a).

plete synthesis requires a light-induced reaction (Mukaida et al., 1993). In addition to the green protochlorophylls (Fig. 12.5; Schoefs, 2000), some carotenoids are present in the seeds and in the oil thereof. Quantitative data on these substances are not available. Lutein is found especially in the pumpkin flesh and the seeds. Matus et al. (1993) found some carotenoids in the defatted seed meal. The main components of the press-residue and the oil were lutein [3,3'-dihydroxy-α-carotene = (3*R*,3'*R*,6'*R*)-β,ε-carotene-3,3'-diol; 52.5%] and β-carotene (β,ε-carotene; 10.1%). In addition to the abovementioned pigments, in small quantities, violaxanthin, luteoxanthin, auroxanthin epimers, lutein epoxide, flavoxanthin, chrysanthemaxanthin, 9(9')-*Z*-lutein, 13(13')-*Z*-lutein, 15-*Z*-lutein, (central-*Z*)-lutein, α-cryptoxanthin, β-cryptoxanthin, and α-carotene (β,ε-carotene) were identified. Matus and co-workers (1993) did not provide quantitative data. These carotenoids are not only found in the seeds and oil but also in the flesh of the fruits (e.g., Azevedo-Meleiro & Rodriguez-Amaya, 2007; Murkovic et al., 2004).

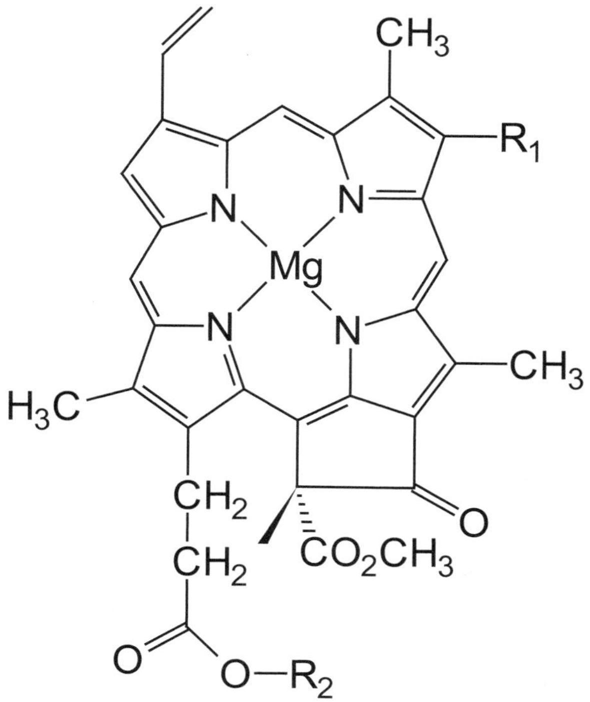

**Fig. 12.5.** Molecular structure of protochlorophyll ($R_1$: ethyl, vinyl; $R_2$: geranylgeraniol, dihydrogeranylgeraniol, tetrahydrogeranylgeraniol, and phytol).

## Lignans

Pumpkin seeds contain some bioactive lignans (see below). The extractability of these lignans in the oil is not known.

# Composition of Pumpkin Press Cake

The press cake contains all substances not soluble in the oil (fiber, protein, and carbohydrates). Due to the technology used (ram press), the remaining oil content in the press cake is 12 to 14% (Murkovic, unpublished results). The press cake is used for animal feed. Earlier work of Garber (1948) showed that the press cake contains up to 8% fat, 65 to 77% protein, 14 to 20% nitrogen-free extractable substances, 1 to 4% of dietary fiber, and 1 to 9% minerals. As salt is added during the roasting process to facilitate the oil extraction, most of it also remains in the press cake. Approximately 93% of the protein in the press cake is digestible (Garber, 1948).

Another group of potentially physiologically active substances is the lignans, especially secoisolariciresinol. The concentration in the seeds was determined to be 3.8 µg/g (Murkovic et al., 2004). However, secoisolariciresinol is destroyed during the roasting process. Adlercreutz and Mazur (1997) found a significantly higher amount of 200 µg/g. This group also found other phytoestrogens, such as genistein and daidzein, in pumpkin seeds, but the concentrations of these isoflavones are several orders of magnitude lower compared to soybean products. The lignan lariciresinol was identified by Sicilia et al. (2003) at trace levels. Experimental evidence in animals shows clear anti-carcinogenic effects of pure lignans in many types of cancer. However, many epidemiological results are conflicting, partly because the determinants of plasma metabolites are very different in different countries. The source of the lignans seems to play a role because other factors in the food obviously participate in the protective effects. The results are promising, but much work is still needed in this area of research (Adlercreutz, 2007).

# Authenticity and Adulteration

Due to the high price of the pumpkin seed oil, adulteration can occur. Erucic acid, which is not present in the pumpkin seed oil, was used earlier to detect adulteration with rapeseed oil. Since most rapeseed oil (called canola oil in the United States and Canada) is currently produced without erucic acid, a new method was developed using the Δ5-sterols as markers of the presence of other oils. In pumpkin seeds, Δ7-sterols dominate the phytosterol fraction with a low content of Δ5-sterols (Mandl et al. 1999; Murkovic et al., 2004). The presence of high amounts of Δ5-sterols indicates a mixture with other plant oils. Since the aroma and color of the pumpkin seed oil are very intense, it is commonly mixed with other oils. This is marketed as "salad oil" which contains up to 50% of other edible plant oils. For identification of the geographical origin of the seeds, a method based on the analysis of rare earth elements was developed by Bandoniene and co-workers (2008).

## Oil Stability

One can not use Styrian pumpkin seed oil for cooking since the aroma changes rather quickly during heating, giving the oil burnt aroma notes. Therefore, the traditional use of the oil is for salad dressings and for adding a typical aroma and color to prepared foods (e.g., soups, scrambled eggs, and ice cream).

Due to its high content of linoleic acid, the stability of pumpkin seed oil is low. Measurements using the Rancimat, which measures the oxidation stability at temperatures above 100°C, showed that the stability is mainly related to the content of linoleic acid. Experiments with added α-tocopherol revealed a pro-oxidative behavior of vitamin E. A statistical evaluation of the stability of 15 different oils (selected for high- and low-linoleic-acid content and high- and low-vitamin-E content) showed a clear relation to the linoleic-acid content, but no relation to the vitamin-E content (Murkovic & Pfannhauser, 2000; Fig. 12.6). Due to the presence of photosensitizers (protochlorophylls), one should store the oil in the dark at cold temperatures and, if possible, under nitrogen to assure a good stability.

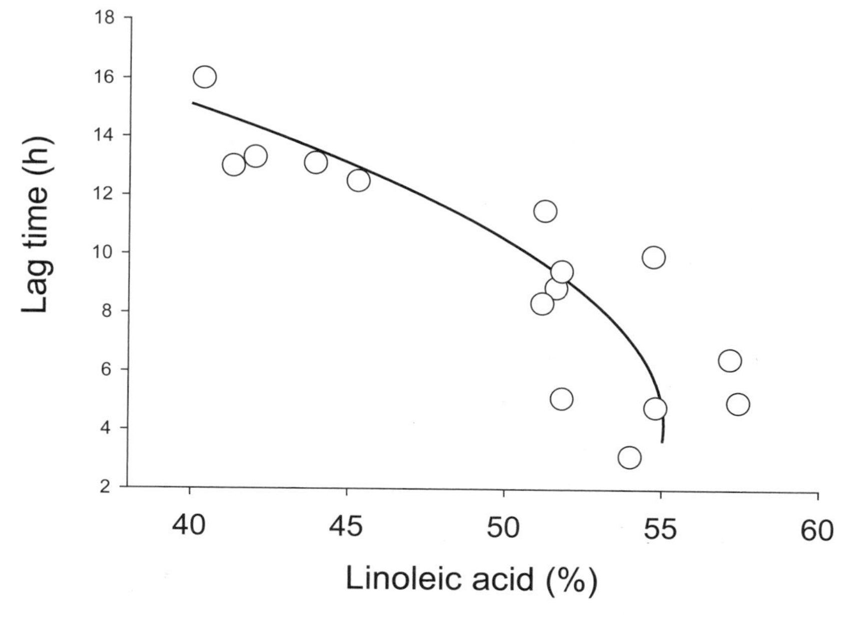


**Fig. 12.6.** Stability of the pumpkin- seed oil in relation to the linoleic-acid content measured in the Rancimat at 110°C.

## Health Aspects of Pumpkin Seed Oil Consumption

Fruhwirth and co-workers (2003) determined the antioxidant activity of Styrian pumpkin seed oil. Since the oil contains vitamin E and carotenoids as lipophilic antioxidants and not-yet-identified more polar antioxidants, the oil could be a dietary source of these physiologically positive substances. Although the vitamin-E content is rather high, it might not suffice to protect the high amounts of unsaturated fatty acids present in the oil. The recommendations of the German, Austrian, and Swiss Societies for Nutrition (DACh, 2000) for the adequate supply of vitamin E are 0.06 mg/g of oleic acid and 0.4 mg/g of linoleic acid. Calculations show that due to the very high content of unsaturated fatty acids, especially linoleic acid, the vitamin-E content is too low to protect the fatty acids against oxidation. Additionally, well-known is that the protochlorophylls present in the oil are photosensitizers (Teppner, 2004). This means that the oil is highly unstable when stored in the light, especially in the sunlight. Because of this photooxidation, the oil very quickly becomes rancid. Therefore, the oil is normally sold in dark-green bottles, and the recommendation is to store the oil at low temperatures in the dark.

In traditional medicine, pumpkin seeds are used to improve the symptoms of benign prostate hyperplasia (BPH). In modern phytotherapy, the enriched polar extracts of pumpkin seeds are used for curing BPH. The biochemical background of the proposed active principle in pumpkin seeds is based on the hypothesis that the inhibition of testosterone 5-α-reductase reduces the conversion of testosterone to dihydrotestosterone (DHT), which is the active male sex hormone. Both, testosterone and DHT produce the androgen-mediated effects; however, DHT is significantly more potent. DHT promotes the development of prostate cells and BPH, and possibly serves as a promoter for prostate cancer. Together with other phytopharmaceuticals, from African plum (*Pygeum africanum*), aspen (*Populus tremula*), dwarf palm (*Serenoa repens*), purple cone flower (*Echinacea purpurea*), rye (*Secale cereale*), South African star grass (*Hipoxis rooperi*), stinging nettle (*Urtica dioica*), and Unicorn root (*Aletrius farinose*), *C. pepo* was studied for its ability to relieve symptoms related to BPH and alter 5-α-reductase activity. Unfortunately, a high placebo effect (40–60%) was observed in several controlled BPH studies (Lowe & Ku, 1996). Additionally, the active principle is normally not known, with the result that the dose of the active ingredient is also unknown, since a standardization of the extract is not possible (Dreikorn et al., 2002). Recently, a placebo-controlled study showed that within 12 months a significant reduction of IPSS (international prostate symptom score) was obtained by administering a polar pumpkin seed extract, whereas the other investigated parameters (maximal urinary flow, $Q_{max}$), quality-of-life (QOL), prostate volume, postvoid residual urine volume (PVR)] did not change (Bach, 2000). Berges and co-workers (2000) published a six-month study by using a dose of 20 mg of sitosterol three times a day along with minor amounts of other phytosterols. In this experiment, the BPH-related parameters (modified Boyarsky symptom score, IPSS, QOL index, $Q_{max}$,

PVR) improved as a result of this treatment. If sitosterol was the active ingredient in this study, the 20 mg of sitosterol would be equivalent to 50 mL of pumpkin- seed oil (calculated from Mandl et al., 1999) or 15 g if the total amount of phytosterols is considered (Murkovic at al., 2004). The evidence for the efficacy of phytotherapeutic agents in the treatment of symptomatic BPH is inconclusive. However, the widespread usage and demand for these agents certainly warrant further well-designed, long-term, placebo-controlled studies (Lowe & Ku, 1996).

Fu et al. (2006) reviewed several other pharmacological activities of pumpkins. The spectrum of activity is comprised of anti-diabetic, anti-hypertensive, anti-tumorigenic, immunomodulatory, anti-bacterial, anti-hypercholesterolemic, intestinal anti-parasitic, anti-inflammatory, and antalgic. The anti-diabetic activity of the fruit pulp and seeds is traced back to the hypoglycemic activity of sugar-removed pumpkin powder. The oil of ungerminated pumpkin seeds improves blood-glucose tolerance (Li et al., 2001). The administration of pumpkin seed oil succeeded in modulating most of the altered parameters affected during arthritis (Fahim et al., 1995). Evidence on anti-tumor activity comes from cell culture and animal experiments which revealed that some of the proteins present in the seeds might be the active principle. Diaz et al. (2004) showed the anthelmintic anti-parasitic activity of pumpkin seeds. Most of the pharmacological activities attributed to pumpkins are related to the fruit pulp. Although some evidence is available on positive pharmacological activities, the beneficial effects on humans remain to be proven.

## Conclusions

Styrian pumpkin seed oil—as it is produced traditionally by first roasting the seeds and then pressing the oil—is a typical product of the southeastern region of Austria. The oil is mainly produced for the local market with a small percentage for worldwide export. With its typical aroma and color, the oil cannot be used for cooking because of its highly reactive substances that deteriorate the aroma during cooking. It is used mainly as salad oil, giving the dressing a dark-green color. The development of new varieties with the aim of higher productivity and improvement of the resistance to zucchini yellow mosaic virus has led to a more stable productivity of the seeds. Due to improvements in the extraction process with the designation of several brands as "PGI" (protected geographical indication) and a continuous effort to improve and standardize the product characteristics (aroma, taste, color, and viscosity), a worldwide marketable product has evolved that reflects the tradition and character of the region.

Pumpkin seeds and pumpkin seed oil are used in folk medicine for curing prostate-related symptoms. Although currently available studies show inconclusive results, when all of their health-promoting substances are taken into account (linoleic acid, vitamin E, lutein, lignans, and phytosterols), pumpkin seeds and pumpkin seed oil certainly contribute to a healthier diet.

## References

Adlercreutz, H. Lignans and human health. *Crit. Rev. Clin. Lab. Sci.* **2007,** *44,* 483–525.

Adlercreutz, H.; W. Mazur. Phyto-oestrogens and Western diseases. *Ann. Med.* **1997,** *29,* 95–120.

Applequist, W.L.; B. Avula; C.T. Schaneberg; Y.-H. Wang; I.A. Kahn. Comparative fatty acid content of seed of four Cucurbita species grown in a common (shared) garden. *J. Food Comp. Analys.* **2006,** *19,* 606–611.

Azevedo-Meleiro, C.H.; D.B. Rodriguez-Amaya. Qualitative and quantitative differences in carotenoid composition among Cucurbita moschata, Cucurbita maxima, and Cucurbita pepo. *J. Agric. Food Chem.* **2007,** *55,* 4027–4033.

Bach, D. Placebokontrollierte Langzeitstudie mit Kürbissamenextrakt bei BPH-bedingten Miktionsbeschwerden. Urologe B **2000**, *40,* 437–443.

Bandoniene, D.; S. Chatzistathis; D. Jöbstl; T. Meisel. Distribution of trace elements: An aid to authentication of speciality oils. *Pflanzliche Lebensmittel–Wein, Obst, Gemüse–Qualität und Sicherheit;* F. Bauer, W. Pfannhauser, Eds.; GÖCh: Graz, 2008; in press.

Berges R.R.; A. Kassen; T. Senge. Treatment of symptomatic benign prostatic hyperplasia with β-sitosterol: an 18 month follow-up. *BJU Int.* **2000,** *85,* 842–845.

Bewley, J.D.; M. Black; P. Halmer. *The Encyclopedia of Seeds: Science, Technology and Uses;* Oxford University Press: Oxford, UK, 2006; pp. 113–117.

Bravi, E.; G. Perretti; L. Montanari. Fatty acids by high-performance liquid chromatography and evaporative light-scattering detector. *J. Chromatogr.* A **2006,** *1134,* 210–214.

Cui, H.; J.B. Loy. Heterosis of seed yield exhibited in hull-less seeded pumpkin. *Cucurbitaceae;* D.N. Maynard, Ed.; ASHS Press: Alexandria, VA, 2002; pp. 323–329.

DACh—Referenzwerte für die Nährstoffzufuhr. 1. Ed.; Umschau/Braus: Frankfurt am Main, Germany, 2000; p. 90.

Diaz, O.; L. Lloja; Z.V. Carbajal. Preclinical studies of cucurbita maxima (pumpkin seeds) a traditional intestinal antiparasitic in rural urban areas. *Rev. Gastroenterol. Peru* **2004**, *24,* 323–327.

Dreikorn, K.; R. Berges; L. Pientka; U. Jonas. Phytotherapy of benign prostate hyperplasia. Current evidence-based evaluation. *Urologe A* **2002,** *41,* 447–451.

Fahim, A.T.; A.A. Abd-el Fattah; A.M. Agha; M.Z. Gad. Effect of pumpkin seed oil on the level of free radical scavengers induced during adjuvant-arthritis in rats. *Pharmacol. Res.* **1995**, *31,* 73–79.

Fruhwirth, G.O.; A. Hermetter. Seeds and oil of the Styrian oil pumpkin: Compounds and biological activities. *Eur. J. Lipid Res. Technol.* **2007,** *109,* 1128–1140.

Fruhwirth, G.O.; T. Wenzl; R. El-'Toukhy; R. Wagner; A. Hermetter. Fluorescence screening of antioxidant capacity in pumpkin seed oils and other natural oils. *Eur. J. Lipid Sci. Technol.* **2003**, *105,* 266–274.

Fu, C.; H. Shi; Q. Li. A review on pharmacological activities and utilization technologies of pumpkin. *Plant Foods Hum. Nutr.* **2006**, *61,* 73–80.

Garber, K. Ölkürbisanbau im norddeutschen Klimagebiet. *Neue Mitt. Landwirtsch.* **1948**, *3,* 165–166.

Garces, R.; C. Sarmiento; M. Manca. Temperature regulation of oleate desaturase in sunflower (Helianthus annuus L.) seeds. *Planta* **1992,** *186,* 461–465.

Lelley, T.; J.B. Loy; M. Murkovic. Hull-less oil seed pumpkin, *Oil Crop Breeding;* J. Vollmann, I. Rajcan, Eds.; Springer, 2009, Chapter 17, in press.

Lelley, T.; G. Stift; M. Pachner. Züchtung virustoleranter Ölkürbissorten mit konventionellen und molekularen Selektionsmethoden. 51. Arbeitstagung der Vereinigung österreichischer Pflanzenzüchter, BAL Gumpenstein, 2000, pp. 111–115.

Li, Q.H.; Z. Tian; T.Y. Tai. Study on the hypoglycemic action of pumpkin extract in diabetic rats. *Acta Nutrmenta Sin* **2001**, *25,* 34–36.

Lowe, F.C.; J.C. Ku. Phytotherapy in treatment of benign prostate hyperplasia: a critical review. *Urology* **1996**, *48,* 12–20.

Mandl, A.; G. Reich; W. Lindner. Detection of adulteration of pumpkin seed oil by analysis of content and composition of specific Δ7-phytosterols. *Eur. Food Res. Technol.* **1999,** *209,* 400–406.

Matus, Z.; P. Molnár; L.G. Szabó. Main carotenoids in pressed seeds (Cucurbitae semen) of oil pumpkin (Cucurbita pepo convar. pepo var. styriaca). *Acta Pharm. Hung.* **1993,** *63,* 247–256.

Mukaida, N.; N. Kawa; Y. Onoue; Y. Nishikawa. Three-dimensional chromatographic analysis of protochlorophylls in the inner seed coats of pumpkin. *Anal. Sci.* **1993**, *9,* 625–629.

Murkovic, M.; A. Hillebrand; S. Draxl; J. Winkler; W. Pfannhauser. Distribution of fatty acids and vitamin E content in pumpkin seeds (Cucurbita pepo L.) in breeding lines. *Acta Hortic.* **1999,** *492,* 47–55.

Murkovic, M.; A. Hillebrand; J. Winkler; W. Pfannhauser. Variability of vitamin E content in pumpkin seeds (Cucurbita pepo L.). *Eur. Food Res. Technol.* **1996a,** *202,* 275–278.

Murkovic, M., A. Hillebrand; J. Winkler; W. Pfannhauser. Variability of fatty acid content in pumpkin seeds (Cucurbita pepo L.). *Eur. Food Res. Technol.* **1996b,** *203,* 216–219.

Murkovic, M.; U. Mülleder; H. Neunteufl. Carotenoid content in different varieties of pumpkins. *J. Food Comp. Anal.* **2004,** *15,* 633–638.

Murkovic, M.; V. Piironen; A. Lampi; T. Kraushofer; G. Sontag. Changes of chemical composition of pumpkin seeds during the roasting process for production of pumpkin seed oil: Part I: non-volatile compounds. *Food Chem.* **2004**, *84,* 367–374.

Murkovic, M.; W. Pfannhauser. Stability of pumpkin seed oil. *Eur. J. Lipid. Sci. Technol.* **2000,** *102,* 607–611.

Rupprechter, A.; E. Höbaus; M. Blass. *Lebensmittelbericht Österreich 2006;* Kastner International: Vienna, Austria, 2006.

Schoefs, B. Pigment analysis of pumpkin seed oil. *Der Ölkürbis (Cucurbita pepo);* M. Murkovic, Ed./publisher: Graz, Austria, 2000; pp. 50–59.

Sicilia, T.; H.G. Niemeyer; D.M. Honig; M. Metzler. Identification and stereochemical characterization of lignans in flaxseed and pumpkin seeds. *J. Agric. Food Chem.* **2003,** *51,* 1181–1188.

Siegmund, B.; M. Murkovic. Changes of chemical composition of pumpkin seeds during the roasting process for production of pumpkin seed oil: Part II: volatile compounds. *Food Chem.* **2004,** *84,* 359–365.

Spangenberg, J.E.; N. Ogric. Authentication of vegetable oils by bulk and molecular carbon isotope analyses with emphasis on olive oil and pumpkin seed oil. *J. Agric. Food Chem.* **2001,** *49,* 1534–1540.

Stevenson, D.G.; F.J. Eller; L. Wang; J.L. Jane; T. Wang; G.E. Inglett. Oil and tocopherol content and composition of pumpkin seed oil in 12 cultivars. *J. Agric. Food Chem.* **2007,** *55,* 4005–4013.

Teppner, H. Notes on *Langenaria* and *Cucurbita* (*Cucurbitaceae*)—review and new contributions. *Phyton.* **2004,** *44,* 245–308.

# 13

# Wheat Germ Oil

**Nurhan T. Dunford**
*Associate Professor, Oklahoma State University, Department of Biosystems and Agricultural Engineering Bioprocessing; and Robert M. Kerr Food & Agricultural Products Center, FAPC Room 103, Stillwater, OK 74078*

## Introduction

Wheat (*Triticum aestivum* L.) is the oldest and the most widespread staple food for humans. Although both maize and rice production are increasing steadily, wheat acreages remain the largest. Worldwide wheat production was about 625 million tons in 2005 (FAO, 2007). The United States ranks third in worldwide wheat production with about 59 million tons per year, behind China and India. Kansas, North Dakota, Montana, Washington, and Oklahoma are the largest wheat-growing states in the United States.

The wheat caryopsis (or the fruit) contains a single seed. Wheat grain consists of endosperm, bran, and germ, which account for ca. 81–84%, 14–16%, and 2– 3% of the grain, respectively (Atwell, 2001). The major portion of a wheat grain at maturity consists of starchy endosperm. Wheat is mainly grown for its endosperm or flour. The embryo, also referred to as germ, serves as an energy source for the grain during germination. Commercial wheat-milling separates the germ from the endosperm. During this process, various other by-products, such as bran and shorts, are also generated.

The reports on the nutritional and health benefits of wheat, specifically bran and germ fractions, date back to the 1920s (Cramer & Mottram, 1927; Plimmer et al., 1931; Voegtlin & Myers, 1919). As is well-established, the beneficial effects of wheat are due to bioactive compounds concentrated in germ and bran (Cureton, 1972a; Kahlon, 1989; Qu et al., 2005; Zhou et al., 2004). Of noteworthy importance is that a majority of the papers on the antioxidant properties of "wheat-bran" extracts published in the scientific literature do not corroborate the presence or absence of wheat germ in bran fractions used for the experiments (Li et al., 2007; Yu et al., 2002, 2003; Yu & Zhou, 2005). While this chapter focuses on wheat germ, bran extracts are also discussed when relevant to the topic.

## Wheat Germ Oil Utilization

Wheat-germ oil (WGO) is a specialty product. Unlike commodity oils such as soybean and canola oils, which are mainly used for their heat-transfer properties during cooking and for providing a pleasant mouth feel in salad dressings, WGO is used for its nutritional value, specifically for its high vitamin-E content. Applications of WGO range from cosmetics, toiletries, pharmaceuticals, and health foods to dietary supplements.

WGO is marketed as a dietary supplement in bottles or in capsules. Reports show that many trainers recommend WGO as a dietary supplement for athletes to enhance endurance and physical fitness (Poiletman & Miller, 1968). This effect was attributed to WGO components called policosanols. WGO is used in food and feed formulations, and it is sometimes mixed with lecithin and cod-liver oil. Feed applications of WGO include diet supplements for farm animals, racehorses, pets, and mink (Kahlon, 1989). WGO is an important source of commercial natural vitamin E.

WGO is used at 0.1–50% concentrations in cosmetic formulations based on the application. Skin moisturizers may contain up to 50% of WGO (Elder, 1980). Betaine from wheat germ is an emulsifier used in cosmetic products for its viscosity building and conditioning properties (Schoenberg, 1985). Amidoamine lactate, sulfosuccinate, amphodiacetate, and amidopropylamine oxide are other surfactants derived from WGO and used in cosmetics, depending on the pH and ingredients in the formulations. WGO is also a potent insect attractant and is used for insect control. The volatile components of WGO are responsible for the aggregation activity of *Trogoderma glabrum* larvae, which is a stored-product pest (Nara et al., 1981). Medium-chain (C13–C16) saturated and unsaturated hydrocarbons, branched hexylbenzene, octanoic acid, gamma-nonalactone, substituted naphthalenes, and cyclic branched ketones are the WGO components which function as insect attractants.

## Oil Extraction and Processing

Mechanical expelling, organic solvent extraction, and supercritical fluid extraction can extract WGO. Mechanical expression and organic solvent extraction are both being used for the commercial extraction of WGO. To our knowledge, supercritical fluid technology is not commercialized for WGO processing in the United States, but small supercritical carbon dioxide (SC-$CO_2$) extraction systems are operating commercially in Asia.

Hexane is commonly used for commercial WGO extraction (Anonymous, 2002). Ethanol and 1,2-dichloroethane are utilized to a lesser extent (Barnes, 1983). Although hot solvents are preferred for vegetable-oil extraction, at least one company reported the use of solvent extraction at temperatures around 38°C for WGO recovery (Barnes, 1983).

Solvent extraction is more efficient than mechanical pressing of WGO. The residual oil content of solvent-defatted wheat germ can be as low as 1% (w/w). Solvent-extracted wheat germ is more stable than the mechanically expressed wheat germ due to its lower lipid content. Pressing recovers only 50% (w/w) of the WGO, and residual wheat germ requires further stabilization to avoid rancidity and to extend shelf life. The mechanical pressing of WGO is successful only when the bran contamination is minimized during the milling operation since the oil content of bran is much lower than that of the germ fraction. Pressed wheat germ is perceived as "natural" and usually preferred by consumers.

Barnes (1982) reviewed the literature on wheat-germ-extract yields using various solvents. Hexane and light-petroleum-ether extraction resulted in yields ranging from 5 to 15% (w/w). This wide range of variation in oil yield was explained by the degree of contamination of germ by bran, which contains only 5% of oil. Diethyl ether, which is expected to extract more relatively polar components, yielded 7–15% of oil. Acetone gave 1.9% more oil yield than light petroleum ether.

Although most edible oils are refined to remove phospholipids (PLs), free fatty acids (FFAs), color compounds and volatile components, WGO is often used in the crude form. Phosphatidylcholine (PC), color, and flavor are desired attributes for the products marketed as "natural" in the health-food stores. However, refining improves the stability of the oil. The FFA content of crude WGO can be quite high, 5–25%, depending on the germ-separation conditions, storage, and oil-extraction method. FFAs contribute to the bitter and soapy flavors in the product; hence, they are removed from WGO by alkali treatment. However, the alkali-deacidification process results in significant losses in oil and, more importantly, in tocopherols. Wang and Johnson (2001) examined the effect of conventional oil-refining processes on the WGO quality. According to this study, the tocopherol content of WGO did not change significantly during degumming, neutralization, and bleaching processes. However, deodorization conditions reduced the tocopherol content significantly. Lower temperature and longer residence time were effective in reducing FFA, peroxide value, and color while retaining tocopherols in WGO during deodorization. Although degumming did not reduce the phosphorous content of the crude oil effectively, phosphorous concentration was reduced at every other stage of WGO refining. Wang and Johnson (2001) suggested that WGO refining should include acid degumming at high temperatures and a high shear for an extended time as compared to that for the typical vegetable oils to maximize PL hydration. Even though PLs have beneficial health effects for humans, they are removed from the crude oil during the degumming process. PL tends to precipitate out in the oil during storage, and has adverse effects on frying operations due to emulsification properties. The neutralization of FFA may require more extensive alkali treatment. WGO bleaching requires more bleaching earth than for typical vegetable-oil refining.

An expired U.S. patent describes the molecular distillation of WGO (Singh & Rice, 1981). Initially, WGO was degummed by using phosphoric acid and water. Bleaching was carried out with activated clay followed by distillation using a centrifugal molecular distillation unit. FFAs were removed at 140–200 °C and below 50 mTorr. In the same patent, a claim was also made that a vitamin-E concentrate was prepared from purified WGO by a second-stage molecular distillation process carried out at 220–300°C and at pressures less than 25 mTorr (Singh & Rice, 1981).

The supercritical fluid extraction of WGO, an alternative method to conventional hexane extraction, was reported by several research groups (Dunford et al., 2003; Eisenmenger, 2005; Gomez & de la Ossa, 2000; Panfili et al., 2003; Taniguchi et al., 1985). WGO solubility in SC-$CO_2$ at 40°C and 200 bar was 0.35% (w/w) (Taniguchi et al., 1985). Oil extracted with SC-$CO_2$ has a lighter color, and contains less phosphorus than the hexane-extracted oil. Although oil-extraction rates from the ground and flaked wheat germ were not significantly different, the utilization of flaked wheat germ is recommended for large- scale SC-$CO_2$ extraction. Ground wheat-germ can be difficult to handle because the small particle size of ground germ causes dust formation. Furthermore, the channeling of the SC-$CO_2$ flow through the ground wheat-germ in the extraction vessel due to compaction may reduce the mass transfer. According to Dunford and Martinez (2003) and Taniguchi et al. (1985), both the α- and β-tocopherol content of SC-$CO_2$-extracted oil were similar to those of hexane-extracted oil. However, Gomez and de la Ossa (2000) reported a higher tocopherol content.

Panfili et al. (2003) characterized the composition of SC-$CO_2$-extracted WGO and defatted cake. According to this study, the FFA content and peroxide value of the oils collected during the initial stages of SC-$CO_2$ extraction (during the first 45 min) were higher than those of the oil fractions collected at the later stages. Similarly, more tocopherols were detected in the oils collected during the first 75 minutes of 3 hours of SC-$CO_2$ extraction. Experiments also indicated that WGO collected during the initial stages of SC-$CO_2$ extraction had a higher tocopherol content (Dunford et al., 2003). The most abundant carotenoid in SC-$CO_2$-extracted WGO was lutein, followed by zeaxanthin and β-carotene. A larger amount of carotenoids was extracted toward the end of SC-$CO_2$ extraction (Panfili et al., 2003).

Studies carried out with liquid and SC-$CO_2$ (50–400 bar) at relatively low temperatures (10–60°C) indicated that pressure had a significant effect on the oil yields, while the effect of temperature was insignificant (Taniguchi et al., 1985). The effect of pressure and temperature on the SC-$CO_2$ extraction yields and WGO composition was also examined at higher pressures and temperatures, 100–550 bar and 40–80°C, respectively (Dunford et al., 2003). Yields of SC-$CO_2$ extracts [(weight loss from the sample during the extraction/initial weight of wheat germ used for extraction) x 100] varied significantly with temperature and pressure in the 2 to 20% (w/w) range. The WGO yield was 11% (w/w) when hot hexane (Soxhlet) was used for extraction. The higher SC-$CO_2$ extraction yield (>11%) indicates that SC-$CO_2$ at high pressures extracted some of the wheat-germ components, which are not soluble in hexane. Moisture in the wheat germ might be co-extracted with oil, resulting in higher extraction yields (Eisenmenger et al., 2006). The highest SC-$CO_2$-extraction yield was obtained at the highest pressure used (550 bar). The temperature dependence of the extract yield was more pronounced at higher temperatures (60 and 80 °C) and the lowest pressure examined in this study (100 bar). This is due to the significant change in SC-$CO_2$ density under those conditions.

Supercritical fluid fractionation (SFF) and the removal of FFA from crude WGO by using a high-pressure packed-column were reported by Eisenmenger (2005). The study demonstrated that the use of the SFF technique can remove FFA from both hexane- and SC-$CO_2$-extracted oil effectively. Tocopherol compositions of extracts (fraction collected from the top of the column) were very low (0.05 mg/g), indicating that tocopherols were retained with triacylglycerides (TAGs) in the raffinate (fraction collected from the bottom of the column). The same study also showed that fatty acid esters of phytosterols were enriched in the raffinate fraction during the deacidification of WGO by using the SFF method.

## Wheat Germ Oil Content

The oil content of wheat germ varies with the variety, the purity, and the extraction method. Pure germ-fractions prepared in the laboratory (dissected by hand) contain a much higher amount of oil (≈15%) than commercially flaked wheat germ (7–11%), because some oil is lost and transferred to flour during the commercial operations (Barnes, 1983; Dunford & Zhang, 2003; Morrison & Hargin, 1981). In general, dissected germ contains both embryo and scutellum. Scutellum contains about two times more oil than does the embryo (Hinton, 1944). The higher oil content in scutellum explains the higher oil content of laboratory-dissected germ. Higher oil yields (25–30%) were obtained when the total acyl-lipid content of wheat germ was determined by acid hydrolysis, which releases more oil from the germ matrix (Morrison et al., 1980). The oil content of wheat germ is lower than that of corn germ and rice bran/germ, about 20% (Dunford, 2005).

## Oil Properties

The specific gravity of WGO varies from 0.925 to 0.938 (Table 13.1) (Firestone, 1999). The typical refractive index of WGO is in the range of 1.469 to 1.483. The iodine and saponification values of the oil are 115–128 and 179–190, respectively. The FFA content of WGO is usually less than 6%. However, the FFA content can be as high as 25% if the germ separation, storage, and oil-extraction conditions are not controlled properly. Solvent-extracted crude oil usually has a lower FFA content than that of the mechanically expelled oil. FFAs are not desirable in the oil due to their contribution to a bitter and soapy flavor in food products; hence, they have to be removed during the edible-oil-refining process. The unsaponifiable fraction of WGO is larger, 2–5%, than that of most other edible oils, 1–1.5%. The composition of the WGO unsaponifiables is discussed later in this chapter.

**Table 13.1. Physicochemical Properties of Wheat Germ Oil (WGO)[1]**

| Property | |
|---|---|
| Specific gravity | 0.928–0.938 (15.5/15.5°C) |
| | 0.925–0.933 (25/25°C) |
| Refractive index | 1.474–1.483 (25°C) |
| | 1.469–1.478 (40°C) |
| Iodine value | 115–128 |
| Saponification value | 179–190 |
| Unsaponifiable (%) | 2–5 |

[1]Adapted from Firestone (1999).

## Composition of Acyl Lipids and Fatty Acid

Significant variations were observed in the fatty acid composition of commercial WGO (Barnes, 1982). These variations were attributed to differences in the varieties of wheat, growth conditions, storage conditions of the germ, method of lipid extraction, and the analysis and post-extraction treatments, such as the removal of fatty acids. WGO is a high-value specialty product. The adulteration of WGO with lower value vegetable oils to lower cost may also affect its fatty acid composition (Barnes, 1982). However, the fatty acid compositions of laboratory-extracted oils (hexane extracts) obtained from wheat germ processed at different milling operations were similar (Barnes, 1982).

Hexane-extracted wheat germ consisted of about 56% of linoleic acid (18:2n-6), which is an essential fatty acid (Table 13.2) (Dunford & Zhang, 2003). The total unsaturated and polyunsaturated fatty acid (PUFA) content of WGO was about 81 and 64%, respectively. As is well-documented, unsaturated fatty acid intake, especially PUFA, reduces coronary heart disease (CHD) (Simopoulos, 1999). Several scientific studies show that n-3 fatty acids have health benefits, such as lowering CHD risk (Hu, 2001). Also suggested is that an n-6/n-3 ratio of 10 or less results in the reduction of fatal CHD risk (Hu, 2001). The n-6/n-3 recommendations of the World Health Organization, Sweden, and Japan are 5–10/1, 5/1, and 2/1, respectively. WGO has very high unsaturated fatty acid and PUFA content and an excellent n-6/n-3 fatty acid ratio (9/1). A high concentration of PUFA is a positive attribute in the functional foods and nutraceutical markets. However, a high content of C18:3 (linolenic acid) fatty acid makes the oil susceptible to oxidative rancidity. The research findings of Dunford and Zhang (2003) demonstrate that no significant change occurred in the fatty acid composition of the oils extracted with various organic solvents (hexane, ethanol, isopropanol, and acetone) even though the extract yield was affected considerably by the solvent type. SC-$CO_2$-extracted WGO had a similar fatty acid content as hexane-extracted oil (Table 13.2) (Eisenmenger, 2005). According to Wang and Johnson (2001), neutralized WGO has a higher palmitic, stearic, and oleic acids content and a lower linoleic and linolenic acids content in comparison to crude and degummed

**Table 13.2. Fatty Acid Composition (%, w/w) of WGO Samples Extracted and Refined Through Various Methods As Compared to Soybean Oil[1]**

| Fatty acid[2] | HE WGO | CR WGO | SC-$CO_2$ WGO | Soybean |
|---|---|---|---|---|
| 16:0 | 16.7[a] | 15.8[b] | 16.8[a] | 10.7[c] |
| 18:0 | 0.77[b] | 0.72[b] | 0.5[c] | 4.56[a] |
| 18:1 | 16.9[b] | 15.8[c] | 13.6[d] | 22.1[a] |
| 18:2 | 57.6[c] | 58.4[b] | 59.7[a] | 54.0[d] |
| 18:3 | 6.4[b] | 6.7[b] | 7.3[a] | 7.2[a] |
| 20:1 | 1.7[a] | 1.6[a] | 1.45[c] | 0.46[c] |

[1]Adapted from Eisenmenger and Dunford (2008).
[2]First number before the semicolon refers to the number of carbon atoms in the fatty acid chain, and second number indicates the number of double bonds on the carbon chain.
[3]Abbreviations: HE WGO: commercially hexane-extracted crude wheat-germ oil (WGO);CR WGO: commercially refined (chemical refining) WGO; SC-$CO_2$ WGO: supercritical-carbon-dioxide–extracted WGO; Soybean: commercially refined soybean oil.WGO means in the same row with the same letter are not significantly different at $P > 0.05$.

oils. This phenomenon was attributed to: (i) the selective hydrolysis of TAG during germ separation and oil extraction and (ii) the removal of PLs, which usually contain more PUFA compared to the neutral oil during the deacidification process.

TAGs compose the major lipid class in WGO (Barnes, 1982). According to Nelson et al. (1963), about 30% of the TAGs consists of 1-palmito-2,3-dilinolein. Trilinolein (≈16%) and 1-palmito-2-linoleo-3-olein (≈12%) are the two other major TAGs present in WGO. Details on the distribution of specific fatty acids in the TAGs of WGO are discussed by Barnes (1982) and Nelson et al. (1963). The diacylglycerol (DAG) content of commercial oils varies from 2 to 11% (Barnes, 1982), whereas the monoacylglycerol (MAG) content of WGO is usually lower (0.1–1.0%). The linoleic-acid content of MAG was significantly lower than that of the other acyl lipids (Barnes & Taylor, 1980).

WGO also contains polar lipids. The amount varies considerably with the extraction method and the solvent used for lipid recovery. According to Hargin and Morrison (1980), dissected wheat-germ's chloroform–methanol extracts contain about 20% of polar lipids. In commercially solvent-extracted crude WGO, of the total acyl lipids 0.3–10% is polar (Barnes & Taylor, 1980), whereas it is only 0.2–1.8% if mechanical expressing is used. Analytical data on WGO PLs are scarce. Of the total PLs in dissected wheat germ, 40–60% is PC, 9–15% is phosphatidylethanolamine (PE), and 13–20% is phosphatidylinositol (PI) (Hargin & Morrison, 1980). Wang and Johnson (2001) examined the effect of processing on the phosphorous content of the oil WGO degumming was very difficult due to the presence of a large amount of nonhydratable PL caused by phospholipase D activity during wheat milling (Wang & Johnson, 2001).

In a recent study, the PL content of commercial and SC-$CO_2$-processed WGO was examined (Table 13.3) (Eisenmenger & Dunford, 2008). Although hexane-extracted crude WGO contained the highest amount of PL among the samples tested in the study, the total PL content of crude WGO was much lower (19.8 mg/g of oil) than the literature values (45–50 mg/g of oil). This was caused by the centrifugation prior to the tests to remove the precipitate formed during cold storage. A significant portion of the PL and wax components was removed with the precipitate. SC-$CO_2$-extracted WGO did not contain any detectable amount of PL because of its low solubility in SC-$CO_2$. The extraction of vegetable oils by SC-$CO_2$ eliminates the need for a degumming step during oil refining. As expected, the PL contents of commercially refined oils were either very low or below the detection levels because these compounds are removed during the degumming step of the refining process. Commercially hexane- extracted WGO contained high amounts of PI + PA (phosphatic acid) (60% of total PLs). Literature reports that the high-PA content in crude vegetable oils may be an indication of poor seed handling and extraction conditions (Wang & Johnson, 2001). PA is a nonhydratable PL, and the separation of this compound by water degumming is very difficult. Other PL components detected in hexane-extracted crude WGO were: 17.7% of PE, 16.7% of PS, and 4.5% of PC, based on the total PL content in the oil (Eisenmenger & Dunford, 2008).

## Nonacylglycerol Components in Wheat-Germ Oil

WGO is relatively rich in unsaponifiable compounds, particularly phytosterols, *n*-alkanols, and tocopherols. Small quantities of triterpenols, carotenoids, and hydrocarbons are also present in the unsaponifiable fraction of the oil. Tocopherols constitute about 18% of the unsaponifiable fraction of WGO, and are currently the most

**Table 13.3. Phospholipid Compositions (mg/g of oil) of WGO Extracted and Refined Through Various Methods[1].**

| Samples | PE[2] | PI + PA[3] | PS[4] | PC[5] |
|---|---|---|---|---|
| HE WGO | 3.5[a] | 12.1[a] | 3.3 | 0.9 |
| CR WGO | 1.9[c] | 0.6[b] | n.d. | n.d. |
| SFE WGO | n.d. | n.d. | n.d. | n.d. |
| Soybean | n.d. | n.d. | n.d. | n.d. |

[1]Adapted from Eisenmenger and Dunford (2008).
[2]PE: phosphatidylethanolamine; [3]PI + PA: phosphatidylinositol and phosphatic acid; [4]PS: phosphatidylserine; [5]PC phosphatidylcholine.
[3]Abbreviations: HE WGO: commercially hexane-extracted crude wheat-germ oil (WGO);CRWGO: commercially refined (chemical refining) WGO; SC-$CO_2$ WGO: supercritical-carbon-dioxide-extracted WGO; Soybean: commercial refined soybean oil. n.d.: not detected. Means in the same column with the same letter are not significantly different at $P > 0.05$.

important unsaponifiable components in WGO. WGO is one of the richest natural sources of α-tocopherols. Natural vitamin E is a fat-soluble vitamin that consists of mainly D-α-tocopherol, with β-, γ-, and δ-forms in lesser quantities (Traber & Packer, 1995). Vitamin E supplements are available as natural and as synthetic vitamin E, which is a mixture of D- and L-α-tocopherol. Also, natural vitamin E has about a two times higher bioavailability than synthetic vitamin E (Burton et al., 1998; Chopra & Bhagavan, 1999). Tocopherols are also known to have antioxidant activity (Azzi & Stocker, 2000). The composition of tocopherol isomers in WGO oil is shown in Table 13.4. WGO is rich in both α- and β-tocopherols. The bran and endosperm of wheat contain less tocopherol than the germ. Hence, the presence of nongerm contamination in wheat germ reduces the tocopherol content. Interestingly, oil from fresh wheat germ contained a similar amount of tocopherol as oil from rancid germ, indicating that conditions causing lipid deterioration did not have any significant adverse effect on the tocopherols (Barnes & Taylor, 1980). A commercial form of vitamin E (α-tocopherol acetate) was detected in some commercial WGO (Barnes, 1982). This is probably due to the synthetic vitamin-E fortification of these commercial oils since no significant amount of α-tocopherol acetate was detected either in any plant-lipid extracts or laboratory-extracted WGO (Barnes, 1982). A recent study showed that both chemical and physical refining lower the total-tocopherol content of WGO significantly, compared to hexane-extracted crude WGO. The total-tocopherol contents of the samples were: 5.7 mg/g in chemically refined, 8 mg/g in physically refined, and 15.1 mg/g in hexane-extracted crude oil (Eisenmenger & Dunford, 2008). The majority of the tocopherols in WGO (90%) were in the form of α-tocopherol, with β-tocopherol as the second most abundant tocopherol. A very small amount of tocotrienols was also detected in the WGO extracted in our laboratories (Table 13.4). According to Barnes (1983), only α- and β-tocopherols are present in the pure dissected wheat germ. The presence of tocotrienols in the germ oil is attributed to the contaminations from endosperm and bran.

The *n*-alkanols, also referred to as policosanol, are a group of high molecular weight primary fatty alcohols present in many plants. Information on the composition of *n*-alkanols in WGO is scarce. The major policosanol component of wheat leaf

**Table 13.4. Tocopherol and Tocotrienol Contents of WGO**

| Tocol (mg/kg) | SC-$CO_2$-extracted WGO | Hexane-extracted WGO |
|---|---|---|
| α-Tocopherol | 1365 | 1377 |
| β-Tocopherol | 998 | 1209 |
| β-Tocotrienol + γ-tocopherol | 24 | 48 |
| δ-Tocopherol | 5 | 5 |
| δ-Tocotrienol | n.d.[1] | 7 |

[1]not detected.
Adapted from Dunford and Martinez (2003).

wax is octacosanol (OC) ($CH_3(CH_2)_{26}CH_2OH$) with a molecular weight of 410.8 (Pollard et al., 1933; Tulloch & Hoffman, 1973). Barnes (1983) reported that unrefined, unadulterated, and solvent-extracted WGO contains 80 mg/kg of OC. A U.S. patent, which expired several years ago, describes the separation of a mixture of OC and triacontanol from WGO unsaponifiables (Levin et al., 1962).

Our laboratory research showed that the policosanol content of the solid fraction which precipitated out of WGO during cold storage was about 17 times higher than that of the clear WGO (oil above the precipitate) (Irmak et al., 2005). Wheat bran contains a significantly higher ($P < 0.05$) amount of policosanol than the germ (Irmak et al., 2005). This might be due to the higher oil content of the wheat germ (about 11%, w/w) than that of the bran (2–3%, w/w). There may be similar amounts of policosanols in the bran and germ, but the high levles of oil in the germ may have cuased the dilution of policosanols in the germ extract. We also showed that the policosanol composition in wheat extracts and milling fractions varied significantly (Irmak et al., 2005). The major policosanol in crude WGO solids was tetracosanol, C24 (34%, w/w, of the total policosanol). Hexacosanol and OC are the two other policosanols present in large quantities in the WGO-solids, 26 and 20% of the total policosanols, respectively.

Phytosterols are a major constituent in the WGO unsaponifiable fraction (about 35%). According to Itoh et al. (1973), WGO contains a significantly higher amount of phytosterols than do the other common commercial oils. Sitosterol (60–70%) and campesterol (20–30%) are the two major phytosterols present in WGO (Anderson et al., 1926; Itoh et al., 1973). The majority of the phytosterols in WGO are present in an esterified form (Kiosseoglou & Boskou, 1987). According to Kiosseoglou and Boskou (1987), the total phytosterol content of WGO is about 3–4%. Esterified sterols constitute 2–3% of the oil. The free- to esterified-phytosterols ratio in the WGO varies between 0.3 and 0.5. Hexane- and SC-$CO_2$- extracted WGO contained similar amounts of total phytosterols (about 3.7 mg/g of oil) (Eisenmenger & Dunford, 2008). A tocopherol-enriched WGO sample that was produced using a proprietary process had a significantly higher total-phytosterol content than both crude and refined WGOs, indicating that the process used for tocopherol concentration also results in phytosterol enrichment in the final product (Eisenmenger & Dunford, 2008). Sitosterol was the major phytosterol (78–85% of the total phytosterols), followed by campesterol and stigmasterol in all WGO samples examined in the study (Eisenmenger & Dunford, 2008).

Cycloartenol, β-amyrin, and 24-methylene cycloartanol are the major triterpenols, which constitute less than 1% of WGO. Hydrocarbons are minor components in the WGO unsaponifiable fraction. According to Kuksis (1964), 50% of the hydrocarbon content was squalene, and the remainder consisted of n-C29 alkanes. The presence of lutein and cryptoxanthin in WGO was first reported in 1935 and 1940, respectively (Drummond et al., 1935; Gisvold, 1940). Recently, Panfili et al. (2003)

reported that petroleum-ether-extracted WGO contained about 25 ppm of lutein, 23 ppm of zeaxanthin, and 8 ppm of β-carotene.

## Health Effects and Antioxidant Properties of Wheat Germ Oil

The nutritional value of wheat germ is very high. It contains three times as much protein of high biological value (26%), seven times as much fat (11%), fifteen times as much sugar (17%), and six times as much mineral content (4%) when compared with flour from the endosperm (Atwell, 2001). Wheat germ is also the richest known source of α-tocopherols (vitamin E) of plant origin. Phytosterols, policosanols, thiamine, riboflavin, and niacin are some of the other health-beneficial bioactive compounds present in wheat germ (Atwell, 2001).

Reportedly, WGO has numerous nutritional and health benefits, such as reducing plasma and liver cholesterol levels, improving physical endurance/fitness, and possibly helping to delay the effects of aging (Kahlon, 1989). Studies carried out with golden hamsters demonstrated that WGO supplementation in diet improved both fertility and successful delivery (Soderwall & Smith, 1962). Similar results were obtained with breeder cows (Mariong, 1962) and sheep (Dukelow & Matalamawki, 1963). Allesandri and co-workers (2006) examined the effect of WGO supplementation on oxidative stress and platelet formation in a double-blinded study including 32 patients with hypercholesterolemia. WGO supplementation was effective in reducing both oxidative stress and platelet formation. A proprietary, fermented, wheat-germ extract inhibits metastatic tumor dissemination and proliferation (Boros et al., 2005; Hidvegi et al.; 1999, Saiko et al., 2007). Unfortunately, these studies lack information on extract composition and its bioactive components.

Borel and co-authors (1990) reported that wheat-germ consumption significantly decreased the intestinal absorption of cholesterol in rats, while wheat bran did not have a significant effect. Although the mechanism of the reduction in cholesterol absorption was not examined, the hypothesis was that phytosterols concentrated in WGO might have been the active compounds. The cholesterol-lowering properties of WGO are reported in other publications (Hirai et al., 1984; Ranhotra et al., 1976). A study carried out in 1951 showed that WGO treatment could alleviate certain neuromuscular disorders (Rabinovitch et al., 1951). The most responsive conditions to WGO therapy were collagen-related diseases, such as dermatomyositis and menopausal muscular dystrophy. The healing effect of WGO was attributed to its high vitamin-E content.

The effect of WGO and its unsaponifiable fraction on both acute and chronic inflammation was examined by Mohamed et al. (2005). Both WGO and WGO unsaponifiables were effective in inhibiting inflammation in rats. Anti-inflammatory activity was dose-dependent. Beneficial effects of WGO on human physical fitness were attributed to its high-policosanol content (Consolazio et al., 1964; Cureton, 1972b). Numerous research studies indicate that a policosanol consumption of 5–20 mg/day is effective in lowering total cholesterol by 17–21%, low-density lipoprotein (LDL) by 21–29%, and in increasing high-density lipoprotein (HDL) by 8–15% (Arruzazabala et al., 1993; Castano et al., 1996; Gouni-Berthold & Berthold, 2002; Torres et al., 1995). Additional beneficial effects of policosanol on smooth-muscle-cell proliferation, reduction in platelet aggregation, and LDL peroxidation (Fraga et al., 1997; Gouni-Berthold & Berthold, 2002; Kato et al., 1995; Valdes et al., 1996) were also reported. Policosanol formulations were used as "antifatigue drugs" (Taylor et al., 2003). According to a study carried out with tail-suspended rats, OC could counteract some of the effects of simulated weightlessness on rats, suggesting that OC-enriched foods might benefit astronauts during space travel (Bai et al., 1997). The majority of animal and clinical studies evaluating the health benefits of policosanols were carried out in Cuba. Reports indicate that cholesterol-lowering properties of policosanols are not reproducible in studies performed in the United States (Varady et al., 2003). A better understanding of the effects of policosanols on disease prevention and treatment requires large-scale independent animal and clinical studies involving various ethnic groups and subjects with different health histories.

WGO is also used as an antioxidant in various fats and oils. The heat stability of lard was extended from 2 hours to over 12 hours by the addition of 1% of WGO (Riemenschneider et al., 1944). Fat degradation was measured by the formation of peroxides during heating. The addition of 10% of WGO can delay the oxidative-polymerization of rapeseed oil during frying (Kurkela & Petro-Turza, 1975). Unsaponifiable matter consists of minor components of TAG oils and fats. These compounds include sterols, tocopherols, hydrophobic pigments and hydrocarbons, and make up 0.5–6% of the oils/fats, depending on the source (Elmadfa, 1995). The addition of unsaponifiable matter (isolated from WGO and other vegetable oils) to frying oils is effective in delaying oil oxidation (Sims et al., 1972). Tocopherol, phytosterol, and squalene, naturally present in WGO, are believed to be the compounds involved in the protection of oils during heating. Malecka (2002) demonstrated that the addition of 0.3% of WGO unsaponifiables to low erucic-acid rapeseed oil was more effective than 0.02% of butylhydroxyanisole (BHA, a synthetic antioxidant widely used for the stabilization of commercial oils) in suppressing peroxide formation during storage at 60°C. In our laboratory, we showed that WGO is effective in increasing induction times (initiation of rapid oil oxidation) of n-3 fatty acid rich and highly unstable fish and flax oils (Fig. 13.1). Induction times of the oils were determined at 8 mL/hour of airflow rate and 97.8°C by using a Rancimat (Model 743; Metrohm Ltd., Westbury, NY) instrument. A higher induction time indicates a higher oil stability. The antioxidant effect of WGO was dose-dependent and varied with oil type.

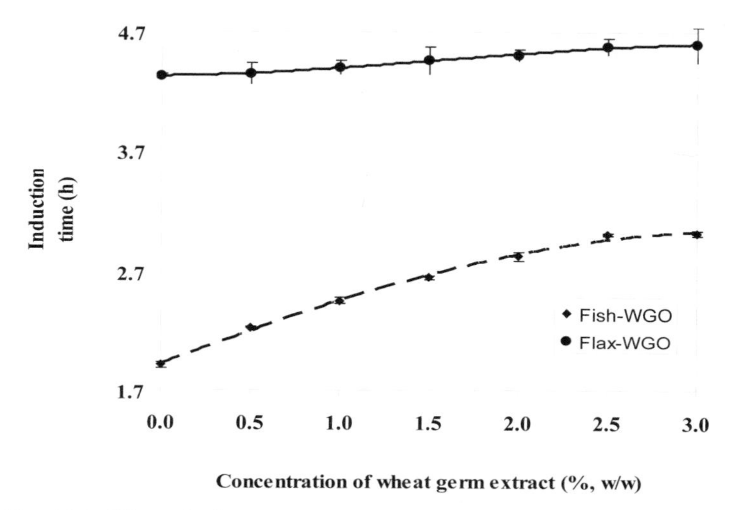


**Fig. 13.1.** Effect of wheat-germ-extract concentration on the oxidative stability of flax and fish oil.

## Conclusions

WGO is a specialty product with very high nutritional value. It is rich in health- beneficial bioactive compounds, such as phytosterols, tocopherols, carotenoids, and omega-3 fatty acids. WGO composition may vary with the variety and growth conditions of wheat, the milling process, germ storage conditions, and oil extraction and refining techniques. Since WGO is a high-value product, adulteration with commodity oils is a serious concern. The presence of very high concentrations of PUFAs and other heat- and light-sensitive compounds, such as tocopherols, makes WGO processing very challenging, and reduces the shelf life of the final product. This, in turn, limits the uses of WGO. The efficacy and economic feasibility of new and advanced processing techniques which operate at low temperatures and minimize oxidation reaction in oils must be evaluated for high-quality WGO production. The development of effective antioxidant systems and other oil-stabilization methods is the key for extending the shelf life of WGO-containing products and for expanding the market share of WGO.

## References

Alessandri, C.; P. Pignatelli; L. Loffredo; L. Lenti; M. Del Ben; R. Carnevale; A. Perrone; D. Ferro;F. Angelico; F. Violi. Alpha-linolenic acid-rich wheat germ oil decreases oxidative stress and cd40 ligand in patients with mild hypercholesterolemia. *Arterioscler. Thromb. Vasc. Biol.* **2006,** *26*, 2577–2578.

Anderson, R.J.; R.L. Shriner; G.O. Burr. The phytosterols of wheat germ oil. *J. Am. Chem. Soc.* **1926,** *48*, 2987–2996.

Anonymous. Isohexane: likely choice for crushers seeking to replace n-hexane. *Inform* **2002,** *13*, 282–286.

Arruzazabala, M.L.; D.R.M. Carbajal; M. Garcia; V. Fraga. Effects of policosanol on platelet aggregation in rats. *Thrombosis Res.* **1993,** *69*, 321–327.

Atwell, W.A. *Wheat Flour;* Eagan Press: St. Paul, MN, 2001.

Azzi, A.; A. Stocker. Vitamin E: non-antioxidant roles. *Prog. Lipid Res.* **2000,** *39*, 231–255.

Bai, S.; L. Xie; C. Liu; Q. Zheng; J. Chen. Effects of octacosanol in food physiological parameters in tail-suspended rats. *Space Medic. Med. Eng.* **1997,** *10*, 450–452.

Barnes, P.J. Lipid composition of wheat germ and wheat germ oil. *Fette Seifen Anstrich.* **1982,** *84*, 256–269.

Barnes, P.J. Wheat germ oil. *Lipid in Cereal Technology;* P.J. Barnes, Ed.; Academic Press: New York, **1983**; pp. 389–400.

Barnes, P.J.; P.W. Taylor. The composition of acyl lipids and tocopherols in wheat germ oils from various sources. *J. Sci. Food Agric.* **1980,** *31*, 997–1006.

Borel, P.; M. Martigne; M. Senft; P. Garzino; H. Lafont; D. Lairon. Effect of wheat bran and wheat germ on the intestinal uptake of oleic acid, monoolein, and cholesterol in the rat. *J. Nutr. Biochem.* **1990,** *1*, 28–33.

Boros, L.G.; M. Nichelatti; Y. Shoenfeld. Fermented wheat germ extract (avemar) in the treatment of cancer and autoimmune diseases. *Ann. N.Y. Acad. Sci.* **2005,** *1051,* 529–542.

Burton, G.W.; M.G. Traber; R.V. Acuff; D.N. Walters; H. Kayden; L. Hughes; K. Ingold. Human plasma and tissue alpha-tocopherol concentrations in response to supplementation with deuterated natural and synthetic vitamin E. *Am. J. Clin. Nutr.* **1998,** *67*, 669–684.

Castano, G.; L. Tula; M. Canetti; M. Morera; R. Mas; J. Illnait; L. Fernandez; J.C. Fernandez. Effects of policosanol in hypertensive patients with type ii hypercholesterolemia. *Curr. Ther. Res.* **1996,** *57*, 691–699.

Chopra, R.K.; H.N. Bhagavan. Relative bioavailabilities of natural and synthetic vitamin E formulations containing mixed tocopherols in human subjects. *Int. J. Vitam. Nutr. Res.* **1999,** *69*, 92–95.

Consolazio, C.F.; L.O. Matous; R.A. Nelson; G.J. Isaac; L.M. Hursh. Effect of octacosanol, wheat germ oil, and vitamin E on performance of swimming rats. *J. Appl. Physiol.* **1964,** *19*, 265–267.

Cramer, W.; J.C. Mottram. On the nutritive value of bread,: with special reference to its content in vitamin B. *Lancet* **1927,** *210*, 1090–1095.

Cureton, T.K. *Influence of Wheat Germ Oil as a Dietary Supplement in a Program of Conditioning Exercises with Middle-Aged Subjects;* Springfield, IL, 1972a.

Cureton, T.K. The physiological effects of wheat germ oil on humans. *Exercise, Forty-two Physical Training Programs Utilizing 894 Humans;* Charles C. Thomas: Springfield, IL, 1972b; p. 525.

Drummond, J.C.; E. Singer; R.J. MacWalter. A study of the unsaponifiable fraction of wheat germ oil with special reference to vitamin E. *Biochem. J.* **1935,** *29*, 456–471.

Dukelow, W.; R. Matalamawki. Effects of ethylene dichloride extracted wheat germ oil on the reproductive efficiency of the sheep. *J. Anim. Sci.* **1963,** *22*, 1137.

Dunford, N.T. Germ oils from various sources. *Bailey's Industrial Oil and Fat Products;* Feridoon Shahidi, Ed.; John Wiley and Sons: NJ, 2005; pp. 195–231.

Dunford, N.T.; J.W. King; G.R. List. Supercritical fluid extraction in food engineering. *Extraction Optimization in Food Engineering;* C. Tzia, G. Liadakis, Eds.; Marcel Dekker, Inc.: NY, 2003; pp. 57–93.

Dunford, N.T.; J. Martinez. *Nutritional Components of Supercritical Carbon Dioxide Extracted Wheat Germ Oil;* 6th International Symposium on Supercritical Fluids, Versailles, France, 2003; pp. 273–278.

Dunford, N.T.; M. Zhang. Pressurized solvent extraction of wheat germ oil. *Food Res. Int.* **2003,** *36*, 905–909.

Eisenmenger, M. Supercritical Fluid Extraction, Fractionation, and Characterization of Wheat Germ Oil. M.S. Thesis, Oklahoma State University, Stillwater, OK, 2005.

Eisenmenger, M.; N.T. Dunford. Bioactive components of commercial and supercritical carbon dioxide processed wheat germ oil. *J. Am. Oil Chem. Soc.* **2008,** *85*, 55–61.

Eisenmenger, M.; N.T. Dunford; F. Eller; S. Taylor;J. Martinez. Pilot scale supercritical carbon dioxide extraction and fractionation of wheat germ oil. *J. Am. Chem. Soc.* **2006,** *10(2)*, 863–868.

Elder, L. Final report on the safety assessment for wheat germ oil. *J. Env. Path. Toxic.* **1980,** *4*, 33–45.

Elmadfa, I. Physiological importance of unsaponifiable components in dietary fats. *Fat Sci. Techn.* **1995,** *97*, 85–90.

FAO (United Nations Food and Agriculture Organization), FAOSTAT. http://faostat.fao.org/site/336/DesktopDefault.aspx?PageID=336 (9/28/2007),

Firestone, D. *Physical and Chemical Characteristics of Oils, Fats and Waxes;* AOCS Press: Champaign, IL, **1999**; p. 152.

Fraga, V.; R. Menendez; A.M. Amor; R.M. Gonzalez; S. Jimenez; R. Mas. Effect of policosanol on in vitro and in vivo rat liver microsomal lipid peroxidation. *Arch. Med. Res.* **1997,** *28*, 355–360.

Gisvold, O.J. Crystalline xanthophyll from wheat germ. *Am. Pharm. Assoc.* **1940,** *29*, 312–313.

Gomez, A.M.; de la Ossa, E.M. Quality of Wheat Germ Oil Extracted by Liquid and Supercritical Carbon Dioxide. *J. Am. Oil Chem. Soc.* **2000,** *77*, 969-974.

Gouni-Berthold, I.; H.K. Berthold. Policosanol: clinical pharmacology and therapeutic significance of a new lipid-lowering agent. *Am. Heart J.* **2002,** *143*, 356–365.

Hargin, K.D.; W.R. Morrison The distribution of acyl lipids in the germ, aleurone, starch, and nonstarch endosperm of four wheat varieties. *J. Sci. Food Agric.* **1980,** *31*, 877–888.

Hidvegi, M.; E. Raso; R. Tomoskozi-Farkas; B. Szende; S. Paku; L. Pronai; J. Bocsi; K. Lapis. MSC, a new benzoquinone-containing natural product with antimetastatic effect. *Cancer Biother. Radiopharm.* **1999,** *14*, 277–289.

Hinton, J.J.C. The chemistry of wheat germ with particular reference to the scutellum. *Biochem. J.* **1944,** *38*, 214.

Hirai, K.; Y. Ohno; T. Nakano; K. Izutant. Effects of dietary fats and phytosterol on serum fatty acid composition and lipoprotein cholesterol in rats. *J. Nutr. Sci. Vitaminol.* **1984,** *30,* 101–112.

Hu, F.B. The balance between ω-6 and ω-3 fatty acids and the risk of coronary heart disease. *Nutrition* **2001,** *17,* 741–742.

Irmak, S.; N.T. Dunford; J. Milligan. Policosanol contents of beeswax, sugar cane and wheat extracts. *Food Chem.* **2005,** *95,* 312–318.

Itoh, T.; T. Tamura; T. Matsumoto. Sterol composition of 19 vegetable oils. *J. Am. Oil Chem. Soc.* **1973,** *50,* 122–125.

Kahlon, T.S. Nutritional implications and uses of wheat and oat kernel oil. *Cereal Foods World* **1989,** *34,* 872–875.

Kato, S.; K.-I. Karino; S. Hasegawa; J. Nagasawa; A. Nagasaki; M. Eguchi; T. Ichinose; K. Tago; H. Okumori. Octacosanol affects lipid metabolism in rats fed on a high-fat diet. *Br. J. Nutr.* **1995,** *73,* 433–441.

Kiosseoglou, B.; D. Boskou. On the level of esterified sterols in cotton seed, tomato seed, wheat germ and safflower. *Oleagineux* **1987,** *42,* 169–170.

Kuksis, A. Hydrocarbon composition of some crude and refined edible seed oils. *Biochemistry* **1964**, *3,* 1086–1093.

Kurkela, R.; M. Petro-Turza. Changes in fatty acid composition of rape-seed oil and a mixture rape-seed oil and wheat germ oil during frying. *Acta Aliment. Acad. Aci. Hung.* **1975,** *4,* 331.

Levin, E.; V.K. Collins; D.S. Varner; J.D. Mosser; G. Wolf. Composition comprising octacosanol, triacontanol, tetracosanol, or hexacosanol, and methods employing same. U.S. Patent 3,031,376 (1962).

Li, W.; M.D. Pickard; T. Beta. Effect of thermal processing on antioxidant properties of purple wheat bran. *Food Chem.* **2007,** *104,* 1080–1086.

Malecka, M. Antioxidant properties of the unsaponifiable matter isolated from tomato seeds, oat grains and wheat germ oil. *Food Chem.* **2002,** *79,* 327–330.

Mariong, B. Effects of wheat germ oil on reproductive efficiency in repeat breeder cows. *J. Dairy Sci.* **1962,** *45,* 904.

Mohamed, D.A.; A.I. Ismael; A.R. Ibrahim. Studying the Anti-Inflammatory and Biochemical Effects of Wheat Germ Oil. *Deutsche Lebensmittel-Rundschau* **2005,** *101,* 66–72.

Morrison, W.R.; K.D. Hargin. Distribution of soft wheat kernel lipids into flour milling fractions. *J. Sci. Food Agric.* **1981,** *32,* 579.

Morrison, W.R.; S.L., T.; K.D. Hargin, K.D. Methods for the quantitative analysis of lipids in cereals and similar tissues. *J. Sci. Food Agric.* **1980,** *31,* 329–340.

Nara, J.M.; R.C. Lindsey; W.E. Burkholder. Analysis of volatile compounds in wheat germ oil responsible for an aggregation response in *Trogoderma glabrum* larvae. *J. Agric. Food Chem.* **1981,** *29,* 68.

Nelson, J.H.; R.L. Glass; W.F. Geddes. The triglycerides and fatty acids of wheat. *Cereal Chem.* **1963,** *40,* 343–351.

Panfili, G.; L. Cinquanta; A. Fratianni; R. Cubadda. Extraction of wheat germ oil by supercritical

$CO_2$: oil and defatted cake characterization. *J. Am. Oil Chem. Soc.* **2003,** *80*, 157–161.

Plimmer, R.H.A.; W.H. Raymond; J. Lowndes. Experiments on nutrition. X. comparative vitamin B1 values of foodstuffs. Cereals 2. *Biochem. J.* **1931,** *25*, 691–704.

Poiletman, R.M.; A.H. Miller. The influence of wheat germ oil on the electrocardiographic T waves of the highly trained athletes. *J. Sports Med. Phys. Fitness* **1968,** *8*, 26–33.

Pollard, A.; A.C. Chibnall; S.H. Piper. CCVII. The isolation of n-octacosanol from wheat wax. *Biochem. J.* **1933,** *27*, 1889.

Qu, H.; R.L. Madl; D.J. Takemoto; R.C. Baybutt; W. Wang. Lignans are involved in the antitumor activity of wheat bran in colon cancer SW480 cells1,2. *J. Nutr.* **2005,** *135*, 598.

Rabinovitch, R.; W.C. Gibson; D. McEachhern. Neuromuscular disorders amenable to wheat germ oil therapy. *J. Neurol. Neurosurg. Psychiat.* **1951,** *14*, 95–100.

Ranhotra, G.S.; R.J. Loewe; L.V. Puyat. Effect of some wheat mill-fractions on blood and liver lipids in cholesterol-fed rats. *Cereal Chem.* **1976,** *53*, 540–548.

Riemenschneider, R.W.; J. Turner; W.C. Ault. Improvement produced in the stability of lard by the addition of vegetable oils. *Oils and Soap (Chicago)* **1944,** *21*, 98–100.

Saiko, P.; M. Ozsvar-Kozma; S. Madlener; A. Bernhaus; A. Lackner; M. Grusch; Z. Horvath; G. Krupitza; W. Jaeger; K. Ammer; M. Fritzer-Szekeres; T. Szekeres. Avemar, a nontoxic fermented wheat germ extract, induces apoptosis and inhibits ribonucleotide reductase in human HL-60 promyelocytic leukemia cells. *Cancer Lett.* **2007,** *250*, 323–328.

Schoenberg, T. Formulating wheat germ surfactants. *Soaps Cosmetics Chem. Specialties* **1985,** *61*, 37.

Simopoulos, A.P. Essential fatty acids in health and chronic disease. *Am. J. Clin. Nutr.* **1999,** *70*, 560S–569S.

Sims, R.J.; J.A. Fioriti; M.J. Kanuk. Sterols additives as polymerization inhibitors for frying oils. *J. Am. Oil Chem. Soc.* **1972,** *49*, 298–301.

Singh, L.; W.K. Rice. Method for producing wheat germ lipid products. U.S. Patent 4,298,622 (1981, 1981).

Soderwall, A.I.; B.C. Smith. Beneficial effect of wheat germ oil on pregnancies in female golden hamsters (*Mesocricetus auratus*, Waterhouse). *Fertility and Sterility* **1962,** *13*, 287–289.

Taniguchi, M.; T. Tsuji; M. Shibata; T. Kobayashi. Extraction of oils from wheat germ with supercritical carbon dioxide. *Agric. Biol. Chem.* **1985,** *49*, 2367–2372.

Taylor, J.C.; L. Prapport; G.B. Lockwood. Octacosanol in human health. *Nutrition* **2003,** *19*, 192–195.

Torres, O.; A.J. Agramonte; J. Illnait; R.M. Ferreiro; L. Fernandez; J.C. Fernandez. Treatment of hypercholesterolemia in NIDDM with policosanol. *Diabetes Care* **1995,** *18*, 393–397.

Traber, M.G.; L. Packer. Vitamin E: beyond antioxidant function. *Am. J. Clin. Nutr.* **1995,** *62*, 1501S–1509S.

Tulloch, A.P.; L.L. Hoffman. Leaf wax of *Triticum Aestivum*. *Phytochemistry* **1973,** *12*, 2217–2223.

Valdes, S.; M.L. Arruzazabala; E. Fernandez. Effect of policosanol on platelet aggregation in healthy volunteers. *Int. J. Clin. Phar. Res.* **1996,** *16*, 67–72.

Varady, K.A.; Y. Wang; P.J.H. Jones. Role of policosanols in the prevention and treatment of cardiovascular disease. *Nutr. Rev.* **2003,** *61*, 376–383.

Voegtlin, C.; C.N. Myers. Distribution of the antineuritic vitamine in the wheat and corn kernel. *Am. J. Phys.* **1919,** *48*, 504–511.

Wang, T.; L.A. Johnson. Refining high-free fatty acid wheat germ oil. *J. Am .Oil Chem. Soc.* **2001,** *78*, 71–76.

Yu, L.; S. Haley; J. Perret; M. Harris. Antioxidant properties of hard winter wheat extracts. *Food Chem.* **2002,** *78*, 457–461.

Yu, L.L.; J. Perret; M. Harris; J. Wilson; S. Haley. Antioxidant properties of bran extracts from "Akron" wheat grown at different locations. *J. Agric. Chem.* **2003,** *51*, 1566–1570.

Yu, L.; K. Zhou. Antioxidant properties of bran extracts from "Platte' wheat grown at different locations. *Food Chem.* **2005,** *90*, 311–316.

Zhou, K.; J.J. Laux; L. Yu. Comparison of Swiss red wheat grain and fractions for their antioxidant properties. *J. Agric. Food Chem.* **2004,** *52*, 1118–1123.

# 14

# Rice Bran Oil

**J. Samuel Godber**
*Department of Food Science, Louisiana State University, Baton Rouge, LA 70803*

## Introduction

Rice bran oil may be the largest underutilized agricultural commodity in the world. It is well-known that rice is one of the major staples of worldwide food consumption. The primary form in which rice is consumed is as white rice, with hull, germ, and bran coat removed. Contained in the bran coat are both the outer aleurone layers of the grain coat, as well as the germ, which is a separate structure of the rice grain but is usually removed along with the bran during milling. Both of these components are rich sources of lipid. Rice bran contains between 15 and 20% lipid, which upon extraction becomes rice bran oil. In 2005 a little less than 2 million metric tons (MMT) of rice bran oil were processed in the world (FAO, 2007). However, if lipids were extracted to produce rice bran oil from all the rice produced worldwide (598 MMT), one could produce approximately 10 MMT of rice bran oil. The difference of 8 MMT represents unutilized rice bran oil that is potentially available. Therefore, only about 20% of the potential world production of rice bran oil is actually being realized.

Questions could be asked such as: what happens to the unutilized rice bran oil, or why isn't it all extracted, or is it even worth extracting, or does it have utility? Unfortunately, and remarkably, the answers to these questions would ultimately reveal that for many, many years the world rice industry squandered a source of oil that is ideally suited to the modern-day concept of healthy oil. Furthermore and even more remarkably, in spite of countless studies that show it to be highly functional from the standpoint of food processing and, even more importantly, that it is one of the most health-promoting forms of vegetable oil, very little current effort exists aimed at recovering and utilizing a greater portion of its unutilized potential.

Numerous monographs written over the years have described the production and utilization of rice bran oil, and, for the most part, the technology associated with rice bran oil production has not changed dramatically over the last few years. Therefore, this chapter does not rehash the development of rice bran oil production and processing, but rather summarizes the current state-of-the-art and updates current developments. A particular focus is directed toward the health-promotion potential of rice bran oil and/or its components. For a more detailed discussion of rice bran oil production and processing, refer to the monographs of Orthoefer (2005), Orthoefer and Eastman (2004), Kao and Luh (1991), and Pillaiyar (1988).

## Overview of Production and Processing of Rice Bran Oil

The production and processing of rice bran oil are similar to many other vegetable oils. In 1982, the World Conference on Oilseeds and Edible Oil Processing was held in the Hague, the Netherlands. The entire second issue of Volume 60 of the *Journal of the American Oil Chemists' Society* was devoted to papers generated from that conference. Virtually every aspect of oil extraction and refining is covered in those proceedings, which represented the state of the art at that time. Therefore, this section provides a brief overview of the extraction and refining process and focuses on aspects of rice bran oil production and processing that are considered unique or peculiar. Rice bran oil is normally obtained from the bran composite that is removed during the milling of brown rice to produce white rice, although some extraction processes were developed that occur during the milling process (to be described later). In actuality, the highest concentration of oil is found in the germ, which represents approximately 2% of the rice kernel. Commercial rice milling practiced in most of the world does not separate the germ from the bran, and in many cases does not completely separate hull from bran. Thus, what is called rice bran usually contains the germ and parts of the hull and endosperm, as well. Stabilized bran is the preferred starting material because lipases in rice bran cause a rapid rise in free fatty acids that markedly reduces oil quality. However, in most of the world the bran is difficult to stabilize, which is the reason that much of the potential oil production is used for nonedible applications (to be discussed later). Bran is stabilized in a number of ways, all usually involving the elevation of temperature that denatures the enzymes responsible for lipid degradation. A popular means to stabilize rice bran is to pass it through an extruder by which the frictional energy generated through the process of forcing the bran through the small orifice of the extruder plate causes an increase in the temperature. Temperatures required for adequate stabilization are generally believed to be above 120°C. Other means of stabilization include pH adjustment using weak acids, microwave heating, and irradiation; however, no other practical methods are utilized commercially. Defatted rice bran becomes a by-product of oil processing. Manufacturers can use defatted rice bran in animal feeds or as an ingredient in human food products, although its utility is not fully realized.

## Extraction

Traditional methods of extraction include both physical and chemical approaches; however, rice bran oil is typically extracted with solvents. The solvent of choice is hexane, although concern for impending regulatory scrutiny of hexane for environmental and toxicological reasons prompted exploration of other approaches to extraction, such as alternative solvents and supercritical fluid extraction (to be discussed later). Although extracting oil from bran is possible through mechanical means via pressing, oil recovery is lower and costs are higher compared to solvent extraction. Pelletizing the fine rice bran particles aids in the percolation of solvent through the bran during solvent extraction, which increases yield. Wet extrusion produces the most satisfactory pellets for solvent extraction, with pellet diameters between 6 and 8 mm (Orthoefer & Eastman, 2004). Alternative solvents that are considered include ethyl acetate, isopropyl alcohol, and acetone, but currently these are not as economical as hexane. Supercritical fluids were evaluated as a potential means to extract oil from rice bran at the pilot scale, but commercial-scale extraction was not undertaken. In principle, this approach offers interesting advantages such as ease of solvent removal and lack of toxicity. Also, the possibility exists to obtain fractionation of oil during the extraction process, which could facilitate refining or production of nutraceuticals. The major obstacles to the use of supercritical fluids are the large capitalization cost of initial start-up and safety concerns due to the high pressures involved. In the future, regulatory actions may stimulate the use of this approach.

### *Developments in Extraction*

The removal of oil from rice bran was accomplished historically using hexane extraction because it is by far the most economical choice currently available. The standard configuration of a rice bran oil extraction facility is one in which rice bran is cooked, pelletized, mixed with solvent (called miscella), usually in a continuous countercurrent extractor, filtered, stripped of solvent by distillation, and then placed in a settling tank to remove the oil foots, which are the solid materials that settle out. The entire process is accomplished under vacuum to minimize solvent losses, for economical and environmental reasons (Pillaiyar, 1988). During the 1970s and early 1980s, the Riviana Rice Mill in Abbeville, Louisiana, operated a plant in which oil was extracted during the milling operation. This process is referred to as the X-M process and was based on technology described in a U.S. patent issued to Wayne (1966). The process had two major advantages: improving rice milling quality and lowering free fatty acids in the oil. However, this approach to rice bran oil extraction was abandoned because it was deemed uneconomical primarily due to inefficient solvent recovery (Kao & Luh, 1991). More recently, because of safety and environmental concerns for the use of hexane in the extraction of edible oil, a fair amount of research was directed at edible-oil extraction by using nontoxic, green approaches. The main avenues include the use of less toxic organic solvents such as isopropanol, supercritical fluid extraction, and enzyme-assisted aqueous extraction.

The potential of supercritical fluid extraction of rice bran oil was suggested by Ramsay et al. (1991) in a position paper espousing the advantages of this "exciting and emerging technology for the food industry." They were able to extract 17.98% oil using supercritical $CO_2$ at 4350 psi (30 MPa) at 35°C compared with 20.21% oil using hexane refluxed in a Soxhlet apparatus for three hours. On the other hand, Garcia et al. (1996) reported that the condition of the highest yield of extract from rice bran in their study was the highest pressure and temperature allowable in their system (28 MPa and 70°C), and the yield was only 16–60% of that obtained by solvent extraction with hexane. The solvation potential of supercritical fluids changes as temperature and pressure change within the supercritical region, which affects the density of the $CO_2$. Definitely consider this feature when comparing data from different studies. It also potentiates the use of supercritical fluids to partition or selectively extract lipid components from lipid-bearing materials. In our laboratory, we found that it was possible to concentrate the oryzanol fraction by optimizing time:temperature:pressure combinations (Xu & Godber, 2000).

Danielski et al. (2005) found that at a constant temperature of 50°C, as pressure increased, extraction efficiency increased due to an increase in density of the $CO_2$. They also found that by coupling supercritical column chromatography to the batch extraction process it was possible to obtain oil with less than 1% free fatty acids, although the final oil would require winterization to remove the high melting point material. An economic analysis indicated that production costs using their system would be around 3.00 EUR/kg, which at that time compared to a market price of 8.50 to 13.00 EUR/kg (Danielski et al., 2005). In actuality, the market price of rice bran oil historically has been much less than the figures stated in this reference; thus, the margin between production cost and market value would be much smaller than indicated.

Sparks et al. (2006) compared supercritical fluid extraction with compressed propane and pressurized hexane relative to rice bran oil extraction efficiency. By using an accelerated solvent extraction system (ASE, Dionex Corp.), which heated hexane to 100°C and produced a pressure of 10.34 MPa, they obtained a yield of 26.1% oil, which is higher than the range normally associated with rice bran. The authors attributed this to the fact that they used bran from parboiled rice, which tends to yield higher lipid extraction rates, presumably due to the migration of lipid from the interior of the rice grain to its surface. Also, possibly, the bran from parboiled rice had a relatively higher concentration of oil because less endosperm is abraded during the milling of parboiled rice, which lessens the dilution effect. Another explanation could be that the increased temperature and pressure produced by the ASE facilitated oil extraction compared with standard Soxhlet extractors. Supercritical $CO_2$ at 35 MPa and 45°C produced a yield of 22.2% oil, and 22.4% was achieved with propane at 0.76 MPa.

Supercritical fluids were also proposed as a means to extract or isolate high-value components from oil refining by-products. Mendes et al. (2005) describe a process in which they isolated sterols and tocols from deodorizer distillate using supercritical $CO_2$.

One commercial drawback to supercritical fluid extraction of oil from rice bran that needs to be overcome is that all of the research to date was accomplished in batch extractors. A continuous extraction system would most likely be required to obtain the product throughput needed to justify commercial scale. A patent issued to Mitchell and Routier (2006) describes a process for continuously extracting oil from solid or liquid oil-bearing materials using supercritical fluids, although no evidence is apparent for the existence of a commercial use of this system.

Johnson and Lusas (1983) provided an overview of the various solvents that were used in the extraction of oil from oil-bearing materials. In the conclusion to this overview, they state that "solvents which pose less health and fire hazards, greater ability to extract neutral oil ... will always be of interest." They provided an overview of the various types of solvents that were employed and indicated that alcohol, especially isopropanol, has the potential to be a replacement for hexane as the solvent of choice in the extraction of oil from oil-bearing material. Thus, a research focus at the Texas A&M Protein Research and Development Center was in the development of isopropyl alcohol as an alternative vegetable oil extraction solvent (Lusas et al., 1991; Watkins et al., 1994). A comparison of isopropyl alcohol with hexane in the extraction of oil and high-value components from rice bran was done in our laboratory (Hu et al., 1996), and hexane tended to extract more total oil than isopropyl alcohol regardless of extraction temperature (60°C versus 40°C). However, isopropyl alcohol extract had a higher concentration of tocols than hexane extract, although the oryzanol concentration was higher in the hexane extract. A more recent innovation in rice bran oil extraction was the use of d-limonene as an alternative solvent with fewer environmental, health, and safety issues (Liu & Mamidipally, 2005).

Enzyme-assisted aqueous extraction refers to the use of enzymes to break down the cell wall and spherosomes to make lipid more readily extractable (Rosenthal et al., 1996). Sharma et al. (2001) applied this approach to the extraction of oil from rice bran. They point out that rice bran presents unique challenges to this approach, but they devised a process that used mixtures of protease, alpha-amylase, and cellulase, resulting in a 77% recovery of oil. A combination of enzyme-assisted extraction with less toxic butanol to facilitate extraction in a three-phase partitioning system was proposed by Gaur et al. ( 2007). They speculated that protease, along with ammonium sulfate, would degrade cell walls and facilitate the release of oil from spherosomes. They were able to achieve an 86% recovery of oil using this approach.

### *Refining*

Following extraction, several processes are undertaken that are collectively referred to as “refining,” to improve the quality of the oil. The refining process may also cause a reduction in nutritionally active compounds such as vitamin E and oryzanol, which may be recovered as by-products of the refining process. The steps that occur in a typical rice bran oil refining process, along with the by-products generated are given in Fig. 14.1. Generally, the first step in the refining process with crude rice bran oil is dewaxing or winterizing, although for other vegetable oils the dewaxing step can be done at different times in the refining process. The reason is that crude rice bran oil can have up to 8% wax depending on the extraction method. Such high levels of wax can have an adverse effect on subsequent refining steps, causing reduced yields, and can cause problems with refining machinery. A common practice is to allow crude oil to sit in settling tanks to allow the higher melting point waxes to settle out. After the removal of the oil, workers can recover these waxes which may have value as a by-product (which is discussed later). The next step is the removal of the phosphatides by water washing, which is referred to as “degumming.” Again, this step is carried out early in the refining process to avoid problems with subsequent refining steps and machinery.

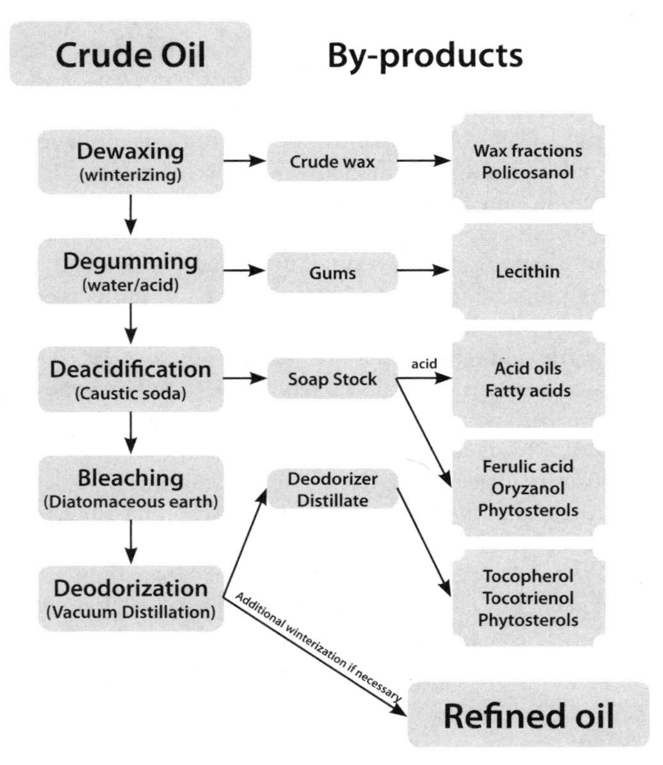


**Fig. 14.1.** Diagram of a typical oil refining process with accompanying by-products.

Perhaps the most critical step in the refining of rice bran oil is the neutralization step. This is because of the propensity for rice bran to undergo hydrolytic rancidity with the accumulation of free fatty acids. High levels of free fatty acids can lead to quality problems related to color and other physical properties. Free fatty acid neutralization is normally accomplished using sodium hydroxide or other caustic compounds such as sodium carbonate. The amount of caustic material is dependent on initial free fatty acid levels. In crude oils with high free fatty acids, the inclusion of physical refining by distillation of free fatty acids may be necessary to increase yields to an acceptable level. A by-product of the neutralization process is soap stock, which is obtained after water washing. In addition to its obvious potential for the production of industrial soaps, soap stock can be an excellent source of a class of compounds called oryzanol, which may have unique health benefits that are discussed later. Furthermore, probably the levels of oryzanol in rice bran oil can be retained at much higher levels if the oil is refined using only physical refining rather than caustic refining (Krishna et al., 2001). Because of the health benefits associated with oryzanol, this approach was advocated as a way to maximize the potential health benefits of rice bran oil. RITO Partnership (Riceland Foods, Inc. and Oilseeds Internaltional, Ltd.) promoted a high oryzanol oil that they call Oryzan™, which is produced by proprietary refining methods. The product specification of Oryzan™ compared with RITO's traditionally refined rice bran oil and crude oil is shown at the RITO Partnership Web site ( http://www.ricebranoil.biz/spec/index.html). The oryzanol content of their crude rice bran oil is given as 1.5% (min.) and that of Oryzan™ as 1.0% (min.), but no level of oryzanol is given for their traditionally refined oils.

The removal of color, primarily chlorophyll and carotenoids, is called bleaching, which is accomplished using clay or diatomaceous earth. Deodorization by countercurrent steam is employed to remove volatile substances. Deodorization also removes some of the tocopherols, tocotrienols, and phytosterols, but these are often fractionated from the deodorizer distillate for use as supplements, nutraceutiacals, and other applications. Other finishing steps could include hydrogenation to reduce unsaturated fat and to harden the oil, and winterization to remove the remaining high-melting-point glycerides and waxes. Winterization would be necessary if the oil was intended for use in salad dressings, which are normally stored at refrigeration temperature.

## Developments in the Refining of Rice Bran Oil

A recent review article (Ghosh, 2007) summarizes some of the key developments in the processing of rice bran oil. The initial consideration mentioned in this review article is the need to remove the fines that become entrained in the oil during the milling of rice bran. The recommended approach is to allow the fines to settle out in settling tanks followed by filtration of the oil. This process removes a certain portion of the wax that also settles out. From that standpoint the review points out that the dewaxing process with rice bran oil is made more challenging by the presence of considerable high-melting-point fatty acids, which necessitates more stringent processing. The rice bran oil refining industry in Japan has generally employed temperature-controlled crystallization (also called winterization) of miscella followed by centrifugation, but this process is not feasible for smaller processing plants found in much of the rest of the world, which continue to depend on settling tanks as the main means to dewax crude oil. Alternative approaches include the use of soap stock along with centrifugation in which wax crystals segregate into the water phase for more consistent separation. Korean patents were granted that describe solvent dewaxing using acetone. A more recent development was the use of membrane technology as a more efficient filtration approach. Various types of plastics and ceramics were employed in this approach to dewaxing (Ghosh, 2007).

In typical vegetable oils, the degumming step tends to be fairly straight-forward and is primarily accomplished through a water-washing step through which the phosphatides hydrate and become insoluble in the oil and are thus easily separated by centrifugation (Haraldsson, 1983). Because crude rice bran oil tends to be higher in phosphatides, additional steps are generally required to reduce the phosphatide content to an acceptable level of less than 1%. Segers et al. (1983) describe a seven-step process for the removal of phosphatides from oil with high initial levels of phosphorus (1000 mg/kg) that is called superdegumming. The process involves initially heating the oil to 70°C, adding a small amount of concentrated food-grade acid, then stirring for 5–30 minutes and subsequent cooling to 25°C. Then, 1–5% water is added and stirred for 2 hours to hydrate the phosphatides, which forms a sludge that centrifugation can separate. Recently, electrolyte degumming was proposed to reduce oil losses during the superdegumming process (Nasirullah, 2005). The use of an aqueous solution of 1.5% potassium chloride and 0.5% sodium chloride in a 95:5 (vol/vol) ratio obtained a phospholipid content of 0.06% in rice bran oil.

Researchers in India, which processes the greatest amount of rice bran oil in the world (FAO, 2007), pursued more efficient processes to refine rice bran oil (Narayana et al., 2002). They note that special attention is needed in the dewaxing, degumming, and deacidification steps. They describe processes aimed at improving these refining steps that include simultaneous dewaxing and degumming, enzymatic degumming, and the removal and separation of glycolipids for potential use in the cosmetics industry. They also promote the use of physical refining as the preferred method of deacidification (Narayana et al., 2002).

Collaboration between the Thai Edible Oil Company and Ghent University in Belgium produced an interesting paper that depicts the effect of chemical refining processes on the minor components of rice bran oil (Van Hoed et al., 2006). These researchers followed a typical refining process used at the Thai Edible Oil Company from crude oil to the final edible oil and assessed the influence of each step of refining on the content of free fatty acids, diglycerides, unsaponifiable matter, total sterols, total tocols, gamma-oryzanol, and individual tocols and phytosterols. The refining steps that they included were neutralization, bleaching, dewaxing, and deodorization. Their findings included expected effects such as the reduction of free fatty acids from 7.53–0.12 g/100g oil and the reduction of gamma-oryzanol from 1.8 g/100 g to 0.4 g/100 g after neutralization. Also, unsaponifiable matter decreased from 5.4 g/100 g in crude oil to 2.7 g/100 g in the final oil product, a 50% reduction. The greatest effect on tocol concentration occurred after deodorization, which was expected, but the reduction was only 20% from 78 mg/100 g to 62 mg/100 g. Also of interest was the fact that the deodorizer distillate only contained 3.2% tocols, whereas the level of sterols was 14.8%. This study provided insight into the nature of phytosterols in rice bran oil and the relative effect of refining. The relative composition of individual phytosterols as a percentage of total phytosterols changed more drastically for phytosterols normally associated with gamma-oryzanol such as 24-methylenecycloartanol. This is an indication that these phytosterols are being lost due to refining, whereas other phytosterols, most notably beta-sitosterol, are retained to a greater extent, most likely because they occur as free sterols. Also, 24-methylenecycloartanol isomerized during refining into new configurations; the most abundant of which was presumed to be cyclobranol, which was not detectable in the crude oil but represented 6.3% of the sterols in fully refined oil.

The deacidification step, often referred to as "refining" even though it is actually a component of the overall process of edible-oil refining, is the most critical step in the overall refining of any vegetable oil. This step is especially important in rice bran oil due to its inherent tendency to undergo hydrolytic rancidity and produce high levels of free fatty acids. In the more commonly used chemical refining (deacidification), an amount of base such as sodium hydroxide calculated to neutralize the free fatty acids in the crude oil is used. After adding water, this produces a soap stock with the free fatty acids that centrifugation can separate and remove.The higher the free fatty acid, the greater the amount of base required. When free fatty acid levels are high, as is often the case with rice bran oil, much higher processing losses are incurred, which greatly reduces yield. A general rule is that as much as ten times the percentage free fatty acids can be lost due to chemical deacidification (Kao and Luh, 1991; Pillaiyar, 1988). This reduction in yield is thought to be due to the entraining of oil in the soap stock that is difficult to overcome through centrifugal separation. Another consequence of the chemical deacidification process is the loss of unsaponifiable components such as gamma-oryzanol, which is undesirable because of the potential health-promoting nature of those compounds. Thus, physical refining was proposed as an alternative deacidification method for rice bran oil. A typical approach to physical refining of vegetable oil is given in Fig. 14.2, adapted from Forster and Harper (1983). The primary difference is the use of vacuum distillation to not only deodorize but also to remove the free fatty acids. This has the advantages of reducing not only oil losses in the soap stock but also the loss of oryzanol and other unsaponifiable components; it also increases the quality of fatty acid recovery and reduces the processing cost associated

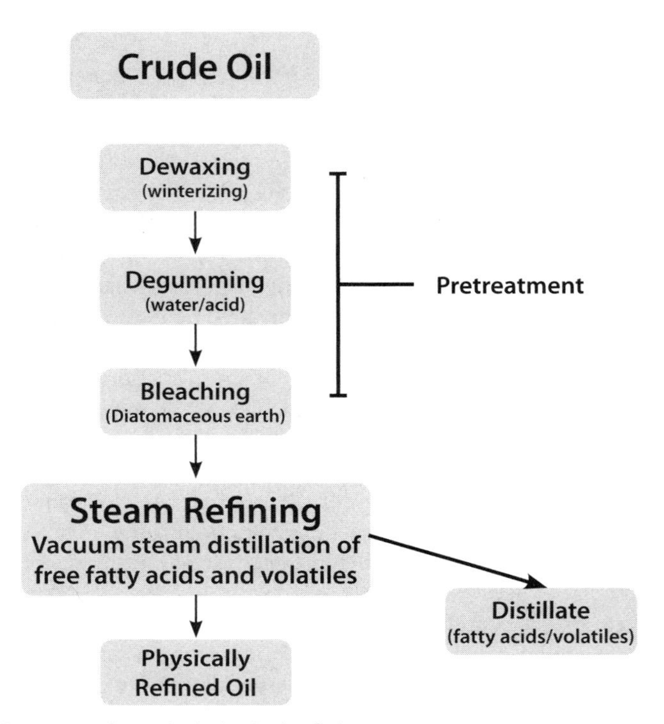


**Fig. 14.2.** Diagram of a typical physical refining process.
(Adapted from Forster and Harper (1983)).

with disposing of the soap stock. However, physical refining does require more care in the pretreatment of the crude oil, especially the removal of phosphatides through degumming, and through bleaching the pigments and metals, especially iron, that could cause darkening of the oil during distillation (Segers, 1983).

Bhosle and Subramanian (2005) provide a review of newer techniques that one could employ in the deacidification of rice bran oil. They discuss biological approaches, primarily related to the reesterification of free fatty acids through microbial fermentation processes, and also mention chemical reesterification. De and Bhattacharyya (1999) describe a process in which rice bran oil is reesterified with monoglyceride to reduce free fatty acids to an acceptable level for further refining. Another approach that they mention is the use of supercritical fluid to remove free fatty acids, employing the selective extraction potential of this approach. Dunford et al. (2003) demonstrated the potential of supercritical $CO_2$ coupled with a pilot-scale packed column to deacidify crude rice bran oil and at the same time enrich the oryzanol concentration of the raffinate fraction threefold. Another approach to deacidification of oil with the purpose of maintaining unsaponifiable material in the refined oil is the use of sodium carbonate rather than the more caustic sodium hydroxide. This approach was first described in the early 1940s as the Clayton soda Ash Refining Process

(Mattikow, 1948). More recently, a patent was issued for the maintenance of higher levels of unsaponifiables in rice bran oil that employs a weak acid salt such as carbonate to precipitate the free fatty acids and separate them by centrifugation (Lawton et al., 2001).

## Edible and Nonedible Applications

Despite the potential abundance of rice bran oil in world commerce, it is considered only a minor player in edible oil applications. It is used most predominately in Japan and India, and to a certain extent in other parts of Southeast Asia, such as Thailand and Indonesia. The majority of rice bran oil produced in much of the world is only suitable for nonedible applications. Because of the high free fatty acid levels of rice bran that is not stabilized immediately after milling, the oil derived from that bran is often used primarily as soap stock. Thus, much of the rice bran oil produced in India, for example, is used to produce soap (Pillaiyar, 1988).

Edible rice bran oil is used in Japan as a staple table oil and is touted as having pleasant sensory characteristics. Rice bran oil is also used as a frying oil, and as an ingredient in salad dressings and mayonnaise and in the making of margarine (Kao & Luh, 1991; McCaskill & Zhang, 1999). The fact that rice bran oil has high levels of antioxidants such as tocols and oryzanol and is less unsaturated than many other vegetable oil, makes it ideal for blending with other oils to improve their stability. Nabisco obtained a patent for blending rice bran oil that was specially formulated to contain high levels of unsaponifiable matter with less stable oils such as soybean oil to increase stability (Taylor et al., 1996). Blending oils to increase stability has become a common approach now that hydrogenation has lost favor due to its production of trans fat. A recent study by Mezouari and Eichner (2007) employed rice bran oil (RO) blended with sunflower oil (SO) as a means to increase stability of frying oils. They attributed the stabilization effect to a decrease in the relative concentration of unsaturated fatty acids, which reduced triglyceride polymer formation. When they added natural antioxidants to simulate levels achieved with a 50:50 blend of RO:SO, they obtained a comparable effect, "remarkably increasing the induction period." Krishna et al. (2005) compared both chemically and physically refined rice bran oil to sunflower oil in terms of frying performance. They found that oil quality as a function of color, peroxide value, and free fatty acids declined after frying Bhujia (a spicy fried Indian snack with a flour-based coating of tubers or vegetables, similar to potato fritters) but that the degradation was not as great with the rice bran oils. They also found that oryzanol from rice bran oil was incorporated into the fried Bhujia product, which could be viewed as increasing the nutritional value of the fried product and possibly its flavor stability.

Rice bran oil is also added to food systems as a means to increase oxidative stability and at the same time to improve the nutritional/health promotion value of the food. We incorporated rice bran oil into restructured beef roasts (Kim et al., 2000;

Kim & Godber, 2001), reducing lipid oxidation during subsequent storage and increasing vitamin E concentration to improve the nutritional value and perception of the beef product. We also used rice bran oil to stabilize dry whole fat milk powder against lipid oxidation during storage (Nanua et al., 2000).

### *Nonedible Applications*

In this age in which much interest is shown in biodiesel, rice bran oil receives scrutiny as a potential feedstock for its production. This is an especially appealing outlet for the vast quantities of industrial-grade rice bran oil. However, a potential problem to overcome is the high free fatty acid levels of this grade of oil. Because an alkali-based transesterification process is needed to produce biodiesel from vegetable oils, high free fatty acids reduce the efficiency of the conversion. Kumar (2007) devised a two-stage conversion process that begins with an esterification step to reduce free fatty acids before the transesterification step.

One can also isolate and utilize components of rice bran oil in numerous different applications. The fact that ferulic acid possesses excellent ultraviolet absorption capacity and acts as an antioxidant makes it a candidate for use in the cosmetic industry. Compton et al. (2007) describe a process in which ferulic acid that was esterified to monoacyl- and diacylglycerols is microencapsulated in starch flakes. This provides a method to carry these lipophilic compounds in an aqueous media while maintaining UV-absorbing characteristics. They contend that one can use this material not only in cosmetic applications but also as a form of natural UV protection in paints, coatings, and insecticidal and herbicidal biocontrol agents. Ferulic acid from rice bran oil soapstock is also used as a substrate for the production of vanillin using *Aspergillus niger* and *Pycnoporus cinnabarinus* (Zheng et al., 2007). The residue left after oil is extracted, sometimes referred to as oil cake, also can be utilized in nonedible applications. Ramachandran et al. (2007) provide a review that describes the application of oil cake from rice bran extraction as a substrate in fermentation processes and as an energy source.

## Major Components of Rice Bran Oil

The lipid material that is extracted from rice bran to make rice bran oil comes primarily from the germ and the outer layers (aleurone and subaleurone) of the rice grain where it exists as lipid droplets or spherosomes. The chapter on rice lipids by Godber and Juliano in the second edition of *Rice: Chemistry and Technology* (2004) provides an extensive discussion of the lipids found in the rice grain. The major constituents of rice bran oil along with their fatty acid composition are listed in Table 14.1 as depicted by Godber and Juliano (2004). Rice bran oil is similar to other vegetable oils with high levels of neutral lipids and lesser amounts of glycolipids and phospholipids. Its fatty acid profile compared with other vegetable oils tends to be higher in

**Table 14.1. Major Lipid Classes of Crude Rice Bran Oil Extracted from Raw Rice Bran and Their Fatty Acid Composition**[a,b]

| | | Fatty acid composition (%) | | | | | | | | |
|---|---|---|---|---|---|---|---|---|---|---|
| Lipid class[c] | wt%[d] | 14:0 | 16:0 | 18:0 | 18:1 | 18:2 | 18:3 | 20:0 | saturated | unsaturated |
| TL | 20.1 | 0.4 | 22.2 | 2.2 | 38.9 | 34.6 | 1.1 | 0.6 | 25.4 | 74.6 |
| NL | 89.2 | 0.4 | 23.4 | 1.9 | 37.2 | 35.3 | 1.1 | 0.7 | 26.4 | 73.6 |
| GL | 6.8 | 0.1 | 27.3 | 0.2 | 36.5 | 35.8 | 0.2 | | 27.6 | 72.4 |
| PL | 4.0 | 0.1 | 22.1 | 0.2 | 38.1 | 39.3 | 0.2 | | 22.4 | 77.6 |

[a]Source: Shin and Godber, 1996.
[b]Values are means of three replicate analyses.
[c]TL, total lipids; NL, neutral lipids (nonpolar lipid and free fatty acids); GL, glycolipids; PL, phospholipids.
[d]TL is wt% of bran; NL, GL, PL are wt% of TL.

**Table 14.2. Fatty-acid Composition as a Percentage of Total Lipid of Rice Bran Oil Compared with that of Corn, Peanut, Soybean, and Cottonseed Oils.**[A]

| Fatty acid | Rice bran | Corn | Peanut | Soybean | Cottonseed |
|---|---|---|---|---|---|
| Myristic (14:0) | 1 | 0 | 0 | 0 | 0 |
| Palmitic (16:0) | 15 | 8 | 7 | 8 | 21 |
| Stearic (18:0) | 2 | 4 | 4 | 4 | 2 |
| Oleic (18:1) | 45 | 46 | 62 | 28 | 29 |
| Linoleic (18:2) | 35 | 42 | 23 | 54 | 45 |
| Linolenic (18:3) | 1 | 0 | 0 | 5 | 3 |
| Arachidic (20:0) | 0 | 0 | 0 | 0 | 1 |

[a]Adapted from Pillaiyar (1988). Values are typical analyses.

oleic acid and lower in linoleic acid, similar to peanut oil in this regard. The major fatty acids of rice bran oil compared to other major vegetable oils are shown in Table 14.2. As with most oils, triacylglycerols make up the bulk of the lipid, although crude oil produced from unstabilized bran has higher levels of free fatty acids, mono- and diacylglycerols due to the active lipases present in the bran. Crude rice bran oil tends to have higher levels of phospholipids, which must be removed through the refining process. Rice bran oil is also unique in that it contains higher levels of wax (3–9%) and unsaponifiable components (2–5%) than other vegetable oils, which can cause processing problems but also may contribute to the unique health benefits attributed to rice bran oil (to be discussed later).

## Minor Components of Rice Bran Oil

One of the most positive features of rice bran oil is its abundance of minor constituents, many of which are suggested to have significant health benefits. In fact, it seems that the scientific community is at odds over which of the many minor constituents in rice bran oil is responsible for its health benefits. The minor constituents most often mentioned as being important include the tocols, oryzanol, phytosterols, and more recently, policosanol. In this section, the nature and abundance of these constituents are discussed, and in a subsequent section, their potential health benefits are reviewed.

## Tocols

The term "tocol" is used to include both the tocopherols and tocotrienols. These two classes of compounds are similar in that they both contain a 6-hydroxychroman ring structure with up to three methyl groups, to which is attached a phytyl tail at the second carbon. They differ in the fact that the phytyl side chain is completely saturated in tocopherols but has three double bonds in a conjugated arrangement in tocotrienols (Fig. 14.3). Crude rice bran oil has one of the highest concentrations of tocols of any vegetable oil and is generally considered one of the best sources of toco-trienols in particular. Table 14.3, which was adapted from Eitenmiller (1997), depicts a comparison of rice bran oil with several other sources of tocols. As shown in this table, at the high end of the range, rice bran has more tocols than any of the other common vegetable oils, followed by palm oil. However, both of these oils generally contain a higher level of tocotrienols and much less α-tocopherol. For that reason, rice bran oil and palm oil are generally not considered good sources of vitamin E because vitamin E activity is greater in the tocopherols, especially α-tocopherol. That is why sunflower oil has the highest vitamin E activity (vitamin E equivalents), even though it has lower total tocols, because almost all of its tocols are α-tocopherol. However, in recent years

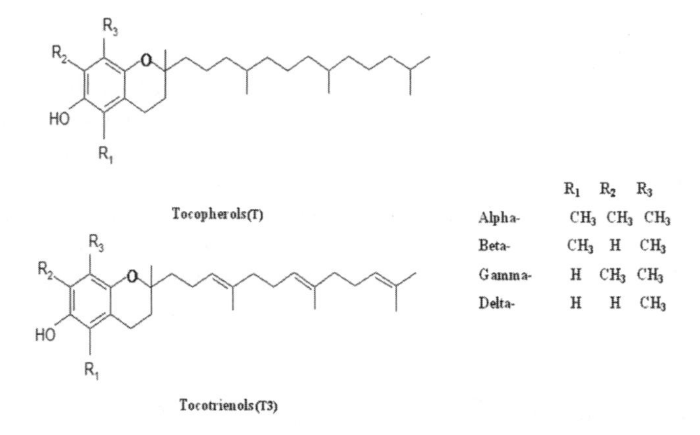


Fig. 14.3. Chemical structures of the tocols.

**Table 14.3. Tocol (Tocopherol (T) and Tocotrienol (T3)) Contents of Rice Bran Oil Compared with Other Common Vegetable Oils in Descending Order of Abundance**

| Oil | Total Tocols mg/100 g | Vit E Equiv. mg/100 g | %T | %T3 | Primary homologs Highest Lowest |
|---|---|---|---|---|---|
| Rice bran | 9-160 | 0.9-41 | 19-49 | 51-81 | γ-T3, α-T3, α-T, β-T, β-T3 |
| Palm | 89-117 | 21-34 | 17-55 | 45-83 | α-T, α-T3, γ-T3, γ-T, δ-T |
| Soybean | 95-116 | 17-20 | 100 | 0 | γ-T, δ-T, α-T |
| Corn | 78-109 | 20-34 | 95 | 5 | γ-T, α-T, δ-T, γ-T3, δ-T3 |
| Canola | 65 | 25 | 100 | 0 | γ-T, α-T, δ-T,α-T3(tr),β-T(tr) |
| Safflower | 49-80 | 41-46 | 100 | 0 | α-T, δ-T, γ-T, β-T |
| Sunflower | 46-67 | 35-63 | 100 | 0 | α-T, γ-T |

Adapted from Eitenmiller (1997).

the potential importance of tocotrienols to health, independent of their vitamin E activity, was recognized (Sen et al., 2005; 2007). As discussed later, the high level of tocotrienols is suggested as a reason for the serum cholesterol-lowering effect of rice bran oil, and recent evidence indicates that tocotrienols have anticancer properties.

## Phytosterols

Plant sterols, also referred to as phytosterols, include a broad class of compounds that were reported to include over 250 different sterol structures (Piironen et al., 2000). They are similar in structure to cholesterol, the more well-known form of sterol found predominately in the animal kingdom, varying primarily in side-chain configurations and small differences in the sterol ring structure. Three classes of sterol structures are found in the plant kingdom based on methylation at the four carbon in the sterol ring structure (Fig. 14.4). The most common form in the plant kingdom is the desmethyl form, with beta-sitosterol as the most often given example. Rice bran oil is relatively unique in that it contains high concentrations of the 4,4-dimethyl form represented by 24-methylenecylcoartanyl. The 4-methyl form, as in citrostadienol, tends to be a minor component of plant lipids, including rice. In addition to differences at the 4-carbon, other differences between phytosterol structures exist that give rise to classifications such as stanols, which are completely saturated, and differences in side-chain structure account for many of the structural differences between different types of phytosterols. The functions of plant sterols appear to be similar to cholesterol in animals in that they act in the cell membranes as structural and surface-active agents and serve as precursors to hormone-like growth factors. A very thorough description of plant sterol synthesis and functionality is given by Piironen et al. (2000).

These phytosterol compounds can exist as either free sterols or esterified to fatty acids or cinnamic acids, such as ferulic acid. Ferulic acid esters of phytosterols are specially classified as gamma-oryzanol, as discussed in the next section. Much more information is available in the literature relative to the oryzanol composition of rice

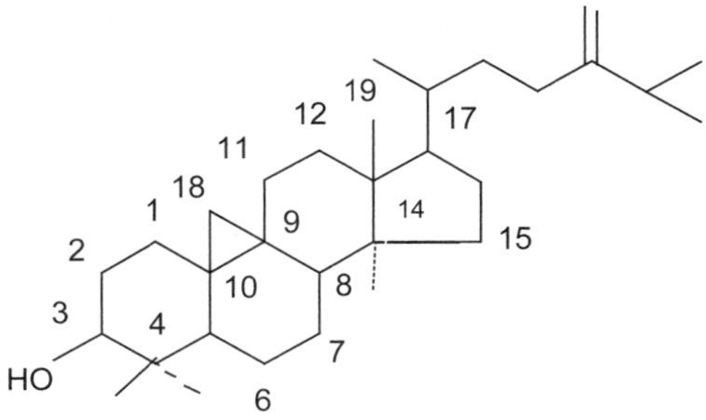

24-Methylenecycloartanol

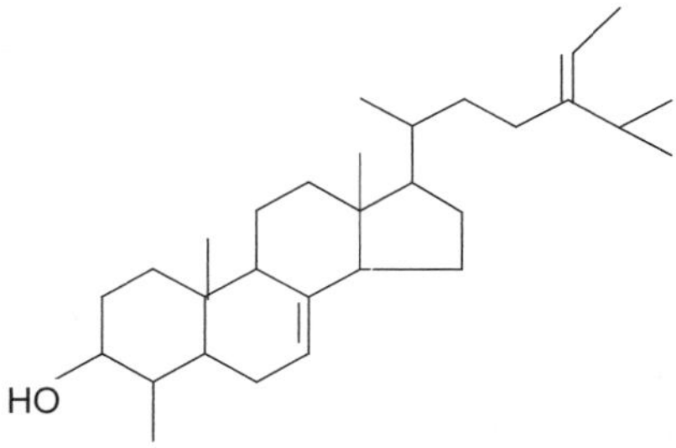

Citrostadienol

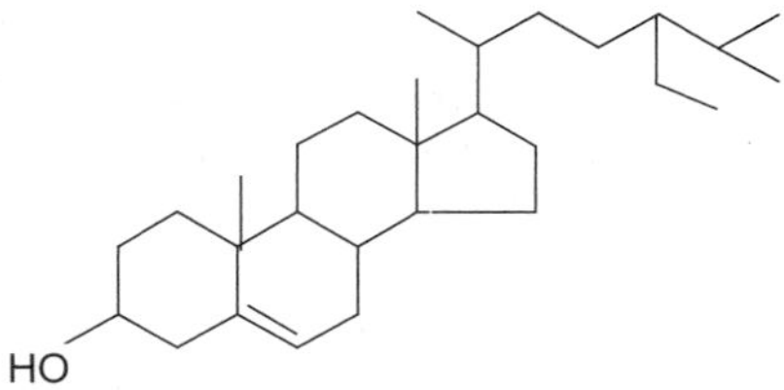

Beta-Sitosterol

**Fig. 14.4.** Basic structures of 4,4-dimethyl- (top), 4-methyl- (middle) and 4-desmethyl- (bottom) sterols found in rice bran oil.

bran oil than on total sterol composition. Phytosterols also occur as steryl glucosides (phytosterol bound to glucose) or as acylated steryl glucosides (phytosterol bound to both glucose and fatty acid). Piironen et al. (2000) indicated that rice bran oil has the highest concentration of phytosterols of the common vegetable oils at 32.25 mg/g in crude oil and 10.55 mg/g in refined oil. More recently, Van Hoed et al. (2006) reported a concentration of 29.4 mg/g total phytosterols in crude oil and 21.5 mg/g, 19.2 mg/g, 19.1 mg/g, and 18.3 mg/g in oil that was neutralized, bleached, de-waxed, and deodorized, respectively. Interestingly, of the total phytosterols in the crude oil, approximately 61% were oryzanols, but only about 16% of the phytosterols in the final oil product were oryzanols.

## Gamma-oryzanol

γ-Oryzanol, as this class of compounds was originally designated, was thought to be a single compound when first isolated in the early 1950s from the soap stock of rice bran oil refining (Seetharamaiah & Prabhakar, 1986). It was since found to be a mixture of steryl and other triterpenyl esters of ferulic acid (4-hydroxy-3-methoxy cinnamic acid). The earliest investigations of this class of compounds classified the major ferulate esters as oryzanol A, B, and C, which were present at 1.5–2.9% of bran oil and had a melting point (m.p.) of 138.5°C (Tsuchiya et al., 1957). They consisted of cycloartenyl ferulate ($C_{40}H_{58}O_4$ m.p. 165.2–166.5°C) (Ohta & Shimizu, 1957), called oryzanol A, and 24-methylenecycloartanyl ferulate ($C_{41}H_{60}O_4$, m.p. 164–164.5°C), termed oryzanol C (Ohta, 1960). Oryzanol B was considered a mixture of oryzanol A and C (Shimizu & Ohta, 1960).

Since this early work in characterizing the oryzanol fraction of rice bran oil, a great deal of interest was generated because of the potential health benefits that it possesses (to be discussed later). Rogers et al. (1993) utilized reverse-phase high-performance liquid chromatography (HPLC) to separate individual ferulates in the oryzanol fraction. They found that cycloartenyl ferulate, 24-methylene cycloartanyl ferulate, and campesteryl ferulate (Fig. 14.5) were the predominant ferulates in oryzanol, along with β-sitosteryl ferulate, stigmasterol ferulate, and cycloartanyl ferulate, which is in basic agreement with the consensus of earlier reports. Norton (1995) found cyclobranyl ferulate, in addition to the ferulates found by Rogers et al. (1993). In our laboratory, Xu and Godber (1999) found a total of ten different ferulates from the oryzanol fraction that included five ferulates that were not previously identified in rice bran, $\Delta^7$-stigmasterol, $\Delta^7$-campesterol, $\Delta^7$-sitostenol, campestanol, and sitostanol. Akihisa et al. (2000) provided insight regarding cis-trans isomerization at the ferulic acid ester linkage and the effect of light on the stereochemistry of that linkage. They also identified several previously unidentified ferulates including ferulic acid esters of 24-methycholesterol, 24-methylenecholesterol, and cycloeucalenol.

Considerable variability was noted in the oryzanol concentration of rice bran oil. Rogers et al. (1993) found that in five commercially available refined rice bran oils the

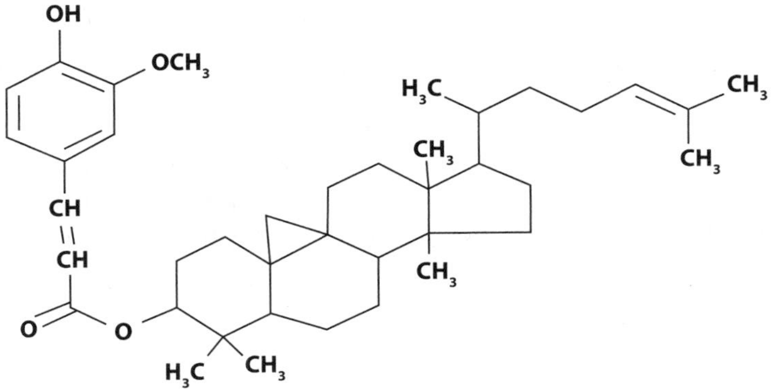

**Cycloartenyl Ferulate**

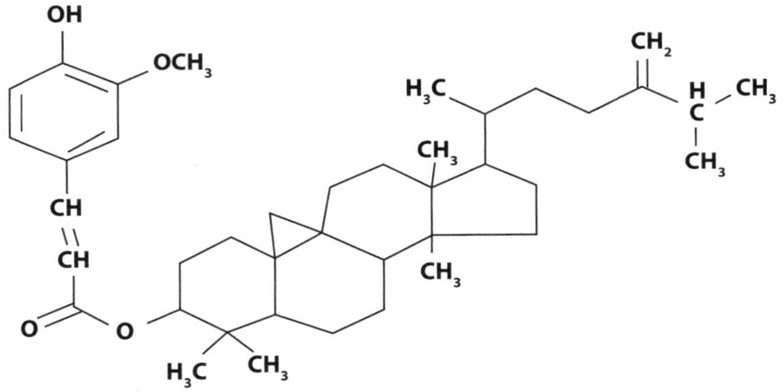

**24-Methylene cycloartanyl ferulate**

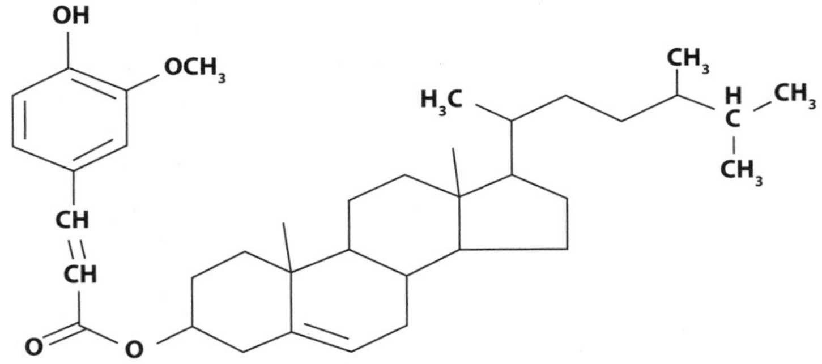

**Campesteryl ferulate**

**Fig. 14.5.** Chemical structures of the three predominate ferulic acid esters in oryzanol.

concentration of oryzanol ranged from 0.115 mg/g to 0.787 mg/g, whereas Norton (1995) found concentrations of 0.4 and 3.0 mg/g in rice bran oils from two different commercial sources, compared to 15.7 mg/g in a crude oil preparation. As mentioned previously, the refining process can have a great impact on the oryzanol concentration. Yoon and Kim (1994) found that a crude oil with 1.61% oryzanol was reduced to 0.77% oryzanol by refining. Xu and Godber (2000) noted that the concentration of oryzanol in a hexane extract of rice bran was 9.8 mg/g without saponification during extraction and 4.6 mg/g with saponification. Krishna et al. (2001) studied the overall influence of refining on the oryzanol content of the oil and found that unrefined oil with 6.8% free fatty acids had 1.86% oryzanol, which was reduced to 1.84% after degumming, to 1.75% after dewaxing, and to 0.10% after alkali deacidification. More recently, the study by Van Hoed et al. (2006) confirmed the drastic effect of alkali deacidification on oryzanol concentration, going from 1.8 g/100 g in crude oil with 7.53% free fatty acids to 0.4 g/100 g in neutralized oil that was reduced to 0.12% free fatty acids.

## Waxes

Waxes are composed of numerous different chemical moieties but are generally thought of in terms of their wax esters. A wax ester is composed of a long-chain fatty acid and a long-chain fatty alcohol. Chain lengths of the fatty acid component generally range from 14 carbons to 24 carbons, with more saturated than unsaturated bonding. The alcohol component contains carbon chain lengths of between 24 and 36, which are almost exclusively saturated. The most notable physical property of wax is a high melting point, which generally makes it solid at room temperature. According to Ito et al. (1983), rice bran oil contains about 90% neutral lipid of which 9.5% is wax esters. These esters were found to be a mixture of 20% alkyl esters and 80% steryl esters. In addition, the wax fraction can contain free fatty acids, free alcohols, aldehydes, and some free hydrocarbons, which complicates its characterization, purification, and analysis. As stated previously, crude rice bran oil has a high concentration of wax material, which can adversely affect the refining process. A common practice in the rice bran oil processing industry is to allow crude oil to sit in settling tanks to allow the wax material to settle out. The composition of waxes from crude rice bran oil tank settlings was studied by Yoon and Rhee (1982) using thin-layer chromatography (TLC) and gas chromatography (GC). They used methyl ethyl ketone to remove the oil and isopropanol to crystallize the wax. They segregated the wax material into soft (m.p. 74°C) and hard waxes (m.p. 79.5°C ) according to their melting points. Hard wax was 64.5% fatty alcohols compared with 51.8% fatty alcohol for soft wax. On the other hand, soft wax had 46.2% fatty acids compared with 33.5% for hard wax. Each had small amounts of other chemical compounds such as hydrocarbons. Recently, Vali et al. (2005) studied a process for the preparation of food-grade rice bran wax and determined its composition. They used as starting materials five different sediments from crude rice bran oil extracted by a Soxhlet method using hexane as the solvent. They obtained almost pure wax esters (>99% purity) after defatting the sediments with hexane and isopropanol, and bleaching with $NaBH_4$ to remove the resinous matter that was mostly free fatty acids, alcohols, and aldehydes. TLC and GC were used for the wax ester analysis. Their results indicate that rice bran wax is mainly a mixture of saturated esters of C22 and C24 fatty acids and C24 to C40 aliphatic alcohols, with C24 and C30 being the predominant fatty acid and fatty alcohol, respectively. The alcohol portion of the wax esters also contained small amounts of branched and odd carbon number fatty alcohols.

## Policosanol

"Policosanol" is a term coined by researchers in Cuba to refer to an extract derived from sugar cane wax that was a mixture of long-chain alcohols. Octacosanol ($CH_3$-$(CH2)_{26}$-$CH_2$-OH, C28) is the predominant alcohol found in sugar cane wax, comprising approximately 63% of the mixture. Other important constituents include triacontanol (13%) and hexacosanol (6%). Minor components include tetracosanol (C24), heptacosanol (C27), nonacosanol (C29), dotriacontanol (C32), and tetratriacontanol (C34). The health benefits of policosanol were primarily evaluated by Cuban scientists. They found that policosanol is effective in improving serum lipids by lowering total cholesterol (TC) and low-density lipoprotein (LDL), while raising high-density lipoprotein (HDL) (Janikula, 2002). They suggested that policosanol performs equal to or better than standard pharmaceutics used to treat abnormalities of cholesterol metabolism such as lovastatin and probucol. Besides improving serum lipid profile, policosanol modifies several other cardiovascular disease risk factors by reducing LDL oxidation, platelet aggregation, and endothelial cell damage (Janikula, 2002). However, recent studies produced contraindicating results compared with the Cuban research (Berthold et al., 2006).

The composition of rice wax is different from sugar cane wax, most notably due to a lower concentration of octacosanol and a higher concentration of triacontanol. A recent study showed similar levels of triglyceride, total cholesterol, and HDL cholesterol in hamsters fed Rice Bran Wax (a policosanol mixture from rice wax with 50% being converted to the corresponding acid; Traco Labs, Chicago, IL), Octa-60 (a policosanol mixture from sugar cane wax; Garuda International, Inc., Lemon Grove, CA), and a control diet (Wang et al., 2003). This research contradicts previous research with policosanols from sugar cane wax, which was shown to effectively lower LDL cholesterol and increase serum HDL cholesterol, thus improving HDL/LDL ratio. No other studies were found in the literature aimed at determining the effectiveness of rice wax policosanols in lowering cholesterol levels. However, a recent study suggested that triacontanol alone was similar to sugar cane wax policosanol in altering cholesterol metabolic enzymes (Singh et al., 2006).

## Flavor and Aroma Compounds

Rice bran oil is characterized as having a slightly bland, nut-like flavor similar to peanut oil. However, because rice bran oil is not viewed as a gourmet-style cooking oil, little research exists as to its specific flavor profile. Aromatic rice varieties have become popular in much of the world, including the United States. The volatile compound responsible for the popcorn-like aroma of aromatic rice varieties is 2-acetyl-1-pyrroline; however, to this author's knowledge, no research exists in which this compound was isolated from rice bran oil.

## Allergic and Toxic Compounds

Rice was promoted as a hypoallergenic commodity for many years; thus assumedly rice bran oil also possesses hypoallergenic properties. This is a reason given for the use of rice bran oil in many cosmetic applications. However, little work was done to verify the hypoallergenic nature of rice bran oil. From a technical standpoint, almost all food allergies are due to specific proteins in food, and highly refined oil does not contain proteins (Hefle, 2008). The exception is cold-pressed oil, which could contain small amounts of proteinaceous residues, but most commercially available rice bran oil is extracted with solvents.

From a general toxicological standpoint, a report published in the *International Journal of Toxicology* in 2006 addressed the safety of cosmetic ingredients derived from rice. The report indicated that undiluted rice bran oil, rice germ oil, and hydrogenated rice bran wax were not irritants in animal skin tests, and were negative in ocular toxicity assays. The Cosmetic Ingredient Review Expert Panel concluded that these rice-derived ingredients are safe as cosmetic ingredients in the practices, uses, and concentrations as described in the report (Anonymous, 2006).

As mentioned previously, because rice bran oil is extracted using hexane, concern exists for residual hexane in the oil, but this concern exists for most commercially produced vegetable oils. Although no maximal levels of residual hexane in vegetable oil are established, good manufacturing practices indicate that one should be careful to minimize its occurrence in crude oil. An international panel of experts determined that hexane at levels found in properly manufactured edible oil has relatively low toxicity (Chipman, 1991). Whether or not it poses a health risk continues to be debated.

Because of the relative stability of rice bran oil during frying or other heat processing, it actually is considered a lower risk oil, relative to the production of toxic heat-generated products, such as hydroperoxides and toxic polymers. Guillen and Ruiz (2005) used $^{1}H$ nuclear magnetic resonance to measure the formation of toxic hydroperoxyalkenals and hydroxyalkenals in corn and sunflower oils during heating. They found that the formation of toxic compounds was higher in sunflower oil than corn oil, which was correlated with higher linoleic acid and lower oleic acid. Rice bran oil is considerably lower than both of these oils in linoleic acid and higher in oleic acid.

From a historical standpoint, two major PCB poisoning events occurred in Japan (in 1968) and China (in 1979) that were both traced to contaminated rice bran oil (Hsu et al., 1985). Because PCBs interact with several functions of the endocrine system, they are called endocrine disruptors (Hsu et al., 2005). Because they persist in the adipose tissue, their effects are long-term, and the consequences of these events were followed for many years (Hsu et al., 2005). Although the source of the contamination was never determined precisely, it could have been from the heat conductors used in the deodorization process (Hsu et al., 1985). Although PCBs are environmental contaminants, which are now prohibited, the stigma of these events potentially taints the reputation of rice bran oil in the Far East.

## Health Benefits of Rice Bran Oil and Oil Constituents

Inhabitants of Japan have used rice bran oil for many years as household cooking oil. The world recognizes the superior general health of the Japanese population and its low level of coronary heart disease in particular. This is often used to characterize the effect of "Western Diets" on Japanese immigrants to the United States (Marmot & Syme, 1976). Possibly, the use of rice bran oil for cooking purposes contributes to this superior health, and Japanese scientists have espoused the potential healthful properties of rice bran oil for a long time (Suzuki & Oshima, 1970). The Indian government initiated a program promoting the use of rice bran oil in their country as a means to reduce their dependence on imported cooking oil (Rukmini & Raghuram, 1991). As part of that program, research was funded to characterize the healthful aspects of rice bran oil. Thus, the work of Rukmini, the lead Indian scientist involved in this research effort, along with her colleagues, is generally regarded as the first well- organized and publicized effort to understand the potential health benefits of rice bran oil (Raghuram & Rukmini, 1995).

Because the primary health benefit ascribed to rice bran oil is its ability to lower serum cholesterol, fairly extensive research was conducted on this property. Yokoyama (2004) provides a summary of the most notable studies aimed at evaluating the cholesterol-lowering capacity of both rice bran and rice bran oil. More recently, Clevidence (2007) presented an interesting overview of available data in human subjects relative to realistic health effects of rice bran and oil at the 2007 Rice Utilization Conference in New Orleans, Louisiana. Her presentation and summary are available at the Web site of the USA Rice Federation (Clevidence, 2007). Both authors explored the various theories for the beneficial role of rice bran oil, or its components, in cholesterol metabolism.

Yokoyama (2004) points out that the Japanese and Indian scientific literature reported the earliest evidence for the cholesterol-lowering capacity of rice bran oil. A study by Suzuki and Oshima (1970) indicated that a 70:30 blend of rice bran oil/safflower oil reduced plasma cholesterol by 26% in healthy young girls. Raghuram et al. (1989) used a study group that they followed during the normal course of their lives for thirty days, with the only intervention being the replacement of their habitual cooking oil with rice bran oil. All subjects had serum cholesterol levels higher than 225 mg/dL. They found that in the group that used rice bran oil (n = 12) in place of their habitual oil a reduction in serum cholesterol from 247.6 to 204 to 182.7 mg/dL occurred after 15 and 30 days, respectively. The control group (n = 9) that continued to use their habitual oil had no change in their serum cholesterol levels. These early studies provided the impetus for a much more extensive evaluation of the cholesterol-lowering effect of rice bran oil. Sugano and Tsuji (1997) provide a classical paper based on their presentation at the VII Asian Conference of Nutrition: Lipid Symposium (held in Beijing, China, October 7–11, 1995), that nicely summarizes the research that was done on the cholesterol-lowering property of rice bran oil. They discussed experimental evidence for the role of fatty acids versus the unsaponifiable fraction, including the tocols, phytosterols, and gamma-oryzanol and its primary constituents. From a mechanistic standpoint, they concurred with the evidence that suggested that the hypocholesterolemic effect of the unsaponifiable material was attributable to increased fecal steroid excretion through interference with cholesterol absorption.

In the United States, work at the USDA-Western Regional Research Center in Albany, California, received national attention and brought focus to rice bran as a potential health-promoting food. These researchers, led by Dr. Robin Saunders, were evaluating the potential of bran from rice to be a serum cholesterol- lowering agent similar to oat bran, which was touted as a new dietary approach to serum cholesterol control. Their studies with hamsters as an experimental model indicated that rice bran was similar to oat bran in serum cholesterol reduction (Kahlon et al., 1990). These researchers noted that the cholesterol-lowering effect of oat bran was attributed to its soluble fiber (primarily beta-glucan), but rice bran was much lower in soluble fiber than oat bran. Thus, an alternative explanation was needed. A hypothesis was developed: because rice bran was also higher in lipid, it or some component therein may be responsible for rice bran's cholesterol-lowering property (Kahlon et al., 1992). From a mechanistic standpoint, their research also focused on the unsaponifiable fraction of rice bran oil and its components as the primary means for the hypocholesterolemic effect (Kahlon et al., 1996). Thus, the unsaponifiable material in rice bran oil received the most attention as having health-promoting potential.

Among the many components in the unsaponifiable fraction, phytosterols (including gamma-oryzanol) and tocotrienols are the components of rice bran oil that were most extensively studied for their serum cholesterol-lowering properties. A common practice is to incorporate phytosterols, or oils that are high in these compounds, into margarine as a means of feeding concentrated sources. In a study specific to rice bran oil, normolipidemic subjects were fed a diet in which margarine provided 2.1g/day of rice bran oil sterols for a three-week period. The researchers observed a 9% reduction in serum cholesterol through this regime (Vissers et al., 2000).

However, the potential that the different forms of phytosterols may affect serum cholesterol levels differently has become an issue. The FDA (2008) granted unqualified health claim status to both sterols and stanols as serum cholesterol-lowering agents; however, a higher consumption of stanols (3.4 g/day) is required to make the health claim than of sterols (1.3 g/day). Although the FDA currently makes no distinction between di-, mono-, and desmethyl forms of phytosterols relative to cholesterol-lowering potential, recent research suggests that the dimethyl form, which is the predominate form of phytosterol in rice bran oil, may not be as effective in preventing cholesterol uptake in the intestine (Trautwein et al., 2002). The potential mechanism for this was evaluated by Moreau and Hicks (2004) and Nystrom et al. (2008) using in vitro assays employing mammalian pancreatic enzymes. They postulated that interference with cholesterol uptake would require that plant sterol esters be hydrolyzed to free sterols. Their results indicate that the ferulate esters of dimethyl sterols were not as effectively hydrolyzed by the mammalian cholesterol esterases that they evaluated (bovine and porcine).

The case for tocotrienols as serum cholesterol-lowering agents was extensively studied by Qureshi et al. (1986, 1996, 1997, 2000) using a variety of sources of tocotrienols, including barley, palm oil, and rice bran. An off-shoot of this work was the development of a supplement that contained a novel form of tocotrienol thought to have superior cholesterol-lowering capacity that was derived from rice bran (Qureshi et al., 1999). The primary avenue by which tocotrienols are thought to lower serum cholesterol is through the reduction of HMG-CoA-reductase activity through a feedback inhibition mechanism.

The most recent well-controlled study with humans evaluating the potential of rice bran oil to lower serum cholesterol was done by Most et al. (2005) at the Pennington Biomedical Research Center, LSU, Baton Rouge, Louisiana. Their study was designed to determine if the cholesterol- lowering effect of rice bran was due to fiber or to its oil component. In one experiment, they compared defatted rice bran with a similar diet containing half the fiber content from a purified source. In a second experiment, they compared a diet in which one-third of the lipid was rice bran oil with a control diet that had a similar level of fat and a fatty acid profile. They found that defatted rice bran had no effect on serum lipids, but rice bran oil caused LDL cholesterol to significantly decrease by 7%, with no change in HDL cholesterol. They stated that rice bran oil, not fiber, was responsible for the cholesterol-lowering effect of full-fat rice bran in mildly hypercholesterolemic individuals, and that this effect was not due to a fatty acid profile, but was most likely due to the unsaponifiable fraction of rice bran oil. Also in 2005, Berger et al. evaluated the effect of rice bran oil with differing levels of gamma-oryzanol (a component of the unsaponifiable fraction) on serum cholesterol of hypercholesterolemic men. They found that rice bran oil, compared with peanut oil, lowered total serum cholesterol by 6.3%, LDL-cholesterol (C) by 10.5%, and LDL-C/HDL-C by 18.9%, but that gamma-oryzanol concentration of the oil did not appear to have an effect (Berger et al., 2005). Cicero and Gaddi (2001) determined that at least 300 mg of gamma-oryzanol was needed in the diet to obtain an LDL-cholesterol-lowering effect.

The research that was done at the USDA-WRRC in Albany, California, (that was mentioned previously as being instrumental in raising the awareness of the cholesterol- lowering potential of rice bran) was done using hamsters as an animal model. Hamsters are commonly used in assessing the mechanism for the serum cholesterol-lowering potential of many different substances because they more closely mimic cholesterol metabolism in humans. Recently, Wilson et al. (2007) conducted a major research project aimed at evaluating the relative effect of several components of rice bran oil on cholesterol metabolism in the hamster. They found that oryzanol had a greater effect on lowering plasma non-HDL cholesterol levels and raising HDL cholesterol levels than ferulic acid. They speculated that this could be due to an increase in fecal excretion of cholesterol and its metabolites. Also using the hamster as an animal model, Ausman et al. (2005) found that physically refined rice bran oil that had 2–4% nontriglyceride components significantly lowered both total serum cholesterol and LDL-C. They also found that the physically refined rice bran oil reduced cholesterol absorption and increased neutral sterol excretion, but not bile acid excretion. They speculated that lipid-lowering by physically refined rice bran oil was due to reduced cholesterol absorption in the gut, but not hepatic cholesterol synthesis, probably due to the nontriglyceride components (e.g., the unsaponifiable fraction). In an effort to identify a potential mechanism for the effect of rice bran oil on diabetes, Chen and Chen (2006) used rats in which Type 2 diabetes was induced using steptozotocin/nicotinamide. They fed the rats diets that contained rice bran oil with varying levels of gamma-oryzanol and gamma tocotrienols. Results indicated that rats fed rice bran oil had greater insulin sensitivity, lower plasma triglycerides, lower serum LDL cholesterol, and increased bile acid excretion than rats fed a control diet. They also found an increase in hepatic LDL-receptors and an increase in HMG-coA-reductase mRNA in rats fed rice bran oil. They speculated that diabetic rats fed rice bran oil diets with high contents of gamma-oryzanol and gamma tocotrienol had suppressed hyperlipidemic and hyperinsulinemic responses due to upregulation of cholesterol synthesis and catabolism.

The in vivo antioxidant activity of microencapsulated gamma-oryzanol was evaluated by Suh et al. (2005). Lard was used as the lipid source, with treatments of fresh lard, heat-treated lard, heat-treated lard with added oryzanol (100 ppm), and heat-treated lard with added microencapsulated oryzanol (100 ppm) fed to rats in a high-cholesterol diet. Results indicated that microencapsulated oryzanol was effective in inhibiting hypercholesterolemia of serum and liver, and also reduced the degree of oxidation of in vivo lipids and cholesterol.

### *Other Health Benefits of Rice Bran Oil or its Components*

The potential of rice bran oil or its constituents in preventing cancer is probably the second most often mentioned potential health benefit after serum cholesterol lowering. A primary reason that rice bran oil is believed to have cancer prevention potential is related to its abundance of antioxidants, again, primarily ascribed to tocols and oryzanol. Wada et al. (2005) found that oral administration of tocotrienols resulted in a significant suppression of liver and lung carcinogenesis in mice. They also found that delta-tocotrienols exerted more significant antiproliferative effects than the other toco-trienols in human hepatocellular carcinoma HepG2 cells. The effect of tocotrienols compared with alpha-tocopherol on cell viability and apoptosis of murine liver cancer cells was studied by Har and Keong (2005). They found that tocotrienols, but not alpha-tocopherol, were able to reduce cancer cell viability, possibly due to early inducement of apoptosis. Iqbal et al. (2004) evaluated the potential of a tocotrienol-rich fraction (TRF) prepared from rice bran oil in the prevention of hepatic lipid peroxidation induced by hepato- carcinogenic agents. Based on their findings, they suggest that long-term intake of TRF could reduce cancer risk by preventing hepatic lipid peroxidation and protein oxidation damage.

Several other means by which rice bran oil components could have potential health benefits are found in the recent literature. Sen et al. (2004) found that tocotrienols, especially the delta homolog, at nanomolar concentrations, protected neurons by an antioxidant-independent mechanism and suggested that tocotrienols fed orally could be a naturally occurring neuroprotective agent. In another investigation (Nagasaka et al., 2007) using macrophages, cycloartenyl ferulate was found to possibly have potential anti-inflammatory properties.

## Conclusion

Rice bran oil represents one of the most underutilized agricultural commodities in the world. Furthermore, its abundance of potentially health-promoting constituents makes rice bran oil one of the most healthful vegetable oils. Difficulties associated with its processing and refining hampered commercial exploitation. However, advancements in appropriate processing and refining techniques lessened the difficulties in producing economically viable products. Continued technological advancement, coupled with tightening world supplies of more traditional vegetable oils such as soybean oils, should potentiate increased commercial interest in rice bran oil. New evidence for health-promoting properties of rice bran oil and its constituents should also drive up consumer demand. A recent announcement by Nutracea, (http://www.reuters.com/article/marketsNews/idINPEK28501520080626?rpc=44), a rice bran and oil processing company headquartered in Phoenix, Arizona, indicating that they are partnering with a Chinese consumer products company to process rice bran oil in China may be a harbinger of the potential future exploitation of rice bran oil.

# References

Akihisa, T.; K. Yasukawa; M. Yamaura; M. Ukiya; Y. Kimura; N. Shimizu; K. Arai. Triterpene alcohol and sterol ferulates from rice bran and their anti-inflammatory effects. *J. Agric. Food Chem.* **2000,** *48*, 2313–2319

Anonymous. Amended final report on the safety assessment of *Oryza sativa* (Rice) bran oil, *Oryza sativa* (Rice) germ oil, rice bran acid, *Oryza sativa* (Rice) bran wax, hydrogenated rice bran wax, *Oryza sativa* (Rice) starch, *Oryza sativa* (Rice) bran, hydrolyzed rice bran extract, hydrolyzed rice bran protein, hydrolyzed rice extract, and hydrolyzed rice protein. *Int. J. Toxicol.* **2006**, *25*, 91–120 Suppl. 2.

Ausman, L.; N. Rong; R. Nicolosi. Hypocholesterolemic effect of physically refined rice bran oil: studies of cholesterol metabolism and early atherosclerosis in hypercholesterolemic hamsters. *J. Nutr. Biochem.* **2005**, *16*, 521–529.

Berger, A.; D. Rein; A. Schafer; I. Monnard; G. Gremaud; P. Lambelet; C. Bertoli. Similar cholesterol-lowering properties of rice bran oil, with varied gamma-oryzanol, in mildly hypercholesterolemic men. *Eur. J. Nutr.,* **2005**, *44*, 163–173.

Berthold, H; S. Unverdorben; R. Degenhardt; M. Bulitta; I. Gouni-Berthold. Effect of policosanol on lipid levels among patients with hypercholesterolemia or combined hyperlipidemia. *JAMA* **2006**, *495*, 2262–2269.

Bhosle, B.; R. Subramanian. New approaches in deacidification of edible oils—a review. *J. Food Eng.* **2005**, *69*, 481–494.

Chen, C.; H. Chen. A rice bran oil diet increases LDL-receptor and HMG-CoA reductase mRNA expressions and insulin sensitivity in rats with strepozotocin/nicotinamide-induced type 2 diabetes. *J. Nutr.* **2006**, *136*, 1472–1476.

Chipman K. Environmental Health Criteria 122: n-Hexane. *WHO,* Geneva, Switzerland, **1991** (accessed online at http://www.inchem.org/documents/ehc/ehc/ehc122.htm).

Cicero A.; A. Gaddi. Rice bran oil and gamma-oryzanol in the treatment of hyperlipoproteinaemias and other conditions. *Phytother. Res.* **2001**, *15*, 277–289.

Clevidence, B. Rice phytonutrients: Do realistic intakes of rice and rice bran oil promote health? **2007** (accessed online at http://www.usarice.com/processing/pdf/ClevidencePhytoEfficacyOnePage.pdf).

Compton, D.; J. Kenar; J. Laszlo; F. Felker. Starch-encapsulated, soy-based, ultraviolet-absorbing composites with feruloylated monoacyl- and diacylglycerol lipids. *Industrial Crops and Products*, **2007**, *25*, 17–23.

Danielski, L.; C. Zetzl; H. Hense; G. Brunner. A process line for the production of raffinated rice oil from rice bran. *J. Supercritical Fluids* **2005**, *34*, 133–141.

De, B; D. Bhattacharyya. Deacidification of high-acid rice bran oil by reesterification with monoglyceride. *J AOCS* **1999**, *76*, 1243–1246.

Dunford, N; J. Teel; J. King. A continuous countercurrent supercritical fluid deacidification process for phytosterol ester fortification in rice bran oil. *Food Res. Int.* **2003**, *36*, 175–181.

Eitenmiller, R. Vitamin E content of fats and oils—Nutritional Implications. *Food Tech.* **1997**, *51(5)*, 78–81.

FAO. FSTAT. **2007** http://faostat.fao.org/site/567/DesktopDefault.aspx?PageID=567.

FDA. Food Labeling Guide, Appendix C: Health Claims, **2008**, http://www.cfsan.fda.gov/~dms/2lg-xc.html

Forster, A.; A. Harper. Physical refining. *J AOCS* **1983**, *60*, 217A–223A.

Garcia, A.; A.D. Lucas; J. Rincon; A. Alvarez; I. Gracia; M.A. Garcia. Supercritical carbon dioxide extraction of fatty and waxy material from rice bran. *J AOCS* **1996**, *73*, 1127–1131.

Gaur, R.; A. Sharma; S.K. Khare; M.N. Gupta. A novel process for extraction of edible oils. Enyzme assisted three phase partitioning (EATPP). *Bioresour. Technol.* **2007**, *98*, 696–699.

Ghosh, M. Review of recent trends in rice bran oil processing. *J AOCS* **2007**, *84*, 315–324.

Godber, J.; B. Juliano. Rice lipids, *Rice Chemistry and Technology*, Third edition; E.T. Champagne, Ed.; AACC: St. Paul, MN, 2004; pp. 163–190.

Guillen M.D.; A. Ruiz. Oxidation process of oils with high content of linoleic acyl groups and formation of toxic hydroperoxy- and hydroxylalkenals. A Study by $^1$H nuclear magnetic resonance. *J. Sci. Food Agric.* **2005**, *85 (14)*, 2413–2420.

Har, C.H.; C.K. Keong. Effects of tocotrienols on cell viability and apoptosis in normal murine liver cells (BNL CL.2) and liver cancer cells (BNL 1ME A.7R.1). *Asia Pacific J. Clin. Nutr.* **2005**, *14*, 374–380.

Haraldsson, G. Degumming, dewaxing and refining. *J AOCS* **1983**, *60*, 203A–209A.

Hefle, S.L. Allergenicity of Edible Oils. Accessed on the Internet, ISEO:Home, **2008,** http://www.iseo.org.

Hsu, P-C.; T-J. Lai; N-W. Guo; G.H. Lambert; Y.L. Guo. Serum hormones in boys prenatally exposed to polychlorinated biphenols and dibenzofurans. *J. Tox. Envir. Health* **2005**, *68*, 1447–1456.

Hsu, S-T.; C-I. Ma; S. K-H. Hsu; S-S. Wu; N. H-M. Hsu; C-C. Yeh; S-B. Wu. Discovery and epidemiology of PCB poisoning in Taiwan: A four-year followup. *Environ. Health Perspect.* **1985**, *59*, 5–10.

Hu, W.; J.H. Wells; T-S. Shin; J.S. Godber. Comparison of isopropanol and hexane for extraction of vitamin E and oryzanol from stabilized rice bran. *J AOCS* **1996**, *73*, 1653–1656.

Iqbal, J.; M. Minhajuddin; Z.H. Beg. Suppression of diethylnitrosamine and 2-acetylaminofluorene-induced hepatocarcinogenesis in rats fed tocotrienol-rich fraction isolated from rice bran oil. *Eur. J. Cancer Prev.* **2004**, *13*, 515–520.

Ito, S.; T. Susuki; Y. Fujino. Wax lipid in rice bran. *Cereal Chem.* **1983**, *60*, 252–253.

Janikula M. Policosanol: a new treatment for cardiovascular disease? *Altern. Med. Rev.* **2002**, *7(3)*, 203–217.

Johnson, L.A.; E.W. Lusas. Comparison of alternative solvents for oils extraction. *J AOCS* **1983**, *60 (2)*, 229–242.

Kahlon, T.S.; F.I. Chow; M.M. Chiu; C.A. Hudson; R.N. Sayre. Cholesterol-lowering by rice bran and rice bran oil unsaponifiable matter in hamsters. *Cereal Chem.* **1996**, *73*, 69–74.

Kahlon, T.S.; R.M. Saunders; F.I. Chow; M.M. Chiu; A.A. Betschart. Influence of rice bran, oat bran and wheat bran on cholesterol and triglycerides in hamsters. *Cereal Chem.* **1990**, *67*,

439–443.

Kahlon, T.S.; R.M. Saunders; R.N. Sayre; F.I. Chow; M.M. Chiu; A.A. Betschart. Cholesterol-lowering effects of rice bran and rice bran oil fractions in hypercholesterolemic hamsters. *Cereal Chem.* **1992**, *69*, 485–489.

Kao, C.; B.S. Luh. Rice oil. *Rice Utilization,* Second edition; B.S. Luh, Ed.; Van Nostrand Reinhold: New York, 1991; Volume II, pp. 295–311.

Kim, J.-S.; J.S. Godber. Oxidative stability and vitamin E levels increased in restructured beef roasts with added rice bran oil. *J. Food Qual.* **2001**, *24*, 17–26.

Kim, J.-S.; J.S. Godber; W. Prinyawiwatkul. Restructured beef roasts containing rice bran oil and fiber influences cholesterol oxidation and nutritional profile. *J. Muscle Foods* **2000**, *11*, 11-127

Krishna, A.G.G.; S. Khatoon; R. Babyltha. Frying performance of processed rice bran oils. *J. Food Lipids* **2005**, *12 (1)*, 1–11.

Krishna, A.G.G.; S. Khatoon; P.M. Shiela; C.V. Sarmandal; T.N. Indira; A. Mishra. Effect of refining crude rice bran oil on the retention of oryzanol in refined oil. *J AOCS* **2001**, *78*, 127–131.

Kumar, N. Production of biodiesel from high FFA rice bran oil and its utilization in a small capacity diesel engine. *J. Sci. Ind. Res.* **2007**, *66*, 399–402.

Lawton, C.W.; R. Nicolosi; S. McCarthy. Refined vegetable oils and extracts thereof. U.S. Patent 6,197,357; 2001.

Liu, S.X.; P.K. Mamidipally. Quality comparison of rice bran oil extracted with d-limonene and hexane. *Cereal Chem.* **2005**, *82*, 209–215.

Lusas, E.W.; L.R. Watkins; S.S. Koseoglu. Isopropyl alcohol to be tested as solvent. *INFORM* **1991**, *2*, 970–976.

Marmot, M.G.; S.L. Syme. Acculturation and coronary heart disease in Japanese-Americans. *Am. J. Epidem.* **1976**, *104*, 225–247.

Mattikow, M. Developments in the refining of oils with sodium carbonate. *J AOCS* **1948**, *25(6)*, 200–203.

McCaskill, D.R.; F. Zhang. Use of rice bran oil in foods. *Food Tech.* **1999**, *53(2)*, 50–53.

Mendes, M.F.; F.L.P. Pessoa; G.V. Coelho; A.M.C. Uller. Recovery of the high aggregated compounds present in the deodorizer distillate of the vegetable oils using supercritical fluids. *J. Supercritical Fluids* **2005**, *34*, 157–162.

Mezouari, S.; K. Eichner. Evaluation of stability of blends of sunflower and rice bran oil. *Eur. J. Lipid Sci. Tech.* **2007**, *109*, 531–535.

Mitchell, A.R.; L.E. Routier, III. Process and system for continuously extracting oil from solid or liquid oil bearing material. U.S. Patent 7,008,528; 2006.

Moreau, R.A.; K.B. Hicks. The *in vitro* hydrolysis of phytosterol conjugates in food matrices by mammalian digestive enzymes. *Lipids* **2004**, *39*, 769–776.

Most, M.M.; R. Tulley; S. Morales; M. Lefevre. Rice bran oil, not fiber, lowers cholesterol in humans. *Am. J. Clin. Nutr.* **2005**, *81*, 64–68.

Nagasaka, R.; C. Chotimarkorn; I.M. Shafiqul; M. Hori; H. Ozaki; H. Ushio. Anti-inflammatory effects of hydroxycinnamic acid derivatives. *Biochem. Biophys. Res. Comm.* **2007**, *358*, 615–619.

Nanua, J.N.; J.U. McGregor; J.S. Godber. Influence of high-oryzanol rice bran oil on the oxidative stability of whole milk powder. *J. Dairy Sci.* **2000**, *83*, 2426–2431.

Narayana, T.; B. Kaimal; S.R. Vali; B.V. Surya; K. Rao; P.P. Chakrabarti; P. Vijayalakshmi; V. Kale; K. Naryana; P. Rani; et al. Origin of problems encountered in rice bran oil processing. *Eur. J. Lipid Sci. Tech.* **2002**, *104*, 203–211.

Nasirullah. Physical refining: Electrolyte degumming of nonhydratable gums from selected vegetable oils. *J. Food Lipids* **2005**, *12*, 103–111.

Norton, R. A. Quantitation of steryl ferulate and $_p$-coumarate esters from corn and rice. *Lipids* **1995**, *30*, 269–274.

Nystrom, L.; R.A. Moreau; A-M. Lampi; K.B. Hicks; V. Piironen. Enzymatic hydrolysis of steryl ferulates and steryl glycosides. *Eur. Food Res. Tech.* **2008**, *227*, 727–733.

Ohta, G. Constituents of rice bran oil. III. Structure of oryzanol-C. *Chem. Pharm. Bull.* **1960**, *8*, 5–9. (Chem. Abstr. 55: 5570, 1961).

Ohta, G.; M. Shimizu. Constituents of rice bran oil. II. Structure of oryzanol-A. *Chem. Pharm. Bull.* **1957**, *5*, 40–44. (Chem. Abstr. 52: 1190, 1958).

Orthoefer, F.T. Rice bran oil. *Bailey's Industrial Oil and Fat Products*, Sixth edition; F. Shahidi, Ed.; Knovel at http://www.knovel.com/web/portal/basic_search/display?_EXT_KNOVEL_DISPLAY_bookid=1432, 2005, Volume 6, p.p. 465-489.

Orthoefer, F.T.; J. Eastman. Rice bran and oil. *Rice Chemistry and Technology*, Third edition; E.T. Champagne, Ed.; AACC: St. Paul, MN, 2004; pp. 569–593.

Piironen, V.; D.G. Lindsay; T.A. Miettinen; J. Toivo; A-M. Lampi. Plant sterols: biosynthesis, biological function and their importance to human nutrition. *J. Sci. Food Agric.* **2000**, *80*, 939–966.

Pillaiyar, P. *Rice Postproduction Manual*; Wiley Eastern Limited: New Delhi, 1988; pp. 374–421.

Qureshi, A.A.; B.A. Bradlow; W.A. Salser; L.D. Brace. Novel tocotrienols of rice bran modulate cardiovascular disease risk parameters of hypercholesterolemic humans. *J. Nutr. Biochem.* **1997**, *8*, 290–298.

Qureshi, A.A.; W.W. Burger; D.M. Peterson; C.E. Elson. The structure of an inhibitor of cholesterol biosynthesis isolated from barley. *J. Biol. Chem.* **1986**, *261*, 10544–10550.

Qureshi, A.A.; R.H. Lane; A.W. Salser. Tocotrienols and tocotrienol-like compounds and method for their use. U.S. Patent 5,919,818; 1999.

Qureshi, A.A.; H. Mo; L. Packer; D.M. Peterson. Isolation and identification of novel tocotrienols from rice bran with hypocholesterolemic, antioxidant, and antitumor properties. *J. Agric. Food Chem.* **2000**, *48*, 3130–3140.

Qureshi, A.A.; B.C. Pearce; R.M. Nor; A. Gapor; D.M. Peterson; C.E. Elson. Dietary α-tocopherol attenuates the impact of γ-tocotrienol on hepatic 3-hydroxy-3-methyglutaryl coenzyme A reductase acitivity in chickens. *J. Nutr.* **1996**, *126*, 389–394.

Raghuram, T.C.; U. Brahmaji Rao; C. Rukmini. Studies on the hypolipidemic effects of rice bran oil in human subjects. *Nutr. Rep. Int.* **1989**, *39*, 889–895.

Raghuram, T.C.; C. Rukmini. Nutritional significance of rice bran oil. *Indian J. Med Res.* **1995**, *102*, 241–244.

Ramachandran, S.; S.K. Singh; C. Larroche; C.R. Soccol; A. Bandey. Oil cakes and their biotechnological applications—A review. *Bioresour. Tech.* **2007**, *98*, 2000–2009.

Ramsay, M.E.; J.T. Hsu; R.A. Novak; W.J. Reightler. Processing rice bran by supercritical fluid extraction. *Food Tech.* **1991**, *5(11)*, 98–104.

Rogers, E. J.; S.M. Rice; R.J. Nicolosi; D.R. Carpenter; C.A. McClelland; L.J. Romanczyk. Identification and quantitation of γ-oryzanol components and simultaneous assessment of tocols in rice bran oil. *J AOCS* **1993**, *70*, 301–307.

Rosenthal, A.; D.L. Pyle; L. Niranjan. Aqueous enzymatic processes for edible oil extraction. *Enzyme Microb. Technol.* **1996**, *19*, 402–420.

Rukmini C.; T.C. Raghuram. Nutritional and biochemical aspects of the hypolipidemic action of rice bran oil: a review. *J. Am. Coll. Nutr.* **1991**, *10*, 593–601.

Seetharamaiah, G.S.; J.V. Prabhakar. Oryzanol content of Indian rice bran oil and its extraction from soap stock. *J. Food Sci. Tech.* **1986**, *23*, 270–273.

Segers, J.C. Pretreatment of edible oils for physical refining. *J AOCS* **1983**, *60*, 214A–216A.

Sen, C.K.; S. Khanna; C. Rink; S. Roy. Tocotrienols: the emerging face of natural vitamin E. *Vitamins and Hormones* **2007**, *76*, 203–261.

Sen, C.K.; S. Khanna; S. Roy. Tocotrienol—The natural vitamin E to defend the nervous system. *Vitamin E and Health*, **2004**, *1031*, 127–142.

Sen, C.K.; S. Khanna; S. Roy. Tocotrienols: Vitamin E beyond tocopherols. *Life Sci.* **2005**, *78(18)*, 2088–2098.

Sharma, A.; S.K. Khare; M.N. Gupta. Enzyme-assisted aqueous extraction of rice bran oil. *J AOCS* **2001**, *78*, 949–951.

Shimizu, M.; G. Ohta. Constituents of rice bran oil. V. Re-examination of oryzanol-B. *Chem. Pharm. Bull.* **1960**, *8*, 108–11.

Shin, T-S.; J.S. Godber. Changes of endogenous antioxidants and fatty acid composition in irradiated rice bran during storage. *J. Agric. Food Chem.* **1996**, *44*, 567–573.

Singh D.K.; L. Li; T.D. Porter. Policosanol inhibits cholesterol synthesis in hepatoma cells AMP-kinase. *JPET* **2006**, *318*, 1020–1026.

Sparks, D.; R. Hernandez; M. Zappi; D. Blackwell; T. Fleming. Extraction of rice bran oil using supercritical carbon dioxide and propane. *J AOCS* **2006**, *83*, 885–891.

Sugano, M.; E. Tsuji. Rice bran oil and cholesterol metabolism. *J. Nutr.* **1997**, *127*, 521S–524S.

Suh, M.H.; S.H. Yoo; P.S. Chang; H.G. Lee. Antioxidant activity of microencapsulated gamma-oryzanol on high cholesterol fed rats. *J. Agric. Food Chem.* **2005**, 53, 9747–9750.

Suzuki, S.; S. Oshima. Influence of blending of edible fats and oils on human serum cholesterol metabolism. *Jap. J. Nutr.* **1970**, *28*, 3–6.

Taylor, J.B.; T.M. Richar; C.L. Wilhelm; M.M. Chrysam; M. Otterburn; G.A. Leveille. Rice Bran Oil Antioxidant. U.S. Patent 5,552,167; 1996.

Trautwein, E.A.; C. Schulz; D. Rieckhoff; A. Kunath-Rau; H.F. Erbersdobler; W.A. de Groot;

G.W. Meijer. Effect of esterified 4-desmethylsterols and –stanols or 4,4'-dimethylsterols on cholesterol and bile acid metabolism in hamsters. *Br. J. Nutr.* **2002**, *87*, 227–237.

Tsuchiya, T; R. Kaneko; O. Okubo. Oryzanol content of rice bran oil. *Tokyo Kogyo Shikensho Hokoku* **1957**, *52*, 1–3.

Vali, S.R.; Y. Ju; T.N.B. Kaimal; Y. Chern. A process for the preparation of food-grade rice bran wax and the determination of its composition. *J AOCS* **2005**, *82*, 57–64.

Van Hoed, V.; G. Depaemelaere; J.V. Ayala; P. Santiwattana; R. Verhe; W. De Greyt. Influence of chemical refining on the major and minor component of rice bran oil. *J AOCS* **2006**, *83*, 315–321.

Vissers M.N.; P.L. Zock; G.W. Meijer; M.B. Katan. Effect of plant sterols from rice bran oil and triterpene alcohols from sheanut oil on serum lipoprotein concentrations in humans. *Am. J. Clin. Nutr.* **2000**, *72*, 1510–1515.

Wada, S.; Y. Satomi; M. Murakoshi; N. Noguchi; T. Yoshikawa; H. Nishino. Tumor suppressive effects of tocotrienol *in vivo* and *in vitro*. *Cancer Lett.* **2005**, *229*, 181–191.

Wang, Y.W.; P.J.H. Jones; I. Pischel; C. Fairow. Effects of policosanols and phytosterols on lipid levels and cholesterol biosynthesis in hamsters. *Lipids* **2003**, *38*, 165–170.

Watkins, L.R.; S.S. Koseoglu; K.C. Rhee; E. Hernandez; M.N. Riaz; W.H. Johnson, Jr.; S.C. Doty. New isopropanol system shows promise. *INFORM* **1994**, *5*, 1245–1253.

Wayne, T.B. Extractive milling of rice in the presence of an organic solvent, U.S. Patent 3,261,690; 1966.

Wilson, T.A.; R.J. Nicolosi; B. Woolfrey; D. Kritchevsky. Rice bran oil and oryzanol reduce plasma lipid and lipoprotein cholesterol concentrations and aortic cholesterol ester accumulation to a greater extent than ferulic acid in hypercholesterolemic hamsters. *J. Nutr. Biochem.* **2007**, *18*, 105–112.

Xu, Z.; J.S. Godber. Purification and identification of components of γ-oryzanol in rice bran oil. *J. Agric. Food Chem.* **1999**, *47*, 2724–2728.

Xu, Z.; J.S. Godber. Comparison of supercritical fluid and solvent extraction methods in extracting γ-oryzanol from rice bran. *J AOCS* **2000**, *77*, 547–551.

Yokoyama, W. Nutritional properties of rice and rice bran. *Rice Chemistry and Technology*, Third edition; E.T. Champagne, Ed.; AACC: St. Paul, MN, 2004; pp. 595–609.

Yoon, S.H.; S.K. Kim. Oxidative stability of high-fatty acid rice bran oil at different stages of refining. *J AOCS* **1994**, *71*, 27–229.

Yoon, S.H.; J.S. Rhee. Composition of waxes from crude rice bran oil. *J AOCS* **1982**, *59*, 561–563.

Zheng, L.R.; P. Zheng; Z.H. Sun; Y.B. Bai; J. Wang; X.F. Guo. Production of vanillin from waste residue of rice bran oil by *Aspergillus niger* and *Pycnoporus cinnabarinus*. *Bioresour. Technol.* **2007**, *98*, 1115–1119.

# 15

# Corn Kernel Oil and Corn Fiber Oil

**Robert A. Moreau[1], Vijay Singh[2], Michael J. Powell[1], and Kevin B. Hicks[1]**
*[1]Eastern Regional Research Center, Agricultural Research Service, U.S. Department of Agriculture, 600 East Mermaid Lane, Wyndmoor, PA 19038, USA; [2]Department of Agricultural and Biological Engineering, University of Illinois at Urbana-Champaign, Urbana, IL 61801, USA*

†Mention of trade names or commercial products in this publication is solely for the purpose of providing specific information, and does not imply recommendation or endorsement by the U.S. Department of Agriculture.

## Introduction

Unlike most edible plant oils that are obtained directly from oil-rich seeds by either pressing or solvent extraction, corn seeds (kernels) have low levels of oil (≈4%), and commercial corn oil is obtained from pressing or extracting the isolated corn germ (embryo), which is an oil-rich portion of the kernel. Commercial corn oil could actually be called "corn germ oil." Most commercial corn oil is obtained from corn germ that is a by-product of the wet milling industry (Moreau, 2002, 2005). The wet milling process was developed in the late 1800s to optimize the production of corn starch from corn kernels. Wet milling efficiently removes the ≈70% of starch in corn kernels, and it produces three major by-products that are usually called "co-products" because the considerable revenues from their sale provide additional profit to the corn wet mill. The first co-product is corn germ, which contains 40–50% of oil (Table 15.1). The corn germ fraction from the wet milling process represents about 5% of the mass of the kernel, and the germ fraction from the dry-milling fraction represents about 10% of the mass of the kernel (Moreau et al., 1999). In our experience, when the corn kernel is carefully dissected with a scalpel, the germ (embryo), hull (pericarp and aleurone cells), and endosperm comprise about 5, 5, and 90%, respectively, of the mass of the kernel. The oil is usually removed by either hexane extraction or mechanical prepressing, followed by hexane extraction (Moreau, 2002, 2005a). The two additional wet milling co-products are corn gluten feed (an animal feed with high levels of fiber and low levels of protein, ≈20%) and corn gluten meal (an animal feed with low levels of fiber and about 60% of protein). Corn gluten feed is obtained by mixing corn fiber with condensed steepwater. Corn oil ranks number 3 in the worldwide production of edible plant oils, with an annual production of 2 million tons, and is surpassed by

**Table 15.1. A Summary of the Properties and Composition of Unrefined Corn Germ Oil, Corn Kernel Oil, and Corn Fiber Oil**

| | Corn germ oil | Corn germ oil | Corn kernel oil | Corn fiber oil | Ref. |
|---|---|---|---|---|---|
| Feedstock (material extracted) | from wet- milled corn germ | from dry- milled corn germ | Ground or flaked whole corn | Corn fiber, a by-product of corn wet milling | |
| Solvent for extraction | hexane | hexane | ethanol | hexane | |
| Oil yield from feedstock | 40–50% | 15–25% | 3–5% | 2–3% | Moreau, 2005 |
| Color | Yellow | Yellow | Red | Amber/Orange | Moreau et al., 2007 |
| Phytosterols (total) | ≈1% | ≈1% | 2–3% | 10–15% | Moreau et al., 1996 |
| Tocopherols | 0.32% | – | 0.15% | 0.08% | Moreau et al., 2005 |
| Tocotrienols | 0.01% | – | 0.04% | 0.05% | Moreau et al., 2005 |
| Lutein and zeaxanthin | 0 | – | 0.21% | 0.05% | Moreau et al., 2007 |

soybean oil and palm oil (Moreau, 2002). The properties of commercial corn oil were studied extensively (Moreau, 2002, 2005a). The remainder of this chapter focuses on two lesser-known types of corn oil which have received much attention in recent years—corn kernel oil and corn fiber oil.

Although well-known for a long time that oil could be obtained directly by extracting it from ground corn, the low levels of oil in most corn kernels (3–5%) are an indication that obtaining corn oil directly from corn kernels may not be economical (Table 15.1). However, if the corn-oil-extraction process could be linked to a second profitable process, then the economics of corn kernel-oil production might be favorable. Hojilla-Evangelista et al. (1992) at Iowa State University developed a "sequential extraction" process which involved an initial extraction of corn oil from flaked corn with ethanol (100%), followed by an extraction of zein protein with 70% of ethanol, and finally an extraction of the remaining proteins and starch. Although the economics of the sequential extraction process were rigorously debated, some recent modifications of the process have improved the economics (Feng et al., 2002; Hojilla-Evangelista et al., 1992). A group at the University of Illinois developed a similar two-step "COPE" (corn oil and protein extraction) process which also focuses on the ethanol extraction of corn oil and zein, but it includes the use of membrane filters to process both products (Cheryan, 2002; Kwiatkowski & Cheryan, 2002). Recently our laboratory demonstrated that corn kernel oil should have health-promoting properties because of its very high levels of lutein and zeaxanthin (valuable for preventing macular degeneration) and its moderate levels of tocopherols, tocotrienols, and phytosterols (Moreau et al., 2007).

Corn fiber oil was first developed in our laboratory in the mid- to late-1990s (Moreau et al., 1996, 1998). The corn fiber co-product from wet milling was known to contain oil (2–3%), and most experts assumed that this small amount of oil was probably attributable to the contamination of the fiber with a small amount of corn germ (Table 15.1). We reported that the chemical composition of corn fiber oil, obtained by extracting wet-milled corn fiber with either hexane or supercritical $CO_2$, was very different from that of commercial corn oil (Table 15.1). Whereas, commercial corn oil contains about 1% of phytosterols (plant sterols), corn fiber oil contains 10–15% of phytosterols (Moreau et al., 1996). During the 1990s, much international interest was shown in phytosterols because of several clinical studies that demonstrated that eating 1–2 grams per day of phytosterols could reduce the levels of low-density lipoprotein cholesterol (LDL–C) in the blood by 10–15%. In 1995 the first phytosterol nutraceutical, a margarine called Benecol, was marketed in Finland by Raisio. In 2000, the United States (U.S.) Food and Drug Administration (FDA) approved the sale of phytosterol-enriched margarines in the United States, followed by the approval of other types of phytosterol-enriched functional foods (Moreau et al., 2003).

# Processing

## Milling/Fractionation of Corn Kernels to Make Corn Kernel Oil

For the production of corn kernel oil, the "sequential extraction process" was developed at Iowa State University, and it involved the flaking of corn before ethanol extraction (Hojilla-Evangelista, 1992). The COPE process was developed at the University of Illinois, and it involved the extraction of ground corn before ethanol extraction (Kwiatkowski & Cheryan, 2002).

## Milling/Fractionation of Corn Fiber and Other Milling Fractions to Make Corn Fiber Oil

For the production of corn fiber oil, our initial papers and patent (Moreau et al., 1996, 1998a) recommended the grinding (milling) of corn fiber to a particle size of 1 micron (20 mesh) or smaller before beginning hexane extraction to achieve optimal oil yields. We observed that grinding to smaller particle sizes such as 40, 60, and 80 mesh resulted in a higher oil yield (up to 3.3% yield at 80 mesh), but smaller particle sizes may be problematic because the higher levels of fines may restrict the flowability of hexane through the particles during hexane extraction. Although common oilseeds are often extruded (using an expander) into collets before hexane extraction, when we made collets from corn fiber our yields of corn fiber oil from the collets were very low, perhaps because the high levels of hemicellulose formed an impermeable gel-like material that impeded the penetration of hexane into the collets (data not shown).

In a comprehensive series of collaborative studies with Dr. Vijay Singh at the University of Illinois, we evaluated various corn-milling fractions and subfractions of corn fiber as feedstocks for the production of corn fiber oil (Table 15.2). During these studies we made an interesting observation about the localization of most of the phytosterols in corn fiber. Corn fiber is comprised of pericarp fiber (the coarse fiber or outer covering of a corn kernel made of dead cell-wall material) and endosperm fiber (a fine fiber or cellular fiber inside the corn kernel). The aleurone layer is also fibrous tissue in a corn kernel, but is structurally part of corn endosperm. Depending on the processing technique used and the soaking/steeping time, the aleurone layer is recovered either with the pericarp or the endosperm fiber (Singh et al., 2001a). In the corn wet milling process, corn fiber is called white fiber, and it consists of a mixture of pericarp and endosperm fiber. In the corn dry-milling process, only pericarp fiber is recovered, which is known as bran, and this fraction represents about 10% of the mass of the kernel (Moreau et al., 1999). New technologies were developed in dry-grind processing for ethanol production that recover pericarp fiber prior to fermentation (Singh et al., 1999; Wahjudi, 2000) or after fermentation from the distillers' dried grains with solubles (DDGS) fraction (Srinivasan et al., 2006; Winkler et al., 2007). DDGS is the material remaining after the fermentation of the starch in whole ground kernels of corn to ethanol. It is dried and is sometimes pelletized and is sold as an animal feed. Our initial research indicated that the levels of oil and levels of total phytosterols were higher in oil obtained from corn white fiber (obtained from the commercial corn wet milling process) than in oil obtained from corn bran (pericarp fiber obtained from the commercial corn dry-milling process) (Moreau et al., 1996, 1999, 2001b).

In subsequent studies, we discovered a possible reason to explain the difference in oil content and phytosterol composition between "white fiber" and "bran fiber." We soaked corn kernels in water for 24 hours to soften the endosperm, and then removed the "pericarp" using a scalpel and forceps. We then separated the outer layer of the pericarp from the inner layer (the aleurone layer, which in most corn accessions is a single layer of cells). We ground both materials (pericarp and aleurone), and extracted the lipids with hexane; we found that most of the oil and most of the phytosterols were in the aleurone layer, and very little was in the pericarp layer (Moreau et al., 2000). We then postulated that the low levels of oil and phytosterols in the corn bran were probably due to the fact that it was mainly pericarp, and the higher levels of oil and phytosterols in white fiber were because it was comprised of both pericarp and aleurone layers (Singh & Moreau, 2003, 2006). Further evidence for this hypothesis was obtained when we found that the levels of oil and phytosterols were much higher in white fiber that was prepared by conventional steeping (with 0.55% of $SO_2$) versus corn fiber that was prepared by steeping in water. We interpreted this to mean that the presence of $SO_2$ during conventional steeping caused the aleurone layer to remain attached to the pericarp in the corn fiber fraction, whereas during water steeping, the aleurone layer remained attached to the endosperm and the fiber was comprised of only pericarp, with little or no aleurone. Another confirmation of this hypothesis is our observation that the levels of oil and phytosterols were low in "Quick Fiber" (an experimental process that involves short — 6 to 12 hours — soaking/steeping of the kernels in water and then coarse grinding and the flotation of the fiber) (Dien et al., 2005).

A tremendous growth has occurred in the U.S. fuel ethanol industry (expanding from about 2 million gallons of ethanol per year in 1998 to a projected 10 million gallons per year in 2008). Currently about 90% of the fuel ethanol is produced by the dry-grind process, which involves fermenting ground corn and produces one major co-product, DDGS, which contains all of the germ and fiber from the original corn kernel. We demonstrated that corn fiber oil can be produced from fiber recovered prior to ethanol fermentation (corn bran and "Quick Fiber") and from corn fiber separated from DDGS (Srinivasan et al., 2005, 2006, 2007, 2008) after fermentation.

In several additional studies, we investigated the effect of various physical and chemical pretreatments of corn fiber to evaluate their effects on the levels of oil and phytosterols that could be extracted from corn fiber (Table 15.2). We concluded that small increases could be achieved by several of these pretreatments.

**Table 15.2. A Comparison of Various Processing Techniques and Their Effect on Corn Fiber Oil Quantity and Quality**

| Fraction | Process used to produce fiber | Oil yield (g oil/100 g fraction) |
|---|---|---|
| White fiber (pericarp + endosperm) | Conventional wet milling | 1.72 |
| Corn fiber (pericarp) | Conventional wet milling | 1.09 |
| Corn fiber (endosperm) | Conventional wet milling | 0.54 |
| Corn bran (pericarp) | Conventional dry milling | 1.32 |
| Quick fiber (pericarp) | Quick fiber process | 1.24 |
| Aleurone layer | Hand-dissected | 5.93 |
| Aleurone layer | Flotation for recovering aleurone layer | 3.3 |
| Corn fiber (pericarp) | Laboratory wet milling process | 0.61 |
| White fiber (pericarp + endosperm) | Conventional wet milling laboratory | 2.09 |
| White fiber (pericarp + endosperm) | Gaseous $SO_2$ wet milling laboratory | 1.98 |
| White fiber (pericarp + endosperm) | Alkali wet milling laboratory | 7.51 |
| White fiber (pericarp + endosperm) | IMDS wet milling laboratory | 1.82 |
| Elusieve fiber | Elusieve process for dry-grinding ethanol | 4.4 |
| Heat-pretreated corn fiber | Conventional wet milling | 1.67 |
| Pretreated white fiber with $H_2SO_4$ and enzymes | Conventional wet milling laboratory | 7.9 |
| White fiber steeped with $NH_4SO_4$ | Conventional wet milling laboratory | 1.73 |

Abbreviation: NM, not measured.

**Table 15.2., cont. A Comparison of Various Processing Techniques and Their Effect on Corn Fiber Oil Quantity and Quality**

| Ferulate phytosterols esters (%) | Total phytosterols (%) | % Change in oil yield compared to control | Reference |
|---|---|---|---|
| 6.75 | 17.67 | Control | Moreau et al., 1996 |
| 5.65 | NM | -36.6 | Moreau et al., 1996 |
| 3.37 | NM | -68.6 | Moreau et al., 1996 |
| 1.50 | NM | -23.3 | Moreau et al., 1996 |
| 4.01 | 12.12 | -27.9 | Singh et al., 1999 |
| 6.31 | 10.19 | 244.8 | Moreau et al., 2000 |
| 4.5 | 7.2 | 91.9 | Singh et al., 2003 |
| 1.72 | 19.64 | -64.5 | Moreau et al., 2000 |
| 5.7 | 15.78 | 21.5 | Singh et al., 2001b |
| 5.89 | 17.13 | 15.1 | Singh et al., 2001b |
| 2.21 | 20.12 | 336.6 | Singh et al., 2001b |
| 3.74 | 8.77 | 5.8 | Singh et al., 2001b |
| 0.37 | 2.55 | 155.8 | Srinivasan et al., 2007 |
| 3 | NM | -2.9 | Moreau et al., 1999 |
| 4.1 | 18.1 | 359.3 | Singh et al., 2003 |
| 4.15 | NM | 0.6 | Singh et al., 2000 |

### Extraction of Corn Kernel Oil

Although extracting corn kernel oil with a variety of solvents and supercritical $CO_2$ is feasible, most of the research on this topic was conducted using 100% of ethanol (Moreau et al., 1996). After the extraction of ground corn or corn flakes, the miscella (extract containing oil and ethanol) is then evaporated by vacuum and heat, and the meal is usually re-extracted with an ethanol/water mixture—70/30, vol/vol—to extract the zein proteins.

Recently, a process was developed to obtain corn kernel oil as a co-product during fuel-ethanol production using the dry-grind method, which involves fermenting ground corn (Anon., 2008). In this new process, corn oil is removed via heating and centrifugation of the "thin stillage," which is the post-fermentation product obtained after the stripping of the ethanol and centrifugation (Anon., 2008). Because it contains high levels of free fatty acids (10–15%), this new type of corn kernel oil is being considered for nonedible applications such as biodiesel production.

### Extraction of Corn Fiber Oil

Although extracting corn fiber oil with other solvents and supercritical $CO_2$ is feasible, most of the research in our laboratory and others was conducted by extracting the ground corn fiber with hexane (Moreau et al., 1996). After the extraction of ground corn fiber, the miscella (extract containing oil and hexane) is then evaporated by vacuum and heat, and the defatted fiber is available for other applications. After the hexane is removed by desolventizing, the ground fiber could be re-extracted with alkaline hydrogen peroxide to obtain corn fiber gum (Doner & Hicks, 1999), or it could be blended with other components (unextracted corn fiber and steepwater) in corn gluten feed, a common co-product of the wet milling of corn (Singh, 2000). A recent U.S. patent (Abbas et al., 2008) describes a "green" process for producing corn fiber oil using ethanol extraction.

## Oil Yields

The yields of corn oil from corn kernels, corn fiber, and various alternative fractions are summarized in Table 15.2.

## Endogenous Hydrolytic Enzymes

Fortunately, the levels of lipases and other lipolytic enzymes are quite low in corn kernels so an attempt to control them during the development of processes to produce corn kernel oil and corn fiber oil has not been necessary. Typically, the levels of free fatty acids in crude (unrefined) corn germ oil, corn kernel oil, and corn fiber oil are about 2–4%. In other studies we noted that corn kernels could be stored for several years, and still be extracted to produce a corn kernel oil with relatively low levels of free fatty acids (<4%). However, if corn kernels were ground and then stored, even at 4°C, a gradual increase occurs in the levels of free fatty acids in the corn kernel oils produced from them (the increase in free fatty acids in ground corn was approximately 1% per month of storage at 4°C) (Parris et al., 2000, 2002).

## Edible and Nonedible Applications

Because both corn kernel oil and corn fiber oil are more costly to extract than conventional corn oil and most other commodity plant oils, the main focus on their utilization was on edible health-promoting applications. An exception is the new corn kernel oil obtained in dry-grind ethanol plants after fermentation which contains high levels of free fatty acids (Anon., 2008). This oil would be very costly to refine, so likely, it will only have nonedible applications, such as for biodiesel production.

## Acyl Lipids (TGs, PLs, etc.) and Fatty Acid Composition

All lipid extracts contain nonpolar lipids (NLs), glycolipids (GLs), and phospholipids (PLs). The proportions of each of these depend on their levels in the plant tissue and on the polarity of the solvent used for extraction. In general, nonpolar solvents, such as hexane, extract most of the NLs but little of the total GLs and PLs. In a recent study we compared the levels of several lipid classes of NLs, GLs, and PLs in the oils resulting from the extraction of ground corn with hexane, methylene chloride, isopropanol, and ethanol (Moreau et al., 2003). As expected, increasing to a mid-polarity solvent—such as methylene chloride—or to a polar solvent—such as ethanol—extracts most of the NLs, but also extracts higher proportions of GLs and PLs.

The major lipid classes in crude (unrefined) corn kernel oil and corn fiber oil are shown in Table 15.3. The most striking difference in the two oils is the much higher levels of the three classes of phytosterols in corn fiber oil, as discussed previously and as documented by the high total phytosterols for corn fiber oil in Table 15.1.

### Fatty Acid Analysis

That commercial corn oil is high (≈55%) in linoleic acid (18:2 n-6) has long been known (Moreau, 2000, 2005). In the 1960s, advertisements for corn oil claimed that it was high in polyunsaturated fatty acids, and thus, it was a very healthy oil. However, many experts now believe that oils highest in oleic acid are probably the healthiest. In addition to oleic acid, commercial corn oil also contains appreciable levels of oleic acid (≈28%), low levels of saturated fatty acids (≈13%), and low levels of linolenic acid (≈1%) (Table 15.4). The low linolenic acid may contribute to the oxidative stability of corn oil. We reported (Moreau et al., 2000) that the fatty acid compositions of corn kernel oil and corn fiber oil are nearly identical to that of commercial corn oil (Table 15.4).

**Table 15.3. Lipid Classes in Unrefined Corn Kernel Oil and Corn Fiber Oil**

| | Ground corn kernels extracted with hexane [a] | Ground corn kernels extracted with ethanol [a] | Corn fiber extracted with hexane[b] | Corn germ (wet- milled) extracted with hexane [a] |
|---|---|---|---|---|
| Nonpolar lipids | | | | |
| SEs | 1.03 ± 0.02 | 0.69 ± 0.03 | 9.05 ± 0.54 | 0.51 ± 0.15 |
| FSs | 0.74 ± 0.92 | 0.81 ± 0.04 | 1.92 ± 0.83 | 0.25 ± 0.04 |
| FPE | 0.38 ± 0.03 | 0.31 ± 0.02 | 6.70 ± 0.25 | 0.03 ± 0.01 |
| FFAs | NR | NR | 2.86 ± 0.04 | NR |
| TAGs | ≈97% | ≈95% | ≈80% | ≈99% |
| Glycolipids | | | | |
| Steryl glycoside | 0 | 0.05 | tr | tr |
| Phospholipids | | | | |
| PE | 0.38 | 1.8 | tr | tr |
| PC | 0.13 | 1.13 | tr | tr |
| Other polar | | | | |
| DFP | 0 | 0.57 ± 0.08 | 0 | 0 |
| CFP | 0 | 0.19 ± 0.02 | 0 | 0 |

Abbreviations: SE, steryl esters; FS, free phytosterols; FPE, ferulate phytosterol esters; FFAs, free fatty acids; TAG, triacylglycerols; PE, phosphatidylethanolamine; PC, phosphatidylcholine; DFP, diferuloylputrescine; CFP, *p*-coumaroylferuloylputrescine; NR, not reported; tr, trace.
[a] From Moreau et al., 2005.
[b] From Moreau et al., 1996.

# Minor (Functional) Lipid Components

## Phytosterols

As noted previously, both corn kernel oil and corn fiber oil contain high levels of total phytosterols, 23% and 10–15%, respectively (Table 15.1). The main phytosterol lipid classes in corn fiber oil are ferulate phytosterol esters (≈7%), phytosterol fatty-acyl

**Table 15.4. Fatty Acids in Crude (Unrefined) Oils Obtained by Extracting Corn Fiber and Corn Germ with Hexane**

| | (Wt %) | |
|---|---|---|
| Fatty acid Abbreviation | Corn fiber oil | Corn germ oil |
| Palmitic 16:0 | 13.8 ± 0.0[a] | 12.5 ± 0.2 |
| Stearic 18:0 | 1.7 ± 0.0 | 1.3 ± 0.0 |
| Oleic 18:1 | 23.8 ± 0.1 | 24.3 ± 0.0 |
| Linoleic 18:2 | 56.4 ± 0.1 | 60.1 ± 0.3 |
| Linolenic 18:3 (n-6) | 2.6 ± 0.0 | 1.0 ± 0.0 |
| Arachidic 20:0 | 0.3 ± 0.0 | 0.3 ± 0.0 |
| Behenic 22:0 | 0.2 ± 0.0 | 0.1 ± 0.0 |
| Lignoceric 24:0 | 0.3 ± 0.0 | 0.3 ± 0.0 |
| Squalene | 0.2 ± 0.0 | 0.2 ± 0.0 |
| Other | 0.6 ± 0.0 | 0.5 ± 0.0 |

[a]From Moreau et al., 2000.

esters, (≈9%), and free phytosterols (≈1%) (Moreau et al., 1997, 1998b, 2001b). These are the highest levels of total phytosterols reported for any natural plant oil. If the total phytosterols in corn oil are hydrolyzed by a basic catalyst, the resulting total phytosterol composition is also quite unique (Table 15.5). The ferulate phytosterol esters in corn fiber oil are similar to those found in rice (which are collectively called "oryzanol"). However, the phytosterol portions of the ferulate phytosterol esters in corn fiber oil are predominantly the 4-desmethylsterols sitostanol and campestanol, whereas in rice-bran oil they are predominantly the 4-dimethylsterols cycloartenol and 24-methylene cycloartenol (Fig. 15.1). The most abundant phytosterol in corn fiber oil is sitostanol (comprising 43% of the total phytosterols), an unusual C29 phytosterol that occurs almost exclusively in cereals. By definition "stanols" are completely saturated phytosterols, so unlike other phytosterols that contain 1 or 2 carbon-carbon double bonds, stanols contain none. The other predominant phytosterols in corn fiber oil are sitosterol (34%), the C-28 stanol called campestanol (15%), and campesterol (5%), followed by lesser amounts of seven other phytosterols (Table 15.5). In commercial corn oil, sitosterol is the predominant phytosterol, and very low levels of sitostanol (<1%) and campestanol (<1%) are present. In corn kernel oil, sitosterol is also the predominant phytosterol, but moderate levels of sitostanol (15%) and campestanol (6%) are present (data not shown).

## Tocopherols/Tocotrienols

Vitamin E is a fat-soluble vitamin that occurs in eight common tocol molecules (vitamers), four tocopherols (α, β, γ, δ), and four tocotrienols (α, β, γ, δ). The ring structures of the tocopherols are identical to the ring structures of the corresponding

**Table 15.5.** Total Sterols (mg/100 g Oil) in Crude (Unrefined) Oils Obtained by Extracting Corn Fiber and Corn Germ with Hexane

| | Campesterol | Campestanol | Stigmasterol | Sitosterol | Sitostanol | Fucosterol | Other |
|---|---|---|---|---|---|---|---|
| Corn fiber oil | 4.9 ± 0.4[a] | 14.7 ± 0.3 | 1.4 ± 0.1 | 34.3± 0.1 | 43.1 ± 0.7 | 1.8 ± 0.0 | Trace |
| Corn germ oil | 22.5 ± 0.1 | 0.0 ± 0.0 | 4.4 ± 0.2 | 71.2± 1.1 | 0.0 ± 0.0 | 1.0 ± 0.5 | Trace |

[a]From Moreau et al., 2000.

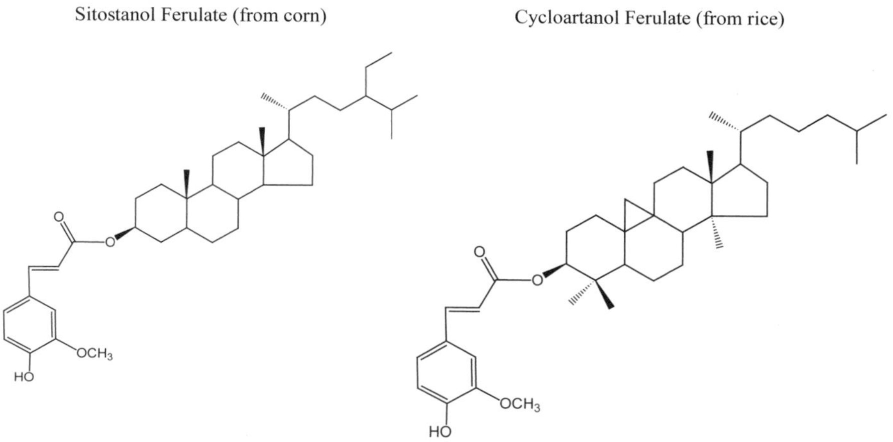

Sitostanol Ferulate (from corn) Cycloartanol Ferulate (from rice)

**Fig. 15.1. The structures of the major ferulate phytosterol esters in corn and rice.**

tocotrienols, but in tocopherols the side chain is saturated and in tocotrienols the side chain contains three carbon-carbon double bonds. Six of the eight tocols occur in corn (α, γ, δ-tocopherols and α, γ, δ-tocotrienols) (Fig. 15.2). Tocols can be quantitatively analyzed using high-performance liquid chromatography (HPLC) with fluorescence detection (Fig. 15.3). The germ portion of most cereals is rich in tocopherols, but contains almost no tocotrienols (Dunford, 2005; Przybylsky, 2006). Commercial corn oil, obtained from corn germ, is consistent with this generalization as it contains high levels of tocopherols (mostly γ-tocopherols), and almost no tocotrienols (Table 15.6). Recently we reported (Moreau et al., 2005) that corn fiber oil differs from commercial corn oil because it contains significant levels of tocotrienols (Table 15.6), probably derived from the aleurone cells. Similarly, corn kernel oil (whose tocol composition should represent a mixture of corn fiber oil and corn germ oil) contained intermediate levels of tocotrienols (Table 15.6), probably derived from the corn fiber portion of the kernel (Moreau et al., 2005).

## Carotenoids/Chlorophylls and Other Pigments

The most popular commodity grain corn is called "yellow dent #2," and the yellow color is caused by high levels of two carotenoids (lutein and zeaxanthin) (Fig. 15.4) in the endosperm (Kale et al., 2007; Moreau et al., 2007). When extracted with a nonpolar solvent such as hexane, very little of these two carotenoids are extracted, because both are xanthophylls and both contain two oxygen molecules, making them more polar in character. However, when ground corn is extracted with a polar solvent like ethanol, the corn kernel oil contains very high levels of lutein and zeaxanthin; the possible health-promoting applications of this feature are discussed in a later section of this chapter (Table 15.7).

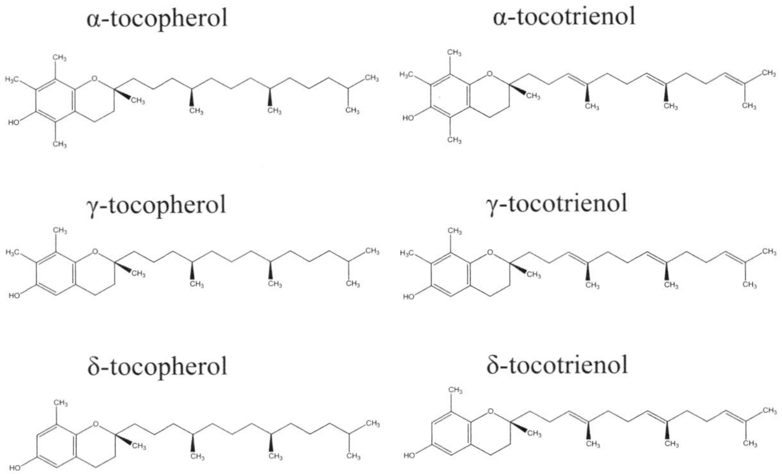

α-tocopherol α-tocotrienol

γ-tocopherol γ-tocotrienol

δ-tocopherol δ-tocotrienol

**Fig. 15.2. The structures of the tocols in corn kernel oil and corn fiber oil.**

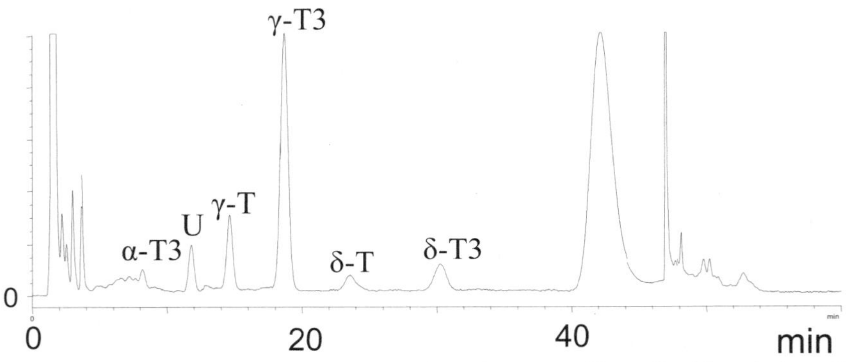


**Fig. 15.3. A chromatogram of the separation of tocols in corn fiber oil. Abbreviations: T, tocopherol; T3, tocotrienol; U, unknown peak.**

## Phenolic Compounds and Natural Products with Antioxidant Properties

Both corn fiber oil and corn kernel oil contain unique phenolic compounds. As mentioned previously, ferulate phytosterol esters are major components of corn fiber oil (≈7%) and minor components of corn kernel oil (≈1%). The ferulate moiety in ferulate phytosterol esters imparts this ester with considerable antioxidant activity (Moreau et al., 2001a). Corn kernel oil, when obtained by ethanol extraction, contains almost 1% of unique phenolics, diferuloylputrescine and *p*-coumaroylputrescine (Fig. 15.5 and Table 15.8).

β-carotene

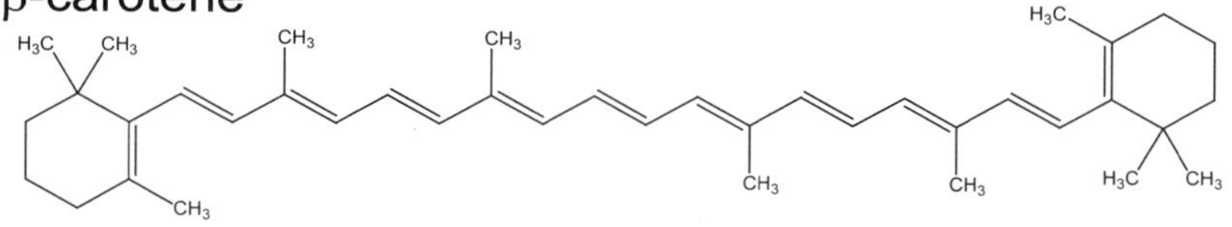

$H_3C$ $CH_3$ $CH_3$ $CH_3$ $H_3C$ $CH_3$ $CH_3$ $CH_3$ $H_3C$ $CH_3$

β-cryptoxanthin

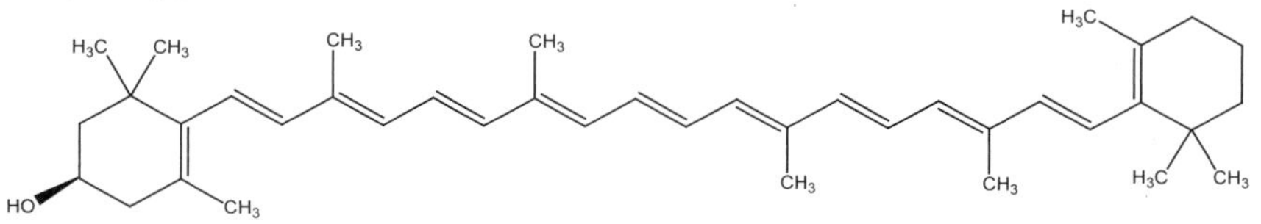

$H_3C$ $CH_3$ $CH_3$ $CH_3$ $H_3C$ HO $CH_3$ $CH_3$ $CH_3$ $H_3C$ $CH_3$

Lutein

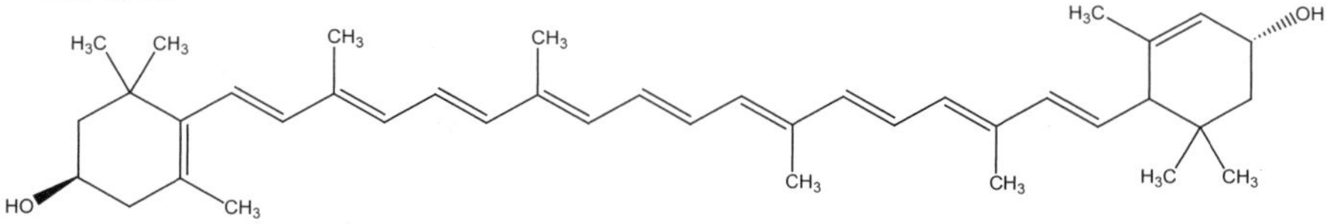

$H_3C$ $CH_3$ $CH_3$ $CH_3$ $H_3C$ OH HO $CH_3$ $CH_3$ $CH_3$ $H_3C$ $CH_3$

Zeaxanthin

$H_3C$ $CH_3$ $CH_3$ $CH_3$ $H_3C$ OH HO $CH_3$ $CH_3$ $CH_3$ $H_3C$ $CH_3$

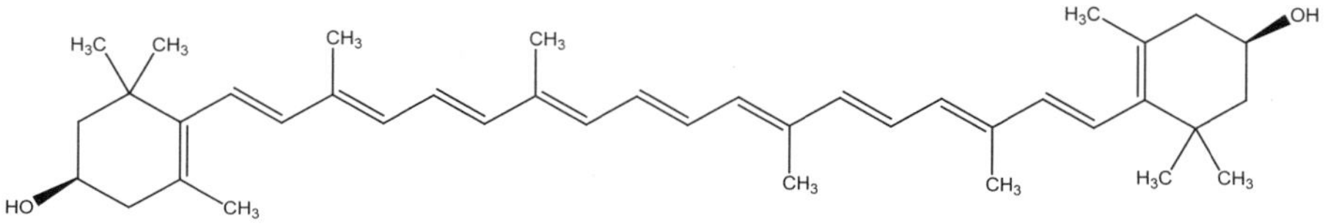

**Fig. 15.4.** Carotenoids in corn kernel oil.

**Table 15.6. Tocopherols and Tocotrienols (mg/kg oil) in Oils Extracted from Ground Corn, Corn Fiber, or Corn Germ**

| Sample, solvent, °C | % Oil extracted with ethanol (e) or hexane (h) | α-T | β-T | γ-T | δ-T | α-T3 | β-T3 | γ-T3 | δ-T3 |
|---|---|---|---|---|---|---|---|---|---|
| Corn kernel oil | 2.70[a] ± 0.06 (e) | 386.7 ± 2.3 | 0 | 1066.7 ± 26.7 | 72.2 ± 3.7 | 138.0 ± 3.5 | 0 | 214.8 ± 2.8 | 0 |
| Corn fiber oil | 2.13 ± 0.02 (h) | 0 | 0 | 79.0 ± 19.5 | 35.7 ± 1.2 | 15.3 ± 2.1 | 0 | 378.3 ± 104.2 | 70 ± 2.7 |
| Corn germ oil | 12.88 ± 0.33 (h) | 218.0 ± 45.2 | 0 | 2756.3 ± 95.8 | 214.8 ± 2.1 | 15.0 ± 3.3 | 0 | 110.7 ± 2.8 | 0 |

[a] Data presented are means ± S.D.

Data are from Moreau and Hicks., 2006.

**Table 15.7.** Carotenoids in Corn kernel Oil, Corn Fiber Oil. and Conventional Corn Oil

| | (μg/g oil) | | | | |
|---|---|---|---|---|---|
| | β-carotene | β-cryptoxanthin | Lutein | Zeaxanthin | Total carotenoids |
| Corn fiber oil (hexane-extracted, unrefined) | 20.3 ± 0.3[c] | 45.7 ± 0.1[b] | 25.0 ± 0.4[b] | 27.3 ± 0.1[b] | 118.3 ± 3.4[c] |
| Corn fiber oil (hexane-extracted, RBD) | 2.2 ± 0.0[d] | 0[c] | 0[c] | 0[c] | 2.2 ± 0.0[d] |
| Ethanol-extracted corn kernel oil (unrefined) | 36.4 ± 1.6[b] | 69.0 ± 0.8[a] | 123.8 ± 4.3[a] | 78.3 ± 6.6[a] | 307.5 ± 13.5[a] |
| Ethanol-extracted corn kernel oil (RBD) | 2.3 ± 0.2[d] | 0[c] | 0[c] | 0[c] | 2.3 ± 0.2[d] |
| Commercial corn oil, Mazola® (RBD) | 3.4 ± 0.1[d] | 0[c] | 0[c] | 0[c] | 3.4 ± 0.1[d] |

[1]Mean ± standard deviation ($n$ = 3). Means in the same column with no letter in common are significantly different ($P < 0.05$) by the Bonferroni least significant difference method [14].
From Moreau et al., 2007.

## Flavor and Aroma Compounds

Commercial corn oil is noted for its mild flavor (Moreau et al., 2002, 2005). No research on flavor or aroma compounds was reported in corn kernel oil or corn fiber oil.

## Allergic and Toxic Compounds

No allergic or toxic compounds were identified in corn kernel oil or corn fiber oil. Two components whose safety is unknown were reported to occur in unrefined ethanol-extracted corn kernel oil, diferuloylputrescine and *p*-coumaroylferuloylputrescine (Moreau et al., 2001a). However, conventional refining, bleaching, and deodorization (RBD) removed all traces of these compounds (unpublished results). Corn fiber oil contains 4–7% phytosterol–ferulate esters (Moreau et al., 1996), whose safety was not established. However, because these esters were hydrolyzed by pancreatic digestive enzymes (Nystrom et al., 2008), likely, they are not toxic, because the two products that are released, phytosterols and ferulic acid, are generally recognized as safe (GRAS).

# Health Benefits of the Corn Kernel Oil, Corn Fiber Oil, and Their Constituents

No study has focussed on the health-promoting properties of corn kernel oil. However, because of its high levels of lutein and zeaxanthin, its consumption may prevent

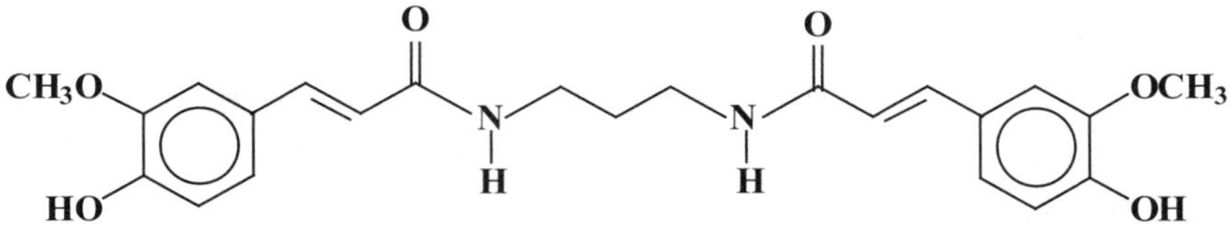

Diferuloylputrescine (DFP)

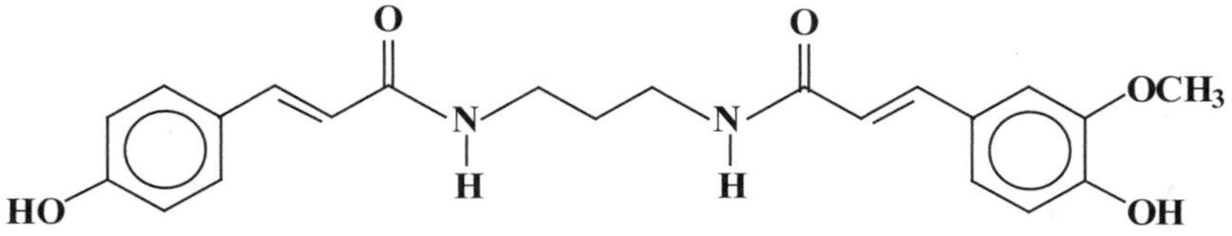

*p*-Coumaroyl-feruloylputrescine (CFP)

**Fig. 15.5. Polyamine conjugates in corn kernel oil.**

**Table 15.8. Polyamine Conjugates in Corn kernel Oil and Corn Fiber Oil**

| | (Wt % of oil) | |
|---|---|---|
| | Diferuolyputrescine | p-Coumaroylputrescine |
| Corn fiber oil (hexane-extracted, unrefined) | 0 | 0 |
| Corn fiber oil (hexane-extracted, RBD) | 0 | 0 |
| Ethanol-extracted corn kernel oil (unrefined) | 0.13 ± 0.02 | 0.46 ± 0.03 |
| Ethanol-extracted corn kernel oil (RBD) | 0 | 0 |
| Commercial Corn Oil, Mazola® (RBD) | 0 | 0 |

[1]Mean ± standard deviation (*n* = 3).
From Moreau et al., 2001a.

the onset of age-related macular degeneration. A daily dosage of two tablespoons of corn kernel oil (≈30 mL) was reported (Moreau et al., 2007) to contain ≈6 mg of lutein + zeaxanthin, which is considered the minimal daily dose to prevent age-related macular degeneration (Bowen et al., 2002; Mozaffarieh et al., 2003).

Many clinical studies demonstrated the serum LDL–C-lowering efficacy of dietary phytosterols (Moreau, 2003, 2005a; Moreau et al., 2002). Because corn fiber oil contains high levels of phytosterols, the main health-promoting property that was-investigated was its ability to lower the levels of serum LDL–C. Wilson et al. (2000)

reported that corn fiber oil had significant cholesterol-lowering activity in hamsters, and caused increased fecal cholesterol excretion. While hamsters fed conventional corn oil showed a 15% reduction in serum LDL–C levels versus a high-cholesterol control diet, hamsters fed corn fiber oil had a 33% reduction in serum LDL–C levels and a 187% increase in the excretion of fecal sterols over the control. Ramjiganesh et al. (2000) reported that corn fiber oil also reduced the levels of plasma LDL and total cholesterol in guinea pigs by decreasing the cholesterol absorption and by increasing the excretion of neutral sterols. Guinea pigs exhibited up to 57% of reductions in plasma LDL–C levels versus control animals fed a high-cholesterol diet without corn fiber oil. Ramjiganesh et al. (2002) also reported that in guinea pigs, corn fiber oil lowered plasma cholesterol by altering hepatic cholesterol metabolism and up-regulating LDL receptors. Recently, Jain et al. (2008) compared the cholesterol-lowering efficacy of corn fiber oil and two of its components—sitostanol and sitostanol ferulate (the major ferulate phytosterol ester). These researchers concluded that corn fiber oil possessed a greater cholesterol-lowering activity than pure sitostanol and its ferulate ester. They also suggested that reductions in the rate of cholesterol absorption induced by corn fiber oil are independent of intestinal enterocyte sterol transporters. To date, no human clinical studies on the effects of corn fiber oil were published.

The antioxidant properties of the ferulate-phytosterol esters in corn fiber oil were also investigated (Nystrom et al., 2008; Wang et al., 2002). The antioxidant and antimelanogenic properties of diferuloylputrescine and *p*-coumaroylferuloylputrescine were studied. These compounds had similar antioxidant properties to ferulic acid and *p*-coumaric acid (Choi et al., 2007). Diferuloylputrescine also significantly inhibited melanin synthesis in B16 melanoma cells. These results suggest that the polyamine conjugates in corn bran may be useful as potential radical scavengers capable of controlling reactive oxygen-mediated pathological disorders, and may be useful as skin-whitening agents in cosmetics. These two polyamine conjugates were reported at total levels of about 1% in unrefined corn kernel oil (Moreau et al., 2007), but they were removed during conventional refining, bleaching, and deodorization (unpublished results).

Although not health-promoting properties, the diferuloylputrescine and *p*-coumaroylferuloylputrescine in corn kernel oil had an interesting biological activity—they inhibit the biosynthesis of alflatoxins by *Aspergillus flavus* (Mellon and Moreau, 2002). Even though they inhibit aflatoxin biosynthesis by this species, they do not inhibit its growth, and thereby are not considered to be toxic to it. This suggests possible agricultural applications for corn kernel oil.

## Other Issues (e.g., Authenticity and Adulteration, etc.)

Because corn kernel oil and corn fiber oil are not yet commercialized, authenticity and adulteration are not concerns. If and when both are commercialized, the unique components in both could potentially be used to monitor their authenticity. For corn kernel oil, the high levels of lutein and zeaxanthin could be used (Moreau et al., 2007). For corn fiber oil, the high levels of ferulate-phytosterol esters and stanols could be used (Moreau et al., 1996).

## Conclusion

Both corn kernel oil and corn fiber oil contain unique health-promoting components that are lacking in commercial corn oil, which is obtained by extracting corn germ. Corn kernel oil contains very high levels of lutein and zeaxanthin; a regular consumption of it may prevent the onset of age-related macular degeneration. Corn fiber oil contains the highest levels of phytosterols of any known natural oil or extract, and a regular consumption of it may reduce the levels of serum LDL- and total cholesterol.

### References

Abbas, C.; A.M. Ramelsberg; K. Beery. Extraction of phytosterols from corn fiber using green solvents. U.S. Patent 7,368,138 (2008).

Anon. GreenShift Corporation's Corn Oil Extraction and Biodiesel Conversion Technologies, 2008, http://www.greenshift.com/pdf/Corn_Oil_Extraction_Presentation-100mmgy.pdf.

Bowen, P.E.; S.M. Herbst-Espinosa; E.A. Hussain. Esterification does not impair lutein bioavailability in humans. *J. Nutr.* **2002,** *132,* 3668–3673.

Choi, S.W.; S.K. Lee; E.O. Kim; J.H. Oh; K.S. Yoon; N. Parris; K.B. Hicks; R.A. Moreau. Antioxidant and antimelanogenic activities of polyamine conjugates from corn bran and related hydroxycinnamic acids. *J. Agric. Food Chem.* **2007**, *55*, 3920–3925.

Dien, B.S.; N. Nagle; V. Singh; R.A. Moreau; M.P. Tucker; N.N. Nichols; D.B. Johnston; M.A. Cotta; K.B. Hicks; Q. Nguyen; R.J. Bothast. Review of processes for producing corn fiber oil and ethanol from "Quick Fiber." *Int. Sugar J.* **2005**, *107*, 187–191.

Dunford, N.T. Germ oils from different sources. *Bailey's Industrial Oil and Fat Products*, 6th ed.; F. Shahidi, Ed.; John Wiley and Sons: New York, 2005; Vol. 3, pp. 195-231.

Feng, F.; D.J. Myers; M.P. Hojilla-Evangelista; K.A. Miller; L.A. Johnson; S.K. Singh. Quality of corn oil obtained by sequential extraction processing. *Cereal Chemistry* **2002,** *79*, 707–709.

Hojilla-Evangelista, M.P.; L.A. Johnston; D.J. Meyers. Sequential extraction processing of flaked whole corn: Alternative corn fractional technology for ethanol production. *Cereal Chem.* **1992**, *69*, 643–647.

Jain, D.; N. Ebine; X. Jia; A. Kassis; C. Marinangeli; M. Fortin; R. Beech; K.B. Hicks; R.A. Moreau; S. Kubow; P.J.H. Jones. Corn fiber oil and sitostanol decrease cholesterol absorption independently of intestinal sterol transporters in hamsters. *J. Nutr. Biochem.* **2008**, *19*, 229–236.

Kale, A.; F. Zhu; M. Cheryan. Separation of high-value products from ethanol extracts of corn by chromatography. *Ind. Crops Prod.* **2007**, *26*, 44–53.

Kwiatkowski, J.R.; M. Cheryan. Extraction of oil from ground corn using ethanol. *J. Am. Oil Chem. Soc.* **2002**, *79*, 825–830.

Mellon, J.E.; R.A. Moreau. Inhibition of aflatoxin biosynthesis in Aspergillius flavus by diferuloylputrescine and p-coumaroylputrescine. *J. Agric. Food Chem.* **2004**, *52*, 6660–6663.

Moreau, R.A. Corn oil. *Vegetable Oils in Food Technology;* F.D. Gunstone, Ed.; Sheffield Academic Press: Sheffield, UK, 2002; pp. 278-296. Moreau, R.A. Phytosterols in functional foods. *Phytosterols as Functional Food Components and Nutraceuticals;* P. Dutta, Ed.; Marcel Dekker: New York, 2003; pp. 317–345.

Moreau, R.A. Phytosterols and phytosterol esters. *Healthful Lipids;* C. Akoh, O-M. Lai, Eds.; AOCS Press: Champaign, IL, 2005a; pp. 335–360. Moreau, R.A. Corn oil. *Bailey's Industrial Oil & Fat Products*, 6th ed.; F. Shahidi, Ed.; Wiley-Interscience: Hoboken, 2005b; Vol. 2, pp. 149–172.

Moreau, R.A.; K.B. Hicks. The composition of corn oil obtained by the alcohol extraction of ground corn. *J. Am. Oil Chem. Soc.* **2005**, *82*, 809–815.

Moreau, R.A.; K.B. Hicks. A reinvestigation of the effect of heat pretreatment of corn fiber on the levels of extractable tocopherols and tocotrienols. *J. Agric. Food Chem.* **2006**, *54*, 8093–8102.

Moreau, R.A.; K.B. Hicks; R.J. Nicolosi; R.A. Norton. Corn fiber oil—its preparation, composition, and use. U.S. Patent 5,843,499 (1998a).

Moreau, R.A.; D.B. Johnston; K.B. Hicks. The influence of moisture content and cooking on the screw pressing and pre-pressing of corn oil from corn germ. *J. Am. Oil Chem. Soc.* **2005**, *82*, 851–854.

Moreau, R.A.; D.B. Johnson; K.B. Hicks. A comparison of the levels of lutein and zeaxanthin in corn germ oil, corn fiber oil, and corn kernel oil. *J. Am. Oil Chem. Soc.* **2007**, *84*, 1039–1044.

Moreau, R.A.; A. Nuñez; V. Singh. Diferuloylputrescine and p-coumaroyl feruloylputrescine, abundant polyamine conjugates in lipid extracts of maize kernels. *Lipids* **2001a,** *36*, 839–844.

Moreau, R.A.; M.J. Powell; K.B. Hicks. Extraction and quantitative analysis of oil from commercial corn fiber. *J. Agric. Food Chem.* **1996,** *44,* 2149–2154.

Moreau, R.A.; M.J. Powell; K.B. Hicks. The occurrence and biological activity of ferulate-phytosterol esters in corn fiber and corn fiber oil. *Physiology, Biochemistry, and Molecular Biology of Plant Lipids;* J.P. Williams, M.U. Khan, N.W. Lem, Eds.; Kluwer Academic Publishers: Dordrecht, The Netherlands, **1997**; pp. 189–191.

Moreau, R.A.; M.J. Powell; K.B. Hicks; R.A. Norton. A comparison of the levels of ferulate-phytosterol esters in corn and other seeds. *Advances in Plant Lipid Research,* Cerdá-Olmedo Sánchez, E. Martínez-Force, Eds.; Universidad de Sevilla: Sevilla, 1998b; pp. 472–474.

Moreau, R.A.; M.J. Powell; V. Singh. Pressurized liquid extraction of polar and nonpolar lipids in corn and oats with hexane, methylene chloride, isopropanol, and ethanol. *J. Am. Oil Chem. Soc.* **2003**, *80*, 1063–1067.

Moreau, R.A.; V. Singh; S.R. Eckhoff; M.J. Powell; K.B. Hicks; R.A. Norton. A comparison of the yield and composition of oil extracted from corn fiber and corn bran. *Cereal Chem.* **1999b**, *76*, 449–451.

Moreau, R.A.; V. Singh; K.B. Hicks. A comparison of oil and phytosterols in the seeds of germplasm accessions of corn, teosinte, and Job's tears. *J. Agric. Food. Chem.* **2001b**, *49*, 3793–3795.

Moreau, R.A.; V. Singh, A. Nunez; K.B. Hicks. Phytosterols in the aleurone layer of corn kernels.

*Biochem. Soc. Trans.* **2000**, *28*, 803–806.

Moreau, R.A.; B. Whitaker; K. Hicks. Phytosterols, phytostanols, and their conjugates in foods: structural diversity, quantitative analysis, and health-promoting uses. *Prog. Lipid Res.* **2002,** *41*, 457–500.

Mozaffarieh, M.; S. Sacu; A. Wedrich. The role of carotenoids, lutein, and zeaxanthin, in protecting against age-related macular degeneration: A review based on controversial evidence. *Nutr.. J.* **2003**, *2,* 1–8.

Nystrom, L.; T.A. Achrenius; A.-M. Lampi; R.A. Moreau; V. Piironen. A comparison of the antioxidant properties of steryl ferulates with tocopherol at high temperatures. *Food Chem.* **2007,** *101*, 947–954.

Nystrom, L.; R.A. Moreau; A.-M. Lamp; K.B. Hicks; V. Piironen. Enzymatic hydrolysis of steryl ferulates and steryl glycosides. *Eur. Food Res. Technol.*, **2008**, *227*, 727–733.

Parris, N.; L.C. Dickey; J.L. Wiles; R.A. Moreau; P.H. Cooke. Enzymatic hydrolysis, grease permeation, and water barrier properties of zein isolate coated paper. *J. Agric. Food Chem.* **2000,** *48,* 890–894.

Parris, N.; L.C. Dickey; M.J. Powell; D.R. Coffin; R.A. Moreau; J.C. Craig. Effect of endogenous triacylglycerol hydrolysates on the mechanical properties of zein films from ground corn. *J. Agric. Food Chem.* **2002** , *50*, 3306–3308.

Ramjiganesh, T.; S. Roy; H.C. Freake; J.C. McIntyre; M.L. Fernandez. Corn fiber oil lowers plasma cholesterol by altering hepatic cholesterol metabolism and up-regulating LDL receptors in guinea pigs. *J. Nutr*. **2002,** *132*, 335–340.

Ramjiganesh, T.; S. Roy; R.J. Nicolosi; T.L. Young; J.C. McIntyre; M.L. Fernandez. Corn husk oil lowers plasma LDL cholesterol concentrations by decreasing cholesterol absorption and altering hepatic cholesterol metabolism in guinea pigs. *J. Nutr. Biochem.* **2000,** *11,* 358–366.

Singh, V.; D.B. Johnston; R.A. Moreau; K.B. Hicks; B.S. Dien; R.J. Bothast. Pretreatment of wet-milled corn fiber to improve recovery of corn fiber oil and phytosterols. *Cereal Chem.* **2003a**, *80*, 118–122.

Singh, V.; R.A. Moreau. Enrichment of oil in corn fiber by size reduction and floatation of aleurone cells. *Cereal Chem.* **2003b,** *80*, 123–125.

Singh, V.; R.A. Moreau. Methods of preparing corn fiber oil and of recovering corn aleurone cells from corn fiber. U.S. Patent 7,115, 295 (2006).

Singh, V.; R.A. Moreau; P.H. Cooke. Effect of corn milling practices on the fate of aleurone layer cells and their unique phytosterols. *Cereal Chem.* **2001a**, *78*, 436–441.

Singh, V.; R.A. Moreau; L.W. Doner; S.R. Eckhoff; K.B. Hicks. Recovery of fiber in the corn dry-grind ethanol process: A feedstock for valuable co-products. *Cereal Chem.* **1999**, *76*, 868–872.

Singh, V.; R.A. Moreau; A.E. Haken; K.B. Hicks; S.R. Eckhoff. Effect of various acids and sulfites in steep solutions on yields and composition of corn fiber and corn fiber oil. *Cereal Chem.* **2000**, *77*, 665–668.

Singh, V.; R.A. Moreau; K.B. Hicks; S.R. Eckhoff. Effect of alternative milling techniques on the yield and composition of corn germ oil and corn fiber oil. *Cereal Chem.* **2001b**, *78*, 46–49.

Srinivasan, R.; R.A. Moreau; C. Parsons; J.D. Lane; V. Singh. Separation of fiber from distillers dried grains (DDG) using sieving and elutriation. *Biomass and Bioenergy* **2008**, *32*, 468–472.

Srinivasan, R.; R.A. Moreau; K.D. Rausch; R.L. Belyea; M.E. Tumbleson; V. Singh. Separation of fiber from distillers dried grains with solubles (DDGS) using sieving and elutriation. *Cereal Chem.* **2005**, *82*, 528–533.

Srinivasan, R.; V. Singh; R.L. Belyea; K.D. Rausch; R.A. Moreau; M.E. Tumbleson. Economics of fiber separation from distillers dried grains with solubles (DDGS) using sieving and elutriation. *Cereal Chem.* **2006**, *83*, 324–330.

Srinivasan, R.; V. Singh; R.A. Moreau; K.D. Rausch. Phytosterol composition and yield of oil from fractions obtained by sieving and elutriation of distillers dried grains with solubles (DDGS). *Cereal Chem.* **2007**, *84*, 626–630.

Wahjudi, J.; L. Xu; P. Wang; V. Singh; P. Buriak; K.D. Rausch; A.J. McAloon; M.E. Tumbleson; S.R. Eckhoff. Quick fiber process: Effect of mash temperature, dry solids, and residual germ on fiber yield and purity. *Cereal Chem.* **2000,** *77*, 640–644.

Wang, T.; K.B. Hicks; R.A. Moreau. Antioxidant activity of phytosterols, oryzanol, and other phytosterol conjugates. *J. Am. Oil Chem. Soc.* **2002**, *79*, 1201–1206.

Wilson, T.A.; A.P. DeSimone; C.A. Romano; R.J. Nicolosi. Corn fiber oil lowers plasma cholesterol levels and increases cholesterol excretion greater than corn oil and similar to diets containing soy sterols and soy stanols in hamsters. *J. Nutr. Biochem.* **2000**, *11*, 443–449.

Winkler, J.K.; K.A. Rennick; F.J. Eller; S. Vaughn. Phytosterol and tocopherol components in extracts of corn distiller's dried grain. *J. Agric. Food Chem.* **2007**, *55*, 6482–6486.

# 16

# Oat Oil

**Kevin Robards[1], Paul Prenzler[1], Danielle Ryan[1] and Afaf Kamal-Eldin[2]**
*[1]School of Agriculture and Wine Sciences, Charles Sturt University, Locked Bag 588, Wagga Wagga 2678, Australia; [2]Department of Food Science, Swedish University of Agricultural Sciences (SLU), Box 7051, SE-750 07, Uppsala, Sweden*

## Introduction

Farmers have historically cultivated oats for numerous uses (e.g., hay, pasture, silage) other than for cash grain (Welch, 1995). About 15 species of oats are grown in cooler regions of the world—primarily Russia, Canada, the United States, Finland, Sweden, Germany, and Poland. The vast majority of the oats grown throughout the world are the cultivated oat, *Avena sativa*. Historically, oats were generally less favored for food use than other grains because of a bland taste and a tendency to undergo spoilage. Nevertheless, oats became a staple in Germany, Ireland, Scotland, and the Scandinavian countries. Scottish settlers introduced oats to the New World in the early seventeenth century (Anon, 2005). The more traditional use of oats as feed for working horses has diminished. Moreover, other cereals have provided better returns to farmers, and worldwide production of oats was in a long-term decline until relatively recently. For example, Russia produced 9.4 million tons in 1998 but only 4.5 million in 2000 (Anon, 2006). Total world production in 2005 was 24.6 million tons (Manunsell, 2007) although only a very small proportion was for human food use and approximately 3 million tons of the oat crop entered world commerce.

The oat crop provides a range of products that are used as animal feeds, in human foods, and as industrial raw materials. These products include crop silage, straw, grain, and grain derivatives. The oat grain is comprised of three main structures, namely, the bran, endosperm, and the germ (embryo). The proximate composition of the grain is given in Table 16.1. Wide variations appear in these data dependent on cultivar, growing conditions, agronomic practices, and even the analytical methodology. As with other grains, starch remains the most abundant component where it constitutes about 35–55% of the dry matter of the entire oat grain. Compared to other cereals, oats are characterized by a lower carbohydrate content (Ozcan et al., 2006), with higher protein and lipid contents. Indeed, oat contains the highest lipid concentration among cereal grains (Gudmundsson & Eliasson, 1989; Peterson & Wood, 1997;

**Table 16.1. Proximate Composition of Oat Grain—Average Values and Ranges (% dry basis)**

| Crude protein | Oil | Carbohydrate (starch+sugars) | Neutral detergent fiber | Ash |
|---|---|---|---|---|
| 10.7<br>(7.2–16.1) | 5.0<br>(1.9–8.0) | 47.8<br>(39.1–57.2) | 31.5<br>(25.5–36.4) | 2.9<br>(2.1–4.1) |

Data compiled from reference (Welch, 1995)

Price & Parsons, 1975; Zhou et al., 1999a). The oat grain is also rich in unsaturated lipids and lipolytic enzymes such as lipases and lipoxygenases, thus accounting for the greater tendency of oats to undergo oxidative spoilage.

The oil content of oat cultivars varies widely (e.g., 3.1–11.6% in over 4,000 entries of world oat collection) (Brown & Craddock, 1972). Oat lines with oil contents outside this range also exist including developed lines with 14.5% and 16.2% oil (Baker & McKenzie, 1972; Branson & Frey, 1989; Frey & Hammond, 1975; Sahasrabudhe, 1979; Schipper & Frey, 1991). However, the most common range is 5–9%. The majority of oat lipids are found in the endosperm, especially in the aleurone and subaleurone cells (White et al., 2006; Youngs, 1986; Youngs et al., 1977). The germ contains the highest concentration of lipids in oats, but as it represents only about 7% of the total kernel weight, its contribution to total kernel oil is minor. Moreover, it is more economical to extract the oil from the whole kernel due to the structural location of the germ within the kernel. The oil is present in oil bodies (Banas et al., 2000; Peterson & Wood, 1997; White et al., 2006). In oilseeds, the oil bodies (0.5–2.5 μm in diameter) are composed of ca. 94–98% triacylglycerols, 0.5–2% phospholipids, and 0.5–3.5% of small basic proteins called oleosins of 15–26 kDa (Tzen et al., 1993). Since oat lipids contain much higher levels of phospholipids and glycolipids than oilseeds, the composition of the oil bodies in oats needs to be studied.

Apart from use as animal feed, oats are used in a range of human foods that include raw oats, rolled oats, oat flakes, oat bran, and oat flour. Oat flour and extracts were among the first antioxidants proposed for the stabilization of lipids and lipid-containing foods (Duve & White, 1991; Emmons & Peterson, 1999; Peters & Musher, 1937). A special fine ground oat flour marketed under such names as Avenex and Aveeno was commercially available and used as a stabilizer in products as diverse as ice cream, fish oil, and cereals (Welch, 1995 ). Oats may provide a useful substitute for wheat products in patients suffering coeliac disease (Hogberg et al., 2004). Oat oil also finds a market but primarily for nonedible applications. Crude oat oil contains a very high level of antioxidants, more than every major oilseed, grain, or grain by-product except wheat germ. Oat oil has excellent shelf-life stability due to the high content of antioxidants unique to oats. Lipids, especially polar lipids, improve loaf volume, grain and texture and delay staling in bread (Erazo-Castrejon et al., 2001). Oat oil with its relatively high content of polar lipids is ideally suited to this application and may ultimately replace existing additives used for this purpose. If this happens, this will open a new market for oat oil which, although commercially available, is currently a speciality oil, with limited food applications.

## Processing

One can extract oat oil from the grain or its pearling fractions by using numerous organic solvents, mainly hexane and petroleum fractions of comparable volatility. Other possible solvents include acetone, methanol, ethanol, 1- and 2-propanols, *tertiary*-butanol, diethyl ether, which one can use separately, in mixtures, or sequentially (Boczewski, 1980; Martin, 1964; Moreau et al., 2003; Potter et al., 1997; Washburn, 1953). These solvents is naturally eliminated by distillation. Another possible route is to use supercritical fluids (e.g. supercritical carbon dioxide) where the solvent naturally separates out the oil upon vessel depressurization at the end of the extraction (Aro et al., 2007; Fors & Eriksson, 1990). The, the obtained oat oils vary significantly

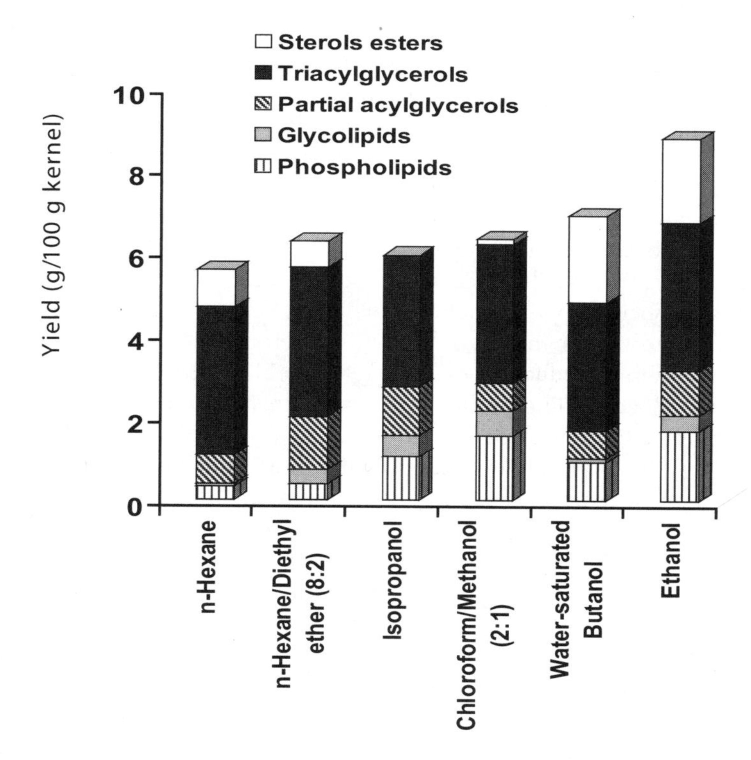


**Fig. 16.1.** Effect of extraction solvent on yield and relative proportions of lipid classes in oats. Data from Sahasrabudhe, 1979.

in their chemical and physical properties, depending on the extraction solvent(s) and temperatures (Fig. 16.1). Oat oils extracted with diethyl ether in a Soxhlet apparatus at 25 degree C (Ozcan et al., 2006), ethanol or carbon dioxide (Ceapro Inc., private communication, 2008) had refractive indices of 1.47plus/minus 0.01 and specific gravity of 0.92 plus/minus 0.01 g/cm3.

The oat grain is rich in enzyme activities that cause lipid peroxidation, including lipases, lipoxygenases, and superoxide dismutases (Giannopolitis & Ries, 1977; Youngs, 1986). These enzymes affect the free fatty acid content and the level of peroxides in the oil, depending on the water activity of the extracted oat, the extraction method, and the temperature of extraction. The level of antioxidants also determines the oxidative status of the extracted oil, including the free and esterified forms of benzoic and cinnamic acids, quinones, flavones, flavonols, chalcones, flavanones, anthocyanidins, and aminophenols as well as tocopherols and tocotrienols (Collins, 1986). Generally oat oil has very high oxidative stability and can stabilize other oils and oil-containing products. The phospholipid fraction seems to be responsible for the greatest part of the stabilization effect (Forssell et al., 1990, 1992).

The color of crude oat oils extracted with different solvents covers the range from light brown, amber brown, tan brown to dark brown and brownish black. The extracted crude oil contains fine protein particles that filtration can remove. One can refine the crude oil in a series of steps, including washing with water to remove some glycolipids and gums, neutralization with alkali to remove free fatty acids, degumming with phosphoric or citric acid to remove phosphatides, alkali neutralization to remove free fatty acids, winterization to remove high-melting triacylglycerols, treatment with activated bleaching earth or activated carbon to remove color, and heating under high vacuum to achieve deodorization (Potter et al., 1997). However, numerous difficulties are associated with the refining of oat oil. For example, the oil contains high levels of free fatty acids and phosphatides, the removal of which is expensive and leads to considerable product losses. U.S. Patent 6,113,908 (Paton et al., 2000) discloses that drying oat pearlings to below 4% moisture allows the endogenous oat lipase to catalyze the esterification of free fatty acids to glycerides and to reduce their level in the oil. Furthermore, pre-extraction of the pearlings with aqueous ethanol was found to selectively remove color and phospholipids and decrease losses on subsequent extraction with hexane or other nonpolar solvents.

Supercritical fluid technologies can also extract and fractionate oat oil. (Aro et al., 2007; Fors & Eriksson, 1990). Extraction with pure carbon dioxide under supercritical conditions (e.g., 70°C, 450 bar and 0.4 L/min) yielded an oil mainly composed of triacylglycerols while the addition of organic modifiers (e.g., ethanol) enables extraction of polar lipids. The two supercritical solvents were used sequentially to remove neutral and polar lipids in separate fractions (Aro et al., 2007). The polar lipid fraction mainly contained digalactosyl diacylglycerol (DGDG) (43%) and phosphatidylcholine (13%).

## Oat Enzymes and Oil Stability

The oat kernel has a high activity of lipase(s) (EC 3.1.1.3) with an optimal pH of 7.5 and temperature of 37.5°C (O'Connor et al., 1992; Sahasrabudhe, 1982). Oat lipase(s) are localized in the aleurone layer as well as the endosperm (Ekstrand et al. 1992; Ekstrand et al. 1993; Hutchinson et al. 1951; Hutchinson et al. 1955; Lehtinen et al. 2003; Urquhart et al. 1983). The indication is that hydrolysis of triacylglycerols by oat lipase(s) does not yield partial glycerides, but is rather a complete hydrolysis to three free fatty acid moieties leading to bitter taste (Liukkonen et al., 1993). Hydrolysis of oat triglycerides during storage or processing increases the susceptibility of the freed fatty acids toward oxidative deterioration by molecular oxygen (Warwick et al., 1979). Enzyme inactivation is generally achieved by subjecting the milled oat products to moist heat (e.g., during extrusion) (Lehtinen & Laakso, 2004). On the other hand, this active lipase in ground oats and moist oat caryopses was reported to contain enough lipase activity to be useful for several applications in fatty acid hydrolysis (Parmar & Hammond, 1994; Piazza et al., 1989; Piazza, 1991; Piazza & Farrell, 1991).

Unlike the lipase activity, the activity of lipoxygenase (EC 1.13.11.12) in oats is weak, and a lipoperoxidase activity (EC 1.11.1) is responsible for the conversion of hydroperoxides to corresponding hydroxy acids (Biermann et al., 1980). The different enzymes involved in the oxidation of oat lipids and the main products of these oxidations are shown in Fig. 16.2. Isomerization and/or further oxidation of these products may form other oxidation products. For example, 3-nonenal is unstable and can isomerize to 2-nonenal or under catalysis by singlet oxygen to pentylfuran. When the kernel is dry and intact, lipid hydrolysis and oxidation reactions are minimal and insignificant. However, when the kernels are disrupted during milling, these reactions are significant and lead to pronounced off-flavor and loss of nutritive value including lowering of tocopherols and tocotrienols. Oils obtained from carefully processed oats may have a free fatty acid content of ca. 5% and a peroxide value of about 5–10 mequiv oxygen/kg oil. At a peroxide value of ca. 20 mequiv oxygen/kg, rancidity is evident.

## Edible and Nonedible Applications

Oat bran and oat oil are normally not consumed as a food supplements but are used in food preparations. For example, the addition of oat oil, especially its polar lipid fraction, to bread formulations increased loaf volume and improved bread appearance and resistance to staling (Erazo-Castrejon et al., 2001). Crude oat oil (and shortening) (at 3%) increased loaf volume by approximately 11% over the zero lipid formulation, while the polar lipid fraction increased loaf volume by nearly the same amount when added at only a 0.5% level. The effect of oat lipids was stronger in breads made of a weak flour (10% protein) than in breads made of a strong flour (14% protein).

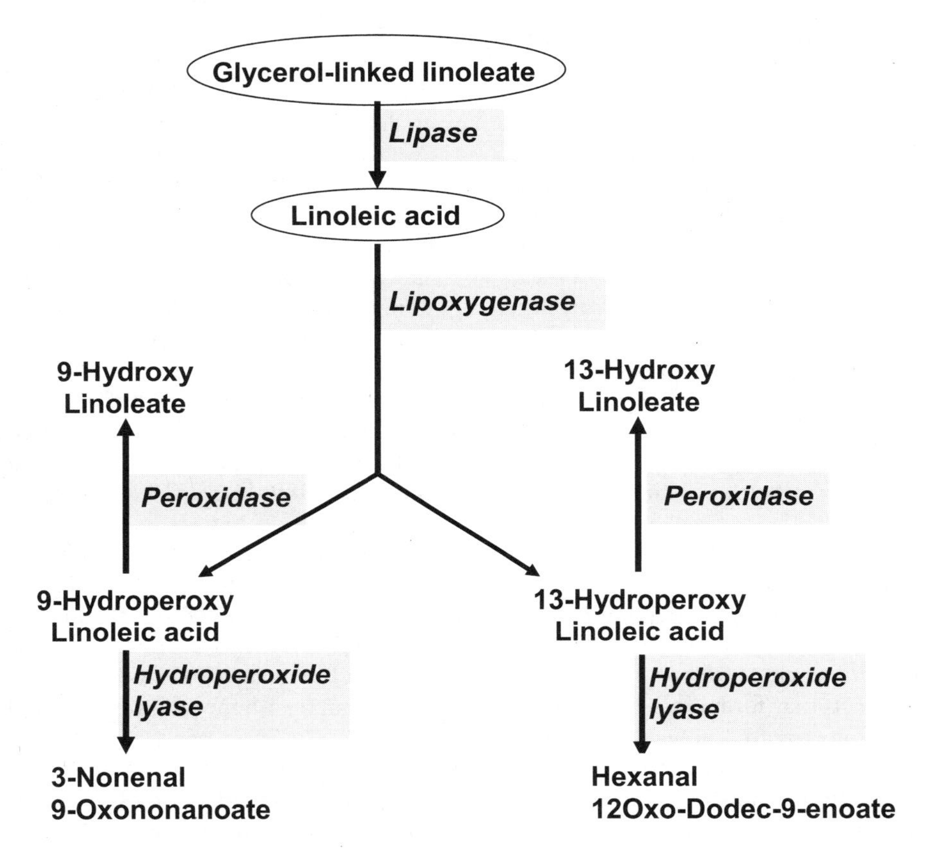


**Fig. 16.2.** Some oat grain oxidation-relevant enzymes and their involvement in the generation of rancid molecular species from triacylglycerol-bound linoleate residues.

These effects of oat oil were attributed to the amphipathic properties of polar lipids that make them able to interact with starch, proteins, and other bread components. Since oat polar lipids may form inclusion complexes with starch, they can modulate the pasting properties of starches leading to increased swelling, peak viscosity, setback and gelatinization temperatures, and freeze-thaw stability (Zhou et al., 1999b). The use of oat oil in bread baking is also promoted as "heart-healthy," since vegetable shortenings traditionally used to increase loaf size contain *trans* fatty acids, the consumption of which are associated with heart disease. Replacing these shortenings with oat oil, known to be free of *trans* fatty acids, therefore offers a healthier alternative for improved bread quality and shelf life (Hardin, 2000).

So far, no standard definition exists for *oat oil*, which poses a difficulty for its potential uses and applications and necessitates proper physicochemical description of the products.

## Cosmetics

The use of oat oil in personal-care products represents a natural extension of the historical use of oatmeal and oatmeal fractions for improving the skin (Aburjai & Natsheh, 2003) in that oatmeal is historically known for its skin soothing and itch- and irritation-relieving properties (Kurtz & Wallo, 2007). Oat oil and its fractions have numerous pharmacological and dermatological properties that make them attractive for pharmaceutical and cosmetic applications (Blom et al., 1996; Dull, 1997). For example, oat oil has the ability to emulsify large quantities of water, thus making it very effective in hydrating and moisturizing epidermal layers (Aburjai & Natsheh, 2003). Phenolic compounds present in the oil provide protection against ultraviolet light as well as providing antioxidant and anti-inflammatory activities; saponins impart a cleansing activity while the phospholipids provide moistening and buffer activities.

Cosmetic formulations can include oat oil and colloidal components of the oat meal (Kurtz & Wallo, 2007). Oat oil is a clear, lightly colored oil with a mild natural odor that is rich in phospholipids and glycolipids (polar lipids), and free of *trans* fatty acids is generally preferred in cosmetic formulations. It has a stabilizing property in emulsions and oils and an excellent shelf-life stability due to the high content of antioxidants found in oats (Anon, undated_b). The oil is claimed to improve the elasticity of the skin and hair and has strong skin-hydrating and moisturizing properties. In skin- and hair-care formulations, a 1–5% concentration is recommended (Anon, 1997–2008 ). Oat oil is available as a viscous liquid, or alternatively one may spray dry (co-dry) it on various solid supports up to a maximal level of 70% by weight (Dull, 1997). The spray-dried product is advantageous because one can easily mix it with other active ingredients in the preparation of packed powder cosmetic formulations (Dull, 1997). Furthermore, one can dose it into colloidal oatmeal formulations to increase oil content and to enhance the moisturization properties of skin-care lotions and creams (Dull, 1997). Various patents relating to the use of oats in cosmetic preparations exist; however, patents relating specifically to oat oil are less common. One particular patent claims that formulations incorporating oat oil compositions have antioxidant and other "dermatologically beneficial properties." The compositions inhibit ultraviolet (UV) irradiation-induced lipid peroxidation, and are thus useful for inhibiting UV irradiation-induced skin damage (Potter et al., 1997).

## Acyl Lipids

Solvent extraction or supercritical fluid extraction can recover oat oil from the grain. The oil yield of the grain is very dependent on the nature of the extracting solvent (Matz, 1991). The efficiency of extractions with pressurized solvents (hexane, methylene chloride, isopropanol, ethanol) of polar and nonpolar oat lipids was examined (Moreau et al., 2003). The effects of solvent polarity and temperature were tested on the recovery of total lipids, triglycerides, glycolipids, and phytosterols. Lipid values obtained by extraction with ethoxyethane (diethyl ether) are often termed "free lipids" to distinguish them from those lipids more or less strongly bound to saccharide or protein components. Normalizing the material extracted with ethoxyethane to 100%, the equivalent yield for material extracted by ethanol was 128% which equaled to values obtained by acid hydrolysis methods (Zhou et al., 1999a). Three categories of lipids were identified (Morrison, 1981, 1988) that are distinguishable experimentally. The internal lipids residing inside native starch granules either in the cavity of the amylose helix or in the spaces between amylose and amylopectin were considered the only true "starch lipids." Such lipids are composed exclusively of monoacyl lipids (free fatty acids and lysophospholipids). Starch surface lipids are artifacts derived from the surrounding proteinaceous matrix of the endosperm, and it was hypothesized that these compounds, which are also monoacyl lipids, formed inclusion complexes with amylose in the surface regions of the granule. The remaining lipids derived from endosperm, aleurone, and germ are termed "nonstarch lipids." The majority are fully acylated (triacylglycerols, diacylglycolipids, and phospholipids) and can reside either in a free state or bound with proteins on the granule surface.

Negligible quantities of the starch lipids are extractable with traditional low-polarity solvents such as chloroform or diethyl ether (Morrison, 1988). Surface and nonstarch lipids, and internal granular lipids, are readily separated by extraction with propan-1-ol/water at ambient temperature or cold water-saturated butan-1-ol (to recover nonstarch and surface lipids) or by refluxing using Soxhlet apparatus to recover internal lipids (Morrison, 1981). Alternatively, cold extraction with chloroform:methanol:water (3:2:1) followed by hot extraction with propan-1-ol was used (Gibinski et al., 1993) to separate the two groups. Hot extraction was significantly more effective in lipid removal as only 6% of the original lipid remained in the starch following hot extraction, whereas 52% remained after cold extraction.

The major fatty acids in oats are palmitic acid (16:0, 14–23%), stearic acid (18:0, 0.5–3.9%), oleic acid (18:1, 29–52%), linoleic acid (18:2, 26–48%), and linolenic acid (18:3, 1–3.5%) (De la Roche et al., 1977; Frey & Hammond, 1975; Karow, 1980; Sahasrabudhe, 1979; Youngs & Puskulcu, 1976; Zhou et al., 1999a). Minor amounts of other fatty acids are also present [e.g., lauric (12:0), dodec-9-enoic (12:1), myristic (14:0), palmitoleic (16:1), arachidic (20:0), eicosenoic (20:1), behenic acid (22:0), erucic (22:1), lignoceric (24:0), and nervonic (24:1)]. Another fatty acid, the oxylipin 15(*R*)-hydroxy-(9*Z*, 12*Z*)-octadecadienoic acid (or avenoleic acid), was also found at very small levels in oat grains as part of a specific glycolipid (Hamberg & Hamberg, 1996). The oil content of oats is highly heritable (Baker & McKenzie, 1972) and is inherited polygenically with additive and nonadditive gene actions (Brown et al., 1974; Frey et al., 1975). Selection for high oil oat cultivars tends to increase the percentage of oleic acid at the expense of that of linoleic acid (Frey & Hammond, 1975; Schipper & Frey, 1991, 1992; Thro & Frey, 1985; Thro et al., 1985).

Oat lipids represent a heterogeneous mixture of acyl lipids, which are classified into neutral and polar lipids. The neutral lipids, mainly triacylglycerols, account for 50–60% of total oat lipids (Sahasrabudhe, 1979). The composition of the triacylglycerols in one sample of oats was as follows: POP (2.3%), PLP (1.8%), POO (8.5%), SOO (0.7%), PLO (11.7%), SLO (1.2%), OOO (13.4%), OOL (26.8%), PLL (6.2%), PLLn (0.6%), OLL (16.3%), OLLn (1.5%), LLL (6.0%), and LLLn (1.2%). Indeed, this composition will vary in cultivars mainly depending on the relative variation in oleate (O) and linoleate (L) residues with minor effects of palmitate (P), stearate (S), and linolenate (Ln) residues.

Oat oil is rich in phospholipids (6–26%) and glycolipids (6–17%) (Alkio et al., 1991; Bedford & Joslyn, 1937; Forssell et al., 1992; Sahasrabudhe, 1979; Youngs et al., 1977; Zhou et al., 1999a). The phospholipids of oats are dominated by lecithin or phospatidylcholine (PC; 1,2-diacyl-*sn*-glycero-3-phospho-1′-choline), which accounts for about 50% of the total (Youngs et al, 1977) followed by phophatidylethanolamine (PE; 1,2-diacyl-*sn*-glycero-3-phospho-1′-ethanolamine) and phosphatidylglycerol (PG; 1,2-diacyl-*sn*-glycero-3-phospho-1′-*sn*-glycerol) (Sahasrabudhe, 1979). Numerous other minor phospholipids were isolated from oats including *N*-acylated glycerophospholipids such as *N*-acylphosphatidylethanolamine (*N*-acyl-PE; 1,2-diacyl-*sn*-glycero-3-phospho-(*N*-acyl)-1′-ethanolamine) and *N*-acyl-phosphatidylglycerol (*N*-acyl-PG; 1,2-diacyl-*sn*-glycero-3-phospho-(3′-acyl)-1′-*sn*-glycerol) (Holmback et al., 2001). Recent reports indicate the importance of *N*-acyl PE as precursors for *N*-acylethanolamines, which in turn play a physiological role during germination of seeds (Chapman et al., 1999) and in defense systems in plants (Tripathy et al., 1999). The major *N*-acylated fatty acids in these phospholipids are 16:0, 18:2, and 18:1 (Holmback et al., 2001).

As mentioned above, the major glycolipid in oat lipids is DGDG (Andersson et al., 1997; Aro et al., 2007; Hauksson et al., 1995). About 65% of avenoleic acid is bound to a galactolipid having the structure 1-[(9′*Z*,12′*Z*)-octadecadienoyl]-2-[15′′*R*-{(9′′′*Z*,12′′′Z)-octadecadienoyloxy}-(9′′*Z*,12′′*Z*)-octadecadienoyl]-3-(α-D-galactopyranosyl-1-6-β-D-galactopyranosyl)-glycerol, which is present in oat kernels at ca. 500 ppm (Hamberg et al., 1998). Whereas Hamberg et al. (1998) reported the occurrence of a DGDG with a third fatty acid (a mono-estolide), Moreau et al. (2008) presented evidence that oat kernels also contain DGDG di-estolides, DGDG tri-estolides, TriGDG, TetraGDG, and several TriGDG and TetraGDG estolides.

## Minor Lipid Components

The nonacyl lipid components of oat oil, also known as the unsaponifiable fraction, contain sterols, hydrocarbons, tocopherols/tocotrienols, phenolic compounds, saponins, carotenoids/chlorophylls, and other pigments, etcetera. The total level of these in the oil is indeed dependent on the oil extraction method. Oils from four

Turkish oat varieties, extracted with diethyl ether, were found to contain about 4% unsaponifiable materials (Ozcan et al. 2006).

Like other cereals, the oat grain is rich in sterols that are present in free, esterified, glycosylated, and acylglycosylated forms. The free and esterified sterols, but not the glycosylated sterols, are usually co-extracted with oil. The range of levels of sterols in the oils from five Swedish cultivars was 0.15–0.35%, the major sterols being β-sitosterol (54%), $\Delta^5$-avenasterol (26%), campesterol (8%), $\Delta^7$-avenasterol (8%), and stigmasterol (4%) (Määttä et al., 1999). The $\Delta^5$- and $\Delta^7$-avenasterols are subject to acid-catalyzed isomerization (e.g., in acid-clay bleaching during the refining of oat oil) (Kamal-Eldin et al., 1998). This concentration of sterols is intermediate between that of low-sterol oils (<0.1%), such as coconut and avocado oils, and oils with a relatively high content up to 0.4%, such as canola and corn oils (Phillips et al., 2002). Plant sterols produce a wide spectrum of biological activities in animals and humans. Comprehensive reviews are available on their diversity, analysis, health-promoting uses (Moreau et al., 2002), metabolism, and potential therapeutic action (Ling & Jones, 1995).

The total vitamin E (tocopherol/tocotrienol) content range is approximately 20–40 ppm (Emmons & Peterson, 1999; Handelman et al., 1999; Peterson, 1995; Peterson & Qureshi, 1993). α-Tocotrienol is generally the major E-vitamer (ca. 65%) followed by α-tocopherol (20–25%) and small amounts of β-tocopherol and β-tocotrienol (10–15%) (Bryngelsson et al., 2002). Analysis of hand-dissected oat kernels showed that α- and γ-tocopherols are concentrated in the germ while the tocotrienols are concentrated in the endosperm and absent from the hull (Peterson, 1995).

No comprehensive or specific study has investigated the rest of the components in the unsaponifiable fraction in oat lipids. However, a large variety of phenolic compounds were isolated and characterized in oat grains. These occur as free compounds, or as soluble conjugated and insoluble bound forms (Ryan et al., In press). Although the distribution between free, soluble, and bound forms varies widely between different cereals, phenolic compounds are rarely found in the free form in cereals; the majority are bound *via* covalent link with cell-wall polysaccharides. In whole oat grains, free phenols accounted for 25% of the total, while the remaining 75% were in bound form (Adom & Liu, 2002). The distribution of ferulic acid between free, soluble, conjugated, and bound forms in oat grains was 0.4%, 1.8%, and 97.8%, respectively (Adom & Liu, 2002; May et al., 2005). During the 1960s, a number of phenolic acid esters were identified as monoesters and α,ω-diols of C26 and C28 alkanols and glycerol and as monoesters of 26- and 28- hydroxyhexacosanoic acid (Daniels et al., 1963; Daniels & Martin, 1961, 1964a, 1964b, 1965a, 1965b, 1968). After alkaline hydrolysis, these are expected to be present in the unsaponifiable fraction as hydroxycinnamic acids, mainly caffeic and ferulic acids. Another group of special antioxidants in oat comprises the avenanthramides (Collins, 1989; Dimberg et al., 1993), avenalumic acid (Collins et al., 1991), and β-truxinic acid (Dimberg et al., 2001). Avenanthramides are found exclusively in oats (*Avena sativa* L.). Avenanthramides-2p (*N*-(4′-hydroxycinnamoyl)-5-hydroxyanthranilic acid), -2c (*N*-(3′,4′-dihydroxycinnamoyl)-5-hydroxyanthranilic acid), and -2f (*N*-(4′-hydroxy-3′-methoxycinnamoyl)-5-hydroxyanthranilic acid) are most commonly investigated since they consistently appear in higher concentrations in extracts (Jastrebova et al., 2006; Peterson et al., 2002; Thomas et al., 2006). In fact, avenanthramide-2c constitutes about one-third of the total avenanthramide content in oat grain (Nie et al., 2004). Avenanthramides constitute by far the major unbound phenols present in oat grains. Nevertheless, total concentrations of avenanthramides in oats are small, 2–289 mg/kg (Ryan et al., 2007). Concentrations of different phenolic antioxidants (Fig. 16.3) in oat oils were not investigated but will depend on the extraction method. No comprehensive study has investigated these other types of components.

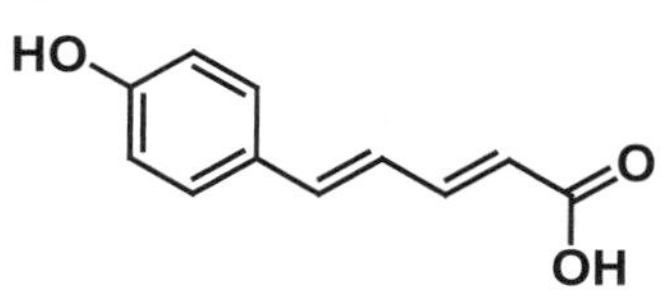

Avenalumic acid

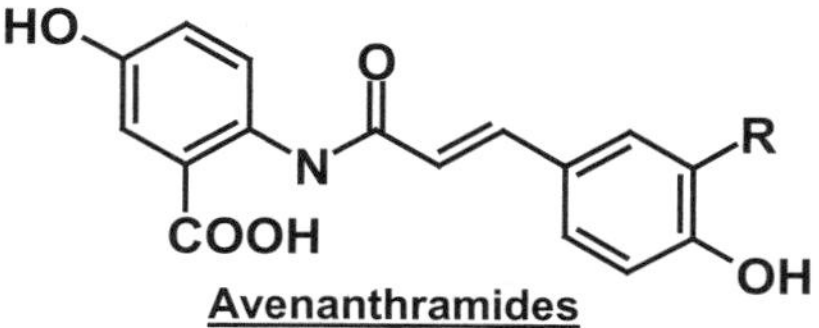

Avenanthramides

R = H (Bp)

R = OH (Bc)

R = $OCH_3$ (Bf)

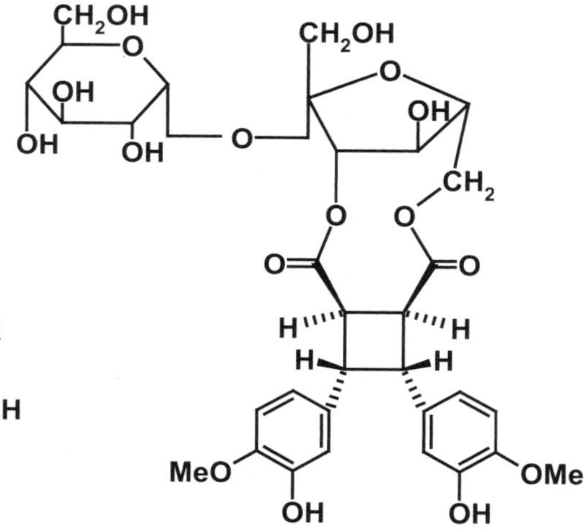

β-Truxinic acid

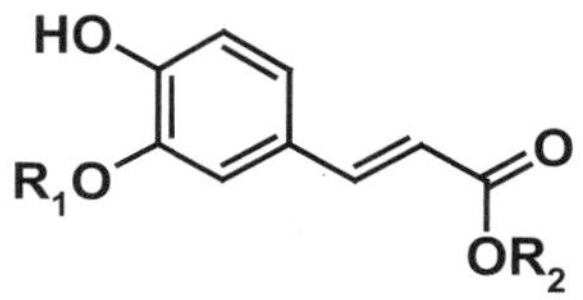

Phenolic acids and their esters

$R_1$ = H (Caffeic acid), or $R_1$ = $CH_3$ (Ferulic acid)

$R_2$ = H (Free Phenolic acids)

$R_2$ = long-chain monoalcohols, diols, or ω-hydroxy acids (Phenolic acid esters)

**Fig. 16.3.** Chemical structures of some unique antioxidant compounds in oat grains (see text for details).

## Flavor and Aroma Compounds

As recently as 2004, the production of oat oil was a minor part of the commercial activity of oats: "Currently, oat oil is not processed as a food oil, primarily for economic reasons, although it might become feasible if high-oil cultivars with good agronomic characteristics are developed" (Peterson, 2004). Consequently, little is published regarding the flavor and aroma of oat oil.

Fors and Schlich have made the most comprehensive analysis of oat oil aroma compounds. (Fors & Schlich, 1989). Although the overall goal of this work was related to oat cereal flavor, it is one of very few studies where sensory analysis has been performed on the oil itself. Moreover, the oil of "crude," that is, nonroasted, oats was examined in the study, which has more relevance to commercial oat oil, which is derived from uncooked oats. Fors' and Schlich's report is not cited for studies of oat oil (ISI Database)—an indication that little in the way of sensory analysis was performed on the aroma and flavor of oat oil.

In the Fors and Schlich (1989) study, eight oils were prepared from four treatments consisting of various combinations of heating/roasting and milling, across two varieties—Magne and Chihuahua—and subjected to gas chromatography–mass spectrometry (GC–MS). GC–olfactometry (GC–O) and analysis by a sensory panel was performed on Magne oils from the four treatments. More than 100 compounds were identified by GC–MS and of those, 75 were quantified (Table 16.2). PCA on the GC–MS data related various chemical classes to the different treatments. For example, heated oil from crude oats was characterized by aldehydes, whereas the unheated oil was characterized by alcohols, ketones, and hydrocarbons. On the other hand, oils that were prepared from roasted oats were characterized by nitrogen or oxygen containing heterocycles.

Interestingly, the sensory panel preferred the oils prepared from roasted oats, and similarities were noted with sesame oil and Swedish crispbread. GC-O analysis revealed notes such as roasted, peanut, butterscotch, sesame seeds, and creamy, caramel-like. Compounds identified and given these descriptors included 2-furfurylalcohol, 2-methylfurfural, 2,6-dimethylpyrazine, and acetylpyrazine. Compounds in the oil from crude oats were given descriptors such as herbaceous, green, pungent, plant-like, spicy, green-house, and moldy from GC–O. Saturated and unsaturated aldehydes were identified as giving these attributes. For example, the presence of hexanal was revealed in the odor of "newly cut grass."

## Allergic and Toxic Compounds

No reports exist in the scientific literature on allergic or toxic reactions to oat oil (ISI database). Searching the Web reveals a large number of sites claiming the low allergy potential of oat products.

**Table 16.2. Oat Oil Volatile Compounds and Associations with Processing (adapted from Fors & Schlich, 1989)**

| Class | Compounds identified | Association of compounds with oil[a] |
|---|---|---|
| Alcohols | butanol, pentanol, heptanol, octanol, l-octen-3-ol | CORC[b] ++ |
| | 3-penten-2-ol | MARC + |
| | hexanol | COUM + |
| Aldehydes | hexanal, (E,Z)-2,4-decadienal, heptanal, (E,E)-2,4-decadienal, 2-nonenal, 2-ethylhexenal, 2-hexenal, 2,4-heptadienal, 2-undecenal, 2,4-nonadienal, 2-Heptenal, nonanal, decenal, decanal, 2-octenal, octanal | CORC ++ |
| Ketones | 2-pentadecanone, dodecadione, cyclic ketone | CORC ++ |
| | 2-heptadecanone | COUM, CORM, MBRM + |
| | Trimethylpentadecanone | |
| N-Heterocycles | Pyridine<br>Pyrazines<br>unsubstituted, 2,3-dimethyl, 2,5-dimethyl-3-ethyl, methyl, 2-ethyl-6-methyl, 2,5-dimethyl, 2-ethyl-5-methyl, 2,6-dimethyl, acetylmethyl, 2-(2-furyl), ethyl, 2-methyl-5-(methylethyl), trimethyl<br>Pyrroles<br>2-acetyl, N-methyl-2-furyl, 1-formyl, l-furfuryl-2-formyl, 5-methyl-2-formyl | MARM ++ |
| | dimethylethylpyrazine<br>2-methyl, dihydrocyclopentapyrazine | MBRM ++ |
| | pentylpyridine | CORC ++ |
| O-Heterocycles | furfural 2-furanylethanone, furanmethanolacetate, hydroxymethylfurfural, 2-methyltetra (or dihydro) furanone, 5-methyl-2-furfural, 2-furanmethanol | MARM ++ |
| | pentylfuran, octylfuran | CORC ++ |
| Hydrocarbons | dodecane, 2-ethyldodecane, 8-methyldecene<br>heneicosane, hexadecene | COUM ++<br>CORM ++ |
| | dodecene, pentadecane, hexadecane, tetradecane | CORC + |
| | nonadecane | COUC + |
| | octadecene | CORM + |

[a] ++ = very good association; + = good association

[b] Processing and variety code: COU = crude oats, unheated; COR = crude oats, roasted; MBR = ground roasted oats, milled before roasting; MAR = whole roasted oats, milled after roasting; M = Magne variety 7.4% lipid; C = Chihuahua variety 8.3% lipid.

## Health Benefits of The Oil and Oil Constituents

For centuries, oats were valued for their medicinal qualities. Ground oat preparations were used on the skin for drying and healing as early as 400 B.C., while seventeenth-century New World immigrants used the grain to relieve stomach aches and other ailments (Anon, 2006 ). More recently, the health benefits of oats were ascribed mainly to the water-soluble mixed linkage (1,3)(1,4)-β-D-glucans (Colleoni-Sirghie et al., 2003), which are the predominant polysaccharide constituents of endosperm cell-walls constituting approximately 85% of the wall in oats (Miller et al., 1995). The health benefits of oats are also often attributed to the presence of various phytochemicals (Zadernowski et al., 1999), including tocols, plant sterols and stanols, and saponins rather than the bulk components. These lipophilic compounds function as antioxidants, and prospective population studies consistently suggest that when consumed in whole foods, antioxidants are associated with significant protection against cancer and cardiovascular disease. The broad range of antioxidant activities from these phytochemicals is probably a significant factor in providing many health benefits. The most abundant antioxidants in oats are Vitamin E (tocols), phytic acid, and phenolic compounds including avenanthramides but flavonoids and sterols are also present (Emmons et al., 1999; Emmons & Peterson, 1999; Peterson, 2001). Depending on the extraction solvent and protocol, these same compounds can be transferred to the oat oil following extraction and thus provide potential health benefits to the oil.

Like palm oil, which has been reported to contain high levels of tocotrienols, the oils of oats, barley, and certain other grains also contain levels of tocotrienols that may be sufficiently high to impart them with health-promoting benefits. Tocols are lipophilic and thus intimately associated with lipid components of the sample matrix. An intrinsic association between the tocols and oil bodies was suggested (White et al., 2006) in which the tocols provide oxidative stability to the membrane and/or oil of oat oil bodies. However, in a more recent trial, tocol and lipid concentrations were not correlated (Peterson et al., 2007). Sterols and tocols are present in oat oil extracted with hexane, but most of the other phenolic antioxidants are not extracted with hexane they are probably partially extracted when oat oil is obtained by extraction with ethanol or other more polar solvents. Tocotrienols were shown to have these abilities: to inhibit hydroxymethylglutaric acid-CoA enzyme activity leading to cholesterol-lowering effects (Parker et al., 1993; Qureshi et al., 1986, 1991, 1989; Raederstorff et al., 2002; Wang et al., 1993), to reduce endothelial expression of adhesion molecules and adhesion to monocytes (Theriault et al., 2002), to minimize atherosclerotic lesions in ApoE-deficient mice (Qureshi et al., 2001), and to inhibit the growth and proliferation of human breast cancer cells (Nesaretnam et al., 1998). Plant sterols are also known for cholesterol-lowering effects even though at much higher doses than may be provided by oat oil. The phenolic compounds in oat oil may contribute to the antioxidant and anti-inflammatory effects of the diet.

The group of Li and co-workers found that the addition of oat lipids to experimental diets induced the formation of three oat-specific I-compounds in liver DNA of female rats (Li et al., 1992; Li & Randerath, 1990; Randerath et al., 1999). I-(endogenous)-compounds represent bulky covalent DNA modifications that increase with age and are detected by $^{32}P$-postlabelling (Randerath et al., 1999). Type-1 I-compounds are modified by age, sex, and diet, where oat oil is a significant inducer, while carcinogens and tumor promoters are significant suppressors (Randerath et al., 1999). Since the connection of I-compounds to cancer is unknown, the pronounced effects of oat lipids may be protective or stimulative to cancer, which deserves elaborated investigations.

## Other Issues

An ISI Web of Science search for oat oil and adultera or authentic yields zero references. The same situation arises when using Biological Abstracts and Biosis previews databases. Such results show that adulteration of oat oil is essentially nonexistent. This is understandable in light of the very low global production of oat oil and the fact that oat oil is not generally consumed as a food product, but rather is used only in food preparations.

## References

Aburjai, T.; F.M. Natsheh. Plants used in cosmetics, *Phytotherapy Res.* **2003**, *17*, 987–1000.

Adom, K.K.; R.H. Liu. Antioxidant activity of grains, *J. Agric. Food Chem.* **2002**, *50*, 6182–6187.

Alkio, M., O. Aaltonen; R. Kervinen; P. Forssell; K. Poutanen. Manufacture of lecithin from oat oil by supercritical extraction. *Proceedings of the 2nd International Symposium on Supercritical Fluids,* M.A. McHugh, Ed.; Boston, MA, 1991 ; 276–278.

Andersson, M.B.O.; M. Demirbuker; L.G. Blomberg. Semi-continuous extraction/purification of lipids by means of supercritical fluids. *J. Chromatogr. A* **1997**, *785*, 337–343.

Anon, Oat Oil (*Avena sativa*), 1997–2008, Accessed on September 26, 2007, http://www.fromnaturewithlove.com/soap/product.asp?product_id=OatOil

Anon, North American Millers Association, 2006, Accessed on September 26, 2007, http://www.namamillers.org/

Anon, Selected Oat Facts, 2007, Accessed on September 26, 2007, http://www.gramene.org/species/avena/oat_facts.html

Anon, Oat Bran and Germ Oil, undated_a, Accessed on September 26, 2007, http://www.florahealth.com/flora/home/Canada/HealthInformation/Encyclopedias/OatBranAndGermOil.htm

Manunsell, C. Oat Services, undated_b, Accessed on September 26, 2007, http://www.oat.co.uk/sectors/oat-oil.asp

Anon. In *The Review of Natural Products,* Fourth ed.; A. DerMarderosian, J.A. Beutler, Eds.; Wolters Kluwer Health: St. Louis, Missouri, 2005; pp. 825–829.

Aro, H.; E. Jarvenpaa; K. Konko; R. Huopalahti; V. Hietaniemi. The characterization of oat lipids produced by supercritical fluid technologies. *J. Cereal Sci.* **2007**, *45,* 116–119.

Baker, R.J.; R.I.H. McKenzie. Heritability of oil content in oats, *Avena sativa* l.. *Crop Sci.* **1972**, *12,* 201–202.

Banas, A.; A. Dahlqvist; H. Debski; P.O. Gummeson; S. Stymne. Accumulation of storage products in oat during kernel development. *Biochem. Soc. Trans.* **2000**, *28,* 705–707.

Bedford, C.L.; M.A. Joslyn. Oat flour and hexane extract of oat flour as antioxidants for shelled walnuts and walnut oil. *Food Res.* **1937**, *2,* 455

Biermann, U.; A. Wittmann; W. Grosch. Occurrence of bitter hydroxy fatty acids in oat and wheat. *Fette Seifen Anstrichmittel* **1980**, *82,* 236–240.

Blom, M.; L. Andersson; A. Carlsson; B. Herslöf; Z. Li; Ĺ. Nilsson. Pharmacokinetics, tissue distribution and metabolism of intravenously administered digalactosyldiacylglycerol and monogalactosyldiacylglycerol in the rat. *J. Liposome Res.* **1996**, *6,* 737–753.

Boczewski, M.P. (Inventor) 1980. Process for the treatment of oats. U.S. Patent 4,220,287.

Branson, C.V.; K.J. Frey. Recurrent selection for groat oil content in oat. *Crop Sci.* **1989**, *29,* 1382–1387.

Brown, C.M.; A.N. Aryeetey; S.N. Dubey. Inheritance and combining ability for oil content in oats (*Avena sativa* L.). *Crop Sci.* **1974**, *14,* 67–69.

Brown, C.M.; J.-C. Craddock, Oil content and groat weight of entries in the world oat collection. *Crop Sci.* **1972**, *12,* 514–515.

Chapman, K.D.; B. Venables; R. Markovic; R.W. Blair; C. Bettinger. n-Acylethanolamines in seeds. Quantification of molecular species and their degradation upon imbibitions. *Plant Physiol.* **1999**, *120,* 1157–1164.

Colleoni-Sirghie, M.; D.B. Fulton; P.J. White. Structural features of water soluble (1,3) (1,4)-beta-D<SMALL CAP>-glucans from high-beta-glucan and traditional oat lines. *Carbohydr. Polym.* **2003**, *54,* 237249.

Collins, F.W. Oat Phenolics: Structure, Occurrence, and Function. *Oats Chemistry and Technology;* F.H. Webster, Ed.; American Association of Cereal Chemists: St. Paul, MN, 1986; pp. 227–295.

Collins, F.W., Oat phenolics—avenanthramides, novel substituted n-cinnamoylanthranilate alkaloids from oat groats and hulls. *J. Agric. Food Chem.* **1989**, *37,* 60–66.

Collins, F.W.; D.C. Mclachlan; B.A. Blackwell. Oat phenolics— avenalumic acids, a new group of bound phenolic-acids from oat groats and hulls. *Cereal Chem.* **1991**, *68,* 184–189.

Daniels, D.G.H.; H.G.C. King; H.F. Martin. Antioxidants in oats: esters of phenolic acids, *J. Sci. Food. Agric.* **1963**, *14,* 385–390.

Daniels, D.G.H.; H.F. Martin. Isolation of a new antioxidant from oats. *Nature* **1961**, *191,* 1302.

Daniels, D.G.H.; H.F. Martin. Antioxidants in oats: light-induced isomerization. *Nature* **1964a**, *203,* 299

Daniels, D.G.H.; H.F. Martin. Structures of two antioxidants isolated from oats. *Chem. Ind.* **1964b**, *50,* 2058.

Daniels, D.G.H.; H.F. Martin. Antioxidants in oats: diferulates of long chain diols. *Chemistry &*

*Industry* **1965a**, *42,* 1763.

Daniels, D.G.H.; H.F. Martin. Antioxidants in oats: monoesters of caffeic and ferulic acids. *J. Sci. Food. Agric.* **1965b**, *18.*

Daniels, D.G.H.; H.F. Martin. Antioxidants in oats: glyceryl esters of caffeic and ferulic acids. *J. Sci. Food. Agric.* **1968**, *19,* 710–712.

De la Roche, I.A.; V.O. Burrows; R.I.H. McKenzie. Variation in lipid composition among strains of oats. *Crop Sci.* **1977**, *17,* 145–148.

Dimberg, L.H.; R.E. Andersson; S. Gohil; S. Bryngelsson; L.N. Lundgren. Identification of a sucrose diester of a substituted beta-truxinic acid in oats. *Phytochemistry* **2001**, *56,* 843–847.

Dimberg, L.H.; O. Theander; H. Lingnert. Avenanthramides—a group of phenolic antioxidants in oats. *Cereal Chem.* **1993**, *70,* 637–641.

Dull, B.J. Oat oil for personal care products. *Cosmetics & Toiletries* **1997**, *112,* 77–81.

Duve, K.J.; P.J. White. Extraction and identification of antioxidants in oats. *J. Am. Oil Chem. Soc.* **1991**, *68,* 365–370.

Ekstrand, B.; I. Gangby; G. Akesson. Lipase activity in oats— distribution, pH-dependence, and heat inactivation. *Cereal Chem.* **1992**, *69,* 379–381.

Ekstrand, B.; I. Gangby; G. Akesson; U. Stollman; H. Lingnert; S. Dahl. Lipase activity and development of rancidity in oats and oat products related to heat-treatment during processing. *J. Cereal Sci.* **1993**, *17,* 247–254.

Emmons, C.L.; D.M. Peterson. Antioxidant activity and phenolic contents of oat groats and hulls. *Cereal Chem.* **1999**, *76,* 902–906.

Emmons, C.L.; D.M. Peterson; G.L. Paul. Antioxidant capacity of oat (*Avena sativa* L.) extracts. 2. *In vitro* antioxidant activity and contents of phenolic and tocol antioxidants. *J. Agric. Food Chem.* **1999**, *47,* 4894–4898.

Erazo-Castrejon, S.V.; D.C. Doehlert; B.L. D'appolonia. Application of oat oil in breadbaking. *Cereal Chem.* **2001**, *78,* 243–248.

Fors, S.M.; C.E. Eriksson. Characterization of oils extracted from oats by supercritical carbon-dioxide. *Food Sci. Technol.-Lebensm. Wiss. Technol.* **1990**, *23,* 390–395.

Fors, S.M.; P. Schlich. Flavor composition of oil obtained from crude and roasted oats, *ACS Symposium Series 409,* American Chemical Society, Washington, D.C., 1989, pp. 121–131.

Forssell, P.; M. Cetin; G. Wirtanen; Y. Malkki. Antioxidative effects of oat oil and its fractions. *Fett Wissenschaft Technol. -Fat Sci. Technol.* **1990**, *92,* 319–321.

Forssell, P.; R. Kervinen; M. Alkio; K. Poutanen. Comparison of methods for separating polar lipids from oat oil. *Fett Wissenschaft Technol. -Fat Sci. Technol.* **1992**, *94,* 355–358.

Frey, K.J.; E.G. Hammond. Genetics, characteristics, and utilization of oil in caryopses of oat species. *J. Am. Oil Chem. Soc.* **1975**, *52,* 358–362.

Frey, K.J.; E.G. Hammond; P.K. Lawrence. Inheritance of oil percentage in interspecific crosses of hexaploid oats. *Crop Sci.* **1975**, *15.*

Giannopolitis, C.N.; S.K. Ries. Superoxide dismutases: I. Occurrence in higher plants. *Plant Physiol.* **1977**, *59,* 309–314.

Gibinski, M.; M. Palasinski; P. Tomasik. Physicochemical properties of defatted oat starch. *Starch-Starke* **1993**, *45*, 354–357.

Gudmundsson, M.; A.C. Eliasson. Some physicochemical properties of oat starches extracted from varieties with different oil content. *Acta Agric. Scand.* **1989**, *39*, 101–111.

Hamberg, M.; G. Hamberg. 15(R)-Hydroxylinoleic acid, an oxylipin from oat seeds. *Phytochemistry* **1996**, *42*, 729–732.

Hamberg, M.; E. Liepinsh; G. Otting; W. Griffiths. Isolation and structure of a new galactolipid from oat seeds. *Lipids* **1998**, *33*, 355–363.

Handelman, G.J.; G. Cao; M.F. Walter; Z.D. Nightingale; G.L. Paul; R.L. Prior; J.B. Blumberg. Antioxidant capacity of oat (*Avena sativa* L.) extracts. 1. Inhibition of low-density lipoprotein oxidation and oxygen radical absorbance capacity. *J. Agric. Food Chem.* **1999**, *47*, 4888–4893.

Hardin, B. Oat Oil Gives Bread a Soft Touch, 2000, Accessed on September 26, 2007, http://www.ars.usda.gov/is/pr/2000/000929.htm

Hauksson, J.B.; M.H.J. Bergqvist; L. Rilfors. Structure of digalactosyldiacylglycerol from oats. *Chem. Phys. Lipids* **1995**, *78*, 97–102.

Hogberg, L.; P. Laurin; K. Falth-Magnusson; C. Grant; E. Grodzinsky; G. Jansson; H. Ascher; L. Browaldh; J.A. Hammersjo; E. Lindberg; U. Myrdal; L. Stenhammar. Oats to children with newly diagnosed coeliac disease: a randomised double blind study. *Gut* **2004**, *53*, 649–654.

Holmback, J.; A.A. Karlsson; K.C. Arnoldsson. Characterization of n-acylphosphatidylethanolamine and acylphosphatidylglycerol in oats. *Lipids* **2001**, *36*, 153–165.

Hutchinson, J.B.; H.F. Martin. The chemical composition of oats. 1. The oil and free fatty acid contents of oats and groats. *J. Agric. Sci.* **1955**, *45*, 411–418.

Hutchinson, J.; H. Martin; T. Moran. Location and destruction of lipase in oats. *Nature* **1951**, *167*, 758759.

Jastrebova, J.; M. Skoglund; J. Nilsson; L.H. Dimberg. Selective and sensitive LC-MS determination of avenanthramides in oats. *Chromatographia* **2006**, *63*, 419–423.

Kamal-Eldin, A.; K. Määttä; J. Toivo; A.M. Lampi; V. Piironen. Acid-catalyzed isomerization of fucosterol and delta(5)-avenasterol. *Lipids* **1998**, *33*, 1073–1077.

Karow, R.S. Oil composition in parental, FI, and F2 populations of two oat (*Avena sativa* L.) crosses. University of Wisconsin; 1980 .

Kurtz, E.S.; W. Wallo. Colloidal oatmeal: history, chemistry and clinical properties, *J. Drugs Dermatol.* **2007**, *6*, 167–170.

Lehtinen, P.; K. Kiiliainen; K. Lehtomaki; S. Laakso. Effect of heat treatment on lipid stability in processed oats. *J. Cereal Sci.* **2003**, *37*, 215–221.

Lehtinen, P.; S. Laakso. Role of lipid reactions in quality of oat products. *Agric. Food Sci.* **2004**, *13*, 88–99.

Li, D.H.; S. Chen; E. Randerath; K. Randerath. Oat lipids-induced covalent DNA modifications (i-compounds) in female Sprague-Dawley rats, as determined by p-32 postlabeling. *Chem.-Biol. Interact.* **1992**, *84*, 229–242.

Li, D.H.; K. Randerath. Association between diet and age-related DNA modifications (i-compounds) in rat-liver and kidney. *Cancer Res.* **1990**, *50*, 3991–3996.

Ling, W.H.; P.J.H. Jones. Dietary phytosterols—a review of metabolism, benefits and side-effects. *Life Sci.* **1995**, *57*, 195–206.

Liukkonen, K.; A. Kaukovirtanorja; S. Laakso. Elimination of lipid hydrolysis in aqueous suspensions of oat flour. *J. Cereal Sci.* **1993**, *17*, 255–265.

Määttä, K.; A.M. Lampi; J. Petterson; B.M. Fogelfors; V. Piironen; A. Kamal-Eldin. Phytosterol content in seven oat cultivars grown at three locations in Sweden. *J. Sci. Food. Agric.* **1999**, *79*, 1021–1027.

Martin, W. Inventor 1964, Producing edible oil from grain. U.S. Patent 3,163,545.

Matz, S.A. *The Chemistry and Technology of Cereals as Food and Feed*, Second edition; AVI: New York, 1991.

May, W.E.; R.M. Mohr; G.R. Lafond; F.C. Stevenson. Oat quality and yield as affected by kernel moisture at swathing. *Canadian J. Plant Sci.* **2005**, *85*, 839–846.

Miller, S.S.; R.G. Fulcher; A. Sen; J.T. Arnason. Oat endosperm cell-walls .1. Isolation, composition, and comparison with other tissues. *Cereal Chem.* **1995**, *72*, 421–427.

Moreau, R.A.; D.C. Doehlert; R. Welti; G. Isaac; M. Roth; P. Tamura; R. Nuñez. The identification of mono-, di-, tri-, and tetragalactosyl-diacylglycerols and their natural estolides in oat kernels. *Lipids* **2008**, In press.

Moreau, R.A.; M.J. Powell; V. Singh. Pressurized liquid extraction of polar and nonpolar lipids in corn and oats with hexane,. methylene chloride, isopropanol, and ethanol. *J. Am. Oil Chem. Soc.* **2003**, *80*, 1063–1067.

Moreau, R.A.; B.D. Whitaker; K.B. Hicks. Phytosterols, phytostanols, and their conjugates in foods: structural diversity, quantitative analysis, and health-promoting uses. *Progr. Lipid Res.* **2002**, *41*, 457–500.

Morrison, W.R. Starch lipids: a reappraisal. *Starch-Staerke* **1981**, *33*, 408–410.

Morrison, W.R. Lipids in cereal starches— a review. *J. Cereal Sci.* **1988**, *8*, 1.15.

Nesaretnam, K.; R. Stephen; R. Dils; P. Darbre. Tocotrienols inhibit the growth of human breast cancer cells irrespective of estrogen receptor status. *Lipids* **1998**, *33*, 461–469.

Nie, L.; M.L. Wise; D.P. Peterson; M. Meydani. Avenanthramides, polyphenols from oats, modulate smooth muscle cell (SMC) proliferation and nitric oxide (NO) production. *Free Radic. Biol. Med.* **2004**, *37*, S40.

O'Connor, J.; H.J. Perry; J.L. Harwood. A comparison of lipase activity in various cereal-grains. *J. Cereal Sci.* **1992**, *16*, 153–163.

Ozcan, M.M.; G. Ozkan; A. Topal. Characteristics of grains and oils of four different oats (*Avena sativa* L.) cultivars growing in Turkey. *Int. J. Food Sci. Nutr.* **2006**, *57*, 345–352.

Parker, R.A.; B.C. Pearce; R.W. Clark; D.A. Gordon; J.J.K. Wright. Tocotrienols regulate cholesterol production in mammalian-cells by posttranscriptional suppression of 3-hydroxy-3-methylglutaryl-coenzyme-A reductase. *J. Biol. Chem.* **1993**, *268*, 11230–11238.

Parmar, S.; E.G. Hammond. Hydrolysis of fats and oils with moist oat caryopses. *J. Am. Oil Chem. Soc.* **1994**, *71*, 881–886.

Peters, F.N.; S. Musher. Oat flour as an antioxidant. *Industrial Engineering Chem.* **1937**, *29*, 146–151.

Peterson, D.M. Oat tocols—concentration and stability in oat products and distribution within the kernel. *Cereal Chem.* **1995**, *72*, 21–24.

Peterson, D.M. Oat antioxidants. *J. Cereal Sci.* **2001**, *33*, 115–129.

Peterson, D.M. Oat—a multifunctional grain. 7th International Oat Conference; Helsinki, Finland; 2004, P. Peltonen-Sainio, M. Topi-Hulmi, Ed.; pp. 21–26 .

Peterson, D.M.; M.J. Hahn; C.L. Emmons. Oat avenanthramides exhibit antioxidant activities *in vitro*. *Food Chem.* **2002**, *79*, 473–478.

Peterson, D.M.; C.M. Jensen; D.L. Hoffman; B. Mannerstedt-Fogelfors. Oat tocols: saponification vs. direct extraction and analysis in high-oil genotypes. *Cereal Chem.* **2007**, *84*, 56–60.

Peterson, D.M.; A.A. Qureshi. Genotype and environment effects on tocols of barley and oats. *Cereal Chem.* **1993**, *70*, 157–162.

Peterson, D.M.; D.F. Wood. Composition and structure of high-oil oat. *J. Cereal Sci.* **1997**, *26*, 121–128.

Phillips, K.M.; D.M. Ruggio; J.I. Toivo; M.A. Swank; A.H. Simpkins. Free and esterified sterol composition of edible oils and fats. *J. Food Comp. Anal.* **2002**, *15*, 123–142.

Piazza, G.J. Generation of polyunsaturated fatty-acids from vegetable-oils using the lipase from ground oat (*Avena-sativa* L.) seeds as a catalyst, *Biotechnol. Lett.* **1991**, *13*, 173–178.

Piazza, G.; A. Bilyk; D. Schwartz; M. Haas. Lipolysis of olive oil and tallow in an emulsifier-free 2-phase system by the lipase from oat seeds (*Avena-sativa* L). *Biotechnol. Lett.* **1989**, *11*, 487–492.

Piazza, G.J.; H.M. Farrell. Generation of ricinoleic acid from castor-oil using the lipase from ground oat (*Avena-sativa* L.) seeds as a catalyst.*Biotechnol. Lett.* **1991**, *13*, 179–184.

Potter, R.C.; J.M. Castro: L.C. Moffatt (Inventors) 1997. Oat oil compositions with useful cosmetic and dermatological properties. U.S. Patent 5620692.

Price, P.B.; J.G. Parsons. Lipids of 7 cereal grains. *J. Am. Oil Chem. Soc.* **1975**, *52*, 490–493.

Qureshi, A.A.; W.C. Burger; D.M. Peterson; C.E. Elson. The structure of an inhibitor of cholesterol biosynthesis isolated from barley. *J. Biol. Chem.* **1986**, *261*, 10544–10550.

Qureshi, A.A.; V. Chaudhary; F.E. Weber; E. Chicoye; N. Qureshi. Effects of brewers grain and other cereals on lipid-metabolism in chickens. *Nutr. Res.* **1991**, *11*, 159–168.

Qureshi, A.A.; D.M. Peterson; C.E. Elson; A.R. Mangels; Z.Z. Din. Stimulation of avian cholesterol-metabolism by alpha-tocopherol. *Nutr. Reports Int.* **1989**, *40*, 993–1001.

Qureshi, A.A.; W.A. Salser; R. Parmar; E.E. Emeson. Novel tocotrienols of rice bran inhibit atherosclerotic lesions in c57bl/6 apoe-deficient mice. *J. Nutr.* **2001**, *131*, 26062618.

Raederstorff, D.; V. Elste; C. Aebischer; P. Weber. Effect of either gamma-tocotrienol or a tocotrienol mixture on the plasma lipid profile in hamsters. *Annals Nutr. Metabolism* **2002**, *46*, 17–23.

Randerath, K.; E. Randerath; G.D. Zhou; D.H. Li. Bulky endogenous DNA modifications (I-compounds)—possible structural origins and functional implications. *Mutation Res—Funda-*

*mental and Molecular Mechanisms of Mutagenesis* **1999**, *424*, 183–194.

Ryan, D.; M. Kendall; K. Robards. Bioactivity of oats as it relates to cardiovascular disease. *Nutr. Res. Rev.* **2007**, *20*, 147–162.

Sahasrabudhe, M.R. Lipid composition of oats (*Avena sativa* L.). *J. Am. Oil Chem. Soc.* **1979**, *56*, 80–84.

Sahasrabudhe, M.R. Measurement of lipase activity in single grains of oat (*Avena sativa* L.). *J. Am. Oil Chem. Soc.* **1982**, *59*, 354–355.

Schipper, H.; K.J. Frey. Observed gains from 3 recurrent selection regimes for increased groat-oil content of oat. *Crop Sci.* **1991**, *31*, 1505–1510.

Schipper, H.; K.J. Frey. Agronomic and bioenergetic consequences of selection for high groat-oil content and high protein yield in oat. *Plant Breeding* **1992**, *108*, 241–249.

Theriault, A.; J.T. Chao; A. Gapor. Tocotrienol is the most effective vitamin E for reducing endothelial expression of adhesion molecules and adhesion to monocytes. *Atherosclerosis* **2002**, *160*, 21–30.

Thomas, S.L.; P. Bonello; P.E. Lipps; M.J. Boehm. Avenacin production in creeping bentgrass (Agrostis stolonifera) and its influence on the host range of gaeumannomyces graminis. *Plant Disease* **2006**, *90*, 33–38.

Thro, A.M.; K.J. Frey. Inheritance of groat-oil content and high oil selection in oats (*Avena sativa* L..), *Euphytica* **1985**, *34*, 251–163.

Thro, A.M.; K.J. Frey; E.G. Hammond. Inheritance of palmitic, oleic, linoleic and linolenic fatty acids in groat oil of oats. *Crop Sci.* **1985**, *25*, 40–44.

Tripathy, S.; B.J. Venables; K.D. Chapman. n-Acylethanolamines in signal transduction of elicitor perception. attenuation of alkalinization response and activation of defense gene expression. *Plant Physiol.* **1999**, *121*, 1299–1308.

Tzen, J.T.C.; Y.Z. Cao; P. Laurent; C. Ratnayake; A.H.C. Huang. Lipids, proteins, and structure of seed oil bodies from diverse species. *Plant Physiol.* **1993**, *101*, 267276.

Urquhart, A.A.; I. Althosaar; G.J. Matlashewski; M.R. Sahasrabudhe. Localization of lipase activity in oat grains and milled oat fractions. *Cereal Chem.* **1983**, *60*, 181–183.

Wang, L.J.; R.K. Newman; C.W. Newman; L.L. Jackson; P.J. Hofer. Tocotrienol and fatty-acid composition of barley oil and their effects on lipid-metabolism. *Plant Foods Human Nutr.* **1993**, *43*, 9–17.

Warwick, M.; W. Farrington; G. Shearer. Changes in total fatty acids and individual lipid classes on prolonged storage of wheat flour. *J. Sci. Food. Agric.* **1979**, *30*, 1131–1138.

Washburn, E.L. (Inventor) 1953, Fractionation of oat oil and use of the same. U.S. Patent 2,636,888.

Welch, R.W. *The Oat Crop.* Chapman and Hall: Great Britain, 1995.

White, D.A.; I.D. Fisk; D.A. Gray. Characterisation of oat (*Avena sativa* L..) oil bodies and intrinsically associated E-vitamers. *J. Cereal Sci.* **2006**, *43*, 244–249.

Youngs, V.L. Oat lipids and lipid-related enzymes. *Oats, Chemistry and Technology;* F.H. Webster, Ed.; American Association of Cereal Chemists: St. Paul, Minnesota, 1986; pp. 205–226.

Youngs, V.L.; M. Puskulcu. Variation in fatty acid composition of oat groats from different cultivars. *Crop Sci.* **1976**, *16,* 881–883.

Youngs, V.L.; M. Puskulcu; R.R. Smith. Oat lipids. 1. Composition and distribution of lipid components in two oat cultivars. *Cereal Chem.* **1977**, *54,* 803–812.

Zadernowski, R.; H. Nowak-Polakowska; A.A. Rashed. The influence of heat treatment on the activity of lipo- and hydrophilic components of oat grain. *J. Food Processing Preservation* **1999**, *23,* 177–191.

Zhou, M.X.; K. Robards; M. Glennie-Holmes; S. Helliwell. Oat lipids. *J. Am. Oil Chem. Soc.* **1999a**, *76,* 159–169.

Zhou, M.X.; K. Robards; M. Glennie-Holmes; S. Helliwell. Effects of oat lipids on groat meal pasting properties. *J. Sci. Food. Agric.* **1999b**, *79,* 585–592.

# 17

# Barley Oil

**Robert A. Moreau**
*Eastern Regional Research Center, Agricultural Research Service, U.S. Department of Agriculture, 600 East Mermaid Lane, Wyndmoor, PA 19038*

†Mention of trade names or commercial products in this publication is solely for the purpose of providing specific information and does not imply recommendation or endorsement by the U.S. Department of Agriculture.

## Introduction

Barley is an ancient domesticated grain. The earliest archaeological evidence for barley cultivation is in the Fertile Crescent region of the Middle East in approximately 8000 BC (Baik & Ullrich, 2008; Newman & Newman, 2006). Barley is fifth among all crops in world production (annual 129-million-metric-ton averages in 2002–2005, behind corn, wheat, rice, and soybeans) (Baik & Ullrich, 2008). Currently, about two-thirds of the world's barley harvest is used for animal feed, one-third for malting, and only about 2% for human food (Baik & Ullrich, 2008). Although most barley cultivars are traditional "hulled" varieties, researchers have developed numerous "hulless" (also called hull-less or naked) cultivars. Like hulled barleys, these hulless varieties have hulls in the field, but the hulls fall off during growth and harvest, resulting in a grain that has lower fiber, and does not need to be pearled before being used in food applications (Ames et al., 2006).

Because the oil content of most hulled and hulless barley cultivars is low (<2%), obtaining oil from whole barley grain is not easy or economical (Moreau et al., 2007a, 2007b). However, when barley is milled by various methods, some of its milling fractions are enriched in oil (up to about 10%), and these fractions are used to produce "barley oil" (Lampi et al., 2004). A recent review (Dunford, 2006) grouped barley oil with other "germ oils" (including corn, wheat, and oats). However, the goals of this chapter are to emphasize the unique composition of barley oil and to describe how it is different from other cereal germ oils. Separate chapters on rice bran oil, wheat germ oil, and oat oil are also included in this book.

Animal-feeding studies were conducted at Montana State University using unrefined barley oil obtained by extracting ground Prowashonupana hulless barley (Wang et al., 1993a) and Azhul hulless barley (Wang et al., 1997a). In the 1990s Miller Brewing Company produced an experimental barley oil that they called "barley-oil extract" and they distributed small amounts to several research groups for evaluation (Babu et al., 1992, Lupton et al., 1994). This barley oil contained very high levels of tocopherols and tocotrienols (up to 0.4%) (Babu et al., 1992). The barley-oil extract was obtained by the hexane extraction of brewer's spent grains, followed by degumming, conventional alkali refining, deodorization, and bleaching (Lupton et al., 1994). The physical and health-promoting properties of barley-oil extract are described in detail in later sections of this chapter.

More recently, my laboratory produced a different form of barley oil which we obtained by extracting oil from the "pearling fines" or "scarification fines," which are produced by abrading barley to remove the outermost layers of the kernel (including pericarp, aleurone, and germ). We characterized some of the physical and chemical properties of this oil (Lampi et al., 2004; Moreau et al., 2007a, 2007b), but its health-promoting properties were not investigated.

Because most barley cultivars contain high levels of β-glucans (similar to oats), most published experimental results on the health-promoting properties of barley focus on the cholesterol-lowering properties of the β-glucans in whole barley and in fractions produced from whole barley. Indeed, in 2006, the U.S. Food and Drug Administration (FDA) announced that manufacturers are allowed to claim that whole-grain barley and barley-containing products reduce the risk of coronary heart disease (CHD). To qualify for the health claim, the barley-containing foods must provide at least 0.75 grams of soluble fiber per serving of the food (FDA, 2006). A cultivar of barley, "Prowashonupana," that contains very high levels of β-glucans, was developed at Montana State University. Its trade name is "Sustagrain®," and it is marketed by ConAgra (ConAgra, 2008). Recently, Cargill launched a commercial barley β-glucan food-ingredient product "Barliv™" (Cargill, 2008). In addition to the health-promoting properties of the β-glucans in barley, this chapter focuses on the evidence that barley lipids also have important health-promoting properties.

## Processing and Properties of Resulting Products

### Milling/Fractionation of Barley Kernels and Flour

Traditional cultivars of barley are "hulled." For most foods that contain barley, producers used "pearled" barley, because removing the hulls improves the taste of the barley. Pearling is an abrasive technology that removes the hull (Fig. 17.1). In addition to removing the hull, pearling or scarification also removes the outer layers of the kernel, including the pericarp, aleurone, and germ (embryo) (Fig. 17.2) (Baik & Ullrich, 2008; Jadhav et al., 1998). In addition to pearling, roller-milling was also investigated as an industrial process to produce barley bran from hulled and hulless varieties of barley (Bhatty, 1993). We recently reported the results of a study where

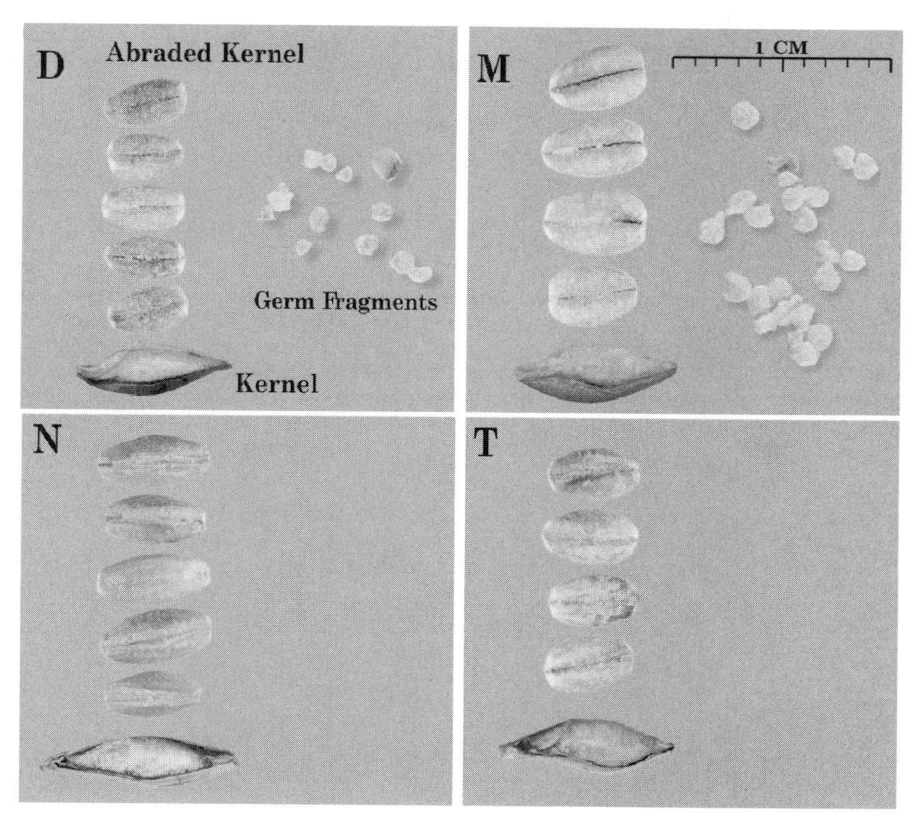


**Fig. 17.1.** Photographs of abraded barley kernels and germ fragments obtained in the largest (U.S. #25) sieve fraction, both after 60 s of scarification. Also included are cross sections of the corresponding intact kernels photographed at the same scale, for comparison. Abbreviations: D, Doyce; M (hull-less), M, Merlin (hull-less); N, Nomini (hulled); T, Thoroughbred (hull-less). Reproduced from Moreau et al., 2007b, with permission.

we scarified hulless barley and separated the scarification fines via sieving. We found that the largest pieces contained the most oil, probably because they contained pieces of the germ (Moreau et al., 2007b). Knuckles and Chiu (1995) investigated the use of air classification to fractionate ground barley into starch-rich and β-glucan-rich (soluble fiber) fractions.

For most animal-feed applications, hulled cultivars of barley are acceptable; however, some advantages result from feeding hulless barley to nonruminants (mainly poultry and swine) because of their lower levels of fiber (Baik & Ullrich, 2008). When used for malting, hulled varieties are preferred, because the presence of the

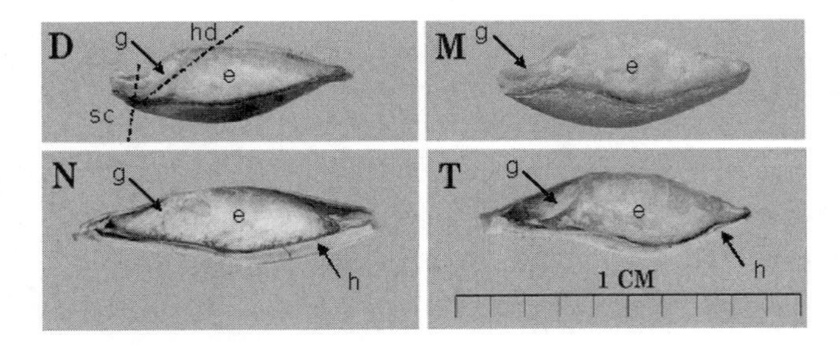


**Fig. 17.2.** Photographs of the cross sections of kernels of the four cultivars of barley used in this study and a comparison of the probable planes of fracture of the germ fragments during scarification (sc) and of the intact germ during hand dissection (hd). Abbreviations: D, M, N, T, see Fig. 17.1; e, endosperm; g, germ; h, hull; sc, probable plane of fracture via scarification which yields the ≈1-mm germ fragments shown in Fig. 17.1; hd, approximate plane of fracture during hand dissection of germ. Reproduced from Moreau et al., 2007b, with permission.

hull improves the filtering efficiency of the mash after brewing (Newman & Newman, 2006). However, many are enthusiastic about using hulless barley cultivars for food applications (Ames et al., 2006). In addition to not requiring pearling to make them acceptable for food applications, the hulless varieties qualify as whole grains, and may contain higher levels of bran-derived phytonutrients than pearled barleys. Hulless cultivars of barley were investigated by processing them via pearling and other abrasive methods (Yeung & Vasanthan, 2001). Zheung et al. (2000) reported that pearling up to 30% of the mass of hulless kernels effectively removed β-glucans from low-β-glucan cultivars but not from high-β-glucan cultivars.

The aleurone layer of barley, wheat, and other grains has long been known to be a rich source of enzymes and phytonutrients and a potential health-promoting food (Bacic & Stone, 1981a, 1981b). My laboratory reported that the aleurone layer from corn is rich in phytosterols and that isolated corn aleurone cells could potentially be valuable as a functional food or as a feedstock to produce a phytosterol-rich functional food (Moreau et al., 2000; Singh et al., 2001; Singh & Moreau, 2006). Bacic and Stone (1981a, 1981b) developed methods to isolate aleurone cells from barley and wheat. They presented some ultrastructural and chemical differences, but they did not identify any obvious phytonutrients. Further studies are needed to investigate the chemical and potential health-promoting properties of "barley aleurone oil."

## Extraction

Almost all of the studies of barley oil and its lipid composition were conducted by extracting the oil and lipids with hexane. These include the three studies from my own lab (Lampi & Moreau, 2004; Moreau et al., 2007a, 2007b) and the barley-oil papers cited in the earlier sections of this chapter. Hexane is the most popular solvent used to extract edible oils, and it extracts the common forms of "nonpolar" or "neutral" lipids, including triacylglycerols, phytosterols, tocopherols, tocotrienols, carotenoids, chlorophyll, hydrocarbons, waxes, free fatty acids, and small amounts of glycolipids and phospholipids. The latter seven types of components are usually removed during oil refining because they are considered undesirable due to off-flavors and/or odors, or they make the oil more susceptible to oxidation.

The extraction of edible oils with supercritical $CO_2$ or with solvents more polar than hexane produces an edible oil that contains more polar lipids and different chemical and health-promoting properties (Moreau et al., 2003; Moreau & Hicks, 2005). Edible oils obtained via pressing or cold-pressing can also contain a greater variety of phytonutrients than those obtained via hexane extraction. Unfortunately, no published studies compare the yields and composition of barley oil with extraction by any method other than hexane extraction. Several patents from the University of Montana identify the processes used to produce barley oil (Goering & Eslick, 1984, 1989, 1991), barley malt oil (Goering & Eslick, 2001), and other products via extraction with ethanol, methanol, or other alcohols.

## Oil Yields

As mentioned previously, the levels of oil in barley kernels are low, and range from about 1.5 – 2% in the most common hulled and hulless barley cultivars (Table 17.1). One barley cultivar that contains higher levels of oil is Prowashonupana (patented with the trade name of Sustagrain®, by ConAgra), which was developed at Montana State University. Although it received attention mostly for its unusually high levels of β-glucans (15–18%) and unusually low levels of starch (21–31%), our laboratory and others also report that it contains unusually high levels of oil (5.1–6.2%) (Andersson et al., 1999a, 1999b,; Moreau et al., 2007a). These properties are attributable to the observation that this cultivar has small kernels and "shrunken endosperms," and the low levels of starch are accompanied by a small endosperm and higher proportions of oil. Oil yields from barley bran were 3.8–5.6%. Oil yields from barley endosperm were 0.5–1.5%. Oil yields from two surface- abrasion methods, pearling fines and scarification fines, ranged from 7.0 – 12.8% and from 2.9 – 9.3%, respectively. Oil yields from barley germ were 7.5 – 17.5% (Table 17.1).

## Endogenous Hydrolytic Enzymes and Their Effects on Storage

Unlike oats, which contain lipases in their kernels (dry kernels, ungerminated) (Parmar & Hammond, 1994; Piazza et al., 1989), lipolytic enzymes were not reported in barley. In my experience with barley (hulled and hulless) lipid extraction and analysis, I have never encountered levels of free fatty acids higher than 2–3%, which is also

**Table 17.1 Oil Yields via Hexane Extraction of Ground Barley and Barley-Processing Fractio**

| Barley cultivar | Hulled or hulless | Material extracted | Oil yield (%) | References |
|---|---|---|---|---|
| Golf | Hulled | Ground kernels | 2.2 | Andersson et al., 1999b |
| High Amylose Glacier | Hulled | Ground kernels | 3.5 | Andersson et al., 1999b |
| SW 906129 | Hulled | Ground kernels | 3.4 | Andersson et al., 1999b |
| Karin | Hulled | Ground kernels | 2.2 | Andersson et al., 1999b |
| Mixed Finnish Sample | Hulled | Ground kernels | 2.1 | Dutta and Appelqvist, 1996 |
| Thoroughbred | Hulled | Ground kernels | 1.8 | Moreau et al., 2007a |
| Nomini | Hulled | Ground kernels | 1.7 | Moreau et al., 2007a |
| SW 8875 | Hulless | Ground kernels | 2.8 | Andersson et al., 1999b |
| Hashonucier | Hulless | Ground kernels | 3.2 | Andersson et al., 1999b |
| Bz 489-30 | Hulless | Ground kernels | 2.9 | Andersson et al., 1999a,b |
| Doyce | Hulless | Ground kernels | 1.6 | Moreau et al., 2007a |
| Merlin | Hulless | Ground kernels | 2.3 | Moreau et al., 2007a |
| Prowashonupana | Hulless | Ground kernels | 6.2 | Andersson et al., 1999a,b |
| Sustagrain® (=Prowashonupana) | Hulless | Ground kernels | 4.7 | Moreau et al., 2007a |
| Doyce | Hulless | Pearling fines | 7.1 | Lampi et al., 2004 |
| Unspecified | Hulled | Pearling flour | 10.7 | Ko et al., 2003 |
| Thoroughbred | Hulled | Scarification fines | 5.3 | Moreau et al., 2007a |
| Nomini | Hulled | Scarification fines | 2.9 | Moreau et al., 2007a |
| Doyce | Hulless | Scarification fines | 7.4 | Moreau et al., 2007a |
| Merlin | Hulless | Scarification fines | 9.3 | Moreau et al., 2007a |
| Merlin | Hulless | "Pearlings" | 12.8 | Fastnaught et al., 2006 |
| Unspecified | Hulled | Bran | 4.57-5.32 | Seog et al., 2002 |
| Scout | Hulless | Bran | 3.8 | Bhatty, 1993 |
| Unspecified | Hulled | Bran | 5.6 | Ko et al., 2003 |
| Thoroughbred | Hulled | Germ | 7.5 | Moreau et al., 2007b |
| Nomini | Hulled | Germ | 17.5 | Moreau et al., 2007b |
| Doyce | Hulless | Germ | 13.3 | Moreau et al., 2007b |
| Merlin | Hulless | Germ | 14.5 | Moreau et al., 2007b |
| Unspecified | Hulled | Germ | 14.7 | Seog et al., 2002 |
| Unspecified | Hulled | Germ | 13.0 | Ko et al., 2003 |
| Thoroughbred | Hulled | Endosperm | 0.5 | Moreau et al., 2007b |
| Nomini | Hulled | Endosperm | 1.2 | Moreau et al., 2007b |
| Doyce | Hulless | Endosperm | 1.5 | Moreau et al., 2007b |
| Merlin | Hulless | Endosperm | 1.5 | Moreau et al., 2007b |
| Unspecified | Hulled | Endosperm | 0.7 | Ko et al., 2003 |
| Thoroughbred | Hulled | Hull | 1.8 | Moreau et al., 2007b |
| Nomini | Hulled | Hull | 3.0 | Moreau et al., 2007b |
| Unspecified | Hulled | Hull | 2.6 | Ko et al., 2003 |

evidence of low levels of lipolytic enzymes in dormant barley kernels (Moreau et al., 2007a, 2007b). Using Merlin, a hulless barley cultivar, Fastnaught et al. (2006) reported a slight increase in free fatty acids during the storage of whole-grain meal at 25 and 37°C, but no increase in free fatty acids when the meal was stored at 4°C. In a recent study, we reported the effect of the storage of milled barley samples for six months at ambient conditions and for three weeks at an elevated temperature and humidity (Liu & Moreau, 2008). Both storage conditions did not affect the oil yields or the concentrations of tocopherols in the milled samples, but both conditions resulted in the measurable degradation of phytosterols and tocotrienols in the milled samples.

## Edible and Nonedible Applications

Because barley oil was only produced on a limited experimental basis and not as a commercial product, only its edible applications have been investigated. As mentioned previously, in recent years barley grain was utilized in the following distribution: about two-thirds for animal feed, about one-third for malting, and about 2% for human-food use. In recent years, especially in Europe, a small amount of barley (<1% of the world production) was used as a feedstock to produce fuel ethanol (Baik & Ullrich, 2008). Because only the starch portion of the barley kernel is currently fermented to ethanol, the remainder of the materials in the kernel is available for feed and other applications. The levels of oil in the "brewer's spent grain" are two- to three-fold higher than the levels in the ground barley, so barley oil is potentially available for extraction and food uses via the extraction of brewer's spent grains (Sosulski et al., 1997; Wang et al., 1997b).

## Acyl Lipids (TGs, PLs, etc.) and Fatty Acid Composition

### Lipid Classes: NLs, GLs, PLs

One can extract, as with most plant materials, the "total lipid" fraction (containing both polar and nonpolar lipids) from barley by using methods such as the Bligh and Dyer (1959) chloroform–methanol extraction method. One can separate the total-lipid fraction into three subfractions by using silica columns or silica solid-phase extraction (Moreau et al., 2008). These three lipid subfractions are: (i) neutral lipids (NLs, comprised of triacylglycerols, phytosterols, phytosteryl fatty acyl esters, free fatty acids, waxes and hydrocarbons such as squalene, and other minor components), (ii) glycolipids (GLs, including steryl glycosides, monogalactosyldiacylglycerol, digalactosyldiacylglycerol, and cerebrocides), and (iii) phospholipids (PLs, including phosphatidylcholine, phosphatidylethanolamine, and phosphatidylinositol). In most of the tissues in barley and other seeds, NLs are the most abundant of these three lipid subclasses (Table 17.2). The proportions of GLs and PLs are smaller, and vary with cultivar and tissue type. When most grain seeds are extracted with hexane, the

**Table 17.2. Total Lipid Subfractions in Total Lipids Extracts of Barley: Cultivar and tissue Total lipid Relative distribution**

| | g/100 g | NL% | GL% | PL% | References |
|---|---|---|---|---|---|
| Riso, whole grain | 4.0 | 75 | 7 | 18 | Bhatty and Rossnagel, 1980 |
| Bonanza, whole grain | 2.0 | 65 | 26 | 9 | |
| Prillar hulless | | | | | Price and Parsons, 1979 |
| Embryo | 19.6 | 75.8 | 6.4 | 17.8 | |
| Bran-endosperm | 2.8 | 64.4 | 12.5 | 23.1 | |
| Hull | 2.2 | 75.9 | 18.2 | 5.9 | |

Abbreviations: NL, neutral lipids; GL, glycolipids; PL, phospholipids.

levels of GLs and PLs in the "oil" are small, usually less than 5% (Moreau et al., 2003; Moreau & Hicks, 2005). However, when extracting with more-polar solvents such as ethanol, methanol, or isopropanol, the levels of GLs and PLs in the "oil" are often higher than 10% (Moreau et al., 2003). Most of the research on barley oil and barley lipids was conducted using hexane extraction.

## Fatty Acid Composition

Several publications reported the fatty-acid analyses of barley kernels and barley oil (Bhatty & Rossnagel, 1980; DeMan & Cauberghe, 1988; Price & Parsons, 1979; Wang et al., 1993a). In each of these studies (Table 17.3), linoleic acid (18:2) was the most abundant fatty acid, followed by oleic acid (18:1). The levels of α-linolenic acid (18:3) ranged from 3.4 to 7% in all of the samples except barley germ oil, where the concentration was higher (9.6%) (Table 17.3). Only one study (Wang et al., 1993a) examined the levels of n-3 (also called ω-3) versus n-6 (also called ω-6) fatty acids in barley, and these researchers reported that all of the linolenic acid (18:3) in barley, 3.4 wt %, was the α-linolenic acid and none was the γ-linolenic acid, an n-6 fatty acid.

# Minor Lipid Components

## Phytosterols

Unlike animals, whose cells contain one predominant sterol, cholesterol, all plant cells contain a variety of sterols, called phytosterols (Moreau et al., 2002; Piironen et al., 2000a). Phytosterols occur in three basic forms in all plant tissues: as free phytosterols (with an unbound OH group), as fatty acyl esters (with the OH esterified to a fatty acid), and as glycosides. In addition, in some cereals, a significant proportion of the phytosterols occurs as esters of hydroxycinnamic acids such as ferulic acid and *p*-coumaric acid (Hakala et al., 2002). In rice kernels, the predominant phytosteryl-hydroxycinnamic acid ester is cycloartenyl-ferulate, commonly called "oryzanol" (Hakala et al., 2002). In corn kernels, the predominant phytosteryl-hydroxycinnamic acid

**Table 17.3. Fatty-Acid Composition of Barley Oil from Various Sources**

| Samples | Cultivar | Hulled or hulless | Fatty acids (relative %) | | | | | | | | | References |
|---|---|---|---|---|---|---|---|---|---|---|---|---|
| | | | 12:0 | 14:0 | 16:0 | 16:1 | 18:0 | 18:1 | 18:2 | 18:3 | 20:0 | |
| Ground barley | Riso | Hulled | nr | nr | 20.6 | nr | 1.0 | 19.1 | 53.5 | 5.7 | nr | Bhatty and Rossnagel, 1980 |
| Ground barley | Bonanza | Hulled | nr | nr | 19.1 | nr | 1.2 | 14.1 | 59.8 | 5.9 | nr | Bhatty and Rossnagel, 1980 |
| Ground barley | Average of three varieties of spring barley | Hulled | nr | nr | 20 | nr | 1 | 14 | 58 | 7 | nr | DeMan and Cauberghe, 1988 |
| Ground barley | Prowashonupana | Hulless | nr | nr | 25.5 | | 1.5% | 19.3 | 50.3 | 3.4 (ω-3) | tr | Wang et al., 1993a |
| Barley germ oil | Prillar | Hulless | nr | 0.2 | 19.5 | tr | 0.6 | 20.4 | 49.7 | 9.6 | tr | Price and Parsons, 1979 |
| Barley bran-endosperm oil | Prillar | Hulless | nr | 0.2 | 20.6 | 0.1 | 1.3 | 16.6 | 56.8 | 4.4 | tr | Price and Parsons, 1979 |
| Barley hull oil | Prillar | Hulless | 1.7 | 5.3 | 26.1 | 2.0 | 5.4 | 21.5 | 30.6 | 5.0 | tr | Price and Parsons, 1979 |

Abbreviation: nr, not reported.

ester is sitostanyl-ferulate (Moreau et al., 1996, 1998). Barley kernels contain little or no phytosteryl-hydroxycinnamic acid esters (Piironen et al., 2000b, 2002). When extracted with hexane, the predominant phytosterols in the barley oil are free phytosterols and phytosteryl fatty acyl esters, at about a 1:1 ratio (Lampi et al., 2004). The levels of total phytosterols in hexane-extracted barley kernel oil range from 3.1–5.6% (Moreau et al., 2007a). The levels of total phytosterols in hexane-extracted barley-scarification-fines oil range from 1.9 – 9.6% (Moreau et al., 2007a).

Two publications (Lampi et al., 2004; Piironen et al., 2002) reported analyses of the levels of individual phytosterols in barley kernels (Table 17.4). As with most plant species, sitosterol (formerly called β-sitosterol) is the predominant phytosterol, followed by stigmasterol and campesterol (Table 17.4). The levels of stanols (saturated phytosterols including campestanol and sitostanol) are very low in barley kernel oil (≈2%) and barley- pearling-fines oil (≈4%) (Table 17.4). Dutta and Appelqvist (1996) reported similar proportions of total sterols (after alkaline and acid hydrolyses) in barley; however, their concentrations of each phytosterol were several- fold higher than those reported in Table 17.4.

## Tocopherols/Tocotrienols in Barley and Barley Oil

Tocopherols and tocotrienols (both are called tocols) are natural forms of vitamin E that occur in many plant tissues (Cahoon et al., 2003). Four tocopherols (α, β, γ, δ) differ from each other in the number and position of the methyl groups on the chromanol ring (Fig. 17.3). Also four tocotrienols are identical to their corresponding tocopherols, except that each has three carbon-carbon double bonds in their phytol chain (Fig. 17.3). Several grains contain very high levels of tocopherols and tocotrienols (Dunford, 2005). We (Moreau et al., 2007b) recently confirmed the previous observations (Falk et al., 2004; Peterson, 1994; Przybylski, 2006) that the tocols in barley germ (embryo) were predominantly tocopherols (97%), whereas those in the endosperm were predominantly tocotrienols (80–90%). Falk et al. (2004) and our lab (Moreau et al., 2007b) recently confirmed that total tocotrienols are more abundant than total tocopherols in barely oil prepared from ground barley kernels. Morrison (1978) is credited with the first report that barley kernels are one of the few plant tissues that contain all eight tocol isomers, and high-performance liquid chromatography (HPLC) with fluorescence detection can separate and quantify them (Fig. 17.4). Others noted that barley kernels contain exceptionally high levels of tocotrienols (Morrison, 1993). The actual levels of tocotrienols in barley kernels are variable, and may be a function of the genotype, the growing conditions of the barley plant, and the storage conditions of the kernels (Morrison, 1993 and personal observation). Peterson and Qureshi (1993) reported that the levels and distribution of tocols (α-tocotrienols was the most abundant tocol in both) were similar in barley and oats, but the total levels of tocols were about two times higher in barley. In most barley cultivars, the levels of α-tocotrienols range from about 20 to 40 mg/kg, and

**Table 17.4. Phytosterols (mg/kg) in Barley**

| Phytosterol | Ground barley, Kustaa (Piironen et al., 2002) | Ground barley, Pokko (Piironen et al., 2002) | Ground barley, Doyce (Lampi et al., 2004) | Pearling fines, Doyce (Lampi et al., 2004) |
|---|---|---|---|---|
| Brassicasterol | 13 ± 7 (2) | tr | nr | nr |
| Campesterol | 150 ± 2 (21) | 192 ± 1 (24) | 181 ± 2 (23) | 391 ± 8 (23) |
| Campestanol | 7 ± 4 (1) | 9 ± 3 (1) | 11 ± 0 (1) | 40 ± 0 (2) |
| Sitosterol | 437 ± 6 (61) | 484 ± 0 (61) | 476 ± 1 (60) | 925 ± 29 (54) |
| Sitostanol | 10 ± 4 (1) | 10 ± 2 (1) | 5 ± 0 (1) | 26 ± 0 (2) |
| Stigmasterol | 24 ± 1 (3) | 36 ± 1 (5) | 39 ± 0 (5) | 120 ± 1 (7) |
| Δ5-Avenasterol | 60 ± 2 (8) | 44 ± 2 (6) | nr | nr |
| Δ7-Avenasterol | 9 ± 1 (1) | 12 ± 9 (2) | nr | nr |
| Other sterols | 5 ± 1 (1) | 13 ± 11 (2) | 86 ± 1 (11) | 221 ± 9 (13) |
| Total sterols | 720 ± 21 (100) | 801 ± 12 (100) | 797 ± 4 (100) | 1732 ± 38 (100) |

Abbreviations: tr, trace; nr, not reported. Values presented are means ± standard deviations. Numbers in parentheses are relative % of each phytosterol.

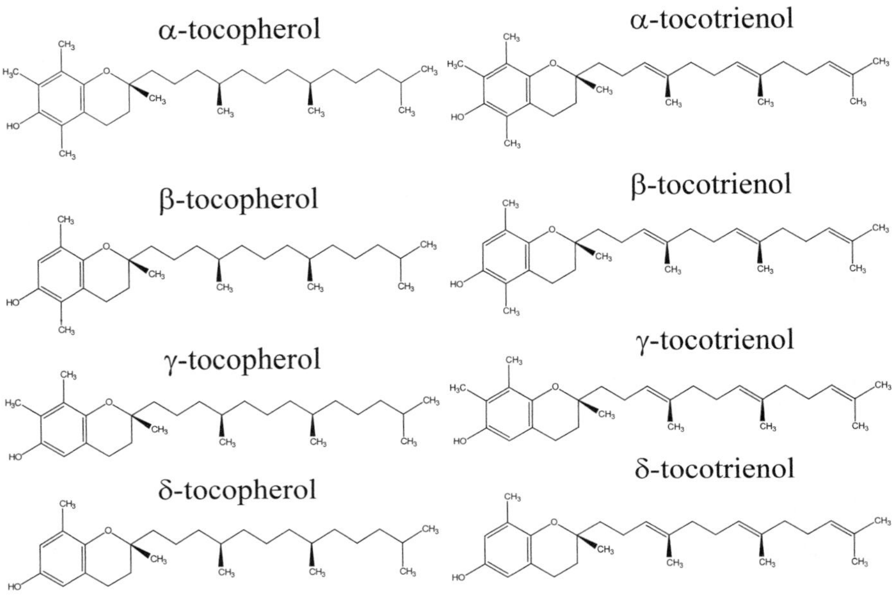

α-tocopherol

α-tocotrienol

β-tocopherol

β-tocotrienol

γ-tocopherol

γ-tocotrienol

δ-tocopherol

δ-tocotrienol

**Fig. 17.3.** The chemical structures of the tocopherols (T) and tocotrienols (T3) in barley oil.

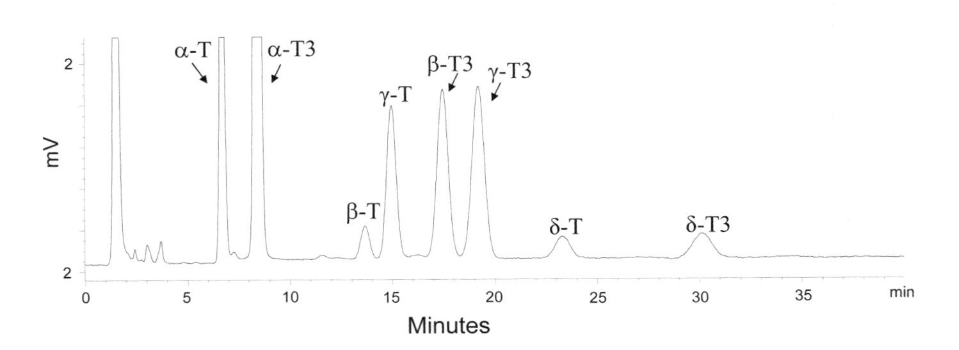


**Fig. 17.4.** High-performance liquid chromatography (HPLC) chromatogram of tocopherols and tocotrienols in barley oil. Exact conditions for HPLC separation and fluorescence detection are described in Moreau et al., 2007b.

the levels of total tocols range from about 40 to 100 mg/kg (Cavallero et al., 2004; Ehrenbgerova et al., 2006; Ko et al., 2003; Moreau et al., 2007a; Peterson, 1994; Piironen et al., 1986; Pryma et al., 2007; Wang et al., 1993b).

Although numerous publications are available on the levels of tocols in barley kernels and barley milling fractions, as noted above, only a few have actually reported the levels of tocols in barley oil (Table 17.5). In each of these reports, barley oil was prepared by extracting ground barley with hexane. However, because the levels of extractable oil in most barley cultivars are low (≈2%), obtaining barley oil by extracting ground barley is costly. When barley kernels are processed or milled, some fractions are enriched in oil. Lampi et al. (2004) reported oil yields of about 7% from the "pearling fines" from Doyce, a hulless variety. Bhatty (1993) reported that the bran from roller-milled barley resulted in ether-extractable oil yields of about 3.6%. We also reported oil yields of 3–9% in the scarification fines of four U.S. barley cultivars (Moreau et al., 2007a). The percentage of tocotrienols/total tocols in barley oil from ground barley ranged from 70–76%, and the percentage of tocotrienols/total tocols in barley oil from scarification fines ranged from 55–70% (Moreau et al., 2007a).

A large variation exists in the levels of tocopherols and tocotrienols in barley oils in the five publications where they were reported (Table 17.5). Babu et al. (1992) reported that barley extract oil, prepared by Miller Brewing Company by extracting brewer's spent grain, contained 0.4% of total tocopherols and tocotrienols. In Table 17.5, the levels of total tocols in the various barley oils ranged from 14 to 8435 mg/kg oil. Interestingly, the levels of total tocols were similar in wheat germ oil and in barley kernel oil; however, the proportions of tocotrienols (%T3) in barley kernel oils were 70–86%, whereas wheat germ oil only contained about 9% of T3. "The proportions of tocotrienols (%T3) in the scarification-fines oils were about 40–60% (Moreau et al., 2007ab).

As mentioned previously, we recently reported that the storage of milled fractions of barley has no effect on the levels of tocopherols in the milled fraction, but considerable degradation of the tocotrienols occurs during both storage conditions (Liu & Moreau, 2008). Among the tocotrienol vitamers, we observed the highest rate of degradation as that of α-tocotrienol (the most abundant tocotrienol in barley).

## Carotenoids/Chlorophylls and Other Pigments

Morrison (1978) reported that barley kernels contain carotene (55 mg/kg) and xanthophylls (41 mg/kg). A recent paper reported that two cultivars of black barley contained high levels of lutein and zeaxanthin (Siebenhandl et al., 2007). When milled and fractionated via sieving, the highest levels of lutein + zeaxanthin for one cultivar (BB6) were in the "shorts" fraction (600–900 μm), whereas with the other cultivar (BB2) the highest levels of lutein + zeaxanthin were in the "middlings" size fraction (180–600 μm).

**Table 17.5.** Tocopherols and Tocotrienols (mg/kg oil) in Barley Oil from Various Sources

| Barley oil extracted from: | Oil yield (%) | αT | βT | γT | δT |
|---|---|---|---|---|---|
| Ground barley Prowashonupana | nr | 142±21 | 6±0 | 104±8 | 1±0 |
| Ground barley Azhull, hulless | nr | 385±18 | 20±2 | 312±18 | 7±1 |
| Bran I | 4.67 | 3.27 | - | 1.23 | - |
| Germ | 14.65 | 90.26 | 1.47 | 18.64 | - |
| Kernels | | | | | |
| Thoroughbred | 1.79±0.05 | 1121±54 | 87±2 | 843±23 | 258±9 |
| Doyce | 1.59±0.05 | 1174±6 | 119±6 | 627±82 | 135±9 |
| Sustagrain | 4.72±0.11 | 372±27 | 59±9 | 88±6 | 32±3 |
| Malted barley | 1.50±0.02 | 1110±8 | 52±3 | 206±8 | 12±2 |
| Germ | | | | | |
| Wheat germ | 9.74±0.03 | 4102±128 | 2000±124 | 44±4 | - |
| Scarification fines | | | | | |
| Thoroughbred | 5.33±0.13 | 1458±91 | 91±10 | 993±164 | 313±18 |
| Doyce | 7.35±0.33 | 1383±15 | 121±2 | 704±27 | 166±10 |
| Sieved Scarification fines, Doyce 60 s | | | | | |
| 0.707-1.410 mm | 11.5±0.2 | 1289±110 | 58±6 | 354±54 | 50±3 |
| 0.287-0.707 | 6.9±1.0 | 1178±172 | 55±11 | 320±69 | 50±9 |
| 0.210-0.287 | 7.1±1.1 | 726±119 | 37±8 | 182±43 | 33±5 |
| < 0.210 | 7.5±0.9 | 583±57 | 31±3 | 139±17 | 28±3 |

Abbreviation: nr, not reported.

| αT3 | βT3 | γT3 | δT3 | Total T | Total T3 | Total tocol | T3(%) T3/T+T3 | References |
|---|---|---|---|---|---|---|---|---|
| 558±22 | - | 85±7 | 5±0 | 253 | 648 | 901 | 71.9 | Wang et al., 1993a |
| 1691±86 | - | 231±15 | 51±3 | 724 | 1973 | 2697 | 73.2 | Wang et al., 1997b |
| 3.47 | 2.73 | 2.66 | 0.57 | 4.50 | 9.43 | 13.93 | 67.7 | Seog et al., 2002 |
| 16.27 | - | 4.67 | - | 110.37 | 20.94 | 131.31 | 16.0 | " |
| | | | | | | | | Moreau et al., 2007a |
| 4249±158 | 693±21 | 1039±44 | 145±9 | 2309 | 6126 | 8435 | 72.6 | |
| 3236±148 | 723±46 | 848±42 | 195±26 | 2055 | 5002 | 7057 | 70.9 | |
| 2521±107 | 395±13 | 664±28 | 77±5 | 551 | 3657 | 4208 | 86.9 | |
| 3978±227 | 614±28 | 864±50 | 246±16 | 1380 | 5702 | 7082 | 80.5 | |
| 66±8 | 396±28 | 170±6 | - | 6146 | 632 | 6778 | 9.3 | Moreau et al., 2007a |
| | | | | | | | | Moreau et al., 2007a |
| 3533±553 | 235±38 | 955±137 | 117±16 | 2855 | 4840 | 7695 | 62.9 | |
| 2364±294 | 225±36 | 593±63 | 123±24 | 2374 | 3305 | 5769 | 57.3 | |
| | | | | | | | | Moreau et al., 2007b |
| 25±2 | 4 ± 0 | 11±4 | 3±1 | 1751 | 43 | 1794 | 2.2 | |
| 210±1 | 8±2 | 43±12 | 8±3 | 1603 | 269 | 1872 | 14.4 | |
| 609±140 | 23±7 | 108±30 | 14±3 | 978 | 754 | 1732 | 43.5 | |
| 515±82 | 21±2 | 92±16 | 13±2 | 781 | 641 | 1422 | 45.1 | |

## Starch Lipids

Starch lipids include those lipids (mostly comprised of lysophospholipids and saturated free fatty acids) that are associated with the starch granules in the kernels of barley and other grains (Kaukovirta-Norja et al., 1997, 1998). Starch lipids are more tightly bound than most lipids, and are commonly extracted with hot butanol or other polar solvent mixtures (Morrison, 1993). Barley oils extracted with hexane probably contain little or no starch lipids, but I am not aware of any studies that actually address this question.

## Ceramides

Ceramides are a common class of plant GLs which contain an amide-bound sphingoid base and no glycerol. Ceramides and other sphingolipids have biological activities, including anticancer properties, maintaining nerve function, and reducing cholesterol absorption (Wehrmüller, 2007). Because of their "intermediate" polarity, most ceramides and other GLs are poorly extracted with hexane, but they are more efficiently extracted with polar solvents such as ethanol, methanol, and isopropanol (Moreau et al., 2003). A recent patent reported a process to produce barley malt oil and ceramides from brewer's spent grains (Goering & Eslick, 2001).

## Phenolic Compounds and Natural Products with Antioxidant Properties

When extracted with hexane, the resulting barley oil most likely will not contain significant levels of phenolic compounds or other antioxidants, since most of them are polar and require more polar solvents for their extraction (Moreau et al., 2005). However, barley kernels contain some common and unique phenolic compounds (including hydroxycinnamic acids, lignans, dehydrodimers, and alkylresorcinols) that may be in barley oils if they are obtained by extraction with ethanol or other polar solvents.

Lignans, such as secoisolariciresinol and matairesionol, are a group of diphenolic compounds that are found in high levels in flax. Some cultivars of barley also have moderate amounts of lignans in their kernels (Nilsson et al., 1997). When consumed, plant lignans are converted to the mammalian lignans enterodiol and entrolactone by gut microflora. Like soy isoflavones, plant lignans possess estrogenic activity. Research is underway to evaluate their phytoestrogenic, anticancer, antioxidant, and serum cholesterol-lowering applications (Przybylski, 2006). Like lignans, dehydrodimers are another group of diphenolic compounds that occur in barley kernels, but their physiological role and biological properties are unknown (Hernanz et al., 2001).

Alkylresorcinols are phenolic compounds that occur in the outer surface of several cereals, including rye and wheat (Przybylski, 2006). Some cultivars of barley also contain high levels of alkylresorcinols (Garcia et al., 1997; Landberg et al., 2008). The physiological role of alkylresorcinols is not known, but because they are localized in the outer layers of kernels, they may protect the kernel from microbes or other threats (Garcia et al., 1997).

In addition to the above types of phenolic compounds, barley also contains high levels of other types of phenolic compounds that can combine to impart barley foods with high levels of total antioxidant activity (Zielinski & Kozlowska, 2000). Some of these phenolic compounds, such as ferulic acid and *p*-coumaric acid, are bound to cell walls and other compounds in the barley kernel. Like other grains, the high concentrations of many of these phenolic compounds in the outer layers of the grain may be the reason for the health-promoting properties of whole grains.

## Flavor and Aroma Compounds

No published reports are available on flavor and aroma compounds.

## Allergic and Toxic Compounds

No published reports are available on allergic and toxic compounds in barley oil.

## Health Benefits of the Oil and Oil Constituents

### General Health-promoting Properties of Barley Kernels

Like oat kernels, barley kernels contain high levels of soluble fiber (β-glucans), and much of the research on the health-promoting properties of barley has focused on mechanisms that involve β-glucans (Gallaher et al., 1992; Lupton et al., 1993; Martinez et al., 1992; Pins & Kaur, 2006; Zhang et al., 1991). In 2006, the FDA amended a previous health claim for oat products, and extended it to include foods that include barley (FDA, 2006): "Based on the available evidence, the FDA concluded that consuming whole-grain barley and dry-milled barley products that provide at least 3 grams of β-glucan fiber per day is effective in lowering blood total and LDL cholesterol and that the cholesterol-lowering effects of β-glucan soluble fiber in dry-milled barley products are comparable [to] that of oat sources of β-glucan soluble fiber now listed in § 101.81(c ) (2) (ii)A)" (FDA, 2006). Although they are generally viewed as health-promoting in foods intended for humans, the high levels of β-glucans in barley restrict the use of barley and barley-processing fractions as ingredients in the feeds for nonruminant animals, such as poultry and swine (Newman et al., 1991). Some researchers investigated the addition of enzymes (such as β-glucanase) to rations to improve the nutritional value of barley (Newman et al., 1991).

## The Cardiovascular Health-promoting Properties of Barley Oil and Its Tocotrienols

In the 1980s, a series of papers by Qureshi et al. (1986) indicated that barley kernels contain lipids that could reduce the levels of serum cholesterol by inhibiting cholesterol biosynthesis. Wang et al. (1993a) reported that barley extract oil (obtained by extracting brewer's spent grains) reduced the levels of total and LDL cholesterol in chicks. However, in a similar study with hamsters, barley flour reduced the levels of serum cholesterol, but barley oil had no effect (Wang et al., 1997a). Robinson and Lupton (1990) reported that barley oil reduced the levels of serum cholesterol in hypercholesterolemic men and women. Lupton et al. (1994) conducted a clinical study with barley oil, and reported that it caused a 7.1% reduction of total cholesterol and a 9.2% reduction of low-density lipoprotein (LDL) cholesterol in humans.

In later papers, Qureshi and colleagues isolated two components in barley oil that inhibited cholesterol biosynthesis (Qureshi et al., 1986). In later papers they demonstrated that the tocotrienols in barley oil were responsible for its ability to inhibit cholesterol biosynthesis, and most likely were also responsible for its ability to reduce serum cholesterol (Qureshi et al., 1991). Parker et al. (1993) presented evidence that tocotrienols inhibit cholesterol biosynthesis by increasing the rate of the degradation of HMG-CoA reductase. Most of the recent studies of the cholesterol-lowering effects of tocotrienols used extracts from palm oil (rich in δ-tocotrienol) and rice- bran oil (Sen et al., 2007). Conducting comparative studies with barley oil or barley-oil tocotrienols will definitely fill a need.

## The Effect of Barley Oil and Its Tocotrienols on Cell Adhesion and Atherosclerosis

Theriault et al. (2002) compared α-tocopherol and α-tocotrienol, and found that the latter was much more effective at inhibiting monocyte cell adhesion. The authors speculated that their results may indicate that α-tocotrienol may effectively protect against atherosclerosis. Because barley oil is rich in α-tocotrienols (Table 17.5), it may also prevent atherosclerosis.

## Immune Response

Babu et al. (1992) reported that barley extract oil (from Miller Brewing Company) depressed the immune response (mitogenic reactivity) in weanling and mature rats. They suggested that the effect may be due to the high levels of tocotrienols in the barley-oil extract.

## Neuroprotection by Tocotrienols

Numerous studies indicate that tocotrienols possess neuroprotective and anticancer properties (Sen et al., 2004, 2007). The authors present evidence that the mechanism of neuroprotection may involve the direct binding of α-tocotrienol to 12-lipoxygenase (Sen et al., 2007). Since barley oil is a rich source of α-tocotrienol, it may also have neuroprotective properties.

### Anticancer Properties of Tocotrienols

Several studies report evidence that some tocotrienols inhibit the growth of the cells of cancers of the breast, prostate, lymphatic system, liver, colon, and skin (Nesaretnam et al., 1998; Sen et al., 2007). As noted above, because barley oil contains all four tocotrienols, and is richest in α-tocotrienol, it may have anticancer properties.

## Conclusion

Barley oil is a "high-linoleate oil" with a fatty-acid composition similar to corn oil and several other "commodity" plant oils. The levels of phytosterols in barley oil are higher than those in most other edible-plant oils. The most unique feature about barley oil is the very high levels of tocotrienols (especially α-tocotrienol). More work is required to evaluate whether the exceptionally high levels of α-tocotrienol barley oil will make it valuable enough to justify the high cost of extracting it from barley kernels or from barley-processing fractions.

## References

Ames, N.; C. Rhymer; B. Rossnagel; M. Therrien; D. Ryland; S. Dua; K. Ross. Utilization of diverse hulless barley properties to maximize food product quality. *Cereal Food World* **2006**, *51*, 23–28.

Andersson, A.A.M.; R. Andersson; K. Autio; P. Aman. Chemical composition and microstructure of two naked waxy barleys. *J. Cereal Sci.* **1999a**, *30*, 183–191.

Andersson, A.A.M.; C. Elfverson; R. Andersson; S. Regner; P. Aman. Chemical and physical characteristics of different barley samples. *J. Sci. Food. Agric.* **1999b**, *79*, 979–986.

Babu, U.S.; M.Y. Jenkins; G.V. Mitchell. Effect of short-term feeding of barley oil extract containing naturally occurring tocotrienols on the immune response in rats. *Annals N.Y. Acad. Sci.* **1992**, *669*, 317–319.

Bacic, A.; B.A. Stone. Isolation and structure of aleurone cell walls from wheat and barley. *Aust. J. Plant Physiol.* **1981a**, *8*, 453–474.

Bacic, A.; B.A. Stone. Chemistry and organization of aleurone cell wall components from wheat and barley. *Aust. J. Plant Physiol.* **1981b**, *8*, 475–495.

Baik, B-K.; S.E. Ullrich. Barley for food: Characteristics, improvement and renewed interest. *J. Cereal Sci.*, **2008**, *48*, 233–242.

Bhatty, R.S. Physicochemical properties of roller-milled barley bran and flour. *Cereal Chem.* **1993**, *70*, 397–402.

Bhatty, R.S.; B.G. Rossnagel. Lipids and fatty acids of Risco 1508 and normal barley. *Cereal*

*Chem.* **1980**, *57*, 382–386.

Bligh, E.G.; W.J. Dyer. A rapid method for total lipid extraction and purification. *Can. J. Biochem. Physiol.* **1959**, *37*, 911–917.

Cahoon, E.B.; S.E. Hall; K.G. Ripp; T.S. Ganzke; W.D. Hitz; S.J. Coughlan. Metabolic redesign of vitamin E biosynthesis in plants for tocotrienols production and increased antioxidant content. *Nature Biotechnol.* **2003**, *21*, 1082–1087.

Cargill, 2008, http://www.cargillhft.com/industry_products_beta.html.

Cavallero, A.; A. Gianinetti; F. Finocchiaro; G. Delogu; A.M. Stanca. Tocols in hulless and hulled barley genotypes grown in contrasting environments. *J. Cereal Sci.* **2004**, *39*, 175–180.

ConAgra, 2008, http://www.conagramills.com/our_products/sustagrain.jsp.

DeMan, W.; N. Cauberghe. Changes in lipid composition in maturing barley kernels. *Phytochemistry* **1988,** *27*, 1639–1642.

Dunford, N.T. Germ oils from different sources. *Edible Oil and Fat Products: Specialty Oils and Oil Products* (in *Bailey's Industrial Oil and Fat Products*) 6th ed.; F. Shahidi, Ed.; John Wiley and Sons, NY, 2005; Vol. 3, pp. 195–231.

Dutta, P.C.; L-Å. Appelqvist. Saturated sterols (stanols) in unhydrogenated and hydrogenated edible vegetable oils and in cereal lipids, *J. Sci. Food Agric.* **1996,** *71*, **383–391.**

Ehrenbergerová, J.; N. Belcrediová; J. Prýma; K. Vaculová; C.W. Newman. Effect of cultivar, year grown, and cropping system on the content of tocopherols and tocotrienols in grains of hulled and hulless barely. *Plant Foods in Human Nutr.* **2006**, *61*, 145–150.

Falk, J.; A. Krahnstover; T.A.W. van der Kooig; M. Schlensog; K. Krupinska. Tocopherol and tocotrienols accumulation during development of caryopses from barley (*Hordeum vulgare* L.). *Phytochemistry* **2004**, *65*, 2977–2985.

Fastnaught, C.E.; P.T. Berglund; A.L. Dudgeon; M. Hadley. Lipid changes during storage of milled hulless barley products. *Cereal Chem.* **2006,** *83*, 424–427.

FDA. (2006). Food labeling health claims; soluble dietary fiber from certain foods and coronary heart disease. http://www.fda.gov/ohrms/dockets/98FR/04p-0512-nfr0001.pdf.

Gallaher, D.D.; P.L. Locket; C.M. Gallaher. Bile acid metabolism in rats fed two levels of corn oil and brans of oat, rye and barley and sugar beet fiber. *J. Nutr.* **1992,** *122*, 473–481.

Garcia, S.; C. Garcia; H. Heinzen; P. Moyna. Chemical basis of the resistance of barley seeds to pathogenic fungi. *Phytochemistry* **1997**, *44*, 415–418.

Goering, K.J.; R.F. Eslick. Processes for production of waxy barley products. U.S. Patent 4,428,967, 1984.

Goering, K.J.; R.F. Eslick. Production of beta-glucan, bran, protein, oil and maltose syrup from waxy barley. U.S. Patent 4,804,545, 1989.

Goering, K.J.; R.F. Eslick. Process for the recovery of products from waxy barley. U.S. Patent 5,013,561, 1991.

Goering, K.J.; R.F. Eslick. Barley malt oil containing vegetable ceramide-associated substances and process for producing the same. U.S. Patent 6,316,032, 2001.

Hakala, P.; A.-M. Lampi; V. Ollilainen; U. Werner; M. Murkovic; K. Wähälä; S. Sarkola; V.

Piironen. Steryl phenolic acid esters in cereals and their milling fractions. *J. Agric. Food. Chem.* **2002**, *50,* 5300–5307.

Hernanz, D.; V. Nunez; A.I. Sancho; C.B. Faulds; G. Williamson; B. Bartolomé; C. Gómez-Cordovés. Hydroxycinnamic acids and ferulic acid dehydrodimers in barley and processed barley. *J. Agric. Food Chem.* **2001**, *49,* 4884–4888.

Jadhav, S.J.; S.E. Lutz; V.M. Ghorpade; D.K. Salunkhe. Barley: Chemistry and value-added processing. *Crit. Rev. Food Sci.* **1998**, *38,* 123–171.

Kaukovirta-Norja, A.R.; P.K. Kotiranta; A-M. Aurola; P.O. Reinkainen; J.E. Olkku; S.V. Laakso. Influence of water processing on the composition, behavior, and oxidizability of barley and malt lipids. *J. Agric. Food Chem.* **1998**, *46,* 1556–1562.

Kaukovirta-Norja, A.; P. Reinkainen; J. Olkku; S. Laakso. Starch lipids of barley and malt. *Cereal Chem.* **1997**, *74,* 773–738.

Knuckles, B.E.; M-C.M. Chiu. β-glucan enrichment of barley fractions by air classification and sieving. *J. Food Sci.* **1995,** *60,* 1070–1074.

Ko, S-N.; C-J. Kim; H. Kim; C-T. Kim; S-H. Chung; B-S. Tae; I-H. Kim. Tocols levels in milling fractions of some cereal grains and soybeans. *J. Am. Oil Chem. Soc.* **2003**, *80,* 585–589.

Lampi, A-M.; R.A. Moreau; V. Piironen; K.B. Hicks. Pearling barley and rye to produce phytosterol-rich fractions. *Lipids* **2004**, *39,* 783–787.

Landberg, R.; A. Kamal-Eldin; M. Salmenkallio-Marttila; X. Rouau; P. Aman. Localization of alkylresorcinols in wheat, rye and barley kernels. *J. Cereal Sci.* **2008**, *48,* 401–406.

Liu, K.S.; R.A. Moreau. Concentrations of functional lipids in abraded fractions of hulless barley and effect of storage. *J. Food Sci.* **2008**, 73, C569–C576.

Lupton, J.R.; J.L. Morin; M.C. Robinson. Barley bran flour accelerates gastrointestinal transit time. *J. Am. Diet Assoc.* **1993**, *93,* 881–885.

Lupton, J.R.; M.C. Robinson; J.L. Morin. Cholesterol-lowering effect of barley bran flour and oil. *J. Am. Diet Assoc.* **1994**, *94,* 65–70.

Martinez, V.M.; R.K. Newman; C.W. Newman. Barley diets with different fat sources have hypocholesterolemic effects in chicks. *J. Nutr.* **1992**, *5,* 1070–1076.

Moreau, R.A.; D.C. Doehlert; R. Welti; G. Isaac;M. Roth; P. Tamura; R. Nuñez. The identification of mono-, di-, tri-, and tetragalactosyl-diacylglycerols and their natural estolides in oat kernels. *Lipids* **2008,** *43,* 533–548.

Moreau, R.A.; R.A. Flores; K.B. Hicks. The composition of functional lipids in hulled and hulless barley, in fractions obtained by scarification, and in barley oil. *Cereal Chem.* **2007a**, *84,* 1–5.

Moreau, R.A.; K.B. Hicks. The composition of corn oil obtained by the alcohol extraction of ground corn. *J. Am. Oil Chem. Soc.* **2005**, *82,* 809–815.

Moreau, R.A.; M.J. Powell; K.B. Hicks. Extraction and quantitative analysis of oil from commercial corn fiber. *J. Agric. Food Chem.* **1996**, *44,* 2149–2154.

Moreau, R.A.; M.J. Powell; K.B. Hicks; R.A. Norton. A comparison of the levels of ferulate-phytosterol esters in corn and other seeds. *Advances in Plant Lipid Research;* J. Sánchez; E. Cerdá-Olmedo; E. Martínez-Force, Eds.; Universidad de Sevilla: Sevilla, 1998; pp. 472–474.

Moreau, R.A.; M.J. Powell; V. Singh. Pressurized liquid extraction of polar and nonpolar lipids in corn and oats with hexane, methylene chloride, isopropanol, and ethanol. *J. Am. Oil Chem. Soc.* **2003**, *80,* 1063–1067.

Moreau, R.A.; V. Singh; A. Nunez; K.B. Hicks. Phytosterols in the aleurone layer of corn kernels. *Biochem. Soc. Trans.* **2000**, *28,* 803–806.

Moreau, R.A.; K.E. Wayns; R.A. Flores; K.B. Hicks. Tocopherols and tocotrienols in barley oil prepared from germ and other fractions from scarification and sieving of hulless barley. *Cereal Chem.* **2007b**, *84,* 587–592.

Moreau, R.A.; B.D. Whitaker; K.B. Hicks. Phytosterols, phytostanols, and their conjugates in foods: structural diversity, quantitative analysis, and health-promoting uses. *Prog. Lipid Res.* **2002**, *41,* 457–500.

Morrison, W.R. Cereal lipids. *Advances in Cereal Sciences and Technology,* Y. Pomeranz, Ed.; American Association of Cereal Chemists: St. Paul, MN, 1978; Vol. II, pp. 221–348.

Morrison, W.R. Barley lipids. *Barley Chemistry and Technology*; A.W. McGregor, R.S. Bhatty, Eds.; American Association of Cereal Chemists: St. Paul, 1993; pp. 199–246.

Nesaretnam, K.; R Stephen; R. Dils; P. Darbre. Tocotrienols inhibit the growth of human breast cancer cells irrespective of estrogen receptor status. *Lipids* **1998**, *33,* 461–469.

Newman, C.W.; R.K. Newman. A brief history of barley foods. *Cereal Foods World* **2006**, *51,* 4–7.

Newman, R.K.; C.W. Newman; P.J. Hofer; A.E. Barnes. Growth and lipid metabolism as affected by feeding of hull-less barleys with and without supplemental β-glucanase. *Plant Foods for Human Nutr.* **1991**, *41,* 371–380.

Nilsson, M.; P. Åman; H. Härkönen; G. Hallmans; K.E. Bach Knudsen; W. Mazur; H. Adlercreutz. Content of nutrients and lignans in rolled milled fractions of rye. *J. Sci. Food Agric.* **1997**, *73,* 143–148.

Parker, R.A.; B.C. Pearce; R.W. Clark; D.A. Gordon; J.J.K. Wright. Tocotrienols regulate cholesterol production in mammalian cells by post-translational suppression of 3-hydroxy-3-methylglutaryl coenzyme A reductase. *J. Biol. Chem.* **1993**, *268,* 11230–11238.

Parmar, S.; E.G. Hammond. Hydrolysis of fats and oils with moist oat caryopses. *J. Am. Oil. Chem. Soc.* **1994,** *71,* 881–886.

Peterson, D.M. Barley tocols: Effects of milling, malting, and mashing. *Cereal Chem.* **1994**, *71,* 42–44.

Peterson, D.M.; A.A. Qureshi. Genotype and environmental effects on tocols of barley and oats. *Cereal Chem.* **1993**, *70,* 150–162.

Piazza,G.; A. Bilyk; D. Schwartz; M. Haas. Lipolysis of olive oil and tallow in an emulsifier free two phase system by the lipase from oat seeds (*Avena sativa*). *Biotechnol. Lett.* **1989**, *11,* 487–492.

Piironen, V.; D.G. Lindsay; T.A. Miettinen; J. Toivo; A-M. Lampi. Plant sterols: biosynthesis, biological function and their importance to human nutrition. *J. Sci. Food Agric.* **2000a**, *80,* 939–966.

Piironen, V.; E.-L. Syvaljo; P. Varo; K. Salminen; P. Koivistoinen. Tocopherols and tocotrienols in cereal products from Finland. *Cereal Chem.* **1986,** *63,* 78–81.

Piironen, V.; J. Toivo; A-M. Lampi. Natural sources of dietary plant sterols. *J. Food Comp. Anal.* Plant sterols in cereals and cereal products. *Cereal Chem.* **2000b,** *79,* 148–154.

Piironen, V.; J. Toivo; A.-M. Lampi. Plant sterols in cereals and cereal products. *Cereal Chem.* **2002**, *79,* 148–154.

Pins, J.J.; H. Kaur. A review of the effects of barley β-glucan on cardiovascular and diabetic risk. *Cereal Foods World* **2006**, *51*, 8–12.

Price, P.G.; H. Parsons. Distribution of lipids in embryonic axis, bran-endosperm, and hull fractions of hulless barley and hulless oat grain. *J. Agric. Food Chem.* **1979**, *27,* 813–815.

Prýma, J.; J. Ehrenbergerová; N. Belcrediová; K. Vaculová. Tocol content in barley. *Acta. Chim. Slov.* **2007**, *54,* 102–105.

Przybylski, R. Cereal grain oils. *Nutraceutical and Specialty Lipids and Their Co-Products;* F. Shahidi, Ed.; CRC Press: Boca Raton, 2006.

Qureshi, A.A.; W.C. Burger; D.M. Peterson; C.E. Elson. The structure of an inhibitor of cholesterol biosynthesis isolated from barley. *J. Biol. Chem.* **1986**, *261*, 10544–10550.

Qureshi, A.A.; N. Qureshi; J.O. Hasler-Rapascz; F.E. Weber; V. Chaudhary; T.D. Crenshaw; A. Gapor; A.S.H. Ong; Y.H. Chong; D.M. Peterson; J. Rapacz. Dietary tocotrienols reduce concentrations of plasma cholesterol, apolipoprotein B, thromboxane B2, and platelet factor 4 in pigs with inherited hyperlipidemias. *Am. J. Clin. Nutr.* **1991**, *53,* 1021S–1026S.

Robinson, M.C.; J.R. Lupton. The effects of barley flour and barley oil on hypercholesterolemic men and women. *J. Am. Diet. Assoc.* **1990**, *90,* 90A–104 (Suppl.).

Sen, C.K.; S. Khanna; C. Rink; S. Roy. Tocotrienols: The emerging face of natural vitamin E. *Vitamins and Hormones* **2007**, *76,* 203–261.

Sen, C.K.; S. Khanna; S. Roy. Tocotrienol: The natural vitamin E to defend the nervous system? *Ann. N.Y. Acad. Sci.* **2004,** *1031,* 127–142.

Seog, H-M.; M-S. Seo; Y-S. Kim; Y-T. Lee. Physicochemical properties of barley bran, germ and broken kernels as pearling by-products. *Food Sci. Biotechnol.* **2002,** *11,* 623–627.

Siebenhandl, S.; H. Grausgruber; N. Pellegrini; D. Del Rio; V. Fogliano; R. Pernice; E. Berghofer. Phytochemical profiles of main antioxidants in different fractions of purple and blue wheat, and black barley. *J. Agric. Food Chem.* **2007**, *55,* 8541–8547.

Singh, V.; R.A. Moreau. Methods of preparing corn fiber oil and of recovering corn aleurone cells from corn fiber. U.S. Patent 7,115,295, 2006.

Singh, V.; R.A. Moreau; P.H. Cooke. Effect of corn milling practices on aleurone layer cells and their unique phytosterols. *Cereal Chem.* **2001**, *78,* 436–441.

Sosulski, K.; S. Wang; W.M. Ingledew; F.W. Sosulski; J. Tang. Preprocessed barley, rye, and triticale as a feedstock for an integrated fuel ethanol-feedlot plant. *Appl. Biochem. Biotechnol. 63–65,* **1997**, 59–70.

Theriault, A.; J.-T. Chao; A. Gapor. Tocotrienol is the most effective vitamin E for reducing endothelial expression of adhesion molecules and adhesion of monocytes. *Atherosclerosis* **2002**, *160,* 21–30.

Wang, L.; S.R. Behr; R.K. Newman; C.W. Newman. Comparative cholesterol-lowering effects of barley beta glucan and barley oil in golden Syrian hamsters. *Nutr. Res.* **1997A,** *17,* 77–88.

Wang, L.; R.K. Newman; C.W. Newman; L.L. Jackson; P.J. Hofer. Tocotrienol and fatty acid composition of barley oil and their effects on lipid metabolism. *Plant Foods Human Nutr.* **1993a** *43,* 9–17.

Wang, S.; K. Sosulski; F. Sosulski; M. Ingledew. Effect of sequential abrasion on starch composition of five cereals for ethanol fermentation. *Food Res. Int.* **1997b**, *30*, 603–609.

Wang, L.; Q. Xue; R.K. Newman; C.W. Newman. Enrichment of tocopherols, tocotrienols, and oil in barley fractions by milling and pearling. *Cereal Chem.* **1993b,** *70,* 499–501.

Wehrmüller, K. Occurrence and biological properties of sphingolipids–A review. *Curr. Nutr. Food Sci.* **2007**, *3*, 161–173.

Yang, B. Natural vitamin E: activities and sources. *Lipid Technol.* **2003**, November, 125–130.

Yeung, J.; T. Vasanthan. Pearling of hull-less barley: product composition and gel color of pearled barley flours as affected by the degree of pearling. *J. Agric. Food Chem.* **2001**, *49*, 331–335.

Zhang, J-X.; E. Lundin; H. Andersson; I. Bosaeus; S. Dahlgren; G. Hallmans; R. Stenling; P. Aman. Brewer's spent grain, serum lipids and fecal sterol excretion in human subjects with ileostomies. *J. Nutr.* **1991**, *121*, 778–784.

Zheung, G.H.; B.G. Rossnagel; R.T. Tyler; R.S. Bhatty. Distribution of β-glucan in the grain of hull-less barley. *Cereal Chem.* **2000,** *77,* 140–144.

Zielinski, H.; H. Kozlowska. Antioxidant activity and total phenolics in selected cereal grains and their different morphological fractions. *J. Agric. Food Chem.* **2000**, *48*, 2008–2016.

# 18

# Parsley, Carrot, and Onion Seed Oils

**Liangli (Lucy) Yu[1] and Junjie (George) Hao[2,3]**
*[1]Department of Nutrition and Food Science, [2]Department of Chemistry and Biochemistry, and [3]Department of Mathematics, University of Maryland, College Park, MD 20742*

## Introduction

Growing evidence indicates the potential of improving human health and the quality of life through improving diet (De Caterina et al., 2006; Griel et al., 2008; Trichopoulou et al., 2006). One of the keys for the development and the production of healthy foods is the development of healthy food ingredients. Edible seed oils represent a group of important food ingredients. Fatty acids, the primary nutritional components in edible oils, are critical in the role that oils may serve in human nutrition and health (De Caterina et al., 2006; Fermandes & West, 2005). The presence of other phytochemicals, such as antioxidants, in oils may also alter their effect on human health (Parry et al., 2005, 2006; Yu et al., 2005). This chapter summarizes the available information about edible parsley, carrot, and onion seed oils with an emphasis on their fatty acid profiles, levels of other beneficial phytochemicals, antioxidant properties, and physical characteristics. In addition, mechanisms by which dietary fatty acids and other phytochemical components in these seed oils may alter the risk of cardiovascular disease are briefly discussed.

## Parsley Seed Oil

Parsley (*Petroselinum crispum*) is an important culinary herb, and is used widely as a flavoring ingredient in China, Mexico, India, South America, and Southeast Asia (Wong & Kitts, 2006). Parsley is also used in folk medicine, and may have antimicrobial activities (Kreydiyyeh & Usta, 2002; Wong & Kitts, 2006). Some studies characterize parsley seed oils and seed extracts by their chemical compositions and possible health-beneficial properties (Gunstone, 1991; Kreydiyyeh & Usta, 2002; Parry et al., 2005, 2006). In 2006, two commercially available cold-pressed parsley seed oils were analyzed for their fatty acid profiles, tocopherol and carotenoid contents, antioxidant properties, oxidative stability, and physical properties (Parry et al., 2006). The commercial cold-pressed parsley seed oils contained approximately 81% of oleic acid

(Table 18.1), along with roughly 11% of linoleic and 3% of palmitic acids according to gas chromatographic (GC) analysis (Parry et al., 2006). This oleic acid concentration was higher than the roughly 68–78% found in olive oil (Okogeri & Tasioula-Margari, 2002). However, Gunstone (1991) reported the presence of petroselinic acid (18:1) along with oleic acid in crude parsley (*Petroselinum sativum*) seed oil prepared with petroleum ether extraction, which was indistinguishable through GC analysis. The crude parsley seed oil contained about 82% of total 18:1 and about 13% of linoleic and 4% of palmitic acids (Gunstone, 1991). Petroselinic and oleic acids, both octadecenoic acids (18:1), were distinguished through $^{13}$C-NMR analysis. In addition, the parsley seed oils were shown to contain about 0.5% of α-linolenic acid, the essential ω-3 fatty acid, which leads to a ω-3 to ω-6 fat ratio of 1:22 (Gunstone, 1991; Parry et al., 2006). Taken together, these data suggest that parsley seed oil may serve as a dietary source of monounsaturated octadecenoic or monounsaturated fatty acids (MUFA). Diets rich in MUFA may have health benefits such as reducing blood-cholesterol levels, modulating immune function, and lowering the risk of atherosclerosis (Griel et al., 2008; Nicolosi et al., 2002; Parker et al., 2003). Multi-biochemical mechanisms may contribute to the biological actions of fatty acids such as altering the expression of genes involved in biosynthesis and the metabolism of fat and cholesterol and the liver LDL receptors. The effects of MUFA and fatty acids in cardiovascular disease and the multi-biochemical mechanisms are found in the two review articles

**Table 18.1. Fatty Acid Composition (g/100 g of fatty acids)**

| Fatty acid | Parsley1[Z] | Parsley2[Z] | Parsley3[Y] | Carrot1[X] | Carrot2[Y] | Carrot3[W] | Onion1[Z] | Onion2[Z] | Onion3[V] |
|---|---|---|---|---|---|---|---|---|---|
| 14:0 | TR | TR | TR | ND | TR | ND | TR | ND | 0.7 |
| 16:0 | 3.1 | 3.1 | 3.5 | 3.71 | 3.8 | 6.7 | 6.4 | 7.1 | 9.1 |
| 16:1 | 0.1 | 0.1 | 0.3 | ND | 0.5 | ND | 0.2 | 0.2 | ND |
| 18:0 | 4.2 | 4.2 | 0.7 | 0.42 | 0.8 | 0.5 | 2.4 | 1.8 | 4.4 |
| 18:1 | 81.0 | 80.9 | 81.9[a] | 82.08 | 80.4[a] | 80.2 | 24.8 | 26 | 34.3 |
| 18:2 | 11.0 | 11.0 | 12.5 | 13.19 | 12.6 | 12.2 | 65.2 | 64 | 44.6 |
| 18:3 | 0.5 | 0.5 | ND | 0.28 | ND | 0.4 | 0.1 | 0.3 | 0.3 |
| 20:0 | 0.1 | 0.1 | 0.3 | 0.33 | 0.2 | ND | 0.3 | 0.3 | ND |
| 20:1 | ND | ND | 0.2 | ND | 0.5 | ND | 0.4 | 0.4 | ND |
| SAT | 7.4 | 7.4 | 4.5 | 4.46 | 4.8 | 7.2 | 9.1 | 9.2 | 14.2 |
| MUFA | 81.1 | 81 | 82.4 | 82.08 | 81.4 | 80.2 | 25.4 | 26.6 | 34.3 |
| PUFA | 11.5 | 11.5 | 12.5 | 13.47 | 12.6 | 12.6 | 65.3 | 64.3 | 44.9 |

[Z] = Parry et al., 2006; [Y] = Gunstone, 1991; [X] = Parker et al., 2003; [W] = Prasad et al., 1987; [V] = Rao, 1994.

[a] Value is total of petroselinic and oleic acids.

ND: not detected; TR: trace; SAT: total saturated fatty acids; MUFA: total monounsaturated fatty acids; PUFA: total polyunsaturated fatty acids.

and one human study published recently (De Caterina et al., 2006; Griel et al., 2008; Trichopoulou et al., 2006). Vegetable oils rich in MUFA are in great demand since a MUFA-rich diet is an alternative to a low-fat diet for those at higher risk for atherosclerosis and cardiovascular disease (Nicolosi et al., 2002; Trichopoulou et al., 2006).

Cold-pressed parsley seed oils contain a total tocopherol concentration of 77.6–80.6 μmol/kg, according to HPLC analysis (Table 18.2) (Parry et al., 2006). The total tocopherol content of the cold-pressed parsley seed oils was higher than that of the cold-pressed cardamom seed oil, but was less than that of 464–1974 μmol/kg found in similarly prepared onion, mullein, roasted pumpkin, and milk thistle seed oils (Parry et al., 2006). Total tocopherol contents are reported in μmol/kg because individual tocopherol isomers have different molecular weights, and it is not accurate to add their isomer contents and report in milligrams per kilograms of oil. Levels of three different vitamers were 29 and 29.5 mg/kg for α-, 2.8 and 4 mg/kg for γ-, and 0.9 and 1.2 mg/kg for δ-tocopherol (Table 18.2). The ratio of α- to γ- to δ-tocopherols in the parsley seed oil was 24.2 and 32.8:3.1 and 3.3:1, suggesting that the percentage of α-tocopherol was much higher than that of the other two tocopherol vitamers in comparison to the vitamers ratio for onion seed oils. The level of α-tocopherol in the parsley seed oil was comparable to that of about 27 mg/kg in mullein and roasted pumpkin seed oils, but was much lower than that found in wheat germ and onion, extra virgin olive, peanut, sunflower, and milk thistle seed oils (Parry et al., 2006). Cold-pressing involves no heat and refining steps during the oil-producing process and may retain more beneficial phytochemicals along with pigments, phospholipids, and volatiles in the cold-pressed oils (Parry et al., 2005, 2006).

High levels of carotenoids were detected in the cold-pressed parsley seed oils (Parry et al., 2006). Total carotenoids in the parsley seed oils ranged from 40.28 to 40.49 μmol/kg, which was comprised of approximately 0.78–0.99 mg/kg of β-carotene, 0.21–0.22 mg/kg of lutein, 20–21 mg/kg of zeaxanthin, and 1.2–1.4 mg/

**Table 18.2. Tocopherol Contents* (Parry et al., 2006)**

| | α-Tocopherol (mg/kg) | γ-Tocopherol (mg/kg) | δ-Tocopherol (mg/kg) | Total tocopherols (μmol/kg) |
|---|---|---|---|---|
| Parsley1 | 29.5 | 2.8 | 0.9 | 77.6 |
| Parsley2 | 29.0 | 4.0 | 1.2 | 80.6 |
| Onion1 | 681.9 | 219.2 | 28.6 | 1973.8 |
| Onion2 | 498.1 | 156.3 | 23.0 | 1762.6 |

* Parsley1 and Parsley2 represent the two cold-pressed parsley seed oil samples; and Onion1 and Onion2 represent the two cold-pressed onion seed oil samples.

kg of cryptoxanthin (Table 18.3) (Parry et al., 2006). The level of total carotenoids found in the parsley seed oils is comparable to that detected in selected cold-pressed berry seed oils (12.5–30.0 μmol/kg) (Parry et al., 2005) and higher than the levels of 0.05–15.80 μmol/kg observed in onion, cardamom, mullein, and milk thistle seed oils, but lower than that of 70.59 μmol/kg of roasted pumpkin seed oils (Parry et al., 2006). Zeaxanthin was the primary carotenoid compound in the cold-pressed parsley seed oil, and its level is greater than that of 5.1–13.6 mg/kg in berry seed oils (Parry et al., 2005) and 1.2–6.4 mg/kg in onion, milk thistle, and mullein seed oils (Parry et al., 2006), but less than 28.5 mg/kg found in the cold-pressed roasted pumpkin seed oil (Parry et al., 2006). These data suggest that parsley seed oil is an excellent dietary source for zeaxanthin and carotenoids in general, but not for tocopherols.

The antioxidant capacity of parsley seed oils was estimated as their scavenging ability against peroxyl (ORAC) and 2,2'-bipyridyl, 2,2-diphenyl-1-picrylhydrazyl (DPPH•) radicals. Total phenolic contents of the parsley seed oils were also examined because phenolic compounds are antioxidants. A higher ORAC value is associated with a stronger free radical scavenging activity or antioxidant capacity. The cold-pressed parsley seed oils had the highest ORAC value of 0.54–1.20 μmol of trolox equivalents (TE)/g oil, indicating its strong scavenging capacity in directly reacting with, and quenching, free peroxyl radicals among the three oils summarized in Table 18.4 (Parry et al., 2006). This value is comparable to that of cardamom seed oil (0.94 mmol of TE/g of oil) and greater than the 0.13 mmol of TE/g for milk thistle seed oil (Parry et al., 2006). This value was much greater than that of 17–78 μmol/g for the selected berry fruit seed oils, and 1.1–27 μmol of TE/g for roasted pumpkin, onion, and mullein seed oils (Parry et al., 2005, 2006). Roughly put, the parsley seed oil had a peroxyl radical scavenging capacity that was two orders of magnitude greater than that of cold-pressed roasted pumpkin seed oil under the same testing conditions. The strong free radical scavenging capacity of parsley seed oil components was also evident in its ability to quench 87–92% of 2,2'-bipyridyl, 2,2-diphenyl-1-picrylhydrazyl radical (DPPH•) in the antioxidant-radical reactions (Table 18.4). This DPPH• scavenging capacity was stronger than that of cold-pressed cardamom, roasted pumpkin, and

**Table 18.3. Carotenoid Contents* (Parry et al., 2006)**

| | β-Carotene (mg/kg) | Lutein (mg/kg) | Zeaxanthin (mg/kg) | Cryptoxanthin (mg/kg) | Total Carotenoids (μmol/kg) |
|---|---|---|---|---|---|
| Parsley1 | 0.78 | 0.22 | 20.40 | 1.43 | 40.28 |
| Parsley2 | 0.99 | 0.21 | 20.55 | 1.20 | 40.49 |
| Onion1 | ND | 0.02 | 1.74 | 0.51 | 4.01 |
| Onion2 | ND | 0.02 | 1.22 | 0.75 | 3.52 |

*Parsley1 and Parsley2 represent the two cold-pressed parsley seed oil samples; and Onion1 and Onion2 represent the two cold-pressed onion seed oil samples.

**Table 18.4. Antioxidant Properties of Oils***

| | ORAC (μmol TE/g) | % DPPH • Remaining[z] | TPC (mg GAE/g) |
|---|---|---|---|
| Pasley1 | 537 | 13.4 | 1.68 |
| Parsley2 | 1097 | 9.2 | 2.27 |
| Carrot | 160 | 37.1 | 1.98 |
| Onion1 | 4.6 | 22.7 | 2.16 |
| Onion2 | 17.5 | 24.2 | 3.35 |

* Parsley1 and Parsley2 represent the two cold-pressed parsley seed oil samples; Carrot stands for the cold-pressed carrot seed oil sample; and Onion1 and Onion2 represent the two cold-pressed onion seed oil samples. ORAC stands for the oxygen radical absorbance capacity, and greater ORAC value is associated with a stronger capacity of the antioxidant to protect lipid oxidation. A lower value of % DPPH• Remaining is associated with a stronger antioxidant activity. TPC means total phenolic content. Data were originally reported in Parry et al., 2006 and Yu et al., 2005.

[z] The final concentration for DPPH• was 100 μM in all antioxidant-radical reactions, and the total volume for each reaction was 2.0 mL. The absorbance at 517 nm of each reaction mixture was estimated at 40 min of reaction for carrot seed oil extract and at 10 min of reaction for parsley and onion seed oil extracts. The final antioxidant concentration was 10.9 mg of oil equivalent per mL for carrot seed oil extract, and 40 mg of oil equivalent per mL for parsley and onion seed oil extracts.

milk thistle seed oils on a per-same-oil-weight basis (Parry et al., 2006). Furthermore, the DPPH• scavenging capacity of parsley seed oil antioxidants showed dose- and time dependence (Fig. 18.1). The total phenolic component (TPC) of parsley seed oil was about 1.7–2.3 mg of gallic acid equivalents (GAE)/g (Table 18.4), which was greater than that of 0.98 mg of GAE/g for roasted pumpkin seed oil and comparable to that of 1.8 and 2.54 mg of GAE/g for cardamom and mullein seed oils, respectively (Parry et al., 2006). This value was also comparable to that of 1.5–2.0 mg of GAE/g for marionberry, blueberry, boysenberry, and red raspberry seed oils (Parry et al., 2005), but was less than the concentration found in milk thistle seed oil (3.1 mg of GAE/g) (Parry et al., 2006). Taking into account the ORAC, % of DPPH• remaining, and TPC values, cold-pressed parsley seed oil is an excellent dietary source for natural antioxidants. Additional research is required to further investigate the health-beneficial properties of parsley seed oil and its components.

Oxidative stability index (OSI) is a measurement of a lipid sample's stability and resistance to oxidation; higher values of OSI reflect greater oxidative stability, and can be associated with longer shelf-life. The two cold-pressed parsley seed oils showed an excellent oxidative stability and did not develop measurable rancidity in 148 or 369 hours under the experimental conditions (Fig. 18.2). The OSI values of these parsley seed oils were 2.2–5.6 and 3.3–7.9 times higher than those of commercial soybean and corn oils, respectively (Parry et al., 2006). The parsley seed oils were more stable than the cold-pressed onion, cardamom, mullein, roasted pumpkin, milk thistle, red

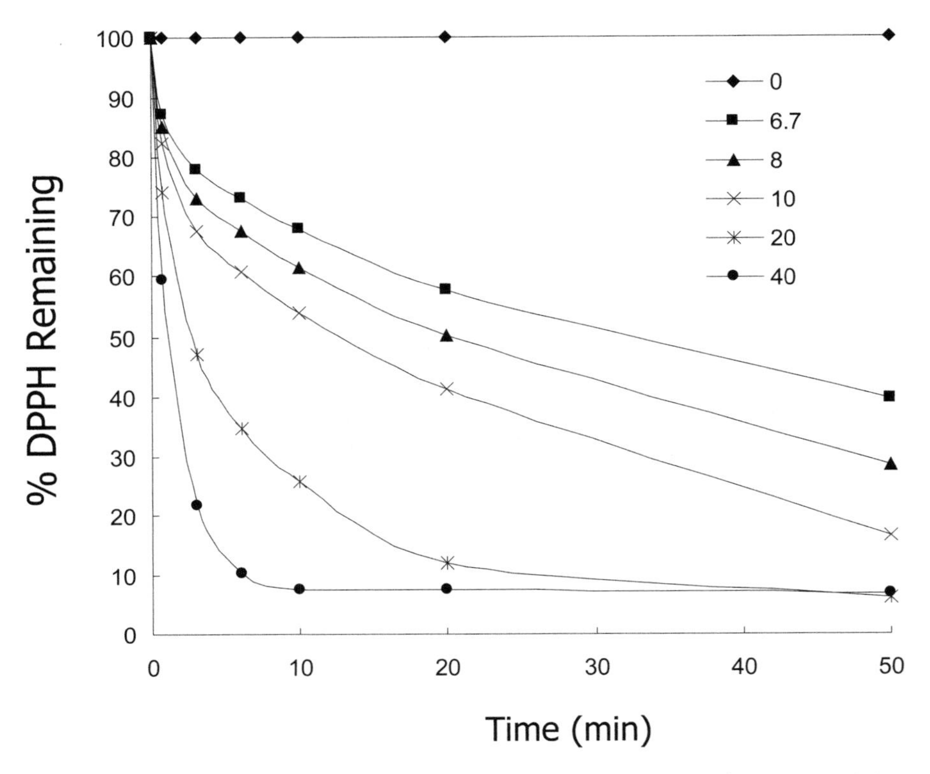


**Fig. 18.1.** Kinetics of the parsley seed oil antioxidants-DPPH˙ reactions. The 100% MeOH extract of cold-pressed parsley seed oil was examined for its capacity to react with DPPH˙. 0, 6.7, 8, 10, 20, and 40 represent the final concentrations of the seed oil extracts at 0, 6.7, 8, 10, 20, and 40 mg of oil equivalents per mL in the antioxidant-radical reaction mixtures, respectively. The initial DPPH˙ concentration was 100 µM in all reaction mixtures.

raspberry, marionberry, boysenberry, cranberry, carrot, and hemp seed oils (Parker et al., 2003; Parry et al., 2005, 2006). The parsley seed oils had density values ranging from 0.981 to 0.985 g/mL (Table 18.5), which was higher than those of mullein, milk thistle, and roasted pumpkin seed oils, and selected berry fruit seed oils (Parry et al., 2005). The refractive index values for the parsley seed oils varied from 1.4858 to 1.4862. The HunterLab color values for the cold-pressed parsley seed oils were also determined and are summarized in Table 18.5.

## Carrot Seed Oil

An examination of the fatty acid composition of carrot seed oil, extracted by either cold-pressing (Parker et al., 2003) or with petroleum ether (Gunstone, 1991), re-

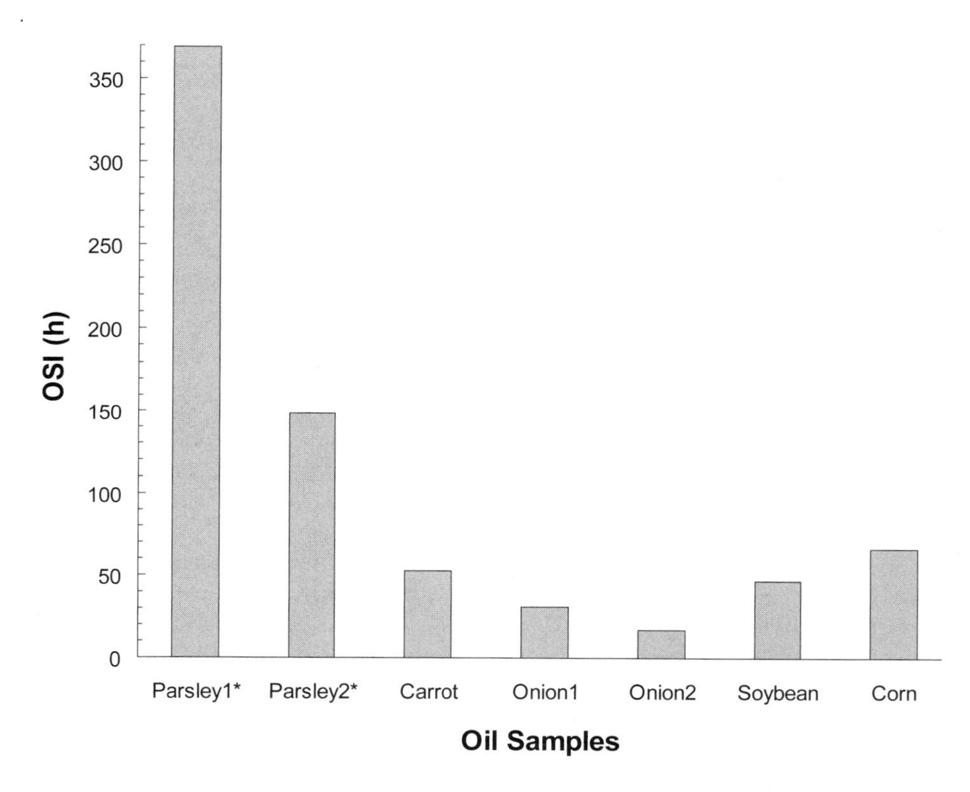


**Fig. 18.2.** Oxidative stability index (OSI) of carrot, parsley, and onion seed oil. The oxidation reactions were performed at 80°C with an air flow of 7 L/hour using a Rancimat instrument (Metrohm Ltd., Herisau, Switzerland). The OSI was defined as the time for an oil sample to develop a measurable rancidity. *Parsley seed oils had no detectable rancidity at 369 or 148 hours under the experimental conditions when the OSI tests were terminated.

vealed that both samples contained predominantly octadecenoic acid (18:1) (Table 18.1). The cold-pressed carrot seed oil contained approximately 82% of oleic acid (18:1n-9), along with roughly 13% of linoleic (18:2n-6) and 4% of palmitic (16:0) acids based on GC analysis (Table 18.1) (Parker et al., 2003). The level of oleic acid was comparable to that of the parsley seed oils discussed previously (Table 18.1) and greater than that found in olive oils (Okogeri & Tasioula-Margari, 2002). A minor amount (0.28%) of α-linolenic acid (18:3n-3) was detected in the cold-pressed carrot seed oil (Parker et al., 2003). In an earlier study conducted in 1991 by Gunstone, however, the carrot seed oil prepared through petroleum ether extraction was shown to contain about 80% of combined petroselinic and oleic (18:1n-11 and 18:1n-9) acids, followed by roughly 13% of linoleic acid and 4% of palmitic acid, while no

**Table 18.5. Physical Properties of Seed Oils***

| | Refractive index $n^{25}_D$ | Density (g/mL) | HunterLab Color | | |
|---|---|---|---|---|---|
| | | | L value | a value | b value |
| Parsley1[Z] | 1.4858 | 0.985 | 1.96 | 2.88 | 1.92 |
| Parsley2[Z] | 1.4862 | 0.981 | 1.22 | 2.70 | 1.44 |
| Carrot[Y] | | | 11.10 | 3.68 | 18.82 |
| Onion1[Z] | 1.4752 | 0.930 | 3.50 | -1.38 | 3.90 |
| Onion2[Z] | 1.4752 | 0.923 | 3.17 | -1.26 | 2.96 |

* Parsley1 and Parsley2 represent the two cold-pressed parsley seed oil samples; Carrot stands for the cold-pressed carrot seed oil sample; and Onion1 and Onion2 represent the two cold-pressed onion seed oil samples. Refractive index and density were determined at ambient temperatures. [Z] = Parry et al., 2006; [Y] = Parker et al., 2003.

α-linolenic acid (18:3n-3) was detected under the experimental conditions (Table 18.1). The distinguishing of petroselinic and oleic acids, both octadecenoic acids, was achieved through $^{13}$C-NMR examination (Gunstone, 1991). The presence of petroselinic acid in carrot (*Daucus carota*) seed oil was also observed in a later study using $^{13}$C-NMR analysis (Lie Ken Jie et al., 1996). In addition, petroselinic and oleic acids accounted for 68.9% and 11.3% of total fatty acids, respectively, in the phospholipids of carrot seeds according to GC-FID analysis, which had a total of 80.2% of 18:1 (Prasad et al., 1987). Prasad and others (1987) also reported 29.1% of phosphatidylcholine, 35.4% of phosphatidylethanolamine, and 23.1% of phosphatidylinositol in the carrot seed phospholipids. These results suggest that carrot seed oil would also serve as a good dietary source of MUFA, and the cold-pressed carrot seed oil may also contain significant levels of phospholipids.

In 2005, Yu and others estimated the antioxidant properties of cold-pressed carrot seed oil as scavenging activity against peroxyl (ORAC), 2,2'-bipyridyl, 2,2-diphenyl-1-picrylhydrazyl (DPPH$^•$), and cation 2,2' azinobis(3-ethylbenzothiazolin-6-sulfonic acid) diammonium salt (ABTS$^{•+}$) radicals, as well as chelating activity against $Fe^{2+}$ and total phenolic content (TPC). Cold-pressed extra virgin carrot seed oil was extracted using 100% of methanol, and the methanol extract was used for DPPH$^•$ scavenging capacity assay. For other antioxidant property estimations, methanol was removed from a known quantity of the extract, and the residue was quantitatively redissolved in dimethyl sulfoxide (DMSO). The carrot seed oil exhibited an ORAC value of 160 μmol of trolox equivalents (TE) per gram, quenched nearly 63% of DPPH$^•$ in a reaction time of 40 minutes, and demonstrated a TPC of approximately 2 mg of GAE/g (Table 18.4). In addition, the carrot seed oil exhibited an ABTS$^{•+}$ scavenging capacity of about 9 μmol of TE/g, and demonstrated chelating activity against $Fe^{2+}$ of roughly 26 mg of 2,2'-bipyridyl, disodium ethylenediaminetetraacetate equivalents (EDTAE) per gram (Yu et al., 2005).

The OSI value of carrot seed oil was roughly 52 hours (Fig. 18.2), which is comparable to a commercial soybean oil (Parker et al., 2003). The cold-pressed carrot seed oil had HunterLab color values of 11.10 for L, 3.68 for a, and 18.82 for b (Table 18.5), appearing in a similar color as carrots. To the best of our knowledge, no other chemical or physical properties of carrot seed oils were evaluated, including tocopherol or carotenoid content, refractive index, or density.

## Onion Seed Oil

Onion is the second most important horticultural crop and widely consumed vegetable worldwide, with an approximate $3–4 billion market in the United States alone (Sellappan & Akoh, 2002; Yang et al., 2004). Commercial onions may come in white, yellow, or red varieties, with or without a sweet taste. Onions contain a number of phytochemicals including flavonoids, organosulfur compounds, and other natural antioxidants, which may contribute to the health benefits associated with onion consumption (Nishimura et al., 2006; Sellappan & Akoh, 2002; Yang et al., 2004).

Recently, cold-pressed commercial onion seed oils were characterized for their chemical composition and antioxidant properties (Parry et al., 2006). The primary fatty acids in the onion seed oils were linoleic (64–65%), oleic (25–26%), palmitic (6–7%), and stearic (close to 2%) acids (Table 18.1). Onion oil had a ratio of PUFA to MUFA to SAT of 7.0–7.2:2.8–2.9:1. Earlier in 1994, India-grown onion (*Allium cepa*) seed oil was prepared by ether extraction and analyzed for its fatty acid profile (Rao, 1994). The onion seed had 23.6% of fat and about 28% of protein on a per-seed-weight basis, along with about 16% (w/w) of fiber and important inorganic elements and vitamins. The onion oil contained around 45% of linoleic and 34% of oleic acids (Table 18.1) and 9.1% and 4.4% of palmitic and stearic acids, respectively. The ratio of PUFA/MUFA/SAT was 3.1:2.4:1 for this oil, which had a lower concentration of PUFA and might be a better choice for applications involving thermal treatments and for food products requiring longer shelf- life. The level of α-linolenic acid (18:3n-3) ranged from 0.1 to 0.3% of total fatty acids according to these studies (Parry et al., 2006; Rao, 1994). In summary, onion seed oil may be a good dietary source of linoleic acid, the essential dietary ω-6 fatty acid, along with a sizeable amount of oleic acid.

The cold-pressed onion seed oils had high levels of tocopherols (Table 18.2). The total tocopherol content in the cold-pressed onion seed oils was in the range of 1.76–1.97 mmol/kg of oil, which was comparable to that of 2.1–2.3 mmol/kg in the cold-pressed red raspberry and boysenberry seed oils (Parry et al., 2005). The onion seed oils had an α-tocopherol level of 498–682 mg/kg, which is equivalent to 460–634 mg/L. This range was greater than that of 89 mg/L found in soybean oil (Cabrini et al., 2001) and 21–151 mg/kg in selected berry seed oils (Parry et al., 2005), comparable to that of 174–578 mg/L for extra virgin olive, corn, peanut, and sunflower seed oils (Parry et al., 2006), and lower than that of 1.330 g/kg in wheat germ oil (Wagner et al., 2004). The cold-pressed onion seed oils were also rich in γ- and δ-tocopherol isomers (Table 18.2). The ratio of α-, γ-, and δ-tocopherols was

21.7–23.8:6.8–7.7:1 for the cold-pressed onion seed oils. These data indicate that onion seed oil may be an excellent dietary source for total, α-, and γ-tocopherols.

The carotenoid content of the cold-pressed onion seed oils was much lower than that detected in the parsley seed oils under the same experimental conditions (Table 18.3). Zeaxanthin was the primary carotenoid compound found in the cold-pressed onion seed oil, along with cryptoxanthin and lutein (Parry et al., 2006). No β-carotene was detected under the experimental conditions. In conclusion, parsley seed oil is a much better dietary source for carotenoids than onion seed oil.

The methanol extracts of the cold-pressed onion seed oils showed antioxidant properties including the scavenging capacity against peroxyl and DPPH radicals, and contained significant amounts of phenolic compounds (Parry et al., 2006). The methanol extracts of the two onion seed oils exhibited an ORAC value of 4.6–17.5 μmol of TE/g oil, which is comparable to that of 13–154 μmol of TE/g detected in common vegetables and fruits on a per dry weight basis (Wang et al., 1996; Wu et al., 2004), but much lower than levels found in carrot and parsley seed oils (Table 18.4), and that found in cold-pressed cardamom, milk thistle, and mullein seed oils (Parry et al., 2006), and wheat grain and bran (51–136 μmol of TE/g) (Zhou et al., 2004a, 2004b). The onion seed oil extracts were able to directly react with and quench DPPH radicals in a dose- and time-dependent manner (Parry et al., 2006), but quenched less DPPH radicals among the three oils discussed in this chapter, although they had a relatively high level of total phenolic contents (Table 18.4). In addition, the two cold-pressed onion seed oils differed significantly in their ORAC and TPC, suggesting the possible variation of antioxidant activities between onion seed oil samples. This difference may be partially explained by the onion genotype, growing conditions, the interaction between genotype and growing conditions, post-harvest treatments, oil-processing conditions, storage conditions, and the age of the oil.

The cold-pressed onion seed oils had OSI values of 17 and 31 hours, which were much lower than those of the parsley and carrot seed oils, as well as those of the corn and soybean oils (Fig. 18.2). This OSI value range was greater than the 13.3 hours for cold-pressed milk thistle seed oil and about 12 hours for hemp seed oil (Parker et al., 2003; Parry et al., 2006) and comparable to that of a little over 20 hours for cold-pressed red raspberry and marionberry seed oils (Parry et al., 2005), but is much lower than that of 45–152 hours for cold-pressed boysenberry, mullein, pumpkin, cranberry, cardamom, and black cumin seed oils (Parry et al., 2005, 2006).

The cold-pressed onion seed oils were also evaluated for their physical properties (Parry et al., 2006). The onion seed oils had density values of 0.923 and 0.930 g/mL and a refractive index value of 1.4752 (Table 18.5). These values are in agreement with $n^{25}_{D}$ of 1.4795 and a density of 0.9230 g/mL at 25° reported previously (El-Hinnawy & El-Tahawi, 1974). The onion seed oils were less dense than the parsley seed oils. The density of the onion seed oils was similar to red raspberry and marionberry seed oils and corn oil (Parry et al., 2005). The HunterLab color values of the onion seed oils are provided in Table 18.5.

In conclusion, parsley and carrot seed oils contain high levels of MUFA, while onion seed oil is highly unsaturated and rich in linoleic acid, the essential n-6 fatty acid. The edible oils rich in MUFA may have greater market potential due to the current consumer trend in reducing total, saturated and *trans* fat, and its relative better oxidative stability during cooking as compared to oils rich in PUFA.

The cold-pressed parsley, carrot, and onion seed oils are commercially available, may contain more beneficial phytochemicals, and may hold currently untapped commercial and nutritional values. These health-beneficial phytochemicals may include, but are not limited to, carotenoids, tocopherols, and natural antioxidants. Cold-pressed onion seed oil may serve as an excellent dietary source for tocopherols and natural phenolics, whereas cold-pressed parsley seed oil is a better source for carotenoids and natural antioxidants. These cold-pressed seed oils may also contain significant levels of pigments, phospholipids, and volatiles, which may alter the overall physicochemical and sensory properties of finished food products. Also, their seeds were not heavily developed by industry and hold little economical value, although parsley, carrot, and onion are found in cuisines from around the world. The development of commercial specialty edible oils from these seeds may enhance the profitability of these vegetable and herb productions.

## References

Cabrini, L.; V. Barzanti; M. Cipollone; D. Fiorentini; G. Grossi; B. Tolomelli; L. Zambonin; L. Landi. Antioxidants and total peroxyl radical-trapping ability of olive and seed oils. *J. Agric. Food Chem.* **2001**, *49*, 6026–6032.

De Caterina, R.; A. Zampolli; S. Del Turco; R. Madonna; M. Massaro. Nutricional mechnisms that influence cardiovascular disease. *Am. J. Clin. Nutr.* **2006**, *83*, 421S–426S.

El-Hinnawy, S.I.; B.S. El-Tahawi. Studies on Egyptian onion seed oil. *Grasas Y Aceites*, **1974**, *25*, 289–291.

Fernandez, M.L.; K. West. Mechanisms by which dietary fatty acids modulate plasma lipids. *J. Nutr.* **2005**, *135*, 2075–2078.

Griel, A.E.; Y. Cao; D.D. Bagshaw; A.M. Cifelli; B. Holub; P.M. Kris-Etherton. A macadamia nut-rich diet reduces total and LDL-cholesterol in mildly hypercholestemic men and women. *J. Nutr.* **2008**, *138*, 761–767.

Gunstone, F.D. The $^{13}$C-NMR spectra of six oils containing petroselinic acid and of aquilegia oil and meadowfoam oil which contain δ5 acids. *Chem. Phys. Lipids* **1991**, *58*, 159–167.

Kreydiyyeh, S.I.; J. Usta. Diuretic effect and mechanism of action of parsley. *J. Ethnopharmacol.* **2002**, *79*, 353–357.

Lie Ken Jie, M.S.F.; C.C. Larm; M.K. Pasha. $^{13}$C Nuclear magnetic resonance spectroscopic analysis of the triacylglycerol composition of Biota orientalis and carrot seed oil. *J. Am. Oil Chem. Soc.* **1996**, *73*, 557–562.

Nicolosi, R.J.; T.A. Wilson; G. Handelman; T. Foxall; J.F. Keaney, Jr.; J.A. Vita. Decreased aortic early atherosclerosis in hypercholesterolemic hamsters fed oleic acid-rich TriSun oil compared to linoleic acid-rich sunflower oil. *J. Nutr. Biochem.* **2002**, *13*, 392–402.

Nishimura, H.; O. Higuchi, K. Tateshita; K. Tomobe; Y. Okuma; Y. Nomura. Antioxidant activity and ameliorative effects of memory impairment of sulfur-containing compounds in *Allium* species. *Biofacors* **2006**, *26*, 135–146.

Okogeri, O.; M. Tasioula-Margari. Changes Occurring in Phenolic Compounds and α-tocopherol of virgin olive oil during storage. *J. Agric. Food Chem.* **2002**, *50*, 1077–1080.

Parker, T.D.; D.A. Adams; K. Zhou; M. Harris; L. Yu. Fatty acid composition and oxidative stability of cold-pressed edible seed oils. *J. Food Sci.* **2003**, *68*, 1240–1243.

Parry, J.; Z. Hao; M. Luther; L. Su; K. Zhou; L. Yu. Characterization of cold-pressed onion, parsley, cardamom, mullien, roasted pumpkin, and milk thistle seed oils. *J. Am Oil. Chem. Soc.* **2006**, *83*, 847–854.

Parry, J.W.; L. Su; M. Luther; K. Zhou; M.P. Yurawecz; P. Whittaker; L. Yu. Fatty acid contents and antioxidant properties of cold-pressed marionberry, boysenberry, red raspberry, and blueberry seed oils. *J. Agric. Food Chem.* **2005**, *53*, 566–573.

Prasad, R.B.N.; Y.N. Rao; S.V. Rao. Phospholipids of palash (*Butea monosperma)*, papaya (*Carica papaya*), jangli badam (*Sterculia foetida*), coriander (*Coriandrum sativum*) and carrot (*Daucus carota*) seeds. *J. Am. Oil Chem. Soc.* **1987**, *64*, 1424–1427.

Rao, P.U. Nutrient composition of some less-familiar oil seeds. *Food Chem.* **1994**, *50*, 379–382.

Sellappan, S.; C.C. Akoh. Flavonoids and antioxidant capacity of George-Grown Vidalia onions. *J. Agric. Food Chem.* **2002**, *50*, 5338–5342.

Trichopoulou, A.; D. Corella; M.A. Martinez-Gonzalez; F. Soriguer; J.M. Ordovas. The Mediterranean diet and cardiovascular epidemiology. *Nutr. Rev.* **2006**, *10*, S13–S19.

Wagner, K.-H.; A. Kamal-Eldin; I. Elmadfa. Gamma-Tocopherol—An underestimated vitamin? *Ann. Nutr. Metab.* **2004**, *48*, 169–188.

Wang, H.; G. Cao; R.L. Prior. Total antioxidant capacity of fruits. *J. Agric. Food Chem.* **1996**, *44*, 701–705.

Wong, P.Y.Y.; D.D. Kitts. Studies on the dual antioxidant and antibacterial properties of parsley (Petroselium crispum) and cilantro (Coriandrum sativum) extracts. *Food Chem.* **2006**, *97*, 505–515.

Wu, X.; L. Gu; J. Holden; D.B. Haytowitz; S.E. Gebhardt; G. Beecher; R.L. Prior. Development of a database for total antioxidant capacity in foods: a preliminary study. *J. Food Com. Anal.* **2004**, *17*, 407–422.

Yang, J.; K.J. Meyers; J. Van Der Heide; R.H. Liu. Varietal difference in phenolic content and antioxidant and antiproliferative activities of onions. *J. Agric. Food Chem.* **2004**, *52*, 6787–6793.

Yu, L.; K. Zhou; J. Parry. Antioxidant properties of cold-pressed black caraway, carrot, cranberry, and hemp seed oils. *Food Chem.* **2005**, *91*, 723–729.

Zhou, K.; J.J. Laux; L. Yu. Comparison of Swiss red wheat grain, and fractions for their antioxidant properties. *J. Agric. Food Chem.* **2004a**, *52*, 1118–1123.

Zhou, K.; L. Su; L. Yu. Phytochemicals and antioxidant properties in wheat bran. *J. Agric. Food Chem.* **2004b**, *52*, 6108–6114.

# 19

# Algal Oils

**Iciar Astiasarán and Diana Ansorena**
*Department of Nutrition and Food Science, Physiology and Toxicology, Faculty of Pharmacy, University of Navarra, Irunlarrea sn 31008, Pamplona (Navarra), Spain*

## Introduction

The industrial culture of microalgae started in the early 1950s and was primarily motivated by the search for new alternative and unconventional protein sources, then for biologically active substances, particularly antibiotics, present in the algae, and also for the use of these microorganisms as generators of renewable energy sources (Spolaore et al., 2006). Microalgae are microscopic algae, usually unicellular or filamentous, with a great biodiversity estimated between 200,000 and several million species (Norton et al., 1996). In the last 30 years, the microalgal biotechnology industry grew and diversified (Spolaore et al., 2006). Successful algal biotechnology mainly depends on choosing the right alga with relevant properties for specific culture conditions and products (Pulz & Gross, 2004).

More recently, microalgae are used to obtain various health-promoting products, such as β-carotene, tocopherols, antioxidant extracts, and fatty acid extracts. The relevance from the nutritional standpoint of the demand for an increase in the dietary supply of long-chain omega-3 fatty acids has intensively developed the application of microalgae, among other microorganisms, for obtaining concentrated sources of certain fatty acids. The oils produced from unicellular organisms (single cell) are considered as "novel foods" (defined as a type of food that does not have a significant history of consumption within the European Union prior to May 1997), so they have to be exhaustively analyzed and studied considering different aspects (composition, stability, and safety) before being commercialized. Nowadays, algae oils from *Crypthecodinium cohnii, Schizochytrium* sp. and *Ulkenia* sp. are allowed to be used for enriching foods or as nutritional supplements under different trade names all over the world. Although the main applications of these oils were initially in infant nutrition, a quick development of the oils for adult consumption is currently underway. Alternatively, one can use the microalgae oil directly as animal feed ingredients to produce eggs, chicken, and pork meat enriched in docosahexaenoic acid (DHA) (Abril & Barclay, 1998; Zelle et al., 2001). Also microalgae oil appears to be an efficient and economically competitive source of renewable biodiesel (Christi, 2007).

## Oil Microalgae Production

The production process of oil microalgae depends on the type of microorganism, but in general includes different steps: biomass production, lipid extraction, and purification.

### *Biomass Production*

The growth of microalgae in batch or in photobioreactors is carried out under highly controlled conditions, allowing the production of uncontaminated strains having a standardized composition with a particular focus on their long-chain polyunsaturated fatty acids (PUFA) content (Molina Grima et al., 1998). Microalgal cultivation presents the advantage to use an indefinitely renewable resource having a negligible environmental impact (Andrich et al., 2005). Many techniques exist for microalgal cultivation, which is divided into two groups: open and closed systems. Open systems are the most traditional ones and use open ponds with water depths of 15–20 cm. They have some problems, such as evaporative losses, diffusion of $CO_2$ to the atmosphere, the permanent threat of contamination and pollution, the large area required, the light limitation in the high layer thickness, and the maintenance of the desired microalgal production. Moreover, this type of system does not permit the control necessary for the production of algae with reproducible composition, and thus to apply GMP (good manufacturing practice) necessary to obtain products to be used for human applications (Pulz, 2001). On the contrary, the closed photobioreactors (PBRs) can be well regularized and show the following benefits: a reduced contamination risk, no $CO_2$ losses, reproducible cultivation conditions, controllable hydrodynamics and temperature, and flexible technical design (Pulz, 2001). However, Mirón et al. (1999) concluded that horizontal tubular photobioreactor technology was too expensive to be used as a system for production of eicosapentaenoic acid (EPA) by *P. tricornutum*. An alternative to photobioreactors to reduce costs is the use of heterotrophic algae grown in conventional fermenters (Apt & Behrens, 1999). With heterotrophic of algae photosynthesis for carbon and energy generation is replaced by supply of glucose or other utilizable carbon source to the medium (Ward & Singh, 2005).

When selecting the best type for commercial applications, Ward and Singh (2005) summarized the characteristics of a good industrial oil producer: the organism must be genetically stable, the strains must be nonpathogenic and toxin-forming, must be able to withstand shear due to impeller mixing and aeration, capable of high growth rates and high rates of product formation, low-cost fermentation media, and finally it has to facilitate cost- effective recovery of the product (oil) to desired specifications. In the case of oil producers the most interesting strains are those which produce high amounts of PUFAs in the form of triglycerides, especially arachidonic acid (ARA), DHA, or EPA. The soil fungus *Mortierella* species was found to be an excellent producer of ARA, producing up to 60% of total fatty acids as ARA, being allowed for human consumption (Ward & Singh, 2005). The algae *Crypthecodinium cohnii* was the first microbial strain used for commercial production of DHA for infant formula because of its low content of EPA (in fact, this dinoflagellate appears to be one of very few organisms that produces DHA as its sole PUFA), whereas *Schizochytrium* and *Ulkenia* strains, the other algae that produce oil for human consumption, were shown to exhibit diverse PUFA production capabilities (Ratledge & Hopkins, 2006a). *Phaeodactylum tricornutum* was considered a potential commercial producer of EPA if grown using a low-expense process developed by Belarbi et al. (2000). PUFA content of algae depend not only on the species, but also on factors related to culture conditions: the composition of the medium, the aeration rate, light intensity, duration of the photoperiod, temperature, and the age of culture (Robles Medina et al., 1998).

### *Oil Extraction and Purification*

The total oil yield of the algal biomass depends on the species, and includes a broad range (<15% to <70% dry weight) (Chisti, 2007). The oil-containing biomass is usually recovered from the culture by filtration or centrifugation to remove the majority of the aqueous material. Then the cells are usually dried and the oil extracted (Fig. 19.1). Mechanical shearing and ultrasonication of the biomass are potentially useful cell disruption methods to increase the lipid yield. Alkali treatment of the microbial biomass can also facilitate solvent extraction of the single-cell oil (Ward & Singh, 2005).

The solvents used have to be inexpensive, volatile (for ready removal later), free from toxic or reactive impurities (to avoid reaction with the lipids), able to form a two-phase system with water (to remove nonlipids), and must be poor extractors of unwanted components. Hexane is the preferred solvent, although one can also use isohexane (Ratledge & Hopkins, 2006a). Also the extraction methods should be fast, efficient, and gentle to reduce degradation of the lipids or fatty acids (Robles Medina et al., 1998). Moist, pressed cells may be directly extracted if these are immediately available from the fermentor. Where a storage period may be involved before the cells can be extracted, then it may be prudent to use spray-dried cells which will be stable much longer than moist cells, even though these will be heat-treated (Ratledge & Hopkins, 2006a).

Oils may be further purified by standard processes of filtration, bleaching, deodorization, and they may be cleaned by chilling and filtering out any solid impurities formed, and antioxidants may be added to prolong shelf life (Ward & Singh, 2005). Robles Medina et al. (1998) published a review on the recovery and purification of microalgae-derived PUFAs detailing the main techniques for concentrating and purifying the different lipid fractions. During all these steps, one must take care to avoid oxidation processes of PUFAs.

Guil-Guerrero et al. (2000) obtained high eicosapentaenoic and arachidonic acid recoveries (around 51 and 40%, respectively) from the red microalga *Porphyridium*

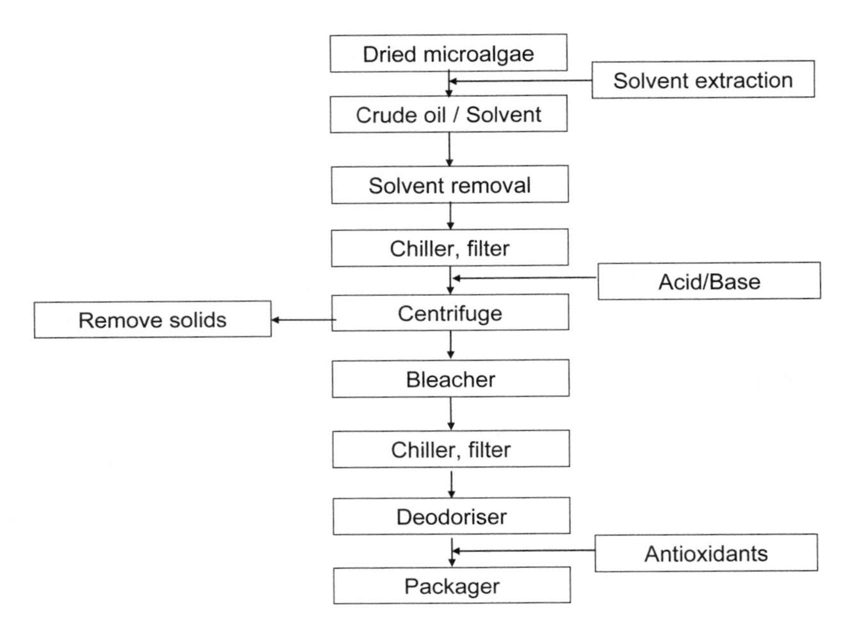


**Fig. 19.1.** Oil extraction and purification schedule from dried microalgae biomass. Source: Final assessment report. Application A428. DHA-rich dried marine microalgae (Schizochytrium sp.) and DHA-rich oil derived from *Schizochytrium* sp. as novel food ingredients. Australia New Zealand Food Authority.

*cruentum* using the method of urea. Sub- and supercritical fluid extractions were used to obtain algal biomass extractions, especially rich in compounds of interest for the food industry such as antioxidants, carotenoids, and diolefines (Herrero et al., 2006; Mendes et al., 1995). Also, this procedure can extract fatty acids (Andrich et al., 2005, 2006), although this method is probably too expensive to use in comparison with the standard solvent extraction protocols. For instance, Mendes et al. (2003) obtaining gamma-linolenic acid (GLA) from *Arthospira (Spirulina) maxima* found that although both supercritical $CO_2$ and *n*-hexane provided similar extraction yields, the supercritical $CO_2$ allowed a higher recovery of GLA-rich oils. The extraction pressure and temperature are critical conditions to achieve optimal yield with supercritical $CO_2$ extraction.

In any case, algal oil should be stored under refrigeration and packed with a minimal exposure to air. Natural antioxidants, such as tocopherols, are partially removed from the oil during the refining steps (especially deodorization), so extra tocopherols added ensure stability of final oil (Ratledge & Hopkins, 2006a).

Microalgal oils containing specific fatty acids (such as the following, produced by Martek Biosciences Corporation: DHASCO-T with 40% DHA; DHASCO-S with

40% DHA, 15% docapentaenoic acid (DPA), and 2.5% EPA; DHA45-TG with a 45% DHA) are now obtained commercially employing strictly controlled processes (Ratledge & Hopkins, 2006b).

DHASCO-T oil is a triglyceride mixture produced in a controlled, closed fermentation process by the microalga *Crypthecodinium cohnii*. This oil contains approximately 40% DHA by weight and no appreciable amount of any other PUFA. This oil was extensively studied in clinical trials in adults, children, and infants and is now used commercially as a source of DHA for infant formula in the United States and around the world. The FDA affirmed this oil as GRAS for use in infant formula when combined with a fungal-derived arachidonic acid ARASCO.

DHASCO-S is a triglyceride mixture produced by the microalga *Shizochytrium* sp. This oil also contains approximately 40% DHA as well as other PUFA, including about 15% of the omega-6 fatty acid, DPA, and 2.5% EPA. The algae producing this oil was used as an animal-feed supplement in applications to enhance the level of DHA in eggs. It was used in numerous clinical studies. The safety of the oil and the originating algae were confirmed GRAS by the FDA for use in fortification at levels up to 1.5 g/day. Both oils have excellent organoleptic qualities, making them useful for food fortification applications (Arterburn et al., 2007).

DHA45-TG or DHActive™ is a refined oil containing a complex mixture of triglycerides derived from the marine microalgae *Ulkenia* sp. This oil contains a number of long-chain fatty acids (C12–C22), with DHA as the major fatty acid (typically 45% of total fatty acids). The extracted oil is colorless to pale yellow, fluid to waxy oil, with a characteristic "bland to fish-like" odor. The FDA confirmed this oil as GRAS, and it enriches various types of foods.

Despite the differences in their composition, these oils, either in capsules or in fortified food, represent safe and equally bioavailable sources of DHA.

## Composition of Microalgae Oils

The PUFA content of algae, and of its oils, depends not only on the species, but also on factors related to culture conditions: the composition of the medium, the aeration rate, light intensity, duration of the photoperiod, temperature, and the age of culture (Robles Medina et al., 1998).

Table 19.1 shows the fatty acid profile of different oils or lipids obtained from the biomass of different species of microalgae both for human consumption and also for aquaculture applications. Great differences in the amounts of the various fatty acids obtained from different microalgae are detected. The commercially available oils are usually good sources of DHA or arachidonic acids. The biomass of *Porphyridium cruentum*, a red microalga, contains significant amounts of EPA and ARA (Rebolloso et al., 2000). Oils obtained from *Parvola lutheri* and *Chroomonas salina* contain significant amounts of EPA and linolenic acid, respectively, and are a common ingredient in fish feed (Volkman et al., 1989).

**Table 19.1. Comparison of Fatty Acid Profiles of Oils Derived from Microalgae (% of Total Fatty Acids)**

| Fatty acid | ABBREV | DHA oil[1] | DHA oil[2] | DHASCO-S[3] | DHASCO oil[4] |
|---|---|---|---|---|---|
| Lauric | 12:0 | - | 0.4 | 0.3 | 4.4 |
| Myristic | 14:0 | 2.7 | 10.1 | 9.1 | 12.7 |
| Palmitic | 16:0 | 32.9 | 23.7 | 22.9 | 9.7 |
| Palmitoleic | 16:1n-7 | - | 1.8 | 0.21 | - |
| Stearic | 18:0 | 1.1 | 0.5 | 0.57 | 1.1 |
| Oleic | 18:1n-9 | | | 1.1 | |
| Vaccenic | 18:1n-7 | - | 0.7 | 0.13 | 27.0 |
| Linoleic | 18:2n-6 | - | - | 0.46 | 1.2 |
| Linolenic | 18:3n-3, n-6 | - | - | 0.1 | - |
| Octadecatetraenoic | 18:4n-3 | - | 0.6 | | - |
| Dihomo-gamma-linolenic and Eicosatetraenoic n-7 | 20:3n-6 20:4n-7 | 1.1 | 2.2 | | - |
| Arachidonic (ARA) | 20:4n-6 | - | 1.8 | 0.51 | - |
| Eicosatetraenoic n-3 | 20:4n-3 | 0.8 | - | | - |
| Eicosapentaenoic n-3 (EPA) | 20:5n-3 | - | 2.6 | 1.25 | 0 |
| Docosatetraenoic | 22:4n-9 | - | 0.6 | | - |
| Docosapentaenoic (DPA) | 22:5n-6 | 11.2 | 13.6 | 15.44 | - |
| Docosahexaenoic (DHA) | 22:6n-3 | 45.6 | 35.0 | 42.41 | 40.0 |

Notes:

1. Derived from *Ulkenia* sp., information taken from Application A522 for assessment as Novel Food (Australia–New Zealand), (average of 3 lots)
2. Derived from *Schizochytrium* sp.; Monsanto derived 1997 analytical data from 5 bench lots.
3. Conchillo et al., 2006.
4. Derived from *Crypthecodinium cohnii*; oil composition data from Martek Home Page, Martek Biosciences Corp., 1996.
5. Abuzaytoun and Shahidi, 2006.
6. Volkman et al., 1989.
7. Rebolloso Fuentes et al., 2000. Only data for the four fatty acids listed were reported. The data in the article were converted to % of total fatty acids by dividing by the total lipid content of 6.53 g/100 g dry biomass.

Source: adapted from Draft Assessment Report. Application A522. DHA-Rich Microalgal Oil from *Ulkenia* sp. as a Novel Food.

**Table 19.1., cont. Comparison of Fatty Acid Profiles of Oils Derived from Microalgae (% of Total Fatty Acids)**

| Fatty acid | DHASCO oil[5] | *Thalassiosira pseudonana*[6] | *Pavlova lutheri*[6] | *Chroomonas salina*[6] | *Porphyridium cruentum*[7] |
|---|---|---|---|---|---|
| Lauric | 6.8 | Trace | 0.3 | Trace | - |
| Myristic | 17.7 | 14.3 | 11.5 | 8.4 | - |
| Palmitic | 12.1 | 11.2 | 21.3 | 14.0 | 24.2 |
| Palmitoleic | 2.2 | 18.0 | 16.8 | 0.6 | - |
| Stearic | - | 0.7 | 1.3 | 0.8 | - |
| Oleic | 15.9 | | | | |
| Vaccenic | - | 0.1 | 1.4 | 3.4 | - |
| Linoleic | 0.65 | 0.4 | 1.5 | 11.1 | 5.7 |
| Linolenic | - | 0.3 | 2.2 | 15.9 | - |
| Octadecatetraenoic | - | 5.3 | 6.0 | 20.6 | - |
| Dihomo-gamma-linolenic and Eicosatetraenoic n-7 | - | - | - | - | - |
| Arachidonic (ARA) | - | 0.3 | Trace | 1.0 | 19.8 |
| Eicosatetraenoic n-3 | - | 0.3 | - | 1.0 | - |
| Eicosapentaenoic n-3 (EPA) | - | 19.3 | 19.7 | 11.4 | 19.4 |
| Docosatetraenoic | - | - | - | - | - |
| Docosapentaenoic (DPA) | - | - | 2.0 | 0.1 | - |
| Docosahexaenoic (DHA) | 41.0 | 3.9 | 9.4 | 5.5 | - |

An interesting observation concerning the PUFA profile deals with the presence of the docosapentaenoic acid (DPA n-6) in oils from *Schizochytrium* and *Ulkenia*, and absent in the oil from *Crypthecodinium cohnii*. This unsusual PUFA initially gave rise to some safety concerns; however, this fatty acid does not interfere with the incorporation of DHA into neural or retinal lipids in either infants or adults, and in fact can even be beneficial in providing additional ARA in the body by retroconversion in situ (Tam et al., 2000).

The unsaponificable fraction of some algae oils contains high natural levels of antioxidants that serve to stabilize the oil during the initial stages of extraction (Ratledge & Hopkins, 2006a). However, during the extraction and purification of oils some of the compounds of this fraction, especially tocopherols, are eliminated (Ratledge & Hopkins, 2006a). In commercial oils, some plant sterols and their derivates were quantitatively analyzed (cholesterol, dehydrocholesterol, stigmasterol, brassicasterol, campesterol, 23-dehydrositosterol) (Draft Assesment report. Application A552; Conchillo et al., 2006).

## Stability of the Oil

One can add algae oils directly to foods; however, their high unsaturation index makes them highly susceptible to oxidation. Their oxidative stability varies widely according to their fatty acid composition, the physical and colloidal states of the lipids, the contents of tocopherols and other antioxidants, and the presence and activity of transition metals, such as iron (Frankel et al., 2002). Numerous processing techniques were developed to stabilize the oils. One of these is microencapsulation, a process by which minute droplets of oil are coated with a stable film. This process increases shelf life, minimizes the fishy taste and odor, and stabilizes the product by reducing oxidation potential (Whelan & Rust, 2006).

Guil-Guerrero et al. (2001) evaluated the stability of the free acids EPA and DHA obtained from the microalgae *Phaeldactylum tricornutum* and the EPA methyl ester and reported that the stability was higher for methyl EPA and that peroxidation is retarded by low-temperature storage and mainly by hexane addition.

A protein-stabilized oil-in-water emulsion at pH 3.0 is a particularly effective carrier for an n-3 PUFA-rich oil because one can easily incorporate it into an aqueous food and it has many physical properties which one can alter to increase oxidative sability (Djordjevic et al., 2004; Hu et al., 2003). In this preemulsified form, algae oil emulsions were used in the preparation of different dairy products: enriched strawberry flavored yogurt (Chee et al., 2005), fluid milk of varying fat contents (Gallaher et al., 2005), and ice cream flavored with either vanilla or strawberry liquid flavoring (Chee et al., 2007). Polysorbate-based emulsions and lecithin-based emulsion were compared to microencapsulated algae oil powder when testing baking performance and quality of white pan bread in which part of the vegetable shortening was substituted by the experimental treatments (Serna-Saldívar et al., 2006). Valencia et al.

(2007) used a soy protein emulsion with algae oil from *Schizochytrium* sp., as a partial substitute for pork back fat in the production of functional dry fermented sausages. They found that the oxidation stability of the products could be guaranteed when antioxidants, such as butylhydroxyanisol (BHA) and butulhydroxytoluene (BHT), were added and the product was stored under vacuum. Lee et al. (2006) also incorporated an emulsion of algae oil into ground turkey, pork sausages, or restructured hams with or without an "antioxidant cocktail" containing citrate, erythorbate, and rosemary. They concluded that cooking resulted in some losses of n-3 PUFAs in fortified meat products and that the "antioxidant cocktail" protected against lipid oxidation during subsequent storage in noncured meat products.

Abuzaytoun and Shahidi (2006) pointed out the lack of published information about the stability of algal oils, especially when the oils were stripped of their minor components. These authors concluded that minor components of algal oils play a major role in their oxidative stability in the dark as well as in the light. They also concluded that the fatty acid composition of the oil and the total tocopherols as well as the type of pigments present contribute to their stability.

## Health Benefits

Most reports of the health-related effects of algae oil have focused on properties associated with their high levels of omega-3 fatty acids, especially their most abundant long-chain polyunsaturated fatty acid, docosahexaenoic acid (DHA). Consumption of long-chain polyunsaturated omega-3 fatty acids was identified as a way to decrease the risk of certain diseases. The first evidences of their health benefits pointed out their capability to reduce the risk of sudden death, through different proposed mechanisms including their effects on fibrinolysis (process wherein a fibrin clot is broken down) and reduction of serum triglyceride levels, platelet activation, and the expression of vascular adhesion molecules (Leaf et al., 2003). Regular consumption of PUFAs also affects several humoral and cellular factors involved in atherogenesis and may prevent atherosclerosis, arrhythmia, thrombosis, and cardiac hypertrophy which make n-3 PUFAs an attractive therapeutic strategy for treating cardiovascular abnormalities in some heart patients (Siddiqui et al., 2008).

Concerning their hypotriglyceridemic role, n-3 fatty acids lower serum triglycerides in a dose-dependent manner especially in persons with hypertriglyceridemia by inhibiting the synthesis of very-low-density lipoprotein (VLDL) cholesterol and triglycerides in the liver (Harris et al., 1997). Two to four grams of (EPA + DHA) per day provided as capsules under a physician's care are recommended by the American Heart Association (americanheart.org, 2007) for that purpose.

Several studies focusing on the serum cholesterol/lipoprotein effects of algal oils were carried out in recent years. During a six-month randomized, double-blind, placebo-controlled, cross over study (the EARLY Study), hyperlipidemic children who were administered six capsules/day of algal-derived DHA (DHASCO oil, 1.2 g/day) showed a positive change in lipoprotein subclasses distribution, decreasing the LDL subclass 3 by 48% compared to placebo capsules (Engler et al., 2005).

A DHA-rich oil from the microalgae *Ulkenia* sp. (Nutrinova$^{R}$–Nutrinova GmbH) was administered to normolipidaemic vegetarians at a dose of 2.28 g oil/day (0.94 g DHA/day) during 8 weeks, resulting in a decrease of both plasma triglyceride levels and also the trigyceride/HDL cholesterol ratio. However, LDL cholesterol increased, leading authors to conclude that the overall effects of that intervention on CHD risk deserve further investigation (Geppert et al., 2006).

Supplementation of three capsules/day (0.7 g DHA) of a refined triacylglycerol derived from *Cryptheconidium cohnii* (DHASCO, Martek B) for three months to 38 healthy volunteers increased DHA in erythrocytes lipids by 58% and significantly lowered diastolic blood pressure, compared to the placebo group; however, it did not influence endothelial function indices or arterial stiffness in the short-term study. The authors concluded that more work is needed to confirm these findings and to investigate further the effects of DHA on cardiac function (Theobald et al., 2007).

In Table 19.2, some large population studies summarizing the implication of omega-3 fatty acids in the decrease of the risk of coronary heart disease are shown. Among them, the GISSI Prevenzione Study (1999) is the largest study to probe the cardiovascular benefits of n-3 PUFAs.

Besides the widely accepted role in improving the cardiovascular risk factors, these fatty acids are also involved in other beneficial health effects. The n-3 PUFAs exhibit anti-inflammatory properties that may play an important role in the modification of subclinical vascular inflammation. The role is generally accepted in the pathogenesis of atherosclerosis, diabetes mellitus, hypertension, metabolic syndrome, and chronic cardiac failure (Fedačko et al., 2007). As a consequence, n-3 PUFAs are used to treat joint pain associated with several inflammatory conditions. A meta-analysis of 17 randomized, controlled trials assessing the pain-relieving effects of omega-3 PUFAs in patients with rheumatoid arthritis or joint pain, secondary to inflammatory bowel disease and dysmenorrhea (menstrual syndromes), pointed out that supplementation with n-3 PUFAs for three to four months causes a reduction in perceived joint pain intensity and also resulted in a reduction in antiinflammatory drug use, among other benefits (Goldberg & Katz, 2007).

Perhaps the most intriguing area for potential benefits of n-3 PUFA is in brain function and mental health. DHA is the major n-3 PUFA constituent in the neuronal membranes, present in approximately 30–40% of the phospholipids of the gray matter of cerebral cortex (Lauritzen et al., 2001). Patients with Alzheimer's disease and other dementias display lower levels of DHA in plasma and brain tissues as compared to age-matched controls (Conquer et al., 2000). Epidemiological studies suggest that high DHA intake might have protective properties against neurodegenerative diseases (Florent-Béchard et al., 2007). Also hypothesized is that a relationship exists between a poor n-3 PUFA status (low n-3 PUFA intake) and an increased risk of depression, evidence arose from supplementation trials that dosages of 0.2–9.6 g EPA + DHA are useful in the treatment of depression (Ruxton et al., 2007).

Lower levels of n-3 fatty acids in blood were also repeatedly associated with various behavioral disorders including attention-deficit/hyperactivity disorder (ADHD), although the exact nature of this relationship is not yet clear and should be further studied (Antalis et al., 2006).

Regarding the evidence related to the beneficial effects of algal oils during pregnancy, research shows some improvements in different functions related to the fetus development. In a review of the beneficial effects of dietary omega-3 fatty acids for pregnant women, Bourre (2007) concluded that the presence of large quantities of EPA and DHA in the diet slightly lengthens pregnancy, and improves its quality, and also ensures an adequate supply of long-chain PUFAs for healthy cerebral and cognitive development of the fetus and newborn. A daily dose of 200 mg of DHA (two capsules of oil) resulted in an increase of the DHA content in human milk, with no appreciable effect on the of the other fatty acids (Fidler et al., 2000).

Clandinin et al. (2005) evaluated the safety and benefits of feeding preterm infants formulas containing docosahexaenoic acid (DHA) and arachidonic acid (ARA). Feeding enriched formulas resulted in enhanced growth and provided better developmental outcomes than unsupplemented formulas. Regarding vision, DHA is important for the retina (the highest DHA contents of all body tissues), the brain, photoreceptors, neurotransmission, rhodopsin activity, the development of rods and cones neuronal connections, and the maturation of cerebral structures (Agostoni & Giovannini, 2001; Neuringer, 2000). In a recent study (Jensen et al., 2005), a 200-mg DHA dose was given to breastfeeding woman for four months after delivery to evaluate visual function and neurodevelopment in breast-fed term infants. Results pointed out that neither the neurodevelopmental indexes of the infants at 12 months of age nor the visual function at 4 or 8 months of age differed from the placebo group. In contrast, the Bayley Psychomotor Development Index was higher at 30 months of age. The importance of the health implications of long-chain n-3 fatty acids supports the Opinion of the Scientific Panel on Dietetic Products, Nutrition and Allergies of the European Food Safety Authority (Request No EFSA-Q-2004-107), that proposes the following nutritional claims:

- "Omega-3 fatty acid source"—The food must contain more than 15% of the Recommended Nutritional Intake (with RNI set at 2 g/day for an adult male) for an adult male of the omega-3 fatty acids concerned per 100 g or 100 mL or 100 kcal.
- "High in omega-3 fatty acids"—The food must contain more than 30% of the Recommended Nutritional Intake for an adult male of the omega-3 fatty acids concerned per 100 g or 100 mL or 100 kcal.

Note that a potential problem with these claims is that they do not distinguish between ALA and long-chain n-3 PUFA, which have different nutritional roles. The

**Table 19.2. Omega-3 Fatty Acid Intake and the Risk of Coronary Heart Disease**

| Study | Study design | Study period | No. and type of subjects | |
|---|---|---|---|---|
| Albert et al. (1998). | RCT, double blind, cohort | 11 years | 20551 male physicians. Subjects stratified by fish intake (serve = 84-112 g fish) | |
| Burr et al. (1989) and Burr et al. (1994) (two articles on same study) | RCT, no blinding | 2 years | 2033 males and females with recent CHD diagnosis, equal group numbers | Received no advice to eat fish |
| | | | | Received advice to eat fish |
| | | | 454 male and female subset of 2033 subjects, equal group numbers | Received no fish oil capsules |
| | | | | Received fish oil capsules |
| GISSI-Prevenzione Investigators (1999) | RCT, double-blinding | 42 months | 11324 males and females with recent CHD diagnosis | Control (placebo) |
| | | | | Omega-3 group |
| | | | | Vitamin E group |
| | | | | Omega-3 and Vitamin E group |
| Hu et al. (2002) | RCT, double blind, cohort | 16 years | 1513 female nurses | Subjects stratified by fish intake (serve = 168-224 g fish) |
| | | | | Subjects stratified by omega-3 fatty acid intake |
| Singh et al. (1997) | RCT, single blinding | 1 year | 360 patients with suspected infarction | Group A – fish oil (n=122) |
| | | | | Group B – mustard oil (n=120) |
| | | | | Group C – placebo (n=118) |

**Table 19.2., cont. Omega-3 Fatty Acid Intake and the Risk of Coronary Heart Disease**

| Intake level | Results: Relative risk total CHD | Results: Relative risk fatal CHD (except where indicated) | Results: Relative risk non-fatal CHD | Significant difference? (P<0.05) |
|---|---|---|---|---|
| < 1 serve / month | - | 1.00 | 1.00 | Yes for trend over intake groups for non-fatal CHD, but not for fatal CHD. |
| 1-3 serves / month | - | 0.96 | 0.64 | |
| 1 serve / week | - | 0.79 | 0.47 | |
| 2-4 serves / week | - | 0.84 | 0.51 | |
| ≥ 5 serves / week | - | 0.81 | 0.39 | |
| 0g fish/week | 1.00 | - | - | No |
| 200-400g fish/week | 0.84 | - | - | |
| 0 g fish oil | - | 9.3 (% incidence of death in group) | - | Yes, between the two groups, although authors indicated there may be selection bias on results |
| 3 g fish oil | - | 3.5 (% incidence of death in group) | - | |
| 0 mg | 1.00 | 1.00 | 1.00 | Yes, the omega-3 group results were significantly lower except for non-fatal CHD. Other groups did not experience a significant lowering in the relative risk of CHD in comparison to the control, except for fatal CHD results. |
| 859-882 mg EPA and DHA / day | 0.80 | 0.70 | 0.96 | |
| 300 mg Vitamin E /day | 0.88 | 0.80 | 1.02 | |
| 859-882 mg EPA and DHA/day + 300 mg Vitamin E /day | 0.88 | 0.80 | 1.01 | |
| < 1 serve / month | 1.00 | 1.00 | 1.00 | Yes, significant trend with increasing intake for total, fatal and non-fatal CHD events |
| 1-3 serves / month | 0.79 | 0.81 | 0.78 | |
| 1 serve / week | 0.71 | 0.66 | 0.74 | |
| 2-4 serves / week | 0.69 | 0.73 | 0.68 | |
| ≥ 5 serves / week | 0.66 | 0.55 | 0.73 | |
| 0.03% of energy intake | 1.00 | 1.00 | 1.00 | Yes, significant trend with increasing intake for total, fatal and non-fatal CHD events |
| 0.05% of energy intake | 0.93 | 0.94 | 0.91 | |
| 0.08% of energy intake | 0.78 | 0.61 | 0.79 | |
| 0.14% of energy intake | 0.68 | 0.41 | 0.66 | |
| 0.24% of energy intake | 0.67 | 0.42 | 0.57 | |
| 1.08 g EPA/day, and 0.72 g DHA/day | 0.70 | 0.52 | 0.51 | Yes, between fish oil and placebo groups, and mustard oil and placebo groups. |
| 20 g/day mustard oil (2.9 g ALA/day) | 0.81 | 0.60 | 0.59 | |
| 0g of omega-3 fatty acids/day | 1.00 | 1.00 | 1.00 | |

***cont. on the next page***

**Table 19.2., cont. Omega-3 Fatty Acid Intake and the Risk of Coronary Heart Disease**

| Study | Study design | Study period | No. and type of subjects | |
|---|---|---|---|---|
| Siscovick et al. (2000) | Case-control, random control selection | 1 month retrospective dietary analysis | 334 CHD case subjects, 493 control subjects | No seafood intake group |
| | | | | Seafood intake group (stratified equally into groups - intake levels are as mean of group) |

RCT: Randomized controlled trial
Source: Draft Assessment Report. Application A522. DHA-Rich Microalgal Oil from *Ulkenia* sp. as a Novel Food

human body can convert dietary α-linolenic acid to EPA and DHA following consumption. However, the extent of this conversion appears to be minimal at normal dietary intakes (around 1%), so it seems that the only effective way to enrich tissue phospholipids with DHA is to consume DHA (Whelan & Rust, 2006). Table 19.3 shows the dietary intakes of n-3 PUFA recommended by various international organizations.

## Safety and Legal Aspects

In relation to the safety of algae oils, two issues must be considered: the absence of contaminants with immediate or long-term toxicological effects and the absence of negative effects due to the intake of high amounts of PUFAs. The microalgae biomass may contain many metabolites, some of which may have pharmaceutical applications as new drugs and others may act as toxins (some toxins were identified in harmful algal blooms) (Berry et al., 2007; Rao et al., 2002; Shimizu, 2003; Yasumoto & Murata, 1993). The processes for oil extraction and purification should eliminate all of these compounds. All the studies carried out with algal oils intended for animal or human feeding must demonstrate the absence of those compounds.

Several toxicological studies including subchronic toxicity studies and genotoxicity studies were carried out with microalgal-derived oils. In the case of oils rich in DHA (Abril et al., 2000, 2003; Arterburn et al., 2000a, b, c; Blum et al., 2007; Boswell et al., 1996; Hammond et al., 2001a, b, c, 2002; Hempenius et al., 1997, 2000; Wibert et al., 1997; Willumsen et al., 1993), no adverse effects were found on mortality, body weight gains, food consumption, or clinical observations (Kroes et al., 2003).

**Table 19.2., cont. Omega-3 Fatty Acid Intake and the Risk of Coronary Heart Disease**

| Intake level | Results: Relative risk total CHD | Results: Relative risk fatal CHD (except where indicated) | Results: Relative risk non-fatal CHD | Significant difference? ($P<0.05$) |
|---|---|---|---|---|
| 0 g omega-3 fatty acids /month | 1.00 | - | - | Yes, significant trend with increasing seafood intake |
| 1 g omega-3 fatty acids /month | 0.90 | - | - | |
| 2.9 g omega-3 fatty acids /month | 0.73 | - | - | |
| 5.5 g omega-3 fatty acids /month | 0.50 | - | - | |
| 13.7 g omega-3 fatty acids /month | 0.38 | - | - | |

In 1993 the U.S. Food and Drug Administration (FDA) identified three possible adverse effects associated with human consumption of omega-3 PUFAs: reduced platelet aggregation, increased low-density lipoprotein cholesterol levels, and reduced glycemic control among diabetics. In 1997, and after an evaluation of several published clinical trials, the FDA concluded that consumption of up to 3 g (DHA + EPA) combined/person/day does not cause a significant risk for increased bleeding time, has no clinically significant effect on glycemic control, and is safe regarding the effect on LDL cholesterol. In 2000 this organization published a new document describing the dietary supplement qualified health claim for omega-3 fatty acids and heart disease, stating DHA from marine algae, as dietary supplements is safe (www.cfsan.fda.gov/~rdb/opa-gras.html). The FDA has approved as "generally recognized as safe" (GRAS), algal oils rich in DHA (*Crypthecodinium cohnii* – GRAS N000041; *Schizochytrium* sp.–GRAS N000137; *Ulkenia* sp.—GRAS N000160), allowing for food fortification with these products.

In Japan, DHA-rich oil is considered a food, and pre-market regulatory permission is not required (Draft Assesment report. Application A552).

In 2002 *Schizochytrium* sp. oil received the approval for incorporation into different foods in Australia and New Zealand, and later on in 2003 it was allowed in Europe to be added to different foods (Commission decision C82003 1790).

The German Competent Authority (Federal Authority for Consumer Protection and Food Safety, Bundesamt für Verbraucherschutz und Lebensmittelsicherheit (BVL)) determined that DHA-rich oil (*Ulkenia* sp.) was substantially equivalent to DHA-rich microalgal oil from *Schizochytrium* sp. and was classified as a novel food under article 5 of the EC Regulation 258/97 and can now be marketed across the EU. The company Nutrinova also placed on the market DHA (from the microalga

**Table 19.3. Recommended Intakes of n-3 PUFA by Various International Organizations**

| Date | Organization | Recommendations |
|---|---|---|
| 2004 | International Society for the Study of Fatty Acids and Lipids | ALA: 0.7% energy<br>EPA +DHA: ≥ 500 mg/d |
| 2004 | Wijendran & Hayes (2004) | ALA: 0.75% energy<br>EPA+ DHA: 0.25% energy |
| 2003 | World Health Organization | Total n-3 PUFA: 1–2% energy |
| 2002 | Food and Nutrition Board | ALA, women: 1.1 g/d of which 10% can be EPA + DHA<br>ALA, men: 1.6 g/d of which 10% can be EPA + DHA |
| 2002 | American Heart Association | Eat fish (fatty fish) at least 2 times per week |
| 2002 | Scientific Advisory Committee on Nutrition (U.K.) | Total n-3 PUFA: >0.2 g/d<br>Eat 2 portions of fish weekly (one being oily) |
| 2001 | Health Councils of the Netherlands | Total n-3 PUFA: 1% energy<br>DHA: 150–200 mg/d |
| 2000 | Simopoulos et al. (2000) | ALA: 2.22 g/d<br>EPA: ≥ 220 mg/d<br>DHA: ≥ 220 mg/d<br>EPA + DHA: 650 mg/d |
| 1999 | British Nutrition Foundation (U.K.) | ALA: 0.2% energy<br>Total n-3 PUFA: 1.25 g/d |

Source: Whelan and Rust (2006).
ALA, α-linolenic acid; DHA, docosahexaenoic acid; EPA, eicosapentaenoic acid; PUFA, polyunsaturated fatty acid

*Ulkenia* sp.) for the same uses following the simplified procedure (Article 5) of the Novel Foods Regulation (EC) No. 258/97. The possibility of additional uses for this oil is under study by the European Commission.

## Nonedible Applications

Microalgal oil can be an efficient source for biodiesel production (Banerjee et al., 2002; Gavrilescu & Chisti, 2005; Sheehan et al., 1998; Xu et al., 2006). Biodiesel is a biodegradable, renewable, and nontoxic fuel that contributes no net carbon dioxide or sulfur to the atmosphere and emits less gaseous pollutants than conventional diesel fuel. It consists of the alkyl esters of fatty acids, and it is receiving considerable attention in recent years.

Various plant oils were demonsrated to be sources of biodiesel (corn, soybean, canola, coconut, oil palm). However, microalgae appear to be the only one with a real potential to displace fossil diesel as a consequence of their oil yield and, the percentage of cropping area needed for sufficient production (Chisti, 2007). Not all algal oils are satisfactory for this application because limitations exist related to a high-yield production, oil content, and the type of fatty acids. The European biodiesel standards limit the contents of FAME with four and more double bonds to a maximum of 1% mol. However, partial catalytic hydrogenation of the oil can effectively reduce the number of double bonds as it is usually done in the technology used in making margarines from vegetable oils.

## Conclusion

Algal oils and fish oils are the main natural sources of omega-3 long chain PUFAs. However, algal oils show certain advantages compared to fish oils such as consistency of composition, sensory properties, and ease of production. As this chapter shows, various possibilities exist for obtaining algae oils with different main omega-3 fatty acids (which increases the possibilities of different specific uses). Also, algal oils lack cholesterol (which is considered to have negative dietary properties) and they lack potential contaminants, which are an important risk in fish oils. Algal oils do not possess odors, in contrast with the fish odor, which accompanies fish oils, limiting its use. Moreover, the production of algae biomass is easier and more environmentally friendly, giving higher yields than in the case of fish oil production.

## References

Abril, R.; W. Barclay. Production of docosahexaenoic acid-enriched poultry eggs and meat using an algae-based feed ingredient. *World Rev. Nutr. Diet* **1998,** *83,* 77–88.

Abril, J.R.; W.R. Barclay; P.G. Abril. Safe use of microalgae (dha gold) in laying hen feed for the production of dha-enriched eggs. *Egg Nutrition and Biotechnology;* J.S. Sim, S. Nakai, W. Guenter, Eds.; CAB International; 2000; pp. 197–202.

Abril, R.; J. Garrett; S.G. Zeller; W.J. Sander; R.W. Mast. Safety assessment of dha-rich microalgae from *schizochytriums sp.*—part v: target animal safety/toxicity study in growing swine. *Regul. Toxicol. Pharm.* **2003,** *37(1),* 73–82.

Abuzaytoun, R.; F. Shahidi. Oxidative stability of algal oils as affected by their minor components. *J. Agric. Food Chem.* **2006,** *18;54(21),* 8253–8260.

Agostoni, C.; M. Giovannini. Cognitive and visual development: influence of differences in breast and formula fed infants. nutr. health. **2001,** *15,* 183–138.

Albert, C.M.; C.H. Hennekens; C.J. O'Donnell; U.A. Ajami; V.J. Carey; W.C. Willett; J.N. Ruskin; J.E. Manson. Fish consumption and risk of sudden cardiac death. *J. Am. Med. Assoc.* **1998,** *279,* 23–28.

Andrich, G.: U. Nesti; F. Venturi; A. Zinnai; R. Fiorentini. Supercritical fluid extraction of bioactive lipids from the microalga *nannochloropsis sp.* europ. *J. Lipid Sci. Technol.* **2005,** *107(6),* 381–386.

Andrich, G.; A. Zinnai; U. Nesti; F. Venturi;R. Fiorentini. Supercritical fluid extraction of oil from microalga *spirulina (arthrospira) platenses. Acta Alimentaria.* **2006,** *35(2),* 195–203.

Antalis, C.J.; L.J. Stevens; M. Campbell; R. Pazdro; K. Ericson; J.R. Burgess. Omega-3 fatty acid status in attention-deficit/hyperactivity disorder. *Prostaglandins Leukot. Essent. Fatty Acids* **2006,** *75(4–5),* 299–308.

Apt, K.E.; P.W. Behrens. Commercial developments in microalgal biotechnology. *J. Phycol.* **1999,** *35(2),* 215–226.

Arterburn, L.M.; K.D. Boswell; S.M. Henwood;D.J. Kyle. A development safety study in rats using DHA-and ARA-rich single-cell oils. *Food Chem. Toxicol.* **2000b,** *38,* 763–771.

Arterburn, L.M.; K.D. Boswell; E. Koskelo; S.L. Kassner; C. Kelly;D.J. Kyle. A Combined subchronic (90-day) toxicity and neurotoxicity study of a single-cell source of docosahexaenoic acid triglyceride (DHASCO© oil). *Food Chem. Toxicol.* **2000a,** *38,* 35–49.

Arterburn, L.M.; K.D. Boswell; T. Lawlor; M.A. Cifone; H. Murli; D.J. Kyle. In vitro genotoxicity testing of ARASCO® and DHASCO® oils. *Food Chem. Toxicol.* **2000c,** *38(11),* 971–976.

Arterburn, L.M.; H.A. Oken; J.P. Hoffman; E. Bailey-Hall; G. Chung; D. Rom; J. Hamersley;D. McCarthy. Bioequivalence of docosahexaenoic acid from different algal oils in capsules and in a DHA-fortified food. *Lipids.* **2007,** *42,* 1011–1024.

Banerjee, A.; R. Sharma; Y. Chisti; U.C. Banerjee. *Botryococcus braunii*: a renewable source of hydrocarbons and other chemicals. *Crit. Rev. Biotechnol.* **2002,** *22,* 245–279.

Belarbi, E.H.; E. Molina; Y. Chisti. A process for high yield and scaleable recovery of high purity eicosapentaenoic acid esters from microalgae and fish oil. *Enzyme Microb. Technol.* **2000,** *26(7),* 516–529.

Berry, J.P.; M. Gantar; P.D.L. Gibbs; M.C. Schmale. The zebrafish (*danio rerio*) embryo as a model system for identification and characterization of developmental toxins from marine and freshwater microalgae. *Comp. Biochem. Phy. C* **2007,** *145(1),* 61–72.

Blum, R.; T. Kiy; S. Tanaka; A.W. Wong;A. Roberts. Genotoxicity and subchronic toxicity studies of dha-rich oil in rats. *Regul. Toxicol. Pharmacol.* **2007,** *49,* 271–284.

Boswell, K.; E.K. Koskelo; L. Carl; S. Glaza; D.J. Hensen; K.D. Williams; D.J. Kyle. Preclinical evaluation of single-cell oils that are highly enriched with arachidonic acid and docosahexaenoic acid. *Food Chem. Toxicol.* **1996,** *34,* 585–593.

Bourre, J.M. dietary omega-3 fatty acids for women. *Biomed. Pharmacother.* **2007,** *61(2-3),* 105–112.

Burr, M.L.; A.M. Fehily; J.F. Gilbert; S. Rogers; R.M. Holliday; P.M. Sweetnam;P.C. Elwood; N.M. Deadman. Effects of changes in fat fish and fibre intakes on death and myocardial reinfarction: diet and reinfarction trial (dart). *Lancet* **1989,** *2(8666),* 757–761.

Burr, M.L.; P.M. Sweetnam; A.M. Fehily; P.C. Elwood. Letters to the editor. *Eur. Heart J.* **1994,** *15,* 1152–1154.

Chee, C.P.; D. Djordjevic; H. Faraji; E.A. Decker; R. Hollender; D.J. McClements; D.G. Peterson; R.F. Roberts; J.N. Coupland. sensory properties of vanilla and strawberry flavored ice cream supplemented with omega-3 fatty acids. *Milchwiss.-Milk Sci. Int.* **2007,** *62(1),* 66–69.

Chee, C.P.; J.J. Gallaher; D. Djordjevic; H. Faraji; D.J. McClements; E.A. Decker; R. Hollender; D.G. Peterson; R.F. Roberts;J.N. Coupland. Chemical and sensory analysis of strawberry flavoured yogurt supplemented with an algae oil emulsion. *J. Dairy Res.* **2005,** *72,* 311–316.

Chisti, Y. Biodiesel from microalgae. *Biotechnol. Advances* **2007,** *25(3),* 294–306.

Clandinin, M.T.; J.E. Van Aerde; K.L. Merkel; C.L. Harris; M.A. Springer; J.W. Hansen;D.A. Diersen-Schade. Growth and development of preterm infants fed infant formulas containing docosahexaenoic acid and arachidonic acid. *J. Pediatr.* **2005,** *146(4),* 461–468.

COMMISSION DECISION of 5 June 2003 authorizing the placing on the market of oil rich in DHA (docosahexaenoic acid) from the microalgae *Schizochytrium sp.* as a novel food ingredient under Regulation (EC) No 258/97 of the European Parliament and of the Council *(notified under document number C(2003) 1790).*

Conchillo, A.; I. Valencia; A. Puente; D. Ansorena;I. Astiasarán. Functional Components in Fish and Algae Oils. Nutrición Hospitalaria **2006,** *21(3),* 369–373.

Conquer, J.A.; M.C. Tierney; J. Zecevic; W.J. Bettger; R.H. Fisher. Fatty acids analysis of blood plasma of patients with alzheimer´s disease, other types of dementia and cognitive impairment. *Lipids* **2000,** *35,* 1305–1312.

Djordjevic, D.; H.J. Kim; D.J. McClements; E.A. Decker. Oxidative stability of whey protein-stabilized oil-in-water emulsions at ph 3: potential v-3 fatty acid delivery systems (part b). *J. Food Sci.* **2004,** *69,* C356–C362.

Engler, M.M.; M.B. Engler; M.J. Malloy; S.M. Paul; K.R. Kulkarni; M.L. Mietus-Snyder. Effect of docosahexaenoic acid on lipoprotein subclasses in hyperlipidemic children (the EARLY study). *Am. J. Cardiol.* **2005,** *95(7),* 869–871.

Fedačko, J.; D. Pella; V. Mechírová; P. Horvath; R. Rybár; P. Varjassyová; V. Vargová. N-3 PUFAS—from dietary supplements to medicines. *Pathophysiology* **2007,** *14(2),* 127–132.

Fidler, N.; T. Sauerwald; A. Pohl; H. Demmelmair;B. Koletzko. Docosahexaenoic acid transfer into human milk after dietary supplementation: a randomized clinical trial. *J. Lipid Res.* **2000,** *41(9),* 1376–1383.

Florent-Béchard, S.; C. Malaplate-Armand; V. Koziel; B. Kriem; J.L. Olivier; T. Pillot; T. Oster. Towards a nutritional approach for prevention of alzheimer's disease: biochemical and cellular aspects. *J. Neurol. Sci.* **2007,** *262(1-2),* 27–36.

Food Nutr. Board, Inst. Med. *Dietary Reference Intakes: Energy, Carbohydrate, Fiber, Fat, Fatty Acids, Cholesterol, Protein, and Amino Acids,* Parts 1 and 2. Natl. Acad. Press: Washington, DC, 2002.

Frankel, E.N.; T. Satue-Gracia; A.S. Meyer; J.B. German. Oxidative stability of fish and algae oils containing long-chain polyunsaturated fatty acids in bulk and in oil-in-water emulsions. *J. Agric. Food Chem.* **2002,** *27,50(7),* 2094–2099.

Gallaher, J.J.; R. Hollender; D.G. Peterson; R.F. Roberts; J.N. Coupland. Effect of composition and antioxidants on the oxidative stability of fluid milk supplemented with an algae oil emulsion. *Int. Dairy J.*. **2005,** *15(4),* 333–341.

Gavrilescu, M.; Y. Chisti. Biotechnology—a sustainable alternative for chemical industry. *Biotechnol. Adv.* **2005,** *23,* 471–499.

Geppert, J.; V. Kraft; H. Demmelmair; B. Koletzko. Microalgal docosahexaenoic acid decreases plasma triacylglycerol in normolipidaemic vegetarians: a randomised trial. *Br. J. Nutr.* **2006,** *95(4),* 779–786.

Goldberg, R.J.; J. Katz. a meta-analysis of the analgesic effects of omega-3 polyunsaturated fatty acid supplementation for inflammatory joint pain. *Pain* **2007,** *129(1-2),* 210–223.

Gruppo Italiano per lo Studio della Sopravvivenza nell'Infarto miocardico. Dietary supplementation with n-3 polyunsaturated fatty acids and vitamin e after myocardial infarction: results of the gissiprevenzione trial. [erratum appears in *Lancet* 2001 Feb 24;357(9256):642]. *Lancet* **1999,** ***354,*** 447–55.

Guil-Guerrero, J.L.; E.H. Belarbi;M.M. Rebolloso-Fuentes. Eicosapentaenoic and arachidonic acids purification from the red microalga *Porphyridium cruentum*. *Bioseparation* **2000,** *9(5),* 299–306.

Guil-Guerrero, J.L.; A. Gimenez-Gimenez; A. Robles-Medina; M.D. Rebolloso-Fuentes; E.H. Belarbi;L. Esteban-Cerdan; E. Molina-Grima. Hexane reduces peroxidation of fatty acids during storage. *Europ. J. Lipid Sci. Technol.* **2001,** *103(5),* 271–278.

Hammond, B.G.; D.A. Mayhew; J.F. Holson; M.D. Nemec; R.W. Mast; W.J. Sander. Safety assessment of dha-rich microalgae from *Schizochytrium sp*—ii. developmental toxicity evaluation in rats and rabbits. *Regul. Toxicol. Pharm.* **2001b,** *33(2),* 205–217.

Hammond, B.G.; D.A. Mayhew; L.D. Kier; R.W. Mast; W.J. Sander. Safety assessment of dha-rich microalgae from *Schizochytrium sp*—iv. mutagenicity studies. *Regul. Toxicol. Pharm.* **2002,** *35(2),* 255–265.

Hammond, B.G.; D.A. Mayhew; M.W. Naylor; F.A. Ruecker; R.W. Mast; W.J. Sander. Safety assessment of dha-rich microalgae from *Schizochytrium sp.*—i. subchronic rat feeding study. *Regul. Toxicol. Pharm.* **2001a,** *33(2),* 192–204.

Hammond, B.G.; D.A. Mayhew; K. Robinson; R.W. Mast; W.J. Sander. Safety assessment of dha-rich microalgae from *Schizochytrium sp*. iii. single-generation rat reproduction study. *Regul. Toxicol. Pharm.* **2001c,** *33(3),* 356–362.

Harris, W.S.; H. Ginsberg; N. Arunakul; N.S. Shachter; S.L. Windsor; M. Adams;L. Berglund; K. Osmundsen. Safety and efficacy of omacor in severe hypertriglyceridemia. *J. Cardiovasc. Risk* **1997,** *4(5-6)* 385–391.

Hempenius, R.A.; B.A.R. Lina; R.C. Haggitt. Evaluation of a subchronic (13-week) oral toxicity study, preceded by an in utero exposure phase, with arachidonic acid oil derived from *Mortierella alpina* in rats. *Food Chem. Toxicol.* **2000,** *38,* 127–139.

Hempenius, R.A.; J.M.H. Van Delft; M. Prinsen; B.A.R. Lina. Preliminary safety assessment of an arachidonic acid-enriched oil derived from *Mortierella alpina*: summary of toxicological data. *Food Chem. Toxicol.* **1997,** *35,* 573–581.

Herrero, M.; A. Cifuentes;E. Ibañez. sub- and supercritical fluid extraction of functional ingredients from different natural sources: plants, food-by-products, algae and microalgae—a review. *Food Chem.* **2006,** *98(1),* 136–148.

http://www.americanheart.org/

Hu, M.; D.J. McClements; E.A. Decker. Impact of whey protein emulsifiers on the oxidative stability of salmon oil-in-water emulsions. *J. Agric. Food Chem.* **2003,** *51,* 1435–1439.

Jensen, C.L.; R.G. Voigt; T.C. Prager; Y.L. Zou; J.K. Fraley; J.C. Rozelle; M.R. Turcich; A.M. Llorente; R.E. Anderson; W.C. Heird. Effects of maternal docosahexaenoic acid intake on visual function and neurodevelopment in breastfed term infants. *Am. J. Clin. Nutr.* **2005,** *82(1),* 125–132.

Kroes, R.; E.J. Schaefer; R.A. Squire; G.M. Williams. A review of the safety of DHA45-oil. *Food Chem. Toxicol.* **2003,** *41(11),* 1433–1446.

Lauritzen, L.; H.S. Hansen; M.H. Jorgensen;K.F. Mickaelsen. The essentiality of long chain n-3 fatty acids in relation to development and function of the brain and retina. *Prog. Lipid Res.* **2001,** *40,* 1–94.

Leaf, A.; J.X. Kang; Y.F. Xiao; G.E. Billman. Clinical prevention of sudden cardiac death by n-3 polyunsaturated fatty acids and mechanism of prevention of arrhythmias by n-3 fish oils. *Circulation* **2003,** *107,* 2646–2652.

Lee, S.; P. Hernandez; D. Djordjevic; H. Faraji;R. Hollender; C. Faustman; E.A. Decker. Effect of antioxidants and cooking on stability of n-3 fatty acids in fortified meat products. *J. Food Sci.* **2006,** *71(3),* C233–C238.

Mendes, R.L.; H.L. Fernandes; J.P. Coelho; E.C. Reis; J.M.S. Cabral; J.M. Novais;A.F. Palabra. Supercritical co2 extraction of carotenoids and other lipids from *Chlorella vulgaris. Food Chem.* **1995b,** *53,* 99–103.

Mendes, R.L.; B.P. Nobre; M.T. Cardoso; A.P. Pereira; A.F. Palabra. Supercritical carbon dioxide extraction of compounds with pharmaceutical importance from microalgae. *Inorganica Chim. Acta* **2003,** *356,* 328–334.

Mirón, A.S.; A.C. Gómez; F.G. Camacho; E.M. Grima; Y. Chisti. Comparative evaluation of compact phobioreactors for large-scale monoculture of microalgae. *J. Biotechnol.* **1999,** *70,* 249–270.

Molina Grima, E.; J.A. Sánchez Pérez; G. García Camacho; A. Robles Medina; A.A. Giménez Giménez; D. López Alonso. The production of polyunsaturated fatty acids by microalgae: from strain selection to product purification. *Process. Biochem.* **1995,** *30,* 711–719.

Neuringer, M. Infant vision and retinal function in studies of dietary long-chain polyunsaturated fatty acids: methods, results, and implications. *Am. J. Clin. Nutr.* **2000,** *71,* 256S–267S.

Norton, T.A.; M. Melkonian; R.A. Andersen. Algal biodiversity. *Phycologia* **1996,** *35,* 308–326.

Pulz, O. Photobioreactors: production systems for phototropic microorganisms. *Appl. Microbiol. Biotechnol.* **2001,** *57,* 287–293.

Pulz, O.; W. Gross. Valuable products from biotechnology of microalgae. *Appl. Microbiol. Biotechnol.* **2004,** *65(6),* 635–648.

Rao, P.V.; N. Gupta; A.S. Bhanskar; R. Jayaraj. Toxins and bioactive compounds from cyanobacteria and their implications on human health. *J. Environ. Biol.* **2002,** *23,* 215–224.

Ratledge, C.; S. Hopkins. Lipids from microbiol sources. *Modifying Lipids for Use in Food;* F.D. Gunstone, Ed.; CRC Press: Boca Raton, FL, 2006a; pp. 81–113.

Ratledge, C.; S. Hopkins. Applications and safety of microbiol oils in food. *Modifying Lipids for Use in Food;* F.D. Gunstone, Ed.; CRC Press: Boca Raton, FL, 2006b; pp. 567–586.

Rebolloso Fuentes, M.M.; G.G. Acién Fernández; J.A. Sánchez Pérez; J.L. Guil Guerrero. Biomass nutrient profiles of the microalga *Porphyridium cruentum. Food Chem.* **2000,** *70(3),* 345–353.

Regulation (EC) No 258/97 of the European Parliament and of the Council of 27 January 1997 concerning novel foods and novel food ingredients. *Official J. L* 043 , 14/02/1997 P. 0001 – 0006.

Robles Medina, A.; E. Molina Grima; A. Gimenez Gimenez; M.J. Ibanez Gonzalez. Downstream processing of algal polyunsaturated fatty acids. *Biotechnol. Adv.* **1998,** *16(3),* 517–580.

Ruxton, C.H.S.; S.C. Reed; M.J.A. Simpson; K.J. Millington. The health benefits of omega-3 polyunsaturated fatty acids: a review of the evidence. *J. Hum. Nutr.* **2007,** *20(3),* 275–285.

Serna-Saldívar, S.O.; R. Zorrilla; C. De la Parra; G. Stagnitti; R. Abril. Effect of DHA containing oils and powders on baking performance and quality of white pan bread. *Plant Foods Hum. Nutr.* **2006,** *61(3),* 121–129.

Sheehan, J.; T. Dunahay; J. Benemann; P. Roessler. (1998) *A Look Back at the U.S. Department of Energy's Aquatic Species Program—Biodiesel from Algae;* National Renewable Energy Laboratory: Golden, CO; Report NREL/TP-580–24190.

Shimizu, Y. Microalgal metabolites. *Curr. Opin. Microbiol.* **2003,** *6(3),* 236—243.

Siddiqui, R.A.; K.A. Harvey; G.P. Zaloga. Modulation of enzymatic activities by n-3 polyunsaturated fatty acids to support cardiovascular health. *J. Nutr. Biochem.* **2008,** *19(7),* 417–437.

Simopoulos, A.P.; A. Leaf; N. Salem Jr. Workshop statement on the essentiality of and recommended dietary intakes for omega-6 and omega-3 fatty acids. *Prostaglandins Leukot. Essent. Fatty Acids* **2000,** *63,* 119–121.

Singh, R.B.; M.A. Niaz; J.P. Sharma; R. Kumar; V. Rastiogi; M. Moshiri. Randomized, double-blind, placebo controlled trial of fish oil and mustard oil in patients with suspected acute myocardial infarction: the indian experiment of infarct survival-4. *Cardiovasc. Drug Ther.* **1997,** *11,* 485–491.

Siscovick, D.; T.E. Raghunathan; I. King;S. Weinmann; V.E. Bovbjerg; L. Kushi; L.A. Cobb; M.K. Copass; B.M. Psaty;R. Lemaitre; B. Retzlaff; R.H. Knopp. Dietary intake of long-chain n-3 polyunsaturated fatty acids and the risk of primary cardiac arrest. *Am. J. Clin. Nutr.* **2000,** *71(1),* 208–212.

Spolaore, P.; C. Joannis-Cassan; E. Duran;A. Isambert. Commercial applications of microalgae. *J. Biosci. Bioeng.* **2006,** *101(2),* 87–96.

Tam, P.S.; R. Umeda-Sawada; T. Yaguchi; K. Akimoto; Y. Kiso; O. Igarashi. The metabolism and distribution of docosapentaenoic acid (n-6) in rats and rat hepatocytes. *Lipids.* **2000,** *35(1),* 71–75.

Theobald, H.E.; A.H. Goodall; N. Sattar; D.C. Talbot; P.J. Chowienczyk; T.A. Sanders.Low-dose docosahexaenoic acid lowers diastolic blood pressure in middle-aged men and women. *J. Nutr.* **2007,** *137(4),* 973–978.

Valencia, I.; D. Ansorena; I. Astiasarán. Development of dry fermented sausages rich in docosahexaenoic acid with oil from the microalgae *Schizochytrium* sp.: influence on nutritional properties, sensorial quality and oxidation stability. *Food Chem.* **2007,** *104(3),* 1087–1096.

Volkman, J.K.; S.W. Jeffrey; P.D. Nichols; G.I. Rogers; C.D. Garland. Fatty acid and lipid composition of 10 species of microalgae used in mariculture. *Exp. Mar. Biol. Ecol.* **1989,** *128,* 345–353.

Ward, O.P.; A. Singh. omega-3/6 fatty acids: alternative sources of production. *Proc. Biochem.* **2005,** *40(12),* 3627–3652.

Whelan, J.; C. Rust. Innovative dietary sources of n-3 fatty acids. *Annu. Rev. Nutr.* **2006,** *26,* 75–103.

Wibert, G.J.; R.A. Burns; D.A. Diersen-Schade;C.M. Kelly. Evaluation of single cell sources of docosahexaenoic acid and arachidonic acid: a 4-week oral safety study in rats. *Food Chem. Toxicol.* **1997,** *35,* 967–974.

Wijendran, V.; K.C. Hayes. Dietary n-6 and n-3 fatty acid balance and cardiovascular health. *Annu. Rev. Nutr.* **2004,** *24,* 597–615.

Willumsen, N.; S. Hexeberg; J. Skorve; M. Lindquist; R.K. Berge. Docosahexaenoic acid shows no triglyceride-lowering effects but increases the peroxisomal fatty acid oxidation in liver of rats. *J. Lipid Res.* **1993,** *34,* 13–22.

Xu, H.; X. Miao; Q. Wu. High quality biodiesel production from a microalga *Chlorella protothecoides* by heterotrophic growth in fermenters. *J. Biotech.* **2006,** *126(4),* 499–507.

Yasumoto, T.; M. Murata. Marine toxins. *Chem. Rev.* **1993,** *93,* 1897–1900.

Zelle, S.; W. Barclay; R. Abril. Production of docosahexaenoic acid from microalgae. *Omega-3 Fatty Acids: Chemistry, Nutrition and Health Effects;* F. Shahidi, J.W. Finley, Eds., American Chemical Society: Washington, 2001; pp. 108–124.

# 20

# Fish Oils

**Jana Pickova**
*Department of Food Science, Swedish University of Agricultural Sciences, P.O. Box 7051, S-750 07 Uppsala, Sweden*

## Introduction

Fish oils, in general, are processed mainly from pelagic (from the open ocean) fatty-fish species; in the modern human diet, they are the major sources of long-chain polyunsaturated fatty acids (LCPUFAs, longer than an 18-carbon chain), especially omega-3 (n-3) fatty acids. In addition, the LCPUFAs in fish oils may also originate from aquaculture fish fed fish meal or via the conversion of shorter n-3 fatty acids from plant oils in fish feeds. The fatty acids eicosapentaenoic acid (EPA, 20:5 Δ 5,8,11,14,17) and docosahexaenoic acid (DHA, 22:6 Δ 4,7,10,13,16,19) are important components of fish oils. Fish-oil PUFAs are sometimes also called highly unsaturated fatty acids (HUFAs), where "highly" means "at least three double bonds". The incorporation of these fatty acids as healthy additives in foods and as formulations in different forms of health-promoting capsules or oils is currently increasing. This makes fish oils extremely valuable, and explains the continuous increase in their prices. This is a new scenario, as fish oils were traditionally a cheap commodity, often hydrogenated to be used in different food applications. Today, larger and larger proportions of these oils are used as healthy additives and ingredients in many value-added (or fortified) food products or health-food capsules, often sold at high prices.

As the global consumer demand for fish continues to climb, especially in affluent developed nations, annual harvests of fish in the wild have remained roughly stable since the mid-1980s, whereas aquaculture has been experiencing a growth rate of around 8–9% per year during the same period. Today, aquaculture continues to expand in almost all world regions. One serious bottleneck in maintaining growth in aquaculture is the limited supply of fishmeal and particularly fish oil used to feed carnivorous cultured species, such as salmonids, flatfish, groupers, and breams. Since fishery harvests have plateaued, the ability to harvest more oils and proteins from this source will not be possible according to the Food and Agriculture Organization (FAO). Aquaculture is crucial to meet the increasing demand for fish and also to support the attempts to maintain wild-fish populations. As aquaculture continues to expand, fish meal and fish oil will become increasingly more expensive. The price of fish oil has exceeded other oils during recent years (Fig. 20.1), and is expected to increase even more (*Oil World*, 2006).

An excellent review on fish oils was published in 2005 by Ackman, and the current chapter attempts to highlight some of the important developments since then. Also, in the same Ackman volume, a chapter on marine mammal oils was published (Shahidi & Zhong, 2005).

## Innovations in the Design of Fish Feeds

Traditionally, the main sources of fish feeds were fish meal and oils from these pelagic fisheries species: anchovy (*Engraulis anchoita*), sprat (*Sprattus sprattus*), herring (*Clupea harringus*), sand eel (*Ammodytes* spp.), and capelin (*Mallotus villosus*). These species are fatty fishes (with a considerable content of fat in their flesh and viscera), but hardly any of them are consumed by humans because their size is very small and they are difficult to keep fresh. Also, the trimmings (processing by-products) of fish from fisheries as well as from aquaculture are often processed to produce fish meals and oils. This part of the by-product utilization from fish to fish oil and meal is undergoing a rapid development. The crude fish oils are produced mostly by cold-pressing and cooking and pressing. The residual content of oils in the meal fractions is often approximately 10%.

The declining wild-living fish populations of many predatory species are increasingly replaced by farmed fish. Demands for fish meal and fish oil are increasing, while production remains stable and cannot be increased on a worldwide basis, due

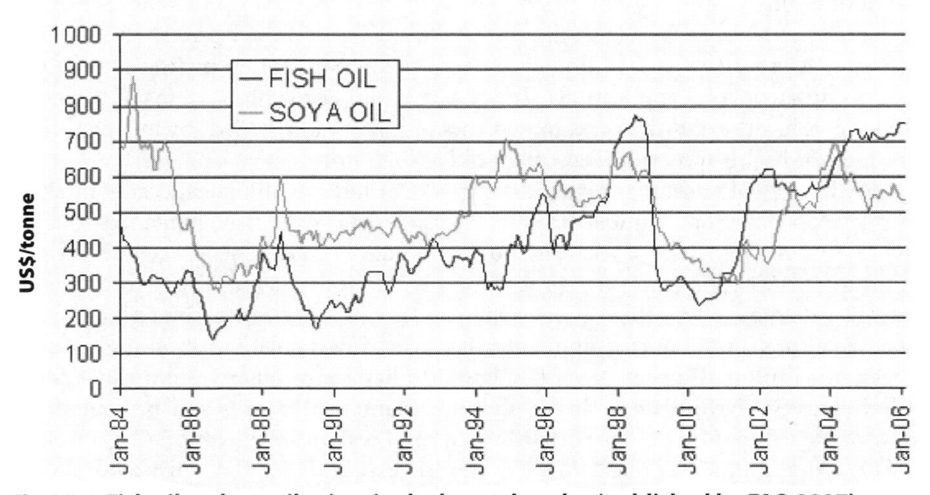


**Fig.20.1. Fish oil and soy oil prices in the latest decades (published by FAO 2007).**

to limited pelagic fish production. During a normal year, approximately 6–7 million tonnes of fish meal and 1.2–1.4 million tonnes of fish oil are produced, and both are mostly used in aquaculture feed. Fish-oil consumption today is about 900,000 tonnes: 783,000 tonnes (87%) is used in aquaculture and the rest for direct human consumption (FAO, 2007). Trials with marine sources other than pelagic fish were performed (e.g., krill and other plankton as well as mezopelagic fish—blue whiting, *Micromesistius poutassou*). Also, in the near future, further utilization of the by-products from fisheries and the farmed-fish industry is needed. Europe, with only a small part of the worldwide production of fish protein and fish oil (15%), is especially dependent on the global pelagic fisheries, being a large consumer of these in predatory temperate- and cold-water farmed fish. The future needs of the aquaculture industry for fish oil are expected to outstrip the current supply within the next decade. In Table 20.1, the nations with the largest industrial fishery harvests are presented.

In general, vegetable sources of protein and oil will likely fill parts of the shortfalls in animal diets and feeds both on land and in water. Important issues to consider when evaluating vegetable resources are their effects on fish growth, health, and welfare, and the final-product's quality. Furthermore, one must consider carefully the impact on human health. Fish feeds are no longer formulated with only the traditional-fish raw materials (oil and meal). Finishing-diet strategies were developed to ensure a high level of the long-chain fatty acids, which are connected to human health (e.g., preventing cardiac, vascular, and mental disorders). When plant-oil-based diets are fed to fish during the growing phase and then replaced by a fish-oil-based diet a few months prior to slaughter, one can restore most of the beneficial fatty acid components of fish in terms of human dietary recommendations. On the other hand, feeding during the period before slaughter means also that a large fish consumes more finishing diet, and thereby requires larger amounts of fish oil.

Vegetable oils are rich in 18-carbon fatty acids, and the most common PUFAs in vegetable oils are linoleic acid (18:2n-6) and α-linolenic acid (ALA, 18:3n-3). Plants of the family *Echium* also contain stearidonic acid 18:4n-3 (Trattner et al., 2008a). The optimal time for restoring these levels of fatty acids in fish-farming activities depends on the inclusion level, the growth rate, and the fish species. One can make predictions of the effects by modeling, as suggested by Jobling (2004) and Robin et al. (2003).

**Table 20.1. Total Fish Oil Production by Country, Statistics (FAO, 2007)**

| Country | Million tonnes | % of Total production |
|---|---|---|
| Peru | 290,422 | 32.4 |
| Chile | 168,923 | 18.9 |
| Denmark | 82,303 | 9.2 |
| United States | 71,523 | 8.0 |
| Japan | 62,741 | 7.0 |
| Total all countries | 895,090 | |

Other n-3 fatty acid sources in larval fish feed include the use of microalgae, which are rich in PUFAs, LCPUFA, and antioxidants. In fact, these microalgae are the main producers of LCPUFAs, which are magnified by the consumers in the different steps of the food chains of the diverse aquatic ecosystems. Some microalgae contain high levels of specific fatty acids [e.g., *Nannochloropsis oculata* produces more than 35% of its fatty acids as EPA, *Phaeodactylum tricornotum* produces almost 25% of EPA, and *Isochrysis galbana* 6% of DHA (unpublished data)].

Krill (*Euphausia superba,* shrimp-like marine invertebrate zooplankton) are a potential new resource, believed to bring new possibilities of increasing the "fish protein and oil" sources to feed production. The drawback with this fishery is the fast degradation of krill after catch. Lipolytic and proteolytic enzymes are abundant in the crustaceans, making this raw material susceptible to the degradation of both lipids and proteins. The solution so far is to process the materials directly onboard ship by using very large and expensive floating factories. Krill-oil content varies largely between seasons, and the fat content is lower than that of most pelagic fish (Olsen et al., 2006). Copepods (*Calanus* spp.) are also harvested since they are the richest source of DHA and EPA, essential for feeding fish larvae. The composition of fatty acids from krill and copepods, among other sources, is presented in Table 20.2.

Seal and whale oils are commercialized mainly as dietary supplements in capsules. Their special role is found in joint inflammation (Arslan et al., 2002). This effect is ascribed to the unique n-3 fatty acid position in the glycerol skeleton (Fig. 20.2). Blubber fatty acid composition from different marine mammals is well-described in the review of Shahidi and Zhong (2005). Also, freshwater organisms (such as microalgae and thereby the whole limnic food- chain) produce DHA and EPA, which is described by Ackman (2002) and Arts et al. (2001), among others.

As mentioned before, the identification of alternative and land-based raw feed materials to partly replace the traditional fish sources is urgently demanded. New fish diets will rely heavily on the use of alternative plant-derived and/or bacteria- and algae-derived ingredients also for carnivorous cold–temperate- water fish species. At present, the cost of all agricultural production is increasing, due in part to a huge demand for cereals and grains as well as oil crops from biofuel production.

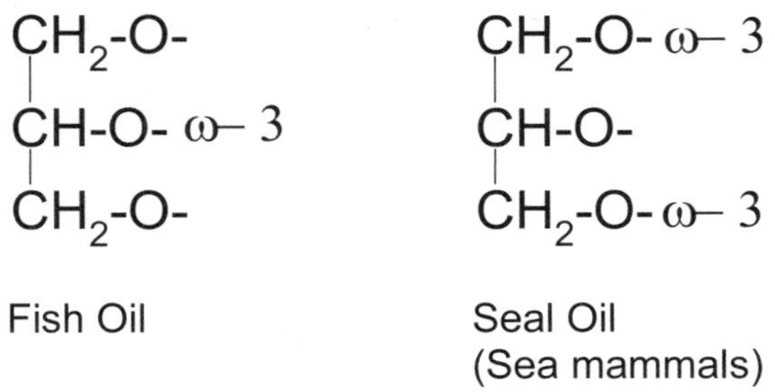

| Fish Oil | Seal Oil (Sea mammals) |
| --- | --- |
| $CH_2$-O- | $CH_2$-O- ω– 3 |
| CH-O- ω– 3 | CH-O- |
| $CH_2$-O- | $CH_2$-O- ω– 3 |

**Fig. 20.2. The position of n-3 fatty acids in fish oils (mainly at *sn*-2 position) versus marine oils from sea mammals or sea blubbers (mainly at *sn*-1, 3 positions) (Fröyland, personal communication)**

The responses of decision makers to the energy crisis may influence the prices of all products, and the competition for grains and oilseeds may be higher. Therefore, not only plant products but also the possible production of other alternatives (single-cell organisms) will be important to our future energy security.

## Microbial Sources of DHA and EPA

Omega-3 oils (rich in DHA and EPA) from microbial sources are currently being produced commercially. The general term for this oil source is "single-cell oil" (SCO). The single-celled organisms producing these fatty acids are fungi, molds and microalgae, which are able to store fat (triacylglycerols). The advantages of these oils are the simple fatty acid profiles (often only a few fatty acids are produced in each culture), the good oxidative stability, and the fact that one can consider them as vegetarian foods. On the other hand, problems with high cost, limited production, and public acceptance (except for algal sources) are still the limiting factors before a large-scale production and sale can take place (Wynn & Ratledge, 2007).

## Extraction and Processing of Fish Oils

Fish oils, produced by a cooking process, are pressed and purified in several steps. The proteins are denatured and removed by pressing. Thereafter, the oil is separated. The main step in the purification of fish oils and the removal of impurities is degumming, where phospholipids, proteins, and carbohydrates (as well as trace metals) are removed. The next step is alkali refining, which removes the free fatty acids, pigments, and residues of the above compounds. Sometimes bleaching is also performed to remove pigments, oxidation products, heavy metals, sulfur compounds, and soaps. One can further purify the oil in a carbon-treatment system to remove dioxins, polychlorinated biphenols (PCBs), furans, and polyaromatic hydrocarbons (PAHs). The oils are often stabilized with $\alpha$-tocopherol. When concentrates of fish oils are processed, the total persistent organic pollutants (POPs) are removed by a "working fluid" process, a new methodology consisting of fatty acid ethyl esters (Breivik & Thorstad, 2005).

## Fatty Acid Composition of Fish Oils

Fish oils differ in their fatty acid composition depending on the species of origin and seasonal variations. Therefore, fishes originating in South America contain other fatty acids than the North Atlantic oils. The fatty acids 20:1n-9 and 22:1n-11 are scarcely present in the South American species, while they can be relatively highly present in the North Atlantic herring and capelin oils, as well as in the liver oils of cod (*Gadus morhua*). The composition of fatty acids is also altered by season, because the fish stocks have a different nutritional status in different seasons. During post-spawning, the oil content is low, and the fishes are usually not harvested until after spawning season. Therefore, the composition of certain oil can vary greatly if all fatty acids are considered. A summary of various oils and the composition of fatty acids are shown in Table 20.2. Tuna oils are especially rich in DHA (approximately 27%) and in EPA (13%) (Nichols, 2007).

## Oxidative Stability and Quality

Fish oils are greatly unsaturated and in need of stabilization with antioxidants to protect against oxidation and deterioration of the product. These changes result in rancidity, reducing shelf life. PUFAs easily undergo oxidation, and one must prevent this by the careful handling, processing, and storage of them (Kamal-Eldin & Yanishlieva, 2002).

The oxidation caused by free radicals originating from the double bonds in the PUFAs may give rise to a production of toxic products when reacting with other compounds such as pigments and proteins in the oils. The oxidative changes are responsible for a series of effects in odor, color, off-flavor, and texture. Bulk oils of today are often stabilized with natural or synthetic antioxidants. The oil quality and oxidative stability are often measured by peroxide value (PV) and anisidine value (AV). PV is a measure of hydroperoxides in the oil analyzed via the oxidation of potassium iodide (AOCS methods and ISO 3960:2001). AV is measured as aldehydes in oils with a para-anisidine reagent (AOCS or ISO 6885: 2006). Other quality parameters of oils include the amount of free fatty acids, *trans*-double-bond measurements, nonsaponifiable components, cholesterol content, and others. Oxidation can also occur in cholesterol, causing the formation of cholesterol oxides, which may be carcinogenic

**Table 20.2. The Composition of Some Selected Fatty Acids (in %) from Raw Product Sources**

| Fatty acids | Krill [a] | Copepods* | South American Anchovy[b] | Menhaden[b] | Herring* | Seal oil[c] | Cod-liver oil* |
|---|---|---|---|---|---|---|---|
| 18:2n-6 | 2.5 | 1 | 1.3 | 1.2 | 1.5 | 1.6 | 3.5 |
| 18:3n-3 | 1.5 | 1.5 | — | — | 0.5 | 0.8 | 0.9 |
| 18:4n-3 | 4 | 4 | — | — | 0.5 | — | — |
| 20:1n-9 | 9 | — | 2 | 1.6 | 12 | 8 | 7 |
| 20:4n-6 | 0.5 | 0.8 | — | — | 0.8 | 0.5 | 0.1 |
| 20:5n-3 | 18 | 17 | 11 | 17 | 5 | 7 | 7.3 |
| 22:1n-11 | 14 | — | 0.6 | 1.2 | 20 | — | 5 |
| 22:5n-3 | 0.5 | tr | 1.9 | 1.6 | 0.5 | 3 | 1.4 |
| 22:6n-3 | 13 | 38 | 9 | 8.8 | 7 | 8 | 11.8 |

* own data; [a] Olsen et al., 2006; [b] Ackman, 2006; [c] Fröyland, personal communication.
—, not detected or not analyzed; tr, trace.

(Pickova & Dutta, 2003). High levels of tocopherols are considered to be valuable attributes of the oil.

## Edible and Nonedible Applications

By tradition, fish oils were utilized first as a source for oils for lighting the streets of European cities. The oil was used in lamps, and made the coastal regions of northern Europe rich. Later, fish oil became widely used in animal feeds, especially for swine and poultry, but this situation has changed dramatically. Today, most of the international fish oil that is produced is used in aquaculture production. As mentioned earlier, one serious problem in the field of aquaculture is finding adequate supplies of oil. New sources of fish or fish-like oils are being investigated to replace and extend the time span for aquaculture development. Also, aquaculture species that are not fully dependent on marine oils are becoming important for the future supply of fish to consumers. The farming of cyprinides, especially carp (*Cyprinus carpio)* and tilapias as well as different freshwater catfishes, is growing.

## Health Effects of Fish Oils

The health effects of fish oils are ascribed especially to the fatty acids EPA (20:5n-3) and DHA (22:6n-3). These fatty acids have a number of beneficial properties in humans. A meta-analysis by Mozaffarian and Rimm (2006) demonstrated the advantage of an intake of these fatty acids in relation to cardiovascular disorders. Beneficial effects on neuro-development during gestation and infancy were also demonstrated. The major conclusion from this study was that the benefits of fish intake exceed the potential risks of the possible harmful effect of pollutants (Mozaffarian & Rimm, 2006). Hites et al. (2004) published an important study of many oil-rich fishes regarding the contents of pollutants. The amount of DHA recommended for daily intake varies between studies. Recommendations on the intake of DHA and EPA are given in several countries, for example, in the United Kingdom (Anonymous, 1994) as well as in the United States (Simopoulos et al., 1999). The need for DHA and EPA is caused by the limited and variable ability of mammals to elongate and desaturate 18-carbon fatty acid chains, both n-6 and n-3. A study showing the desaturases involved in this conversion was published by Voss et al. (1991). Today, the importance of DHA as a nutrient that is vital in gene expression and thereby a factor in lipid metabolism is recognized (Massaro et al., 2008). Recommended dosages of fish oils as foods and as additives or supplements in different foods are being developed. The measured levels of omega-3 in foods are often used in the marketing and labeling of functional food products rich in omega-3 fatty acids.

Health claims can relate to the contents of nutritionally beneficial compounds or disease-risk reduction claims. The omega-3 health claims can be divided into two different claims: omega-3 PUFAs (ALA-rich PUFA oils) and long-chain omega-3 PU-

FA-containing oils (LC omega-3-rich PUFA oils, e,g., fish oils), the latter containing mostly EPA 20:5n-3 and DHA 22:6n-3. The legislation related to the omega-3 oil's fatty acid content and composition is still under revision, and will likely continue to undergo changes as new research studies are published and scrutinized.

Seafood is the richest and most efficient source of dietary very long chain n-3 fatty acids. World aquaculture will constitute more than one-half of the world's seafood production within only a few years (FAO, 2007). Therefore, aquaculture is the world's fastest-growing food- production sector, bringing great potential for food supply, enhanced trade, and economic benefits. Nordic farming of cod is a brand-new business, but the production is expected to exceed 100,000 metric tonnes within a decade—corresponding to a tenfold increase. The use of high-energy feeds is of major importance for the development of cost-effective fish farming and fast- growing fish.

Fish meal and fish oil are limited resources. The importance of plant seeds (rich in protein and fat) in fish feeds is increasing. Grains and other plant products are cost-efficient in fish feed, and are considered more sustainable than fish meal and fish oil. Grains and other seeds may contain biologically active compounds that affect nutrient utilization and possibly fish health, positively or negatively, by their potential modulating effects on the immune system, as well as an interaction with the fat metabolism. As a consequence, the rearing of salmon, cod, and other high-value fish is increasing, but this is also limited by the availability of aquaculture feeds based on marine products, and alternative sources of nutrients are desperately needed for fish farming. Such sources have the advantage, beyond reducing the direct pressure on wild-fish resources, to add to the human-food bank rather than only transforming one human-food source to another. Different vegetable sources of oils and proteins are being tested with good results, except for the amounts of EPA and DHA in the fish (Pickova & Mørkøre, 2007).

Recently shown was that the addition of sesame lignans to different vegetable oils increased the percentages of DHA in the white muscle of rainbow trout by approximately 30–40% by increasing the desaturation index (22:6/18:3) (Trattner et al., 2008a). The conclusion was that the supplementation of fish feed with n-3 fatty acids (e.g., from rapeseed oil, linseed oil, or mixed oils) and sesame lignans increases the proportions of highly unsaturated n-3 fatty acids, especially DHA. It also affects the expression of a few nuclear receptors (Trattner et al., 2008b). Recent developments in the field of nutrigenomics and metabonomics are expected to promote our understanding of gene–nutrient interactions. Of special importance is the effect of n-3 fatty acids (including ALA, EPA and DHA and lipid-soluble synergists) on nuclear receptors including peroxisome proliferator-activated receptors (PPARs), retinoid X receptors (RXRs), and sterol regulatory element-binding proteins (SREBPs). These receptors and proteins control the expression of genes involved in glucose and lipid homeostasis, inflammation, cell proliferation and differentiation, as well as immune response (Clark et al., 2000). The n-3 fatty acids and certain bioactive phytochemicals, such as sesame-oil lignans, are examples of naturally occurring potent ligands for these receptors (Kleveland, 2007; Trattner et al.; 2008a). Mixtures of sesamin and EPA/DHA increase the expression of PPAR and the activity of peroxisomal and liver enzymes involved in fatty acid β-oxidation (Arachchige et al., 2006).

Pharmaceuticals based on fish oils are still rare; the largest product so far is Omacor (Hjaltason & Haraldsson, 2006), containing approximately 50% of EPA and 35% of DHA. It is documented that Omacor lowers blood triacylglycerols (up to 45%), and one can use it together with statins. In the European Union (EU), Omacor is approved for use following a myocardial infarction. In the United States, a petition was filed with the Food and Drug Administration (FDA) to permit the use of Omacor as a prescription drug for the treatment of hypertryglycerolemia (Hjaltason & Haraldsson, 2006). Other products containing EPA and DHA are commercialized in Asia (Hjaltason & Haraldsson, 2006).

## Toxic Substances

For oils from areas where high loads of pollutants such as PCBs and dioxins are discovered, purification by different techniques is often performed to ensure nontoxic levels in the produced oils. Within the EU, regulations and legislation on the acceptable levels of contaminants in fish and fish oils were given. These values are presented separately for dioxin-like PCBs, furans and dioxins, which at present have a threshold of 10 pg/g of oil (European Commission, 2006b), while PAHs are set to a limit of 2 μg/kg of oil, measured as the amount of the benzo[a]pyrene compound. The limit of polychlorinated dibenzo-para-dioxins (PCDDs) and polychlorinated dibenzo-furans (PCDF) toxic equivalents is set to 2 pg/g of oil. The limit of mercury is set at 0.5 g/kg in fishery products (European Commission, 2006a).

## Genetically Modified Organisms

In the future, the need for nutritional fatty acids will grow, and the development of genetically modified organisms (GMOs) producing "marine" fatty acids is a likely scenario. Several international research groups are engineering new genes in rapeseed and linseed to enable the plants to synthesize EPA and DHA. This development is expensive and time-consuming, and many obstacles remain. Applications for these new oils include direct human consumption and health additives. For the time being, the EU has still not decided whether it will approve the use of GMOs for daily human consumption. The expected increase of aquaculture production is another of the markets and applications for these oils (Napier et al., 2005).

### References

Ackman, R.G. Freshwater fish lipids—an overlooked source of beneficial long-chain n-3 fatty acids. *Eur. J. Lip. Sci. Technol.* **2002,** *104,* 253–254.

Ackman, R.G. Fish oils. *Bailey's Industrial Oil and Fat Products;* 6th ed.; Six-volume set. F. Shahidi, Ed.; John Wiley & Sons, 2005; pp. 279–317.

Ackman, R.G. Marine lipids and omega-3 fatty acids. *Handbook of Functional Lipids;* C.C. Akoh, Ed.; CRC Press, Taylor and Francis Group: Boca Raton, FL, 2006; pp. 311–324.

Anonymous. Department of Health, UK. *Nutritional aspects of cardiovascular disease.* Report on Health and Social Subjects No. 46; London: United Kingdom. Her Majesty's stationary office. 1994.

Arachchige, P.G.; Y. Takahashi; T. Ide. Dietary sesamin and docosahexaenoic and eicosapentaenoic acids synergistically increase the gene expression of enzymes involved in hepatic peroxisomal fatty acid oxidation in rats. *Metabolism* **2006,** *55,* 381–390.

Arslan, G.; L.A. Brunborg; L. Frøyland; J.G. Brun; M. Valen; A. Berstad. Effects of duodenal seal oil administration in patients with inflammatory bowel disease. *Lipids* **2002,** *37,* 935–940.

Arts, M.T.; R.G. Ackman; B.J. Holub. "Essential fatty acids" in aquatic ecosystems: a crucial link between diet and human health and evolution. *Can. J. Fisheries Aquat. Sci.* **2001,** *58 (1),* 122–137.

Bimbo, A.P. Processing of marine oils. *Long-Chain Omega-3 Specialty Oils*; H. Breivik,Ed.; The Oily Press, PJ Barnes & Associates: Bridgewater, England, 2007; pp. 43–110.

Breivik, H.; O. Thorstad. Removal of organic environmental pollutants from fish oil by short path distillation. *Lipid Technol.* **2005,** *17,* 55–58.

Clark, R.B.; D. Bishop-Bailey; T. Estrada-Hernandez; T. Hla; L. Puddington; S.J. Padula. The nuclear receptor PPARα mediates inhibition of helper T cell responses. *J. Immunol.* **2000,** *164,* 1364–1371.

European Commission 2006a. European Commission Regulation EC, 1881/2006 setting maximum levels for certain contaminants in foodstuffs. *Off. J. EU.* L364/5, 20 December 2006.

European Commission 2006b. European Commission Regulation EC, 1883/2006 on methods of sampling and analysis for the official control of dioxins, and dioxin-like PCBs in foods. *Off. J. EU.* L364/32, 20 December 2006.

FAO, 2007. http://www.faostat.org. Accessed Nov. 2007.

Fröyland, Livar. NIFES, Bergen Norway. Personal correspondence.

Hites, R.A.; J.A. Foran; D.O. Carpenter;M.C. Hamilton; B.A. Knuth; S.J. Schwager. Global assessment of organic contaminants in farmed salmon. *Science* **2004,** *303,* 226.

Hjaltason, B.; G. Haraldsson. Fish oils and lipids from marine sources. *Modifying Lipids for Use in Food*; F.D. Gunstone, Ed.; Woodhead Publishing Ltd.: Cambridge, 2006; pp. 56–79.

ISO, International Standards Organization. ISO 3960:2001.

ISO, International Standards Organization. ISO 6885:2006.

Jobling, M. "Finishing" feeds for carnivorous fish and the fatty acid dilution model. *Aquaculture Res.* **2004,** *35,* 706–709.

Kamal-Eldin, A.; N.V. Yanishlieva. n-3 Fatty acids for human nutrition: stability considerations. *Eur. J. Lipid Sci. Technol.* **2002,** *104,* 825–836.

Kleveland, E.J. Gene expression studies in Atlantic salmon (*Salmo salar* L.). *Effects of 3-Thia Fatty*

*Acids and Characterization of Scavenger Receptor Class B, Type 1.* Faculty of Mathematics and Natural Sciences University of Oslo; 2007. ISSN 1501-7710 No. 585.

Massaro, M.; E. Scoditti; M.A. Carluccio; R. De Catarina. Basic mechanisms behind the effects of n-3 fatty acids on cardiovascular disease. *Prostaglandins Leukotrienes and Essential Fatty Acids,* **2008**, *79,* 3–5, 109–115.

Massaro, M.; E. Scoditti; M.A. Carluccio; M.R. Montinari; R. De Catarina. Omega-3 fatty acids, inflammation and angiogenesis: Nutrigenomic effects as an explanation for anti-atherogenic and anti-inflammatory effects of fish and fish oils. *J. Nutrigenetics Nutrigenomics* **2008,** *1,* 4–23.

Mozaffarian, D.; E.B. Rimm. Fish intake, contaminants, and human health. Evaluating the risks and the benefits. *JAMA* **2006,** *296,* 1885–1899.

Napier, J.; F. Beaudoin; O. Sayanova. Reverse engineering of long chain polyunsaturated fatty acid synthesis into transgenic plants. *Eur. J.Lipid Sci. Technol.* **2005,** *107,* 249–255.

Nichols, P.D. Fish oil sources. *Long-chain Omega-3 Specialty Oils;* H. Breivik, Ed.; The Oily Press, P.J. Barnes & Associates: Bridgewater, England, 2007; pp. 23–42.

Oil World, 2006; www.oilworld.biz, accessed Dec. 2007.

Official methods and recommended practices of American Oil Chemists' Society, Cd 18-90, Cd 8-53, and Cd 8b-90.

Olsen, R.E.; J. Suontama; E. Langmyhr;H. Mundheim; E. Ringo; W. Melle; M.K. Malde; G.-I. Hemre. The replacement of fish meal with Antarctic krill, *Euphausia superba* in diets for Atlantic salmon, *Salmo salar. Aquaculture Nutr.* **2006,** *12,* 280–290.

Pickova, J.; P.C. Dutta. Cholesterol oxidation products in some samples of herring, fish roe, fish meal and fish oil. *J. Am. Oil Chem. Soc.* **2003,** *80,* 993–996.

Pickova, J.; T. Mørkøre. Alternate oils in fish feeds. *Eur. J. Lipid Sci.Technol.* **2007,** *109,* 256–263.

Robin, J.H.; C. Regost; J. Arzel; S.J. Kaushik. Fatty acid profile of fish following a change in dietary fatty acid source: model of fatty acid composition with a dilution hypothesis. *Aquaculture* **2003,** *225,* 283–293.

Shahidi, F.; Y. Zhong. Marine mammal oils. *Bailey's Industrial Oil and Fat Products,* 6th ed.;F. Shahidi, Ed.; John Wiley & Sons, 2005; pp. 259–278.

Simopoulos, A.P.; A. Leaf; N. Salem, Jr. Workshop on the essentiality of and recommended dietary intakes for omega-6 and omega-3 fatty acids. *J. Am. College Nutr.* **1999,** *18,* 487–489.

Tacon, A.G.J. Salmon aquaculture dialogue: State of information on salmon aquaculture feed and the environment. *Int. Aquafeed* **2005,** *8,* 22–37.

Trattner, S.; A. Kamal-Eldin; E Brännäs; A. Moazzami; A. Zlabek; P. Larsson; B. Ruyter; T. Gjøen; J. Pickova. Sesamin supplementation increases white muscle docosahexaenoic acid (DHA) levels in rainbow trout (*Oncorhynchus mykiss*) fed high alpha-linolenic acid (ALA) containing vegetable oil: Metabolic actions. *Lipids* **2008a,** *43,* 989–997.

Trattner, S.; B. Ruyter; T.K. Østbye; T. Gjøe; V. Zlabek; A. Kamal-Eldin; J. Pickova. Sesamin increases alpha-linolenic acid conversion to docosahexaenoic acid in Atlantic salmon (*Salmo salar* L.) hepatocytes: Role of altered gene expression. *Lipids* **2008b**, *43*, 999–1008.

Voss, A.; M. Reinhart; S. Sankarappa; H. Sprecher. The metabolism of 7,10,13,16,19-docosapentaenoic acid to 4,7,10,13,16,19-docosahexaenoic acid in rat liver is independent of a 4-desaturase. *J. of Bio. Chem.* 266, **1991**, *30*, 19995–20000

Wynn J.P.; C. Ratledge. Microbial oils: production, processing and markets for specialty long-chain omega-3 polyunsaturated fatty acids. *Long-chain Omega-3 Specialty Oils;* H. Breivik, Ed.; The Oily Press, P.J. Barnes & Associates: Bridgewater, England, 2007; pp. 43–76.

# 21

# Butter, Butter Oil, and Ghee

**Bhavbhuti M. Mehta**
*Assistant Professor, Dairy Chemistry Department, Sheth M.C. College of Dairy Science, Anand Agricultural University, Anand-388 110, Gujarat, India*

## Introduction

The art of butter-making has a long history. References are made in ancient writings to the use of butter as a food in India between 2000 B.C. and 1400 B.C., and according to the conventional rules of the Hindu of the eighth and ninth centuries B.C., milk, honey, and butter were used in their ceremonial feasts. The importance of butter is well-documented. "Butter and honey shall he eat that he know to refuse the evil and choose the good" (Isaiah 7:15). The butter used in the early years of recorded history was churned directly from milk, and in many parts of India, butter for the preparation of "ghee" is still made by the direct churning of milk. Until the middle of the nineteenth century, factory butter-making was unknown; most butter was made on the farm from cream obtained by gravity creaming. However, with the development of the centrifugal cream separator, fat test, butter churn, artificial refrigeration etcetera, factory butter-making developed rapidly. Depending on the manufacturing process, three main types of butter exist, each having a specific flavor: (i) sour-cream butter, obtained from cream inoculated with starter cultures; (ii) sweet-cream butter, derived from unfermented cream; and (iii) acidified-cream butter, produced with sweet cream, to which lactic acid and flavor concentrates are then added. Multiple fat products—including butter oils, ghee (clarified milk fat), anhydrous butterfat, butterfat–vegetable oil blends, and fractionated butterfats—are manufactured around the world today. In the past, converting milk fat to butter was the primary preservation technique. Today, the preferred preservation method involves the processing of fat to the anhydrous butter oil/ghee state and then hermetically packaging it under nitrogen to substantially increase the shelf life and reduce the incidence of degradation (McDowall, 1953; Solanky, 1990; Sukumar De, 1991).

"Butter oil" is the term used commercially in the United States to describe fat recovered from butter, and also to describe fat obtained directly from cream by de-emulsification and direct centrifugation. The terms "milk fat," "anhydrous milk fat," "dry butter fat," and "dehydrated butter fat" are used synonymously with butter oil (McDowall, 1953; Sukumar De, 1991).

In India, the preservation of milk and milk products is primarily achieved by heat-induced desiccation. Ghee is obtained by the clarification of milk fat at a high temperature. Ghee is almost completely anhydrous milk fat, and no similar product exists in other countries. It is the most ubiquitous indigenous milk product, and is prominent in the hierarchy of the Indian diet. Being a rich source of energy, fat-soluble vitamins and essential fatty acids, and due to its long shelf life at room temperature (20 – 40°C), a major portion of ghee is used for culinary purposes, and the remainder is used in confectionary items that are consumed on formal occasions such as religious ceremonies (Ganguli & Jain, 1972).

## Processing of Butter, Butter Oil, and Ghee

### Butter

In simplest terms, butter may be considered as a solid emulsion of fat globules, liquid as well as crystallized fat, water, and air. Butter may be defined as a fat concentrate which is obtained by churning cream, gathering the fat into a compact mass, and then working it. It contains slightly more than 80% of fat, which is partly crystallized. Many kinds of butter are found in the market including: pasteurized-cream butter, ripened-cream butter (*makkhan*), unripened-cream butter, salted butter, unsalted butter, sweet-cream butter, sour-cream butter, creamery butter, and cold-storage butter (Solanky, 1990; Sukumar De, 1991).

The butter-making process, whether by a batch or a continuous method, consists of the following steps (Thakar, 1990):

- Preparation of cream
- Destabilization and breakdown of the oil-in-water (o/w) emulsion
- Aggregation and concentration of the fat particles
- Formation of a stable water-in-oil (w/o) emulsion
- Packaging
- Storage and distribution of the product

The flow chart for conventional butter manufacturing is shown in Fig. 21.1.

The cream—after standardization for fat (and acidity if necessary), pasteurization, cooling, and aging—is transferred to the butter churn. Optimal temperatures for the aging of cream (5–10°C for 10–12 hours, generally overnight) are normally lower than that needed for efficient churning. The aged cream is also susceptible to damage if mishandled, which may result in blocked pipelines and excessive fat losses in the buttermilk. The efficiency of churning or churnability is measured in terms of the time required to produce butter granules and by the loss of fat in the buttermilk. Hence, the temperature of churning is adjusted to produce butter in a reasonable

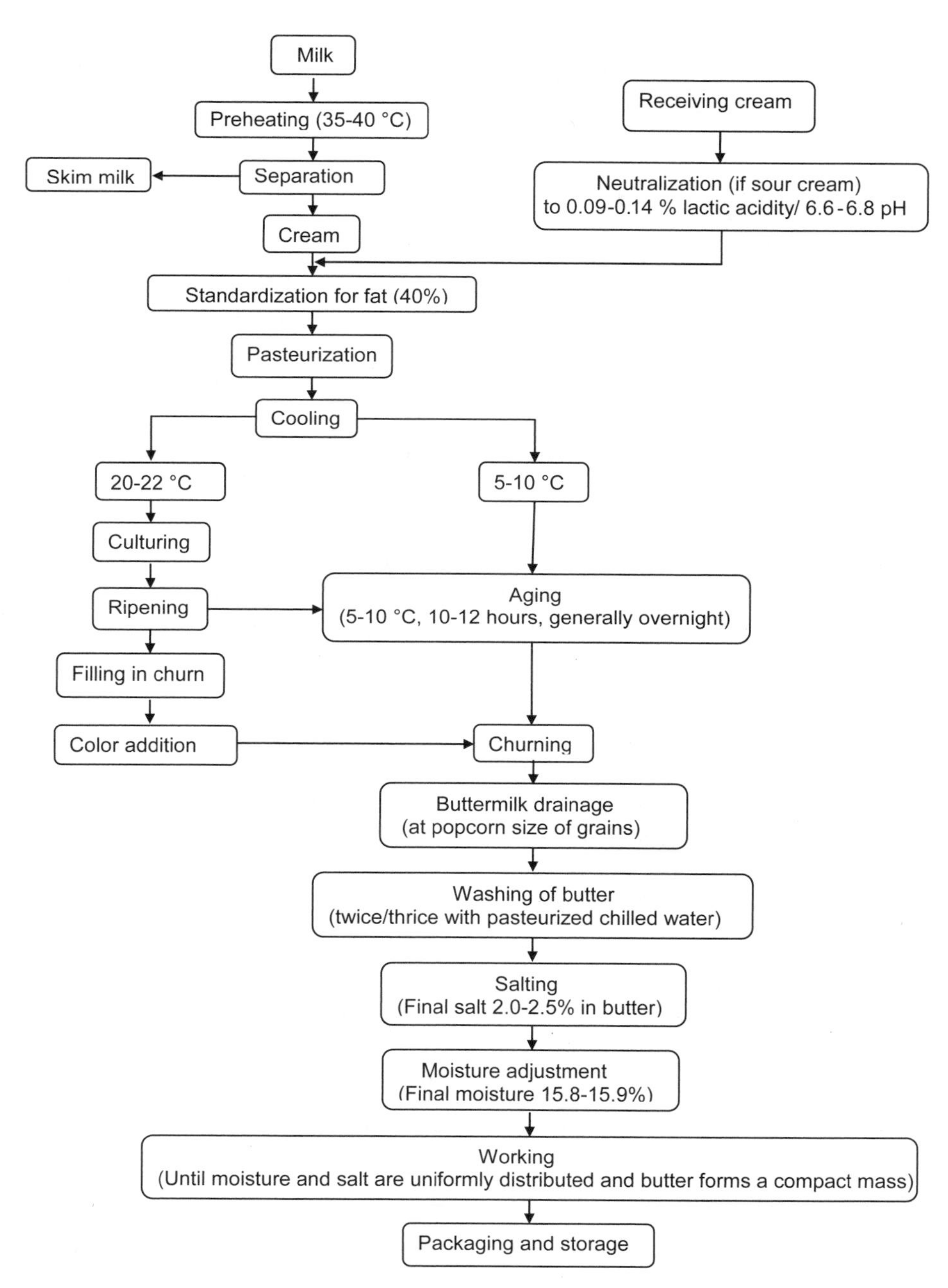


**Fig. 21.1. Flow chart for conventional butter manufacturing.**

period of time and to prevent the excessive losses of fat in buttermilk that occur if the temperature is too high. Usually the churning of cream is carried out at a temperature of 9–11°C (in summer) or 11–15°C (in winter) for a batch method (Sukumar De, 1991; Thakar, 1990).

### Cream Ripening

Ripening (fermentation) of cream is carried out (in ripened cream butter) to produce butter with a pleasing, pronounced characteristic flavor and aroma and to minimize the fat losses in buttermilk. The starter culture containing lactic-acid producers such as *Streptococcus lactis* and/or *S. cremoris* together with aroma (diacetyl) producers such as *S. diacetilactis*, *Leuconostoc citrovorum*, and *L. dextranicum*, in correct proportions, is added to the standardized, pasteurized, and cooled (20–22°C) cream at 0.5–2.0%. After being thoroughly mixed, the cream is incubated at 21°C for 15–16 hours. The typical flavor of butter from ripened cream is mainly the effect of diacetyl (also known as 2,3-butanedione, as per nomenclature of the International Union of Pure and Applied Chemistry), and, to a smaller extent, of acetic and propionic acids. No diacetyl is present in sweet cream. The flavor intensity in butter depends on its diacetyl content (Hunziker, 1940), as shown in Table 21.1. The normal diacetyl content of ripened cream butter is on average 2.5 ppm and rarely over 4 ppm. Diacetyl is produced from acetyl–methyl–carbinol. One way to improve the flavor of butter is to increase the citric-acid content of cream or milk before fermenting begins (Sukumar De, 1991).

Cream ripening is expensive, time-consuming, and exacting. Further, most of the flavoring substances enter into the buttermilk and wash water, and are lost to the butter. Hence, the use of starter, starter distillate, or synthetic-flavor compounds, which are mixed with sweet cream during the working process, imparts the characteristic flavor of ripened cream butter to the finished product. However, such butter has a somewhat harsh unnatural aroma, and lacks the pleasing, mellow, uniformly blended aroma of ripened cream butter (Sukumar De, 1991).

### Butter Color

Prior to the churning process, yellow color is sometimes added to the butter. This is done to maintain the uniformity of yellow color in butter throughout the year for consumer satisfaction. The amount of standard color added varies from 0 – 250 mL

**Table 21.1. Correlation of Flavor Intensity of Butter with Its Diacetyl (2,3-Butanedione) Content**

| Diacetyl content in butter | Flavor intensity |
|---|---|
| None | Flavorless |
| 0.2–0.6 ppm | Mild flavor |
| 0.7–1.5 ppm | Full flavor |

Source: Hunziker (1940).

or more per 100 kg of butterfat. The butter color should preferably be added to the cream in the churn. Generally, either annatto or β-carotene is used to color butter. Annatto is obtained from the seeds of the annatto plant (*Bixa orellana* L.). The yellowish-red coloring substances are extracted from the seeds by dissolving them in neutral oil (such as castor, groundnut, and sesame). The annatto plant is of tropical origin. β-Carotene is extracted from carrots and other β-carotene-rich vegetable matter. The yellow color from β-carotene has a slight greenish tint. Its use is growing because it increases vitamin-A potency (Satyanarayana et al., 2003; Sukumar De, 1991).

### Churning Process

The churning of cream consists of agitation at a suitable temperature until the fat globules adhere, forming larger and larger masses, and until a relatively complete separation of fat and serum occurs. During churning, air is beaten into the cream, and is dispersed into small bubbles. The fat globules touch these bubbles, often spreading part of their membrane substances and some of their liquid fat over the air–water interface, and become attached to the bubbles; one bubble catches several globules. This resembles flotation, although in true flotation the foam is collected. In the churning process, however, the air bubbles keep moving through the liquid and collide with each other. They thus coalesce, and in this way their surface area diminishes. As a consequence, the adhering fat globules are driven toward one another. The liquid fat then acts as a sticking agent, and the fat globules are clumped together. In this way, small fat clumps are formed (Walstra et al., 1999).

The clumps, in turn, participate in the churning process, resulting in still larger clumps. When the clumps become larger, a direct collision between them increasingly occurs, and the clumps now grow without the air bubbles any longer playing an important part. Flotation thus predominates initially, and simple mechanical clumping occurs later on. In addition, more and more liquid fat and surface layer material are released (they first spread over the air bubbles and partly desorb as soon as the bubbles coalesce); this is called colloidal fat, and it consists of tiny liquid-fat globules and membrane remnants. Toward the end of the churning process, little foam is left, presumably because too few fat globules remain to cover the air bubbles and thereby stabilize these bubbles (Walstra et al., 1999).

The churning time (i.e., the time required to destabilize the fat globules) depends on the following factors (Sukumar De, 1991; Sukumar De & Mathur, 1968; Thakar, 1990):

- Size of the fat globules
- Season of the year, stage of lactation, and feeding conditions
- Churning temperature and conditions of cooling of cream
- Fat content of cream

- Load of cream in the churn, speed of rotation of churn, and design of the churn
- Acidity of cream

Cream with large fat globules churns more readily than cream with small fat globules. The reduction in the size of the fat globules with advancing lactation causes an increase in the churning time. Cream from the milk of animals very near the end of lactation gives serious churning difficulties due to the formation of soaps by the action of lipases on the fat. During winter, a larger proportion of hard fat (higher melting point) is present in cream (due to the type of feed given to the animals), whereas in summer relatively soft fat is present in cream. The feeding of hay, cottonseed, and some other agricultural materials may make the butter harder, while the feeds rich in vegetable oils tend to decrease the hardness of butter. Low-fat cream churns less readily due to the presence of more entrapped serum, which sometimes prevents the contact and cohesion of fat globules. Cream with very high fat content is also too viscous to churn readily, and it tends to cling to the wall of the churn. Hence, it does not receive sufficient agitation and mixing.

The temperature of cream at churning will influence the churning of cream. When the cream is to be churned soon after cooling (i.e., without aging), it should be cooled to 5–8°C below the usual churning temperature. If the time of holding is 1–2 hours, the temperature should be 2–4°C below the normal churning temperature. In most butter factories, holding the cream overnight and churning it on the following morning is more convenient. This intermediate period between cooling and churning allows crystallization of the fat to reach equilibrium, and hence permits a uniformity in the churning conditions (Thakar, 1990).

Overloading the churn results in delayed churning due to the formation of froth to such an extent that it completely fills the free space above the cream, and thereby does not allow any mixing.

### *Draining of Buttermilk*

After producing butter granules of approximately the size of a peanut, the buttermilk should be drained from the churn. The draining of the buttermilk at smaller butter-granule sizes results in heavy fat losses in buttermilk, whereas overchurning to larger butter granules results in more entrainment of buttermilk within these granules, making the subsequent washing of the butter more difficult. The fat contained in the buttermilk is derived from three main sources, namely, fat globules that are too small to churn out, fat globules entrained in the curd particles, and fat in the form of butter granules that are sufficiently small to pass through the buttermilk strainer (McKay & Larsen, 1939; Walker-Tisdale & Robinson, 1919).

### Washing the Butter

Washing the butter (with pasteurized chilled water) enables the removal of free buttermilk from the butter granules, thereby reducing the curd content of the butter, and it helps to control the temperature of the granules for the subsequent working process. The amount of buttermilk solids retained (after washing treatment) within the granules depends on granule size and serum content in the cream. The water should be allowed to remain in contact for 5–10 minutes to allow time for uniform cooling throughout the granules. The wash water should be free from iron, odorless, and bacteriologically sterile. Unwashed butter may contain 1.1–1.5% of curd compared to 0.6–1.0% found normally in washed butter. The washing stage can be eliminated also if good- quality pasteurized cream is used for butter-making (Thakar, 1990).

### Salting the Butter

The quantity of salt to be added is based on an estimate of the amount in finished butter expected from the churn. The salt used for the salting of butter should be of good quality, free from foreign insoluble matter, and bacteriologically sterile. The addition of salt may be done by sprinkling the calculated quantity of salt (dry salting), followed by the sprinkling of water (wet salting) on the butter granules in the churn. Alternatively, the required quantity of salt is dissolved in water (brine salting) and then sprinkled on butter granules. After salting, the churn is run at a slow speed for about 10 revolutions (Thakar, 1990).

### Adjusting the Moisture Content

After the addition of salt, working is continued until the free moisture is absorbed by the butter and the churn is dry. No appreciable quantity of free moisture should appear on a trier plug taken from the butter in the churn. A sample is taken for the first moisture test, and based on the results, an additional quantity of water may be added. The first moisture test sometimes is higher than desired. This may be due to the churning of very thin cream, churning at high temperature, churning to too small granules, or faulty judgment in controlling the draining of the churn. Then the necessity may arise to drain the buttermilk for a longer period when cream is too thin or when butter is soft. When difficulty in reducing the moisture content to a desired level is expected, the draining of the granules may be facilitated by the use of cold wash water (Thakar, 1990). On completion of the working stage, the churn is unloaded and the butter is taken in the trolley for packaging.

### Continuous Butter-Making

Continuous methods of butter-making may be divided into three main types (Sukumar De, 1991):

**Fritz process:**
It involves the use of high-speed beaters to form butter grains within a few seconds. Buttermilk is drained away, and the resulting grains, washed or unwashed, are worked in a kneading section. Contimab, Westfalia, Silkeborg, and some machines from other manufacturers utilize this principle.

**Concentration or inversion process:**
It involves a system whereby cream of 30–40% of fat is concentrated in a special separator to 80–82% of fat according to the fat content required in butter. After standardization, the concentrated butter fat is cooled and worked. Alfa-Laval and Meleshin processes utilize this principle.

**Emulsification process:**
It involves the concentration of 30–40% of fat cream; during concentration, the emulsion is broken and the standardization of fat, water, and salt content is carried out, followed by re-emulsification, cooling, and working. Machines utilizing this principle are Creamery Package and Cherry Burrel.

## Butter Oil

Butter oil refers to the fat-concentrate obtained mainly from butter or cream by the removal of practically all the water and nonfat solids. The terms "milk fat," "anhydrous milk fat," "dry butter fat," and "dehydrated butter fat" are used synonymously with butter oil. It can be prepared by the following ways (Jana, 1990; Sukumar De, 1991):

- Alfa-Laval process (using cream as raw material): Cream having 30–40 % of fat is first warmed at 55–58°C and passed through a pre-concentrator. This is, in effect, a hermetic separator, which concentrates the cream up to a fat level of 70–75%. The high-fat cream is then passed into a centrifixator (used for homogenization of the cream), a specially designed paring chamber, for phase inversion into a water-in-oil emulsion. The next concentrator separates out the milk fat, which is washed, reseparated, and then passed into a vacuum chamber at 80–90°C for the removal of residual moisture. It is subsequently cooled to 20–26°C for packaging.
- Using white butter as raw material: The butter is first melted into a pumpable form, at around 50°C, passed through a plate heat exchanger to raise the temperature to 70–80°C, and then held in a sealed tank for a few minutes. This helps the emulsion to break. The butter is then pumped into a separator, where the butter serum is removed and the butterfat is washed with hot water and subsequently reseparated. This fat is then passed through to a vacuum chamber for the removal of the residual moisture. It is then cooled and stored for further use.

- Westfalia process: In this process, cream having 40% of fat is placed in a cream concentrator that removes the nonfat-milk solids portion and increases the fat to 75% in the cream, followed by homogenization to break the membrane around the fat globules. The fat is free of the solids and most of the aqueous phase in the centrifuge. The semi-dried oil is washed with water, centrifuged, and finally vacuum-dried. After cooling to 20°C, it is packaged.

Butter oil should be cooled and crystallized under careful control to form a large number of fine crystals. The desired result may be obtained by rapidly supercooling to 13–18°C and stirring the mass during forced crystallization (by adding 5–15% of finely crystalline fat from a previous batch); this yields a smooth homogeneous mass which does not separate into solid and liquid layers on standing (Sukumar De, 1991).

## Ghee

Ghee is an anhydrous milk fat. It occupies a prominent place in the Indian diet. In India, ghee is produced from the cream of both dairy cows (*Bos indicus* and *Bos taurus)* and buffalo (*Bubalus bubalis,* the domestic Asian water buffalo). The milk of water buffalo has a higher fat content than the milk from most dairy cows, and its lipid composition is also different as will be noted in a later part of the chapter. Ghee is manufactured by the direct heating of cream or butter churned from fresh or ripened cream or *dahi* obtained by the fermentation of milk with bacteria native to milk or selected starter cultures (Srinivasan, 1976).

The different methods of ghee manufacture (Pandya & Sharma, 2002) are:

- *Desi* or indigenous or traditional method
- Creamery-butter method
- Direct-cream method
- Pre-stratification method
- Continuous method

### *Desi or Indigenous or Traditional Method*

Of all the methods, ghee produced by the traditional method still represents a major share of ghee production. Simple technology, inexpensive equipment, small-scale operation, and superior organoleptic quality of ghee could be some of the reasons for this. The principle of ghee manufacture by this method involves: (i) fermentation (by lactic-acid bacteria, *S. lactis*) of primary raw material (i.e., milk); (ii) a mechanical process to gather milk fat in a concentrated form; and (iii) heating of the fat concentrate at a specified range of temperatures to remove moisture and to induce the interaction of milk fat with fermented residues of nonfat-milk solids. The characteristic aroma, flavor, and taste of ghee depend on the first and third steps (Pandya & Sharma, 2002).

### *Creamery-butter Method*

The making of ghee by the creamy-butter method is the usual industrial practice. In this process, the milk is first heated to around 40°C and separated through a centrifugal cream separator. The cream thus obtained is pasteurized, cooled, aged, and converted into butter. Sometimes, when such butter is immediately converted into ghee, the pasteurization of cream is eliminated. To further improve the flavor of the final product, the cream is sometimes ripened by using lactose-fermenting starter culture, and the churning is carried out in the usual manner. Butter is then clarified at temperatures ranging from 110–140°C. The ghee residue is removed either by filtration or through a ghee clarifier.

### *Direct-cream Method*

The direct-cream method is another way of obtaining ghee from milk. The ghee is obtained by direct-heat clarification of the cream. The whole process is simply a four-step procedure. The steps include: (i) dilution of cream to the original volume of milk with water and reseparating the same, (ii) washing the cream with ordinary or acidified water, (iii) use of cream with a high-fat percentage, and (iv) ripening of cream. All of these attempts are mainly aimed at trying to increase the recovery of fat in ghee and to improve the flavor, which otherwise is mild and milky in the case of direct-cream ghee.

### *Pre-stratification Method*

When the butterfat is heated to around 80°C and left undisturbed, distinct layers form. The lower layer has the highest specific gravity, and is made up of the serum portion of the butter. The top layer is mostly fat, and the curd particles generally form an intermediate stratum. Based on this expected separation, the pre-stratification method was developed. Apart from the shortening of the period of clarification and saving energy, the yield of ghee also increases by about 8% by this method.

### *Continuous Method*

Dairy plants have tried to modify, scale up, and adopt the traditional batch process for commercial production. Although large quantities of ghee are made by this process, a need exists for the development of a continuous process. One of the ways is to adopt the established process for producing butter oil involving the centrifugal separation of moisture, followed by final dehydration under vacuum.

### Butter Oil versus Ghee

The main differences between butter oil and ghee were given by Ganguli and Jain (1972). Butter oil has a bland flavor, whereas ghee has a pleasing flavor. Ghee has less moisture, contains more protein solids, and differs in fatty acid and phospholipid as compared to butter oil. Butter oil is prepared by melting butter at a temperature not exceeding 80°C, whereas ghee is manufactured at 100–140°C. Butter oil can be reconstituted with skim-milk powder, whereas ghee can not be.

## Defects in Butter, Butter Oil, and Ghee

### Butter

Butter is subject to, and possesses or develops, one or more of a great variety of defects. The defects that may be found in butter are of diverse nature and origin. According to their general nature, these defects are classified under three main groups: flavor and aroma; body and texture; and color.

#### *Defects in Flavor and Aroma*

Most consumers agree that butter should possess a clean fresh "buttery" flavor, which is a composite of the natural basic flavor of fresh butter fat and of the effect of the unique physical structure of butter on the flavor sensation. However, commercial butter varies widely in its final flavor because of differences in the character of the other flavors naturally present with, or imposed upon, this basic flavor and because of variation in physical properties.

Flavor defects, more than any other faults of butter, affect butter consumption adversely, penalizing the entire industry, while their absence makes for increased consumption and thus provides an essential stimulus for an active and prosperous commerce in butter.

The possible flavor defects of butter are numerous and of great variety. They may be due to direct absorption by cream or butter of objectionable flavors and odors from other substances or products, or to bacterial fermentation, or to enzymatic or chemical reactions, or to combinations of two or more of these causes. They may have originated in the cream during the production, handling and transits, or were produced during manufacturing, or developed in butter after manufacturing, or they may be the result of a combination of conditions, such as the quality of cream, method of manufacture, and conditions of storage. The same off-flavors may be produced independently by several different factors, or combinations of factors, and the same factor may cause several different off-flavors. The flavor defects in butter may be grouped as:

- Flavor defects most likely due to off-flavor in the cream include: feed, weed, cowy and barny, unclean, musty, malty, bitter (rancid), stale, yeasty, cheesy, metallic flavors, etcetera.
- Flavor defects of fresh butter due to faulty methods in manufacture include: flat, high acid and sour, coarse, cooked or scorched, neutralizer, oily or oily metallic flavors, etcetera.
- Flavor defects that may develop after manufacture include: surface taint, cheesy, putrid, rancid, tallowy (oxidized) flavors, etcetera.

### *Defects in Body and Texture*

Body and texture refer chiefly to the physical properties of butter. A physically ideal butter is one that has a firm, compact body that will stand up well under unfavorable temperature conditions, and is free from visible moisture and that has a waxy, plastic and spreadable texture, free from objectionable crumbliness, stickiness, salviness, greasiness, mealiness, and grittiness.

The factors and combinations of factors that determine or influence the body and texture of butter are: composition of butter fat, structure of the fat globules, rate of fat crystallization in cream and butter, amount of liquid fat, and size of fat crystals in the butter. Some of these factors are, in turn, related to certain phases in the process of manufacture, such as the intensity of cream cooling, temperature of wash water, manner and intensity of working, and temperature at which the butter is held immediately after manufacture. The examples are crumbly, sticky, weak (soft), gritty, gummy body and leaky, greasy and oily, grainy, lumpy, open, coarse, short texture, etcetera.

### *Defects in Color*

The ideal color of butter ranges between a straw color and a golden- yellow color, according to market requirement. It must be uniform from churning to churning, and must be the same shade of intensity throughout the body of the butter. A change in the color of butter may suggest to the consumer that the butter is not genuine or that it is of inferior quality. A lack of uniformity in color of different churnings of butter also may lead to streakiness and other color defects.

Defects in color of freshly made butter are mainly due to faults in manufacturing technique, especially to the uneven incorporation of water and salt. To avoid any color defect, one must dissolve salt properly and disperse brine droplets evenly in the butter before working is complete. Uneven color may be due to the use of too low a churning temperature in conjunction with too short a period of working. It may also be due to the use of too high a temperature of working, allowing salt crystals to become surrounded by fat and so become difficult to dissolve. The solution of these crystals is completed after the butter is made, giving spotted color. Uneven color also may be due to an uneven incorporation of added color, and certain color defects are due to microbial action or chemical action on artificial butter colors. The examples are pale or dull, high color, mottled or spotted, wavy or streaky, white specks, bleached, green discoloration, brown or black specks, pink color, etcetera.

### Butter Oil and Ghee

Due to low-moisture content, ghee/butter oil has a better shelf life than that of butter. However, it undergoes deterioration which affects not only the economic value of the product but also spoils its appetizing flavor, making it unpalatable and sometimes toxic (Rangappa & Achaya, 1974; Varshney, 1997).

Ghee deteriorates through the process of rancidification which may occur through hydrolytic and/or oxidative routes (Kuchroo & Narayanan, 1973). Hydrolytic rancidity is the secondary pathway in which fatty acids are released from triglycerides of milk fat in the presence of moisture or lipolytic enzymes. However, it can be prevented to a great extent by proper heat treatment of raw materials before the manufacture of ghee to inactivate the lipolytic enzymes (Rangappa & Achaya, 1974). Oxidative rancidity is the major pathway by which ghee undergoes deterioration. This is referred to as autoxidation because the rate of oxidation increases as the reaction proceeds under usual processing and storage conditions.

Several researchers have devoted considerable effort to improve the stability of ghee against autoxidation through developing specific feeds for milch (domestic milk-producing) animals (Hagrass et al., 1983; Rama Murthy & Narayanan, 1972; Tandon, 1977), altering processing parameters (Rama Murthy et al., 1968; Singh et al., 1979), using proper packaging materials and storage conditions (Amr, 1990a; Chauhan & Wadhwa, 1987), adding milk components (Bhatia et al., 1978; Gupta Sudha et al., 1977; Megha, 1981; Santha & Narayanan, 1979b), adding synthetic antioxidants (Chatterjee, 1977; Kuchroo & Narayanan, 1972), incorporating natural antioxidants from edible-plant materials, spices and condiments, aromatic herbs, *Amla* (Indian gooseberry) juice, Betel, curry and drumstick leaves, mango-seed kernel, tomato-seed powder, onion-skin extract, *Tulsi* (*Ocimum sanctum* L.) leaves, Ragi (*Eleusine coracana* L.), and other natural materials (Ahmad et al., 1960; Amr, 1990b; Mehta, 2006; Parmar & Sharma, 1986; Sethi & Aggarwal, 1952). The prolonged consumption of synthetic antioxidants may cause health hazards such as teratogenic, carcinogenic, and mutagenic effects in experimental animals and primates (Hathway, 1966; Heijden et al., 1986; Maeura et al., 1984).

## Chemical Composition

The esters of fatty acids which are soluble in nonpolar organic solvents and insoluble in aqueous liquids are known as lipids/lipine/lipoid. Generally the terms "lipid" and "fat" are used interchangeably, but fat consists mostly of a mixture of neutral triglycerides, whereas "lipid" is a broader term. Fats which are liquid at room temperature are called oils (Jenness & Patton, 1959).

Fats are one of the most important constituents of milk as they play a significant role in milk and milk products from the perspectives of economics, nutrition, flavor, and physical properties. Fats are a primary source of energy in the diet. In addition to their fatty-acid components, fats appear to perform a number of vital functions in the body which are quite unrelated to their action as energy-bearing materials. Edible fats have certain important non-nutritional functions such as their use in baked goods, frying, contribution to flavor, and palatability of foods. These functions are derived principally from the distinctive physical properties of fats. Fat absorbs odors very easily, and therefore readily takes up off-flavor if stored near substances with a strong odor (Walstra & Jenness, 1984).

Milk fat is composed of a number of different lipid classes, but triacylglycerols represent about 97–98% of the total lipid. Phospholipids, cholesterol and cholesterol esters, diacylglycerols, monoacylglycerols, and free fatty acids (FFAs) make up the remaining 2–3% (Bitman & Wood, 1990). In addition, trace amounts are present of ether lipids, hydrocarbons, fat-soluble vitamins, flavor compounds (lactones and hydroxyl- and keto-acids), and compounds introduced from feed such as β-ionone and gossypol, which have physiological properties (Parodi, 1999). The structure and composition of the typical milk-fat globule are exceedingly complex. The globule is typically 2–3 μ in diameter with a 90 Å thick membrane surrounding a 98–99% triglyceride core (Hettinga, 2005; Mulder & Walstra, 1974).

On the basis of saponification, two classes of milk fat exist:

i. Saponifiable matter which is the major part, comprising about 99.5% of the total. It consists of various glycerides (mainly triglycerides), phospholipids, waxes, cholesterol esters, etcetera and

ii. Unsaponifiable matter which, after saponification with the alkali and extraction with suitable solvent, remain nonvolatile on drying at 80°C. It is comprised of cholesterol, lanosterol, hydrocarbon– squaline, fat-soluble vitamins A, D, and E, pigments/carotenoids, aldehydes and ketones, etcetera.

The composition of milk lipids is presented in Table 21.2 (Christie, 1995) along with the total concentration of fat in the milk. Each lipid class consists of many different kinds of molecules with varying component fatty acids. The glycerides are compounds in which 1, 2, or 3 –OH groups of glycerol have reacted with one or more fatty acids to form esters. By all estimates, about 500 separate fatty acids were detected in milk lipids. Three positions of attachment of fatty acid are designated as α, β, and α' or 1, 2, and 3, resulting in various positional isomers. The positional isomers may show differences in physical property (e.g., melting point) and in chemical property (e.g., rate of hydrolysis). The type of fatty acid attached at a certain position also influences the

**Table 21.2. Composition of Milk Lipids**

| | Amount (wt% of total lipid) | | |
|---|---|---|---|
| Lipid class | Cow milk (3.3–4.7% fat) | Buffalo milk (4.7–8.9% fat) | Human milk (3.8–4.0% fat) |
| Triacylglycerol | 97.5 | 98.6 | 98.2 |
| Diacylglycerol | 0.36 | — | 0.7 |
| Monoacylglycerol | 0.027 | — | Trace |
| Cholesterol ester | Trace | 0.1 | Trace |
| Cholesterol | 0.31 | 0.3 | 0.25 |
| Free fatty acids | 0.027 | 0.5 | 0.4 |
| Phospholipids | 0.6 | 0.5 | 0.26 |

Source: Christie (1995).

properties of the fat. Thus, the fatty-acid composition of triglycerides is an important factor in determining lipid properties (Joshi, 1990; Kurtz, 1974).

The composition of the milk-fat globule membrane (Table 21.3) is quite different from milk fat (Bracco et al., 1972; Brunner, 1969). Approximately 60% of the lipids in the milk-fat globule membrane are triglycerides, much less than in the parent milk fat (Hettinga, 2005). Milk phospholipids are associated mainly with the fat-globule membrane. Phospholipids constitute about 1% of total milk lipids. The major phospholipid classes are phosphatidylcholine, phosphatidylethanolamine, phosphatidylinositol, phosphatidylserine, and sphingomyelin. Phospholipids do not contain short-chain fatty acids, but they contain higher concentrations of long-chain and polyunsaturated fatty acids (PUFAs) than triacylglycerols (Parodi, 2004).

Milk fat contains many different fatty acids (Table 21.4), including a considerable concentration of short-chain fatty acids. The concentration of long-chain saturated fatty acids is higher in buffalo (domestic Asian water buffalo) milk fat which make it harder than the cow-milk fat; the melting point of buffalo fat is also high. As observed from Table 21.4, short-chain fatty acids are abundant. These are important in contributing flavor, particularly to butter. The concentration of butyric acid in milk fat (3–4%) is higher than in all other types of natural fats (Joshi, 1990; Walstra, 1983).

Unsaturated fatty acids and fat containing unsaturated fatty acids have a tendency to undergo various additional types of reactions. Most vegetable or plant oils contain a large proportion of unsaturated fatty acids such as oleic, linoleic, and linolenic acids with low melting points, and so they are liquid at room temperature and will be softer at low temperature. On the other hand, animal-body fats contain a high proportion of saturated fatty acids such as palmitic and stearic acids of high melting points, and hence are solid or semisolid at room temperature. The melting points of triglycerides in milk fat range from –40°C to 72°C; however, 37°C is considered the average melting point (Joshi, 1990; Mortensen, 1983).

**Table 21.3. Composition of Lipids from Milk-Fat Globule Membrane**

| Lipid component | Wt% of membrane lipids |
|---|---|
| Carotenoids (pigment) | 0.45 |
| Squalene | 0.61 |
| Cholesterol ester | 0.79 |
| Triglycerides | 53.4 |
| Free fatty acids | 6.3[a] |
| Cholesterol | 5.2 |
| Diglycerides | 8.1 |
| Monoglycerides | 4.7 |
| Phospholipids | 20.4 |

[a] Contained some triglycerides.
Sources: Bracco et al. (1972) and Brunner (1969).

**Table 21.4. Fatty-Acid Composition (Mole %) of Milk Fat**

| Sr. no. | Fatty acid | Abbreviation | Cow milk | Buffalo milk |
|---|---|---|---|---|
| 1. | Butyric | $C_{4:0}$ | 3.3 | 3.6 |
| 2. | Caproic | $C_{6:0}$ | 1.6 | 1.6 |
| 3. | Caprylic | $C_{8:0}$ | 1.3 | 1.1 |
| 4. | Capric | $C_{10:0}$ | 3.0 | 1.9 |
| 5. | Lauric | $C_{12:0}$ | 3.1 | 2.0 |
| 6. | Myristic | $C_{14:0}$ | 9.5 | 8.7 |
| 7. | Palmitic | $C_{16:0}$ | 26.3 | 30.4 |
| 8. | Stearic | $C_{18:0}$ | 14.6 | 10.1 |
| 9. | Oleic | $C_{18:1}$ | 29.8 | 28.7 |
| 10. | Linoleic | $C_{18:2}$ | 2.4 | 2.5 |
| 11. | Linolenic | $C_{18:3}$ | 0.8 | 2.5 |

Sources: Consolidated from various sources.

"Conjugated linoleic acid (CLA)" is a collective term for a series of conjugated dienoic positional and geometrical isomers of linoleic acid ($C_{18:2}$), which are found in relative abundance in the milk of ruminants compared with other foods (Chin et al., 1992; Lawson et al., 2001). The biochemical nomenclature for linoleic acid (*cis*-9, *cis*-12, octadecadienoic acid) designates this fatty acid as an 18-carbon ("octa-deca") fatty acid containing two double bonds ("di-en"), specifies the location of the double bonds (the 9 and 12 carbon atoms), and identifies the double bonds as being in a *cis*-isomeric configuration. This structural configuration results in two single bonds separating the double bonds. CLA is formed when reactions shift the location of one or both of the double bonds of linoleic acid in such a manner that the two double bonds are no longer separated by two single bonds. Unlike linoleic acid, which is a single unique molecule, several dozen different CLA isomers are possible, depending on which double bonds are relocated and the resultant isomeric reconfigurations (Bell & Kennelly, 2001; Kelly, 2001; & MacDonald, 2000).

CLA is formed as an intermediate product in the ruminal digestion of dietary fat. Forages and grains fed to dairy cows are characterized by a high percentage of PUFAs such as linoleic acid ($C_{18:2}$) and linolenic acid ($C_{18:3}$). The rumen bacteria, *Butyrivibrio fibrisolvens*, is capable of converting linoleic acid to CLA in a process known as biohydrogenation. The two double bonds in linoleic acid shift from their normal position at the 9th and 12th carbon to the 8/10, 9/11, 10/12, or 11/13 positions. Each of these double bonds can be in a *cis* or *trans* configuration, giving a range of possible CLA types, or isomers. The term "conjugated linoleic acid" actually refers to this whole group of 18 carbon conjugated fatty acids. The *cis*-9, *trans*-11 (*c*-9, *t*-11 CLA) is the principal dietary form of CLA found in a ruminant product exhibiting biological activity, and accounts for 73–94% of total CLA in milk and dairy products of ruminant origin (Bell & Kennelly, 2001; Kelly, 2001; MacDonald, 2000; Parodi, 2004). The chemical formula of *cis*-9, *trans*-11 CLA is shown below:

$$CH_3\text{-}(CH_2)_5\text{-}CH{=}\underset{\substack{\uparrow \\ trans}}{C}H\text{-}CH{=}\underset{\substack{\uparrow \\ cis}}{C}H\text{-}(CH_2)_7\text{-}COOH$$

Conjugated linoleic acid (*cis*-9, *trans*-11)

## Chemical Aspects of Butter and Butter Oil

Butter is a concentration of only one of the three main constituents (protein, fat, and carbohydrates/milk sugar) of milk—the fat. The other two constituents, protein and milk sugar, are present, but only in the proportion carried into the butter by buttermilk retained after churning. The fat carries with it into the butter the fat-soluble minor constituents of milk, the yellow pigment β-carotene, and its transformation products vitamins A and D. The curd is the dried residue from the buttermilk included in the butter during churning. It consists of the nonfatty constituents of the buttermilk such as casein, albumin, and milk sugar and a small amount of minerals. The curd content varies to some extent according to the conditions of manufacture, degree of dilution of the cream with water churning, the size of grain formed in the churn before the buttermilk is run off, and the thoroughness of the washing with chilled water, but it is not often higher than 1.0% (except with butter from nonwashed granules, when it may be as high as 1.50%). The normal diacetyl content in ripened cream butter is on average 2.5 ppm. The typical composition of table butter (creamery butter) and butter oil is presented in Table 21.5 and Table 21.6, respectively.

**Table 21.5. Composition of Butter (Creamery Butter)**

| Constituents | Wt% |
|---|---|
| Fat (minimum) | 80.0 |
| Moisture (maximum) | 16.0 |
| Salt (maximum) | 3.0 |
| Curd (maximum) | 1.0 |

**Table 21.6. Chemical Composition of Butter Oil**

| Constituent | Wt% |
|---|---|
| Butterfat | 99.5-99.8 |
| Moisture | 0.1-0.3 |
| Acidity (oleic) | 0.2-0.5 |
| Peroxide value | 0.0-0.1 |

Source: McDowall (1953).

## Chemical Aspects of Ghee

Chemically ghee is a complex mixture of glycerides (usually mixed), FFAs, phospholipids, sterols, sterol esters, fat- soluble vitamins (A, D, E, K), tocopherol, carbonyls, hydrocarbons, carotenoids (only in ghee derived from cow milk), small amounts of charred casein and traces of calcium, phosphorus, iron, etcetera. It contains not more than 0.3% of moisture. Glycerides constitute about 98% of the total material. Of the ≈2% remaining constituents of ghee, sterols (mostly cholesterol) comprise 0.5% (Ganguli & Jain, 1972; Sharma, 1981). The concentration of some major and minor constituents of ghee is shown in Table 21.7, and the fatty-acid composition of ghee is shown in Table 21.8.

Ghee residue is a charred (burnt) light- to dark-brown residue which is obtained on the cloth strainer after the ghee (prepared by any of the methods given above) is filtered. It is a by-product of the ghee industry. Essentially, it contains heat-denatured milk proteins, caramelized lactose, and varying amounts of entrapped fat, besides some minerals, phospholipids, and water. It is a rich source of the characteristic flavoring compounds such as FFAs, carbonyls, and lactones present in ghee (Santha & Narayanan, 1979a; Sharma, 1980; Srinivasan & Anantakrishnan, 1964). The antioxidant properties of ghee residue were attributed to the presence of phospholipids, α-tocopherol, vitamin A, amino acids, proteins with free sulfuryl groups, and protein–carbohydrate interaction products (Lal et al., 1984; Santha & Narayanan, 1978). Ghee residue may be used to enhance the flavor and quality of bland products such as vegetable fats and butter oil, in addition to its antioxygenic properties. The incorporation of ghee residue at a level of 15–20% in butter oil prior to heating at 120°C for 3 minutes enhanced the shelf life of butter oil, making it comparable to that of ghee (Wadhwa et al., 1991). Ghee residue is also used for direct consumption, as a spread on bread sandwiches, and for the preparation of toffees and other sweets (Sukumar De, 1991).

**Table 21.7. Concentration of Some Major and Minor Constituents of Ghee**

| Component | Cow ghee | Buffalo ghee |
|---|---|---|
| Saponifiable constituents | | |
| Triglycerides* | | |
| Short-chain (%) | 37.6 | 45.3 |
| Long-chain (%) | 62.4 | 54.7 |
| Trisaturated (%) | 39.0 | 40.7 |
| High-melting (%) | 4.9 | 8.7 |
| Partial glycerides* | | |
| Diglycerides (%) | 4.3 | 4.5 |
| Monoglycerides (%) | 0.7 | 0.6 |
| Phospholipid | 38.0 | 42.5 |
| Unsaponifiable constituents | | |
| Total cholesterol (mg%) | 330.0 | 275.0 |
| Lanosterol (mg%) | 9.32 | 8.27 |
| Lutein (μg/g) | 4.2 | 3.1 |
| Squalene (μg/g) | 59.2 | 62.4 |
| Carotene (μg/g) | 7.2 | 0.0 |
| Vitamin A (μg/g) | 9.2 | 9.5 |
| Vitamin E (μg/g) | 30.5 | 26.4 |
| Ubiquinone (μg/g) | 5.0 | 6.5 |

* Based on the percentage of total glycerides.
Source: Sharma (1981).

## Flavor Compounds in Butter and Butter Oil

Depending on the manufacturing process, three main types of butter exist, each having specific flavor. Sour-cream butter, obtained from cream inoculated with starter cultures; sweet-cream butter, derived from nonfermented cream; acidified-cream butter, produced with sweet cream, that is acidulated in a subsequent step with lactic acid and flavor concentrates (Mallia et al., 2008).

The flavor compounds of butter and butter oil primarily originate from the cream used to make it. When cream undergoes various processes such as pasteurization, sterilization, and ultra-high temperature (UHT), the processed cream has different odorants such as diacetyl (buttery), 2-pentanone (carrot-like), 2-heptonone (dairy-like), 3-hydroxy-2-butanone (buttery), dimethyl trisulfide (cabbage-like), acetic acid (acidic), furfural (caramel-like), and butanoic acid (cheese-like) (Pionneir & Hugelshofer, 2006). The aroma of cream is mainly due to the contribution from the aqueous phase of milk and from the fat-globule membrane (Badings & Neeter, 1980), while butter aroma is primarily derived from the volatile compounds present in the fat fraction. The odor-active compounds in sweet-cream butter (SwCB), sour-cream butter (SoCB), and butter oil (BO) are mentioned in Table 21.9, and the chemical structures of some of the potent odor-active compounds of butter and butter oil are in Fig. 21.2 (Mallia et al., 2008).

**Table 21.8. Fatty-Acid Composition of Cow and Buffalo Ghee (Mole%)**

| Fatty acid | Abbreviation | Western cow | Indian cow | Buffalo |
|---|---|---|---|---|
| Saturated | | | | |
| Butyric | $C_{4:0}$ | 9.6 | 8.8 | 11.4 |
| Caproic | $C_{6:0}$ | 3.5 | 3.5 | 3.1 |
| Caprylic | $C_{8:0}$ | 1.8 | 2.2 | 1.0 |
| Capric | $C_{10:0}$ | 3.1 | 3.0 | 1.6 |
| Lauric | $C_{12:0}$ | 3.6 | 3.8 | 2.6 |
| Myristic | $C_{14:0}$ | 9.5 | 9.9 | 10.6 |
| Palmitic | $C_{16:0}$ | 23.4 | 26.1 | 30.3 |
| Stearic | $C_{18:0}$ | 9.2 | 9.1 | 10.5 |
| High-saturated | $C_{20-26}$ | 0.8 | 1.0 | 0.7 |
| Unsaturated | | | | |
| Lower unsaturated | $C_{10-14:1}$ | 1.8 | 1.8 | 1.0 |
| Hexadecenoic | $C_{16:1}$ | 3.6 | 2.8 | 3.6 |
| Oleic | $C_{18:1}$ | 26.2 | 24.7 | 21.6 |
| Unsaturated polyethenoid | $C_{18-22:2+}$ | 3.9 | 3.5 | 2.0 |
| Values of some fat constants | | | | |
| Reichert value | | 30 | 27 | 32 |
| Polenske value | | 2.5 | 1.8 | 1.6 |
| Iodine value | | 36 | 35 | 30 |

Sources: Consolidated from various sources.

## Ghee Flavor

The chemistry of ghee flavor is very complex. More than 100 flavor compounds responsible for ghee flavor were identified using gas-liquid chromatography (GLC) (Wadhwa & Jain, 1990). Various flavor components, namely FFAs, carbonyls, lactones, reducing substances etcetera, constitute the ghee flavor. Among these, carbonyls and lactones play a major role in imparting typical ghee flavor. FFAs $C_6$–$C_{12}$, although present in a very low concentration (0.4–1.0 mg/g) and accounting for only 5–10% of total fatty acids, contribute significantly to ghee flavor. The average FFA level of cow ghee is higher than buffalo ghee (Pandya & Sharma, 2002). About 50% of the flavoring compounds in ghee are carbonyls (Wadhwa, 1998). The proportion of various carbonyls is probably the most important factor that creates the characteristic ghee flavor. Like carbonyls, a definite proportion of lactones and a balanced

**Table 21.9. Odor-Active Compounds in Sweet Cream Butter (Swcb), Sour Cream Butter (Socb) and Butter Oil (BO) as Determined by Gas Chromatography-Olfactometry**

| | | Concentration (μg/kg butter) | | |
|---|---|---|---|---|
| Compound | Odor quality | SwCB | SoCB | BO |
| Acids | | | | |
| Butanoic acid | Buttery, sweaty, cheesy, rancid | 192 | 4480 | nq |
| Hexanoic acid | Pungent, musty, cheesy, acrid | 732 | 1840 | nq |
| Aldehydes | | | | |
| 2-Methylbutanal | Chocolate, fruity | 4.9 | — | — |
| 3-Methylbutanal | Chocolate | 11.9 | — | — |
| Hexanal | Green, fatty | 29 | nq | nq |
| Nonanal | Waxy, fatty, floral | 43 | — | — |
| (*E*)-2-Nonenal | Green, fatty, tallowy | 10 | nq | 6.75 |
| (*Z*)-2-Nonenal | Green, fatty | nq | nq | 0.2 |
| (*Z*)-4-Heptenal | Green, fatty, creamy, biscuit-like | nq | nq | 0.3 |
| Ketones | | | | |
| 2,3-Butanedione* | Buttery | 6.6 | 620 | nq |
| 1-Hexen-3-one | Vegetable-like, metallic | 0.004 | — | nq |
| 1-Octen-3-one | Mushroom-like | 0.58 | nq | 1.1 |
| Lactones | | | | |
| δ -Hexalactone | Creamy, chocolate, sweet aromatic | 47.9 | — | — |
| δ -Octalactone | Coconut-like, peach | 72.8 | — | nq |
| δ -Decalactone | Coconut-like, peach | 1193 | 5000 | nq |
| γ -Dodecalactone | Peach | 441 | — | — |
| (*Z*)-6-Dodeceno-γ-lactone | Peach | — | 260 | nq |
| Sulfur-containing compounds | | | | |
| Dimethyl sulfide | Corn-like, fresh pumpkin | 20 | — | — |
| Dimethyl trisulfide | Garlic, sulfury | 17.4 | — | — |
| Nitrogen-containing compound | | | | |
| 3-Methyl-1 H-indole (Skatole) | Mothball, fecal | 12.6 | nq | nq |

nq = compound detected but not quantified.
* 2,3-Butanedione is also called Diacetyl
Modified from Mallia et al., 2008.

ratio are responsible for the normal pleasing flavor. Apparently, lactones are only one component of ghee flavor, and two others (FFAs and carbonyls) also play a dominant role. The average concentration of flavor components in cow and buffalo ghee is given in Table 21.10 (Pandya & Sharma, 2002).

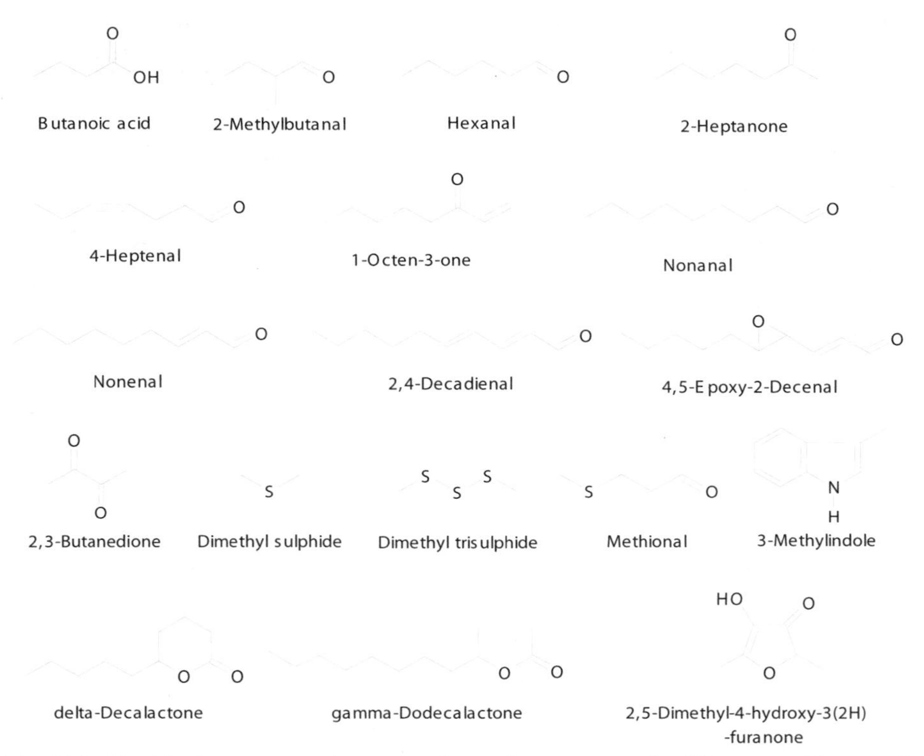


**Fig. 21.2. Chemical structures of potent odor-active compounds of butter and butter oil. Note that Diacetyl is a common name for 2,3-Butanedione.**

**Table 21.10. Carbonyl* Flavor Components of Cow and Buffalo Ghee**

| Flavor compound | Cow ghee | Buffalo ghee |
|---|---|---|
| Total carbonyls (μmoles/g) | 7.200 | 8.640 |
| Volatile carbonyls (μmoles/g) | 0.330 | 0.260 |
| Headspace carbonyls (μmoles/g) | 0.035 | 0.027 |

* Gas-stripped carbonyls
Source: Pandya and Sharma (2002).

The metabolic activity of starter bacteria on various milk/cream/butter constituents such as lactose, citrates, and glucose may lead to the enhancement of ghee-flavor compound production (especially carbonyls and FFAs). The compounds produced during the ripening process are incorporated into the final product upon moisture removal during the clarification process. The incorporation is facilitated by the acidity of ripened cream/butter. In addition to this, the lower pH produced by the activity of starter organisms helps to increase the intensity of chemical reactions such as caramelization during heat clarification. The ripening of cream prior to churning into butter and the subsequent clarification of this butter into ghee yield a ghee with superior flavor (Pandya & Sharma, 2002).

## Health Effects

Compared to other fats and oils, milk fat is easily digestible. The digestibility of milk fat is 99%, while that of natural palm is 91%. The excellent digestibility of milk fat is due to the dispersion of fat globules in the aqueous phase of milk forming an emulsion. These are absorbed directly, unlike other dietary fats that have to be emulsified by bile, pancreatic enzymes, and intestinal lipases before they can pass through the intestinal wall. Also, milk fat is rich in short- and medium-chain fatty acids, which are more easily absorbed than long- chain fatty acids. The ester bonds involving short-chain fatty acids are more easily cleaved by lipases. The easy digestibility of milk fat makes it a valuable dietary constituent for the treatment of diseases of the stomach, intestine, liver, gall bladder, kidney, and disorders of fat digestion (Haug et al., 2007; Kansal, 1994; Miller et al., 1995).

In infant and child nutrition, milk fat is of immense benefit. It helps to meet their energy requirements by increasing the energy density of the diet. A sufficient fat supply is essential for thriving babies, causes a rosy and smooth skin, and also increases resistance to bacterial infections. (Kansal, 1995). Short- and medium-chain fatty acids with 4–12 carbon atoms, which occur in a relatively high concentration in milk fat, have antibacterial and fungicidal activity against gram-negative bacteria and certain molds. Butyric acid is a well-known modulator of gene function, and may also play a role in cancer prevention (German, 1999). Caprylic and capric acids may have antiviral activities, and caprylic acid has delayed tumor growth (Thormar et al., 1994). Lauric acid may have antiviral and antibacterial functions (Sun, 2002), and might act as an anticaries and antiplaque agent. A protective effect against human tooth decay was ascribed, in part, to the adsorption of milk fat onto the enamel surface and, in part, to the antimicrobial effect of milk fatty acids (Kansal, 1995; Schuster et al., 1980). Milk fat has several bioprotective molecules (e.g., sphingomyelins, butyric acid, myristic acid, β-carotenes, and fat-soluble vitamins) that have the potential to inhibit the processes of carcinogenesis and atherogenesis and to stimulate the immune system. While butyric acid has anticancer properties, myristic acid may help the body fight infection by activating macrophages (Haug et al., 2007; Parodi, 1996; Szakaly et al., 2001).

Coronary heart disease (CHD), the most common and serious form of cardiovascular diseases (CVDs), is the leading cause of death. Milk fat was often implicated in CHD because of its cholesterol content and the composition of its fatty acids. However, to judge the implication of milk fat in the development of CHD solely on the basis of its fatty-acid composition and cholesterol content is incorrect. The average cholesterol content in cow and buffalo milk is only 2.8 and 1.9 mg/g of fat, respectively. Moreover, humans only absorb about 10–14% of dietary cholesterol; thus, only 20–40 mg of cholesterol will be absorbed from 50 g of dietary milk fat or from 1 L of milk. On the other hand, the body itself synthesizes cholesterol (1–4 g daily) in much higher amounts than what is absorbed from the diet (Bartov et al., 1973; Grundy et al., 1969; Kansal, 1994; Quintao et al., 1971; Tanaka & Portman, 1977). A common belief is that ghee, because of its saturated fatty acids, increases the levels of serum cholesterol. Therefore, more PUFA should be in the diet. Studies show that the reduction in the serum cholesterol levels by PUFA was attributed partly to a redistribution of cholesterol into other body tissues. An increased excretion of cholesterol in the form of bile acids may accelerate the formation of gall stones (Renner, 1983). A high intake of PUFA reduces the conversion of linoleic acid to arachidonic acid and hence to growth factors. An excessive intake of PUFA increases the vitamin-E requirements. The oxidation products of PUFA such as peroxides may cause an alteration in the membrane of blood cells. Further, a diet containing predominantly either oleic or linoleic acid results in LDL enriched with these fatty acids and thus more susceptibility to oxidation. According to Hodgson et al. (1993), the chances of atherosclerosis are higher when the linoleic-acid content of adipose tissue is higher. The increased excretion of bile acids due to a PUFA-rich diet may lead to increased colonization of bile- degrading bacteria in the intestine—which are thought to be carcinogenic.

Aneja and Murthy (1990) reported that depending upon the method of preparation of ghee, the CLA contents of ghee increases from 5–6 µg CLA per mg (milligram) fat to as high as 25–28 µg CLA per mg (milligram) fat in both buffalo- and cow-ghee. CLA content could be increased to as high as 25–28 µg of fat. Both butter and ghee consumption improve the CLA content in human breast milk. Fogerty et al. (1998) noted that breast milk from women of the Hare Krishna religious sect contained twice as much CLA as milk from conventional Australian mothers (1.12% versus 0.58%). This difference was attributed to the large amount of butter and ghee consumed by Hare Krishna women. The CLA content of milk fat from commercial dairy products such as milk, butter, cheese, and cultured products ranging from 0.1–2.9% of fat for different countries and regions were recently compiled by Parodi (2003). The CLA content of various milk and dairy products is given in Table 21.11.

CLA has an anticarcinogenic activity, and inhibits the growth of melanoma, leukemia, mesothelioma, and glioblastoma together with breast, prostate, colon, and

**Table 21.11. Total Conjugated Linoleic Acid (CLA) Content in Milk and Dairy Products***

| Food | Total CLA content (mole% of fatty acids) |
|---|---|
| Whole milk | 0.25–1.80 |
| Sour cream | 0.46–0.75 |
| Butter | 0.47–1.19 |
| Cow ghee | 0.60–2.10 |
| Buffalo ghee | 0.50–1.90 |

*Sources: Compiled from various sources.

ovarian cancer-cell lines (Dhiman et al., 2005; Haug et al., 2007; Lin et al., 1995; McBean, 2000; Pandya & Kanawjia, 2002; Pariza, 1991; Parodi, 1994; Visonneau et al., 1996). Other research reports indicate that CLA may have other physiological effects, in addition to the cancer-fighting properties. Some of these effects include: a role in reducing atherosclerosis (Lee et al., 1994; MacDonald, 2000; Nicolosi et al., 1997), a benefit for controlling type II diabetes treatments (Bauman et al., 2001; Houseknecht et al., 1998; Kelly, 2001), a reduction in body-fat accretion (Belury, 2002; Khanal & Olson, 2004), an increase in the body protein in growing animals (Park et al., 1997; Whigham et al., 2000), benefits in bone formation (Belury, 2002; Watkins et al., 1999), and modulating the immune system (Bassaganya-Riera et al., 2002; Bell & Kennelly, 2001).

## Nutritional Aspects of Rancidity

Oxidized fat contains several classes of compounds which have toxic effects. These toxic compounds may be categorized as lipid peroxides, hydroxyl fatty acids, carbonyl compounds such as malonaldehyde, cyclic monomers, dimers, polymers, polycyclic aromatic compounds, and oxidized sterols (Addis, 1986; Alexander, 1986; Logani & Davies, 1980; Nawar, 1985). An acute effect of consuming oxidized fat is diarrhea. The other groups of compounds responsible for acute adverse biological effects are secondary lipid-oxidation products. Tissue congestion (an excessive accumulation of fluids in a body part), fatty degeneration, and neurosis were more severe in mice dosed with autoxidized methyl linoleate containing secondary oxidation products than with methyl- linoleate hydroperoxides (Alexander, 1986). The chronic effects of consuming oxidized fat were summarized by Sanders (1989) which include diarrhea, poor growth rate, myopathy, hepatomegaly, steatites, yellow-fat disease, hemolytic anemia, and a secondary deficiency of vitamins A and E. The long-term effects associated with the consumption of oxidized fats are the initiation and the promotion of tumor growth. These are mainly because of the presence of fatty-acid hydroperoxides, which are mutagenic (MacGregor et al., 1985), and malonaldehydes, which are both mutagenic and carcinogenic (Shamberger et al., 1974). A new point of concern is the involvement of lipid peroxides in atherogenesis leading to atherosclerosis. Further, the oxidation products of cholesterol are also atherogenic in experimental animals (Imai et al., 1981). Oxidized cholesterol may be carcinogenic or may promote tumor growth (Alexander, 1986). Possibly, the high incidence of CHD among Asian Indian men of Indian descent in Britain may be related to their intake of oxidized cholesterol in ghee (Jacobson, 1987).

## Methods of Chemical Analysis

Readers who are interested in official and unofficial methods for the chemical analysis of butter, butter oil, and ghee should refer to Sharma (1990).

### *Detection of Adulteration of Milk Fat*

Because milk fat is the costliest of all cooking fats, it is widely adulterated with cheaper fats, known as adulterants. The main adulterants include both vegetable oils and fats and animal-body fats (such as lard and tallow).

The methods used in the detection of foreign fats in milk fat fall under the following categories:

**Physical Methods**

- Refractive index and refractive dispersion
- Melting point, solidification point, and opacity profile
- Microscopic examination
- Cryoscopic examination
- Critical temperature of dissolution
- Examination of fluorescence and luminescence under the light of different wave lengths
- Differential thermal analysis
- Spectroscopic analysis, including ultraviolet and infrared spectroscopy.

**Chemical Methods**

- Identification of principal fatty acids—Reichert, Polenske, Kirschner Values, etcetera
- Measurement of unsaturated fatty acids
- Enzymatic hydrolysis
- Complexes with urea
- Chromatography—paper, column, TLC, GC, HPLC
- Fractionation of glycerides and glycerides' structure, Bomer value
- Unsaponifiable matter and study of sterols.

**Other Methods**

- Correlation between various constants—Reichert and iodine value, Reichert and saponification values, etcetera
- Detection of added tracer substances—Baudouin test, Halphan Test, etcetera

## Conclusion

Butter, butter oil, and ghee are considered to be fat-rich dairy products. These products provide energy, fat-soluble vitamins (A, D, E, and K), and essential fatty acids. They have anticarcinogenic, antitumor, anticaries, antibacterial, and fungicidal properties. The physicochemical properties of butter, butter oil, and ghee mainly depend on the initial milk-fat composition, the fatty-acid profile, as well as different types of processing methods. The fat-rich dairy products are susceptible to hydrolytic and oxidative changes, and these decrease the nutritive values of the products. The enhanced level of cholesterol in serum is linked to CVDs. CVDs are also associated with changes in lifestyle— the use of other types of animal fats rather than fat-rich dairy products (e.g., butter, butter oil, ghee). These products have a cholesterol-reducing ability. If an individual consumes fat-rich dairy products in moderation and reduces total fat intake by making other personal food choices (e.g., using low-fat salad dressing and eating fewer snack foods, such as potato chips), this combination of changes may improve overall cardiovascular heart health. In conclusion, the gourmet applications of butter, butter oil, and ghee are acknowledged in most parts of the world. The health-promoting properties of CLA and other components in these products were also demonstrated by numerous clinical studies. However, more research is needed to answer the question of whether the health benefits of these products outweigh the possible health risks associated with consuming the moderate amounts of saturated fats and cholesterol that they contain.

## References

Addis, P.B. Occurrence of lipid oxidation products in foods. *Food Chem. Toxic.* **1986**, *24*, 1021–1030.

Ahmad, I.; M.K.N. Karimullah; M.K. Saeed. A comparative study of *amla* products and synthetic antioxidant for edible fats. *Pak. J. Sci. Res.* **1960**, *12*, 71.

Alexander, J.C. Heated and oxidized fat. *Dietary fat and Cancer;* (D.F. Burte, A.E. Rogers, C. Milton, Eds.; Aleen R. Liss: New York, 1986; p. 185.

Amr, A.S. Storage of sheep Samna packed in traditional and modern packing materials. *Ecology Food Nutr.* **1990a**, *24*, 289.

Amr, A.S. Role of some aromatic herbs on extending the stability of sheep ghee during accelerated storage. *Egyptian J. Dairy Sci.* **1990b**, *18*, 335–344.

Aneja, R.P.; T.N. Murthy. Conjugated linoleic acid content of Indian curds and ghee. *Indian J. Dairy Sci.* **1990**, *43*, 231–238.

Badings, H.T.; R. Neeter. Recent advances in the study of aroma compounds of milk and dairy products. *Neth. Milk Dairy J.* **1980,** *34*, 9–30.

Bartov, I.; R. Reiser; G.R. Henderson. Hypercholesterolemic effect in the female rat of egg yolk. *J. Nutr.* **1973**, *103*, 1400–1405.

Bassaganya-Riera, J.; R. Hontecillas; M.J. Wannemuehler. Nutritional impact of conjugated linoleic acid: a model functional food ingredient. *In Vitro Cell. Dev. Biol.* **2002**, *38*, 241–246.

Bauman, D.E.; B.A. Corl; L.H. Baumgard; J.M. Griinari. Conjugated linoleic acid (CLA) and the dairy cow. *Recent Advances in Animal Nutrition;* P.C. Garnsworthy, P.C. Wisemen, Eds.; Nottingham University Press: Nottingham, UK, 2001; pp. 221–250.

Bell, J.A.; J.J. Kennelly. Conjugated linoleic acid enriched milk: a designer milk with potential. *Adv. Dairy Technol.* **2001**, *13*, 213–228.

Belury, M.A. Dietary conjugated linoleic acid in health: Physiological effects and mechanisms of action. *Annu. Rev. Nutr.* **2002**, *22*, 505–531.

Bhatia, I.S.; N. Kaur; P.S. Sukhija. Role of seed phosphatides as antioxidants for ghee (Butter fat). *J. Sci. Food Agric.* **1978**, *29*, 747–752.

Bitman, J.; D.L. Wood. Changes in milk fat phospholipids during lactation. *J. Dairy Sci.* **1990**, *73*, 1208–1216.

Bracco, U.; J. Hidalgo; H. Bohren. Lipid composition of the fat globule membrane of human and bovine milk. *J. Dairy Sci.* **1972**, *55*, 165–172.

Brunner, J.R. *Structural and Functional Aspects of Lipoproteins in Living Systems;* E. Tria, A.M. Scann, Eds.; Academic Press, Inc.: London, 1969; p. 545.

Chatterjee, K.L. Shelf life of ghee produced and stored under commercial condition. *Indian Dairyman* **1977**, *29*, 797–803.

Chauhan, P.; B.K. Wadhwa. Comparative evaluation of ghee in tin and polythene package during storage. *J. Food Processing and Preservation* **1987**, *11*, 25–30.

Chin, S.F.; W. Liu; J.M. Storkson; Y.L. Ha; M.W. Pariza. Dietary sources of conjugated dienoic isomers of linoleic acid, a newly recognized class of anticarcinogens. *J. Food. Comp. Anal.* **1992**, *5*, 185–197.

Christie, W.W. Composition and structure of milk lipids. *Advanced Dairy Chemistry*, 2nd ed.; P.F. Fox, Ed.; Chapman & Hall: London, 1995; Vol. 2, p. 136.

Dhiman, T.R.; S.H. Nam; A.L. Ure. Factors affecting conjugated linoleic acid content in milk, meat. *Crit. Rev. Food Sci.Nutr.* **2005**, *45*, 463–482.

Fogerty, A.C.; G.L. Ford; D. Svornos.Octadeca-9,11-dienoic acid in food stuffs and in the lipids of human blood and breast milk. *Nutr. Reports Int.* **1998**, *38*, 937–942.

Ganguli, N.C.; M.K. Jain. Ghee: its chemistry, processing and technology. *J. Dairy Sci.* **1972,** *56*, 19–25.

German, J.B. Butyric acid: a role in cancer prevention. *Nutr. Bull.* **1999**, *24*, 293–299.

Grundy, S.M.; E.H. Ahrens; J. Davingnon.The interaction of cholesterol absorption and cholesterol synthesis in man. *J. Lipid Res.* **1969**, *10*, 304–315.

Gupta, Sudha; P.S. Sukhija; I.S. Bhatis. Role of amino acids as antioxidants for ghee. *Indian J. Dairy Sci.* **1977**, *30*, 319–324.

Hagrass, A.E.A.; A.A. Asker; S.H. Hafez; A.E. Shehata. Properties and stability of samna made from Freisan cow milk as affected by stage of lactation and seasonal variation. *Annals. Agric. Sci.* **1983,** *28*, 1493.

Hathway, D.E. Metabolic fate in animals of hindered phenolic antioxidants in relation to their safety evaluation and antioxidants function. *Adv. Food Res.* **1966,** *15*, 1–56.

Haug, A.; A.T. Hostmark; O.M. Harstad. Bovine milk in human nutrition—a review. *Lipids in Health and Dis.* **2007**, *6*, 25–40.

Heijden, V.E.A.; D.J.C.M. Vansen; J.J.T.W.A. Stirk. Toxicology of gallates. A review and evaluation, *Food Chem. Toxicol.* **1986**, *24*, 1067–1070.

Hettinga, D. Butter. Bailey's Industrial Oil and Fat Products, 6th ed.; S. Fereidoon, Ed.; John Wiley & Sons, Inc.: New York, **2005;** Vol. 6, pp. 1–59.

Hodgson, J.M.; M.L. Wahlqvist; J.A. Boxall; N.D. Balazs. Can linoleic acid contribute to coronary artery disease? *Am. J. Clin. Nutr.* **1993**, *58*, 228–234.

Houseknecht, K.L.; J.P. Vanden Heuvel; S.Y. Moya-Camarena; C.P, Portocarrerro;L.W. Peck; K.P. Nickel; M.A. Belury. Conjugated linoleic acid normalizes impaired glucose tolerance in the Zucker diabetic fatty fa/fa rat. *Biochem. Biophys. Res. Commun.* **1998**, *244*, 678–682.

Hunziker, O.F. *The Butter Industry,* 3rd ed.; Products Corp.: Chicago, 1940, pp. 1–23.

Imai, H.; N.T. Werthessen; V. Subramanayam; P.W. Kequesne; A.H. Soloway;M.C. Kanisawa. Angiotoxicity of sterols and possible precursors. *Science* **1981**, *207*, 651–653.

Jacobson, M.S. Cholesterol oxides in Indian ghee. Possible cause of unexplained risk of atherosclerosis in immigrant population. *Lancet* **1987,** *2*, 656–658.

Jana, A. Quality Requirements of Manufacture of Fat Rich Dairy Products for Use in Recombined Dairy Products. In a compendium of lectures delivered at refresher course on "Technology of Fat Rich Dairy Product" organized by SMC College of Dairy Science, Gujarat Agricultural University, Gujarat, India, between 16 to 30 June, 1990, pp. 188–199.

Jenness, R.; S. Patton. *Principles of Dairy Chemistry;* John Wiley & Sons, Inc.: New York, 1959.

Joshi, N.S. Physico-chemical Characteristics of Fat Relevant to Manufacture of Fat Rich Dairy Products. In a compendium of lectures delivered at refresher course on "Technology of Fat Rich Dairy Product" organized by SMC College of Dairy Science, Gujarat Agricultural University, Gujarat, India between 16 to 30 June, 1990, pp. 1–27.

Kansal, V.K. Milk fat and human health. *Indian Dairyman* **1994**, *46*, 345–350.

Kansal, V.K. Advances in milk fat and its role in human health. *Indian Dairyman* **1995**, *47*, 20–27.

Kelly, G.S. Conjugated linoleic acid: a review. *Alternative Med. Rev.* **2001**, *6*, 367–382.

Khanal, R.C.; K.C. Olson. Factors affecting conjugated linoleic acid (CLA) content in milk, meat and egg: A review. *Pakistan J. Nutr.* **2004**, *3,* 82–98.

Kuchroo, T.K.; K.M. Narayanan. Effect of addition of antioxidants on the oxidative stability of buffalo ghee. *Indian J. Dairy Sci.* **1972**, *25,* 228–232.

Kuchroo, T.K.; K.M. Narayanan. Preservation of ghee. *Indian Dairyman* **1973**, *25*, 405–409.

Kurtz, F.E. The lipids of milk: composition and properties. *Fundamentals of Dairy Chemistry;* B.H. Webb, A.H. Johnson; J.A. Alford, Eds.; The AVI Publishing Company Inc.: Westport, Connecticut, 1974; p. 125.

Lal, D.; T. Rai; I.M. Santha; K.M. Narayanan. Standardization of a method for transfer of phospholipids from ghee-residue to ghee. *Indian J. Anim. Sci.* **1984**, *54,* 29–33.

Lawson, R.E.; A.R. Moss; D. Ian Givens. The role of dairy products in supplying conjugated linoleic acid to man's diet: a review. *Nutr. Res. Rev.* **2001**, *14*, 153–172.

Lee, K.N.; D. Kritchevsky; M.W. Pariza.Conjugated linoleic acid and atherosclerosis in rabbits. *Atherosclerosis* **1994**, *108*, 19–25.

Lin, H.; T.D. Boylston; M.J. Chang; L.O. Luedecke; T.D. Shultz. Survey of the conjugated linoleic acid contents of Dairy Products. *J. Dairy Sci.* **1995**, *78*, 2358–2365.

Logani, M.K.; R.E. Davies. Lipid oxidation: Biological effects and antioxidants—a review. *Lipids* **1980**, *15*, 485–495.

MacDonald, H.B. Conjugated linoleic acid and disease prevention: A review of current knowledge. *J. Am. College of Nutr.* **2000**, *19*, 111S–118S.

MacGregor, J.T.; R.E. Wilson; W.E. Neff; E.N. Frankel. Mutagenicity tests on lipid oxidation products in *Salmonella typhimurium*; monohydroperoxide and secondary oxidation products of methyl-linoleate and methyl-linolenate. *Food Chem. Toxicol* **1985**, *23*, 1041–1047.

Maeura, Y.; J.L. Weisherger; G. Williams. Dose development reduction of N-2 fluoryl acetamide liver cancer and enhancement of bladder cancer in rats by butylated hydroxyl toluene. *Cancer Res.* **1984**, *44*, 1604–1610.

Mallia, S.; F. Escher; H. Schlichtherle-Cerny. Aroma- active compounds of butter: a review. *Eur. Food Res. Technol.* **2008**, *226*, 315–325.

McBean, L.D. Emerging health benefits of CLA. *The Dairy Council Digest* **2000**, *71*, 19–24.

McDowall, F.H. *The Butter Maker's Manual;* New Zealand University Press, University House: Wellington, New Zealand, 1953; Vols. I, II.

McKay, G.L.; C. Larsen. *Principles and Practice of Butter Making;* John Wiley & Sons: New York, 1939.

Megha, A.V. Effect of Casein, Lactose and Casein-Lactose Mixture on Oxidative Stability of Ghee. M.Sc. Thesis, Gujarat Agril. Uni.: Sardar Krushinagar, India, 1981.

Mehta, B.M. Ragi (*Eleusine coracana* L.) —a natural antioxidant for Ghee (butter oil). *Int. J. Food Sci.Technol.* **2006**, *41*, 86–89.

Miller, G.D.; J.K. Jarvie; L.D. McBean. *Handbook of Dairy Foods and Nutrition;* CRC Press, Inc.: Boca Raton, Florida, 1995.

Mortensen, B.K. Physical properties and modification of milk. *Development in Dairy Chemistry;* P.F. Fox, Ed.; Applied Science Published Ltd., London, 1983; Vol. 2, pp. 159–194.

Mulder, H.; P. Walstra. *The Milk Fat Globule;* Centre for Agricultural Publishing and Documentation, Pudoc: Wageningen, Netherlands, 1974, p. 1.

Nawar, W.W. Lipids. *Food Chemistry, 2nd ed.;* O.R. Fennema, Ed.; Marcel Dekker Inc.: New York, 1985; p. 139.

Nicolosi, R.J.; E.J. Rogers; D. Kritchevsky; J.A. Scimeca; P.J. Huth. Dietary conjugated linoleic acid reduces plasma lipoproteins and early aortic atherosclerosis in hypercholesterolemic hamsters. *Artery* **1997**, *22*, 266–277.

Pandya, N.C.; S.K. Kanawjia. Ghee: A traditional nutraceutical. *Indian Dairyman* **2002**, *54*, 67–75.

Pandya, A.J.; R.S. Sharma. Ghee—Its Chemistry, Technology and Nutrition—An Overview. National seminar on role of pure ghee in health and nutrition exploding myths, Jointly organized by Indian Dairy Association, Gujarat chapter, Anand, GCMMF, Anand and J.S. Ayurveda College, Anand on 13–14 June, 2002, 1–14.

Pariza, M.W. CLA: A new cancer inhibitor in dairy products. *IDF Bull.* **1991,** *257*, 29–30.

Park, Y.; K.J. Albright; W. Liu; J.M. Storkson; M.E. Cook; M.W. Pariza. Effects of conjugated linoleic acid on body composition in mice. *Lipids* **1997**, *32*, 853–858.

Parmar, S.S.; R.S. Sharma. Use of mango seed kernels in enhancing oxidative stability of ghee. *Asian J. Dairy Res.* **1986**, *5*, 91–99.

Parodi, P.W. Conjugated linoleic acid: an anticarcinogenic fatty acid present in milk fat. *Aust. J. Dairy Technol.* **1994**, *49*, 93–97.

Parodi, P.W. Milk fat components: possible chemo preventive agents for cancer and other diseases. *Aust. J. Dairy Technol.* **1996**, *51*, 24–32.

Parodi, P.W. Conjugated linoleic acid and other anticarcinogenic agents of bovine milk fat. *J. Dairy Sci.* **1999**, *82*, 1339–1349.

Parodi, P.W. Conjugated linoleic acid in food. *Advances in Conjugated Linoleic Acid Research;* J.L. Sebedio, W.W. Christies, R.O. Adlof, Eds.; AOCS Press: Champaign, Illinois, 2003; Vol. 2, pp. 101–122.

Parodi, P.W. Milk fat in human nutrition. *J. Dairy Technol.* **2004**, *59*, 3–59.

Pionneir, E.; D. Hugelshofer. *Flavor Science: Recent Advances and Trends;* W.L.P. Bredie, M.A. Petersen, Eds.; Elsevier: Amsterdam, 2006: pp. 233–236.

Quintao, E.; S.M. Grundy; E.H. Ahrens.Effect of dietary cholesterol on the regulation of total body cholesterol in man. *J. Lipid Res.* **1971**, *12*, 233–247.

Rama Murthy, M.K.; K.M. Narayanan. Polyunsaturated fatty acids of buffalo and cow milk fat. *Milchwissenschaft* **1972,** *27*, 695.

Rama Murthy, M.K.; K.M. Narayanan;V.R. Bhalerao. Effects of phospholipids on the keeping quality of ghee. *Ind. J. Dairy Sci.* **1968,** *21*, 62–65.

Rangappa, K.S.; K.T. Achaya.*Indian Dairy Products*; Asia Publishing House: New Delhi, 1974; p. 327.

Renner, E. Milk and Dairy Products in Human Nutrition; Volkswirtschaftlicher Verlag: Muenchen (Germany, F.R.), 1983; p. 450.

Sanders, T.A.B. Nutritional aspects of rancidity. *Rancidity in Foods;* J.C. Allen, R.J. Hamilton, Eds.; Elsevier Applied Science Publishers: London, 1989; p. 125.

Santha, I.M.; K.M. Narayanan. Antioxidant properties of ghee residue as affected by temperature of clarification and method of preparation of ghee. *Indian J. Anim. Sci.* **1978**, *48,* 266–271.

Santha, I.M.; K.M. Naryanan. Changes taking place in proteins during the conversion of butter/cream to ghee. *Indian J. Dairy Sci.* **1979a**, *32*, 68–74.

Santha, I.M.; K.M. Narayanan. Studies on the constituents responsible for the antioxidant properties of ghee residue. *Indian J. Anim. Sci.* **1979b**, *49,* 37–41.

Satyanarayana, A.; P.G. Rao; D.G. Rao. Chemistry, processing and toxicology of Annatto (*Bixa orellana* L.). *J. Food Sci. Technol.* **2003**, *40*, 131–141.

Schuster, G.S.; T.R. Dirksen; A.E. Ciarlone; G.W. Burnett; M.T. Reynolds; M.T. Lankford. Anticaries and antiplaque potential of free-fatty acids in vitro and in vivo. *Pharmacol. Ther. Dent.* **1980**, *5*, 25–33.

Sethi, S.C.; J.S. Aggarwal. Stabilization of edible fats by spices and condiments. *J. Sci. Ind. Res.* **1952**, *11B*, 468–470.

Shamberger, R.J.; T.L. Andrione; E.C. Willis. Antioxidants and cancer. III. Initiating activity of malonaldehyde as a carcinogen. *J. Natl. Cancer Inst.* **1974**, *53*, 1771–1773.

Sharma, R.S. Ghee residue: yield, composition and uses. *Dairy Guide* **1980**, *6*, 21–24.

Sharma, R.S. Ghee: A resume of recent researches. *J. Food Sci. Technol.* **1981**, *18*, 70–77.

Sharma, R.S. In Line Testing of Fat Rich Dairy Products for Quality Assurance: The Chemical Aspect; In a compendium of lectures delivered at a refresher course on "Technology of Fat Rich Dairy Product" organized by SMC College of Dairy Science, Gujarat Agricultural University, Gujarat, India between 16 to 30 June, 1990, pp. 298–322.

Singh, S.; B.P. Ram; S.K. Mittal. Effect of phospholipids and methods of manufacture on flavor and keeping quality of ghee. *Indian J. Dairy Sci.* **1979**, *32*, 161–166.

Solanky, M.J. Handling and Processing of Milk and Cream Intended for Butter Making, In a compendium of lectures delivered at refresher course on "Technology of Fat Rich Dairy Product" organized by SMC College of Dairy Science, Gujarat Agricultural University, Gujarat, India between 16 to 30 June, 1990, pp. 61–73.

Srinivasan, M.R. Ghee making in the tropical countries and possibilities of its industrial production. *Indian Dairyman* **1976,** *28,* 279–283.

Srinivasan, M.R.; C.P. Anantakrishnan. *Milk Products of India;* ICAR: New Delhi, 1964.

Sukumar, De. *Outlines of Dairy Technology;* Oxford University Press, YMCA Library Building: New Delhi, India, 1991; pp. 143–173.

Sukumar, De; B.N. Mathur. Some investigation on the churning efficiency of Indian Creams. *Indian Dairyman* **1968**, *20*, 351–353.

Sun, C.Q.; C.J. O'Conner; A.M. Roberton.The antimicrobial properties of milk fat after partial hydrolysis by calf pregastric lipase. *Chem. Biol. Interact.* **2002**, *140*, 185–198.

Szakaly, S.; B. Schaffer; P. Horn; C.S. Sarudi; Z. Szakaly; J. Dohy. Nutritional value of milk based on the latest research findings. *Tejgazdasag* **2001**, *61*, 1–10.

Tanaka, N.; O.W. Portman. Effect of type of dietary fat and cholesterol absorption rate in squirrel monkeys. *J. Nutr.* **1977**, *107*, 814–821.

Tandon, R.N. Effect of feeding cotton seed to milch animals on the opacity pattern of ghee and changes in its physicochemical constants on storage. *Indian J. Dairy Sci.* **1977**, *30*, 341–343.

Thakar, P.N. Manufacture of Table Butter by Batch and Continuous Butter Making Method, In a compendium of lectures delivered at refresher course on "Technology of Fat Rich Dairy Product" organized by SMC College of Dairy Science, Gujarat Agricultural University, Gujarat, India between 16 to 30 June, 1990, pp. 74–88.

Thormar, H.; E.E. Isaacs; K.S. Kim; H.R. Brown. Interaction of visna virus and other enveloped viruses by free fatty acids and monoglycerides. *Ann NY Acad. Sci.* **1994**, *724*, 465–471.

Varshney, N.N. Dairy professional's ready reckons. *Dairy India,* 5th ed.; P.R. Gupta, Ed.; P.R. Gupta Printers: New Delhi, 1997; p. 364.

Visonneau, S.; A. Cesano; S.A. Tepper; J. Scimeca; D. Santoli; D. Kritchersky. Effect of different concentrations of conjugated linoleic acid (CLA) on tumor cell growth *in vitro. FASEB J.* **1996**, *10*, 182 (Abstr.).

Wadhwa, B.K. Chemistry of ghee flavor and flavor simulation studies. *Advances in Traditional Dairy Products.* In compendium brought for a short course by N.D.R.I., Karnal, India, 1998, pp. 86–91.

Wadhwa, B.K.; M.K. Jain. Chemistry of ghee flavor—a review. *Indian J. Dairy Sci.* **1990**, *43*, 601–607.

Wadhwa, B.K.; K. Surinder; M.K. Jain. Enhancement of the shelf life of flavored butter oil by natural antioxidants. *Indian J. of Dairy Sci.* **1991**, *44*, 119–121.

Walker-Tisdale, C.W.; T.R. Robinson. *Practical Buttermaking: A Treatise for Butter Makers and Students;* Kessinger Publishing Company, 1919.

Walstra, P. Physical chemistry of milk fat globules. *Developments in Dairy Chemistry;* P.F. Fox, Ed.; Applied Science Publishers Ltd.: London, 1983; Vol. 2, p. 119.

Walstra, P.; T.J. Geurts; A. Noomen; A. Jellema; M.A.J.S. van Boekel.Butter. *Dairy Technology—Principals of Milk Properties and Processes;* Marcel Dekker Inc.: New York, 1999; pp. 485–515.

Walstra, P.; R. Jenness. *Dairy Chemistry and Physics;* John Wiley & Sons, Inc.: New York, 1984; pp. 58–97.

Watkins, B.A.; Y. Li; M.F. Seifert. Bone metabolism and dietary conjugated linoleic acid. *Advances in Conjugated Linoleic Acid Research;* M.P. Yurawecz, M.M. Mossoba, J.K.G. Kramer, M.W. Pariza, G.J. Nelson,Eds.; AOAC Press: Illinois, Chapter 25, 1999; Vol. 1, pp. 327–339.

Whigham, L.D.; M.E. Cook; R.L. Atkinson. Conjugated linoleic acid: implications for human health. *Pharmacol. Res.* **2000**, *42*, 503–510.

# Contributors

**Diana Ansorena**, Department of Nutrition and Food Science, Physiology and Toxicology, Faculty of Pharmacy, University of Navarra, Irunlarrea sn 31008, Pamplona (Navarra), Spain.

**Ramón Aparicio-Ruiz**, Instituto de la Grasa (CSIC), Padre García Tejero, 4, 41012, Sevilla, Spain.

**Iciar Astiasarán**, Department of Nutrition and Food Science, Physiology and Toxicology, Faculty of Pharmacy, University of Navarra, Irunlarrea sn 31008, Pamplona (Navarra), Spain.

**D. Ed Barre**, Cape Breton University, P.O. Box 5300, Sydney, Nova Scotia, Canada B1P-6L2.

**Cherie Bulley**, HortResearch, The Horticulture & Food Research Institute of New Zealand Limited, Auckland, New Zealand.

**J.C. Callaway**, Finola ky, PL 236, Kuopio, FI-70101 Finland, www.finola.com; Departments of Pharmaceutical Chemistry and Neurobiology, University of Kuopio, FI-70211 Kuopio, Finland.

**Nurhan T. Dunford**, Associate Professor, Oklahoma State University, Department of Biosystems and Agricultural Engineering Bioprocessing; and Robert M. Kerr Food & Agricultural Products Center, FAPC Room 103, Stillwater, OK 74078, USA.

**Laurence Eyres**, Oil and Fats Group, N.Z. Institute of Chemistry, Auckland, New Zealand.

**Kelley C. Fitzpatrick**, FLAX CANADA 2015, 465-167 Lombard Ave., Winnipeg, Manitoba R3B 0T6, Canada.

**Ellen Friel**, Diageo Baileys Global Supply, Nangor House, Nangor Road, Dublin 12, Ireland.

**Diego L. García-González**, Instituto de la Grasa (CSIC), Padre García Tejero, 4, 41012, Sevilla, Spain.

**J. Samuel Godber**, Department of Food Science, Louisiana State University, Baton Rouge, LA 70803, USA.

**Clifford Hall III**, Department of Cereal and Food Sciences, North Dakota State

University, 210 Harris Hall, Fargo, North Dakota 58105, USA.
**Junjie (George) Hao**, Department of Chemistry and Biochemistry, and Department of Mathematics, University of Maryland, College Park, MD 20742.
**Marina Heinonen**, Department of Applied Chemistry and Microbiology, P.O. Box 27, Latokartanonkaari 11, FI-00014 University of Helsinki, Finland.
**Kevin B. Hicks**, Eastern Regional Research Center, Agricultural Research Service, U.S. Department of Agriculture, 600 East Mermaid Lane, Wyndmoor, PA 19038, USA.
**Afaf Kamal-Eldin**, Department of Food Science, Swedish University of Agricultural Sciences, Box 7051, 750 07 Uppsala, Sweden.
**Anna-Maija Lampi**, Department of Applied Chemistry and Microbiology, P.O. Box 27, Latokartanonkaari 11, FI-00014 University of Helsinki, Finland.
**Cynthia Lund**, HortResearch, The Horticulture & Food Research Institute of New Zealand Limited, Auckland, New Zealand.
**Tony McGhie**, HortResearch, The Horticulture & Food Research Institute of New Zealand Limited, Auckland, New Zealand.
**Bhavbhuti M. Mehta**, Assistant Professor, Dairy Chemistry Department, Sheth M.C. College of Dairy Science, Anand Agricultural University, Anand-388 110, Gujarat, India.
**Ali Moazzami**, Department of Food Science, Swedish University of Agricultural Sciences, 750 07 Uppsala, Sweden.
**Robert A. Moreau**, Eastern Regional Research Center, Agricultural Research Service, United States Department of Agriculture, 600 East Mermaid Lane, Wyndmoor, Pennsylvania 19038, USA.
**Michael Murkovic**, Graz University of Technology, Institute for Food Chemistry and Technology Petersgasse 12/2, A-8010 Graz, Austria.
**Shane Olsson**, HortResearch, The Horticulture & Food Research Institute of New Zealand Limited, Auckland, New Zealand.
**David W. Pate**, Centre for Phytochemistry and Pharmacology, Southern Cross University, Lismore, NSW 2480, Australia.
**Jana Pickova**, Department of Food Science, Swedish University of Agricultural Sciences, P.O. Box 7051, S-750 07 Uppsala, Sweden.
**Michael J. Powell**, Eastern Regional Research Center, Agricultural Research Service, U.S. Department of Agriculture, 600 East Mermaid Lane, Wyndmoor, PA 19038, USA.
**Paul Prenzler**, School of Agricultural and Wine Sciences, Charles Sturt University, Locked Bag 588, Wagga Wagga 2678, Australia.
**Mohamed Fawzy Ramadan**, Biochemistry Department, Faculty of Agriculture, Zagazig University, Zagazig 44511, Egypt.
**Cecilia Requejo-Jackman**, HortResearch, The Horticulture & Food Research Institute of New Zealand Limited, Auckland, New Zealand.

**Kevin Robards**, School of Agricultural and Wine Sciences, Charles Sturt University, Locked Bag 588, Wagga Wagga 2678, Australia.
**Danielle Ryan**, School of Agricultural and Wine Sciences, Charles Sturt University, Locked Bag 588, Wagga Wagga 2678, Australia.
**Vijay Singh**, Department of Agricultural and Biological Engineering, University of Illinois at Urbana-Champaign, Urbana, IL 61801, USA.
**Mindy Wang**, HortResearch, The Horticulture & Food Research Institute of New Zealand Limited, Auckland, New Zealand.
**Yan Wang**, Institute of Food Nutrition and Human Health, Massey University, Albany, Auckland, New Zealand.
**Allan Woolf**, HortResearch, The Horticulture & Food Research Institute of New Zealand Limited, Auckland, New Zealand.
**Marie Wong**, Institute of Food Nutrition and Human Health, Massey University, Albany, Auckland, New Zealand.
**Liangli (Lucy) Yu**, Department of Nutrition and Food Science.
**Haiyan Zhong**, Faculty of Food Science and Engineering, Central South University of Forestry and Technology, Changsha 410004, Hunan, P. R. China.

# Index

## C

## H

# I

## P

## Q



## R

## S

## T

## U